INNOVATIONS IN NONLINEAR ACOUSTICS

To learn more about the AIP Conference Proceedings, including the Conference Proceedings Series, please visit the webpage **http://proceedings.aip.org/proceedings**

INNOVATIONS IN NONLINEAR ACOUSTICS

ISNA17

17th International Symposium on Nonlinear Acoustics

including the

International Sonic Boom Forum

State College, Pennsylvania   18 – 22 July 2005

17th ISNA

International Symposium on Nonlinear Acoustics

EDITORS
Anthony A. Atchley
Victor W. Sparrow
Robert M. Keolian
The Pennsylvania State University
University Park, Pennsylvania

SPONSORING ORGANIZATIONS
Graduate Program in Acoustics, The Pennsylvania State University
The Acoustical Society of America - (ASA)
The International Commission for Acoustics - (ICA)

◎ CD-ROM INCLUDED

AIP
75 Years of Service

Melville, New York, 2006
AIP CONFERENCE PROCEEDINGS ■ VOLUME 838

Editors:

Anthony A. Atchley
Victor W. Sparrow
Robert M. Keolian

The Pennsylvania State University
Graduate Program in Acoustics
University Park, PA 16802
U.S.A.

E-mail: atchley@psu.edu
        vws1@psu.edu
        keolian@psu.edu

The articles on pp. 59 – 62, 540 – 543, and 659 – 662, were authored by U.S. Government employees and are not covered by the below mentioned copyright.

L.C. Catalog Card No. 2006925986
ISBN 0-7354-0330-9
ISSN 0094-243X

Printed in the United States of America

CONTENTS

v

SECTION 2

NONLINEAR ACOUSTICS IN SOLIDS

SECTION 3

ELASTIC-WAVE EFFECTS ON FLUIDS IN POROUS MEDIA

SECTION 4

HIGH INTENSITY FOCUSED ULTRASOUND
IN MEDICINE AND BIOLOGY

SECTION 5

HARMONIC IMAGING IN DIAGNOSTIC ULTRASOUND

SECTION 6

SHOCK WAVE THERAPY

SECTION 7

NONLINEAR ACOUSTICS IN MEDICINE AND BIOLOGY

SECTION 8

THERMOACOUSTICS

SECTION 9

SOUND BEAMS, RESONATORS, AND STREAMING

SECTION 10

BUBBLES, PARTICLES, AND FLOWS

SECTION 11

INFRASOUND, PROPAGATION, SHOCKS, AND NOISE

SECTION 12

SONIC BOOM

PREFACE

Most, if not all, of us have had the experience of looking at a photograph of the participants of a scientific conference from the early 1900's and, recognizing many now familiar names, wondering what it must have been like to be a part of it. What exciting discoveries were being discussed for the first time? What were these people really like? Times have changed and scientific conferences have evolved considerably over the intervening decades. However, one thing has not changed. Conferences and symposia still play a vital role in the advancement of science and engineering. They provide a forum for the exchange of ideas and cultures and provide a sense of continuity and community. Faced with the ever increasing number of conferences that one could or should attend, it is fair to ask whether a symposium series such as the International Symposium on Nonlinear Acoustics (ISNA) is necessary. Does it play a unique and important enough role to justify the resources required to sustain it? Based on the experience gained through organizing and hosting the 17th ISNA (ISNA17), we believe that the answer to this question is a resounding yes! While it is true that topics presented at ISNA17 are found at other conferences, no single venue captures the diversity of the field of nonlinear acoustics as does ISNA. It brings together researchers from disparate fields as no other forum does. It provides a place to renew old friendships and make new ones. And it provides an opportunity to collect the latest results into a single volume, a snapshot in time from which the breadth, the depth, and the interconnectivity of the field of nonlinear acoustics can be appreciated.

The 17th International Symposium on Nonlinear Acoustics was held 18-22 July at the Penn Stater Conference Center Hotel on the campus of The Pennsylvania State University in State College, Pennsylvania, USA, with 178 participants. Following in the footsteps of the previous ISNAs the scope of the symposium covered nonlinear acoustic phenomena in solids, liquids and gases. Nineteen technical sessions were held. Sessions devoted to special topics and consisting of invited and contributed papers included Nonclassical Nonlinear Acoustics of Solids and NDE Applications (organized by P.A. Johnson, L.A. Ostrovsky, and I. Solodov), Elastic Wave Effects on Fluids in Porous Media (P.M. Roberts and I.B. Esipov), the Science and Application of High Intensity Focused Ultrasound in Medicine and Biology (R.A. Roy), Shock Wave Therapy (M. Bailey), Infrasound (K.A. Naugolnykh and A.J. Bedard), Harmonic Imaging in Diagnostic Ultrasound (R.O. Cleveland), and Thermoacoustics (R.M. Keolian). General sessions consisting of contributed papers included Sound Beams, Resonators and Streaming (chaired by B.O. Enflo), Bubbles, Particles, and Flows (P.L. Marston), Propagation, Shocks and Noise (H. Hobæk Nonlinear Acoustics in Solids (V. Espinosa), and Nonlinear Acoustics in Medicine and Biology (Yu.A. Ilinskii and E.A. Zabolotskaya), as well as a poster session.

In addition, ISNA17 included the International Sonic Boom Forum (ISBF) co-organized by V.W. Sparrow and F. Coulouvrat. This set of special sessions encapsulated increasing international interest in building and operating small supersonic jets with low-amplitude, shaped boom signatures, and timing of the ISBF occurred when many research programs were beginning to address the technical feasibility and design for such aircraft. The purpose of holding the ISBF was to foster technical communication and exchange between university, industry, and government scientists, engineers, and executives interested in sonic booms. This Forum provided a timely and unique opportunity to communicate on sonic boom and get information on the latest research advances and progress toward possible community acceptance of shaped sonic boom. Although not represented in these proceedings, the participation of Peter Coen of NASA Langley Research Center as one of the ISBF Plenary speakers and of all the ISBF panelists is greatly appreciated. The panelists included Akira Murakami, Institute of Space Technology and Aeronautics (JAXA), Laurette Fisher, Federal Aviation Administration; Kenneth Orth, Consultant to Gulfstream; Gerard Duval, retired Concorde pilot for Air France; Thierry Auger, Airbus France; Sam Bruner, Raytheon; Nicolas Heron, Dassault Aviation; Tom Hartmann, Lockheed-Martin Aeronautics; and Richard G. Smith, NetJets Inc. The ISBF organizers also appreciate Gulfstream Aerospace Corporation temporarily locating their portable sonic boom simulator, Supersonic Acoustic Signature Simulator Generation II, at ISNA17 for demonstration to interested ISBF and ISNA17 attendees.

ISNA17 and ISBF gratefully acknowledge the financial support of the Penn State's Graduate Program in Acoustics, the Acoustical Society of America (ASA), and the International Commission on Acoustics (ICA). We are particularly indebted to the Bioresponse to Ultrasound/Bioresponse to Vibration and the Physical Acoustics Technical Committees of the ASA for being strong advocates for ISNA17 and ISBF within the ASA. The support from the ICA allowed us to defray the costs of participation for some of the international participants. We are also indebted to Chriss Schultz and Kathy Liebrum of Penn State's Office of Conferences and Institutes. We very much appreciate assistance from Karen Brooks, Carolyn Smith and Christine Popovich of the Penn State Graduate Program in Acoustics who managed to fit the demands of ISNA into their already busy schedules. A special thanks goes to Karen for long hours spent managing transportation between the airport, hotels, dormitories, conference venue and any other place where participants wanted to go. We also appreciate the assistance of the students of the Graduate Program in Acoustics with audio visual and other logistical support.

Finally, we express our gratitude to all of the participants in ISNA17 and to the ISNA International Organizing Committee for giving us the opportunity to host the international nonlinear acoustics community for a brief time.

A. A. Atchley
V. W. Sparrow
R. M. Keolian

INTERNATIONAL ORGANIZING COMMITTEE

H. Hobaek (Norway, General Secretary)
P. Blanc-Benon (France)
M. A. Breazeale (USA)
L. A. Crum (USA)
X. F. Gong (China)
M. F. Hamilton (USA)
P. A. Johnson (USA)
T. Kamakura (Japan)
V. K. Kedrinskii (Russia)
W. Lauterborn (Germany)
W. G. Mayer (USA)
K. A. Naugolnykh (USA/Russia)
L. A. Ostrovsky (USA/Russia)
O. V. Rudenko (Russia)
E. A. Zabolotskaya (USA)

SPONSORING ORGANIZATIONS

PENNSTATE

GRADUATE PROGRAM IN ACOUSTICS

xvii

PREVIOUS INTERNATIONAL SYMPOSIA ON NONLINEAR ACOUSTICS

1st ISNA
27 May 1968
New London, Connecticut, USA
R.H. Mellen, Chair

2nd ISNA
10-11 November 1969
Austin, Texas, USA
T.G. Muir, Chair

3rd ISNA
1-2 April 1971
Birmingham, UK
H.O. Berktay, Chair

4th ISNA
18-19 April 1972
Buffalo, New York, USA
D.T. Blackstock, Chair

5th ISNA
20-22 August 1973
Copenhagen, Denmark
L. Bjørnø, Chair

6th ISNA
8-10 July 1975
Moscow, USSR
R.V. Khokhlov, Chair

7th ISNA
19-21 August 1976
Blacksburg, Virginia, USA
A.H. Nayfeh, Chair

8th ISNA
3-6 July 1978
Paris, France
A. Zarembowitch, Chair

9th ISNA
20-24 July 1981
Leeds, UK
D.G. Crighton, Chair

10th ISNA
24-28 July 1984
Kobe, Japan
A. Nakamura, Chair

11th ISNA
24-28 August 1987
Novosibirsk, USSR
V.K. Kedrinskii, Chair

12th ISNA
27-31 August 1990
Austin, Texas, USA
M.F. Hamilton and D.T. Blackstock, Chairs

13th ISNA
28 June-2 July 1993
Bergen, Norway
H. Hobæk, Chair

14th ISNA
17-21 June 1996
Nanjing, China
R.J. Wei, Chair

15th ISNA
1-4 September 1999
Göttingen, Germany
W. Lauterborn, Chair

16th ISNA
19-23 August 2002
Moscow, Russia
O.V. Rudenko, Chair

SECTION 1
NONCLASSICAL NONLINEAR ACOUSTICS
OF SOLIDS AND NDE APPLICATIONS

Nonclassical Nonlinear Acoustics in Solids: Methods, Applications, and the State of the Art

Paul A. Johnson[1] and Lev Ostrovsky[2]

[1]*Los Alamos National Laboratory of the University of California*
[2]*ZelTech/NOAA ETL*

Abstract. We outline the area of "nonclassical" nonlinear acoustics that deals with media exhibiting anomalously strong elastic nonlinearity, including hysteresis, conditioning, and slow dynamics. The entire range of behaviors we term nonlinear, *nonlinear, nonequilibrium dynamics (NND)*. The link between the diverse materials that exhibit NND appears to be the presence of "damage" at many scales, ranging from order 10^{-8} m to 10^{-1} m at least. The "damage" may be distributed as in granular materials or isolated, as in a sample with a single crack. As to the precise physical origin of the behavior, it is relatively clear in granular media for instance (Hertz-Mindlin interaction physics); however, in other materials the origin is less clear. We believe it to be due to shear sliding, related to crack and possibly dislocation dynamics as well as other, as yet unidentified physical processes. Because the origin of the behavior is related to damage, damage diagnostics in solids, nonlinear nondestructive evaluation, follows naturally. New areas of research and application have appeared recently. For instance, NND plays a significant role in earthquake strong ground motion and potentially to earthquake source dynamics. Medical applications are in development as well as nonlinear imaging methods as well as applications to locating landmines.

Keywords: Nonclassical nonlinear acoustics, nonlinear elasticity, nonlinear nonequilibrium dynamics, nonlinear imaging, slow dynamics.
PACS: 61.41.+-j, 62.20.Dc, 62.20.Mk,91.60.x, 91.60.Fe, 91.60.Lj, 43.25.Dc, 43.25.Ed, $3.25.Gf

INTRODUCTION

Over the last two decades, studies of nonlinear dynamics in materials, known as *anomalous, nonclassical*, and more recently *nonlinear nonequilibrium* [1] that include rock, damaged materials, some ceramics, sintered metals, granular media, etc., have increased markedly (e.g., see reviews in [2,3]). *Nonlinear, nonequilibrium dynamics,* outlined in Figure 1, appears even at small strain amplitudes ($\sim 10^{-6}$) under ambient conditions, and are manifest in different manners. Specifically, when the material is disturbed by a wave, the period-average elastic modulus decreases. We call this *nonlinear fast dynamics.* Following this, it takes tens of minutes to hours to return to its equilibrium state. This is called *slow dynamics* [4,5]. Further, *conditioning* takes place during nonlinear fast dynamics; conditioning and slow dynamics provide a complexity that is not observed in materials whose nonlinearity is due to lattice anharmonicity (see below). Nonlinear, nonequilibrium dynamics is due to mechanically "soft" inclusions (soft matter) in a "hard" matrix (e.g., [2,3]). For instance, a crack in a solid will induce nonequilibrium dynamics, but a void will not; a sandstone exhibits such behavior due to distributed soft inclusions, also known as the bond system, but a bar of aluminium does not. Numerous experimental methods and theory have been developed to interrogate the elastic response in solids. Here we provide a brief overview of the state of the art, based on the perspective of our research groups and those with which we are associated, in hope that it will stimulate a more broad discussion.

Fundamentally, classical elastic/acoustic nonlinearity implies that the stress-strain relation, the equation of state (EOS) is nonlinear. In particular, the one-dimensional stress (σ) - strain (ε) can be described by

$$\sigma = K_o \varepsilon (1 + \beta \varepsilon + \delta \varepsilon^2 + ...), \qquad (1)$$

where K_0 is the linear modulus, and β and δ are the first- and second- order classical nonlinear parameters; e.g., β is normally of order 1-10 in value. At low dynamic wave amplitudes (strains of less than order 10^{-6}) under ambient conditions, there is evidence that most solids behave in a manner according to the above equation [1]; however, in materials such as rock, β and δ may be much larger than in the classical cases. Further, at ambient conditions, for wave amplitudes above approximately 10^{-6} strain, the material EOS is thought to be hysteretic. A hysteretic EOS can be represented as

$$\sigma = K_o \varepsilon [1 + \beta \varepsilon + \hat{\alpha}(\varepsilon)], \qquad (2)$$

where $\hat{\alpha}(\varepsilon)$ is a hysteretic nonlinear coefficient, depending on the sign of the strain time derivative, $\partial \varepsilon / \partial t$. Specific forms of $\hat{\alpha}(\varepsilon)$ can be retrieved from experiments or, in simple cases, derived from phenomenological models such as the P-M space model [6]. Eq. (2) can be used for practical estimates, especially for NDE applications, but it does not capture the important slow dynamics or conditioning behaviors [5,7] as outlined in Figure 1. As previously noted, slow dynamics is manifest by material relaxation, where its elastic modulus returns to its rest level, K_0, as the logarithm of time, after a large amplitude disturbance. The two aspects of conditioning as we view them are as follows: if a rock sample is driven at a fixed amplitude and frequency for a period of time, (a) the modulus will immediately decrease with the onset of the wave, but then continue to decrease slowly to a new equilibrium value as long as the drive is maintained [5]. Furthermore, one can imagine that (b), slow dynamics is initiated through each pressure cycle of an excitation wave. It is easy to imagine this if the frequency is very low. As wave pressure increases in the first wave oscillation, slow dynamics is initiated. Slow dynamical recovery commences as the wave passes through its dilatational cycle. If the drive is *cw* the process is repeated through each wave cycle. As frequency is increased, eventually the wave oscillation may be too fast for the material to respond and the slow dynamics may be "locked". Aspect (a) has been verified; aspect (b) has yet to be verified experimentally, however. Conditioning is a relatively small effect in most materials. Despite its shortcomings, simple forms of Eq. (2) have been applied broadly to describe the material elastic nonlinearity. The rate-dependent effect of conditioning appears to have only a minor influence on estimates of α [7].

There are numerous methods by which to measure the nonlinear response, including resonance and application of harmonics and wave modulation. These are well described elsewhere in this volume and in the literature [2,8,9,10,11]. In the following we will briefly describe nonlinear imaging and some new areas of research, in addition to addressing the origin of nonlinear behaviours.

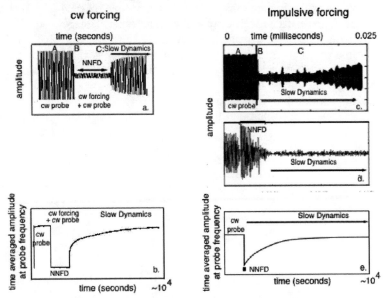

FIGURE 1. *Nonlinear, nonequilibrium dynamics for two types of forcing.* The figure illustrates the fully nonlinear, nonequilibrium behaviour that includes *nonlinear fast dynamics, conditioning,* and *slow dynamics.* Figures (a) and (b) show how nonlinear, nonequilibrium dynamics is manifest in the case when a low-amplitude, continuous wave (*cw*) probe wave is input into a sample in the presence of a large-amplitude vibration. One sees in (a) the undisturbed probe wave (time A) and the corresponding time-average amplitude of the *cw* probe signal in (b) ["*cw* probe"]. At the time the vibration is turned on the signal amplitude sharply decreases (time B). Until it is turned off, nonlinear fast dynamics, including conditioning, take place (denoted by *NNFD* in (a) and (b)). As soon as the large-amplitude wave is terminated, one sees in (a) and (b) an instantaneous, partial recovery of the amplitude, and then a longer term recovery that is linear with the logarithm of time where slow dynamics is the sole process acting in the system. Figures (c-e) show the situation where the sample is disturbed by an impact, such as a tap, in the presence of the probe (time B in (c)). Figure (d) shows a zoom of (c) where one can observe the onset of the tap-induced vibration and its ring down (*NNFD* in (d-e)). After the vibration energy has dissipated, slow dynamics is the sole process operating in the system, the onset of which is shown in (c-d), and the long term behaviour is seen in (e) [12].

NONLINEAR IMAGING

Nonlinear elastic wave imaging is in its infancy in solids. One method is described in [13]. In this method, a low frequency, continuous wave (*cw*) excitation is applied to the specimen simultaneously with a group of high frequency tone-bursts. The tone-bursts, in the presence of the vibration, will exhibit nonlinear effects and these can be used for imaging. In an experiment with a small steel plate, the wave scattering from a hole created by drilling, and a crack created by cyclic loading, were measured. By windowing the high-frequency pulse signal in time, the method provides the means to isolate a nonlinear scatterer in time (and in space, if material velocity is known). It has not been demonstrated in three-dimensions, however.

New methods exploiting the focusing properties of time reversal (e.g. [14]) together with the elastic nonlinear properties of cracks (and other damage features) are being developed for application to damage location and imaging (e.g., [15,16]). A significant aspect of time reversal in regards to elastic nonlinearity is that it provides one the ability to focus an ultrasonic wave, regardless of the position of the initial source and of the heterogeneity of the medium in which the wave propagates. By scanning the surface using a laser vibrometer in tandem with TR focusing at each scan point, then analyzing for nonlinear response at that point, one can determine if damage exists in

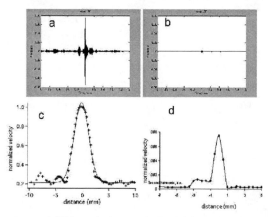

FIGURE 2. Nonlinear time reversal (TR) results in a cracked solid. In the experiment, a time –signal is emitted from a transducer and detected elsewhere (on or adjacent to the crack) using a laser vibrometer. The detected signal is then time reversed, and then re-emitted at the source location. The focused signal is again detected on or adjacent to the crack (a) The time-reversed and focused signal, bandpass-filtered around the second harmonic, detected above a crack; (b) the time reversed and focused signal located 30 mm away from the crack. The fact that a signal is observed over the crack indicates that the crack is strongly nonlinear as expected. By scanning along the crack, it can be imaged: (c) spatial distribution of the fundamental and (d), the second harmonic TR signal detected along a line that includes the crack, normed to the maximum amplitude. The crack was about 2mm in diameter.

the scanned area. The feasibility of this technique was evaluated in an experiment where the time reversal focusing was conducted along a single line scan in the glass sample with and without damage present (Figure 2). This experiment demonstrates the feasibility of the application of the method for crack imaging (see [15,16] for details). The general approach of nonlinear time reversal is being applied elsewhere including to location of landmines [17]. Other methods of nonlinear imaging are in development as well, in particular one using nonlinear modal analysis [18].

GRANULAR MEDIA AND EARTH PROCESSES

Granular media is another member of the large class exhibiting nonequilibrium dynamics. As such, granular media are an interesting material from the perspective of earth processes, particularly strong ground motion and earthquake physics, and because it is a much simpler system to understand than many others of the class. One can apply Hertz-Mindlin theory (or some variation thereof) in order to understand its elastic behavior (see, e.g., [2]).

In earthquake strong-ground-motion, broad-frequency band waves propagate from hypocentral depths of order 10 km to the earth's surface. Sediments at the surface, composed of granular media, can respond by ringing at their resonance modes. The shear modes are particularly dangerous because, if they couple into a building or other structures, damage or failure of the structure can take place.

Predicting the elastic linear and nonlinear behavior of near-surface sediments during an earthquake is a large field of study in itself. Some years ago, it was demonstrated that significant elastic nonlinear behavior may take place during large earthquakes manifesting itself by changes in surface-layer resonances. For instance, a ~75% decrease in fundamental-mode resonance frequency of the uppermost layer was observed at one site in the Los Angeles Basin during the 1994 Parkfield, California earthquake [19]. The magnitude of this change came as a significant surprise to the seismic community.

Currently, research in applying active, large-vibrator sources to *in situ* characterization of the near surface layers is being attempted. In one recent experiment at Garner Valley California, located near the San Andreas Fault southeast of Los Angeles, a significant nonlinear response of a near-surface layer was observed. A decrease in resonance frequency of order 25% over a strain interval of approximately $10^{-6} - 5 \times 10^{-4}$ was observed. Slow dynamics appeared to be present as well, and were clearly verified in a follow-on experiment conducted at a site just outside of Austin, Texas [20]. Note that the results indicate that full nonlinear, nonequilibrium dynamics takes place over frequencies from order 1 Hz to hundreds of kHz if we compare them to laboratory experiments.

Recently, it was speculated that a phenomenon known as *dynamic earthquake triggering* [21] could be due, at least in part, to the effects of nonlinear, nonequilibrium dynamics [22]. Normally, an earthquake exhibits precursors known as foreshocks, followed by a main shock (the magnitude of which is reported for an earthquake—the associated smaller earthquakes are not normally reported to the public), followed by aftershocks. Under certain, and apparently rare conditions [21], some of the aftershocks can take place at hundreds of kilometers from a main shock at the time or soon after the seismic wave from the main shock impinges on a distant fault. This is the phenomenon of *dynamic earthquake triggering*. It has been a puzzle as to why dynamic triggering takes place because wave strains at these distances tend to be order 10^{-7}-10^{-6} and is it difficult to understand how such small strains could be responsible for this phenomenon.

It has been proposed that, if the conditions are right, triggering may be due to nonlinear softening and weakening of the fault core known as *fault gouge*, which is granular in nature. Gouge is created by a fault as it progressively slips over the history of the fault, creating a thick granular layer between adjacent fault blocks. The conceptual model is that the dynamic wave temporarily reduces the core modulus. The modulus reduction is accompanied by a material strength reduction sufficient to induce fault slip, thereby triggering earthquakes [22]. Recent seismic wave observations from a suite of earthquakes support the proposed mechanism [23]. Much recent experimental work in granular media related to this work has been conducted at the University of Marne-la-Vallee (France), and numerous other studies in granular media been conducted by the group at the University of Le Mans (e.g., [24]).

ON MECHANISMS OF NONLINEAR BEHAVIOR

At present, some understanding has been achieved regarding the origins of the complex behavior of materials described above. As mentioned, the corresponding class of materials possesses soft inclusions in a harder matrix. More specifically, grain contacts, cracks and other defects can contribute into strong nonlinearity. Some theoretical models have been developed for such media (see, e.g., [2,3]). However, many aspects of such a behavior remain to be understood. One of them is the presence of hysteresis than can be described by semi-phenomenological models such as Granato-Lucke model known for metals, and the P-M Space model mentioned above. Still, physical mechanisms of hysteresis, especially in such materials as rock, are not well understood.

These issues are being progressively addressed by various groups. There is now evidence that there are clear regimes of elastic behavior (e.g., [1]). In the lowest amplitude regime, the materials behave linearly--there is no modulus dependence on strain amplitude. In the next, intermediate regime, the materials act as classical nonlinear oscillators that can be described by Landau theory (up to strains of roughly $1-3 \times 10^{-6}$ at ambient conditions), and above this, nonlinear nonequilibrium dynamics emerges. This last and most interesting regime occurring at larger strains remains to be carefully understood in terms of what is physically taking place, and to develop a verifiable physics-based theory that describes all aspects of it. The original P-M Space theory does not account for conditioning or slow dynamics (e.g., [6]). The variations of the P-M space theory that include conditioning and slow dynamics, although effective at modeling observed behaviors, are *ad hoc*, based on thermal fluctuations (e.g., [25]). Other models such as a recently proposed ratchet model [26] are as yet to be verified by experiment. Models indicating that thermal heating and diffusion are the source of nonequilibrium dynamics are questionable if one invokes three-dimensional thermal diffusion, which we believe is the case [27], as shown in [28].

Some have suggested that fluids are responsible for nonequilibrium dynamics. Van Den Abeele et al. [29] have shown that fluids act to modify the internal forces in porous media and thereby influence the nonlinear behavior; fluids appear not to be fundamentally responsible for the behavior. In any case, some materials in the class are dry with no means for fluid penetration (e.g., gray iron, alumina ceramic are two examples described in [7]).

A very important issue that has not been explored experimentally is whether shear sliding is the fundamental cause, at least in some cases, of nonlinear, nonequilibrium dynamics. Many controlled experiments have been conducted with longitudinal or bulk modes. There is evidence suggesting the nonlinear response to shear motions is larger than to the bulk mode but experiments aimed at isolating shear from other motions have not been methodically conducted to our knowledge. Such experiments would aid tremendously in helping verify such a hypothesis and developing a theory.

One suggested approach to addressing a generalized, physics-based model is to start with specific systems and look for similarities in the underlying physics. For instance, in granular media, Hertz-Mindlin theory can be applied as noted above; for cracks in metals, crack dynamics and dislocation-point defect interaction may be applied. Incorporating in slow dynamics adds significant complexity. Such an approach could presumably evolve to a consistent theory of nonclassical behavior (a generalized nonlinear, nonequilibrium theory) for all or most of the materials in the class.

An additional problem is to understand acoustic wave propagation in materials and resonators with hysteresis, if only phenomenologically. This problem has been addressed for traveling waves in cases of relatively simple, symmetric hysteretic loops (see, e.g., [2]). Even more difficult is the description of waves in resonators such as bars in which most of experiments have been performed. Recently, some models have appeared [30,31]. To compare these models with experiments and modify them if necessary, much of work is still needed.

ACKNOWLEDGMENTS

Institutional support (LDRD) and the Office of Basic Energy Science, Engineering and Geoscience funded this work.

REFERENCES

1. TenCate, J., Pasqualini, D. Habib, S., Heitmann, K., Higdon, D. and Johnson, P., *Phys. Rev. Lett.* **93**, 06551-06555 (2004).

2. Ostrovsky, L. and Johnson, P., *Rivista del Nuovo Cimenta* **24**, 1-46. (2001).

3. Guyer, R. A., and Johnson, P. A., *Physics Today* **5**, 30-35 (1999).

4. Johnson, P. A., Zinszner, B., and Rasolofosaon, P. N .J., *J. Geophys. Res.* **10**, 11553-11564 (1996).

5. TenCate, J. and Shankland T., *Geophys. Res. Lett.* **23**, 3019-3022 (1996).

6. Guyer, R. A., McCall, K. R., Boitnott, G. N., Hilbert, L. B., and Plona, T. J., *J. Geophys. Res.* **102**, 5281-5293 (1997).

7. Johnson, P. and Sutin, A., *J. Acoust. Soc Am.* **117**, 124-130 (2005).

8. Johnson, P., *Materials World, the J. Inst. Materials* **7**, 544-546 (1999).

9. Van Den Abeele, K. E.-A., Johnson, P. A. and Sutin, A. *Research on NonDestructive Evaluation* **12**, 17-30 (2000).

10. Van Den Abeele, K. E.-A., Carmeliet, J., TenCate, J. and Johnson P. A., *Research on NonDestructive Evaluation* **12**, 31-43 (2000).

11. Zheng, Y., Maev, R.Gr., and Solodov, I.Yu., *Canadian Journal of Physics* **77**, 927-967 (1999).

12. Johnson, P., "Nonequilibrium nonlinear-dynamics in solids: state of the art", *in The Universality of Nonclassical Nonlinearity, with Applications to NDE and Ultrasonics*, edited by P. P. Delsanto, and S. Hirshekorn, New York: Springer, *in press* (2005).

13. Kazakov, V. V., Sutin, A. and Johnson, P. A., *Appl. Phys. Letters* **81**, 646-648 (2002).

14. Fink, M., *Physics Today* **5**, 34-40 (1997).

15. Sutin, A. and Johnson, P. A., QNDE 2004, July 25- 30 at the Colorado School of Mines in Golden Colorado, *in press* (2005).

16. Ulrich, T. J., Johnson, P. A. and Sutin, A., *J. Acoust. Soc. Am.*, *in review* (2005).

17. Sutin, A. Johnson, P., TenCate, J. Sarvazyan, A., *in Detection and Remediation Technologies for Mines and Minelike Targets X*, edited by R. Harmon, J. Broach and J. Holloway, *Proc. of SPIE*, **5794**, 706-716 (2005).

18. Van Den Abeele, K., Pers. Comm. (2004).

19. Field, E. H., Johnson, P. A., Beresnev, I. and Zeng, Y., 1997, *Nature* **39**, 599-602 (1997).

20. Pearce, F., Bodin, P. Brackman, T., Lawrence, Z., Gomberg, J., Steidl, J., Menq, F., Guyer, R., Stokoe, K., and Johnson, P. A., *Eos Trans. American Geophysical Union* **85**, Fall Meeting Supplement, Abstract S42A-04 (2004).

21. Gomberg, J., Reasenberg, P. A., Bodin P. and Harris R.A., *Nature* **41**, 462-466 (2001).

22. Johnson, P.A., and Jia, X, *Nature, in press* (July 2005)

23. Gomberg, J. and Johnson, P. A., *Nature, in press* (July 2005).

24. Tournat, V, Gusev, V. and Castagnede, B., *Physics Letters A* **326**, 340-345 (2004).

25. Van Den Abeele, K., Carmeliet, J., Johnson, P. and Zinzsner, B. *J. Geophys. Res.* **107**, 10,1029-10,1040 (2002).

26. Vakhnenko, O. O., Vakhnenko, V. O., Shankland, T. J. and TenCate. J. A., *Physical Review E.* **70**, 015602R-1-015602-R-4 (2004)

27. Zaitsev, V., Gusev, V. and Castagnede, B., *Phys. Rev. Lett.* **84**, 159-162 (2002).

28. Pasqualini, D., *in preparation* (2005).

29. Gusev, V., J. Acoust. Soc. Am. **107**, 3047-3058 (2000); Gusev, V., Wave Motion **42**, 97-108 (2005).

30. Ostrovsky, L. A., *J. Acoust. Soc. Amer.* **116**, 3348-3353 (2004).

Fundamental and Nonclassical Nonlinearity in Crystals

M. A. Breazeale and I.V. Ostrovskii

Jamie Whitten National Center for Physical Acoustics and Physics Department
University of Mississippi, University, MS 38677

Abstract. Fundamental nonlinearity results from the fact that the interatomic potential function is not a parabola. It determines the third order elastic constants. The means of measuring fundamental nonlinearity in cubic crystals is discussed, as well as conclusions that can be drawn from the measurements. Since many cubic crystals had been measured, an attempt was made to expand the measurements to crystals of other symmetries by measuring $LiNbO_3$. Elasticity is not responsible for nonlinearity in $LiNbO_3$, but $LiNbO_3$ appears to produce a nonclassical nonlinearity. Structural inhomogeneity of ferroelectric $LiNbO_3$ may be the cause. Among the phenomena discussed are non-exponential echo patterns which cannot be explained by nonparallelism of the sample surfaces and acoustic memory; unusual temperature dependence of ultrasonic attenuation and its temperature hysteresis; amplitude characteristics of ultrasonic attenuation.

Keywords: Nonclassical, Nonlinear, Ultrasound, Attenuation, Hysteresis, Memory.
PACS: 43.25.Dc; 43.25.Ba ; 77.80.Dj

1. FUNDAMENTAL NONLINEARITY IN CRYSTALS.

The elastic potential energy in a crystal can be written as a power series in the strains η:

$$\Phi(\eta) = \frac{1}{2!}\sum_{ijkl} C_{ijkl}\eta_{ij}\eta_{kl} + \frac{1}{3!}\sum_{ijklmn} C_{ijklmn}\eta_{ij}\eta_{kl}\eta_{mn} + \ldots \quad (1)$$

This equation defines both second order elastic constants and third order elastic constants. The coefficients C_{ijkl} are the ordinary (second order) elastic constants which can be determined from sound velocity. The coefficients C_{ijklmn} are the third order elastic constants which are determined by the nonlinearity of the medium.

One can present the nonlinear wave equation in a similarly simple manner. Using the appropriate form of Lagrange's equations and specializing to a specific orientation of the coordinates with respect to the wave propagation direction, the nonlinear wave equation can be written

$$\rho_o \frac{\partial^2 u}{\partial a^2} = K_2 \frac{\partial^2 u}{\partial a^2} + (3K_2 + K_3)\frac{\partial u}{\partial a}\frac{\partial^2 u}{\partial a^2} \quad (2)$$

where the values of K_2 and K_3 for the principal directions are listed in Table I. This wave equation is written in this form to show the effect of elastic nonlinearity since the last term is the first nonlinear term. The ratio of the coefficients in this term to those in the linear term is called the nonlinearity parameter. It can be written

$$\beta = -\left(\frac{K_3}{K_2} + 3\right) \quad (3)$$

CP838, *Innovations in Nonlinear Acoustics: 17th International Symposium on Nonlinear Acoustics*,
edited by A. A. Atchley, V. W. Sparrow, and R. M. Keolian
© 2006 American Institute of Physics 0-7354-0330-9/06/$23.00

(The negative sign is written for convenience. In this way β is a positive quantity.) With values of the nonlinearity parameter plus measurements of sound velocity dependence on pressure it is possible to determine all six third order elastic constants. With only nonlinearity parameters, however, we are able to draw useful conclusions.

TABLE 1). Expressions for the linear combinations K_2 and K_3 in cubic crystals.

Substance	K_2	K_3
Cubic Crystal		
[100]	C_{11}	C_{111}
[110]	$1/2[C_{11}+C_{12}+2C_{44}]$	$1/4[C_{111}+3C_{112}+12C_{166}]$
[111]	$1/3[C_{11}+2C_{12}+4C_{44}]$	$1/9\,[C_{111}+6C_{112}+12C_{144}$ $+23C_{166}+2C_{123}+16C_{456}\,]$

Nonlinearity parameters of a number of cubic crystals have been measured. They have been found to be essentially independent of temperature, and their magnitudes do not vary greatly for the three principal directions in a cubic crystal. If one evaluates the average nonlinearity parameter in the [100] direction he gets the results shown in Table II, where results from measurement of unusual isotropic materials have been added. This table is analogous to Table I of Cantrell [1] who investigated crystalline structure dependence of acoustic nonlinearity parameters.

TABLE 2). Average nonlinearity parameters in cubic crystals.

Material	Bonding	β_{av}
Zincblende	Covalent	2.2
Flourite	Ionic	3.8
FCC	Metallic	5.6
FCC (Inert gas)	Van der Waals	6.4
BCC	Metallic	8.2
NaCl	Ionic	14.6
$YBa_2Cu_3O_{7-\delta}$ Ceramic	Isotropic	14.3
Fused Silica	Isotropic	-3.4

From these results one can conclude that fundamental (elastic) nonlinearity parameters of cubic crystals are typically in the range between 2 and 16. Nonlinearity parameters outside this range probably come from other effects. After measurement of cubic crystals we tried the trigonal crystal lithium niobate. Since sono-luminescence has been observed in this crystal [2], we assumed that it would exhibit nonlinear effects as well. The [100] axis is a pure mode direction; however, when we measured in this direction we found a nonlinearity parameter of 2.1. This means that in this direction the elastic nonlinearity is essentially negligible. Other measurements are necessary.

2. NONCLASSICAL NONLINEARITY IN LITHIUM NIOBATE.

2.1. Non-exponential Echo Patterns and Acoustoelectric Memory In Lithium Niobate.

The fundamental physical properties of lithium niobate ($LiNbO_3$) continue to be of interest [3-6] both because of industrial applications and of fundamental physics.

LiNbO$_3$ is a ferroelectric material that exhibits several unusual properties. Difficulty with measurement of ultrasonic attenuation below 100MHz has been experienced [7], but recent observations about acoustical memory may lead to a correct interpretation of acoustical losses in the medium as well as a better understanding of the acoustical nonlinear response of the lithium niobate medium. Observations of acoustical energy redistribution that occurs in LiNbO$_3$ single crystals, an effect labeled as acoustical memory were reported [8].

The non-exponential echo trains were observed in three samples of Lithium Niobate and reported. The frequency of the transducers used in these experiments was in the range from 15 to 30 MHz. One transducer was used as both the emitter and receiver of the acoustical signal. Three samples of single crystal LiNbO$_3$ with different wedge angles were prepared for the measurements. The results below were performed with acoustical longitudinal waves propagating along the piezoelectrically active direction, the [001] axis. The samples were cylindrically shaped with lengths 2.4 and 3.9 cm, and diameters 4.5 cm (samples LNO-2P and LNO-2U) and 2.5 cm (LNO-3P). Initial experiments were performed using a 39° rotated Y-cut 30 MHz LiNbO$_3$ transducer bonded to the top surface of the sample. A capacitive receiver below the sample detected the resulting signal. Sinusoidal tone bursts of approximately 5 μs in duration were generated by a synthesizer/function generator. Experimental results were verified by eliminating the capacitive receiver and using the transducer as both the emitter and receiver. As a check, a second transducer was bonded to the opposite face of the sample and a standard 2 transducer setup was used. The experiments were then repeated using 30 MHz quartz transducers. Typical non-exponential patterns are shown in Fig. 1. Two

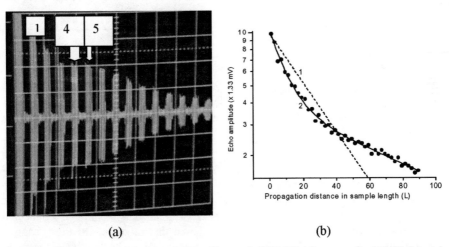

(a)   (b)

FIGURE 1. Non-exponential echo train from the sample LNO-2P at frequency f = 16 MHz (a), and LNO-2U (L=3.9 cm) at f=29.5MHz at room temperature (b). Step in propagation distance in (b) is 2L, 1- one-exponential and 2- two-exponential decay simulation.

echoes, number 4 and 5 are of the same amplitude in the oscillogram (a), and strong non-exponential decay is demonstrated in panel (b), which must be simulated by adding the second order exponential decay function (line 2). Other behavior has been

13

observed at other frequencies. If non-exponential behavior is not strong, one-exponent decay simulation (4) can be used:

$$A = A_o e^{-\alpha L} \qquad (4)$$

where A is the measured amplitude, L is the propagation distance, and α is the absorption coefficient. In many cases, the non-exponential echo trains are too far from the exponential behavior, like in Fig. 1-b. Than first order exponential simulation gives a significant error. It was found that fitting these data with a double exponential curve of the form

$$A = A_0 \left(e^{-\alpha_1 L} + e^{-\alpha_2 L} \right) \qquad (5)$$

greatly reduced the deviation in the fit. The data in Fig.1 were taken in a one-transducer setup. In our experiments at frequencies of the megahertz range, we obtained unusual echo trains shown in Fig. 2. We name the second group of echoes "Acoustical Memory". Memory depends on frequency and temperature [9]. We noted

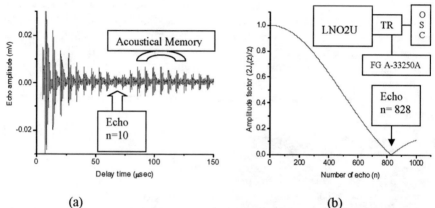

(a)   (b)

FIGURE 2.. Oscillogram of the Acoustic Memory type echo train taken from the sample LNO-2U (wedge angle $\theta = 0.45'$) at f = 4.6 MHz (a), and computer simulation by the function ($2\mathbf{J_1}$ (z) / z) of the experimental data of fig. 2-a. Insert in (b) shows the experimental setup: TR- transducer, OSC-oscilloscope LeCroy 9400, FG- Function Generator Agilent-33250A . The **n** is the number of echoes.

that the time position of the memory signal does not depend on input amplitude and that the amplitude of the memory signal varies linearly with the input amplitude. The maximum of the memory envelope may shift to a different position (delay time) at a different frequency.

An alternative explanation of occurrence of acoustical memory is connected with nonparallelism of a sample. It has been shown that in some situations, sample non-parallelism can affect the observed echo train and cause deviations from an expected exponential decay [10]. For a sample with a given wedge angle θ, the decay train is modulated by the factor

$$\frac{2J_1(z)}{z} \quad \text{with} \quad z = \frac{d\omega\theta n}{V} , \qquad (6)$$

14

where $J_1(z)$ is the Bessel function of the first kind, d is the diameter of the transducer, ω is the angular frequency of the acoustic wave, θ is the wedge angle, V is the velocity of the acoustic wave, and n is the number of the echo (essentially the distance the acoustic pulse has traveled). Thus, for comparison of the experimental data with the theoretical predictions of multiple diffraction due to non-parallelism [10] , the wedge angles in our samples were carefully measured with a Nikon 6B auto collimator. This angle is found to be 0.5 minute (1.45 x 10^{-4} rad) for sample LNO-2P and 0.45 minutes (1.3 x 10^{-4} rad) for sample LNO-2U. The third sample LNO-3P has the wedge angle of 3 minutes (8.7 x 10^{-4} rad).

Experimental observation and relevant computer simulation are shown in Fig. 2-b. An enormous difference between experimentally observed first minimum amplitude echo of n = 10 and wedge theory prediction n = 828 leads to the conclusion that nonparallelism of the sample cannot explain the effect of Acoustical Memory. We assume this effect is a nonclassical nonlinear phenomenon, which arises from the real crystalline structure of a ferroelectric with its various internal substructures.

The explanation of acoustical memory in terms of ferroelectric crystal internal micro substructures is supported by the frequency dependencies of memory amplitude and peculiarities in dielectric properties [8]. Here we can add a certain correlation with piezoelectric properties. In the references [13-16] it was reported that ferroelectric substructures of the type of periodically poled ferroelectric domain arrays posses their own resonance frequencies, which were much higher than fundamental resonance frequency of a crystal-host, and they can be measured from the rf-admittance of the crystal. Actually a periodic array of inversely poled ferroelectric domains has a resonance frequency that matches a domain length. We made experiments with the sample LNO-2U and find that it has a high frequency region with resonance-type rf-admittance (4.4 to 5.4 MHz) and a fundamental frequency near 150 KHz – Fig. 3. In this frequency region the acoustical memory was measured as well. These results are shown in Fig. 2-a.

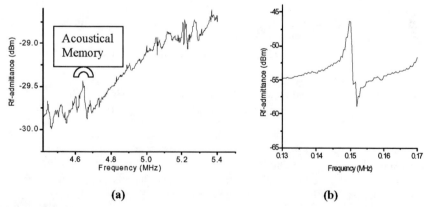

(a)   (b)

FIGURE 3. Rf-admittance in MHz frequency range (a) and in fundamental frequency (b) from the sample LNO-2U at room temperature.

The frequency range with additional resonances in admittance that matches the measurements of the acoustical memory is shown in Fig. 3-a by the semi–circle that identifies acoustic memory group of bursts in Fig. 2-a. The resonances in Fig. 3-a, where acoustical memory is detected, are not harmonics of the fundamental. The results of Fig.3 can be explained by the internal substructures mechanism of acoustical memory.

Acoustical memory was found to be sensitive to temperature. Increasing temperature can lead to significant suppression of the acoustical memory signal. For this resonance, it was necessary to consider the effect that thermal expansion of the sample might have on the sample nonparallelism. For thermal expansion in $LiNbO_3$ along the x and y directions above room temperature, the thermal expansion coefficient $\gamma_{x,y} = 16.7x10^{-6}$ K^{-1}. The change in the wedge angle θ is caused by non-equal changes in sample length. After heating the sample by some ΔT, the new wedge angle is be given by

$$\theta'(T_0 + \Delta T) = \frac{\Delta L_0 + \gamma_z \Delta T \Delta L_0}{D_0 + \gamma_{x,y} \Delta T D_0} = \theta_0 \left(\frac{1 + \gamma_z \Delta T}{1 + \gamma_{x,y} \Delta T} \right). \tag{7}$$

Where ΔL_0 is the biggest difference between sample lengths measured on opposite sides of the diameter D_0. We note that the temperature expansion coefficients are themselves a function of temperature over the range of temperatures used. Maximum values were chosen. It was found that the change in the wedge angle over the temperature range investigated in these experiments was negligible. Under heating by $\Delta T = 60°C$ one gets:

$$\theta' (T_0 + 60°C) = 0.9994 \theta_0(T_0). \tag{8}$$

This result means that for our experimental conditions the heating changes the wedge angle by the factor of 0.0006, which means 1.449×10^{-4} rad instead of 1.45×10^{-4} rad for LNO-2P sample, and 8.695×10^{-4} rad instead of 8.7×10^{-4} rad for LNO-3P sample. Computer simulation by the wedge angle theory [10] turned out to be insensitive to such a small variation of non-parallelism for our experimental conditions. Thus, higher temperatures practically can not affect a wedge angle, but can influence acoustical memory itself. This is an additional argument that memory is a nonclassical nonlinear phenomenon, which is connected to the internal structures of the ferroelectric samples.

2.2. Nonclassical Nonlinear Properties and Hysteresis of Ultrasonic Attenuation in Lithium Niobate.

For measurements of the temperature dependence of the absorption coefficient, the sample was placed in an oven and the temperature was adjusted in small increments. Once a stable temperature was reached in the oven, the sample was left for 20 min to achieve equilibrium. Measurements of the absorption coefficients were made at different temperatures under equilibrium conditions. Figure 4-a shows typical results with decreasing attenuation when temperature increases. A strong hysteresis in both components α_1 and α_2 of a two-exponential simulation is measured. Since the temperature was changed in these experiments, we checked for a possible variation in the transducer's fundamental frequency due to thermal expansion effects over the entire temperature range used (25 - 85°C). The amplitudes of the echo train were a maximum at 17.4 ± 0.05 MHz for the whole temperature range. Our observations are in agreement with the theoretical temperature dependence of a LiNbO₃ transducer.

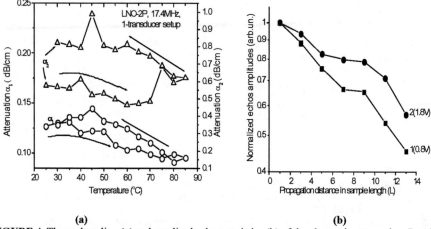

(a)

(b)

FIGURE 4. Thermal cycling (a) and amplitude characteristics (b) of the ultrasonic attenuation. Panel (a) shows data taken from the sample LNO-2P at 17.4 MHz, and panel (b) – from LNO-3P at f = 33.7 MHz; amplitude of first echo is 0.8 Volts for plot 1 and 1.8 Volts for plot 2.

The coefficient of thermal expansion of LiNbO₃ along the z direction is given by γ_z = 2.0 x 10^{-6} K^{-1} (for 25°C < T < 100°C). Thus, the shift in the fundamental frequency is given by the expression

$$f(T)= f_o(1 + \gamma \Delta T) = 17.4 \ (1 + 2.0 \ x \ 10^{-6} \ \Delta T) \ MHz \qquad (9)$$

For the heating by ΔT = 60°C, the fundamental frequency shift is very small:

$$\Delta f = f(T)- f_o =2.09 \ KHz \quad , \qquad (10)$$

which is within the limits of experimental uncertainty. Another peculiarity of ultrasound attenuation is shown in Fig. 4-b. Attenuation is higher for smaller acoustic amplitude. This is clear from the larger decrease in the first 4 echoes shown in Fig.4-b.

2.3. Discussion and Conclusions.

Measurement of the nonlinearity parameters resulting from interatomic forces in cubic crystals gives a range of nonlinearity parameters between 2 and 16. Nonlinearity parameters outside this range probably result from other effects.

Measurement of attenuation in LiNbO₃ produced some surprises. Careful measurement of the temperature dependence of the attenuation revealed that as the temperature increases the attenuation may decrease. This is the opposite of attenuation behavior of materials previously measured. There is a temperature hysteresis of ultrasonic attenuation. Finally, we observed smaller attenuation at larger amplitudes. This means that a nonclassical nonlinear effect must be responsible.

In order to explain the presence of this unusual behavior, we introduced the concept of Acoustical Memory, which can explain some peculiarities of the observed nonclassical nonlinear phenomena.

We associated the nonclassical nonlinearity in lithium niobate with the presence of the substructures inside the samples of ferroelectric LiNbO$_3$. These may be a) ferroelectric micro domains with their inter-domain boundaries, b) small angle boundaries, c) dislocations with their networks especially near the small angle boundaries and inter-domain walls, d) groups of electrically charged and neutral point defects, which in turn are attracted by linear and extended imperfections like internal boundaries, e) inclusions of other than main chemical phases including non-polar niobium-rich phase of LiNb$_3$O$_8$ [17], etc. We believe, therefore, that the observed nonlinearity in lithium niobate is not fundamental. It does not depend on the interatomic potential function, but depends more on the presence of crystal substructures and crystal lattice imperfection. It must be nonclassical, and cannot be calculated from the interatomic potential function.

REFERENCES

1. John H. Cantrell, *Proceedings of the IEEE Ultrasonics Symposiu*m, Vol. 1, 425-428 (1987).
2. I.V. Ostrovskii and P. Das, *Appl. Phys. Lett.*, **70** 167-169 (1979)
3. E. V. Podivilov, B. I. Sturman, G. F. Calvo, F. Agullo-Lopez, M. Carrascosa and V. Pruneri, *Physical Review* B **62**, 13182-13187 (2000).
4. N. G. R. Broderick, G. W. Ross, H. L. Offerhaus, D. J. Richardson and D. C. Hanna, *Physical Review Letters* **84**, 4345-4348 (2000).
5. U. T. Schwarz and Max Maier, *Physical Review* B **55**, 11041-11044 (1997).
6. B.B. Ped'ko, I.L. Kislova, T.R. Volk, and D.V. Isakov, *Izvestiya Akademii Nauk Seriya Fizicheskaya (Bulletin of the Russian Academy of Sciences / Physics)*, **64**, 1145-1153 (2000).
7. M. F. Lewis and C. L. West, *Properties of Lithium Niobate,* Emis Datareview Series, 82-83, (1989).
8. M. McPherson, I. Ostrovskii, and M.A. Breazeale, *Physical Review Letters* **89**, 155506-1-3 (2002).
9. M. A. Breazeale, I.V. Ostrovskii, M. S. McPherson. Jour. Appl. Phys. **96**, 2990-2994 (2004).
10. Rohn Truel and William Oates, J. Acoust. Soc. Am. 35, 1382 (1963). Rohn Truel, Charles Elbaum, and Bruce B. Chick, Ultrasonic Methods in solid State Physics (Academic, New York, 1969).
11. V. Ermolov, Jyri Stor-Pellinen and Mauri Luukkala., *J. Phys. D: Appl. Phys.* **30** 1734-1740 (1997).
12. Ya. I. Lepikh. *Semiconductor Physics, Quant. Electronics & Optoelectronics*, **3**, 308-310 (2000).
13. V.V. Antipov et al. *Sov. Phys. Crystallogr.* **30**, 428 (1985).
14. Yong-Feng Chen, Shi-Ning Zhu, Yong-Yuan Zhu et al. *Appl. Phys. Lett.* **70**, 592 (1997).
15. G. V.Golenishchev-Kutuzov et al. *Physics-Uspehi*, **43**, 647-662 (2000).
16. I. V. Ostrovskii, A. B. Nadtochiy. *Appl. Phys. Lett.* **86**, 222902-1-3 (2005).
17. V. Ya. Shur, R.K. Route, M.M. Fejer, R.L. Byer et al. *Appl. Phys. Lett.* 80 (1037-1039).

Localizing Nonclassical Nonlinearity in Geological Materials with Neutron Scattering Experiments

Timothy W. Darling[a,b], James A. TenCate[a], Sven Vogel[a], Thomas Proffen[a], Katie Page[a], Christine Herrera[a], Aaron M. Covington[b], Erik Emmons[b]

[a]Los Alamos National Laboratory, Los Alamos NM USA 87545
[b]Physics Dept. University of Nevada, Reno NV 89557

Abstract. Deviations from linear elasticity in consolidated rocks have been measured by acoustic methods and by quasistatic application of stress since the early 1900s. We have used elastic neutron scattering to probe a large volume (on a grain-size scale) of a variety of consolidated rocks under dynamic and quasi-static conditions. Differences between macroscopic *external* strain and average *atomic-scale* strain were determined. Nonlinear behavior is intimately related to strain magnitudes so the *microscopic* distribution of strain is important. Our results indicate that a very small volume ($< 5\%$) of the rock is responsible for almost all of the nonlinear response of the bulk rock. We discuss our results and possible future experiments.

Keywords: neutron scattering, nonlinear, nonclassical, hysteresis, rocks.
PACS: 43.20.Jr, 43.25.Dc, 43.25.Ts

INTRODUCTION

Non-Hooke's Law behavior, hysteresis, and end-point memory seen in quasi-static cyclic stress-strain measurements [1]—and similar behavior in the elastic constants measured by resonance peak shifts—demonstrate the inherent nonlinearity of many rocks, even at strains as low as 10^{-7} [2]. In consolidated rocks the internal distribution of strain occurs at grain size scales where the details of crystal anisotropy and the nature of interfaces and porosity come into play. Using acoustic measurements or quasi-static stress-strain techniques provides only an *average* strain value over the volume of a relatively large sample. Unfortunately, most microscopic measurement techniques do not have sufficient penetrating capability to study the interior volume of a large sample; neutrons, however, penetrate deeply into materials. Elastic scattering—both coherent (Bragg) and diffuse—of thermal neutrons can provide the average atomic lattice spacing in a large gauge volume. It is also fortuitous that the grains which make up a rock are randomly oriented and so the rock satisfies the conditions needed for powder diffraction techniques. Bragg scattering measurements done on beamlines of the Lujan Center (http://www.lansce.lanl.gov/lujan/index.html) at the Los Alamos Neutron Science CEnter (LANSCE) can provide lattice spacings with accuracies of a few parts in 10^5; differences in those spacings gives an atomic

CP838, *Innovations in Nonlinear Acoustics: 17th International Symposium on Nonlinear Acoustics*,
edited by A. A. Atchley, V. W. Sparrow, and R. M. Keolian
© 2006 American Institute of Physics 0-7354-0330-9/06/$23.00

level strain. This accuracy is insufficient for acoustic strains (usually 10^{-9} to 10^{-6}), but will resolve stress and thermal induced strains. We have done experiments on three beamlines at the Lujan Center. Two involve the measurement of average atomic strains in the crystalline (grain volume) material for comparison with macroscopic properties, and one which uses all the elastic scattering data in a model including non-crystalline scattering sources.

EXPERIMENTS AT SMARTS, HIPPO AND NPDF

SMARTS beamline – Quasi-Static Stress-Strain Measurements

Samples for measurements in SMARTS were cylinders of 13.4 mm diameter and 26.2 mm long, mounted between the conical anvils of a horizontal Instron compression test machine. The neutron beam is collimated and hits the sample at an

angle of 45°, Fig. 1. Lattice planes with normals parallel to the applied stress diffract neutrons to detector 2, giving the lattice strain for the independent planes collinear with the stress.

A strain gauge measures the overall length change. Figure 2 shows results from two of our samples. An Arkansas novaculite (left side)—a dense quartzite—shows that the behavior of the independent **a** and **c** planes (data points, dotted lines, respectively) is rather similar to the macroscopic data (continuous line). It is linear—the initial curvature is an effect due to asperities—and of similar slope, with the anisotropy of the lattice separating

FIGURE 1. SMARTS geometry and sample (sandstone) mounted between conical anvils.

the independent planes. On the other hand, the Berea sandstone (right side) shows that while the overall macroscopic sample behavior is *nonlinear and hysteretic* with a large

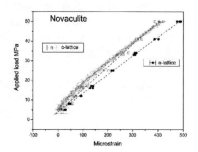
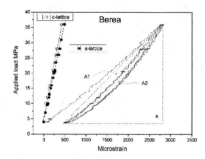

FIGURE 2. Neutron and strain gauge stress-strain curves for a novaculite and Berea sandstone.

20

strain, the bulk of the crystal volume is *linear* and has a small strain. This is typical of the response of most sandstones, even at quite low strains.

TABLE 1. SMARTS samples and results, Summary and Conclusions

Sample	Mineral	Porosity %	Atomic/macro strain %	a-slope/c-slope %
Novaculite	100% Quartz	0.2±0.2	106±10	85±4
Fontainebleau	99% Quartz	24 ± 0.5	21±2	98±4
Berea sandstone	85% Quartz	20±2	20±2	89±4
Meule sandstone	70% Quartz	21±2	18±2	90±4
Carrara marble	100% Calcite	0.4±0.2	66±6	160±4
Lavoux limestone	100% Calcite	22±0.5	35±3	162±4

In these experiments the applied stress must be constant through any plane in the sample for equilibrium. The measured strains in a homogeneous medium should also be constant; this is borne out by the narrow line widths of the diffraction peaks. It is clear from Fig. 2 that for the novaculite (1) the strain in the independent lattice spacings (**a** and **c**) is very similar to the total strain from the strain gauge measurement; (2) that the response is linear and (3) the **a** and **c** lattice planes have different slopes, appropriate for the anisotropy of quartz crystals which have an **a:c** slope ratio of 82%. Novaculite also has very small grains and almost no porosity, so the entire surface of each grain is in contact with a neighbor. *In contrast,* the sandstone is macro-mechanically very nonlinear and experiences a non-recoverable strain before the cyclic behavior starts. On the other hand, the lattice strains show only a fraction of the total macroscopic strain—about 20% seems common in sandstones—and they *are entirely linear*. The line widths are still too narrow to show a strain variation, and the slopes are now closer, being equal for the case of the Fontainebleau sandstone. Where is the remaining 80% of the strain? How can the anisotropy of the quartz be vanishing? The neutron data represents scattering from the crystalline volume, but volumes smaller than a few percent will have insufficient statistics to show up in the spectrum, and amorphous (incoherent scattering) volumes will not produce peaks. The roughly 20% porosity in the sandstones means that individual grains only make small area contacts with neighboring grains. This necessitates a stress focusing—*the tiny (invisible to neutrons) volumes near the contacts are responsible for both the majority of the macroscopic strain and for the nonlinear effects.* Further discussion of our work and additional results may be found in Ref. [3].

HIPPO beamline – Dynamic Experiments with Temperature Changes

HIPPO is an excellent instrument for measurements which require watching lattice structure evolve in time because it is very near the neutron source and there are many detectors. Counting/statistics on this beamline make watching *dynamic* changes (on order of a few 10s of minutes) relatively easy. HIPPO was chosen to watch lattice changes while simultaneously measuring the macroscopic modulus during rapid temperature changes. To this end we built a special aluminum sample holder and container mounted to a Peltier stage (for heating and cooling) to hold long, finger-sized cylindrical-rod samples. See Fig. 3. The sample holder was constructed in such a way as to be sure the entire sample experienced the same temperature changes and still

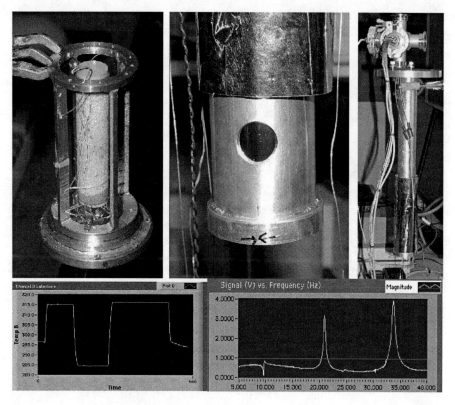

FIGURE 3. HIPPO sample holder (top photos), sample temperature vs time (lower left), and resonance response spectrum (lower right) to monitor sample modulus during temperature changes.

permitted neutrons to get in and scatter out without interacting with the container (via round, thin mylar "windows" like the one shown in the figure). Small piezoelectric transducers were mounted at both ends of the sample which was held suspended in place with thin dental floss. Repeated acoustic resonance sweeps on the sample were made to measure elastic modulus changes with temperature.

A typical temperature protocol (with time) is shown in the lower left of Fig. 3. At each measurement time, a single macroscopic modulus sweep (i.e., a resonance spectrum of the sample like shown in the lower right hand figure) and crystalline lattice information—obtained with neutron scattering—were obtained.

To test the system, we measured resonance frequency and lattice spacing of a steel sample during a rapid temperature change. See Fig. 4. No surprises were seen; the lattice spacing tracked the temperature, the resonance frequencies (and thus the modulus) tracked the temperature too. The same experiment (with more temperature cycles) was then performed with a Fontainebleau sandstone. In Fig 5. we show the (scaled to fit) variation of (left) the resonance frequency and (right) the unit cell volume—derived from the neutron data—compared to the temperature changes.

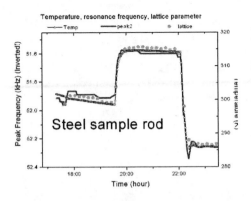

FIGURE 4. A steel sample responds essentially instantly in every way to a temperature change.

In the sandstone the resonance does *not* match the temperature, Fig. 5 left, (the inversion is OK) and there are significant differences between cycles indicating an irreversible effect, like the first leg of the stress cycles in Fig. 1. A gradual change at constant temperature is seen within each cycle, with an unusual downward "spike" seen at *both* upward and downward temperature changes. In contrast, the neutron data (Fig. 5 right) tell us that the overwhelming bulk of the crystal lattice responds *immediately* to a temperature change, at least on the scale of a minute. The change in the lattice corresponds well with the known thermal expansion of crystalline quartz.

In this high frequency (~25 kHz) measurement, once again we find that a volume which is not affecting the coherent elastic scattering of neutrons is determining the bulk response of the material. These rapid thermal strains are about 300 micro-strains in the crystal grains and are comparable to the isothermal quasi-static strains we measured on SMARTS. In this case the heating and cooling cycles may be gradually driving water from the rock, giving the long term variability. The "unmeasured" volume giving these nonlinear responses must again be either very small (<2% of volume) or incoherently scattering the neutrons (disordered) or both.

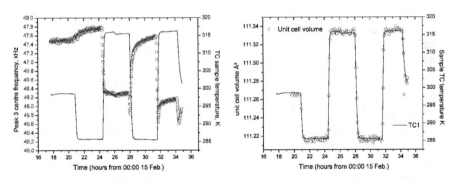

FIGURE 5. (Left) The macroscopic sandstone elasticity (symbols) does not match temperature change (solid line), while (Right) the bulk of the crystalline material (symbols) follows temperature accurately.

NPDF beamline – Searching for Disorder in Crystalline Rocks

Incoherent scattering shows up as a broad background in a scattering spectrum. The analyses used for lattice spacing and strain determination ignore this contribution and concentrate instead on the Bragg peaks. Pair Distribution Function (PDF) analysis uses this data to determine (essentially) the number of *pairs of atoms* with a given separation. A crystal has long-range order so the separations of atoms across tens of angstroms will be well determined (and repeated) giving sharp peaks in the PDF. A liquid or glass will show that only the distances to the nearest neighbors are repeated, so no peaks will be seen at large separations. In Fig. 6. we show PDF plots for six nearly pure silica samples. With the exception of the crushed crystal quartz, all are consolidated materials cored into cylindrical samples.

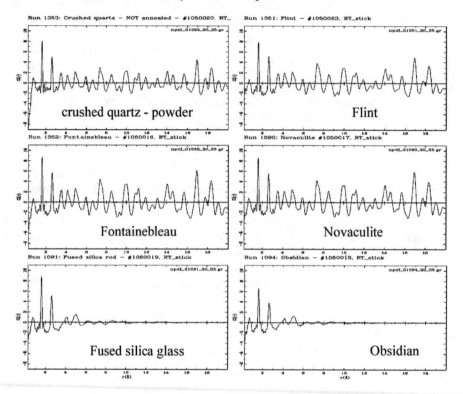

FIGURE 6. PDF plots for a variety of (almost) pure silica materials.

Despite the different forms, the upper four plots indicate a strongly crystalline nature, shown by the long range (up to 20Å) peaks. The commercial silica glass and the natural volcanic glass show no peaks beyond 10 Å. The two sharp peaks common to all the plots correspond to the Si-O and O-O bond lengths of the rigid SiO_4 tetrahedral units which exist even in the "disordered" glassy silicates. Given a good model of the crystal lattice and scattering lengths, we can calculate the expected PDF spectrum for a crystal—it is essentially a complex counting problem—and compare it to the data. When we make this comparison for Fontainebleau sandstone, Fig. 7, left plot, we note that if all the long range peaks are fit well, then there is an unexpected discrepancy in the two short-range amplitudes. It seems the data has too many short-range bonds to correspond with the expected number of long-range bonds. If we add to the model a contribution from our empirical glass spectrum, we find that we can fit the lower peaks without affecting the longer range peaks. We suggest this is evidence for an amorphous component within the sandstones, with a volume fraction in the range 5-10%. Further discussion of our PDF research can be found in Ref. [4].

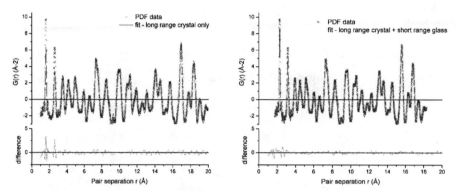

FIGURE 7. Fontainebleau sandstone PDF data with a crystalline model fit (left) and a crystal-plus-glass model fit (right).

CONCLUSIONS AND FUTURE DIRECTIONS

Neutron scattering experiments have been done at three beamlines at the Lujan Center at LANSCE using a time-of-flight spallation neutron source to pinpoint the sources of mechanical and acoustic nonlinearity in porous rocks. Comparing results from various silica rocks enabled us to demonstrate that all of the nonlinear behavior observed is associated with a tiny fraction of the rock volume, which we think is the volume near the limited area contacts in porous granular rocks, where stress focusing occurs. We also have evidence that amorphous material exists in many of the sandstones: high stresses are known to amorphize crystalline quartz [5] so this material may be localized near the contacts and contribute to the nonlinear properties. The 3-D network of contacts is apparent in Fig. 8 (left) but the thin section (right) does

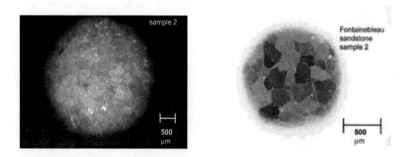

FIGURE 8. Fontainebleau, in 3-D (left) and a thin section under crossed polarizers (right).

not convey the point-like nature of the contacts well. We note that experienced petrographers generally do *not* see evidence for significant quantities (e.g., 5-10%) of amorphous material in these thin sections. The test of our hypothesis will involve microscopic examination of the contacts. We have made preliminary measurements of Raman scattering, Fig. 9, from crystalline and glassy silicates and sandstone. We will soon have available a confocal Raman microscope with resolution sufficient to examine grain and contact regions separately to provide a definitive answer on the contributions of glassy SiO_2 to nonlinear behavior in sandstones.

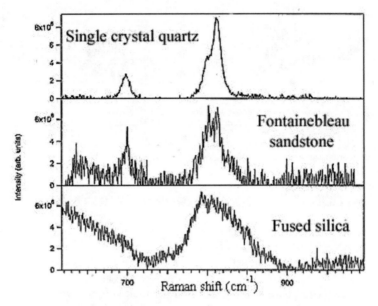

FIGURE 9. Raman scattering results for three different forms of pure SiO_2

ACKNOWLEDGMENTS

This work was primarily supported by Laboratory Directed Research and Development program funds at the Los Alamos National Laboratory. JT was partly funded by the DOE's Office of Science, Basic Energy Sciences. KP and CH were funded on the NASA-USRP summer student program. The Raman measurements at UNR were supported by DOE/NNSA under UNR grant DE-FC52-01NV14050.

REFERENCES

1. One of the earliest measurements is F. D. Adams and E. G. Coker (1906), An investigation into the elastic constants of rocks, more especially with reference to cubic compressibility, Publ. 46, Carnegie Inst. of Washington, Washington, D. C.
2. J. A. TenCate et al., *Phys. Rev. Lett.*, 93, 065501 (2004).
3. T. W. Darling et al., *Geophys Res. Lett.*, 31, L16604, (2004).
4. K. Page et al., *Geophys. Res. Lett.*, 31, L24606, (2004).
5. R. J. Hemley, A. P. Jephcoat, H. K. Mao, L. C. Ming & M. H. Manghnani, *Nature* 334, 52 - 54 (07 July 1988); doi:10.1038/334052a0

Broken Symmetry in the Elastic Response to Temperature of Consolidated Granular Media

TJ Ulrich

University of Nevada, Reno, Physics Department MS 220, Reno, NV, 89557

Abstract. When subjected to externally applied forces consolidated granular media (CGM), take a Berea sandstone as example, are elastically soft, unusually nonlinear, and have hysteresis with end point memory. In response to a variety of transient external disturbances CGM exhibit slow dynamics, e.g. $log(t)$ recovery of the strain following a step change in applied pressure. These elastic properties have led to a bricks (sand grains) and mortar (bond system) picture to describe the physics of the system. Because the grains are thermally anisotropic, temperature drives the bond system altogether differently than applied stress. Consequently temperature provides the means to probe new features in the elastic response of CGM. I describe an experiment/analysis in which the temperature, used to probe the elastic state of a CGM, reveals unusual behavior. The elastic state of CGM at fixed applied stress and temperature, is a function of the applied stress protocol and the temperature protocol. Working at constant stress I find that all aspects of the elastic response to temperature exhibit behavior which presents a broad range of time scales, i.e. slow dynamics, and the response to a transient temperature disturbance is asymmetric in the sign of $\Delta T(t)$.

Keywords: Resonant Ultrasound Spectroscopy, elasticity, rocks, thermal expansion, jamming, non-linear elasticity, slow dynamics
PACS: 62.20.-x, 61.43.Gt, 62.20.Dc, 62.40.+i, 65.40.De, 43.25.Dc, 81.40.Jj, 91.60.-x

INTRODUCTION

The goal of this research was to study the elasticity of weakly consolidated granular materials using temperature, rather than pressure, to induce strain in the material. Three timescales have been identified and studied in this research. These timescales range from a short timescale of minutes to an intermediate one of hours and finally to the long timescale on the order of weeks (or longer). The experiments investigating these timescales were conducted during a 7 month time frame where the sample was isolated in a 1 atm dry He atmosphere and carfully temperature controlled. In this 7 month time frame, the sample was held at several baseline temperatures, at which the experiments were performed. The experiments presented here are described in the following section

EXPERIMENT

The sample, a 1 X 1 X 4 cm block of Berea sandstone, was subjected to several temperature protocols at various baseline temperatures, designated T_0. When the baseline temperature was changed, the sample was always given 7 days to equilibrate at that temperature before further temperature changes were made. The sample sat at T_0 for 48 hours between each temperature protocol (see fig. 1), as well. The elastic state of the sample was monitored using a Resonant Ultrasound Spectroscopy (RUS) system

CP838, *Innovations in Nonlinear Acoustics: 17th International Symposium on Nonlinear Acoustics*,
edited by A. A. Atchley, V. W. Sparrow, and R. M. Keolian
© 2006 American Institute of Physics 0-7354-0330-9/06/$23.00

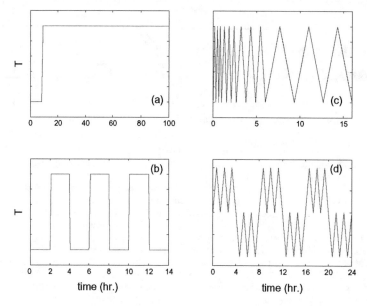

FIGURE 1. The four temperature protocols used, (a) Step, (b) Plateau, (c) Chirp, (d) AC. The times labeled on the axes are typical values for the given protocols. The temperature range for each experiment varied from $\Delta T = 1$ to 20 K. Exact values of time and temperature can be found in the results for the experiments presented here.

from DRS Intl. to measure a resonance frequency of the sample. The integrity of the experimental apparatus and procedure was verified on Ti alloy standards[1]. Further experimental details, results and speculations can be found in [2, 3].

The temperature protocols ($T(t)$) used in this study are shown in fig. 1. The protocols were developed to explore different timescales of the thermomechanical equilibrium process due to the applied thermal strain ($\varepsilon(T,t)$) at a constant stress ($\sigma_0 = 1$ atm). The "step" protocol (fig. 1 (a)) involved simply changing from one baseline temperature T_0 to another and measuring the change in modulus (i.e., the resonance frequency) as a function of time. The "plateau" protocol (fig. 1 (b)) explores the reproducibility of the "step" protocol as well as the symmetry, or lack thereof, between increasing or decreasing temperature steps. Both the "step" and "plateau" protocols investigate a slow time evolution of the elastic modulus. To explore the realm of the faster elastic changes in response to $T(t)$, the "chirp" and "AC" protocols were designed. The "chirp" protocol (fig. 1 (c)) makes use of a variable rate of change of temperature $\dot{T}$, while the "AC" protocol maintains a constant $\dot{T}$ and implements a $T(t)$ containing cycles that are periodic on multiple timescales. Each of the protocols also had an analagous "reciprocal" data set where $T(t) \rightarrow -T(t)$ to investigate the possibility that a difference may arise if the protocol involved a ΔT of opposite sign.

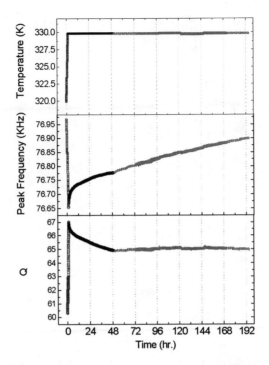

FIGURE 2. Results from a "step" protocol (see fig. 1 (a)). (a) Measured temperature profile, (b) resonance frequency response and (c) inverse attenuation Q. Note that despite a well controlled constant temperature, the resonance frequency is continually evolving, indicating that an equilibrium condition has not yet been reached. The evolution of the frequency can be shown to be logarthimic in time as seen in fig. 3 (c).

RESULTS

"Step" Protocol

The results of the "step" protocol experiments, shown in fig. 2, provide the first indication of extremely slow changes in the elastic behavior due to an induced thermal strain. The temperature profile in this experiment illustrates the high precision to which the temperature was controlled in this and all other experiments. It is quite obvious that atleast two different reaction timescales exist in the material when looking at the resonance frequency measurement (fig. 2 (b)), i.e., the fast elastic response during the temperature change and the slow "recovery" at constant temperature.

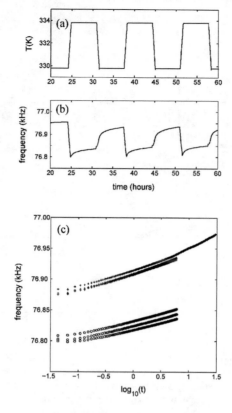

FIGURE 3. Results from a "plateau" protocol experiment (see fig. 1 (b)). (a) Measured temperature profile and (b) resonance frequency response in time. (c) The frequency response at the constant T portions of the temperature profile are shown shifted in time to $t = 0$ when T = constant and plotted in semilog space (i.e., $\log(t)$).

"Plateau" Protocol

Further exploration of the long time "recovery", and how that "recovery" is manifest after a decrease, as opposed to an increase, in temperature is done using the "plateau" protocol. Figure 3 presents the measured temperature profile (fig. 3 (a)) and the frequency response on both linear and logarithmic timescales (fig. 3 (b) & (c)). The succession of up and down temperature changes, each followed by holding T constant, highlights the phenomenon of an always upward shifting resonance frequency after a temperature change, regardless of the sign of ΔT. This translates to an elastic stiffening of the sample over time. Also apparent in the "plateau" experiments is the assymetry of the initial frequency response due to the sign of ΔT. This asymmetry gives rise to the need to investigate the shorter timescales involved in the elasticity changes, thus the "chirp" and "AC" protocols were developed.

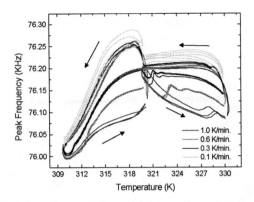

FIGURE 4. Results from two "chirp" protocol experiments (see fig. 1 (c)). Both experiments originate at a baseline temperature $T = 320$ K. The arrows above/below the loops indicate the direction of the temperature change. The different line weights indicate the rate of temperature change as denoted in the legend. Notice the asymmetry in the hysteresis loops below and above the baseline temperature of 320 K.

"Chirp" Protocol

Systematically varying the rate $\dot{T}$ for a constant temperature change ΔT brought about hysteretic elastic behavior and a means to analyze this behavior. Figure 4 shows the hysteretic elastic behavior for the various $\dot{T}$'s used. There is a striking asymmetry in the response to $\Delta T > 0$ and $\Delta T < 0$ and in the shape of the hysteresis loops as well. The shrinking of the hysteresis loops can be quantified by calculating the area enclosed as well as the locations of the loop end points. These values are presented graphically in fig. 5. From fig. 5 it can be seen that the hysteresis is predominantly, if not entirely, due to the fact that the sample is never in an equilibrium state (due to the long timescales of elastic changes that are apparent from the "step" and "plateau" experiments) as the hysteresis approaches approximately zero as $\dot{T} \rightarrow 0$. The asymmetry of the frequency shift is currently unexplained. It is also of note that a decreasing temperature gives rise to a decrease in the resonance frequency. This is not only counter-intuitive, but also alludes to an instability of some kind.

"AC" Protocol

The processing of the data from the "AC" protocol experiments begins with the normalization of the vectors $|f\rangle$, $|T\rangle$ and $|\dot{T}\rangle$ such that

$$\begin{aligned} \langle f|f\rangle &= 1 \\ \langle T|T\rangle &= 1 \\ \langle \dot{T}|\dot{T}\rangle &= 1. \end{aligned} \tag{1}$$

The vectors $|T\rangle$ and $|\dot{T}\rangle$ are also orthogonalized, such that $\langle \dot{T}|T\rangle = \langle T|\dot{T}\rangle = 0$. Fortunately, the experimental temperature profile is already largely orthogonal to the rate of

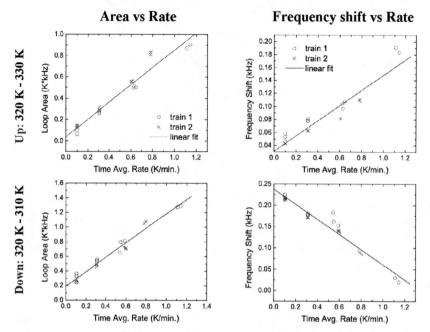

FIGURE 5. Analyzing the hysteresis loops shown in fig. 4 provides information about the area of the loops and the frequency shift (difference in loop end points). (a) and (b) show the loop area as a function of rate for both $\Delta T > 0$ (red) and $\Delta T < 0$ (blue), respectively. Notice that when extrapolated to an infinitely slow rate, the loop area minimizes to approximately zero. (c) and (d) display the degree of frequency shift as a function of rate. The frequency shift extrapolates to a finite, but non-zero, value.

change.

I postulate that the frequency vector can be described by an orthonormal basis set whose most important components are temperature and rate (i.e., the experimental controls). Thus

$$|f\rangle = a|T\rangle + b|\dot{T}\rangle + c|R\rangle, \tag{2}$$

where $|R\rangle$ represents the sum of all residual influences unaccounted for in the $|T\rangle$ and $|\dot{T}\rangle$ vectors and a and b are parameters whose values indicate the importance of temperature and rate respectively on the frequency response. To obtain information on multiple timescales, the data is filtered with low and high pass filters using the two periods apparent in the temperature protocol as defining times for the filters (fig. 6). Furthermore, the data sets were broken into blocks where $\Delta T > 0$ and $\Delta T < 0$, thus each of he "AC" experiments contains 6 data blocks. The projection analysis described above is then employed for both the high and low pass filtered data (fig. 7) and for each data block.

With the data being orthonormalized, and providing that the basis vectors $|T\rangle$ and $|\dot{T}\rangle$ describe the data completely, the values of a and b found from the projection analysis

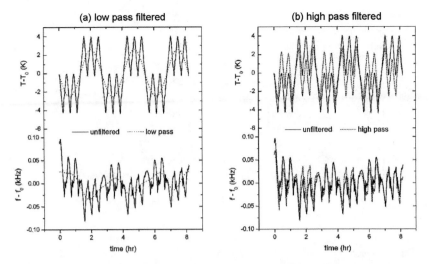

FIGURE 6. The measured temperature profile for an "AC" protocol experiment(top of (a) and (b)) and the frequency response (bottom of (a) and (b)). The data was filtered with low and high pass filters for use with the projection analysis described in this paper. Note, the data shown in (a) and (b) are identical data sets separated only to clearly display the results of the filtering.

should lie on a circle of radius 1 in *a-b* space as

$$a^2 + b^2 = \langle f|f \rangle = 1. \tag{3}$$

Figure 7 displays the results of the projection analysis from an "AC" experiment with a unit circle overlaid for perspective. It is readily apparent that on the intermediate timescale (the low pass filtered data) that the elastic changes are directly related the the temeprature as $|a| \to 1$ for all data blocks, while the faster timescale (from the high pass filtered data) displays an asymmetry between the blocks $\Delta T > 0$ and $\Delta T < 0$. The analysis suggests that the contraction of the sample due to $\Delta T < 0$ responds more to the rate at which the temperature change is occuring than the temperature itself. It also appears that there may be another influence that is unaccounted for, as the results do not lie completely on the unit circle.

SUMMARY

The description above is in terms of the resonance frequency of a particular mode of the sample, a Young's mode. This is surrogate for the "elastic state of the sample". In all that I have done I see evidence for slow time evolution of the elastic state. This makes itself known in the "chirp" experiment in the appearance of hysteretic behavior. When I probe this hysteresis with the T-chirp temperature protocols I see that most of the hysteresis goes away as $t \to \infty$. Because of the $log(t)$ recovery at the longest times we can not rule out the possibility of a small residual hysteresis in the elastic state driven by temperature.

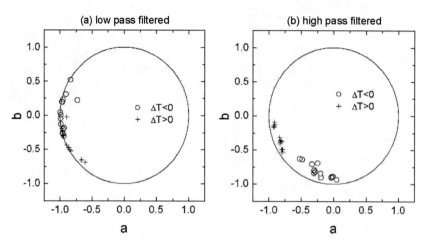

FIGURE 7. Results from the projection analysis performed on the (a) low pass filtered data and (b) high pass filtered data taken from an "AC" protocol experiment (see fig. 1 (d)).

Long time scales have been seen in these systems in the past, so what I report is not that surprising. More surprising is the strong asymmetry in response to the sign of a temperature change. This confirms a result seen earlier by TenCate et al [4]. Apparently there is an important difference in the ability of the bond system to respond to the sign of temperature. When asked to "expand", in response to forces in the grains, the bond system moves relatively easily. When asked to "contract" it has a hard time. We take expanding and contracting to be associated with un-crowding and crowding. There is the suggestion from recent neutron work[5] for an important amorphous component in the structure of a sandstone. Possibly this component mediates the grain/grain contacts so that an encounter between grains is an encounter between disordered molecular tangles which each carries.

ACKNOWLEDGMENTS

I would like to thank Katherine McCall, Robert Guyer, Jim TenCate and Tim Darling for their comments, input and assistance on this project. This work was funded by the NSF geophysics program.

REFERENCES

1. T. J. Ulrich, D. Chandra, and A. Imam, "RUS on Ti alloys," in *Proceedings of the World Conference on Ti 2003, Hamburg, Germany*, 2003, p. 2222.
2. T. J. Ulrich, K. R. McCall, and R. A. Guyer, *submitted to Phys. Rev. Lett.* (2005).
3. T. J. Ulrich, *University of Nevada, Reno, Doctoral Dissertaion* (2004).
4. J. A. TenCate, D. E. Smith, and R. A. Guyer, *Phys. Rev. Lett.*, **85**, 1020 (2000).
5. T. W. Darling, J. A. TenCate, S. C. Vogel, and T. E. Proffen, *Geophys. Res. Lett.* (to be published).

Nonlinear Acoustic NDE Using Complete Nonclassical Spectra

I. Solodov, K. Pfleiderer, and G. Busse

*Institute for Polymer Testing and Polymer Science (IKP) - Nondestructive Testing - (ZFP),
Stuttgart University, Stuttgart 70569, Germany*

Abstract. In the presence of cracked defects, material nonlinear response is determined by local contact dynamics which strongly depends on the amplitude of acoustic wave. At moderate driving amplitude, the contact acoustic nonlinearity suggests a fully deterministic scenario of higher harmonic generation and/or wave modulation. Unlike classical nonlinear materials, these effects feature much higher efficiency, specific dynamic characteristics, modulated spectra, and unconventional acoustic waveform distortion. At higher excitation, the contact vibrations acquire a dynamic instability which is a forerunner of transition to chaos. Such a dynamics is interpreted on the basis of nonlinear resonance phenomena for a defect conceived as a set of coupled oscillators. It is shown to result in a decay of external excitation into either a combination frequency pair or a subharmonic mode. For higher-order contact nonlinearity, the nonlinear spectrum expands considerably to include the ultra-subharmonic and ultra-frequency pair modes. Experiments show that even a moderate acoustic excitation of realistic cracked defects gives rise to the instability vibration modes which exhibit threshold behavior and distinctive hysteretic dynamics. All the modes–contributors to such non-classical nonlinear spectra display a high localization in the areas of nonlinear contacts and visualize readily various fractured defects in solids. The case studies presented include Hi-tech and constructional materials and demonstrate their applicability for defect-selective imaging in nonlinear NDE.

Keywords: nonlinear NDE, higher harmonics, ultra-subharmonics, ultra-frequency pairs, nonlinear acoustic imaging.
PACS: 43.25.Dc; 62.30.+d; 68.35.Ja; 62.20.Mk.

INTRODUCTION

Nonlinear acoustics was advocated as a tool for non-destructive evaluation (NDE) in early 60`s after an increase in material nonlinearity due to a mechanical impact was observed by Krasil`nikov et al. [1]. However, the classical approach to acoustic nonlinear NDE was based on the assumption of a low local nonlinearity and relied on accumulation of the nonlinearity along the wave propagation distance. As a result, all classical nonlinear NDE applications were focused around measurements of the material *nonlinearity distributed* over some extended lengths. Even in the presence of

CP838, *Innovations in Nonlinear Acoustics: 17th International Symposium on Nonlinear Acoustics*,
edited by A. A. Atchley, V. W. Sparrow, and R. M. Keolian
© 2006 American Institute of Physics 0-7354-0330-9/06/$23.00

a localized damage an overall material nonlinearity parameter averaged over a total area changed insignificantly. Though in some cases these changes could be used for detection of damage but were unable to provide information on the location of defects.

Another implication of the weakly nonlinear wave propagation was a negligible waveform distortion giving rise solely to the second harmonic (or generation of sum- and difference-frequency waves) whose amplitude was normally down to $\leq 10^{-2}$ of that for fundamental waves. Hence, a number of precautions hardly-accomplished in general NDE-practice should have been taken to provide an extremely low higher harmonic level in an intense acoustic source, a lack of dispersion in propagating waves, etc. The latter classical requirement has also confined a selection of types of acoustic waves applicable for nonlinear NDE proposing an alternative of the longitudinal or surface waves only.

Over the last decade, however, a significant progress has been achieved in making nonlinear NDE a less critical and more practically oriented methodology [2, 3]. It is concerned with an anomalously high local nonlinear response of damaged area caused by strongly nonlinear vibrations of defect fragments driven by an acoustic wave [4]. In this case, the intact material outside the defect is no more than a "linear carrier" of the acoustic wave while the *nonlinearity* is *localized* in the defect area. A scanning methodology used on the receiver side then enables not only to reveal the presence but also to locate and visualize the damage [5]. Such an approach also lifts the main classical requirements to the equipment for nonlinear NDE. The higher harmonic "noise" in the source has now a little impact on the contrast of nonlinear image that simplifies greatly the excitation. Velocity dispersion can be considered as a favorable factor, which reduces nonlinear background outside the defect and highly dispersive Lamb modes can be used for nonlinear NDE. This enables to reduce the driving frequency and hence to test a much wider range of materials and components.

The high local nonlinearity of the vibrating defect strongly modifies its nonlinear frequency response. At moderate driving amplitude, the classical spectral pattern of the fundamental frequency and its second harmonic, changes for a wide nonlinear spectrum of the higher-order ultra-harmonics. For higher levels of excitation, the spectrum of local oscillations demonstrates a qualitative departure from the classical collection: successive bifurcations of period doubling and instability eventually result in a chaotic-like spectrum. At the 16th ISNA [6], we reported on the results of experimental simulations of such nonclassical spectra extended from a set of higher-order harmonics to a quasi-continuum. This paper is concerned with their applications for "on-the-spot" control and imaging of realistic flaws in materials and components.

HIGHER HARMONIC MODE

The fractured defects (micro-cracks, debondings, delaminations, impact and fatigue damages) are normally supported by internal compression stresses. As soon as the amplitude of an acoustic wave exceeds the static stress it causes vibrations of an intermittent contact between the defect fragments. Clapping and/or rubbing of the micro-asperities provide a highly nonlinear contact dynamics due to a strong modulation of its stiffness [6].

Clapping results in asymmetrical modulation of the contact stiffness: it is higher for compression (C) and lower for the contact extension ($C - \Delta C$). Such a bimodular behaviour of a pre-stressed contact driven by a harmonic acoustic strain $\varepsilon(t) = \varepsilon_0 \cos \nu t$ is similar to a "mechanical diode" and results in a pulse-type modulation of its stiffness $\Delta C(t)$. Since $\Delta C(t)$ is a pulse-type periodic function of the driving frequency ν, the spectrum of the stress induced in the damaged area ($\Delta C(t) \cdot \varepsilon(t)$) contains a number of its higher harmonics $n\nu$ (both odd and even orders) whose amplitudes are modulated by the *sinc*-envelope function. The depth of stiffness modulation ($\Delta C / C$) can generally be as high as ~1 (for weakly-stressed contacts) thus providing a very efficient higher harmonic generation by clapping defects [6].

The dynamics of the damaged area driven by shear traction results in friction controlled contact between its micro-contacts [6]. In this case, the stiffness modulation is caused by the transition between stick- and slide phases: higher contact stiffness provided by the static friction in the stick phase drops substantially as the contact surfaces start sliding. Such an abrupt transition takes place twice for the period of acoustic excitation (symmetric stiffness modulation). Therefore, the contact stiffness modulation $\Delta C(t)$ is a 2ν-pulse-type function which comprises its higher harmonics $2n\nu$. As a result, the spectrum of nonlinear shear vibrations of the defect ($\Delta C(t) \cdot \varepsilon(t)$) contains a number of odd harmonics of the driving frequency. In classical terms, both of the above mechanisms are higher-order nonlinearities.

Apparently, the variety of the higher harmonic spectra in realistic materials is not confined to the two basic nonlinear responses shown above to be characteristic of the intermittent contacts. However, a prevalence of any of the two mechanisms in the measured nonlinear response for a given acoustic excitation may cast light on the structure of the material and the type of defects it contains.

The experimental methodology used for nonlinear NDE includes an intense CW acoustic excitation of 20 kHz-frequency combined with a fast and remote scanning laser vibrometry [5]. After a 2D-scan and FFT of the signal received, the C-scan images of the sample area are obtained for any spectral line within the frequency bandwidth of 1 MHz.

The clapping mechanism of higher harmonic generation is typical for a number of mechanical units which use gear boxes and push-pull components. In many cases, the presence of clapping may indicate a faulty part of an instrument. Fig. 1 illustrates a feasibility of detection of the clapping components in a commercial cutting tool. The 3d harmonic image of the motor rotation frequency clearly locates position of the worm-gear transmission (dark area) which moves in a nonlinear (clapping) way.

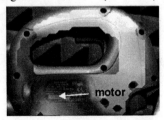 

FIGURE 1. Fundamental (left) and 3d harmonic (right) images of saw-cutting tool.

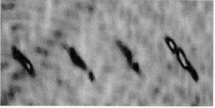

FIGURE 2. 4th harmonic images of impacts in CFRP-plate: (left) impacted side; (right) reverse side.

An example of damage detection using higher harmonic imaging is shown in Fig. 2. The 150x70x1mm3 sample of a multi-ply (+ 45°; - 45°) carbon fibre - reinforced composite contains a set of four impacts. The images in Fig. 2 illustrate development of impact-induced damage in composites: on the impacted side it "imprints" a point-like area of the tapping tool only (left), while the damage transforms into a line delamination along a ply of 45°-fibres on the reverse side (right). Interestingly, that some sections of the 45°-fibre structure are also seen in the nonlinear images which is an indication of a weak fibre-matrix bonding in the material.

The mechanism of friction (thermoelastic) nonlinearity was found to prevail in wood which is a natural fibre-reinforced composite [7]. In intact wood, the nonlinear spectrum averaged over the specimen surface exhibits an evident odd harmonic domination (Fig. 3, (left)). However, due to the strong material inhomogeneity caused by the annual rings, a local nonlinear response of wood is also expected to be spatially inhomogeneous with a maximum nonlinear output in areas of the highest compliancy where peak strains are developed. A typical higher harmonic C-scan of the LR-plane (cut along the trunk) of spruce specimen is shown in Fig. 3, (right). One can see the wavy distribution of local nonlinearity in the radial direction with maxima located in the earlywood area close to the latewood/earlywood transition interface. Therefore, the local nonlinear response indicates that the most load-vulnerable part of wood is formed early in the growing season when the thin-walled earlywood cells appear.

Since the odd harmonic domination is a characteristic of intact wood, any deviation from such a spectrum indicate the presence of clapping defects responsible for even harmonic generation (Fig. 4, (left). Such a "clapping-selective" imaging is shown in Fig. 4, (right) for simulated delaminations between a decorative oak veneer lamina and a particleboard substrate (12x6.5x1cm). The strips of the delaminated areas are clearly indicated by a sharp local increase in the 4th harmonic amplitude (dark strips) due to the clapping mechanism. It is worthwhile noting, that acoustic NDE of particleboard

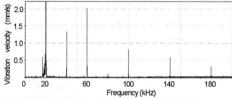

FIGURE 3. Higher harmonic mode in intact wood (spruce): left - nonlinear spectrum (20 kHz excitation); right – 3d harmonic distribution in LR-plane.

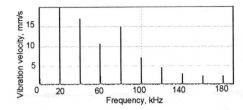

FIGURE 4. Higher harmonic mode in wood with delaminations: left - nonlinear spectrum (20 kHz excitation); right – 4th harmonic image of simulated stripes of delaminations.

normally fails due to the high damping in thick specimens while the nonlinear response is virtually independent of the specimen thickness.

ULTRA-SUBHARMONIC AND ULTRA-FREQUENCY PAIR MODES

To illustrate a feasibility of the new nonlinear vibration modes, we assume that the damaged area exhibits both resonance and nonlinear properties and can be identified as a nonlinear oscillator with s degrees of freedom [8]. The equations of motion of the oscillator driven by an acoustic excitation in normal coordinates (Q_α) takes the form:

$$\ddot{Q}_\alpha + \omega_\alpha^2 Q_\alpha = f_0 \cos \nu t + F_\alpha^{NL}, \tag{1}$$

where $\alpha = 1, 2, \ldots s$, ω_α - are the normal frequencies.

In the first approximation of the perturbation theory $F_\alpha^{NL} = 0$ and the solution to (1) is: $Q_\alpha^{(1)} = A_\alpha \cos \omega_\alpha t + B_\alpha \cos \nu t$, where A_α, B_α are constants.

This expression is used in the second-order equations $\ddot{Q}_\alpha^{(2)} + \omega_\alpha^2 Q_\alpha^{(2)} = F_\alpha^{NL}(Q^{(1)})$ to yield: $F_\alpha^{NL} \sim \sum_{\beta\gamma} (A_\beta \cos \omega_\beta t + B_\beta \cos \nu t)(A_\gamma \cos \omega_\gamma t + B_\gamma \cos \nu t)$ with combination frequency components: $\sum_{\beta\gamma} [F_{\beta\gamma}^{(2)} \cos(\omega_\beta \pm \omega_\gamma) + F_{\nu\beta}^{(2)} \cos(\nu \pm \omega_\beta)]$. The last term causes a resonance increase in $Q_\alpha^{(2)}$ if $\nu \pm \omega_\beta \approx \omega_\alpha$, while a similar resonance of $Q_\beta^{(2)}$ is achieved when $\nu \pm \omega_\alpha \approx \omega_\beta$. Thus, the driving force of frequency ν provides a simultaneous resonance growth for the pair of normal modes if: $\omega_\alpha + \omega_\beta \approx \nu$, i.e. a resonance decay into a pair of "phonons" of combination frequencies..

The resonance increase of the spectral components $\omega_{\alpha,\beta}$ then expands the nonlinear spectrum due to the higher order terms in the driving force (assuming clapping or friction force nonlinearities). After account for the N th-order terms, the spectrum acquires the following frequency components:

$$F^{NL}(\omega) \sim \sum_{m,n,p} F_{mnp}(n\nu + m\omega_\alpha + p\omega_\beta), \tag{2}$$

where $m + n + p = N - 1$.

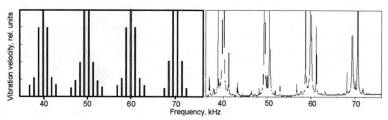

FIGURE 5. Ultra-frequency pair spectra: calculated (left) and measured (right).

Besides the ultra-harmonics $n\nu$, the nonlinear spectrum (2) comprises ultra-frequency pairs (UFP) $(n\nu + m\omega_\alpha + p\omega_\beta)$ and $(n\nu + p\omega_\alpha + m\omega_\beta)$. They are separated by $|m - p|\Delta$ $(\Delta = \omega_\beta - \omega_\alpha)$ and centered around $[n\nu + (m + p)\nu / 2]$. For odd values of $m + p$, the latter term describes the spectrum of ultra-subharmonics (USB) (of the second order). The UFP spectrum changes for the USB-spectrum when a single degree of freedom of a nonlinear defect is taken into account ($\omega_\alpha = \omega_\beta = \omega$). A part of the UFP-spectrum calculated from (3) for $F_{mnp} = 1$, $N = 8$ is shown in Fig. 5 (left). A comparison with the UFP-spectrum measured for a glass fibre-reinforced composite with a 9.5J- impact damage (right) justifies the model developed above.

The dynamics of the decay into UFP or USB is totally different from the classical power-law dependences and features a threshold amplitude and frequency instability (avalanche-like "jumps") and hysteresis [9].

Fig. 6 (left) provides direct experimental evidence of the spectral transformations beyond the subharmonic instability threshold for a delamination area in a C/C-SiC-composite. As the driving amplitude of 20-kHz excitation increases above the threshold for the subharmonic mode (≈ 0.5 μm in Fig. 6), another instability threshold gives rise to an avalanche-like energy-decay into the UFP-components (at ≈ 1 μm-drive). A further increase of the input results in widening of the UFP-lines into quasi-continuous frequency bands which are the forerunners of the transition to chaos. The results of our experiments on the frequency and dynamic nonlinear responses of the fractured flaws are summarized schematically in Fig. 6 (right) for a defect represented by a pair of coupled oscillators (normal frequencies ω_1 and ω_2).

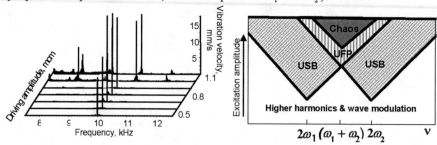

FIGURE 6. Dynamics of nonclassical nonlinear spectra.

FIGURE 7. Defect selective nonlinear NDE: 4th harmonic image (left); 11th suharmonic image (right).

Since the subharmonic mode results from the nonlinear resonance in acoustic wave-defect interaction it may be sensitive to the input frequency, while the higher harmonics, normally, are generated invariably. This can be a basis for a frequency selective nonlinear NDE: for a given driving frequency only "resonant" defects can be discerned in the subharmonic mode. Such a case is illustrated in Fig. 7 for the multi-ply plate of CFRP foam laminate with two artificial defects. Both of them, the edge delamination (left) and a thermal impact (right) are detected in the higher harmonic mode but the ultra-subharmonics are sensitive only to the resonance impact area.

Fatigue loads in metals (rotors, turbines, etc.) cause minute cracks of micro-meter scale which gradually develop into major cracking and initiate an abrupt material fracture. The linear ultrasound is virtually unable to detect the fatigued crack at the stage of their development. The examples of the nonlinear imaging of fatigue induced micro-flaws and degradation of metal micro-structure using the USB- modes are given in Figs. 8 and 9. Fig. 8 is concerned with fatigue cracking produced by cyclic loading in Ni-base super-alloys. Such a crack of less than 2 mm length with average distance between the edges of only $\approx 5\mu$m (left) is clearly detected in the USB- $(15\nu/2)$ image (right). The traditional linear NDE by using slanted ultrasonic reflection failed to work with such a small crack.

The USB-mode of nonlinear NDE of even smaller defects (dislocations, grain boundaries, etc) which are the forerunners of micro-cracking is illustrated in Fig. 9. A steel cylindrical sample (diameter 0.8 cm; length 6.5 cm) was subjected to a 6%-plastic tensile deformation. Some necking initiated in its central part (Fig. 9, right) indicated that the strain was mostly concentrated in its central part. A strong standing wave pattern with no indication of any local nonlinearity was observed in an intact sample (left). On the contrary, the USB-image (right) clearly detects a cluster of defects localized in the neighborhood of the necking area.

 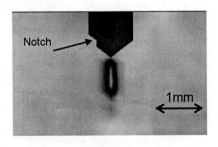

FIGURE 8. Nonlinear USB-NDE of fatigue crack (left) in Ni-base super-alloy: $15\nu/2$ -image (right).

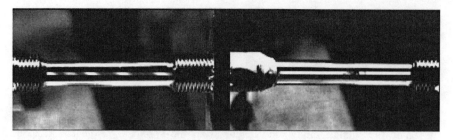

FIGURE 9. Nonlinear NDE of micro-damage induced by plastic strain in steel: 3d harmonic standing wave in intact sample (left); $43\nu/2$-USB image of damage at 6% plastic strain (right).

CONCLUSIONS

At moderate amplitude of an acoustic excitation, the nonlinear vibration pattern of defect fragments develops on a fully deterministic scenario of the higher harmonic generation and/or wave modulation. Unlike their classical counterparts, these effects feature much higher efficiency, specific dynamic characteristics, modulated spectra, and result in an unconventional waveform distortion. For higher levels of excitation, the contact vibrations acquire a dynamic instability which is a forerunner of transition to chaos. The ultimate nonlinear spectrum expands considerably to include the ultra-subharmonic and ultra-frequency pair modes. All the modes contributing to such nonclassical nonlinear spectra display a high localization in the areas of nonlinear contacts and, thus, can visualize readily various fractured defects in solids. Numerous case studies demonstrate their applicability for nonlinear NDE and defect-selective imaging in various materials. Multiple frequency components comprising complete nonclassical nonlinear spectra in imperfect materials provide abundant information on properties and location of defects that enables to improve reliability and quality of nonlinear acoustic NDE.

REFERENCES

1. Gedroits, A.A., and Krasil'nikov, V.A., *Sov. Phys. JETP* **16**, 1122-1131 (1963).
2. Krohn, N., Dillenz, A., and Busse, G., "Zerstoerungsfreie Pruefung formadaptiver Strukturen" in *Verbundwerkstoffe und Werkstoffverbunde*, edited by K. Schulte and K.U. Kainer, Weinheim: Wiley-VCH, 1999, pp. 602-607.
3. Johnson, P.A., *Materials World* **9**, 544-546 (1999).
4. Solodov, I.Yu., *Ultrasonics* **36**, 383-390 (1998).
5. Krohn, N., Pfleiderer, K., Stoessel, R., Solodov, I., and Busse, G., "Nonlinear acoustic imaging: Fundamentals, methodology and NDE-applications" in *Acoustical Imaging, Vol.* 27, edited by Arnold, W. and Hirserkorn, S., Dordrecht: Kluwer Academic/Plenum Publishers, 2004, pp. 91-98.
6. Ballad, E.M., Korshak, B.A., Solodov, I.Yu., Krohn, N., and Busse, G., "Local nonlinear and parametric effects for non-bonded contacts in solids", in *Nonlinear Acoustics at the Beginning of the 21ˢᵗ Century*, edited by O. Rudenko and O. Sapozhnikov, Moscow, 2002, pp. 727-734.
7. Solodov, I., Pfleiderer, K., and Busse, G, *Holzforschung* **58**, 504-510 (2004).
8. Solodov, I., Wackerl, J., Pfleiderer, K., and Busse, G., *Appl. Phys. Letters* **84**, 5386-5388 (2004).
9. Solodov, I.Yu., and Korshak, B.A., *Phys. Rev. Letters.*, **88**, 014303, 1-3 (2002).

Nonlinear NDE of Concrete Mechanical Properties

Iosif E. Shkolnik

Department of Civil and Environmental Engineering, Wayne State University, Detroit, MI 48202, USA

Abstract. Obtained theoretical relationship shows that the strength of concrete increases if the nonlinear parameter decreases. Experimental data proved that modulus of elasticity, ultrasound pulse velocity and nonlinear parameter are independent characteristics of concrete. Two nondestructive patent methods based on the measurement of resonant frequency shift and phase shift are described. These nonlinear nondestructive methods can be used when conventional acoustic methods are not applicable for evaluating strength of concrete.

The relationship between static and dynamic modulus is obtained from the thermofluctuation theory and nonlinear equation of state of concrete. Corresponding relationship shows that the ratio of the static to the dynamic modulus of elasticity depends on the strength of concrete, its temperature, ratio and rate of loading, and that dynamic modulus is greater than static modulus of elasticity.

Comparative study illustrates substantial agreement between obtained relationships and existing experimental results as well as general equations given in standards.

Presented data illustrate the potential of the nonlinear approach, and indicate a new direction for nonlinear nondestructive methods of evaluating mechanical properties of concrete.

INTRODUCTION

Beginning with experiments by James Bernoulli, and until present careful investigations of different materials (stone, concrete, wood, glass, metal etc.) show the nonlinear dependence between stress and strain even at infinitesimal values of deformation [1].

The presence of the nonlinear dependence between stress and strain has been derived from the basic statements of the modern theory of solids, and is stipulated by the nature of the interatomic forces [2].

Different flaws in structure (dislocations, micropores, and microcrevices) have a significant influence on the nonlinear behavior of concrete under loading, and predetermine its modulus of elasticity and strength [3]. For instance, the examination of concrete specimens subjected to increasing uniaxial compression, using microscopes and X-ray technique, showed that bond cracks exist in concrete even before any loading. Starting at 30 percent of ultimate load, bond cracks increase rapidly in length, width and number [4].

Dominant influence on nonlinearity of deformation such as flaws in the structure of concrete like micropores, microcracks and other discontinuities (defects) is established

CP838, Innovations in Nonlinear Acoustics: 17th International Symposium on Nonlinear Acoustics,
edited by A. A. Atchley, V. W. Sparrow, and R. M. Keolian
© 2006 American Institute of Physics 0-7354-0330-9/06/$23.00

from the results of ultrasonic, strain and gamma radiation measurements on saturated and dried to constant weight specimens [5]. It was recognized that in dry specimens of concrete the velocities of longitudinal and shear waves (or dynamic modulus of elasticity) increase at stress level below (25-30) percent of ultimate load, and start to decrease above this level. In general the changes of modulus of elasticity of concrete, or nonlinear parameters, are typically 2-3 orders of magnitude greater in comparison with those of crystalline or polycrystalline materials. It shows a high sensitivity of the nonlinear parameter to the element of structure, which dimensions are much smaller than the wavelength at dynamic tests of concrete [6, 7, 8]. Simultaneously such elements predetermine the nonlinear behavior of a real material and have a significant influence on its strength.

This paper presents a nonlinear approach to unify the nondestructive methods for evaluating modulus of elasticity at quasi-static loading, and strength of concrete.

BASIC STATEMENTS

As it was mentioned above, concrete subjected to increasing uniaxial compression reflects the reaction of its structure on different values of external forces and indicates the nonlinearity of deformation. It means that the fracture of concrete during loading does not occur as a critical event. It is rather the result of a process of nucleation, accumulation, and development of damage elements. In accordance with thermofluctuation theory, the fracture is a kinetic or time dependent phenomenon [9].

At the atomic-molecular level this process is controlled by events of breaking of interatomic bonds by thermal fluctuations, and can be described as [10]:

$$\sigma = \sigma_{max} - (2g/mL^2)kT\ln(\tau/\tau_0) \qquad (1)$$

where σ is the average stress; σ_{max}- the strength for ideal solid; g- anharmonic coefficient, m- constant of elasticity; L is the distance between particles in the state of equilibrium; k is Boltzmann's constant; T- temperature; τ - time to failure of a solid under stress σ; τ_0–the period of interatomic vibration (~10^{-13}second); kT=404*10^{-23}J (2.44kJ/mole) at the usual temperature of test (T=293K). Thus, if g=0, the effect of temperature and time vanishes.

At the macroscopic level the fracture of stressed concrete proceeds by a process that develops in the body from the moment of load application, while the fracture itself is the final act in this process. For the simplest case of stress growth at constant rate $\dot{\sigma}$ it was obtained the value of strength (σ_r) for concrete taking into account the principle of summation of damages under loading [11, 12]:

$$\sigma_r = (kT/\gamma)\ln(\dot{\sigma}/\dot{\sigma}_o) \qquad (2)$$

where γ is the structurally sensitive coefficient, $\dot{\sigma}$ is the rate of loading. For a number of different cases of fracture of concrete (different concretes, different forms of stress, simply different investigations in different countries) was found a fairly reliable value of $\log\dot{\sigma}_o$= -8.5($\dot{\sigma}_o$ in MPa/s). This value corresponds to the stable value of activation energy for concrete $U_o \sim 120kJ/mole$, which is close to the dissociation energy of the silicon-oxygen bonds in accordance with hydrolytic mechanism.

For the development of the nonlinear approach the most important is the condition that within a wide range of strains the relationship between stress (σ) and strain (ε) of different materials, including those that are brittle, can be described by one and the same quadratic parabola [1]. This nonlinear function 1) demonstrates qualitatively and quantitatively the response from the largest (10^{-4}) to the smallest deformations (10^{-6}), and 2) the value of modulus of elasticity extrapolated to zero stress (E_o) closely agrees with the dynamic modulus determined from longitudinal vibrations with amplitudes of 10^{-6}. Therefore the dependence between stress acting on the body and strain can be presented as

$$\sigma = E_o\varepsilon - b\varepsilon^2 \qquad (3),$$

where constant "b" is the nonlinear parameter. Eq. (3) clearly illustrates that the modulus of elasticity linearly proportional to the strain

$$E = d\sigma/d\varepsilon = E_o - 2b\varepsilon \qquad (4)$$

Corresponding value of the nonlinear parameter "b" can be written as

$$b = (E_o - E)/2\varepsilon = E_o(1 - E/E_o)/2\varepsilon \qquad (5)$$

$$b/E_o = (E_o - E)/2E_o\varepsilon \sim dE/2d\sigma \qquad (6)$$

By assuming equal distribution of external load, all interatomic connections are strained equally, and the average stress σ acting on the body is equal to the stress σ_a acting on the atom. Under the action of stress σ_a, which is less than the ultimate value, the potential energy $U(\sigma_a)$ is proportional to the volume of atom (W_a), and in accordance with (3) becomes [8]:

$$U(\sigma_a) = U_o - W_a\sigma_a b/E_o \qquad (7)$$

where U_o, W_a is the initial potential energy, and volume of atom ($W_a \sim 10^{-2}nm^3$), respectively.

On the other hand, it was shown [9] that the potential energy decreases linearly with applied stress as

$$U(\sigma) = U_o - \gamma\sigma \qquad (8)$$

Then from (6), (7) and (8) follows

$$\gamma = W_a b/E_o = W_a(E_o - E)/(2\varepsilon E_o) \sim W_a(E_o - E)/2\sigma \sim W_a \, dE/2d\sigma \qquad (9)$$

or $\quad b = \gamma E_o/W_a = (kT/\sigma_r)(E_o/W_a)\ln(\dot\sigma/\dot\sigma_o) \qquad (10)$

45

One can see from Eq.(3) and Eq. (10) the adequate equation of state is

$$\sigma = E_o\varepsilon - (kT/W_a)(E_o/\sigma_r)[\ln(\dot\sigma/\dot\sigma_o)]\varepsilon^2 \tag{10a}$$

Thus, equations (9, 10) indicate the relationship between the structurally sensitive coefficient "γ" (the thermofluctuation theory), and the nonlinear parameter "b" (the original equation of state). Substituting Eq. (9) into Eq. (2), and taking into account the values of kT, $\dot\sigma_o$, W_a, the formulas for modulus of elasticity E, depending on stress σ, can be written as:

$$\frac{E}{E_o} = 1 - \frac{kT}{W_a}\frac{2\sigma}{E_o\sigma_r}\ln\frac{\dot\sigma}{\dot\sigma_o} = 1 - 808\frac{\sigma}{E_o\sigma_r}(\ln\dot\sigma + 19.57) \tag{11a}$$

$$E = E_o - \frac{kT}{W_a}\frac{2\sigma}{\sigma_r}\ln\frac{\dot\sigma}{\dot\sigma_o} = E_o - 808\frac{\sigma}{\sigma_r}(\ln\dot\sigma + 19.57) \tag{11b}$$

$$\frac{E}{E_o} = \frac{E}{E + 808\dfrac{\sigma}{\sigma_r}(\ln\dot\sigma + 19.57)} \tag{11c}$$

where $\sigma, \dot\sigma$ in MPa, MPa/s, respectively. Thus, Eqs. (11a-11c) confirm that the value of modulus of elasticity E depends on the temperature T, level of stress (σ/σ_r), and rate of loading ($\ln\dot\sigma$). It can be seen that the static modulus of elasticity at any level of loading has to be less than the initial modulus of elasticity (or dynamic modulus) E_o. Also the ratio of the static modulus of elasticity to initial modulus of elasticity (or dynamic modulus) increases, when E_o increases-Eq. (11a). These conclusions correspond in general to existing experimental data for concrete.

Taking into account Eq. (9), the Eq. (2) for strength can be reduced to the form:

$$\sigma_r = \frac{kT}{W_a}\frac{2}{\dfrac{dE}{d\sigma}}\ln\frac{\dot\sigma}{\dot\sigma_o} = 808\frac{1}{\dfrac{dE}{d\sigma}}(\ln\dot\sigma + 19.57) \tag{11d}$$

Based on the thermofluctuation theory and nonlinear deformation, Eqs. (11a-11d) acquires a physical explanation for concrete.

EVALUATION OF MODULUS OF ELASTICITY

In order to make a proper comparison between quasi-static and dynamic tests of the modulus of elasticity of concrete the values of strains (stresses) have to be specified. One can see also from Eqs. (11a)- (11c) that the modulus of elasticity are strain-rate sensitive. But this dependence is logarithmic, i.e. small in comparison with linear dependence. Therefore, the influence of the level of stress (strain) on the modulus of elasticity must be taken into consideration first.

Table 1 shows the results of compressive stress-strain measurements on concrete cylinders 15 by 30cm for modulus E, and of dynamic tests for E_o [13]. The values of the stresses less than σ_{min} indicate that the stress-strain curve coincides with dynamic (resonance) tests for E_o. The range of corresponding strains is $(1-4)*10^{-5}$ (ε_{min}), the ratio ($\sigma_{min}/\sigma_r = \sigma_{min}/f_c'$), where f_c' –the strength, is small ($\sigma_{min}/f_c'=0.01-0.04$), and the ratio E/E_o is high: $E/E_o=0.98-0.99$ at $\dot{\sigma}=0.24$MPa/s (Eq. 11a). At the stresses σ_{max} the actual relationship "stress-strain" falls off from linear. In accordance with experimental data the ratio of modulus of elasticity E/E_o (Eq. 11a) decreases ($E/E_o=0.96-0.89$), if the ratio of stress to compressive strength increases ($\sigma_{min}/f_c'=0.10-0.23$). The relative value of the nonlinear parameter b/E_o (Eq. (6), Table 1, last column) is calculated using Eq. (11b) for the difference (E_o-E), and $E_o\varepsilon_{max} \sim \sigma_{max}$. Thus the nonlinear parameter "b" is 2 orders of magnitude greater than Young's Modulus E_o of concrete.

TABLE 1. Experimental data [13] and the ratio of the modulus of elasticity by Eq. (11a).

f_c' MPa	σ_{min} MPa	E_o, MPa	ε_{min} x10^5	σ_{min}/f_c' ratio	E/E_o Eq. (11a)	σ_{max} MPa	σ_{max}/f_c' ratio	E/E_o Eq. (11a)	b/E_o Eqs.(6)
53	1.38	36544	3.77	0.03	0.99	6.90	0.13	0.95	132
47	0.69	36544	1.89	0.01	0.99	6.90	0.15	0.94	159
53	1.38	39302	3.51	0.03	0.99	5.52	0.10	0.96	142
46	1.38	37923	3.64	0.03	0.99	6.90	0.15	0.94	165
34	1.38	35165	3.92	0.04	0.98	6.90	0.20	0.92	204
24	0.69	31717	2.17	0.03	0.99	5.52	0.23	0.89	316

Experimental correlations (12-14) illustrate the relationship between the static and dynamic modulus for different concrete mixes and cured conditions [14]:

$$E= 1.033E_o - 7.242 \text{ (GPa)}, R^2= 0.996 \qquad (12),$$

$$E= 0.966E_o - 4.914 \text{ (GPa)}, R^2= 0.981 \qquad (13),$$

$$E= 1.005E_o - 6.577 \text{ (GPa)}, R^2= 0.980 \qquad (14),$$

R^2 is the coefficient of determination. One can see Eqs. (12-14) is similar to each other and theoretical Eq. (11b). It is important to note that the average value of the first term is equal one, and the average value of the second term equal 6244 MPa, which coincides with the counterpart in Eq. (11b) at $\sigma/\sigma_r= 0.426$, $\dot{\sigma}=0.24$ MPa/s. The supporting data show that Eq. (11b) can be used to predict the static modulus of elasticity for aged concrete ranging between (1-28) days using the results of dynamic tests.

Equation (15)

$$E/E_o=1.25-23.75/(E*10^{-3}+19) \text{ (MPa)} \qquad (15)$$

is derived from the general relation E $=1.25E_o-19$ (GPa) given in BS 8110: Part 2:1985 for concrete at age of 28 days [15]. The ratio of static to dynamic modulus of elasticity, Eq. (11c), calculated at constant values $\bar{\sigma}/\sigma_r = \sigma/f_c =0.4$, $\dot{\sigma} =0.24$MPa/s. It can be seen (Table 2) that the error between the basic Eq.(15) and Eq. (11c) does not exceed 10% for the static modulus of elasticity in the range $(3-5)*10^4$MPa.

TABLE 2. Empirical and predicted values of E/E_o

E, MPa	E/E_o Eq. (15)	E/E_o Eq. (11c)	Error, %
30000	0.77	0.84	-9
34500	0.81	0.85	-6
38000	0.83	0.87	-4
41400	0.86	0.88	-2
44800	0.88	0.88	0
48300	0.90	0.89	1
50000	0.91	0.90	1

EVALUATION OF STRENGTH

To investigate directly the relationship between strength (f_c') and nonlinear parameter $(dE/d\sigma)$ mix proportions of concrete having different water-cement ratios, (w/c), were prepared. During the loading the ratio of normal stress to corresponding strain, i.e. modulus of elasticity, was measured within strain range (10-1000) $\mu\varepsilon$. Then the nonlinear parameter $dE/d\sigma$ as a slope between the modulus of elasticity and stress has been found from the data having a correlation coefficient at least -0.9 [16]. Table 3 shows the average values of the experimental compressive strength, and predicted results using Eq.(11d) at $\dot{\sigma}=0.24$ MPa/s. It is seen in Table 3 that the average relative error of predicted strength does not exceed 10% for each w/c ratio. This data proved that

Table 3. Water-cement ratio, nonlinear parameters and mechanical properties of concrete

W/C	E, MPa	$dE/d\sigma$	f_c', MPa Experiment	f_c', MPa Eq. (5d)	$b=(E*dE/2d\sigma)10^4$ MPa, from Eq. (6)
0.5	31768	375	40	39	596.250
0.45	36340	279	49	53	506.385
0.4	35020	325	47	47	568.750
0.35	38335	273	53	55	522.795

modulus of elasticity decreases linearly with increase in the strain (stress) at quasi-static loading: Eq. (4). On the other hand, detailed observations on the effect of the magnitude of dynamic loading on the flexural resonance frequency showed that the fundamental frequency "always decreased with increase in the vibratory strain" [17]. For concrete beams with dimensions 8 by15 by 83cm the relative resonance frequency shift was about -2% in the range (5-75) microstrain. Similar data were obtained using the longitudinal resonance technique [6]. For concrete prisms with dimensions 10 by 10 by 30cm the relative resonance frequency shift was -0.5% in the range (0.1-7) microstrain. Takings into account these results as well as the equation of state (3), and the

48

relationship between modulus of elasticity and resonance frequency, the nonlinear parameter (b/E_o) can be estimated from Eq.(6) as: $b/E_o=0.5dE/d\sigma{\sim}df/fd\varepsilon$ (f is the resonance frequency; ε is the amplitude of excitation). Thus, $df/f{\sim}(b/E_o)d\varepsilon$, i.e. the relative change of resonance frequency is linearly proportional to the changes of strain, and the nonlinear parameter b/E_o is the slope. For beams and prisms b/E_o equals:~285, and ~725, respectively. It is important that fundamental researches of rocks illustrate the linear dependence between the decrease of resonance frequency and the strain [18, 19].

Two nondestructive methods based on the measurement of resonance frequency shift or phase shift as nonlinear parameters were developed to evaluate the strength [6, 7, 20, 21, 22]. Fig. 1 illustrates the phase difference for two levels of excitation amplitude on concrete specimens with constant ultrasound pulse velocities respectively at age 3 days and 28 days. The obtained results proved that nonlinear parameters were different for high performance concrete specimens with constant ultrasound pulse velocities. It confirms that the nonlinear parameters are more sensitive to the structure of concrete than ultrasound velocities (linear parameters). It was shown also that the more the phase shift the less the strength. A nonlinear (hyperbolic) relation between the cube strength (Y) and the phase shift (x) produced the equation $Y=55.362x^{-0.1544}$ in the investigated range of strength (coefficient of determination $R^2=0.7291$).

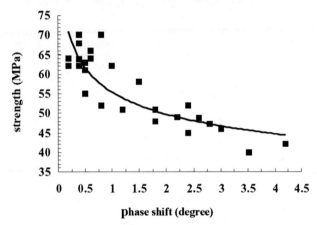

FIGURE 1. Strength- phase shift relation for concrete

CONCLUSION

Presented data illustrate the potential of the nonlinear approach, and indicate a new direction for NDE the static modulus of elasticity and strength of concrete.

ACKNOWLEDGMENTS

I wish to acknowledge the dedicated pursuit of "Quality" instruction in the field of Science, and Mathematics by the Carman-Ainsworth Board of Education (Flint, MI, USA).

REFERENSES

1. J. F. Bell, *"Mechanics of solids. The experimental foundations of solid mechanics"*, Springler-Verlag. Berlin, Heilderberg, New York, 1984.
2. Y. I. Frenkel, *"Introduction in the theory of metals"*, OG.Leningrad-Moscow, 1948 (in Russian)
3. Shkolnik I.E., "Effect of nonlinear response of concrete on its elastic modulus and strength" in *Journal Cement and Concrete Composites*, Elsevier Science Publishers Ltd, 27(7-8), 2005, pp. 747-757.
4. T. T. C. Hsu , F. O. Slate, G. M. Sturman, G. Winter, "Microcracking of plain concrete and shape of the stress-strain curve" in *Journal of the American Concrete Institute*. Detroit, 1963; 60(2), pp. 209-224.
5. Y.A . Nilender, G.Y. Pochtovik, I. E. Shkolnik IE. "The Relation of the Velocity of Elastic wave Propagation on Internal Energy Losses in Concrete under Loading" in *Concrete and Ferroconcrete;* 1969; No.7, pp.7-8, (in Russian).
6. L. K. Zarembo, V. A. Krasilnikov, I. E. Shkolnik., "About nonlinear acoustical defectoscopiya of brittle inhomogeneous materials and evaluation their strength" in *Defectoscopiya*. Academe of Science,USSR, 1989, No. 10, pp. 16-23 (in Russian).
7. I. E. Shkolnik, L. K. Zarembo, V. A. Krasilnikov, "On structural nonlinear diagnostics of solid and estimation of their Strength" in *Proceedings of the 12th International Symposium on Nonlinear Acoustics* edited by M. Hamilton et al., USA, Elsevier Science Publishers Ltd, 1990, pp. 589-594.
8. I. E. Shkolnik, " *Evaluation of dynamic strength of concrete from results of static tests"* in Journal of Engineering Mechanics, USA, 1996, 122(12), pp. 135-138.
9. V. R. Regel, A. I. Slutzker, E. E. Tomashevsky, *"The Kinetic Nature of the Strength of Solids"*, Nauka. Moscow, 1974. (in Russian).
10. S. N. Zhurkov, V. A. Petrov , *Rep. of USSR AS* **239,** 1316-1319 (1978). (in Russian).
11. V. S. Kuksenko, A. I. Slutzker, I. E. Shkolnik, Estimating the dynamic strength of concrete. *Tech. Phys. Lett.* American Institute of Physics; **19(2)**, 127-129 (1993).
12. V. I. Betekhtin, V.S. Kuksenko, A. I. Slutzker, I. E. Shkolnik, "Fracture kinetics and the dynamic strength of concrete". *Phys. Solid State.* American Institute of Physics, 1416-1421, (1994).
13. T. C. Powers, "Measuring Young's modulus of elasticity by means of sonic vibrations" in *Proceedings of the Forty first Annual Meeting of the American Society for Testing Materials,* 1938, 38(2), pp. 460-470.
14. D. Yuan, S. Nazarian, D. Zhang , "Use of Stress Wave Techniques to Monitor and Predict Concrete Strength Development" in *Proceedings of the International Symposium NDT CE.* 2003.
15. A. M. Neville, J. J. Brooks, "Concrete *Technology",* Longman Scientific & Technical, New York 1987.
16. I. E. Shkolnik, H. M. Aktan, R. Birgul, "Nonlinear nondestructive methods for evaluating strength of concrete" in *Proceedings of the International Symposium NDT CE.* 2003.
17. E. N. Gatfield, "A method of studying the effect of vibratory stress, including fatigue, on concrete in flexure", in *Magazine of Concrete Research,* 1965, 17, pp. 211-216.
18. R. A. Guyer, J. TenCate, P. Johnson, *Phys. Rev. Letters* **82,** 3280-3283 (1999).
19. L. A. Ostrovsky, P. A. Johnson, "Dynamic nonlinear elasticity in geomaterials" in *Rivista del Nuovo Cimento* 24, 2001, pp. 1-46.
20. I. E. Shkolnik, " Nondestructive Testing of concrete: new aspects", in *Nondestr. Test. Eval.,*1993, 10, pp. 351-358.
21. I. E. Shkolnik, V. A. Yurovsky, V. Y. Fishman, L.K. Zarembo, V. A. Krasilnikov, "A method of evaluation the mechanical properties of constructions", Patent No. 1770889, 1992. (Russia).
22. I. E. Shkolnik, T. M. Cameron, "Nonlinear Acoustic Methods for Strength Testing of Materials" in *Proceedings of the 14th International Symposium on Nonlinear Acoustics. Nonlinear Acoustics in Perspective,* edited by R. J. Wei, Nanjing, 1996, pp. 316-320.

V.A.Robsman: Nonlinear Testing and Building Industry

Oleg V.Rudenko

Faculty of Physics, Moscow State University, 119992 Moscow, Russia
Blekinge Institute of Technology, S-371 79 Karlskrona, Sweden

Abstract. This talk is devoted to the memory of outstanding scientist and engineer Vadim A. Robsman who died in January 2005. Dr.Robsman was the Honored Builder of Russia. He developed and applied new methods of nondestructive testing of buildings, bridges, power plants and other building units. At the same time, he published works on fundamental problems of acoustics and nonlinear dynamics. In particular, he suggested a new equation of the 4-th order continuing the series of basic equations of nonlinear wave theory (Burgers Eq.: 2-nd order, Korteveg - de Vries Eq.: 3-rd order) and found exact solutions for high-intensity waves in scattering media.

Keywords: Nonlinear nondestructive testing, cracks, huge nonlinearity.
PACS: 43.25.Gf, 43.25.Dc

INTRODUCTION

During 1970-1981 V.Robsman worked at the Department of Geography of Moscow State University and concerned himself with the fundamental problems of mathematics, physics and geology [1-3]. R.V.Khokhlov awaked his interest in nonlinear problems. After 1981 V.Robsman worked as Chief scientist at Transport Building Institute (Moscow). Just before catastrophic earthquake in Spitak (Armenia) he constructed the tunnel to re-direct a river Arpa to the shoaling lake Sevan. Being near the epicenter of this earthquake which caused the death of several tens thousands of people, V.Robsman participated in the elimination of destructions. He invented new methods of acoustic testing with the aim to determine which of the buildings could be reconstructed and which of them should be demolished. V.Robsman observed nonlinear spectral transformations of waves propagating in beams and overlaps. Later he proposed empirical laws connecting nonlinear transformations with the loss of strength [4]. During last 15 years V.Robsman applied his methods for testing of about 30 bridges, tens of viaducts, more than 20 power stations, many tunnels and subways. At the same time, he published works on fundamental problems of nonlinear dynamics [5, 6]. In particular, he suggested [6] a new equation of the 4-th order and found exact solutions. During his work at Chernobyl nuclear power plant V.Robsman got the great dose of ionizing radiation. He was seriously ill during 6 last years, but continued his researches and acoustic inspections of new buildings and transport constructions in Moscow. His last work [7] was published several days before his death

CP838, *Innovations in Nonlinear Acoustics: 17th International Symposium on Nonlinear Acoustics*,
edited by A. A. Atchley, V. W. Sparrow, and R. M. Keolian
© 2006 American Institute of Physics 0-7354-0330-9/06/$23.00

NONLINEARITY OF BUILDING MATERIALS

V.Robsman took part in the elimination of the consequences of the earthquake in Armenia. He proposed methods of acoustic nondestructive testing of the load-carrying structures of damaged buildings. By combining active sounding of beams and floors by acoustic signals and shock pulses with recording the spectra of acoustic emissions under quasi-static loads, Robsman observed nonlinear distortions of wave spectra. He understood that the formation of isolated cracks, their growth, collective behavior (interaction), and subsequent coalescence are the origin of increasing nonlinear distortions. When the external force action on a structure was increased, the nonlinear effects became more pronounced. On the basis of full-scale testing, Robsman developed empirical criteria for the relation between the nonlinear distortions of spectra and the loss of strength in a structure. Later, owing to the close cooperation with the Department of Acoustics of the Moscow State University, the effects observed by Robsman were give a more elaborate theoretical explanation on the basis of the ideas and results of nonlinear acoustics.

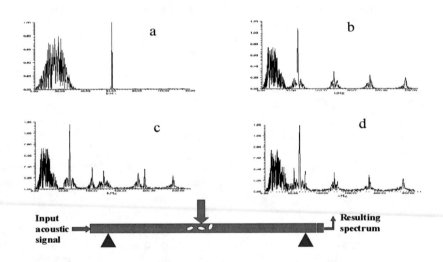

FIGURE 1. Nonlinear interaction of spectra of wide-band noise and harmonic wave at different stages of the development of cracks in loaded concrete beam: a) no cracks, b) single untied micro-cracks up to 10-15% of loss of strength, c) at the formation of systems of cracks (20-25% loss of strength), d) just before sudden destruction

In the last fifteen years, the methods of nonlinear nondestructive testing were put to practice by Robsman and his colleagues in transportation, power industry, and civil engineering. NDT was carried out for many bridges, highway overpasses, power stations, tunnels and subways. The work carried out in zones of seismic hazard made it possible to restore tens of structures damaged by earthquakes and to provide for the seismic safety of residential areas and power stations. In Moscow, nondestructive testing was used to monitor different stages of construction of the third transport ring, as well as in the restoration of some of the ancient buildings.

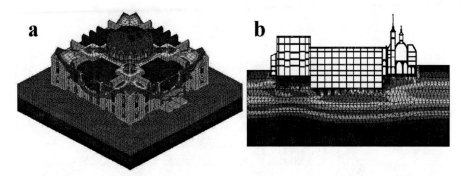

FIGURE 2. a) Computer model of deformation in the brickwork of Travel palace of Peter the Great. Model takes into account results of nonlinear testing of defects and damages. b) Modeling of force interaction between newly designed frame-house and old church (historic monument) in the center of Moscow. To form database for this finite-element problem, the strength of church was investigated by NDT method

NONLINEAR WAVE PROBLEMS

Robsman not only worked on the problems of construction industry but also obtained some fundamental results. He proposed and solved exactly a new nonlinear equation with a fourth-order derivative that describes intense waves in a scattering medium [6]

$$\frac{\partial p}{\partial x} - \frac{\varepsilon}{c^3 \rho} p \frac{\partial p}{\partial \tau} = -\beta \frac{\partial^4 p}{\partial \tau^4} \quad , \qquad \beta = \frac{8\langle \mu^2 \rangle a^3}{c^4}. \tag{1}$$

For turbulent media $\langle \mu^2 \rangle$ is the mean square of fluctuations of the refractive index, and a is the correlation radius. Shock wave in such media is non-monotonic; as distinct from Burgers equation, it contains fading oscillations. Although [6] was published only recently, references to the new model as the "Robsman equation" have already appeared in the mathematical literature.

In the paper that appeared one month before his death [7], Robsman solved the problem of nonlinear waves in a hysteretic medium. The formulation of the problem is directly related to the NDT of piles used in the construction of the third transport ring of Moscow. In solving this applied problem, Robsman generalized the well-known Mandel'shtam--Leontovich theory to the case of relaxation of the nonlinear "internal parameter". The form of single nonlinear in such medium is described by the equation

$$\left(\frac{\partial}{\partial \tau} + \frac{1}{T_R} \right) \left[\frac{\partial p'}{\partial x} - \frac{\varepsilon}{c_1^3 \rho_1} p' \frac{\partial p'}{\partial \tau} \right] = -\frac{\varepsilon}{2 c_1^3 \rho_1 T_R} \frac{\partial}{\partial \tau} [p' - p'_m(x)]^2 \tag{2}$$

53

where T_R is relaxation time for nonlinearity, and $p'_m(x)$ is the peak pressure of pulse wave. The profile of triangular pulse in hysteretic soil strongly differs from analogous profile in usual media.

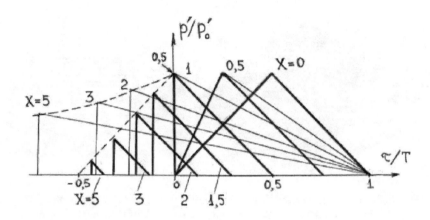

FIGURE 3. Transformation of monopolar pulse in hysteretic (thick solid curves) and usual (thin curves) media

At the reflection of pulse from the pile-ground interface self-blooming phenomena is observed: due to compacting of soil the reflected pulse disappears, and the wave energy completely penetrates to the ground.

REFERENCES

1. V.A.Robsman. *Moscow University Geography Bulletin* (1972).

2. V.A.Robsman, M.Sh.Shikhsaidov. *Phys.Solid State* **30**, No.8 (1988).

3. M.V.Kas'yan, G.N.Nikogosyan, V.A.Robsman. *Doklady-Geophysics.* No.3 (1989).

4. A.V.Kashko, V.A.Robsman. In the Book: *Testing ensuring earthquake-resistance of structures.* Yerevan, Armenia (1990) (in Russian).

5. Landa P.S., Firsov G.I., Robsman V.A. *J.Tech.Phys.*.**37**, No.3-4 (1993).

6. O.V.Rudenko, V.A.Robsman. *Doklady Physics* **7**, No.6 (2002).

7. O.V.Rudenko, V.A.Robsman. *Acoustical Physics.***50**, No.6 (2004).

Acoustoelectric Nonlinearity in Periodically Poled Lithium Tantalite.

I.V. Ostrovskii and A.B. Nadtochiy.

Department of Physics and Astronomy, University of Mississippi, University, MS 38677

Abstract. Some new nonclassical nonlinear acoustic phenomena in ferroelectric lithium niobate are reported last years. They are a) thermal hysteresis of nonlinear ultrasonic attenuation, b) acoustic memory, which is a strong non-exponential echo train that may not be explained by a nonparallelism of a sample, c) acousto-domain interaction that was detected by X-Ray technique and consists of a reorientation of the crystal lattice substructures under nonlinear action of ultrasound, d) new class of crystal sonoluminescence, which is generated at ferroelectric surface by the acoustic vibrations. In these phenomena, the involvement of the ferroelectric domain boundaries is experimentally identified by acoustically induced evolution of the X-ray diffraction patterns, chemical etching of the crystal surfaces, temperature characteristics of the effects observed, and by the changes in acousto-electric properties of a sample.
In this work, we report our experiments with lithium tantalite vibrators of two different types: a) two-dimensional multi domain array of the inversely poled ferroelectric domains in a y-rotated cut, and b) regular vibrator made of a single crystal. We measure the amplitude characteristics of rf-admittance (Y) of the samples in Megahertz frequency (f) range at room temperature. A strong nonlinearity is detected. The distortions in the Y(f) dependencies occur under increasing of ultrasonic excitation. These distortions are much stronger in multi domain sample with a number of so called domain walls. The experimental results allow making a conclusion on nonlinear vibrations of the domain walls in an acoustic field, and to discuss their contribution to nonlinear nonclassical phenomena.

Keywords: Periodically, Poled, Ferroelectric, Nonclassical, Nonlinear,
PACS: PACS: 43.25.Dc; 77.80.Dj

1. TWO-DIMENSIONAL PERIODICALLY POLED FERROELECTRIC RESONATOR.

Main areas of research in periodically poled (PP) ferroelectrics are the fabrication of inversely poled micro-domains in bulk crystals, which are sometime referred to as one dimensional acoustic super-lattice, and their applications for ultrasonic transducers [1-3]. Different ferroelectric crystals including $LiNbO_3$ and $LiTaO_3$ with one-dimensional periodically poled domain structures have been reported [3-5]. Scanning Electron Microscopy was employed to show a significant interaction of the surface acoustic waves with the periodically poled domains [6], which results in some nonlinear effects [7]. The domain walls and associated complexes of crystal lattice defects are responsible for a new type of nonlinear ultrasonic attenuation [8] and non-exponential echo patterns [9] in lithium niobate. Last years a fabrication of the ferroelectric domains at the nanoscale is of great interest [10]. We can not refer to any publication on acoustoelectric properties of multidomain ferroelectric structures in

CP838, *Innovations in Nonlinear Acoustics: 17th International Symposium on Nonlinear Acoustics*, edited by A. A. Atchley, V. W. Sparrow, and R. M. Keolian
© 2006 American Institute of Physics 0-7354-0330-9/06/$23.00

two-dimensional ferroelectric resonators. This work is devoted to theoretical calculations, computer simulations and experimental verification of the nonclassical nonlinear acousto-electric properties of the multidomain periodically poled ferroelectric structures in the thin resonators made of 3m-symmetry crystals.

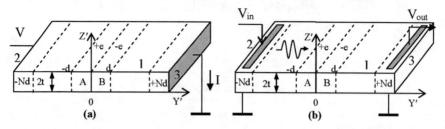

FIGURE 1. Experimental setup for the rf- admittance measurements (a) and acoustic vibration measurements (b). 1–LiTaO$_3$ plate with (2Nd)-length , 2 and 3 input and output electrodes, respectively.

We consider a ferroelectric plate consisting of N pairs of inversely poled domains as shown in Fig. 1. Axis Y' is parallel to wave vector $\vec{k}$, and Z'-axis is parallel to vector-normal $\vec{n}$. An inverse polarization in two adjacent domains is symbolized by opposite signs of the piezoelectric coefficients in the adjacent domains, that is $+e$ for A-type domains and $-e$ for B-type domains.

By solving the equations of motion, electrodynamics and boundary conditions including equalities of the stress tensor and acoustical displacements at the interfaces between the adjacent inversely poled domains, we derive the expression for rf-admittance $Y(\omega)$ of the vibrator:

$$Y(\omega) = \frac{i\omega A \varepsilon}{2L}\left[1 - \frac{T(N)K^2}{((k' + ik'')L/4)}\tan((k' + ik'')L/4)\right]^{-1} , \quad (1)$$

where A is an electrode area, wave vector $k = k' + ik''$, $K^2 = e^2/(c\varepsilon)$, and $T(N)$ is a trigonometric function. Actually, $T(N)$ is a sum of the trigonometric functions that represent all the domains, nevertheless the resonance in $Y(\omega)$ is at the frequency that corresponds to the length of only two inversely poled domains ($2d$). For the case of two inversely poled domains (N=1), $T(N)= 1$ at any frequency, and for arbitrary N the function $T(N)$ is about one at domain resonance/anti-resonance frequency range. The equation (1) yields an anti-resonance frequency $\omega_a = (\pi V/d)$ that is twice higher than for a single crystal vibrator [11] of the same length L and same acoustic wave speed V. To characterize a perfection of a material, a quality factor Q can be used, Q=($k'/2k''$). One can determine Q from the experimental measurements of rf-admittance $Y(\omega)$:

$$Q = (f_R / \Delta f), \quad (2)$$

where f_R is a resonance frequency corresponding to maximum magnitude of the rf-admittance (Y_M), and Δf is a frequency range within which $Y \geq 0.707\, Y_M$.

2. EXPERIMENTAL RESULTS.

We measure Y(f, A), Q(A) (setup in Fig.1a) and the echo trains (setup in Fig.1b) from a standard 0.35 mm thick lithium tantalite chip of $42°y$-rotated cut with a sequence of the periodically poled domains of 1 mm thick and periodicity of 2 mm. Three samples have width 20mm and number of domain pairs N from 7 to 12. The measurements are done in a digital mode at room temperature. The rf-admittance is measured with the help of spectrum analyzer Advantest-R3131A. The typical experimental result on Y(f) are shown in Fig. 2.

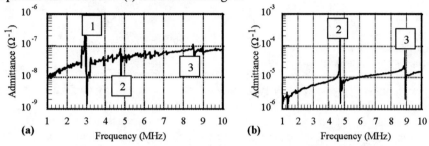

(a)   (b)

FIGURE 2. Rf-Admittance Y(f) in the PP sample LT4a3 with N=7, L=14mm (a) and single-domain sample LT10b1a, L=12.3mm (b). 1 – domain resonance (along Y′ axis), 2 – fundamental shear wave resonance along thickness, 3 – fundamental longitudinal wave resonance along thickness.

The domain resonance at 2.87-2.947 MHz (1 in fig. 2a) exists only in PP sample LT4a3, and the usual resonances are observed for both PP and single domain (SD) crystals. By using the data of Fig. 2 taken under different levels of applied rf-voltage, we determine the amplitude dependencies of Q for PP and SD crystals – Fig. 3. The strong nonlinearity and hysteresis are observed for PP crystal (panel (a)). An amplitude induced change in Q-factor reaches 30% for PP-crystal versus only 7% for

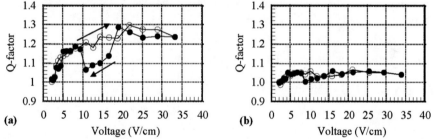

(a)   (b)

FIGURE 3. Amplitude dependencies of Q-factor in the periodically poled sample LT4a3 at domain resonance frequency of 2.87 MHz (a) and in the single domain sample LT10a3 at longitudinal resonance frequency of 0.82 MHz (b).

SD-sample. Since PP-domains bring to SD-crystal the inter-domain walls, which are a strong imperfection, we may conclude observed in Fig. 3a nonlinearity and hysteresis are connected to the domain walls. To clarify the physical processes involved, we examine the oscillogram of rf-burst propagation in two types of crystals. Typical results are shown in Fig. 4. *Unexpected new result consists of the secondary vibrations occurring in PP- crystal and denoted as S1, S2, S3 in Fig. 4a. These S-signals follow the main bursts A1, A2, A3.* No S-vibrations are detected from the SD-crystal- Fig. 4b.

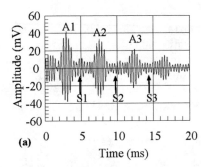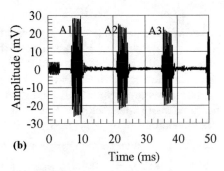

FIGURE 4. Wave propagation in PP LTO plate, sample LT4a3, rf 2.87 MHz (a) and wave propagation in Single-domain LTO plate, sample LT4a, rf 2.2 MHz (b).

In conclusion, we observe nonclassical nonlinearity in periodically poled lithium tantalite chips of SAW-type cut, which consists in a) amplitude induced distortions in the Y(f), b) amplitude dependent electro-mechanical quality factor Q and its hysteresis, and c) observation of the secondary vibrations that are complimentary signals to the initially introduced in periodically poled crystal. Since these peculiarities are absent or suppressed in a single domain crystal, we assume the inter-domain walls along with their surrounding including point and linear defects are responsible for the S-vibrations and other observed nonlinearities.

Experiment shows that as acoustic amplitude is higher, a relative amplitude of S-vibrations is increasing. That is why Q-factor of the domain resonance is increasing in Fig. 3a. The S-bursts can be responsible for "Acoustic Memory" [9].

Our Finite Element Method computations reveal the acousto-electric "domain resonance" [12], and *the secondary S-vibrations occurring only in multidomain ferroelectric*. The computer simulation is pretty close to the experimental data presented. Two dimensional PP-ferroelectrics at nanoscale are prospective devices for communication in the Gigahertz frequency range of 5 to 10 GHz and beyond.

ACKNOWLEDGEMENTS.

Authors want to thank Dr. Mack Breazeale for very useful discussions.

This work is made possible in part due to the research grant "Acousto-electric phenomena in multidomain ferroelectrics", UM, 2004/5.

REFERENCES.

1. Yong-yuan Zhu, Nai-ben Ming. *J. Appl. Phys.* **72**, 904 (1992).
2. O. Yu. Serdobolskaya, G.P. Morozova. *Ferroelectrics*, **208-209**, 395 (1998).
3. Yong-feng Chen, Shi-Ning Zhu, Yong-Yuan Zhu, Nai-ben Ming, Bio-Bing Jin, Ri-Xing Wu. *Appl. Phys. Lett.* **70**, 592 (1997).
4. A. Feisst, P. Koidl, *Appl. Phys. Lett.*, **47**, 1125 (1985).
5. V.V. Antipov et al. *Sov. Phys. Crystallogr.*, **30**, 428 (1985).
6. D.V. Roshchupkin, M. Brunel, *IEEE Trans. on UFFC.*, **41**, 512 (1994).
7. Golenishchev-Kutuzov A., Golenishchev-Kutuzov V., Kalimullin R. and Batanova N. *Ferroelectrics.*, **285**, 321 (2003).
8. Mack A. Breazeale, Igor V. Ostrovskii, Michael S. McPherson. *Jour. Appl. Phys.*, **96**, 2990 (2004).
9. M.S. McPherson, Igor Ostrovskii, and M. Breazeale. *Phys. Rev. Lett.*, **89**, 115506 (2002).
10. S. Kim, V. Gopalan, K. Kikamura, Y. Furukawa. *J. Appl. Phys.*, **90**, 2949, (2001).
11. Mason W.P. Physical acoustics, V. 1, pt.A, ch.3. Academic Press, New York (1964).
12. I.V. Ostrovskii, A.B. Nadtochiy., *Appl. Phys. Lett.*, **86**, 222902 (2005).

Experimental Investigation Of Excitation Techniques for Nonlinear Acoustic Mine Detection

Douglas Fenneman and Brad Libbey

U.S. Army RDECOM CERDEC Night Vision and Electronic Sensors Directorate
10221 Burbeck Road
Fort Belvoir, Virginia 22060

Abstract. Recently, a nonlinear acoustic landmine detection technique has been investigated [1, 2]. In this detection scheme, two tones are broadcast near a buried mine. The acoustic energy couples into the soil and causes both the mine and the surrounding soil to vibrate. Vibration sensors are then used to measure intermodulation frequencies generated in the vicinity of the mine. A comparison of the intermodulation effects above a mine to those at an off target location identifies the position of the buried target. Research indicates that the contrast ratios measured using nonlinear detection exceed those obtained using linear detection. However, the velocities produced at the intermodulation frequencies are significantly less than the velocities measured using linear detection.

The objective is to maximize the on/off target contrast ratio obtained in conjunction with non-linear acoustic mine detection. This goal will be pursued by experimentally examining excitation methods and the resultant surface velocities at nonlinearly generated frequencies. Specifically, the impact of increasing the primary amplitudes on soil surface velocity will be examined.

Keywords: nonlinear, acoustics, landmine detection
PACS: 43.25.Ts, 43.25.Zx, 43.40.Ga

INTRODUCTION

Acoustic mine detection techniques have been under investigation for a number of years [3, 4]. These detection methods employ airborne and/or seismic excitation to induce structural vibration in soil and emplaced mines. Landmines possess unique mechanical properties (e.g. compliant diaphragms flexible casings, air pockets, etc.) that distinguish the vibration response of soil above a buried mine from the vibration response at an off target location. A thorough comparison of velocity measurements over an area of interest is used to identify the position of a buried target.

Nonlinear acoustic mine detection was initially investigated by researchers at the Stevens Institute of Technology [1]. In this detection scenario, two tones, f_1 and f_2, are broadcast in the vicinity of a buried mine. Donskoy theorizes separation at the mine/soil interface due to relatively weak bonding between the soil particles and the mine's top plate. This nonlinear mechanism leads to the generation of combination frequencies (e.g. $|af_1 \pm bf_2|$ where a and b are integers) distinct from the excitation. In this formulation, the nonlinear effects above a buried mine are a function of the relative compliance of the soil compared to the buried mine. Donskoy's model, however, does not account for the nonlinear nature of the soil itself. Researchers at the Georgia Institute of Technology, for

CP838, *Innovations in Nonlinear Acoustics: 17th International Symposium on Nonlinear Acoustics*,
edited by A. A. Atchley, V. W. Sparrow, and R. M. Keolian
2006 American Institute of Physics 0-7354-0330-9/06/$23.00

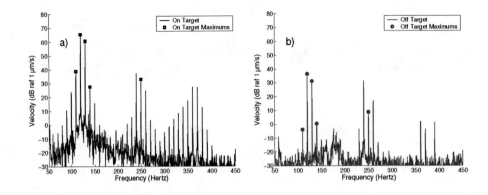

FIGURE 1. Comparison of a) on target and b) off target velocity spectra

example, have observed soil nonlinearities in the presence of displacements as small as $1\mu m$ [5]. These findings suggest that it is important to consider the nonlinear nature of the soil in conjunction with nonlinear acoustic mine detection techniques. The present research then is concerned with quantifying the crucial difference between the nonlinear response inherent to soil and nonlinearities produced in the presence of a buried mine.

EXPERIMENTAL FINDINGS

Introduction

The motivation underlying this research is to explore the difference between the nonlinear response of soil and nonlinearities generated above a buried target. To this end, experiments were conducted in which the amplitude of the nonlinear effects were tracked as a function of increasing drive amplitude. Specifically, one of the primary frequencies, f_2, was maintained at a constant level, while the amplitude of the other primary, f_1 was incremented. The response of the soil at representative combination frequencies was then studied as a function of increasing drive amplitude.

The experimental set-up consisted of a concrete rectangle measuring 57 by 57 by 23 cm filled with sifted, white sand. A VS1.6 anti-tank mine was buried at a depth of 2.5 cm in the geometric center of the rectangle and two loudspeakers were positioned above the soil surface. A geophone (Input Output SM-11) was employed to measure the velocity of the soil.

Nonlinear Testing

Prior to the start of nonlinear testing, a preliminary linear experiment was performed. The loudspeakers were driven in parallel using a swept sine excitation. This sweep identified a strong mine/soil resonance at 125 Hz. The primary frequencies, f_1=120

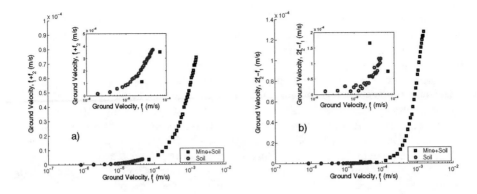

FIGURE 2. Nonlinear amplitude as a function of drive level at a) sum frequency f_1+f_2 and b) inter-modulation frequency $2f_2-f_1$. Insets provide zoomed in view of off target measurements.

Hz and f_2=130 Hz, were chosen above and below resonance to take advantage of the increased velocities near resonance. The soil was then ensonified at the primaries and the resultant ground motion was measured. A representative on/off target comparison is displayed in Figure 1. Observe that the VS1.6 is detectable linearly and nonlinearly. Contrast ratios of 29 dB and 30 dB are observed at the two primaries f_1 and f_2, respectively. At the combination frequencies, f_1+f_2, $2f_2-f_1$, and $2f_1-f_2$, contrast ratios of 24 dB, 27 dB, and 43 dB are observed, respectively.

After identifying the resonance frequency of the mine/soil system, the amplitude of the second primary, f_2, was maintained at its maximum value, while the amplitude of the other primary, f_1 was incremented. These measurements were performed in the presence of the VS1.6 and after removal of the target. In the presence of the VS1.6, soil velocities measured at f_2 were on the order of 2 mm/s. Off target soil velocities at f_2 were observed to be on the order of 80 μm/s.

The results of this amplitude study are presented in Figure 2. Soil velocity trends at the sum frequency, f_1+f_2, and a representative intermodulation frequency, $2f_2-f_1$, are displayed. It is interesting to observe that the amplitude of the nonlinear effects increase linearly at the sum frequency and quadratically at the combination frequency, $2f_2-f_1$, with increasing drive level. This trend is observed both in the presence of a buried target and after its removal. It is equally noteworthy that the nonlinear response of the soil at high drive levels is comparable in amplitude to the response of the mine/soil system at low drive levels.

Discussion

The nonlinear response at the soil surface in the presence and absence of a buried target is observed to increase for the combination frequencies considered. It has been shown that the nonlinear interaction of the mine/soil system "softens" when subjected to increasing excitation levels[2]. The fact that the nonlinear effects increased linearly and quadratically for the intermodulation frequencies considered suggest that the mine/soil system remained at or near resonance for these measurements. Additional increases in drivel level will effectively detune the mine/soil system. As the resonance frequency shifts downward, the primaries should be updated to reflect the resonance of the mine/soil system in order to properly quantify nonlinear effects.

The experimental results indicate the utility of nonlinear acoustic detection techniques. Contrast ratios observed at the combination frequencies of interest are sufficient for target detection. At the maximum excitation level, on/off target ratios of 26 dB and 40 dB are recorded at f_1+f_2 and $2f_2-f_1$, respectively. It is important to keep in mind, however, that these findings were obtained under ideal conditions. The impact of a number realistic field conditions, for example, the introduction of soil inhomogeneities, on the nonlinear response of the soil is subject to ongoing investigation.

Nonlinear discrimination requires the amplitude of the nonlinear effects above a buried mine to be large compared to off target locations. If the discrepancy between on and off target velocities is minimal, it is important to define criteria (e.g. thresholds) to differentiate nonlinear effects generated as a result of the mine/soil system and nonlinearities inherent to the soil.

CONCLUSIONS

The experimental results indicate that the nonlinear nature of soil must be considered in conjunction with nonlinear landmine detection. Additional research is required to improve understanding of the nonlinear mechanism governing the mine/soil system that will lead to robust detection techniques. An investigation is also recommended to experimentally quantify contrast ratios that are observed in a realistic environment.

REFERENCES

1. D. Donskoy, A. Ekimov, N. Sedunov, and M. Tsionskiy, *The Journal of the Acoustical Society of America*, **111**, 2705–2714 (2002).
2. M. Korman, and J. Sabatier, *The Journal of the Acoustical Society of America*, **116**, 3354–3369 (2004).
3. J. Sabatier, and N. Xiang, "Laser-Doppler Based Acoustic-to-Seismic Detection of Buried Mines," in *Detection and Remediation Technologies for Mines and Minelike Targets VI*, edited by A. C. Dubey, J. F. Harvey, J. T. Broach, and R. E. Dugan, 3710, International Society for Optical Engineering, 1999, pp. 215–222.
4. W. Scott Jr., and J. Martin, "Experimental investigation of the acousto-electromagnetic sensor for locating land mines," in *Detection and Remediation Technologies for Mines and Minelike Targets VI*, edited by A. C. Dubey, J. F. Harvey, J. T. Broach, and R. E. Dugan, 3710, International Society for Optical Engineering, 1999, pp. 204–214.
5. G. Larson, J. Martin, W. Scott Jr., and G. McCall, "Environmental factors that impact the performance of a seismic landmine detection system," in *Detection and Remediation Technologies for Mines and Minelike Targets VI*, edited by A. C. Dubey, J. F. Harvey, J. T. Broach, and V. George, 4394, International Society for Optical Engineering, 2001, pp. 563–574.

Nonlinear Acoustic Experiments Involving Landmine Detection: Connections with Mesoscopic Elasticity and Slow Dynamics in Geomaterials

Murray S. Korman

Physics Dept. U.S. Naval Academy, Annapolis, Maryland 21402, and U.S. Army RDECOM CERDEC, Night Vision and Electronic Sensors Directorate, 10221 Burbeck Road, Fort Belvoir, Virginia 22060

James M. Sabatier

National Center for Physical Acoustics, University of Mississippi, University, Mississippi 38677

Abstract. The vibration interaction between the top-plate of a buried VS 2.2 plastic, anti-tank landmine and the soil above it appears to exhibit similar characteristics to the nonlinear mesoscopic/nanoscale effects that are observed in geomaterials like rocks or granular materials. [J. Acoust. Soc. Am. **116**, 3354-3369 (2004)]. When airborne sound at two primary frequencies f_1 and f_2 (closely spaced near resonance) undergo acoustic-to-seismic coupling, (A/S), interactions with the mine and soil generate combination frequencies $|n f_1 \pm m f_2|$ which affect the surface vibration velocity. Profiles at f_1, f_2, $f_1 - (f_2 - f_1)$ and $f_2 + (f_2 - f_1)$ exhibit single peaks whereas other combination frequencies may involve higher order modes. A family of increasing amplitude tuning curves, involving the surface vibration over the landmine, exhibits a linear relationship between the peak particle velocity and corresponding resonant frequency. Subsequent decreasing amplitude tuning curves exhibit hysteresis effects. New experiments for a buried VS 1.6 anti-tank landmine and a "plastic drum head" mine simulant behave similarly. Slow dynamics explains the amplitude difference in tuning curves for first sweeping upward and then downward through resonance, provided the soil modulus drops after periods of high strain. [Support by U.S. Army RDECOM CERDEC, NVESD, Fort Belvoir, VA.]

Keywords: nonlinear, acoustic, landmine detection
PACS: 43.25.Ts, 43.40.Fz, 43.40.Ga

INTRODUCTION

It has been established that buried plastic anti-tank mines (e.g. VS 1.6 and VS 2.2) and smaller plastic anti-personnel mines (e.g. VS 50), can be detected in soils and roadways with very low rates of false alarms using linear A/S coupling techniques.[1,2] Donskoy[3,4] has introduced a nonlinear acoustic technique for buried landmine detection. Here, nonmetallic landmines are buried in soil and a two tone excitation (f_1 and f_2) provides the airborne sound which through A/S coupling ultimately interacts

CP838, *Innovations in Nonlinear Acoustics: 17th International Symposium on Nonlinear Acoustics*,
edited by A. A. Atchley, V. W. Sparrow, and R. M. Keolian
© 2006 American Institute of Physics 0-7354-0330-9/06/$23.00

with the top surface of the mine. Donskoy's results show significant soil surface vibration amplitude at the so-called combination frequencies $|mf_1 \pm nf_2|$, where m and n are integers, indicating that the nonlinear coupling is strong. In some cases the nonlinear technique improves the "on the mine" to "off the mine" contrast ratio and eliminates certain types of false alarms.

Recent measurements of the surface vibration tuning curve response (over the mine) exhibit a decrease in resonant frequency with increasing acoustic drive amplitude and hysteresis effects.[5,6] These effects are characteristic of geophysical materials which exhibit *mesoscopic/nanoscale nonlinear elasticity* undergoing vibration near a resonance.[7,8]

EXPERIMENTAL RESULTS

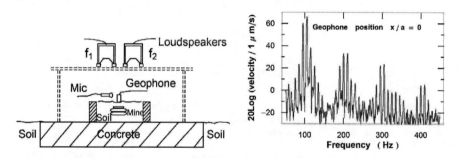

FIGURE 1. Experimental setup for the inert VS 1.6 landmine buried 5.0 cm deep in sifted loess soil. FIGURE2. Spectral geophone response for the two-tone excitation at $f_1=95$ and $f_2=105$ Hz.

Experiments are performed by placing the primary acoustic tones $f_1 = 95$ and $f_2 = 105$ Hz on either side of the strong resonant peak 100 Hz which was found from tuning curve results that will be presented later in this paper.

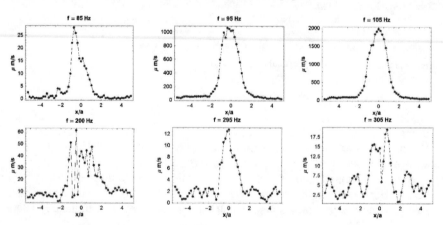

FIGURE 3. Profiles of the soil surface particle velocity at a particular frequency component are obtained from the spectrum of the geophone response to the two-tone excitation at each position x/a .

64

The profiles shown in Figure 3 were selected from a larger set and correspond to f_1-(f_2-f_1) = 85, f_1=95, f_2=105, f_+=200, $2f_1+f_2$=295, and $2f_2+f_1$=305 Hz. The latter three suggest that higher order mode shapes of the top plate are involved. Let the profile coordinate x run from into the page to out of the page (in Figure 1) across the center of the mine. The top plate radius of the VS 1.6 is a = 7.2 cm.

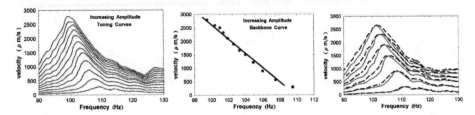

FIGURE 4. VS 1.6 tuning curve results in loess soil at x/a=0; (a) sweeping up and then incrementing the drive amplitude, (b) the corresponding back-bone curve, and (c) first sweeping up, then sweeping down (broken curve) and then incrementing the drive amplitude. In (c) any sweep time is 26 sec, with a step size of 1/8 Hz and time increment of 65 ms between steps.

Tuning curve experiments are performed when the 12 inch subwoofers (shown in Figure 1) are connected with a tee and driven with an amplified swept sinusoidal signal. The geophone response was recorded by the Agilent 35670A dynamic spectrum analyzer operating in the swept sine mode. In Figure 4(b) the back-bone curve represents the locus of points involving the resonant frequency (obtained from the horizontal tangent) and associated peak particle velocity. This linear behavior cannot be predicted from classical nonlinearity[9] but is characteristic of the mesoscopic nonlinear behavior that is found in geomaterials.[8]

An attempt was made to observe the effects of *slow dynamics* behavior from tuning curve measurements (involving the VS 1.6 buried landmine) that might be similar to the effects reported by J.A. TenCate, E. Smith and R.A. Guyer in a variety of sandstones, limestones and concrete.[10] (See also Ref. 11). Figure 4(c) shows a subset of results where one first sweeps upward through resonance from 80 to 130 Hz, then immediately downward through resonance. Next, the acoustic drive amplitude is incremented and the sweeping up and down process is repeated. The results for low amplitude up/down sweeps show little deviation as expected, however for higher drive amplitudes the up/down tuning curves are notably different in shape.

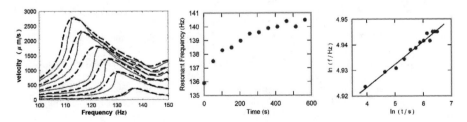

FIGURE 5. VS 1.6 tuning curve results in white sand at x/a=0; (a) first sweeping up, then sweeping down and then incrementing the amplitude, (b) and (c) recovery after large acoustic strain is removed.

When the VS 1.6 is buried at 2.5 cm deep in fine sifted white sand, the bending of the tuning curves with drive amplitude (Figure 5) is much greater than in the loess soil case. It is possible, as in slow dynamic behavior in geomaterials, that the differences in the shape of the tuning curve (first swept upward, then downward) in Figure 5(a) indicates a persistent modulus reduction following periods of high strain, which shifts the resonance downward for a finite time after each passage through resonance. The sand recovers from a high acoustic strain at a much slower rate than the loess soil. Figures 5(b,c) show the time dependence of resonance at a low strain, sometime after a high strain has been removed.

ACKNOWLEDGEMENTS

This work is supported by the U.S. Army RDECOM CERDEC, NVESD, Fort Belvoir, Virginia. The authors wish to express their gratitude to Kelly Sherbondy, Charles Amazeen, Douglas Fenneman and Brad Libbey for all their support at NVESD. Special thanks go to Paul Johnson for enlightening discussions on mesoscopic nonlinear elasticity. MSK wishes to acknowledge support from the Physics Dept. at the U.S. Naval Academy and by the Naval Academy Research Council.

REFERENCES

1. J.M. Sabatier and N. Xiang, "An investigation of a system that uses acoustic-to-seismic coupling to detect buried anti-tank landmines," IEEE Trans. Geoscience and Remote Sensing **39**, 1146-1154 (2001).
2. N. Xiang and J.M. Sabatier, "An experimental study on antipersonnel landmine detection using acoustic-to-seismic coupling," J. Acoust. Soc. Am. **113**, 1333-1341 (2003).
3. D.M. Donskoy, "Nonlinear vibro-acoustic technique for land mine detection," in *Detection and Remediation Technologies for Mines and Minelike Targets III*, ed. by A. C. Dubey, J. F. Harvey and J. T. Broach, SPIE Proc. **3392**, 211-217 (1998); "Detection and discrimination of nonmetallic land mines," in *Detection and Remediation Technologies for Mines and Minelike Targets IV*, ed. by A. C. Dubey, J. F. Harvey, J. T. Broach and R. E. Dugan, SPIE Proc. **3710**, 239-246 (1999).
4. D.M. Donskoy, A. Ekimov, N. Sedunov, and M. Tsionskiy, "Nonlinear seismo-acoustic land mine detection and discrimination," J. Acoust. Soc. Am. **111**, 2705-2714 (2002).
5. M.S. Korman and J.M. Sabatier, "Nonlinear tuning curve vibration response of a buried land mine," in *Detection and Remediation Technologies for Mines and Minelike Targets VIII*, ed. by R.S. Harmon, J. H. Holloway, Jr., and J. T. Broach, SPIE Proceedings, **5089**, 476-486 (2003).
6. M.S. Korman and J.M. Sabatier, "Nonlinear acoustic techniques for landmine detection," J. Acoust. Soc. Am. **116**, 3354-3369 (2004).
7. R.A. Guyer and P.A. Johnson, "The astonishing case of mesoscopic elastic nonlinearity," *Physics Today* **52**, 30-35 (1999).
8. L.A. Ostrovsky and P.A. Johnson, "Dynamic nonlinear elasticity in geomaterials," Rivista Del Nuovo Cimento, **24**, serie 4, No. 7, 1-46 (2001).
9. R.T. Beyer, *Nonlinear Acoustics* (Naval Ship Systems Command, Department of the Navy, 1974).
10. J.A. TenCate, E. Smith and R.A. Guyer, "Universal Slow Dynamics in Granular Solids," Phys Rev.. Letters, **85**, No. 5, p. 1021, Fig. 1 (31 July 2000).
11. J.A. TenCate and T.J. Shankland, "Slow dynamics in the nonlinear response of Berea sandstone," Geophys. Res. Lett. Vol. **23** 3019-3022 (1996).

Experimental study of nonlinear acoustic effects in unconsolidated granular materials

V. Tournat[*], V.E. Gusev[†] and B. Castagnède[*]

[*]Laboratory of Acoustics at University of Maine UMR-CNRS 6613, University of Maine, Av. Olivier Messiaen, 72085 Le Mans, France.
[†]Laboratory of Solid State Physics UMR-CNRS 6087, University of Maine, Av. Olivier Messiaen, 72085 Le Mans, France.

Abstract. Experimental results of ultrasonic wave propagation through unconsolidated granular materials are reported. Influence of the applied static pressure and the propagation distance on the linear and nonlinear acoustic properties is studied. The aim of this work is to understand better the linear and nonlinear acoustic propagation through granular assemblages when the wavelength is of the order of the bead diameter. In particular, the role of the so-called force chains is currently not well understood, as well as the influence of the contact distribution of the medium. In particular, the self-demodulation of shear wave packets with variation of the excitation amplitude is documented for different configurations of granular media (propagation distance, static pressure etc ...). Modeling is currently in progress and should allow together with these results to understand better the acoustic propagation through granular assemblages.

Keywords: Nonlinear acoustics, granular media, multiple scattering, self-demodulation
PACS: 43.25.+y, 45.70.-n, 43.20.Fn, 43.25.Lj

INTRODUCTION

Complex media in acoustics often exhibit several physical features at the same time: velocity dispersion, frequency dependent absorption, scattering, nonlinearity ... The study of regimes where several of these features play an important role on wave propagation at the same time can be difficult to achieve. For instance, in a complex medium where the linear properties of absorption are not well-known, it is much more difficult to study quantitatively the nonlinear absorption properties. In this paper, the nonlinear self-demodulation of shear wave packets is studied in a strong multiple scattering regime. The multiple scattering manifests itself in the attenuation of the primary waves but also in the structure of the acoustic field, both of first importance in the self-demodulation process [1, 2]. Consequently, the physical features considered here are closely linked, which makes traditionnally information difficult to extract. By different experiments, we try in the following to analyze these features selectively.

EXPERIMENTAL SETUP

The granular medium considered here is composed of glass beads of 2 mm in diameter. It is contained in a cylindrical box of 20 cm diameter. Two shear transducers (wideband centered on 100 kHz), are located at the top and at the bottom of the container (see Fig.

CP838, *Innovations in Nonlinear Acoustics: 17th International Symposium on Nonlinear Acoustics*,
edited by A. A. Atchley, V. W. Sparrow, and R. M. Keolian
© 2006 American Institute of Physics 0-7354-0330-9/06/$23.00

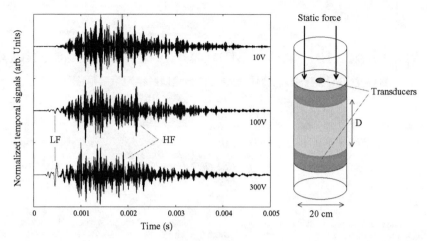

FIGURE 1. Left: temporal signals obtained after a propagation distance of 10.5 cm with a static stress of 300 MPa for three different excitation amplitudes. Signals are normalized to their corresponding excitation amplitude. Right: Schematics of the experimental configuration.

1). A static pressure can be applied on the granular medium and can be varied from several MPa to 300 MPa. The propagation distance D can also be varied. The emitted signals are Gaussian wave packets (sine wave amplitude modulated with a Gaussian function) having 6-8 wave periods and centered on 100 kHz, which is of the order of the cut-off frequency of the granular medium [3]. Consequently, strong multiple scattering is expected in this configuration. Before amplification with a linear amplifier, signals are high-pass filtered above 50 kHz to avoid direct low-frequency generation in the medium.

EXPERIMENTAL RESULTS

In Fig. 1, the presented signals are typical for multiple scattering media, and have already been observed in a very similar form in granular materials [3, 4]. They can be divided into two different contributions, a high frequency (HF) contribution, which corresponds to the generated frequency, and a low frequency (LF) contribution, which corresponds to a frequency which is not generated by the transducer but is much lower. The HF part is commonly associated with the incoherent contribution due to mutiple propagation pathes through the disordered grain and contact network. It is often denoted as the "coda" and its intensity shape can de fitted by the solution of a diffusion equation [5]. We show in this article that the LF part of the signal, often denoted as the coherent part of the total signal, is in our experimental conditions, a self-demodulated signal. It corresponds to a nonlinearly generated wave resulting from the interactions of the different frequency components of the primary emitted wave [1]. The amplitude dependence of the LF contribution is clearly shown in Fig. 1 where the HF part behaves almost linearly (with a slight amplitude saturation shown in the following) while the LF part amplitude increases nonlinearly with excitation amplitude.

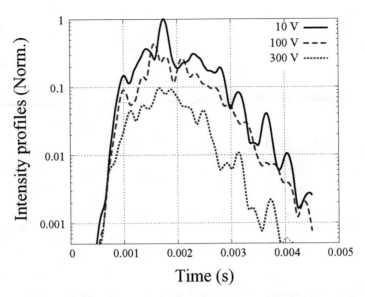

FIGURE 2. Temporal representation of the intensities of the HF contributions of signals shown in Fig. 1, normalized to their respective excitation intensity. These profiles are obtained by integrating the squared temporal signal windowed by a sliding Hanning function and by associating the obtained value to the central time of the window.

A visible amplitude saturation effect can be seen on the intensity profiles of the HF contribution as shown in Fig. 2. The interesting feature in terms of granular medium characterization is that the amplitude saturation is not the same as a function of delay time. At latest delay times, the amplitude saturation is higher than at the middle delay times, and is almost absent at earliest delay times. Theoretical interpretations in terms of amplitude dependent medium parameters are currently in progress. But the important feature described above may also give information on characteristic times in the medium that would explain the time dependence of the observed saturation mechanism.

A classical representation to illustrate the self-demodulation effect is shown in Fig. 3(a). The relative contributions of the HF and LF parts of the signal clearly evolves with propagation distance : ratios of HF energy over LF energy are 0.3 for D=12.5 cm, 10 for D=10.5 cm and 200 for D=6.5 cm. These relative changes are mainly due to the strong HF energy decrease with distance more than the LF energy variation.

Finally, in Fig. 3(b), the evolution of the signal energy with static pressure is plotted for two excitation amplitudes. HF and LF parts are separated by filtering. While HF contribution is strongly dependent on the static pressure (through scattering and absorption), the LF part does not exhibit any clear dependence. The decrease of the medium nonlinearity with the increase of static pressure is just compensated by the decrease in the attenuation of the HF part. A LF linear signal containing frequencies close to the self-demodulated signals has also been generated and monitored as a function of the applied static pressure. As seen in Fig. 3(b), the LF linear signal energy is increasing with static pressure mainly due to the decrease of dissipation (scattering is low for these low

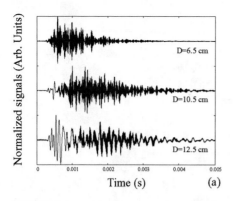 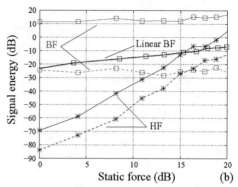

FIGURE 3. (a) Signals received after propagation through different thicknesses of granular medium, for identical static pressure of 300 MPa and excitation amplitude 300 V. (b) Signal energy dynamics as a function of the applied static pressure for two excitation amplitudes. The HF and LF contributions are separated by filtering. The energy of a reference LF linear signal is also plotted for comparison.

frequencies). The difference in the energy dynamics as a function of the static pressure for the LF linear signal and the self-demodulated signal may allow to characterize the relative roles of absorption and scattering in the acoustic attenuation on the nonlinear self-demodulation effect.

CONCLUSIONS

In this paper, we have shown that under some experimental conditions, two different contributions can be identified in the mutiple scattered signal that are recorded after propagation through granular materials: a HF contribution which is incoherent and a LF contribution which is coherent and nonlinear (a self-demodulated signal). Specific features of each of these contributions are exhibited: amplitude saturation effect and strong signal energy dependence on the applied static pressure concerning the HF contribution, independence of the signal energy on the applied static pressure for the LF contribution.

The results presented here are expected together with the currently developed modeling to provide a better understanding of the nonlinear acoustic properties of granular materials, shown to be appropriate to probe the contact properties inside granular media.

REFERENCES

1. V. Tournat, V.E. Gusev and B. Castagnède, *Phys. Rev. E*, **66**, 041303 (2002).
2. V. Tournat, B. Castagnède, V.E. Gusev and P. Béquin, *C.R. Mecanique*, **331**, 119-125 (2003).
3. X. Jia, C. Caroli and B. Velicky, *Phys. Rev. Lett.*, **82**, 1863-1866 (1999).
4. X. Jia, *Phys. Rev. Lett.*, **93**, 154303 (2004).
5. J.H. Page, H.P. Schriemer, A.E. Bailey and D.A. Weitz, *Phys. Rev. E*, **52**, 3106 (1995).

Nonlinear Resonant Ultrasound Spectroscopy (NRUS) Applied to Damage Assessment in Bone

M. Muller[1], J. A. Tencate[2], T. W. Darling[2], A. Sutin[3], R. A. Guyer[4], M. Talmant[1], P. Laugier[1], and P. A. Johnson[2]

[1]Laboratoire d'Imagerie Paramétrique, CNRS, Université Paris VI, France
[2]University of California, Los Alamos National Laboratory, Los Alamos, New Mexico
[3]Stevens Institute of Technology, Hoboken, New Jersey
[4]Department of Physics, University of Massachussets, Amherst, Massachussets

Abstract. This study shows for the first time the feasibility of Nonlinear Resonant Ultrasound Spectroscopy (NRUS) techniques for damage characterization in bone. Two diaphysis of bovine bone were subjected to a progressive damage experiment. Fatigue damage was progressively induced in the samples by mechanical testing in 11 steps. At each damage step, the nonlinear elastic parameter was measured using NRUS. For independent assessment of damage, high energy X-ray CT imaging was performed, but only helped in the detection of the prominent cracks. As the amount of damage accumulation increased, a corresponding increase in the nonlinear response was observed. The measured nonlinear response is much more sensitive than the change in modulus. The results suggest that NRUS could be a potential tool for micro-damage assessment in bone. Further work has to be carried out for a better understanding of the physical nature of damaged bone, and for the ultimate goal of in vivo implementation of the technique.

Keywords: Nonlinear Resonant Ultrasound Spectroscopy, bone, microcracks
PACS: 43.80.Qf, 43.80.Ev, 43.25.Ba, 43.25.Gf

INTRODUCTION

Several studies have revealed a strong correlation between bone micro-damage accumulation and bone fragility (1), suggesting the importance of micro-damage assessment for fracture risk prediction. However, in vivo micro-damage has remained relatively poorly documented due to the lack of non-invasive techniques for its in vivo assessment. It is therefore crucial to develop a damage assessment technique that could be used in vivo.

Accumulation of damage in bone leads to a stress-strain behaviour that is correspondingly more nonlinear than in healthy bone (2), which may be a manifestation of a softening of the bone, due to a larger crack density. It is now well known that damaged materials display a characteristic, elastic-nonlinear behaviour (3). Nonlinear Resonant Ultrasound Spectroscopy (NRUS) is a technique exploiting this behaviour and has proved to be valuable for damage detection, because of its high sensitivity (4). The objective of this study was to explore for the first time the potential of Nonlinear Resonant Ultrasound Spectroscopy to assess progressively induced bone damage.

CP838, *Innovations in Nonlinear Acoustics: 17th International Symposium on Nonlinear Acoustics*,
edited by A. A. Atchley, V. W. Sparrow, and R. M. Keolian
© 2006 American Institute of Physics 0-7354-0330-9/06/$23.00

NONCLASSICAL BEHAVIOUR

When materials are subjected to strains above roughly 10^{-6}, it is currently thought that macro and micro-cracks, as soft mesoscopic structural features in a rigid matrix, are responsible for a characteristic nonlinear response related to the presence of strain memory and hysteresis in the stress-strain relation. One of the manifestations of this behaviour is a shift of the resonance frequency as the excitation level increases. This phenomenon can be understood as a softening of the material when the excitation level increases, that becomes more pronounced with accumulated damage. A phenomenological model (5) has been developed to describe this specific response. In this formalism, the frequency shift can be expressed as a function of the average strain applied to the material over a cycle (Eq.1 where f is the resonance frequency and f_0 is the resonance frequency at the lowest, linear drive level). The parameter α is called the nonlinear hysteretic parameter and it correlates directly to accumulated damage.

$$\frac{\Delta f}{f_0} = \frac{f - f_0}{f_0} \approx \alpha \Delta \varepsilon \quad (1)$$

Using Eq.1, we can extract α from the frequency shift as a function of strain, as the sample is progressively damaged and becomes more elastically nonlinear.

EXPERIMENTAL PROCEDURE

The study was carried out on two fresh bovine femur specimens with soft tissue and marrow removed. The two samples (termed B1 and B2) were progressively damaged in 11 steps by compressional fatigue cycling in an INSTRON 5569 press.

At each damage step, NRUS experiments were performed on the samples: each sample was probed using a step-sweep in frequency around one of its eigenmodes and the process was repeated at gradually increasing drive levels. An analytical simulation based on a unidimensional propagation model helped in the selection of an eigenmode of the bone sample which would be well separated from adjacent modes. Particle velocities related to the radial displacement close to the top of the bone sample were measured using a laser vibrometer.

High energy X-ray, 3-dimensional computed tomography (CT) was performed in order to have an independent assessment of damage in the samples (Hytec Inc, Los Alamos, NM, USA). The pixel size of the CT images was 127µm. The common length of bone microcracks is in a range of 5µm to 500µm so this resolution only allowed the detection of the larger cracks. Nonetheless, imaging the samples at each step allowed us to control that we were actually inducing macro-damage in the samples.

RESULTS

Fig. 1 shows resonance curves, obtained by measuring the radial velocity close to the top of sample B1, and corresponding CT images, for damage steps #0 and #9. The resonant peak shift was more significant in the progressively damaged stages (right) than in the undamaged sample (left). α was extracted according to Eq. 1.

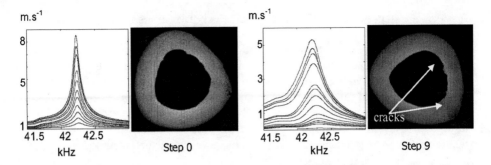

Figure 1. Example of resonance curves, obtained around the 42kHz eigenmode of the system for damage steps 0 (Left) and 9 (Right). The radial velocity measured close to the top of sample B1 is plotted as a function of frequency.

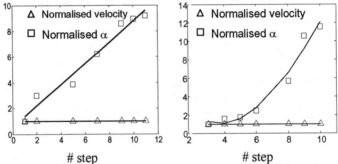

Figure 2. Normalised nonlinear parameter α (squares) and normalised velocity (triangles) as a function of damage step in the two samples (left: sample B1, right: sample B2).

Fig. 2 shows the behaviour of the speed of sound in the sample (derived from the linear resonance frequency f_0 at each step) and α as a function of damage step for the two samples. As damage accumulates, the speed of sound is almost constant (although a slight decrease can be observed in sample B2 when damage increases). On the other hand, α dramatically increases with the accumulated damage. The difference between linear and nonlinear measurements is remarkable.

DISCUSSION

Fig. 1 shows a resonance frequency shift in both the "intact" and damaged samples. This was expected since healthy bone contains continually-healing microcracks. Moreover, intact bone by itself could be classified as a nonlinear mesoscopic elastic material, because of its heterogeneous mesoscopic structure. The behaviour of α and the speed of sound (SOS) as a function of damage is illustrated in Fig. 2. Clearly, the nonlinear parameter α measured by NRUS is much more sensitive to damage than the speed of sound, related to the Young's modulus of the material, which stays nearly constant during the entire progressive damage experiment. α begins to increase at the first damage step, even though macro-damage is not yet discernable on the X-ray CT images.

An independent quantitative evaluation of micro-damage (higher resolution imaging like micro-CT for example) would be useful here to determine a semi-quantitative relation between the measured nonlinear parameter α and micro-damage. Another limitation of this study is the small number of data points. Indeed, the number of damage steps is not sufficient to correctly understand the shape of the curves shown in Fig. 2. In any case, the trend is clearly demonstrated in this work: there is a strong relation between micro-damage and the nonlinear hysteretic parameter α.

CONCLUSION

This is the first study showing the ability of NRUS to detect damage in bone. A progressive damage experiment has been conducted on two samples of bovine bone. The increasing amount of damage mechanically induced in the sample leads to an increased shift in the resonance frequency with wave amplitude, indicating a progressively more nonlinear behaviour of the sample, as damage accumulates. This work is a preliminary study and some aspects, such as the fact that only the radial displacement is sensitive to damage, are not well understood yet. For an accurate quantification of damage in human bone, those experiments must be repeated on human bone, with a larger number of data points. Further work will have to be carried on for the *in vivo* application of the technique to see if it may be viable as diagnostic tool for skeletal status assessment.

ACKNOWLEDGMENTS

This work was supported by Institutional Support (LDRD) and by the Institute of Geophysics and Planetary Physics at Los Alamos, and the Centre National pour la Recherche Scientifique (CNRS, France). The authors would like to acknowledge P. Zysset for helpful discussions and comments, and Hytec Incorporated (www.hytecinc.com) for the imaging facility.

REFERENCES

1. P. Zioupos, *J Microsc* **201**, 270-8, (2001)
2. D. R. Carter, et al., *Acta Orthop Scand* **52**, 481-90, (1981)
3. J. Ostrovsky, *La Rivista del Nuovo Cimento dellà Socità Italiana de Fisica* **24**, (2001)
4. K. V. D. Abeele, et al., *Research on non destructive evaluation* **12**, 31-42, (2000)
5. R. A. Guyer and K. R. McCall, *Physical Review Letters* **74**, 3491-94, (1995)

Nonlinear Acoustic and Ultrasonic NDT of Aeronautical Components

Koen Van Den Abeele[a)], Tomasz Katkowski[a,b)], and Christophe Mattei[c)]

[a)] K.U.Leuven Campus Kortrijk, Interdisciplinary Research Center
E. Sabbelaan 53, B-8500 Kortrijk, Belgium
[b)] University of Gdańsk, Institute of Experimental Physics, ul. Wita Stowsza 57, 80-952 Gdańsk, Poland
[c)] CSM Materialteknik AB, Gelbgjutaregatan, SE-581 13 Linköping, Sweden

Abstract. In response to the demand for innovative microdamage inspection systems, with high sensitivity and undoubted accuracy, we are currently investigating the use and robustness of several acoustic and ultrasonic NDT techniques based on Nonlinear Elastic Wave Spectroscopy (NEWS) for the characterization of microdamage in aeronautical components. In this report, we illustrate the results of an amplitude dependent analysis of the resonance behaviour, both in time (signal reverberation) and in frequency (sweep) domain. The technique is applied to intact and damaged samples of Carbon Fiber Reinforced Plastics (CFRP) composites after thermal loading or mechanical fatigue. The method shows a considerable gain in sensitivity and an incontestable interpretation of the results for nonlinear signatures in comparison with the linear characteristics. For highly fatigued samples, slow dynamical effects are observed.

Keywords: CFRP, Nonlinear Resonance Spectroscopy, Microdamage Diagnostics, Nondestructive Testing, Slow Dynamics.
PACS: 43.25.+x, 43.25.Gf, 43.25.Dc, 43.25.Ba

NONLINEAR RESONANCE TESTING PROCEDURE AND ANALYSIS

A fully non-contact measurement system has been developed to excite and measure the response of precut samples of CFRP. The nominal size of the samples is 120 by 12 by 4 mm^3. The beams are hung up by nylon wires from a frame. A low distortion, high frequency speaker with a focusing cone (opening = 20 mm) is positioned near the middle of the beam facing the 120 mm by 12 mm surface. Typical frequencies of excitation range from 1kHz to 2kHz. On the back side, the response is measured using a laser vibrometer in a point near the left or right edge of the sample. The driving of the speaker and the acquisition of the response is fully automated and controlled through LabVIEW. Two types of "single mode nonlinear resonance acoustic spectroscopy (SIMONRAS)" experiments were performed: a time domain measurement (TD-SIMONRAS) [1,2] and a frequency domain (FD-SIMONRAS) process [3-5]. In TD-SIMONRAS, the sample is excited at a frequency near its fundamental flexural vibration mode, till a steady state is reached (duration of 1000 periods). Then the excitation is stopped, and the response of the reverberation of the sample (with time) is recorded. To achieve a high accuracy, we implemented a variable dynamic range acquisition procedure based on an automated feedback of the

CP838, *Innovations in Nonlinear Acoustics: 17th International Symposium on Nonlinear Acoustics*,
edited by A. A. Atchley, V. W. Sparrow, and R. M. Keolian
© 2006 American Institute of Physics 0-7354-0330-9/06/$23.00

instantaneous amplitude response. Doing so, the dynamic range is adjusted each 20ms, and the signals are averaged 10 times. In the frequency domain method, FD-SIMONRAS, several traditional "discrete" frequency sweeps, encompassing the fundamental flexural mode frequency, are performed at gradually increasing excitation, and the response of the fundamental component in steady state is recorded as function of the drive frequency and drive level.

We analyze the "composed" signals obtained in the TD-SIMONRAS experiment by fitting small moving time windows of the signal (converted to strains) with an exponentially decaying sine function: $\varepsilon_A e^{-\alpha t} \sin(2\pi f t + \phi)$. The output of this analysis procedure yield the evolution of the frequency (f) and damping (α) as function of the strain amplitude ε_A in the decaying signal. Similarly, we analyze the FD-SIMONRAS resonance curves by determining the coordinates of each resonance peak using polynomial fitting, and we save the resonance frequency as function of the converted strain amplitude at the resonance peak. The typical analysis of both experiments is shown in Figure 1a. In order to quantify the degree of nonlinearity, we determine the linear proportionality coefficient γ between the resonance frequency shift and the strain amplitude, $\Delta f/f_o = \gamma \varepsilon_A$.

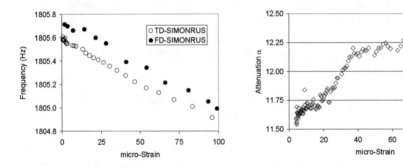

FIGURE 1. a) Typical analysis of the strain amplitude dependent frequency using TD-SIMONRAS and FD-SIMONRAS; b) Typical analysis of the strain dependent attenuation using TD-SIMONRAS.

RESULTS FOR THERMALLY DAMAGED CFRP

Figure 2 shows the results of the nonlinearity coefficient γ for one untreated reference sample (displayed 4 times) and 20 samples of CFRP which have been subjected to thermal treatments at various temperatures (240-300°C) and different exposure times (15-60 minutes). All measurements were performed at room temperature long after the treatment. We clearly observe a gradual increase of the nonlinear parameter as function of exposure temperature and time (up to a factor 10). We attribute the increase in nonlinearity to the global increase of the amount of soft bonding system of the material (elastically soft constituent within a rigid matrix) which originated as a result of the thermal treatment. On the other hand, we discovered that the linear (or low excitation) value of attenuation in the samples has decreased with increasing temperature. At first this seemed unexpected, since damage normally is reflected in a higher attenuation. However, in this case, we believe that the decrease in the linear attenuation property corresponds to the evaporation of fluids and the physico-chemical transformation of the rigid material components rather than to the alteration of the mechanical properties of the interaction zone between them. The latter effect can be observed through the fact that the attenuation characteristic is increasing with driving amplitude (cfr. Figure 1b).

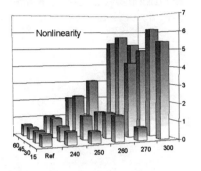 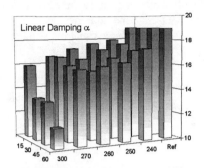

FIGURE 2. a) Nonlinearity coefficients obtained by TD-SIMONRAS for thermally exposed CFRP samples. b) Linear attenuation characteristic for the same set of samples.

RESULTS FOR FATIGUED CFRP

Following the tests on thermally damaged CFRP samples, we performed additional tests on similar CFRP samples after being subjected to fatigue loading. Figure 3a compares the TD-SIMONRAS analysis for 4 samples: a reference sample and 3 samples with increasing number of fatigue cycles (17000, 25000 and 80000 cycles). Up to 25000 fatigue cycles the behavior is similar to the results obtained in the case of thermally damaged samples, i.e., a linear reduction of the frequency with amplitude. For a larger number of fatigue cycles, higher order effects on the nonlinearity are present, causing an apparent saturation of the frequency reduction at high amplitudes. If we take the tangent of the $f(\varepsilon)$ curve as the measure of nonlinearity (first order), we notice a substantial increase of the nonlinearity coefficient γ relative to the reference sample, from a factor 28 at 17000 cycles, 43 at 25000 cycles, to a factor 1300 at 80000 cycles. In contrast with the observations on thermally damaged samples, we now also measured an increase of the linear damping factor, as expected (the low amplitude α changes from 16 to 19 with increasing number of cycles). Indeed, for fatigued samples, there is no evaporation of fluids, and no physico-chemical transformation of the rigid material constituents due to the mechanical loading. All of the linear (and nonlinear) attenuation increase is attributed to the increase of the soft bonding system.

For all thermally damaged samples and all moderately fatigued CFRP samples (less than 25000 cycles), we observed consistent results between FD and TD-SIMONRAS, independent of the initial excitation amplitude. However, once the fatigue level is high, we systematically observe a difference between the FD-SIMONRAS results obtained at increasing levels of excitation and the results for TD-SIMONRAS for similar initial levels of the excitation. On the one hand, Figure 3b clearly shows that the highest amplitude analysis of the TD-SIMONRAS (gray circles, connected with lines) corresponds quite well with the results of the FD-SIMONRAS (black squares) technique at the same excitation level, but on the other hand it is obvious that the recovery of the resonance frequency after being excited at high amplitude is not immediate. Moreover the recovery is dependent on the highest excitation/response level. This is a clear evidence of a slow dynamical effect [6-7] and can be attributed to the fact that the material memorizes temporarily the modulus reduction which it experienced at large dynamic strains.

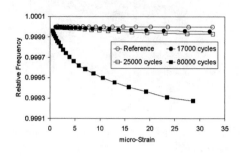 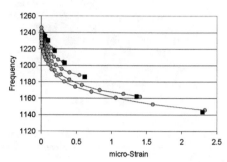

FIGURE 3. a) Analysis of TD-SIMONRAS for fatigued CFRP samples as function of the number of fatigue cycles. The frequency is normalized to the linear value of the resonance frequency for each sample b) Observation of Slow Dynamical effects in fatigue damaged samples.

ACKNOWLEDGMENTS

The authors gratefully acknowledge the support of the Flemish Fund for Scientific Research. (G.0206.02 and G.0257.02), the provisions of the European Science Foundation Programme NATEMIS and the European FP5 and FP6 Grants DIAS (EVK4-CT-2002-00080) and AERONEWS (AST-CT-2003-502927).

REFERENCES

1. Van Den Abeele K., Campos-Pozuelo C., Gallego-Juarez J., Windels F., Bollen B., "Analysis of the nonlinear reverberation of titanium alloys Fatigued at high amplitude ultrasonic vibration" in *Proc. Forum Acusticum 2002* (3st Convention of the EAA,. Acta Acustica), Sevilla, September 2002, publication on CD
2. Van Den Abeele K. and De Visscher J., *Cement and Concrete Research* **30/9**, 1453-1464 (2000)
3. Johnson P.A., Zinszner B. and Rasolofosaon P.N.J., *J. Geophys. Res* **101**, 11553-11564 (1996)
4. Van Den Abeele K., Carmeliet J., TenCate J.A. and Johnson P.A., *Res. Nondestr. Eval.* **12/1**, 31-42 (2000)
5. Van Den Abeele K., Van De Velde K., and Carmeliet J., *Polymer Composites* **22(4)**, 555-567 (2001)
6. TenCate, J.A., and T. J. Shankland, *Geophys. Res. Lett.* **23(21)**, 3019-3022 (1996)
7. Johnson P.A. and Sutin A., *J. Acoust. Soc. Am.* **117(1)**, 124-130 (2005)

Nonlinear Parameter Measurement for Nondestructive Evaluation of Solids : Calibrated Phase Modulation Method

O. Bou Matar[a], M. Vila[b], F. Vander Meulen[b], L. Haumesser[b] , J. Fortineau[b], T. Goursolle[b] , S. Dos Santos[b]

[a]IEMN UMR8520 CNRS, Cité Scientifique - Avenue Poincaré, BP 69, 59652 Villeneuve d'Ascq Cedex, France
[b]Laboratoire UltraSons Signaux et Instrumentation de L'Université François Rabelais /GIP Ultrasons, FRE2448 CNRS, EIVL, Rue de la Chocolaterie, BP 3410, 41034 Blois Cedex, France

Abstract. A phase modulation method is used for Nondestructive Evaluation of solid samples. We take advantage of a self reciprocity calibration procedure to achieve quantitative measurements of the nonlinear parameters of homogeneous solids samples made of fused silica, aluminum, steel and glass, as well as inhomogeneous ones (cracked glass and quenched steel).

Keywords: NDE, calibration, phase modulation, nonlinear parameter.
PACS: 43.25.Ba, 43.25.Dc, 43.25.Zx

INTRODUCTION

It is now confirmed that nonlinear methods of damage detection turn out to be more sensitive to damage-related structural alterations that any known method based on measurements of linear parameters. In order to complete the variety of nonlinear methods for nonlinear acoustic parameter measurements, a phase modulation method, using the parametric interaction of a low frequency (LF) wave (1,5 MHz) and a high frequency wave (15 MHz), has been improved for sample characterization in contact [1]. The main achievement is the calibration procedure based on the self-reciprocity principle extended and validated in the time domain: the absolute value of the LF particle velocity needs to be known for quantitative determination of the sample nonlinear parameter.

We present the set-up, which allows the evaluation of the LF particle velocity through electrical measurements.. Nonlinear parameters of fused silica, aluminum, steel and glass samples will be presented and advantages of this method for any kind of homogeneous material with parallel interfaces, will be highlighted. Furthermore, results on cracked glass samples will be used in order to investigate the influence of cracks in the phase modulation process.

CP838, *Innovations in Nonlinear Acoustics: 17th International Symposium on Nonlinear Acoustics*, edited by A. A. Atchley, V. W. Sparrow, and R. M. Keolian
© 2006 American Institute of Physics 0-7354-0330-9/06/$23.00

EXPERIMENTAL METHOD DESCRIPTION

The sample is placed between two contact transducers. A high frequency (HF) tone burst of frequency $f_{HF} = \omega_{HF}/2\pi$ comprised between 15 and 20 MHz is inserted in the material from the right interface of the sample(figure 1a). A low frequency (LF) pulse of central frequency equal to one- tenth of the HF wave one, is applied at the other face in the opposite direction so that the two waves propagate on a distance d collinearly during the back propagation towards the HF frequency transducer.. The nonlinear parameter is proportional to the phase modulation index $\Delta\Phi$ of the high frequency wave and the particle velocity v_{LF} at the LF transducer radiating area:

$$\beta = \frac{\Delta\Phi}{v_{LF}} \frac{c_{HF}^2}{\omega_{HF}d} e^{\alpha_{HF}d} \tag{1}$$

Phase modulation measurement

The experimental setup include three function generators. Two of them send electrical excitations to the transducers and the last one to synchronize the propagation of the two waves in the sample. A 60dB power amplifier is added to evaluate the phase modulation index at various excitation levels of the LF generator. Another amplifier is needed for the tone burst treatment in the case of attenuating material, to ensure sufficient signal to noise ratio. The signal is received by the HF transducer, filtered and transmitted through a diplexer to a numerical scope recording and further processing. The high pass filter is crucial to avoid wave mixing in the electronic part of the receiving system. In the same manner, test have been conducted to verify that no parametric phase modulation occurred in the HF transducer.

Self-reciprocity calibration procedure

The HF transducer is removed to calibrate the LF transducer (figure 1b). The LF transducers is laid in contact with the sample using an aqueous gel, excited by a pulse delivered by a HP 3314A generator and amplified by an ENI amplifier. The Thévenin-voltage source in figure 1b is equivalent to both the generator and the amplifier. A x10 voltage probe and a 60 MHz current probe are used to measure the excitation pulses V_{in} and I_{in}, and the receiving echo signals V_{out} and I_{out}. Using the self-reciprocity technique, the LF velocity v_{LF} is calculated in time domain by:

$$v_{LF}(t) = FFT^{-1}\left[H_v(f)I_{out}(f)e^{\alpha_{LF}d} \right] \tag{2}$$

where:
$$H_v(f) = \sqrt{-\frac{1}{2AZ_0} \frac{E_{in}}{I_{out}} \frac{1}{D(f)}} \; e^{jk_{LF}d} \tag{3}$$

with A as the effective transducer radiating area, Z_0 as the acoustic impedance of the media, E_{in} as the Thévenin-equivalent voltage source of the amplifier, I_{out} as the output electrical current, α_{LF} and k_{LF} as the attenuation and wave number of the LF wave into the sample, respectively, and $D(f)$ as a diffraction correction for transmitter.

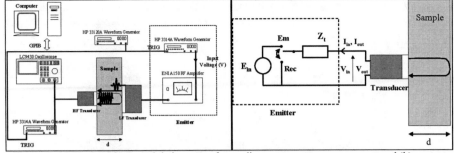

Figure 1. (a) Contact phase modulation setup for nonlinear parameter measurement and (b) contact self-reciprocity calibration setup.

RESULTS AND DISCUSSION

The samples studied can be classified as undamaged or damaged. The undamaged samples are homogeneous plates, made of fused silica, aluminum, glass and XC48 steel. Two kinds of damaged samples are investigated: two XC48 steel samples quench processed, and cracked glass. The three steel samples are identified by their HRC hardness values: HRC = 10 (undamaged sample), 33 and 64. The values of the nonlinear parameters measured from various samples, are summarized in Table 1. The results agree with referenced ones for undamaged samples implying the method is well suited for homogeneous materials with parallel plane sides. Furthermore, it is sensitive to the sign of the nonlinear parameter as can be seen on the fused silica sample (figure 2). However, it appears that the method becomes difficult to implement in the case of highly attenuated samples. In order to maintain the SNR for the HF wave, frequencies of both signals have to be turned down making the condition $f_{HF} = 10 f_{LF}$ and the thickness of the sample as the restricting factors to the method. Concerning the steel samples, it can be noted that the nonlinear parameter value decreases as a function of HRC hardness values (figure 3.b). Phase modulation results are shown for healthy glass (figure 4.a). Because of the inhomogeneity in the distribution of cracks, no value of β can be given. But the measured β in cracked samples (figure 4.b) are about an order of magnitude higher than in the healthy sample. A promising feature to note is the shape distortion of the phase modulation index in the cracked glass samples.

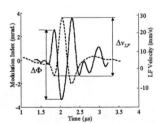 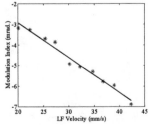

Figure 2. (a) Time evolution of phase modulation extracted from the HF wave (solid line), and LF velocity (dashed line) in fused silica. (b) Modulation index of the HF signal for fused silica versus the amplitude of the LF transducer surface velocity.

81

TABLE 1. Nonlinear parameter values for many plate samples.

Material	β measured		
Fused silica	-6		
Aluminum	8		
Glass	7		
Steel	β (HRC10) = 7.8	β (HRC33) = 6.3	β (HRC64) =1.7

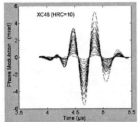 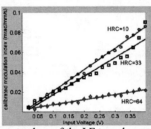

Figure 3. (a) Phase modulation in XC48 steel for different values of the LF transducer surface velocity. (b) Same as Figure 2b., but now the slope include calibration. Nonlinear parameter is evaluated for different HRC hardness values obtained after quenching process.

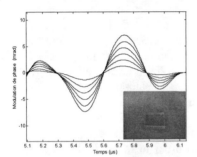 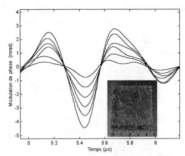

Figure 4. Phase modulation in (a) healthy and (b) cracked glass for different values of the LF transducer surface velocity. Note the shape distortion of the phase modulation behavior between samples.

ACKNOWLEDGMENTS

The authors would like to thank Marion Bailly for technical assistance in the experimental setup design. This work was supported by the EC Sixth Framework Program AERONEWS: Specific Targeted Research FP6-502927 (http://www.kulak.be/AERONEWS).

REFERENCES

1. Vila M., Vander Meulen F., Dos Santos S., Haumesser L. and Bou Matar O. *Contact phase modulation method for acoustic nonlinear parameter measurement in solid* Ultrasonics 2004, V.42(1-9) pp. 1061–1065.
2. Barrière C. and Royer D. *Diffraction effects in the parametric interaction of acoustic waves : application to measurements of the nonlinearity parameter B/A in liquids* IEEE, Trans. on UFFC2001, V.48(6) pp. 1706–1715.
3. J.K. Na, J.H. Cantrell, and W.T. Yost. *Linear and nonlinear ultrasonic properties of fatigued 410cb stainless steel.* Rev. Prog. QNDE, 15: pp. 1347–1351, 1996.

High Amplitude Studies of Natural Dome Salt and Hydrostatically Pressed Salt

I. Lucas and B. R. Tittmann

The Pennsylvania State University
Department of Engineering Science and Mechanics
212 Earth-Engineering Sciences Bldg.
University Park, PA 16802

Abstract. Techniques and results are described for obtaining the onset of nonlinearity of forced resonance vibrations in natural dome salt and hydrostatically pressed salt at ~500 Hz. The technique is centered on the use of an inertially loaded vibrating bar apparatus. The onset of nonlinearity was observed at ~10^{-6} strain. The attenuation (Q^{-1}) was observed to increase while the resonant frequency decreased. Concomitant with the nonlinear behavior were asymmetries in the resonance wave and the observation of a memory effect when the samples were subjected to a large number of cycles.

Keywords: Nonlinear, ultrasonics, salt, attenuation, Q-factor, resonance frequency, forced vibration resonance, memory effect.
PACS: 43.25+y

INTRODUCTION

Salt domes are formed when a thick bed of halite found a depth intrudes vertically into surrounding rock. Since the density of salt is less than that of the surrounding material, it moves upward toward the surface forming large bulbous domes from 1-10 km across and at as far down as 6.5 km.

The rock salt in these domes is impermeable so pockets can form where petroleum and natural gas collect. A major source of the petroleum produced is from salt domes in the Gulf of Mexico. Other uses of salt dome pockets include storage of oil, gas, and hazardous waste. For locating the pockets by seismic exploration the linear and nonlinear acoustic properties must be known. This paper presents preliminary data on some of these properties.

EXPERIMENTAL APPROACH

The natural dome salt (Texas) samples were cylindrically shaped with diameter of ~2 cm and length of ~20 cm. The material was homogeneous and isotropic and low in moisture. The stress relaxation occurs over small distances relative to the sample dimensions.

CP838, *Innovations in Nonlinear Acoustics: 17th International Symposium on Nonlinear Acoustics*,
edited by A. A. Atchley, V. W. Sparrow, and R. M. Keolian
© 2006 American Institute of Physics 0-7354-0330-9/06/$23.00

The halite samples were received after hydrostatic pressing. They were found to be prone to moisture uptake and had to be dried at 120°C for several days to remove the moisture. Substantial differences were found for both the dielectric constants and loss factor between dry and moist (25% humidity) salt between 0.1 and 100 kHz.

Figure 1 shows the inertially loaded resonant bar apparatus described in detail previously [1-7]. The use of non-contact electromagnetic transducers allowed the resonant vibrations to be driven in either flexure or torsion. The design features included a stiff clamp-arm-rotor assembly to minimize bending moments in any parts other than the sample and rigid mechanical coupling between sample and rotor-arm assembly. The maximum torsion and flexure strain amplitudes were $\varepsilon_{max} = \dfrac{2r\theta}{L}$ and $\varepsilon_{max} = \dfrac{\delta L}{L}$ respectively where L is the sample length and r is the sample radius. θ is the torsion angle and δL is the change in the sample length. The sample cylinders were clad and sealed with soft metal. The apparatus was inserted into a pressure chamber which was used to apply effective pressure in the range of 0 to ~70 MPa.

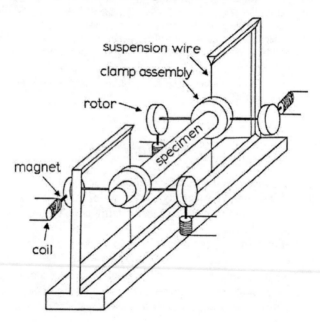

FIGURE 1. Schematic illustration of the inertially loaded resonant bar apparatus. Resonant vibrations are driven in either flexure or torsion using noncontact electromagnetic transducers.

RESULTS

For natural dome salt the Q-factor and modulus showed a strain amplitude dependence above 2×10^{-6} and at all effective pressures to ~70 MPa. For pressed salt (Halite 336B-IOB), weakly nonlinear behavior persisted to very low amplitudes (below 10^{-8}) corresponding to stresses of ~0.01 bar. Under ambient pressure conditions the attenuation was sensitive to the amount of moisture absorbed. The Q^{-1} and resonant frequency for flexure as a function of strain amplitude for Halite 336B-IOB is shown in Figure 2 for three different effective pressures. The graphs show some of the features described above.

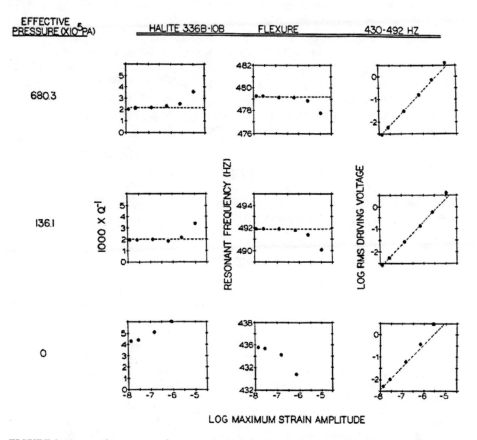

FIGURE 2. Attenuation, resonant frequency, and log driving voltage plotted as a function of log maximum vibration amplitude for extensional waves (flexure) at 430-492 Hz, and at several effective pressures.

The natural dome salt samples were subjected to repeated forced torsional resonance cycles up to ~2.5×10^5 over a time period up to ~6×10^3 seconds under 0.7 MPa effective pressure. For low strain amplitudes there was negligible change in peak vibration amplitude with time but for successively increased strain amplitudes (nonlinear regime) the peak vibration amplitude was found to decrease successively more with time. This effect is apparently due to damage accumulation perhaps in the form of increasing dislocation density, grain boundary weakening and microvoid formation. These effects were found to reverse in comparable periods of time.

CONCLUSION

The results show that the force swept resonance technique appears satisfactory for measuring the Q of a linear anelastic material and for defining the transition from linear to nonlinear behavior. For natural dome salt the transition occurs at strains of 2 x 10^{-6}. The linear anelastic Q for dry specimens was ~500 in flexure and ~1000 in torsion independent of pressure up to 6.8×10^7 Pa and very slightly dependent on frequency between ~80 Hz and ~480 Hz. The behavior of pressed salt contrasted with that of natural dome salt. Weakly nonlinear behavior persisted to very low amplitudes (below 10^{-8}) corresponding to stresses of approximately 0.01 bars. Strongly nonlinear behavior was observed at strain amplitudes higher than 10^{-6}. Under ambient pressure conditions attenuation in pressed salt was sensitive to the amount of moisture adsorbed either on the surface or within the interior of the specimen.

ACKNOWLEDGEMENT

The authors gratefully acknowledge the contributions of Larry Bivens and James Bulou. The work was carried out under AFOSR contract number F49620-82-C-0015.

REFERENCES

1. Tittmann, B. R., "Studies of Absorption in Salt, Final Report for the Period Dec. 1, 1981 through Nov. 30, 1982." Prepared for Air Force Office of Scientific Research under contract number F49620-82-C=0015 (1983).
2. Clark, V. A. Tittmann, B. R., and Spencer, T. W., "Effect of Volatiles on Attenuation (Q^{-1}) and Velocity in Sedimentary Rocks," *J. of Geoph. Research*, No. B10, #80B0180, 85:5190-5198 (1980).
3. Tittmann, B. R., Nadler, H., Clark, V. A., Ahlberg, L. A., and Spencer, T.W., "Frequency Dependence of Seismic Dissipation in Saturated Rocks," *Geophys. Research Lett.* No. 1, 8:36-38 (1981).
4. Bulau, J. R., Tittmann, B. R., Abdel-Gawad, M., and Salvado, C., "The Role of Aqueous Fluids in the Internal Friction of Rocks," *J. Geophys. Res.*, 89:4207-4212 (1984).
5. Tittmann, B. R., "Characterization of Porous Media with Elastic Waves," in *Ultrasonic Methods in Evaluation of Inhomogeneous Materials*, NATO Workshop on Wave Propagation and Inhomogeneous Media, Traponi, Italy (1985). Edited by A. Alippi and W. Mayer, NATO Series E:Applied Sciences No. 126:301-319 (1987).
6. Tittmann, B. R., "Internal Friction Measurements and their Implications in Seismic Q Structure Models of the Crust," Geophysical Monograph 20, "The Structure and Physical Properties of the Earth." (John G. Heacock, Editor), *American Geophysical Union*, pp. 197-215 (1978).
7. Tittmann, B. R., Ahlberg, L., Nadler, H., Curnow, J., Smith, T., and Cohen, E.R., "Internal Friction Quality-Factor Q Under Confining Pressure," *Proc. Lunar Science Conf. 8th*, 3:1209-1234 (1977).

Three Nonlinear NDE Techniques On Three Diverse Objects

Kristian C.E Haller, and Claes M. Hedberg

Blekinge Institute of Technology, 371 79 Karlskrona, Sweden

Abstract. Non-Destructive Evaluation has been carried out on three different test objects, with three different methods based on exhibits of slow dynamics and nonlinear effects. The three diverse objects were cast iron, ceramic semi-conductors on circuit boards, and rubber. The three approaches were Higher Harmonics detection (HH), Nonlinear Wave Modulation Spectroscopy (NWMS), and Slow Dynamics (SD). For all of the objects the three approaches were tried. The results showed that for each of the objects, a different method worked the best. The cast iron worked best with nonlinear wave modulation, the ceramic semi-conductors worked well with the higher harmonics detection, while the rubber showed best results with slow dynamics.

Keywords: Nondestructive testing, nonlinear acoustics, higher harmonics, nonlinear wave modulation spectroscopy, slow dynamics.
PACS: 43.25.Dc

INTRODUCTION

Sensitive nonlinear methods are in strong progressive development. High sensitivity gives the opportunity to find defects and characterize the material characterization in a very careful manner. The Higher Harmonic (HH) [1] method has been used in evaluation for some time. Nonlinear Wave Modulation Spectroscopy (NWMS) [2] and Slow Dynamics (SD) [3],[4] have been introduced in later years. In this work, all three of these methods have been applied to diverse objects to evaluate the objects damage and to determine what method is best to use. Three diverse objects were examined and they are from different industrial areas.

One object is a cast iron steel part of a large cutting tool. First an intact undamaged part was tested to find out if the response was nonlinear from pre-existing cracks and defects. A damage was introduced at the top of the cutting edge and then the result of the new measurement is compared to the one for the undamaged.

Ceramic semiconductors used for capacitive loads are very common in electrical components and are manufactured as standard components to be machine mounted on printed circuits. These semiconductors sometimes break partially and can give intermittent errors. The ceramic semiconductors were bent to introduce cracks and defects but since they are brittle they tend to break completely (instead of partly cracking). A solution was chosen where the piezo ceramics are bonded to the printed circuit. Damage is introduced while the ceramic semiconductors are still held in place by its soldering spots.

CP838, *Innovations in Nonlinear Acoustics: 17th International Symposium on Nonlinear Acoustics,*
edited by A. A. Atchley, V. W. Sparrow, and R. M. Keolian
© 2006 American Institute of Physics 0-7354-0330-9/06/$23.00

Rubber for cable transmission sealing through walls is manufactured in blocks and then cut to desired size and shape. Before the cut operation one need to determine the quality and homogeneity of the block. Two different batches from the production were tested, one of good quality and another less good.

EXPERIMENTAL METHODOLOGY AND RESULTS

These experiments are all conducted using commercially available piezoceramic transducers (PZT) glued with epoxy. One PZT is acting as an actuator to generate the acoustic wave in the specimen and the other is acting as a pickup of the acoustical wave.

For the higher harmonics measurement a single sinusoid acoustic wave is sent to the specimen by the signal generator through one PZT. The signal is distorted by cracks and other defects on its way to the PZT acting as a sensor. In the recorded frequency spectrum higher harmonics are detected to indicating the amount of cracks or defects.

For the Nonlinear Wave Modulation Spectroscopy (NWMS), a single frequency signal is repeated as for (HH), but now with an addition of some of the specimen's own resonant frequencies, lower in frequency than the single high frequency signal. This is accomplished by hitting the object with a test hammer. In the recorded frequency spectrum sidebands appear beside the single frequency if defects and cracks are present.

The Slow Dynamics time signal method (SD) is the monitoring of the resonance frequency after a material is conditioned by a transient force. If the material is damaged, the resonance frequency is found to drop instantaneously and then the resonance frequency is slowly recovering - logarithmically with time - to its initial value.

All methods were applied on all objects for evaluation. For each object the best suited method is presented, with results from the other methods briefly commented.

For the cast iron cutting tool the NWMS method was best, see Figure 1. The material itself shows nonlinear response in the undamaged state, indicated by all three methods. This was an expected result because casting iron is a complicated process and inhomogeneities and inclusions can be problematic for this process. Calculating the sideband energy and comparing the undamaged and damaged states showed a high difference in values. The HH showed differences in number of harmonics and in their amplitude, but was not as clear as NWMS. The difference in SD indicated defect, but was not large.

FIGURE 1. Nonlinear Wave Modulation Spectroscopy results for cast iron shown as sideband energy for undamaged in the right column, and damaged in the left column.

The ceramic-semiconductors mounted to the printed circuit board showed the HH worked best. Higher harmonics was present even when the semiconductors not was damaged. This indicates that either the board itself gives nonlinear response or some component mounted was damaged, but testing a larger number of boards indicates the nonlinear response arises from the board. When damaging the semi-conductors the number of multiple harmonics increases also the amplitude of the harmonics increases, Figure 2. The results from NWMS was indicating defects but not as strong as the HH. Also SD was indicating defects but not as strong as HH.

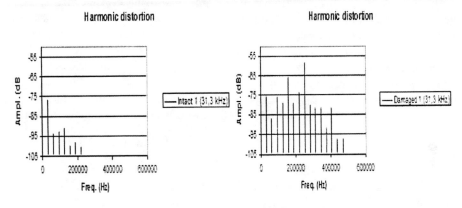

FIGURE 2. Higher Harmonics results of an undamaged circuit board ceramics on the left, and a damaged one on the right.

For the rubber bricks the SD method was the only one to indicate damage and quality differences, Figure 3. NWMS gave no measurable differences in sideband energy. No clear differences in HH could give results to conclude material quality differences.

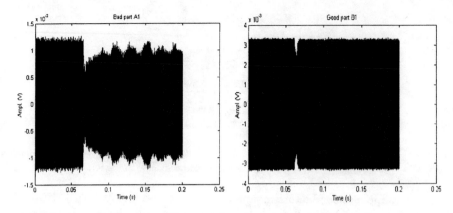

FIGURE 3. Slow Dynamics time signal results of good quality rubber on the left, and bad quality rubber on the right.

DISCUSSION

The method of Higher Harmonics (HH) was succesfully applied to circuit boards for detection of damage of ceramic semi-conductors. For circuit boards the method with no impacts is convenient when one would like to be careful with the components. The Nonlinear Wave Modulation Spectroscopy (NWMS) method worked very good with the cast iron. Rubber was an unexperienced area for the authors before conducting these experiment. Remarkably only Slow Dynamics (SD) could detect differences. It is known SD is a very sensitive method, but the other methods gave no indication to separate between the different samples. This is the first time we have encountered the case where SD yields high response, while the fast dynamics methods (NWMS and HH) would not indicate any change with damage.

ACKNOWLEDGEMENTS

This work is financed by the grant "Nonlinear nondestructive evaluation of material conditions - resonance and pulse techniques" from Vetenskapsrådet (Sweden). We would like to thank the TANGO-Verkstad for project support. The European Science Foundation programme NATEMIS (Nonlinear Acoustic Techniques for Micro-Scale Damage Diagnostics) is also acknowledged.

REFERENCES

1. Buck,O., Morris,W.L. and Richardson,J.M., *Appl.Phys.Lett.* **33**. 371-372 (1978).
2. Van Den Abeele, K.E.-A., Johnson, P.A. and Sutin, A., *Res.Nondestr.Eval.*, **12**, 17 - 30 (2001).
3. R.A.Guyer and P.A.Johnson, *Physics Today*, April 1999, 30-36.
4. TenCate, J.A., and Shankland, T.J., *Geophys.Res.Lett.* **23**(21), 3019-3022 (1996).

Multiscale Approach For Simulating Nonlinear Wave Propagation In Materials with Localized Microdamage

Sigfried Vanaverbeke and Koen Van Den Abeele

K. U. Leuven Campus Kortrijk, Interdisciplinary Research Center
Nonlinear Acoustics laboratory, E. –Sabbelaan 53, 8500 Kortrijk, Belgium

Abstract. A multiscale model for the simulation of two-dimensional nonlinear wave propagation in microcracked materials exhibiting hysteretic nonlinearity is presented. We use trigger-like elements with a two state nonlinear stress-strain relation to simulate microcracks at the microlevel. A generalized Preisach space approach, based on the eigenstress-eigenstrain formulation, upscales the microscopic state relation to the mesoscopic level. The macroscopic response of the sample to an arbitrary excitation signal is then predicted using a staggered grid Elastodynamic Finite Integration Technique (EFIT) formalism. We apply the model to investigate spectral changes of a pulsed signal traversing a localized microdamaged region with hysteretic nonlinearity in a plate, and to study the influence of a superficial region with hysteretic nonlinearity on the nonlinear Rayleigh wave propagation.

Keywords: nonlinear wave propagation, nonlinearity, micocracks, hysteresis, multiscale models
PACS: 43.25.+y, 43.25.+x

INTRODUCTION

Nonlinearity and hysteresis in the quasi-static stress-strain relation are instantly recognizable phenomena observed in all damaged solids. It is reasonable to assume that nonlinearity has important consequences for the dynamics of microcracked solids as well. Several experiments conducted in the last decade have provided ample evidence for this [1-3]. Including nonlinearity, and in particular hysteresis, in numerical models for wave propagation presents significant challenges. Over the last decade, several researchers have tried to incorporate hysteretic stress-strain relations into existing codes for linear wave propagation, mostly in the case of one-dimensional simulations [4-7], but recently the extension to two dimensional versions [8-9] has gained interest because of its appropriateness in the field of non-destructive testing. In this contribution, we extend the previously reported 1D EFIT multiscale model proposed by Van Den Abeele et al. [7] by applying a generalization of the Preisach-Mayergoyz-(PM)-space model for multidimensional hysteretic nonlinearity [10]. We illustrate the model by results of two-dimensional wave propagation for nonlinear Rayleigh waves and in plane pulse propagation in a plate with localized damage.

CP838, Innovations in Nonlinear Acoustics: 17th International Symposium on Nonlinear Acoustics,
edited by A. A. Atchley, V. W. Sparrow, and R. M. Keolian

91

NUMERICAL MODEL

The implementation of the nonlinear 1D EFIT multiscale model has been described in detail by Van Den Abeele et al. [7]. EFIT, originally introduced by Fellinger et al. [11] and intensively applied by Schubert [12], basically implements a discretisation of the stress rate equation and of the Cauchy equation of motion in terms of a staggered grid formulation in both space and time. In our previous one-dimensional model, we introduced hysteretic nonlinearity in the stress-strain relation at the level of each cell in the numerical grid by a multiscale approach: each material element can be thought of as being composed of a statistical ensemble of trigger like microscopic features with elementary two-state hysteresis operators (e.g. microcracks, inhomogeneities). Subsequently, the elastic modulus at the level of a cell (which we consider as the mesoscopic level) is updated at each time step by means of the PM-space model. The macroscopic response of the sample for any excitation signal is then predicted using EFIT.

The extension of the 1D model to two dimensions is based on the suggestion by Helbig [10] to reformulate the elastic tensor in terms of its associated eigensystem. Briefly, our procedure consists in rewriting the stress rate equations in terms of volumetric, deviatoric and shear stresses and strains and defining new corresponding moduli. As a first approximation, we assume that the moduli only depend on the actual value and the history of each corresponding eigenstress component. The rate equations can then be treated as scalar equations and the moduli are updated independently by attributing a scalar PM-space to each eigenstress/eigenstrain pair. Once the eigenstress components are updated, the Cartesian stresses are recalculated and we proceed to the next step. For simplicity, we will only consider the effects of volumetric hysteresis in this paper. The approach is described in more detail in [8], and will be the subject of an upcoming paper.

SIMULATIONS

As a first example, we consider an aluminium plate in the absence of linear attenuation. The dimensions of the plate are 300 by 500 mm (Fig. 1a) and rigid boundary conditions are applied at the edges. The medium is linearly elastic except for a zone where damage is modelled by imposing a 80 % reduction of the shear modulus and a hysteretic modulus-stress relation with uniform PM-space density and hysteretic strength parameter $\hat{\gamma} = 10^{-3}$ [7], distributed over a square region of 30 by 30 mm in the middle of the plate. The excitation signal (an apodized pulse with a center frequency of 200 kHz) is introduced as an externally applied stress with a 2D spatial Gaussian distribution applied over a region of 10×50 mm centered at ($x_s = 125$ mm, $y_s = 150$ mm). Figure 1b shows the T_{xx} stress component at the receiver position ($x_r = 375$ mm, $y_r = 150$ mm) over a timescale of 0.225 ms for a source amplitude $A = 2 \cdot 10^5$ Pa.

It is not easy to visually discern the nonlinearity from the normalized time signals of the simulations with and without hysteresis. We therefore use Fourier analysis and wavelet transforms to analyze the signals in the frequency domain. Fig. 1c shows the

Fourier transforms of the stress signals with hysteresis at $A=5\cdot10^4$ Pa and $A=2\cdot10^5$ Pa together with the linear signal at $5\cdot10^4$ Pa for comparison. Only the first arrival of the longitudinal wave was considered in the FFT and it was isolated by apodizing the full signal with a Gaussian with maximum at $t=5.48\cdot10^{-5}$ s. One clearly observes the generation of odd harmonics with a quadratic dependence on the fundamental amplitude. The wavelet transform of the stress signal at $A=2\cdot10^5$ Pa is shown in Fig. 1d and provides a time resolved picture of the harmonic content of the signal. Again the odd harmonics are evident at 600 and 1000 kHz.

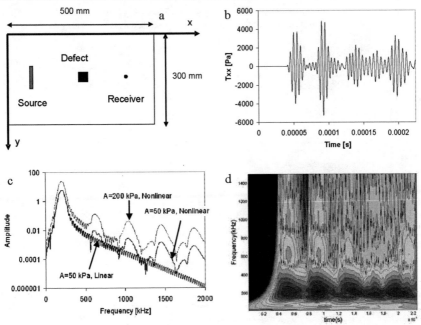

FIGURE 1. a) Geometry of the simulations; b) Received stress signal for $A=2\cdot10^5$ Pa.; c) FFT of stress signals including hysteresis at $A=5\cdot10^4$ Pa and $2\cdot10^5$ Pa and a linear signal at $A=5\cdot10^4$ Pa; d) wavelet map of the stress signal at $A=2\cdot10^5$ Pa.

In the second example we consider a solid medium measuring 120 by 40 mm with a density of 1000 kg/m^3 and longitudinal and shear velocities of 2000 and 1155 ms^{-1}, respectively (Fig. 2a). The medium has a free upper surface and is surrounded by three absorbing boundary layers to simulate an infinite halfspace. A source at the upper left corner generates a Rayleigh wave (on top of the usual bulk waves) which propagates towards the receiver positions situated at the surface at 20 and 100 mm from the source. A microdamaged surficial zone, centered at 60 mm from the source, has a smooth distribution for $\hat{\gamma}$, decaying from the surface, with a horizontal width of 12 mm. Using our multiscale model, we analyzed both the Rayleigh wave speed and the harmonic content of the signals. The main conclusion is that (1) the presence of hysteretic nonlinearity induces no measurable extra effect on the linear wave speed and (2) the amplitude of the harmonics of the Rayleigh wave is highly sensitive to the nonlinearity and depth of the near surface layers. A typical result for the second conclusion is shown in Fig. 2b, in which the third harmonic of the Rayleigh wave's vertical particle velocity at the second receiver is plotted as a function of the excitation frequency for different depths of the damaged zone (in mm) and at a fixed excitation level. The resulting frequency dependence is cubic and the strength of the third harmonic reduces with decreasing depth of the microdamaged zone. These results encourage further research on the use of nonlinear Rayleigh wave propagation for "harmonic" depth profiling and microdamage inspection in non-destructive testing application, e.g. characterizations of natural building stones and historical monuments.

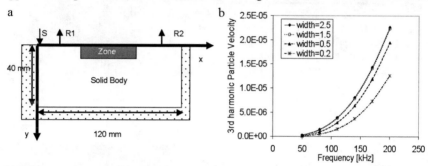

FIGURE 2. a) Geometry of the 2D Rayleigh wave propagation simulations; b) Frequency dependence of the third harmonic vertical particle velocity component in the Rayleigh wave for different depths of the microdamaged zone as a function of the depth.

ACKNOWLEDGMENTS

The authors gratefully acknowledge the support of the Flemish Fund for Scientific Research. (G.0206.02 and G.0257.02), the provisions of the European Science Foundation Programme NATEMIS and the European FP5 and FP6 Grants DIAS (EVK4-CT-2002-00080) and AERONEWS (AST-CT-2003-502927).

REFERENCES

1. Nazarov V.E., Ostrovskii L.A., Soustova I.A. and Sutin, A.M., *Sov. Phys. Acoust.* **34**, 284 (1988).
2. Nagy P.B., and Adler L., in *Rev.QNDE* **11**, D.O. Thompson, D.E. Chimenti (Eds.), 2025 (1992).
3. Johnson P.A., Zinszner B. and Rasolofosaon P.N.J., *J. Geophys. Res.* **101**, 11553 (1996).
4. Nazarov V.E., Radostin A.V., Ostrovskii L.A., and Soustova I.A., *Acoust. Phys.* **49**, 344, 2003.
5. Scalerandi M., Agostini V., Delsanto PP, Van Den Abeele K., Johnson P.A., *J.Acoust.Soc.Am.* **113**, 3049, 2003.
6. Gusev V. and Aleshin V., *J. Acoust. Soc. Am.* **112**, 2666, 2002.
7. Van Den Abeele K., Schubert F., Aleshin V., Windels F., and Carmeliet J., *Ultrasonics* **42**, 1017, 2004.
8. Van Den Abeele K. And Vanaverbeke S., in *The universality of nonclassical nonlinearity with applications to NDE and Ultrasonics,*, Chapter 12, in press.
9. Hirsekorn M., Gliozzi A., Nobili M., Van Den Abeele K., in *The universality of nonclassical nonlinearity with applications to NDE and Ultrasonics,* Chapter 18, in press.
10. Helbig K. and Rasolofosaon P.N.J., *Anisotropy 2000, Soc. of Expl. Geophys.*, pp383-398 (2001).
11. Fellinger P., Marklein R., Langenberg K.J., and Klaholz S., *Wave Motion* **21**, 47, 1995.
12. Schubert F. and Koehler B., in: *Nondestructive Characterization of Materials VIII*, Plenum, 567, 1998

Pseudospectral simulation of elastic waves propagation in heterogeneous nonlinear hysteretic medium

O. Bou Matar[1], S. Dos Santos[2], J. Fortineau[2], T. Goursolle[2], L. Haumesser[2], F. Vander Meulen[2]

[1] Joint European Laboratory LEMAC
IEMN - DOAE UMR CNRS 8520, Ecole Centrale de Lille, B.P.48, 59651 Villeneuve d'Ascq Cedex
[2] Laboratoire UltraSons Signaux et Instrumentation de L'Université François Rabelais /GIP
Ultrasons, FRE2448 CNRS, EIVL, Rue de la Chocolaterie, BP 3410, 41034 Blois Cedex, France

Abstract. A pseudo-spectral time domain algorithm (PSTD) has been developed for solving elastic wave equation in nonlinear hysteretic heterogeneous solids. The hysteretic nonlinearity is introduced owing to a PM space model, based on a multiscale approach. 1D resonant bar simulations and 2D time reversal experiments in a plate with fixed boundaries are presented.

Keywords: Pseudospectral simulation, nonlinear hysteretic solid.
PACS: 43.25.Dc, 43.25.Gf

INTRODUCTION

There is a growing interest for nondestructive testing methods based on nonlinear acoustic effects in solids. It has been shown that micro-inhomogeneities such as cracks lead to an anomalously high level of "nonclassical" nonlinearity[1].

A PSTD algorithm developed for solving elastic wave equation in nonlinear hysteretic heterogeneous solids is described. Rod resonance curves simulated in 1D are presented, including conditioning effects and nonlinear attenuation. Finally, 2D extension based on a Kelvin notation method[2] is described and results of pulse propagation and time reversal experiment in an aluminium plate are presented.

PSEUDOSPECTRAL WAVE SOLVER

A 1D pseudo-spectral time domain algorithm (PSTD) has been developed for solving elastic wave equation in nonlinear hysteretic heterogeneous solids using FFTs for calculation of the spatial differential operator on staggered grid[3]. The solver uses a staggered fourth order Adams – Bashforth method, by which stress and particle velocity are update at alternating half time steps, to integrate forward in time. To circumvent wraparound inherent to FFT-based pseudo-spectral simulation, a novel implementation of Perfectly Matched Layers (PML) boundary condition named convolutional PML (CPML) has been used.

CP838, *Innovations in Nonlinear Acoustics: 17th International Symposium on Nonlinear Acoustics,*
edited by A. A. Atchley, V. W. Sparrow, and R. M. Keolian
© 2006 American Institute of Physics 0-7354-0330-9/06/$23.00

1D SIMULATIONS

Governing equations

Consider a heterogeneous medium in which compressional wave propagation is modeled by the following one-dimensional hyperbolic system:

$$\frac{\partial v}{\partial t} = \frac{1}{\rho_0}\frac{\partial \tau}{\partial z} \tag{1}$$

$$\frac{\partial \sigma}{\partial t} = K_r(t,\sigma)(1 + L\tau(t,\sigma))\frac{\partial v}{\partial x} + \sum_{l=1}^{L} r_l \tag{2}$$

$$\frac{\partial r_l}{\partial t} = -\frac{r_l}{\tau_{\sigma l}} + K_r\left(\sum_{l=1}^{L}\frac{\tau}{\tau_{\sigma l}}e^{-t/\tau_\sigma}\right)\frac{\partial v}{\partial x} \tag{3}$$

where ρ_0 and K_r are respectively the density and the bulk modulus of the material, v is the particle velocity, σ is the longitudinal stress, r_l are relaxation variables with amplitude τ and relaxation time $\tau_{\sigma l}$. As shown, linear attenuation is introduced in simulations with a multi-relaxation method. The hysteretic nonlinearity is considered owing to a PM space model, based on a multiscale approach[4]. In this model, the bulk modulus is calculated by summation of the strain contribution of a numerous number of elementary hysteretic unit (HEU) described by two stresses P_C and P_O, corresponding to the transition between two states (open and close) when the stress is increased or decreased, respectively. For each cell of the calculation grid N_0 hysteretic units are considered with different values of the two stresses characteristic. This representation is commonly termed "PM-space" and can be described mathematically by its density distribution. Two differents hysteretic elementary units (HEU) are considered: inelastic (or rectangular) two states elements[1,4], and elastic (or triangular) two states elements[5]. Finally, nonlinear attenuation is introduced with a stress dependent relaxation amplitude: $\tau(t) = \tau_0(1 + \alpha. \text{ % open elements})$, where α is a constant.

Resonant bar simulations in nonlinear hysteretic media

A 25 cm long bar of rock (ρ_0 = 2600 kg/m^3, K_r = 10 Gpa, Q factor = 80) with stress free ends is considered. A source is located at one end of the bar, and a receiver is attached on the other end. Resonance sweeps are simulated for source amplitude between 50 and 51200 Pa (Figure 1a.). Hysteretic nonlinearity is introduced with rectangular HEU giving contribution to the strain, when all elements are closed, equals to 10^{-4}. The resonance frequency linearly decreases with increasing stress source amplitude, corresponding to a softening of the rod. The other caracteristic behavior of hysteretic nonlinearity described by the simulations is the quadradic dependence of the third harmonic strain on the fundamental strain (Figure 1b). To describe conditioning effets triangular HEU need to be used. The resonance curve obtained at low amplitude, after a large amplitude prestressing process, is shifted in frequency in comparison with the same low amplitude resonance curve before prestressing. Moreover, the resonance curves obtained in the case of increasing or decreasing frequencies, show slightly different results, as observed in experimental measurements (Figure 2b.).

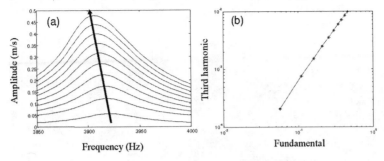

FIGURE 1. (a) Fundamental frequency resonance curves for increasing source amplitude and (b) third harmonic strain as a function of the fundamental strain for a nonlinear hysteretic bar.

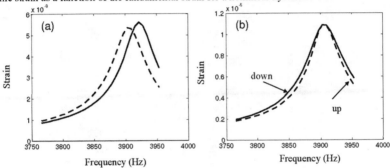

FIGURE 2. (a) Comparison of resonance curves at low amplitude ($\sigma = 1$ Pa) before (solid line) and after (dashed line) a high amplitude ($\sigma = 512$ Pa) conditioning process. (b) Comparison of the strain resonance curve obtained for increasing (dashed line) and decreasing (solid line) frequency sweeping.

2D SIMULATIONS

Propagation in 2D transverse isotropic solid, where the propagation of P-S_V waves can be decoupled from the propagation of S_H wave, is considered. The 2D elastic wave solver developed is based on a Kelvin notation method[2]. Using this notation, the 3 eigenvectors of the elastic constants tensor (in 2D) correspond to 3 eigenstress / eigenstrain vectors. These vectors represent directions where applied stress and created strain are in the same direction. It is then possible to use for each of these 3 directions a scalar "PM space" model similar to the one used for 1D simulations. Each PM space associated to the different directions modifies the associate eigenstiffness. Then, with these new eigenstiffnesses, "effective" elastic coefficients are calculated at each time step. Simulation of the propagation of a burst sent inside an aluminium plate containing a nonlinear hysteretic small area is shown on Figure 3. If the received signal is simply time reversed and sent back in the plate by the receiver, the emitted wave returns to the source as shown on Figure 4a. Now, if one of the harmonics (here the third) of the incident wave is selectively time reversed and sent back in the plate, then a pulse, at the chosen frequency, propagates backward to the small area of strong nonlinearity (Figure 4b.). This proscess is the basis of a new imaging method of localised defects, now experimentally studied in our laboratory.

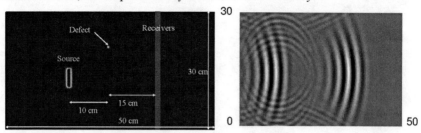

FIGURE 3. Set-up of the numerical experiment for 2D pulse propagation in an aluminium plate, and obtained particle velocity at $t = 50$ μs.

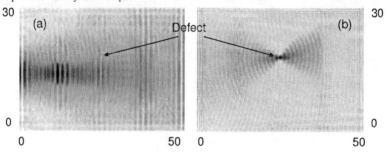

FIGURE 4. Maximum amplitude obtained, in all points of the plate, in the case of (a) time reversal of the entire received signal, and (b) time reveral of only the third harmonic component of the signal.

ACKNOWLEDGMENTS

This work was supported by the EC Sixth Framework Program AERONEWS: Specific Targeted Research FP6-502927 (http://www.kulak.be/AERONEWS).

REFERENCES

1. Guyer R.A. and Johnson P.A., *Physics Today*, 30-36 (1999).
2. Helbig, K. and Rasolofosaon, P.N.J., "A theoretical paradigm for describing hysteresis and nonlinear elasticity in arbitrary anisotropic rocks," in *Proceedings of the Nineth Int. Workshop on Seismic Anisotropy*, Society of Exploration Geopysicistry, Tulsa, 2000.
3. Bou Matar, O., Preobrazhensky, V., and Pernod, P., "2D Axisymmetric numerical simulation of supercritical phase conjugation of ultrasound in active solid media," accepted for publication in J. Acoust. Soc. Am.
4. Van Den Abeele, K., Schubert, F., Aleshin, V., Windels, F. and Carmeliet, J., *Ultrasonics* **42**, 1017-1024 (2004).
5. Scalerendi, M., Agostini, V., Delsanto, P.P., Van Den Abeele, K. and Johnson, P., *J. Acoust. Soc. Am.* **113**, 3049-3059 (2003).

Acoustic Nonlinearity of Porous Viscoelastic Medium

Zheng Fan, Jian Ma, Bin Liang, Zhemin Zhu, Jianchun Cheng

Institute of Acoustics and State Key Lab of Modern Acoustics, Nanjing University, Nanjing 210093, P. R. China

Abstract. An approach based on the equivalent medium method is used to describe the propagation of acoustic longitudinal wave in porous viscoelastic medium. Combining the wave equation for the porous medium and the dynamic equation for micropores and using the perturbation method, equations for fundamental and second harmonic are obtained. Treating medium containing micropores as a homogeneous medium, the effective nonlinearity parameter is defined, and a formula for it is achieved. The frequency dependence of effective nonlinear parameter is studied in detail, and its low frequency limit consists with Ostrovsky's static results.

Keywords: rubberlike medium, porous, nonlinearity
PACS: 43.25.Cb

INTRODUCTION

The nonlinear acoustic wave propagation in the porous medium has attracted much attention in recent years. Many experiments show that very strong elastic nonlinearity may be detected in porous media, such as soil, rocks and sediment, etc.

Moreover, there exists a class of 'weakly compressible' media which could be called rubberlike material. L.A.Ostrovsky [1] has well studied the high nonlinearity of such rubberlike porous material at very low frequency, which is far below the resonant frequency of the micropores. Also he proposed a concept of equivalent nonlinear parameter derived from the nonlinear Hooke' law to estimate the nonlinearity of the rubberlike porous medium, which is validated by the relative experiment.

In this paper, we extended the theoretical model to estimate the nonlinear property of the rubberlike porous medium at higher frequency around the resonance of micropores. At the resonant frequency, microbubbles oscillate vigorously in an external ultrasonic field. These nonlinear oscillating micropores behave as secondary sources for ultrasound. Because of the nonlinear nature of the oscillation, higher harmonics, particularly the second harmonic component, become significant. Consequently, the nonlinearity of the rubberlike media is dramatically enhanced. Furthermore, a general study of frequency dependence of the nonlinearity is needed, since the nonlinear oscillation of the micropores is also determined by the frequency of the excitation ultrasound. At the very low frequency range, the presented results agree very well with Ostrovsky's work, which has been confirmed by experimental data.

CP838, *Innovations in Nonlinear Acoustics: 17ᵗʰ International Symposium on Nonlinear Acoustics*,
edited by A. A. Atchley, V. W. Sparrow, and R. M. Keolian
© 2006 American Institute of Physics 0-7354-0330-9/06/$23.00

THEORY

Nonlinear Waves In Porous Rubberlike Medium

If the amount of pores per wavelength is large enough, the porous medium can be viewed as equivalent continuous medium, the "averaged Hooke's law" relating stress σ and strain ε for a longitudinal deformation can be described as: [1]

$$\sigma = (K + \frac{4}{3}\mu)(\varepsilon - NU) \tag{1}$$

where K and μ refer to bulk and shear modulus of the viscoelastic medium, which is complex respectively. U is defined as the volume variation of an individual bubble. Substituting equ(1) to the wave equation of pure viscoelastic medium, we obtain the one-dimensional longitudinal wave equation of porous rubberlike medium as:

$$\frac{\partial^2 \sigma}{\partial x^2} - \frac{1}{c_0^2(1+j\eta)}\frac{\partial^2 \sigma}{\partial t^2} = \rho_0 N \frac{\partial^2 U}{\partial t^2} \tag{2}$$

Here η refers to the viscosity of the pure medium.

Nonlinear Pore Dynamic

To solve Eq. (2), another equation describing the oscillation of the individual pore in ultrasound field is required. L.A.Ostrovsky derived a Rayleigh-Plesset-like equation to describe the oscillation of the pores. In this article, we use the same equation but add an additional term of viscosity into it for completeness. The equation therefore could be expressed as

$$\ddot{U} + \omega_0^2 U + \upsilon\dot{U} = GU^2 + q(2U\ddot{U} + \dot{U}^2) + e\sigma \tag{3}$$

Here, the resonant frequency $\omega_0 = 2\sqrt{\dfrac{\mu}{\rho r_0^2}}$ and r_0 is the static radius of the cavity,

$\mu_s = \dfrac{\eta_M}{\omega}$ describes the shear viscosity of the medium which has a great influence in the pore dynamics.

Second Harmonic

Combining Eq(3) with eq(2) and using perturbation approach, it gets:

$$\frac{\partial^2 \sigma_2}{\partial x^2} - \frac{1}{c_0^2(1+j\eta)}\frac{\partial^2 \sigma_2}{\partial t^2} = \frac{\rho_0 Ne}{1 - \dfrac{1}{4z^2} - j\dfrac{1}{2Q}}\sigma_2 - \rho_0 ND\sigma_1^2 \tag{4}$$

Where $D = \dfrac{(G - 3q\omega^2)e^2}{(1 - \dfrac{1}{z^2} - j\dfrac{1}{Q})^2(1 - \dfrac{1}{4z^2} - j\dfrac{1}{2Q})\omega^4}$, and σ_1 is the solution of fundamental

wave. Solving this inhomogeneous differential equation with the boundary condition $x = 0$, $\sigma_2 = 0$, we obtain:

$$\sigma_2 = \frac{\rho_0 ND}{(4\bar{k_1}^2 - \bar{k_2}^2)} \sigma_{1A}^2 [\exp(-2j\bar{k}_1 x) - \exp(-j\bar{k}_2 x)] \exp j(2\omega t)$$

(5)

When the x is small, the approximation $2\alpha_1 - \alpha_2 \ll 1, 2k_1 - k_2 \ll 1$ can be made to simplify the solution, [3] thus the amplitude of the second harmonic is expressed:

$$\sigma_{2A} = \frac{\rho_0 ND \sigma_{1A}^2 x}{\sqrt{(2k_1 + k_2)^2 + (2\alpha_1 + \alpha_2)^2}}$$

(6)

Equivalent Acoustic Nonlinear Parameter

For nonporous nondissipative elastic media, the one-dimensional wave equation of stress tensor can be written as [2]:

$$\sigma = \sigma_{1A} \exp[j(\omega t - kx)] + \frac{\Gamma \omega \sigma_{1A}^2 xj}{c^3 \rho} \exp[2j(\omega t - kx)]$$

(7)

And the second harmonic wave could be express as $\sigma_{2A} = \frac{\Gamma \omega x}{c^3 \rho} \sigma_{1A}^2$, which is quite

similar with eq(6) in formulation. So the concept of equivalent nonlinear parameter Γ_e could be presented by treating the second harmonic field of porous medium as a nonporous elastic medium with equivalent nonlinear parameter as:

$$\Gamma_e = \frac{\rho c^3 ND}{\omega \sqrt{(2k_1 + k_2)^2 + (2\alpha_1 + \alpha_2)^2}}$$

(8)

NUMERICAL RESULTS

Figure 1 shows the frequency dependence of the equivalent acoustic nonlinearity parameter of the rubberlike medium containing micropores based on equation (8). Two medium are used for calculation: silicon and plastizole. From the figure we can see, the effective nonlinear parameter is much larger by including the effect of nonlinear oscillation of the pores compared to the low frequency results. It is interesting to find that the spectrum of effective nonlinear parameter reaches its maximum at about the half of the resonant frequency, which means the second harmonic is most effectively generated at this frequency.

To explain this phenomenon, we need to study the nature of the nonlinear oscillation of individual pores, since it is the source of the high nonlinearity of the mixture medium. The nonlinear oscillation equation, eq.(3), could be solved by fourth Runge-Kutta method, the second harmonic of each driven sound pressure can be obtained, which is shown in figure 2. It is quite clearly to see that, under the excitation of external ultrasound, for the individual pore, it could most effectively generate the second harmonic at the half of its resonant frequency. Thus, it can be concluded that the nonlinearity of mixture medium is, to great extent, determined by the nature of the oscillation of the individual pores.

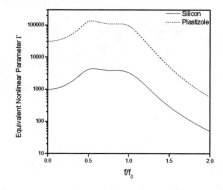 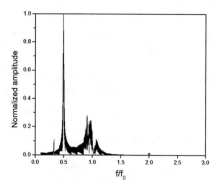

FIGURE 1. This is Frequency spectra of equivalent acoustic nonlinearity parameter

FIGURE 2. Frequency spectra of second harmonic amplitude of a single pore oscillation

Table I shows that our predictions of effective nonlinear parameter at low frequency agrees very well with Ostrovsky's static results, which means the presented effective nonlinear parameter could estimate both the static and dynamic effect of the micropores on the nonlinearity of the whole porous media.

TABLE 1 Comparison between our calculation and Ostrovsky's static results					
Material	Young's modulus	K/μ	Optimal Volume Fraction	Ostrovsky's static results	Our prediction
Silicone	1.80×10^8	7225	1.85×10^{-4}	1016.1	985.9
Plastizole	3.0×10^4	2.1×10^5	6.31×10^{-6}	29673	30776

CONCLUSION

In this paper, an effective medium method is used to describe the nonlinearity of the porous rubberlike medium, which is due to the nonlinear nature of pore oscillation. A concept of equivalent nonlinear parameter Γ is proposed to estimate the nonlinear property. It is found that the effective nonlinear parameter spectrum always reaches its maximum at a frequency that is close to half of the pores' resonant frequency. At low frequency, our prediction agrees well with Ostrovsky's counterpart results.

REFERENCES

1. L.A.Ostrovsky, Wave progresses in media with strong acoustic nonlinearity, J,Acoust,soc, Am., 90(6), 3332-3337 ,1991

2. D.M.Donskoy, etal , "Nonlinear acoustic waves in porous media in the context of Biot's theory", J.Acoust,Soc,Am, 102(5), 2521-2528,1997
3. Junru Wu and Zhemin Zhu, "Measurements of the effective nonlinearity parameter B/A of water containing trapped cylindrical bubbles", J,Acoust,soc, Am,89(6),2634-2639,1991.

Physical constitutive equations for nonlinear acoustics of materials with internal contacts

Vladislav Aleshin, Koen Van Den Abeele

K.U.Leuven Campus Kortrijk, E. Sabbelaan 53, B-8500 Kortrijk, Belgium

Abstract. A physical theory for the acoustics of microdamaged materials is presented based on the analysis of internal contact roughness. As a start, we adopt the Whitehouse and Archard contact mechanics yielding a decomposition of continuously roughs surfaces into a set of spherical asperities of different sizes and heights. Coupled with a model for the force-displacement relationship for microcontacts, taking into account adhesion hysteresis, it provides a description for an individual intergranular contact. We upscale this model using continuum mechanics of solids with cracks and derive a set of equations for the evolution of acoustical waves. We use this model to predict results for nonlinear acoustical experiments and derive amplitude dependences of resonant frequency and harmonics, and slow dynamics phenomena.

Keywords: nonclassical nonlinearity, adhesion hysteresis, contact mechanics, roughness statistics, microdamaged solids.
PACS: 43.25.+y; 62.20.Dc; 62.20.Qp; 62.65.+k

INTRODUCTION

The history of nonclassical nonlinearity in acoustics started at the beginning of the 1980s when it was noted that for many media, such as geomaterials, composites, ceramics, and polycrystalline metals, the response to an acoustical excitation deviated from the one expected by the classical nonlinear acoustics theory. This was attributed to a unifying feature of this class of materials, namely the presence of internal contacts, existing as an inherent microstructure or as a result of damage or fatigue. Static loading tests revealed hysteresis in stress-strain relation, and as a consequence the hypothesis of hysteretic acoustical equations of state naturally appeared. This enabled to successfully resolve many of the observed discrepancies but raised an important question about the underlying physical mechanisms of hysteresis. Over the years, researchers have attributed hysteresis to one or more of the following factors: sliding friction in internal cracks [1], hysteretic adhesion interaction of asperities [2], double-well potentials of cracks [3]. The present study concerns with the adhesion hysteresis mechanism, and constructs the nonlinear constitutive relations using contact mechanics of rough surfaces, known interaction forces for individual asperities and continuum mechanics equations to upscale these micro- and mesoscopic models to the macroscopic level. On the latter level, we solve the conventional wave equation for a resonant bar with nonclassical nonlinearity and present some results for the amplitude dependences of the resonant frequency shift and the generated higher harmonics.

CP838, *Innovations in Nonlinear Acoustics: 17th International Symposium on Nonlinear Acoustics*,
edited by A. A. Atchley, V. W. Sparrow, and R. M. Keolian
© 2006 American Institute of Physics 0-7354-0330-9/06/$23.00

MACROLEVEL: ELASTISITY OF SOLIDS WITH CRACKS

We consider an individual crack (slit) of aperture a in an infinite volume of intact material subjected to a constant stress σ on the infinity. Following a reasoning similar as in reference [3], we can write the force balance equation (deduced from the Lagrangian) in the following manner

$$\sigma \cos^2 \theta = F(a) + T(a),$$ (1)

where θ is the angle between the direction of the stress and the normal vector to the crack's plane, $F(a)$ is the force of the contact interaction, and $T(a)$ describes the elastic reaction of surrounding material layers on a change of aperture a, preventing it from becoming too large or too small. The form of $T(a)$ for a thin circular crack of radius R is [4]:

$$T(a) = \frac{3}{2\pi} \frac{E}{1-v^2} \frac{1}{R}(a-A)$$ (2)

with E the Young's modulus, and v the Poisson's ratio of the intact material. Assuming a set of such non-interacting cracks (as an idealization of a complicated network of interacting contacts), we can write the purely geometrical relation

$$\varepsilon = \sigma/E + S_V \int_0^{\pi/2} (a-A)\cos^2 \theta \sin \theta d\theta$$ (3)

for the strain ε of such a configuration. Here σ/E corresponds to the strain of the intact material (classical nonlinearity of the intact material is neglected) and the integral represents the contribution from an ensemble of cracks (contacts) with apertures $a=a(\theta,\sigma)$. The integral is written for an isotropic crack orientation. Here A is the aperture in the absence (virtual) of the contact interaction force, i.e. when $F=0$.

In the next paragraph, we derive the form of the contact force $F(a)$ from contact mechanics.

MESOLEVEL: CONTACT MECHANICS

Upon classifying the different approaches in contact mechanics, one notes that most of them use spheres or other simple geometrically shaped units as an approximation of individual asperities, to transform a complicated continuous roughness profile into an equivalent set of simple shapes with differing parameters. To model the contact interaction we adopted the decomposition procedure of Whitehouse and Archard [5] which explicitly produces a combined distribution of summits and curvatures of individual asperities from the statistics of a continuous roughness. For a random stationary profile $z(x)$ with a Gaussian distribution of heights z:

$$\varphi(z) = \frac{1}{\sqrt{2\pi s^2}} \exp\left(-\tfrac{1}{2} z^2 / s^2\right)$$ (4)

and an exponential correlation function

$$K(\beta) = \frac{1}{L} \int_{-L/2}^{L/2} z(x) z(x+\beta) dx = \exp\left(-\frac{\beta}{\beta^*}\right),$$ (5)

the decomposed distribution reads

105

$$g(z,C) = \frac{3}{10\pi\sqrt{2}l^2} \exp\left(-\frac{1}{2}z^2\right) \exp\left[-\left(z-\frac{1}{2}C\right)^2\right] \mathrm{erf}\left(\frac{1}{2}C\right). \tag{6}$$

Here C is the normalized curvature, connected to the accrual radius r of an asperity as $r = l^2/(Cs)$ with s the standard deviation of heights and l the sampling length. The sampling length l is a key notion in Whitehouse-Archard's contact mechanics and it is taken to be comparable to the correlation radius β^* (see. Eq. 5): $l = 2.3\beta^*$, such that $K(l) = 0.1$ and the intervals separated by this length l can be considered as uncorrelated. For each of the samples l, the continuous roughness is substituted by a single z-measurement. The array of discretized z-data is analyzed statistically [5] and finally leads to the derivation of the combined distribution of peaks and curvatures (Eq. 6).

The contact force for this distribution is consequently given by

$$F(a) = \int_a^\infty f(z-a)\,dz \int_0^\infty g(z,C)\,dC. \tag{7}$$

In the next paragraph, we consider the theoretical expression for the microscopic force $f(\delta)$ for a single asperity deformed by the displacement δ.

MICROLEVEL: AHDESION HYSTERESIS

One of the most frequently used models for a microcontact force is the Johnson-Kendall-Roberts-Sperling (JKRS) [6] mechanics which represents a modification of the Hertz contact theory taking into account the surface energy γ. This force is usually expressed in its parametric form

$$f = \left(b^3 - b\sqrt{6b}\right)f_0, \quad \delta = \left(b^2 - \frac{2}{3}\sqrt{6b}\right)\delta_0, \tag{8}$$

with the normalization constants $\delta_0 = \left(3\pi\gamma\left(1-\nu^2\right)/(2E)\right)^{\frac{2}{3}} r^{\frac{1}{3}}$, $f_0 = \pi r \gamma$. Figure 1 shows that it is possible to split the JKRS interaction force into a hysteretic f_H (purely rectangular) and a non-hysteretic f_{NH} (purely Hertzian) component. The sum $f_{NH} + f_{NH}$ is plotted in Fig. 1 with a thin line. The directions of hysteretic switching are indicated by the arrows. For decreasing displacement, a neck is formed at negative δ which brakes at $\delta \approx -\delta_0$. For displacements δ increasing from large negative values, the neck is not formed, and the contact remains absent till $\delta = 0$ where it suddenly appears due to the adhesion force.

SOLUTION AND RESULTS

The calculation of the double integral in Eq. 7 gives: $F(a) = F_{NH}(a) + F_H(a)$, where

$$F_{NH}(a) = \frac{\sqrt{2}}{5\pi} \frac{E}{1-\nu^2} \frac{s}{l} \varphi(a) \tag{9}$$

and $\varphi(a)$ is a known function (see [5]). The hysteretic force can be expressed as

$$F_H(a) = -\iint_\Omega \rho(a_o, a_c)\,da_o\,da_c \tag{10}$$

where a_c and a_o substitute z and C: $z = a_c$, $C = \lambda/(a_o - a_c)^3$, and Ω is the area in the space (a_o, a_c) corresponding to the elements in contact. The function $\rho(a_o, a_c)$ is usually called the Preisach density (see. Fig. 2). Each pair of (a_o, a_c) characterizes an individual hysteretic element, i.e. an asperity with a definite summit coordinate z and curvature C. The procedure for determining the area Ω is described in detail, for instance, in [7].

Applying this formulation of the hysteretic and non-hysteretic forces to the resonant bar simulation we have found the results for the harmonic generation and the resonance frequency shift as shown in Figs.3 and 4.

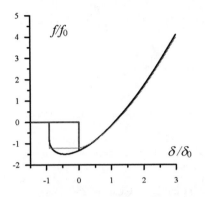

FIGURE 1. JKRS force-displacement relation.

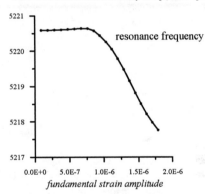

FIGURE 2. Preisach density in aperture space.

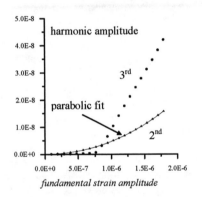

FIGURES 3,4. Resonant bar simulation: harmonic's amplitudes and resonant frequency shift for the proposed model of nonlinearity.

REFERENCES

1. Walsh, J. B., *J. Geophys. Res.* **71**, 2591-2599 (1966); Mavko, *J. Geophys. Res.* **84**, 4769 (1979).
2. Sharma, M. M., and Tutuncu, A. N., *Geoph. Res. Lett.* **21(21)**, 2323-2326 (1994); Pecorari, C., *J. Acoust. Soc. Am.* **116(4)**, 1938-1947 (2004).
3. Aleshin,V., and Van Den Abeele, K.,*J. Mech. Phys. Solids* **53(4)**, 795-824 (2005).
4. Sneddon, I., *Fourier Series* (Routledge & Kegan, New York, 1951).
5. Whitehouse, D.J., and Archard, J. F., *Proc. R. Soc. London, Ser. A* **316(1524)**, 97-121 (1970).
6. Johnson, K.L., Kendall, K., and Roberts, A. D., *Proc. R. Soc. London, Ser.A* **324**, 301 (1971).
7. Aleshin, V.V., Gusev, V.E., and Zaitsev, V.Yu., *J. Comput. Acoust.*, **12(3)**, 319-354 (2004).

This research was supported by the Foundation for Scientific research, Flanders, Belgium (grants G.0206.02, and G.0257.02), and by the European Union (ESF-PESC program NATEMIS, and STREP FP6-502927 AERONEWS).

Acoustic Lockin-Thermography of Imperfect Solids for Advanced Defect Selective NDE

G. Riegert, Th. Zweschper, A. Gleiter, I. Solodov, and G. Busse

Institute for Polymer Testing and Polymer Science (IKP) - Nondestructive Testing - (ZFP), Stuttgart University, Stuttgart 70569, Germany

Abstract. A propagation of an intense acoustic wave in imperfect solids results in a local temperature increase in a defect area. To raise a sensitivity of detection of such faint heat sources, the lock-in approach was applied by using a low-frequency amplitude modulation of acoustic excitation (Ultrasound-Lockin-Thermography). However, for a single carrier frequency, the temperature distribution is often affected by standing wave pattern in the specimen that impedes acoustic thermography of defects. Alternatively, a wide-band time varying modulation enables to avoid resonances of the specimen and minimize the standing wave impact. A burst-phase acoustic thermography uses short acoustic bursts to derive phase angle images from a temperature image sequence at various frequencies. It, therefore, combines a high sensitivity of the lock-in counterpart with a wide bandwidth of Burst-Phase-Thermography. A number of case studies concerned with various flaws (delaminations, cracks, voids, non-bonds, etc.) in a wide range of materials are reported and demonstrate a high reliability of defect recognition. A striking similarity found between the results of acoustic lockin-thermography and higher harmonic acoustic nonlinear NDE suggests nonlinear acoustic contribution into the mechanism of acoustic thermography.

Keywords: acoustic thermography, ultrasound-lockin-thermography, nonlinear vibrometry
PACS: 43.25.Ed; 81.70.Cv; 43.35.Zc.

INTRODUCTION

Defect selective inspection improves a reliability of defect detection and allows for automated non-destructive evaluation (NDE) of materials and components. The ultrasound activated thermography and nonlinear vibrometry are the examples of defect selective methods which are promising for a wide range of NDE- applications. The selectivity of the ultrasound thermography based on a locality of ultrasonic wave-defect interaction: acoustic waves, used for sample excitation, are converted into heat only in the areas of defects due to a local increase of mechanical losses. The produced heat depends on the frequency of the elastic waves and normally a reasonable sensitivity for infrared detection is achieved in the kHz-frequency range [1].

The simplest way of implementing ultrasound thermography is to apply a short ultrasonic burst (or pulse) to the sample while an infrared camera records a temperature image sequence, then the image of highest contrast is evaluated. The disadvantage of this method of transient ultrasound thermography is a poor signal-to-noise ratio with images highly affected by reflections and surface features. The lockin-technique solves these problems.

CP838, *Innovations in Nonlinear Acoustics: 17th International Symposium on Nonlinear Acoustics*,
edited by A. A. Atchley, V. W. Sparrow, and R. M. Keolian
© 2006 American Institute of Physics 0-7354-0330-9/06/$23.00

Ultrasound activation also leads to elastic nonlinear behaviour of the defects. As a result, the higher harmonics (and other nonlinear spectral components) are generated in the defect area and can be detected with a laser scanning vibrometer. In this paper, the results of the ultrasound lockin-thermography are presented and compared with the data obtained by using nonlinear vibrometry.

ULTRASOUND ACTIVATED LOCKIN-THERMOGRAPHY

Ultrasound-Lockin-Thermography (ULT) uses a periodical heat generation by amplitude modulated elastic waves excited in a sample (Figure 1). This results in the defects pulsating at the modulation (lockin) frequency and emitting thermal waves. The local temperature modulation on the surface above the defects is detected by an infrared camera which records a sequence of temperature images over several modulation periods. A Fourier transform at the modulation frequency ("lockin-frequency") provides an image of the amplitude and phase of the thermal wave field [2, 3]. These images have a highly improved signal-to-noise ratio as compared to single temperature images because all images of the temperature image stack are used in the Fourier transformation. Artefacts like inhomogeneity of the coefficient of infrared emission or spurious temperature distribution are reduced by the use of phase angle for imaging.

For faster measurements the sample excitation with a short ultrasound burst can also be applied [4]. Such an Ultrasound-Burst-Phase Thermography (UBP) uses the whole sequence of temperature images obtained during cooling of the sample for Fourier evaluation in contrast to the transient thermography where only one temperature image is processed. The UBP provides similar information to the lockin method but operates much faster. Since a short ultrasonic pulse comprises a number of frequency components a single UBP measurement replaces several ULT measurements at different lockin-frequencies [5].

However, the ULT which uses ultrasound of a fixed carrier frequency also suffers from some disadvantages. In particular, heating outside a defect area increases considerably if the carrier frequency is close to a resonance frequency of the sample. In this case, a strong ultrasonic standing wave pattern reduces a contrast of (or makes impossible) the ULT-imaging of defects. An improved ULT-version uses frequency modulated ultrasound in addition to the amplitude modulation by the lockin-frequency [6].

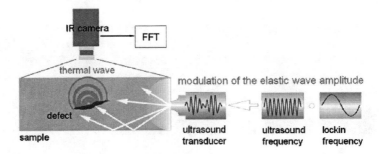

FIGURE 1. Setup of Ultrasound-Lockin-Thermography.

The standing wave pattern is reduced substantially and an enhancement in the contrast and resolution of the images area is obtained.

NONLINEAR VIBROMETRY

The nonlinear approach to NDE is concerned with nonlinear material response, which is related to the frequency changes of the input acoustic signal. These spectral changes are caused by anomalously nonlinear local dynamics of defects of various scale and nature: from large-scale structural debonds or faulty joints in the finished components to micro-scale cracks, grains, delaminations in materials. By monitoring the local nonlinearity, the nonlinear NDE modes appeal directly to the vulnerable (faulty) areas within a material or a product. As a result, such a defect selective NDE is extremely sensitive to a wide-range of imperfections and can deliver information about the disruptions of the material structure, defects in components, integrity, quality and lifetime of the manufactured products [7].

CASE STUDIES OF DEFECT SELECTIVE NDE

NDE of impacts in carbon fibre reinforced plastics (CFRP) is of mayor importance for many areas of applications since such impacts leave nearly no visible trace on the surface though the damage within the material might be critical.

A model sample (quasi-isotropic setup 4.3±0.1 mm thickness) with a 16 J-impact damage was inspected with Ultrasound-Burst-Phase thermography (Figure 2). Two different excitation modes were used: 20 kHz-burst (duration 300 ms; estimated strain amplitude $\varepsilon \sim 3 \ 10^{-4}$) and frequency modulated (3.77 Hz-frequency modulation in the range 15-25 kHz) burst (duration 500 ms; $\varepsilon \sim 2 \ 10^{-4}$). The lower ultrasound amplitude in this mode was due to the limitation of the output amplifier; to compensate it the burst length was increased. For comparison, the image of the highest contrast out of the sequence of images obtained during cooling of the sample after the burst application is also shown (Figure 2, right).

This example demonstrates the potential of lockin-evaluation as compared to single temperature shots out of the temperature image sequence. In the temperature image of the highest amplitude contrast of the cooling sequence (Figure 2, right), the image

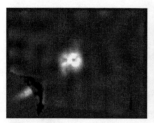

FIGURE 2. NDE of 16 J-impact damage in CFRP plate. UBP using 20 kHz-excitation, phase image at 0.3 Hz lockin-frequency (left) and frequency modulation at the same lockin-frequency (middle). Image of the highest contrast 1.6 s after frequency modulated burst (right).

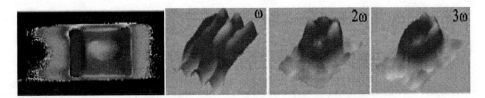

FIGURE 3. Defect selective NDE of delamination in CFRP sample with embedded piezo-ceramic actuator. Left: actuator excited ULT, phase image at 0.03 Hz; right: nonlinear vibrometry: linear, second and third harmonic images.

quality is low due to temperature gradients produced by the ultrasonic excitation and the defect is virtually not distinguishable. The spurious temperature gradient is diminished in the phase UBP-images and the defect is clearly detected in both cases due to an increased signal-noise-ratio after Fourier transformation of the cooling sequence of images.

The phase image with a single carrier frequency excitation reveals a standing wave pattern which could cause problems in detection of smaller defects (Figure 2, left). In the frequency modulation mode, the standing waves are reduced dramatically (Figure 2, middle). This example demonstrates a defect selective character of ultrasound activated thermography: the heat is produced most efficiently only in the defect areas so that they are solely discerned in the ULT-images on the background of a "cool" intact material.

The results of a comparative NDE using both techniques are presented in Fig. 3. It is concerned with a detection of a delamination in a CFRP structure (2x30x200 mm) with an embedded piezoceramic actuator ("smart structure") which is used for excitation of 50 kHz ultrasound. The ULT-image (left) corresponds to 100 V input while the higher acoustic harmonic generation was observed already at 5-10 V input (right).

The linear image observed at the excitation frequency (Fig. 3) detects only a standing wave pattern in the sample. On the contrary, the higher harmonics are located exclusively in the delamination area with the amplitudes of about 20 dB above the background. A similar result is clearly seen in the ULT-image (left) that demonstrates a defect selective capability of both techniques.

REFERENCES

1. Mignogna, R.B., Green Jr., R.E., Henneke, E.G., and Reifsnider, K.L., *Ultrasonics* **7**, 159-163 (1981).
2 Rantala, J., Wu, D., and Busse, G., *Research in Nondestructive Evaluation* **7**, 215-218 (1996).
3 Krapez, J.-C., Taillade, F., Gardette, G., Fenou, B., Gouyon, R., and Balageas, D. *Journée Thermografie quantitative de la Soc. Fr. des Thermiciens* (1999).
4 Maldague, X., and Marinetti, S., *J. Appl. Phys.* **79**, 2694-2698 (1996).
5 Dillenz, A., Zweschper, Th., and Busse, G., *Insight* **42**, 815-817 (2000).
6 Zweschper, Th., Dillenz, A., Riegert, G., Scherling, D., and Busse, G., *Insight* **45**, 178-182 (2003).
7 Krohn, N., Pfleiderer, K., Solodov, I.Yu., and Busse, G., "Nonlinear vibro-acoustic imaging for non-destructive flaw detection", Nonlinear Acoustics at the Beginning of the 21st Century, edited by O. Rudenko and O. Sapozhnikov, 2002, pp. 779-786.

Acoustic Emission as Large Cracked Foundation Response on Static and Dynamic Loading

Nora Vilchinska

Energolaboratory, 56/58 Ventspils Street, Riga, LV-1046, Latvia

Abstract. Large scale long time experiment is carried out in situ on foundations of hydro power station located on soft soil. Loading assessment and analyses of recorded response are developed. Foundation load may bee varied from quasi static to strong motion regimes. AE spectra configuration is the same in quasi static and slow dynamic and its energy carried frequency show frequency shift by loading-unloading. AE spectra are high sensitive to static load changes. By regime strong motion in some measurement points fracture opening are observe. Development of AE as the sum of the nonlinear resonance frequencies of fractals and cracks under dynamic loads is discussed. Experiment results future applicability may bee in non destructive nonlinear testing of large objects in-situ and in modeling fractured media response to long time dynamic with goal to estimate dynamic loading time and amplitude limits for intensification of filtration of fluids through porous media.

INTRODUCTION

The non-linear methods of investigating the earth's crust were rapidly developed in the recent decades; soft soil, rocks, concrete cracked and fractured massifs were under research. The non-linear methods of analysis of response require a medium with a clearly expressed non-linearity. The requirement of the practice of forecast and analysis of the response of cracked media to dynamic loading - this is the main stimulus of concentration of attention to intensive studies of non-linear processes [1] in these media. Information about the non-linear phenomena in geophysics was accumulated from the last quarter of the past century [3, 4] during the study of the propagation of waves in the earth's crust, in the grainy media, with the vibration action to the earth's crust and with studies of strong earthquakes. Concrete belongs to the class of non-linear materials, beginning from a certain stage of hardening [6]. Non-linear wave methods for the examination of damage in materials are the new frontier of acoustical non-destructive testing [6,8]. As practice shows, in the process of operation concrete is cracked and it becomes an ever more non-linear material. The non-linearity of concrete plus the non-linearity of soil [2] under it not only complicates the task, but also considerably increases practical interest in it, that also made us use the non-destructive testing methods of non-linear acoustics. They are the only and unique methods, which are able to clear up the complex spectral picture of the

CP838, *Innovations in Nonlinear Acoustics: 17ᵗʰ International Symposium on Nonlinear Acoustics*,
edited by A. A. Atchley, V. W. Sparrow, and R. M. Keolian

response. The evaluation of changes in the foundation after the boosting regimes and recommendations regarding the selection of the saving regimes with the daily operation - these are the primary tasks of this investigation.

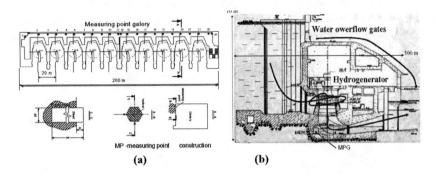

(a)   (b)

FIGURE 1. There are 23 MPs (measurement points) in the MPG (measurement point gallery); their locations and configuration are shown on Fig. 1(a). Cross-section of the dam along the axis of a hydraulic turbine generator in the flow direction, Fig.(1b).

Measurements

Measurement location.

The work was carried out investigating foundation concrete structures of an operational hydroelectric power station, site and setup description was published in[2], in short form it is seen in Fig. 1.

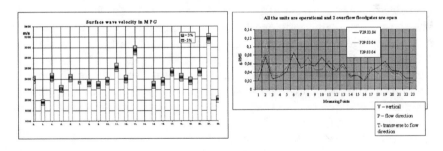

(a)   (b)
FIGURE 2 (a) Surface wave propagation velocity along the MPG. **(b)** Response RMS by overflow in 23 MP

The surface wave propagation velocity along the gallery is distributed as given in Figure 2(a). Vertical polarized Rayleigh wave propagation velocity is measured between MPs and it is within 1850-3200 m/s. The latter reflects the fractures inside the concrete body (the less is the wave velocity, the higher is the crack concentration). Propagating impulse spectral changes allow to assess the fracture size. Measurements are made in silence.

Equipment.

The following equipment was used during the work: 8 accelerometers manufactured by Wilcoxon Research, a SONY 8-channel digital data recorder, type PC208A, an 8-channel data analysis software PCscan MKII and a specialised 8-channel spectrum analysis programme. In some cases, a one-channel data collector-analyser CMVA55 and vibration sensor manufactured by SKF Condition Monitoring were used, allowing carrying out the signal analyses in situ. The data analyses of all kinds were aimed at determining vibration acceleration.

Load assessment

Dynamic load assessment is made for future analysis of response spectra within the framework of non-linear elasticity.

Quasi-static realization: the power station foundation is 200 m long and it is based on soft soils and preloaded with total vertical stress approximately 350 kPa. In the basement of the foundation, there are silty-clayey and clayey-silty soils with non-uniform settlement and non-uniform relaxation time. All hydroelectric generators are idle. Static stressed state of the foundation changes with the change of the previous history of grouping of working hydroelectric units and duration of their work. The possible explanation for that phenomenon: the impact of vibrations of the working hydroelectric unit acts on the bearing capacity of weak soil directly under the unit, as well as the difference in consolidation time for clayey sand and sandy clay. That means that the relaxation processes take place in a different way, and the process of consolidation of weak soils is at a different stage under each unit, causing slow and weakly changing stress in the body of the foundation. Thus, in the "silent" regime, the conditions of "quasi-static" loading of the dam foundation due to relaxation processes are complied   with.
• response RMS in MP = 0.001 – 0.004 g

Slow dynamic realization: Working some hydroelectric generators.
• response RMS in MP = 0.002 – 0.02 g

Strong motion realization : Working 10 hydroelectric generators + water overflow
• response RMS in MP = 0.02 – 0.08 g
• assessment of the emitted and dissipated energy: 25-73 MJ

Response RMS of the foundation by overflow

RMS vibration accelerations by overflow (Fig.2b) (0.02 – 0.08) g, corresponding to: a strong or very strong earthquake (Force 5-7) on the seismic scale of the Institute of the Earth Physics (Moscow), slight to medium on the Richter scale, Force 3-4 on the Mercalli scale.

Based on approximate estimates, the maximum lost energy, which is, consequently, 1% from the generated energy, the possible seismic event comprises 25-73 MJ with all the units in operation plus water discharge that corresponds to a slight to medium earthquake. The bulk of the energy was emitted in the high frequency range – (1- 4) kHz, thus ensuring that the event is local. There still remained a possibility of the excitation of the medium by high-frequency energy, which later is emitted in the low-frequency range.

Measurement results and analysis
What do we measure?

Two little simple experiments on the samples (diameter 7 cm, length 40 cm) of foundation concrete give an answer to that question. The sensor is glued on the one end (cross section) of the specimen, see Figure 3.

Experiment 1

Compression of a sample in metal clamps acts from two sides in the middle of the specimen length; The response contains the low-frequency part - sample oscillations and the high-frequency part. It is shown in Fig. 3 (b). The high-frequency area is caused by friction of metal about a granular surface of concrete, that is acoustic emission (AE). AE frequencies depend on the pressure on a sample and on amplitudes of sample low-frequency oscillations.

Experiment 2

Two samples are in contact with the end of a rough surface. They are squeezed manually. Response spectra are shown in Fig. 3(a). The dynamic contact metal - concrete is absent, see Fig. 4. The high-frequency area is caused by friction granular surfaces of concrete against concrete that is acoustic emission (AE). AE frequencies depend on the pressure between the samples and on amplitudes of sample low-frequency oscillations.

(a)

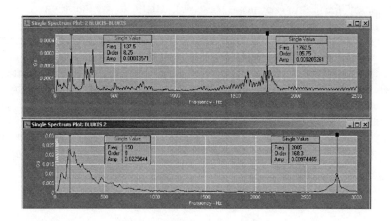

(b)

FIGURE 3 (a) Response spectra: two samples are in contact with the end of rough surface.(b). Response spectra: compression of a sample-pressure is transferred through metal tips and acts from two sides in the middle of specimen length. In both cases the gauge is glued on one end face of one specimen

FIGURE 4 The collector-analyzer CMVA55 was used. Sensor is glued on one end of the specimen

We measure the wide range response spectra. The response contains the low frequency part – the part of foundation oscillations and the high-frequency part, caused by friction of metal over concrete against the granularar surface concrete, that is acoustic emission (AE).

Detailed response analysis in MPs

Figure 5(a) gives a comparison of spectra V17 of the response in silence before (18-Mar -04) and after (14-May-04) floods of 2004, the same forV18 on fig.6. Fig. 5 (b), (c) show V17 response spectra development from 2003 to 2005 in two frequency bands. Fig.5 (d) shows the response strength growth. By the application of maximum loading 16-Mar-2004, resonance occurred. Real response in MP, Fig. 5, and response in laboratory experiment, Fig. 3 (a, b) are similar, only in situ energy-carrying frequencies are in the range (2000– 5000) Hz, but low-frequency oscillations (0.3 – 1000) Hz exist at low amplitudes. It is explained simply: a laboratory sample is extracted from a borehole in solid concrete, MPs in situ are located in cracked concrete and close to cracks. High signal level is obtained from there, the source of which is friction of crack edges.

Response analysis simultaneously in 23 MPs.

Informatively, V,P,T spectra from 23 MP considerably expands and deepens the research. It is possible to track the resonances arising in iron rods of Ferro-concrete, and their distribution in space, on distance some tens meter. In a concrete body, it is possible to single out volume into some hundreds cubic meters, which oscillates as a single whole on the nonlinear resonant frequency. Volume border cracks on three dimensions in space create it. For great volumes it is required to enter significant energy into system that these oscillations in general have occurred.

Summary

The purpose of the publication is to show an opportunity of a method of auscultation of an object with the subsequent analysis of the recorded signals, applying latest developments in the theory of non-linear dynamic elasticity and in acoustic emission. The simple and non-destructive mode of detailed research and monitoring of so big an object needs future development. It is necessary to have 3D visualisation of accumulated spectra and detailed laboratory experiments for the exact assessment of the interdependence of the emitted frequencies of the material and the applied stress for the object under investigation.

(a)   **(c)**

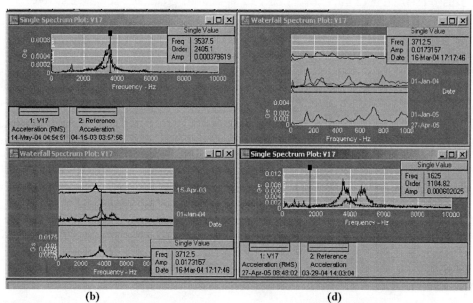

(b)   **(d)**

FIGURE 5. (a) Comparison of spectra of the response in silence before (18-Mar -04) and after (14-May-04) the floods of 2004. (b), (c) Response development in V17 (MP17). 7 measurements were made: 1) 15-Apr -03 - silence, 2) 27-May-03 8 HG worked, 3) 18-Mar-04 - silence, 4) 29-Mar-04 –10 HG + 2 overflow, 5) 1-Apr-2004 (in the report it is 16-Mar-2004 but it is operator's mistake) -10 HG + 3 overflow (it is the maximum dynamic load), 6)14-May-04 - silence, 7) 27-May-05 - 10 HG worked, (d) Response spectra on maximum load, 2005 and equivalent load + overflow, 2004.

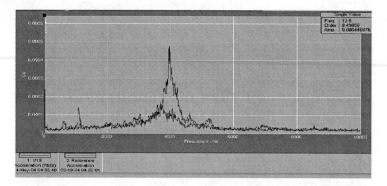

Figure 6. V18 till and after floods 2004 Analyze in details two silences /High pressure drop down as results of strong oscillations. Q-factor is changed. Additional 2 max. in low frequency range are seen

ACKNOWLEDGMENTS

This work was supported by Energolaboratory. All measurements were carried out by engineers of the Energolaboratory. In the processing of results software of the Laboratory was used. Energolaboratory is an industrial laboratory. Thanks for financial support and for the opportunities given me to carry out this research

REFERENCES

1. *Nonlinear Acoustics at the Beginning of the 21st Century.* Volume 1 and 2. Editors O.V. Rudenko and O.A. Sapozhhnikov. Moscow 2002.
2. N.Vilchinska and Dz.Slapjums *Proseadings of ISNA-16" Nonlinear Acoustics at the Beginning of the 21st Century.* Volume 2. Editors O.V. Rudenko and O.A. Sapozhhnikov. Moscow, (2002)
3. N.Vilchinska and V.Nikolaevskiy, Acoustical emission and spectrum of seismic signals. *Solid Earth Physics* (Transactions of Russian Academy of Sciences), # 5, pp.91 – 100, (1984)
4. L.A.Ostrovsky and P.A. Johnson. *La Rivista del Nuovo Cimento* V 24, Serie 4, numero 7, 1-45 pp., Bologna (2001)
5. N.Vilchinska and V.Nikolaevskiy *AC SU 1827655 A1*, Priorities 31.01.91. Published 15.07.93. № 26 (1993)
6. M. Skalerandi, P..P. Delsanto, and P. A. Johnson. *J. Phys. D. Appl. Phys.* **36**, 288-293 (2003)
7. R.A. Guyer,, P..A. Johnson and J.N. TenCate *Phys. Rev. Lett.* **82**, 3280-3283, (1999)
8. J.-Y Kim.,V.A.Yakovlev., S.I.Rokhlin. *J. Acoust. Soc. Am.* **115,** Issue 5.pp.1961-1972 (2004)

Soft-ratchet modeling of slow dynamics in the nonlinear resonant response of sedimentary rocks

Oleksiy O. Vakhnenko*, Vyacheslav O. Vakhnenko[†],
Thomas J. Shankland[¶], and James A. TenCate[¶]

*Department of Quantum Electronics, Bogolyubov Institute for Theoretical Physics
14-B Metrologichna Street, Kyïv 03143, Ukraïna
[†]Department of Dynamics of Nonhomogeneous Media, Institute of Geophysics,
63-B Bohdan Khmel'nyts'kyy Street, Kyïv 01054, Ukraïna
[¶]Earth and Environmental Sciences Division, Los Alamos National Laboratory, Los Alamos,
New Mexico 87545, USA

Abstract. We propose a closed-form scheme that reproduces a wide class of nonlinear and hysteretic effects exhibited by sedimentary rocks in longitudinal bar resonance. In particular, we correctly describe: hysteretic behavior of a resonance curve on both its upward and downward slopes; linear softening of resonant frequency with increase of driving level; gradual (almost logarithmic) recovery (increase) of resonance frequency after large dynamical strains; and temporal relaxation of response amplitude at fixed frequency. Further, we are able to describe how water saturation enhances hysteresis and simultaneously decreases quality factor. The basic ingredients of the original bar system are assumed to be two coupled subsystems, namely, an elastic subsystem sensitive to the concentration of intergrain defects, and a kinetic subsystem of intergrain defects supporting an asymmetric response to an alternating internal stress.

INTRODUCTION

Sedimentary rocks, particularly sandstones, are distinguished by their grain structure in which each grain is much harder than the intergrain cementation material [1]. The peculiarities of grain and pore structures give rise to a variety of remarkable nonlinear mechanical properties demonstrated by rocks, both at quasistatic and alternating dynamic loading [1-4]. Thus, the hysteresis earlier established for the stress-strain relation in samples subjected to quasistatic loading-unloading cycles has also been discovered for the relation between acceleration amplitude and driving frequency in bar-shaped samples subjected to an alternating external drive that is frequency-swept through resonance. At strong drive levels there is an unusual, almost linear decrease of resonant frequency with strain amplitude, and there are long-term relaxation phenomena such as nearly logarithmic recovery (increase) of resonant frequency after the large conditioning drive has been removed.

In this paper we present a short sketch of a model [5, 6] for explaining numerous experimental observations seen in forced longitudinal oscillations of sandstone bars. According to our theory a broad set of experimental data can be understood as various aspects of the same internally consistent pattern [5, 6].

CP838, *Innovations in Nonlinear Acoustics: 17th International Symposium on Nonlinear Acoustics*,
edited by A. A. Atchley, V. W. Sparrow, and R. M. Keolian
© 2006 American Institute of Physics 0-7354-0330-9/06/$23.00

SOFT-RATCHET MODEL

A reliable probing method widely applied in resonant bar experiments is to drive a horizontally suspended cylindrical sample with a piezoelectric force transducer cemented between one end of the sample and a massive backload, and to simultaneously measure the sample response with a low-mass accelerometer attached to the opposite end of the bar [2, 4].

The evolution equation for the field of bar longitudinal displacements u as applied to above experimental configuration is assumed to be

$$\rho\frac{\partial^2 u}{\partial t^2} = \frac{\partial \sigma}{\partial x} + \frac{\partial}{\partial x}\left[\frac{\partial \Im}{\partial(\partial^2 u/\partial x\partial t)}\right]. \tag{1}$$

Here we use the Stokes internal friction associated with the dissipative function $\Im = (\gamma/2)\left[\partial^2 u/\partial x\partial t\right]^2$. The quantities ρ and γ are, respectively, mean density of sandstone and coefficient of internal friction. The stress-strain relation $(\sigma - \partial u/\partial x)$ we adopt in the form

$$\sigma = \frac{E\,\text{sech}\,\eta}{(r-a)\left[\cosh\eta\,\partial u/\partial x + 1\right]^{a+1}} - \frac{E\,\text{sech}\,\eta}{(r-a)\left[\cosh\eta\,\partial u/\partial x + 1\right]^{r+1}}, \tag{2}$$

which for $r > a > 0$ allows us to suppress the bar compressibility at strains $\partial u/\partial x$ tending toward $+0 - \text{sech}\,\eta$. Thus, the parameter $\cosh\eta$ is assigned for a typical distance between the centers of neighboring grains divided by the typical thickness of intergrain cementation contact. The indirect effect of strain on Young's modulus, as mediated by the concentration c of ruptured intergrain cohesive bonds, is incorporated in our theory as the main source of all non-trivial phenomena.

We introduce a phenomenological relationship between defect concentration c and Young's modulus E. Intuition suggests that E must be some monotonically decreasing function of c, which can be expanded in a power series with respect to a small deviation of c from its unstrained equilibrium value c_0. To lowest informative approximation we have

$$E = (1 - c/c_{cr})E_+. \tag{3}$$

Here c_{cr} and E_+ are the critical concentration of defects and the maximum possible value of Young's modulus, respectively.

The equilibrium concentration of defects c_σ associated with a stress σ is given by

$$c_\sigma = c_0 \exp(\upsilon\sigma/kT), \tag{4}$$

where the parameter $\upsilon > 0$ characterizes the intensity of dilatation. Although formula (4) should supposedly be applicable to the ensemble of microscopic defects in crystals, it was derived in the framework of continuum thermodynamic theory that does not actually need any specification of either the typical size of an elementary defect or the particular structure of the crystalline matrix. For this reason we believe it should also work for an ensemble of mesoscopic defects in consolidated materials, provided that for a single defect we understand some elementary rupture of intergrain cohesion. The

121

approximate functional dependence of c_0 on temperature T and water saturation s based on experimental data was treated in [5, 6].

In order to achieve reliable consistency between theory and experiment we have used the concept of blended kinetics, which finds more-or-less natural physical justification in consolidated materials [6]. The idea presents the actual concentration of defects c as some reasonable superposition of constituent concentrations g, where each particular g obeys rather simple kinetics. Specifically, we take the constituent concentration g to be governed by the kinetic equation:

$$\partial g / \partial t = - [\mu \theta (g - g_\sigma) + v \theta (g_\sigma - g)](g - g_\sigma). \qquad (5)$$

Here $\mu = \mu_0 \exp(-U/kT)$ and $v = v_0 \exp(-W/kT)$ are the rates of defect annihilation and defect creation, respectively, and $\theta(z)$ designates the Heaviside step function. A huge disparity $v_0 \gg \mu_0$ between the priming rates (attack frequencies) v_0 and μ_0 is assumed, notwithstanding the native cohesive properties of cementation material.

Typical resonant response experiments [1, 2, 4] correspond to forced longitudinal vibration of a bar, which we associate with the boundary conditions:

$$u(x = 0|t) = D(t)\cos(\varphi + \int_0^t d\tau \, \omega(\tau)), \qquad \sigma(x = L|t) + \gamma \frac{\partial^2 u}{\partial x \, \partial t}(x = L|t) = 0, \qquad (6)$$

where L is sample length, and $D(t)$ is driving amplitude. The initial conditions are

$$u(x|t = 0) = 0, \qquad \frac{\partial u}{\partial t}(x|t = 0) = 0, \qquad g(x|t = 0) = c_0. \qquad (7)$$

COMPUTERIZED REPLICAS OF ACTUAL EXPERIMENTS

Computer modeling of nonlinear and slow dynamics effects was performed in the vicinity of the resonance frequency $f_0(2)$, which we choose to be the second frequency ($l = 2$) in the fundamental set,

$$f_0(l) = \frac{2l-1}{4L} \sqrt{(1 - c_0 / c_{cr}) E_+ / \rho} \qquad (l = 1, 2, 3, ...). \qquad (8)$$

Figure 1 shows typical resonance curves, i.e., dependences of response amplitudes R (calculated at $x = L$) on drive frequency $f = \omega / 2\pi$, at successively higher drive amplitudes D. Solid lines correspond to conditioned resonance curves calculated after two frequency sweeps were performed at each driving level in order to achieve repeatable hysteretic curves. The dashed line illustrates an unconditioned curve obtained without any preliminary conditioning. Arrows on the three highest curves indicate sweep directions. To improve the illustration, results of the computer simulations were adapted to experimental conditions appropriate to the data obtained by TenCate and Shankland for Berea sandstone [2]. In particular, $L = 0.3\,\text{m}$, $f_0(2) = 3920\,\text{Hz}$, $vE_+ / k \cosh \eta = 275\,\text{K}$, $\cosh \eta = 2300$, $r = 4$, $a = 2$.

The shift of resonance frequency as a function of drive amplitude D was found to follow the almost linear dependence typical of materials with nonclassical nonlinear response, i.e., materials that possess all the basic features of slow dynamics (see [5, 6] for more details).

122

Figure 2 shows the gradual recovery of resonance frequency f_r to its maximum limiting value f_0 after the bar has been subjected to high amplitude conditioning and conditioning was stopped. We clearly see the very wide time interval $10 \le (t - t_c)/t_0 \le 1000$ of logarithmic recovery of the resonant frequency, in complete agreement with experimental results [4]. Here t_c is the moment when conditioning switches off and $t_0 = 1\mathrm{s}$ is the time scaling constant. Curves $j = 1, 2, 3$ on Figure 2 correspond to successively high water saturations $s_j = 0.05(2j - 1)$.

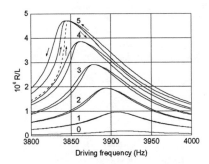

FIGURE 1. Resonance curves $j = 0, 1, 2, 3, 4, 5$ at successively higher driving amplitudes $D_j = 3.8(j + 0.2\delta_{j0})10^{-8} L$. The time to sweep back and forth within the frequency interval $3700 - 4100\mathrm{Hz}$ is chosen to be $120\,\mathrm{s}$.

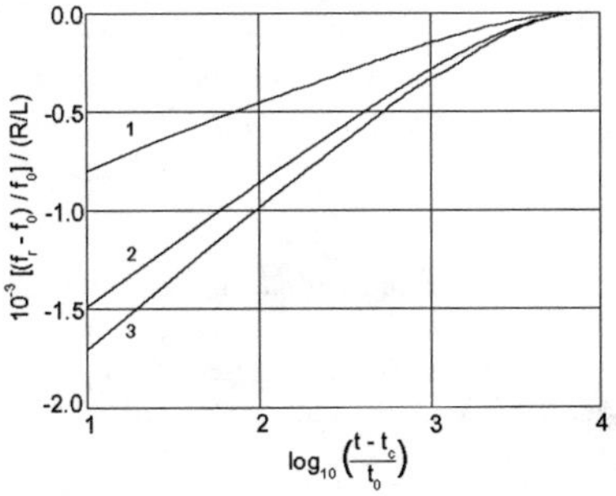

FIGURE 2. Time-dependent recovery of resonant frequency f_r to its asymptotic value f_0. The frequency shift $f_r - f_0$ is normalized by both the asymptotic frequency f_0 and the unitless response amplitude R/L attained at conditioning resonance.

ACKNOWLEDGMENTS

This research was supported by the Science and Technology Center in Ukraine under Grant No. 1747. O.O.V. acknowledges support from the National Academy of Sciences of Ukraine, Grant No. 0102U002332. J.A.T. and T.J.S. thank the Geosciences Research Program, Office of Basic Energy Sciences of the US Department of Energy for sustained assistance.

REFERENCES

1. R. A. Guyer, and P. A. Johnson , *Phys. Today* **52**, 30 (1999).
2. J. A. TenCate, and T. J. Shankland, *Geophys. Res. Lett.* **23**, 3019 (1996).
3. K. E.-A. Van Den Abeele, J. Carmeliet, P. A. Johnson, and B. Zinszner, *J. Geophys. Res., [Solid Earth]* **107**, 2121 (2002).
4. J. A. TenCate, E. Smith, and R. A. Guyer, *Phys. Rev. Lett.* **85**, 1020 (2000).
5. O. O. Vakhnenko, V. O. Vakhnenko, T. J. Shankland, and J. A. TenCate, *Phys. Rev. E* **70**, 015602(R) (2004).
6. O. O. Vakhnenko, V. O. Vakhnenko, and T. J. Shankland, *Phys. Rev. B* **71**, 174103 (2005).

Dynamical realization of end-point memory in consolidated materials

Vyacheslav O. Vakhnenko*, Oleksiy O. Vakhnenko†,
James A. TenCate¶, and Thomas J. Shankland¶

*Department of Dynamics of Nonhomogeneous Media, Institute of Geophysics,
63-B Bohdan Khmel'nyts'kyy Street, Kyïv 01054, Ukraïna
† Department of Quantum Electronics, Bogolyubov Institute for Theoretical Physics,
14-B Metrologichna Street, Kyïv 03143, Ukraïna
¶Earth and Environmental Sciences Division, Los Alamos National Laboratory, Los Alamos,
New Mexico 87545, USA

Abstract. Starting with a soft-ratchet model of slow dynamics in nonlinear resonant response of sedimentary rocks we predict the dynamical realization of end-point memory in resonating bar experiments with a cyclic frequency protocol. The effect we describe and simulate is defined as the memory of previous maximum amplitude of alternating stress and manifested in the form of small hysteretic loops inside the big hysteretic loop on the resonance curve. It is most clearly pronounced in the vicinity of bar resonant frequency. These theoretical findings are confirmed experimentally.

INTRODUCTION

Sedimentary rocks are prototypical of a class of materials that exhibit unusual elastic properties. In particular, they possess hysteresis and discrete memory [1-4]. For exploration of their equation of state, both quasi-static [1-3] and dynamic measurements [4, 5] have been used. Dynamic experiments may contain more information than quasi-static measurements. However, a description of the dynamic processes, and in particular, finding the equation of state, is a more difficult problem.

In modeling experiments on the longitudinal vibrational resonance of bar-shaped sedimentary rocks, we have proposed a closed system of equations for describing these processes [6-8]. Moreover, we have predicted the phenomenon of hysteresis with end-point memory in its essentially dynamical hypostasis [7]. In this paper we present recent experimental measurements that confirm above prediction.

PRINCIPAL PHYSICAL FOUNDATIONS OF MODEL

We here restrict the presentation of the theory only by its principal physical foundations and send the reader to our recent papers [6, 7] for details. First of all we believe that the number of intergrain defects determines the strength properties of consolidated medium. On the other hand, during dynamical loading the number of defects in a sample is changed (increased or decreased) relaxing continuously toward its dynamically driven would-be equilibrium value. Arguing a substantial difference between the typical rates of defect creation and defect annihilation, we come to a

CP838, *Innovations in Nonlinear Acoustics: 17th International Symposium on Nonlinear Acoustics*,
edited by A. A. Atchley, V. W. Sparrow, and R. M. Keolian
© 2006 American Institute of Physics 0-7354-0330-9/06/$23.00

physical mechanism that breaks the symmetry of system response to an alternating external drive and acts as a sort of soft ratchet or leaky diode. The formalization of these and other basic ideas gives rise to a model system [6-8] that enables us to describe correctly a wide class of experimental facts concerning the unusual dynamical behaviour of such mesoscopically inhomogeneous media as sandstones [4-8], as well as to forecast a new essentially dynamical form of end-point memory [7].

DYNAMICAL REALIZATION OF END-POINT MEMORY: THEORETICAL PREDICTION

The question of whether an effect similar to the end-point (discrete) memory that is observed in quasi-static experiments with a multiply-reversed loading-unloading protocol [1-3] could also be seen in resonating bar experiments with a multiply-reversed frequency protocol has been risen in [7] and firstly was examined

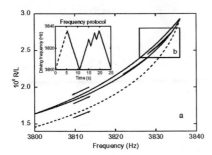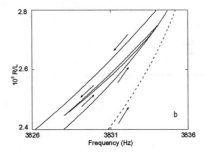

FIGURE 1. Manifestation of end-point memory in dynamic response with a multiply-reversed frequency protocol. R is the response amplitude, L is the length of the bar.

theoretically. The graphical results of this investigation are presented in Figure 1 (see also Fig. 16 in [7], where the model constants are given). One of the features of dynamical end-point memory, defined here as the memory of the previous maximum amplitude of alternating stress, is seen as small loops inside the big loop. The starting and final points of each small loop coincide, which is typical of end-point memory. According to our theory when producing an extremely small inner loop on conditioned (solid-line) curves the chance to find it closed diminishes in proportion to its linear size, being lower on the downward going curve and higher on the upper part of upward going curve. The reason for such behavior is the existence of a threshold stress amplitude (depending on previous history) that must be surmounted in order for the kinetics of the slow subsystem to be switched from defect annihilation at lower amplitudes to defect creation at higher amplitudes. This restriction can be substantially relaxed provided the linear size of the inner loop becomes comparable with that of the big outer loop. In contrast, when dealing with unconditioned (dashed-line) curve, a closed inner loop can be produced anywhere without any restrictions on its smallness (not shown).

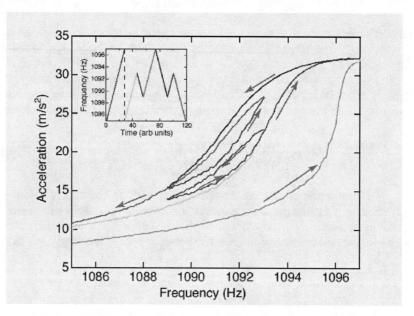

FIGURE 2. The low frequency sides of experimental resonance curves for Fontainebleau sandstone.

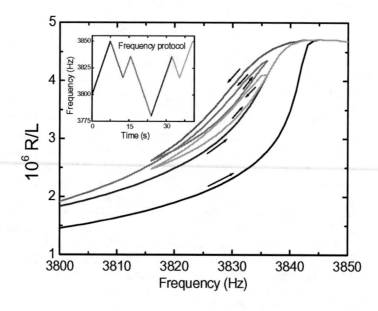

FIGURE 3. The low frequency sides of the resonance curves calculated for Berea sandstone.

DYNAMICAL REALIZATION OF END-POINT MEMORY: EXPERIMENTAL CONFIRMATION

Following the theoretical results, shown in Fig. 1, we performed experimental measurements to verify our prediction. The sample bar was a Fontainebleau sandstone and the drive level produced a calculated strain of about $2 \cdot 10^{-6}$ at the peak. Figure 2 shows the low frequency sides of resonance curves that correspond to the frequency protocol given on inset of Fig. 2. We clearly see that the beginning and end of each inner loop coincide, i.e., a major feature of end-point memory.

The experimental results for the Fontainebleau sandstone shown in Fig. 2 were simulated by using existing model equations (including a state equation) [6-8] and corrected (as compared with Fig. 1) constants for Berea sandstone. We note the good qualitative agreement between the experimental (Fig. 2) and the theoretical (Fig. 3) curves suggesting that our physical model is appropriate for both sandstones.

ACKNOWLEDGMENTS

This research was supported by the Science and Technology Center in Ukraine under Grant No. 1747. O.O.V. acknowledges support from the National Academy of Sciences of Ukraine, Grant No. 0102U002332. J.A.T. and T.J.S. thank the Geosciences Research Program, Office of Basic Energy Sciences of the US Department of Energy for sustained assistance.

REFERENCES

1. G. N. Boitnott, "Fundamental observations concerning hysteresis in the deformation of intact and jointed rock with applications to nonlinear attenuation in the near source region", in *Proceedings of the Numerical Modeling for Underground Nuclear Test Monitoring Symposium, Los Alamos Natl. Lab. Rev., LA-UR-93-3839*, 1993, p. 121.
2. L. B. Hilbert, Jr., T. K. Hwong, N. G. W. Cook, K. T. Nihei, and L. R. Myer, "Effects of strain amplitude on the static and dynamic nonlinear deformation of Berea sandstone", in *Rock Mechanics Models and Measurements Challenges from Industry,* edited by P. P. Nelson and S. E. Laubach. Rotterdam, Netherlands, 1994, p. 497.
3. T. J. Plona, and J. M. Cook, "Effects of stress cycles on static and dynamic Young's moduli in Castlegate sandstone", in *Rock Mechanics,* edited by J. J. K. Daemen and R. A. Schultz, Proceedings of the 35th U.S. Symposium. A. A. Balkema, Rotterdam, Netherlands, 1995, p. 155.
4. J. A. TenCate, and T. J. Shankland, *Geophys. Res. Lett.* **23**, 3019 (1996).
5. J. A. TenCate, and T. J. Shankland, "Slow dynamics and nonlinear response at low strains in berea sanstone", in *Proceedings of the 16th International Congress on Acoustics and 135th Meeting of the Acoustical Society of America,* edited by P. A. Kuhl and L. A. Crum. New York: American Institute of Physics, **3**, 1998, p. 1565.
6. O. O. Vakhnenko, V. O. Vakhnenko, T. J. Shankland, and J. A. TenCate, *Phys. Rev. E* **70**, 015602(R) (2004).
7. O. O. Vakhnenko, V. O. Vakhnenko, and T. J. Shankland, *Phys. Rev. B* **71**, 174103 (2005).
8. O. O. Vakhnenko, V. O. Vakhnenko, T. J. Shankland, and J. A. TenCate, "Soft-ratchet modeling of slow dynamics in the nonlinear resonant response of sedimentary rocks", in this *Proceedings*.

SECTION 2
NONLINEAR ACOUSTICS IN SOLIDS

Excitable behavior of ultrasound in a magnetoacoustic resonator

V.J. Sánchez–Morcillo, J. Redondo,
J. Martínez–Mora, V. Espinosa and F. Camarena

*Departamento de Física Aplicada, Universidad Politécnica de Valencia,
Ctra. Natzaret-Oliva S/N 46730 Grao de Gandia, (Spain)*

Abstract. The dynamical aspects of ultrasound generated by a parametric process in magnetostrictive ceramics are considered. The model takes into account both magnetic and magnetoacoustic nonlinearities. The numerical analysis shows the existence of self-pulsing dynamics of the fields. It is also demonstrated the existence of homoclinic bifurcations and, as a consequence, the excitable character of the resonator under given conditions.

Keywords: Magnetostriction, parametric generation, self-pulsing, excitability.

INTRODUCTION

The parametric generation of ultrasound in magnetostrictive ceramics is an active topic of research, and the basis of important applications related with the wave phase conjugation properties of these materials[1]. Physically, the process occurs because a time-varying external magnetic field modulates the sound velocity in the material. In this work we analize the dynamical behaviour of this system in the presence of magnetic nonlinearity. It is shown that a cubic magnetic nonlinearity is responsible for the appearance of new effects not reported before in this system, such as self-pulsing dynamics, and spiking and excitable behaviour, related to the existence of homoclinic bifurcations.

MODEL, STATIONARY SOLUTIONS AND STABILITY

The physical system consists in a electric RLC circuit, driven by an external *ac* source at frequency 2ω and variable amplitude ε. The inductance coil, with density of turns n, transverse section S and length L, contains a magnetostrictive ceramic material in the from of a rod with paralel and flat end surfaces, which acts as an acoustical resonator and is the origin of nonlinearities in the system when the driving is high enough. The theoretical model for this system has been derived in Ref. 2 and next it is shortly reviewed.

CP838, *Innovations in Nonlinear Acoustics: 17th International Symposium on Nonlinear Acoustics*,
edited by A. A. Atchley, V. W. Sparrow, and R. M. Keolian

The model consider the coupled dynamics of the acoustic and magnetic fields. The acoustic part consists in a wave equation for the displacement u, driven by a magnetic field

$$\frac{1}{v^2}\frac{\partial^2 u}{\partial t^2} - \nabla^2 u = \alpha H u \tag{1}$$

where $\alpha = (2k)^2 (\partial \ln v / \partial H)$ is the coupling constant. The magnetic field H with frequency 2ω has an amplitude proportional to the current in the circuit. For the RLC circuit, the electric variables (potentials) obey

$$N\frac{d}{dt}\int B dS + IR + \frac{q}{C} = \mathcal{E}\cos(2\omega t) \tag{2}$$

where $B = \mu_0 (H + M)$. Two kind of nonlinearities are present in the first term in Eq. (2), the nonlinear magnetic susceptibility of the material and the magnetoacoustic interaction. The first is introduced by assuming that, for weak fields, the magnetization is described by $M = \chi^{(1)} H + \frac{1}{6}\chi^{(3)} H^3$. On the other side, H is the magnetic field resulting from three contributions, a static (bias) field H_0, an alternating field $H_q(t) = H_q \cos(2\omega t)$, and an acoustically induced field $H_{int}(t) = -\alpha \langle u^2 \rangle$, where the brackets denote average over the material volume[1]. Assuming weak losses (high reflectivity at resonator walls), the charge in the capacitor and the acoustic displacement can be assumed to be of the form of quasi-harmonic waves, i.e., $q(t) = Q(t)e^{i2\omega t} + c.c.$ and $u(\vec{r},t) = U(t)g(r_\perp)\sin(kz)e^{i\omega t} + c.c$, where Q and U are slowly varying functions of time. After some algebra the following system is obtained

$$\frac{dX}{d\tau} = P - X + Y^2 + i\eta X|Y|^2, \qquad \frac{dY}{d\tau} = -\gamma(Y - XY^*) \tag{3}$$

where the parameters are defined as

$$P = \frac{v^2 n \alpha}{8\gamma_Q \gamma_U \omega L}\mathcal{E}, \quad \gamma = \frac{\gamma_U}{\gamma_Q}, \quad \tau = \gamma_Q t, \quad \eta = \frac{4\omega H_0 \gamma_U}{\alpha v^2}\frac{\chi^{(3)}}{\chi^{(1)}} \tag{4}$$

and X and Y are proportional to Q and U respectively.

Eqs. (3) possess the stationary solution $|\overline{X}| = 1$, $|\overline{Y}|^2 = \dfrac{1 \pm \sqrt{P^2(1+\eta^2) - \eta^2}}{1+\eta^2}$, which

shows bistability in the acoustic field between $P = 1$ (threshold) and $P = \eta/\sqrt{1+\eta^2}$.

The stability of the stationary solutions has been also analyzed, by considering the evolution of the system under small perturbations, inserting solutions of the form $X_i(t) = \overline{X}_i + \delta X_i(t)$, and linearizing the resulting system. This leads to

$\delta X_i(t) \propto \exp(\lambda t)$, where λ are the eigenvalues of the linear matrix. From this study[3] results that the solutions experience a Hopf (self-pulsing) bifurcation when

$$4\eta^4(\gamma-1)\left|\overline{Y}\right|^8 + 2\eta^2(\gamma+1)^2\left|\overline{Y}\right|^6 + \eta^2(4\gamma^3-3\gamma-3)\left|\overline{Y}\right|^4 + 2(\gamma+1)^2(2\gamma+1)\left|\overline{Y}\right|^2 = (2\gamma+1)^2$$

SELF-PULSING AND EXCITABLE REGIMES

The numerical integration of Eqs. (3) was performed in order to demonstrate the existence of dynamical solutions. For typical experimental conditions, $\gamma_Q < \gamma_U$, and consequently $\gamma < 1$. We take, according to Ref. 3, the value $\gamma = 0.1$ for numerical integration, although we note that similar results are obtained for a smaller loss ratio. For this value, the linear stability analysis predicts that the solutions display temporal dynamics when the nonlinearity coefficient take values in the range $0 < \eta < 0.82$.

In Fig. 1(a) the bifurcation diagram as computed numerically for $\gamma = 0.1$ and $\eta = 0.7$ is shown. Dashed lines correspond to the analytical solutions, and the existence of the backward (inverted) Hopf bifurcation of the upper branch is observed at $P = 1.87$. Dynamical states exist for pump values below this critical point. For higher pump values, the solution is stationary, and the dynamics decay to a fixed point. The upper and lower branches with open circles in Fig. 1(a) correspond to maximum and minimum values of ultrasonic amplitudes in the oscillating regime.

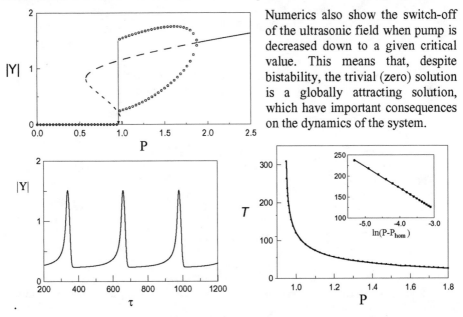

Numerics also show the switch-off of the ultrasonic field when pump is decreased down to a given critical value. This means that, despite bistability, the trivial (zero) solution is a globally attracting solution, which have important consequences on the dynamics of the system.

FIGURE 1. Self-pulsing dynamics of ultrasound. (a) Bifurcation diagram for $\gamma = 0.1$ and $\eta = 0.7$. (b) Temporal evolution for P = 1 (c). Interspike period as a funcion of the pump.

133

Close to the threshold, the period of oscillation increase and tends to infinity [Figs 2(b) and (c)]. This behavior signals the presence of a homoclinic bifurcation.

For pump values below the homoclinic bifurcation point, the system possess a property called excitability, characterized by (a) perturbations of the rest state beyond a certain threshold induce a large amplitude response before coming back to the rest state; otherwise the system smoothly relaxes to the rest (b) above threshold, the amplitude of response is independent of the amplitude of the perturbation, but the response time decreases with it, and (c) there exist a refractory time during which no further excitation is possible. These properties, which are characteristic of several biological problems (e.g. the behaviour of action potentials in neurons) and some laser models, are demonstrated to exist in the acoustic system described by Eqs. (3).

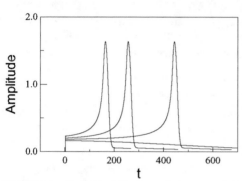

FIGURE 2. Excitable behaviour of ultrasonic field for $P = 0.9$. The response of the system to four perturbations with different amplitudes is shown.

Figure 2 shows the response on the resonator in the excitable regime for several perturbation amplitudes at $t = 0$. Note that the perturbation below the excitability threshold decay, but those above threshold produce a pulse at different times, according to the definitions given above. In this sense, the behaviour of the system in the excitable regime has an all-or-nothing character.

Based in the previous experimental work on this system[1,4], we could estimate the numerical values of the different parameters appearing in Eqs. (3). This analysis suggest that, under typical operation conditions, the nonlinearity parameter η has a magnitude of order 1, and consequently the prediced effects could be observed on these systems.

ACKNOWLEDGMENTS

The work was financially supported by the CICYT of the Spanish Government, under the project BFM2002-04369-C04-04.

REFERENCES

1. A. Brysev, L. Krutyansky and V. Preobrazhensky, Phys. Uspekhi 41, 793--805 (1998)
2. V.N. Streltsov, BRAS Physics Supplement 61, 228-230 (1997)
3. V.J. Sánchez-Morcillo, J. Redondo, J. Martínez-Mora, V. Espinosa and F. Camarena, arxiv.org/pdf/nlin.CD/0502034
4. A. Brysev, P. Pernod and V. Preobrazhensky, Ultrasonics 38, 834-837 (2000)

Acoustoelectric Harmonic Generation in a Photoconductive Piezoelectric Semiconductor

W. Arthur, R. E. Kumon, F. Severin, and R. Gr. Maev

*Centre for Imaging Research and Advanced Materials Characterization,
Department of Physics, University of Windsor, 401 Sunset Ave., Windsor, Ontario N9B 3P4 Canada*

Abstract. Piezoelectric semiconductors can exhibit harmonic generation because of nonlinear interactions between the acoustic and electric fields in the solid. To observe this effect, longitudinal waves were excited in crystalline cadmium sulfide. Because CdS is highly photosensitive, its conductivity can be changed by several orders of magnitude by varying the applied light level. The velocity and attenuation at 4.1 MHz and 8.2 MHz were measured and shown to be strong functions of conductivity in the range 0.001-0.010 $\Omega^{-1}\text{m}^{-1}$. Fitting the resulting data with linearized theory yielded values for the piezoelectric stress and elastic moduli that were consistent with literature values. Also, harmonic generation resulting from excitation at 4.1 MHz was measured. The amplitudes of the second to fifth harmonics exhibited oscillatory behavior as a function of conductivity, particularly in the aforementioned range. Finally, the second harmonic amplitude was also measured at multiple propagation distances, and the conductivity of its peak amplitude was shown to shift to lower conductivity at further distances.

Keywords: Acoustoelectric effect, Harmonic generation, Piezoelectric semiconductor, Photoconductivity, Cadmium sulfide
PACS: 43.25.Dc, 72.50.+b, 77.65.Dq, 72.40.+w

INTRODUCTION

When appropriately polarized acoustic waves propagate in specific directions in piezoelectric semiconductors, the modulated electric field induced by the wave causes conduction electrons to bunch together. This *acoustoelectric effect* involves nonlinear coupling of the electric field and charge density, which results in an internal dc electric field as well as higher electric and acoustic harmonics. Most of the theoretical and experimental work in nonlinear acoustics on this topic has been limited to second harmonic generation [1–3] or parametric interactions [4,5] or has included the effects of an applied external electric field [1–6]. The latter can give rise to amplification and other effects, as has been described in several reviews [7–9] and tends to result in exponential growth of certain modes over all others.

In this paper, we describe experiments on the generation of higher harmonics in the low frequency regime ($\omega\tau \ll 1$, where ω is angular frequency and τ is the electronic relaxation time) and without application of an external electric field. Longitudinal bulk waves were excited along the hexagonal axis of undoped cadmium sulfide, a direction that was chosen because of its high piezoelectric activity.

CP838, Innovations in Nonlinear Acoustics: 17th International Symposium on Nonlinear Acoustics,
edited by A. A. Atchley, V. W. Sparrow, and R. M. Keolian

THEORY

It can be shown that the propagation of longitudinal bulk waves in a piezoelectric semiconductor in one dimension can be described by [3]:

$$\frac{\partial^2 T}{\partial z^2} - \frac{\rho_m}{c^D}\frac{\partial^2 T}{\partial t^2} - \frac{\rho_m}{c^D}\kappa\sqrt{\frac{c}{\varepsilon}}\frac{\partial^2 D}{\partial t^2} = 0,$$ (1)

$$\frac{\partial D}{\partial t} + \frac{\mu q n_0}{\varepsilon^T}\left(D - \sqrt{\frac{\varepsilon}{c}}\kappa T\right) - D_n\frac{\partial^2 D}{\partial z^2} = \frac{\mu}{\varepsilon^T}\left(D - \sqrt{\frac{\varepsilon}{c}}\kappa T\right)\frac{\partial D}{\partial z},$$ (2)

where z is distance, t is time, T is stress, ρ_m is mass density, $c^D = c(1+\kappa^2)$, c is stiffness modulus, $\kappa^2 = e^2/c\varepsilon$, e is piezoelectric modulus, ε is dielectric permittivity, D is electric displacement, μ is mobility, q is magnitude of electric charge, n_0 is equilibrium electron number density, $\varepsilon^T = \varepsilon(1+\kappa^2)$, and D_n is diffusion coefficient. In a quasilinear solution with fundamental $T_1(z=0) = T_1^0\exp(i\omega t - q_1 z)$ and second harmonic $T_2(z=0)=0$, it was shown that the second harmonic of the acoustic stress has the form

$$T_2 = \frac{F}{q_2 - 2q_1}\left(e^{-i2q_1 z} - e^{-iq_2 z}\right)e^{i2\omega t},$$ (3)

where q_j is wavenumber of the jth harmonic and F is given in [3]. Qualitatively, the amplitude of the second harmonic exhibits oscillations as a function of conductivity, and the nature of these oscillations evolves as a function of distance [2–5]. However, experimental data corroborating this result are limited [2,4] and higher harmonics do not appear to have been measured.

EXPERIMENT

Figure 1 shows the experimental apparatus. The CdS crystal was placed in a temperature-controlled bath, and acoustic transducers were placed on opposite faces. The conductivity of the crystal was varied by changing the illumination using an Interlux DC1100 digitally-controlled light source with dc power to enhance its stability. A photometer was utilized to monitor and ensure consistent illumination. To generate, receive, and process the acoustic signal, a Ritec SNAP-1-30 system was employed along with a LeCroy LT342L Digital Oscilloscope. For the velocity and

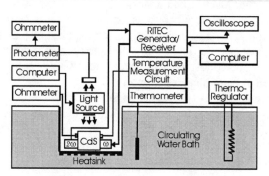

FIGURE 1. Experimental apparatus.

attenuation measurements, 8 cycle, 4.1 MHz and 8.2 MHz signals were generated with 3.5 MHz and 7.5 MHz Xactex transducers, respectively, each operating in reflection mode. For measurement of the harmonics, a transmission setup was used with the fundamental of 4.1 MHz generated with the 3.5 MHz Xactex transducer, the second harmonic detected with a 7.5 MHz Aerotech transducer, and higher harmonics detected with a 15 MHz Aerotech transducer. A 5 MHz low-pass filter assured that no higher harmonics entered the crystal, while a 5 MHz high-pass filter eliminated the fundamental frequency from entering the Ritec system. The amplitudes of the higher harmonics were measured relative to the second harmonic and were appropriately scaled with respect to the sensitivity of the 15 MHz transducer.

RESULTS AND DISCUSSION

Figure 2 shows the resulting velocity and attenuation measurements as a function of conductivity, along with theoretical curves [10] fitted by setting $c = 88.7$ GPa, $e = 0.515$ C/m^2, and the light penetration depth $d_p = 140$ μm, the latter quantity used to compute the conductivity from resistance measurements. These values are consistent with those reported by Hudson and White [10] and Bube [11]. In addition, a constant was added to the attenuation to account for attenuation that was not a result of the acoustoelectric effect. The velocity and attenuation are not independent, and compromises were made in the fitting to favor the velocity data, which were deemed more accurate. The reasons for the significant underestimation of the attenuation at 8.2 MHz are not entirely clear but may be at least partially attributed to the generation of higher harmonics, which is not accounted for in the linear theory of Ref. 10.

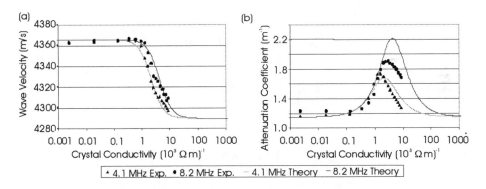

FIGURE 2. (a) Velocity and (b) attenuation of the fundamental frequencies 4.1 MHz and 8.2 MHz as a function of crystal conductivity.

Figure 3(a) shows amplitude of the second harmonic at 8.2 MHz as a function of conductivity. The successive curves show the first received pulse of the second harmonic as well as three consecutive echoes. As can be seen, the amplitude of the first pulse passes through a maximum around 0.009 Ω^{-1} m^{-1}, but the maxima of subsequent echoes shifts to lower conductivities as the pulse travels further. Initial theoretical calculations which account for only the second harmonic [Eq. (3)] appear to qualitatively correspond with these observations, Figure 3(b) shows the second (8.2 MHz), third (12.3 MHz), fourth (16.4 MHz), and fifth (20.5 MHz) harmonics as a function of conductivity. The second harmonic has a minimum near 0.009 Ω^{-1} m^{-1}, while the other harmonics have multiple minima in the range 0.005–0.009 Ω^{-1} m^{-1}.

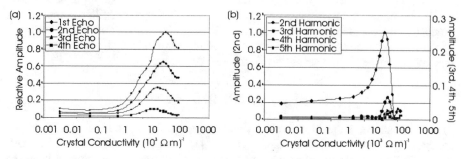

FIGURE 3. Relative amplitudes of (a) consecutive echoes of the second harmonic and (b) amplitudes of the second to fifth harmonics of a 4.1 MHz fundamental.

CONCLUSION

Measurements of the velocity of longitudinal bulk waves along the hexagonal axis of CdS have been shown to be consistent with known linearized theory. The amplitudes of the second through fifth harmonics of a 4.1 MHz fundamental have been measured and show oscillations of the sort predicted by quasilinear theory for the second harmonic component.

ACKNOWLEDGMENTS

This work was supported by the Industrial Research Chair in Applied Solid State Physics and Material Characterization, which is funded by NSERC, DaimlerChrysler, and the University of Windsor. We would also like to acknowledge useful discussions with V. G. Mozhaev.

REFERENCES

1. H. Kroger, *Appl. Phys. Lett.* **4**, 190-192 (1964).
2. B. Tell, *Phys. Rev.* **136**, A772-A775 (1964).
3. V. G. Mozhaev and I. Yu. Solodov, *Vestnik Moskovskogo Universiteta Fizika* **35**, 46-53 (1980).
4. R. Mauro and W. C. Wang, *Phys. Rev. B* **1**, 683-687 (1970).
5. A. K. Ganguly and E. M. Conwell, *Phys. Lett.* **29A**, 221-3 (1969); *Phys. Rev. B* **4**, 2535-58 (1971).
6. P. K. Tien, *Phys. Rev.* **171**, 970-986 (1968).
7. J. H. McFee, "Transmission and Amplification of Acoustic Waves in Piezoelectric Semiconductors," in *Physical Acoustics* **4A**, edited by W. P. Mason, New York: Academic Press, 1966, pp. 1-45.
8. V. L. Gurevich, *Soviet Phys. Semicond.* **2**, 1299-1325 (1969).
9. H. Kuzmany, *Phys. Stat. Sol. A* **25**, 9-67 (1974).
10. A. R. Hutson and D. L. White, *J. Appl. Phys.* **33**, 40-47 (1962).
11. R. H. Bube, *Photoelectronic Properties of Semiconductors*, New York: Cambridge U.P., 1992, p. 8.

Time-domain evolution equation for dispersive nonlinear Rayleigh waves

Won-Suk Ohm[*,†] and Mark F. Hamilton[*]

[*]Department of Mechanical Engineering, The University of Texas at Austin, 1 University Station C2200, Austin, Texas 78712-0292
[†]Acoustical Standards, Institute for National Measurement Standards, National Research Council, M-36, 1200 Montreal Road, Ottawa, Ontario, Canada K1A 0R6

Abstract. Korteweg-de Vries (KdV) and Benjamin-Ono (BO) equations are often cited as time-domain models for nonlinear Rayleigh waves in a dispersive medium such as a coated substrate. However, these canonical equations do not account for the nonlocal nonlinearity that is a hallmark of surface acoustic waves. In this paper, a time-domain evolution equation derived for nonlinear Rayleigh waves in nondispersive isotropic media [Hamilton, Il'insky, and Zabolotskaya, J. Acoust. Soc. Am. **97**, 891–897 (1995)] is extended to include dispersion due to thin-film coating. It is demonstrated that a class of soliton-like solutions resembling Mexican hats satisfies the evolution equation for the case of KdV-type dispersion, a result obtained previously by Eckl et al. [Phys. Rev. E **70**, 046604 (2004)] from a frequency-domain form of the evolution equation.

Keywords: Rayleigh wave, nonlinear, thin film, time domain, pole expansion, soliton
PACS: 43.25.Fe, 43.25.Rq

INTRODUCTION

Theoretical studies of nonlinear Rayleigh waves are based predominantly on evolution equations in the frequency domain. This is primarily because evolution equations in the time domain contain awkward convolution integrals, often in the form of Hilbert transforms, that account for the nonlocal nonlinearity of surface acoustic waves (SAWs) [1]. In contrast, evolution equations in the frequency domain are expressed as coupled first-order differential equations that are well suited for numerical integration. Eckl et al. [2] recently investigated solitary SAWs in substrates coated with thin films, and they inferred the existence of surface acoustic solitons from frequency-domain simulations of solitary SAW interaction. Although the study offers strong computational evidence of SAW solitons, a mathematically rigorous solution should resort to analytical techniques such as the inverse scattering transform (IST) [3] that require a time-domain evolution equation. Solitons are formally defined within the framework of the IST as solutions of an evolution equation corresponding to discrete eigenvalues of the associated scattering problem.

The purpose of the present paper is to approach the problem of solitary SAW propagation using a time-domain evolution equation. As a preliminary step towards application of the IST to the case of nonlinear Rayleigh waves in thin-film systems, it is demonstrated that a class of exact analytic solutions, equivalent to those obtained previously by Eckl et al. [2], is developed by means of pole expansion.

CP838, *Innovations in Nonlinear Acoustics: 17th International Symposium on Nonlinear Acoustics*, edited by A. A. Atchley, V. W. Sparrow, and R. M. Keolian
© 2006 American Institute of Physics 0-7354-0330-9/06/$23.00

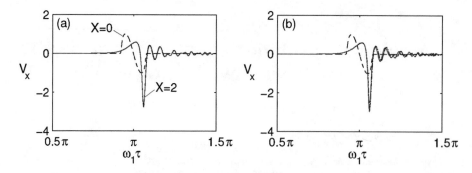

FIGURE 1. Evolution of a Rayleigh-wave pulse on a gold-plated fused-quartz substrate. Dimensionless horizontal velocity V_x at the surface is shown in retarded time τ: dashed line, source waveform; dash-dot line, full model [5]; solid line, simplified model [Eqs. (1) and (2)]. In frame (a), both BO and KdV dispersion terms are used in the simplified model, whereas in frame (b), only the BO term is retained.

EVOLUTION EQUATION

To obtain a time-domain evolution equation, we start with a system of dimensionless spectral equations in Ref. 1, augmented here with a two-term Taylor expansion approximating an arbitrary dispersion law:

$$\frac{dV_n}{dX} + i(D_{BO}n^2 - D_{KdV}n^3)V_n = \gamma\left(-\frac{n}{4}\sum_{m=1}^{n-1}V_m V_{n-m} + \frac{n^2}{2}\sum_{m=n+1}^{\infty}\frac{1}{m}V_m V_{m-n}^*\right). \quad (1)$$

Here, V_n is the spectral amplitude of the nth harmonic, X propagation distance, and $\gamma = \pm 1$ the sign of nonlinearity. The coefficients $D_{BO,KdV}$ represent strength of dispersion relative to nonlinearity. Given a dispersion relation $c = c(\omega)$, the coefficients are

$$D_{BO} = \frac{\bar{x}\omega_1^2 c_0'}{c_0^2}, \qquad D_{KdV} = \frac{\bar{x}\omega_1^3 (c_0')^2}{c_0^3} - \frac{\bar{x}\omega_1^3 c_0''}{2c_0^2}, \quad (2)$$

where $\bar{x}$ characterizes the shock formation distance, ω_1 is repetition frequency, $c_0 = c(0)$ the propagation speed at zero frequency, $c_0' = (dc/d\omega)_{\omega=0}$, and $c_0'' = (d^2c/d\omega^2)_{\omega=0}$. See Ref. 4 for details. Equations (1) and (2) constitute a simplified version of the complete spectral model for nonlinear Rayleigh waves in coated substrates [5]. Although reduction of the full model is undertaken in both nonlinearity and dispersion, the simplified model describes waveform distortion with great accuracy. Figure 1 shows the evolution of a bipolar pulse (dashed line) in a gold-plated fused-quartz substrate. Horizontal velocity waveforms computed with the complete (dash-dot line) and simplified (solid line) models are almost indistinguishable in Fig. 1(a). The error associated with discarding the higher-order KdV dispersion term in the simplified model is relatively insignificant [compare Figs. 1(a) and 1(b)]. However, as suggested by Eckl *et al.* [2], there may be a rare film-substrate material combination that renders the KdV dispersion term much more pronounced than the BO term ($|D_{KdV}|/|D_{BO}| \gg 1$). For this type of material pair, soliton-like solutions exist as shown in the next section.

The time-domain evolution equation obtained from Eq. (1) is written in terms of dimensionless horizontal displacement at the surface, $d = \sum_n |n|^{-1} V_n(X) e^{-in\theta}$ [1]:

$$\frac{\partial d}{\partial X} = \frac{\gamma}{8} \frac{\partial^2 d^2}{\partial \theta^2} + \frac{\gamma}{4} \mathscr{H} \left[d \mathscr{H} \left[\frac{\partial^2 d}{\partial \theta^2} \right] \right] + D_{\text{BO}} \mathscr{H} \left[\frac{\partial^2 d}{\partial \theta^2} \right] + D_{\text{KdV}} \frac{\partial^3 d}{\partial \theta^3}, \qquad (3)$$

where θ is dimensionless retarded time and $\mathscr{H}$ denotes the Hilbert transform. Note that the greatest departure of Eq. (3) from such canonical equations as the BO and KdV equations is the presence of the nonlocal nonlinear term involving the Hilbert transform. In the linear limit ($\gamma \to 0$), Eq. (3) reduces to dispersion models of the BO and KdV equations [hence the designations "BO" and "KdV" in Eqs. (1) and (3)].

ANALYTIC SOLITARY-WAVE SOLUTIONS

One direct method for finding soliton solutions of a certain class of evolution equations is by pole expansion [3]. Because Eq. (3) without the nonlocal nonlinear term resembles the BO equation at the leading order in dispersion, it is reasonable to assume a trial solution composed of the Lorentzian function (the BO soliton) and its powers,

$$\frac{\partial d}{\partial \theta} = \sum_{n=1}^{N} \hat{B}_n \left[\frac{1}{(\theta - aX)^2 + b^2} \right]^n = \sum_{n=1}^{N} B_n \left(\frac{i}{\theta - \theta_1} + \frac{-i}{\theta - \theta_1^*} \right)^n, \qquad (4)$$

where $\partial d / \partial \theta$ is dimensionless horizontal velocity, and the Lorentzian function is expanded in terms of poles at $\theta_1, \theta_1^* = aX \pm ib$. Following substitution of the ansatz into Eq. (3), partial fraction expansion is performed noting that $\mathscr{H}[1/(\theta - \theta_1)^n] = -i/(\theta - \theta_1)^n$, $\text{Im}\,\theta_1 > 0$. Preventing terms other than poles from appearing in the resulting equation and matching coefficients of successive powers of $1/(\theta - \theta_1)$ [or $1/(\theta - \theta_1^*)$] produces a series of algebraic equations to be solved for the unknown constants B_n, a, and b. Although Eq. (4) is a seemingly natural outgrowth of the BO soliton, it satisfies Eq. (3) only if $D_{\text{BO}} = 0$. It turns out that Eq. (3) with $D_{\text{BO}} = 0$ can be shown to be related to the third member of the BO hierarchy known as the BO_3 equation [2].

The only solitary-wave solution in the form of Eq. (4) is thus found to require $D_{\text{BO}} = 0$ in Eq. (3), and it is given by

$$\frac{\partial d}{\partial \theta} = 48\gamma D_{\text{KdV}} \left\{ \frac{-1}{(\theta - aX)^2 + b^2} + \frac{2b^2}{[(\theta - aX)^2 + b^2]^2} \right\}, \qquad (5)$$

where constants a and b scale with a free parameter B_1 according to

$$a = -\frac{B_1^2}{192 D_{\text{KdV}}}, \qquad b = \frac{24 D_{\text{KdV}}}{B_1}. \qquad (6)$$

Eckl et al. [2] arrived at the same solution via frequency-domain analysis. Equation (5) represents a one-parameter (B_1) family of solitary-wave solutions of Eq. (3) with the KdV-type dispersion, whose scaling law is given by Eqs. (6). These solitary waves have

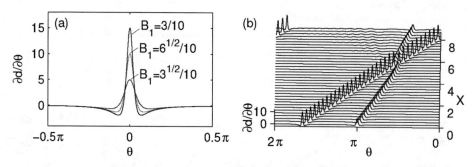

FIGURE 2. A family of solitary Rayleigh waves for the case of positive nonlinearity ($\gamma = 1$) and loading dispersion ($D_{KdV} = 0.0005$): (a) solitary waves corresponding to different values of B_1, and (b) collision of two solitary waves with $B_1 = \sqrt{3}/10$ and $B_1 = \sqrt{6}/10$.

tripolar shapes as illustrated in Fig. 2(a). Their polarity is determined by the sign of the product γD_{KdV} [see Eq. (5)], while the rate a at which they advance or recede in retarded time depends on dispersion and the amplitude of the wave [Eqs. (6)]. For example, for the case of loading dispersion ($D_{KdV} > 0$) considered in Fig. 2(a), all solitary waves propagate at speeds greater than the nondispersive Rayleigh-wave speed, with taller and narrower pulses traveling faster than shorter and wider ones.

Collision of two solitary pulses from Fig. 2(a) is shown in Fig. 2(b). Propagation of the pulses was simulated via numerical solution of Eq. (1) with $D_{BO} = 0$. While Eqs. (1) and (3) are equivalent models, we chose the former to solve numerically as a matter of convenience. The solitary waves interact with each other in a soliton-like fashion, apart from the fact that small oscillations are shed behind the smaller pulse following collision. The cause of the oscillations is not clear at the time of this writing.

Solitary-wave solutions of Eq. (3) with BO-type dispersion ($D_{BO} \neq 0$), which is the more usual combination for surface waves, have been obtained numerically [2,4], but none was shown to survive collision. Given the current state of knowledge, it appears most likely that surface acoustic solitons will be identified in a rather unusual substrate-film combination that gives rise to KdV-like dispersion.

REFERENCES

1. M. F. Hamilton, Yu. A. Il'insky, and E. A. Zabolotskaya, J. Acoust. Soc. Am. **97**, 891–897 (1995).
2. C. Eckl, A. S. Kovalev, A. P. Mayer, A. M. Lomonosov, and P. Hess, Phys. Rev. E **70**, 046604 (2004).
3. M. J. Ablowitz and H. Segur, *Solitons and the Inverse Scattering Transform* (SIAM, Philadelphia, 1981).
4. W.-S. Ohm, "Effects of dispersion on nonlinear surface acoustic waves in substrates laminated with films," Ph.D. Dissertation (The University of Texas at Austin, 2001).
5. W.-S. Ohm and M. F. Hamilton, J. Acoust. Soc. Am. **115**, 2798–2806 (2004).

Application of Acoustical Activity for Control of Light Intensity

F.R. Akhmedzhanov

Department of Physics, Samarkand State University,
15 University blvd., Samarkand, 703005, Uzbekistan

Abstract. The proposed new technology is differed from all the existing technologies for measuring and control of light intensity. In the proposed method is used the Bragg light diffraction on the hypersonic wave in gyrotropic crystals. The light intensity can be changed by modification of generator frequency by which the acoustic wave is excited. The increase or decrease of light intensity is provided electronically, fluently and with high accuracy. Using previously the standard calibration, it is possible to measure and change the light intensity to the necessary value.

Keywords: Acoustic wave, Bragg diffraction, control, light intensity, measuring.

PACS: 42.79.Jq

INTRODUCTION

In this work the possibility of application of the acoustical activity phenomenon [1-10] for control of light intensity has been investigated. The new technology is different from all the existing technologies for measuring and control of light intensity.

The proposed method is based on using the Bragg light diffraction on the hypersonic acoustic wave in gyrotropic crystals [10-12]. The light intensity can be changed by modification of generator frequency by which the acoustic wave is excited. As a result, the increase or decrease of the light intensity is provided electronically, fluently and with high accuracy because the frequency of electromagnetic vibrations can be determined very precisely. Using the standard calibration it is possible to measure and to change the light intensity to the necessary value.

To understand the proposed technology below is briefly described the phenomenon of acoustical activity without theoretical details.

INTRODUCTION TO THE CONCEPT OF ACOUSTICAL ACTIVITY IN CRYSTALS

A number of authors [1-7] have independently introduced the concept of the acoustical activity. It is well known that the acoustical activity is the mechanical analogue of optical activity. Its simplest manifestation is the progressive rotation of the polarization plane of a transverse acoustic wave propagating in a crystal along some special directions.

CP838, *Innovations in Nonlinear Acoustics: 17th International Symposium on Nonlinear Acoustics,*
edited by A. A. Atchley, V. W. Sparrow, and R. M. Keolian
© 2006 American Institute of Physics 0-7354-0330-9/06/$23.00

Although the phenomenon of the acoustical activity has been discovered relatively long ago, its direct experimental investigation is considerably more difficult as compared with the optical analogue. Because of that, the acoustical activity has been experimentally investigated in a limited number of crystals [8-12]. More complicated situation than that occurs at the optical activity may arise when the propagation direction of the acoustic wave does not coincide with acoustic axis [10, 11]. This probably explains the relatively slow progress of such investigations. However, the situation is bond to improve with continuing progress in experimental technique. [12].

Unlike the optical activity, the acoustical activity can occur in all of the 21-acentric crystal point groups of the symmetry [11]. Moreover, the acoustical activity has been explained in terms of first order spatial dispersion: a strain ε created at time τ at a point r' gives the rise to a stress δ at another point r at a later moment of time. Thus, taking into account the spatial dispersion the complex elastic constants C_{ijkl} are equal [11]:

$$C_{ijkl}(\omega, q) = C_{ijkl}(\omega) + i \cdot \gamma_{ijklm}(\omega) \cdot \mathbf{q} \cdot \mathbf{m}, \qquad (1)$$

where γ_{ijklm} is the tensor of acoustical activity, ω is the frequency of acoustic wave, $\mathbf{m}$ is the normal vector and $\mathbf{q}$ is the wave vector of acoustic wave.

BRIEF DESCRIPTION OF THE EXPERIMENTAL METHOD FOR CONTROL OF LIGHT INTENSITY

In present work acoustical activity and Bragg light diffraction on hypersonic waves in ferroelectrics crystal $LiNbO_3$ have been used for control and measuring of the light intensity. The applied sample of $LiNbO_3$ was cut from optically clear single crystal and oriented along the crystallographic axis of the third order with the accuracy of 10'. Piezoelectric transducers of Lithium Niobate of appropriate cuts are used to excite the plane-polarized transverse acoustic waves in the frequency range from 900 MHz to 1.1 GHz. The block diagram of acoustic-optical system for control of light intensity is shown in Fig. 1.

Measurements of the dependence of the diffracted light intensity from the distance to the piezoelectric transducer along the direction of the acoustic wave propagation have been carried out in automatic regime by using the computer, which worked under the control of the program adapted specially for solving this problem.

The values of the specific rotation of polarization vector (β) of the transverse acoustic waves are estimated from the measurements of the dependence of the scattering light intensity (I) from the distance (z) to the piezotransducer along the direction of the acoustic wave propagation. In common case, the light intensity is the function from some parameters including the attenuation coefficient of acoustic wave (α) and the specific rotation of polarization vector (β) [12]:

$$I = I_0 \cdot [\exp(-\alpha \cdot z)] \cdot \cos^2(\beta \cdot z + \varphi), \qquad (2)$$

where I_0 is the laser light intensity; φ is initial phase of acoustic wave.

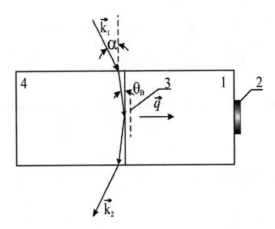

FIGURE 1. Block diagram of acoustic-optical system for control of light intensity. The observation of diffracted light beam is shown for the time moment when the acoustic pulse is moved through the area of the acousto-optical interaction.

Here: 1 – sample of LiNbO₃ crystal, 2– piezotransducer, 3 – acoustic wave front, 4 – back of sample; k_1 and k_2 are wave vectors of incident and diffracted light, accordingly, q is wave vector of acoustic wave, α and θ_B are outside and inside angles of the incident light beam.

In the experiment, it is possible to obtain different values of the diffracted light intensity in the range from zero to the maximum intensity of the diffracted light. These intensities can be used for simultaneous determination of the specific rotation of polarization vector and the attenuation coefficient of the acoustic wave by the appropriate model equation. Some experimental results for pure LiNbO₃ crystal obtained at the frequency 1.0 GHz are shown in Fig. 2.

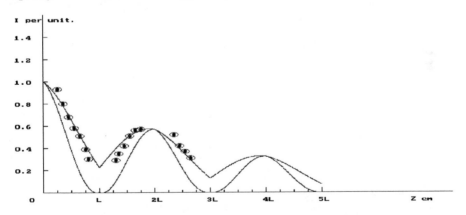

FIGURE 2. Dependence of Bragg diffraction light intensity on the distance from piezotransducer along the direction of propagation of transverse acoustic wave with frequency 1.0 GHz. The length of real sample is L. The points are the experimental results; the first solid curve is the calculation by equation (2). The second solid curve is the calculation for abstract sample with the specific rotation coefficient equal $\pi/2L$.

How it can see from the Fig. (2), the intensity of diffracted light is changed while the transverse acoustic wave is propagated along the crystal. Alternatively, we can observe the variation of the light intensity by observing the diffracted light in some point of the crystal if the frequency of acoustic wave will be changed when the acoustic wave is exited.

REFERENCES

1. Andronov, A. A., *Proceedings of the Higher Educational Institutions, SU, Radio Physics Series* **3**, 645 648 (1960).
2. Portigal, D. L., and Burstein, E., *Phys. Rev.* **170**, 673-679 (1968).
3. Kumaraswamu, K., and Krishnamurthy, N., *J. Acoust. Soc. Am.* **72**, 418-420 (1983).
4. Bhagwat, K. V., and Subramanian, R., *Phys. Rev. B* **33**, 5795-5800 (1986).
5. Li, Yin-yuan, and Chen, Liang, *Phys. Rev. B* **36**, 9507-9513 (1987).
6. Srinivasan, T. P., *J. Phys. C: Solid State Phys.* **21**, 4207-4219 (1988).
7. Bhagwat K. V., and Subramanian, R., *Acta Cryst.* **A44**, 551-554 (1988).
8. Pine, A. S., *Phys Rev. B* **2**, 2049-2054 (1970).
9. Lin, Q., Tao, F., Zhang, T. Y., Niu, S. W., Gou, C., and Shi, Z. J., *Solid State Communications* **54**, 803-806 (1985).
10. Bryzhina, M. F., and Esayan, S. Kh., *Solid State Physics (SU)* **20**, 2628-2636 (1978).
11. Lyamov, V. E., *Polarization Effects and Anisotropy of Acoustic Waves*, Moscow: Moscow University, 1983, pp. 24-36.
12. Akhmedzhanov, F. R., "Acoustical activity in $La_3Ga_5SiO_{14}$ and $Nd_3Ga_5SiO_{14}$ crystals" in *The 14th Symposium on Thermophysical Properties - 2000*, Colorado: Colorado University, 2000, p. 227.

On Absence of Decay of Bulk Solitary Waves in Elastic Wave Guides

Alexander M. Samsonov, Galina V. Dreiden, Irina V. Semenova

The Ioffe Institute of the Russian Academy of Sciences, St. Petersburg, 194021 Russia[1]

Abstract. Propagation of bulk solitary density waves is studied in lengthy elastic wave guides, which have strong dispersion and are made of polymers with remarkable linear dissipation, that leads to decay of any of linear or shock waves at short distances. Nonlinear elasticity of materials results in generation of strain solitons even under short-run and reversible (elastic) loading. Theoretical and experimental research has been performed to prove the existence of long bulk strain solitary waves produced by a laser-induced impact in nonlinearly elastic isotropic wave guides. New experiments in lengthy bars (over 0.5 m) confirm that bulk solitons do not reveal any considerable amplitude decay and shape transformation, while linear or shock waves disappear at much shorter distance completely. Decrement values are small, indeed, and estimated first for elastic nonlinear strain waves dissipation in polymeric bars.

Keywords: Soliton, Dissipation, Nonlinear Elasticity, Longitudinal Wave.
PACS: 42.65.Tg, *43.25.–x, 05.45.–a, 62.20.Dc

INTRODUCTION

How far may a bulk solitary elastic wave propagate in isotropic wave guides without decay? In theory a soliton is powerful and lossless, contrary to 'common experience' with most of bulk longitudinal waves in mechanics of real solids. Any of *linear* elastic or shock waves in polymeric materials disappear completely at very short distances, that allows to prefer polymers for experiments with *nonlinear* long solitary waves in solids. Nonlinear elasticity of these materials results in generation of strain solitons even under short-run and reversible (elastic) loading, while decay of linear waves helps to detect solitons in solids. Mathematical theory of long nonlinear waves in elastic wave guides was developed and resulted in successful experiments in generation of strain solitons in real solids [1]. In experiments with short and thin wave guides (1x1x15 cm), we were not able to prove a lossless propagation of bulk strain solitons observed [2]. Now we demonstrate new experimental results on soliton behavior in lengthy polymeric wave guides (up to 60 cm). Both polystyrene (PS) and plexiglas (PMMA) were used for wave guides. The materials were selected due to unique combination of non-linear elasticity and optical properties allowing the translucent recording of wave patterns. The first generation of solitary nonlinear

[1] E-mail samsonov@math.ioffe.ru, fax +7-812-2471017

CP838, *Innovations in Nonlinear Acoustics: 17th International Symposium on Nonlinear Acoustics*,
edited by A. A. Atchley, V. W. Sparrow, and R. M. Keolian
© 2006 American Institute of Physics 0-7354-0330-9/06/$23.00

longitudinal strain wave (the soliton) was made in various 1D and 2D elastic wave guides by means of laser driven shock wave. The holographic interferometry used for nonlinear wave recognition [1,2] involves the laser generation and optical recording of the waves. The apparatus consists of a channel to produce the strain wave in a solid from a weak shock wave, which is induced by laser pulse evaporation of a metallic target placed nearby an entrance of the wave guide, a synchronizer and a holographic interferometer. The set-up allows to record a wave pattern in and outside the transparent wave guide due to the density variation caused by wave, which leads to the shifts of the carrier fringes on the resulting holographic interferogram. We use a 57 cm long PS bar and 60 cm long PMMA bar, both having 1x1 cm cross sections.

MOTIVATION, METHODOLOGY AND RESULTS

We have performed numerous experiments on the generation and propagation of bulk strain solitary waves in PS and PMMA wave guides. Firstly we certified the observed wave as a genuine bulk solitary wave. It keeps the shape permanent, has no long wave of opposite sign behind, has no recognizable decay on distances up to 30 length scale units, and its amplitude and width are in correct proportion to the wave speed. However the problem on the density soliton decay when propagating along a lengthy bar made of *real isotropic material* remained unsolved.

Physical concepts of nonlinear elastic wave decay in glassy polymers seem to be insufficient so far. Most of studies were devoted to shock attenuation in polymers at short distances up to several millimeters. Polymers are among best materials for shock and vibration damping [7], however references to elastic wave dissipation in bulky polymers are infrequent. PMMA was named in [5] as the material with the highest loss modulus and high damping capacity value, in [6] it was mentioned that even for isotropic materials theoretical formuli for decay rate of density waves are often far from being confirmed by experiments, moreover for low frequencies the frequency dependence is negligible as a rule.

We proved now that even at very long distances bulk solitons do not show any *considerable* decrease of amplitude or shape transformation when propagating in an isotropic wave guide, and are able to estimate now the decrements of nonlinear dissipation.

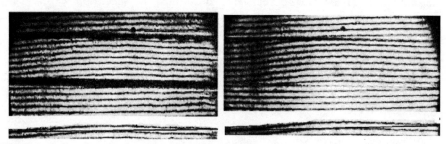

FIGURE 1. Bulk solitons in the PMMA bar at the distance 345-400 mm (left), and in the PS bar at 340-395 mm (right) from the entrance cross section of a bar.

Evaluation of Nonlinear Dissipation Decrement in Polymeric Wave Guides

The dissipation logarithmic decrement is estimated for non-linear bulk waves via an elementary formula:

$$\alpha = [1/(x_2 - x_1)]\log(\Delta K_1 / \Delta K_2) \qquad (1)$$

where x_1, x_2 are points of measurement (i.e., where the solitary wave amplitude approaches its maximum), and ΔK_i is the fringe shift measured in the picture made at the distance interval i, see Fig.1, where the fringe shift, representing a soliton, was extracted and plotted below the observation images for convenience. Note that ΔK is proportional to wave amplitude at x_i with a coefficient depending on the wave guide parameters. Following linear wave theory [3] one may calculate $\alpha=0.17$ cm^{-1} for PS and $\alpha=0.25$ cm^{-1} for PMMA, therefore an amplitude declines in e times on 6 cm in PS and 4 cm in PMMA respectively.

Our main results for solitary wave decay in PS and PMMA bars are shown in Tables 1 and 2 resp. The values of logarithmic decrement of dissipation are calculated in (1) using parameters of the solitons shown in Fig.1, and the amplitude estimation:

$$A_{max} = \frac{\Delta K \lambda_0}{h(1 - v)(n_i - 1)} \qquad (2)$$

where h is a bar thickness, ΔK is the fringe shift, v is the Poisson ratio, n_i is the refraction index of an i-th bar material, $\lambda_0 = 6.94*10^{-5}$ cm is the laser light wave length. For both PS and PMMA $v=0.34$, $h=1$ cm, while $n_{PS} =1.59$, and $n_{PMMA} =1.49$.

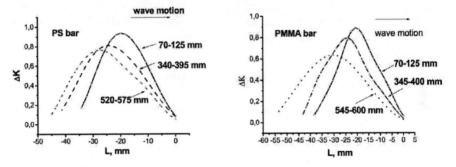

FIGURE 2. Bulk solitons in 3 different distances demonstrate experimental evidence of small dissipation in the PS (left) and PMMA (right) bars. The holographic fringe shift ΔK vs the solitary wave width is shown.

TABLE 1. Disspation Decrement Evaluation for Bulk Solitons in Polystyrene Bar.

Distance (mm)	Fringe shift (ΔK)	Max strain amplitude, 10^{-4}	Width (mm)	$\alpha 10^{-3}$, cm^{-1} (and distance for e times decay)	"Mass" $\int A dx$	"Energy" $\int A^2 dx$
1: 70-125	1.00	1.78	34.2	1-2: **4.3** (232.6)	20.8	15.84
2: 340-395	0.89	1.58	41.4	1-3: **4.8** (208.3)	21.2	14.05
3: 520-575	0.81	1.44	45.0	2-3: **5.5** (181.8)	21.3	12.52
averaged α	and	distance for	e times decay:	**4.9 (204.8)**	δ_{13}=+2%	δ_{13}=-11 %

149

TABLE 2. Disspation Decrement Evaluation for Bulk Solitons in PMMA Bar

Distance (mm)	Fringe shift (ΔK)	Max strain amplitude, 10^{-4}	Width (mm)	$\alpha 10^{-3}$, cm^{-1} (and distance for e times decay)	"Mass" $\int A dx$	"Energy" $\int A^2 dx$
1: 70-125	1.00	2.15	38.4	1-2: **4.9** (204.1)	18.9	12.23
2: 345-400	0.88	1.89	43.2	1-3: **8.6** (116.3)	19.1	11.26
3: 545-600	0.65	1.39	54.4	2-3: **12.6** (79.4)	20.6	10.07
averaged α	and	distance for	e times decay:	**8.7 (114.94)**	δ_{13}=+1%	δ_{13}=-8 %

We conclude that variations in values of α for solitary waves may be provided by nonlinearity of dissipation. Conventional values of α estimated for linear waves are well distinguished from those calculated for bulk elastic solitons, see Table 3.

TABLE 3. Disspation Decrements for Linear and Non-linear Waves

	linear waves: α, cm^{-1}	solitary waves : α, cm^{-1}	$\alpha_{lin}/\alpha_{solitons}$
PS	**0.17** (decay in e times at **6 cm**)	**0.005** (decay in e times at **205 cm**)	34
PMMA	**0.25** (decay in e times at **4 cm**)	**0.009** (decay in e times at **115 cm**)	28

We show that while moving along a thin (1 cm) bar over half a meter the bulk soliton does not exhibit a *remarkable* decay in both amplitude and shape, that is in excellent agreement with the general theory of solitons. On the other hand the dissipation decrement for density solitons was calculated first, that allows now to estimate energy losses for bulk solitons in lengthy wave guides. The solitary wave propagation in a polymeric bar is unusually stable in comparison with any conventional bulk elastic wave in polymers. Long nonlinear solitary strain waves exhibit extremely low decay, which is now evaluated, and may be of considerable importance to study the time evolution of elastic strains, the impact loading of materials with radiation and/or wear resistance, the nonlinear elastic parameters of materials etc.

REFERENCES

1. Samsonov, Alexander M., *Strain Solitons in Solids and How to Construct Them*. Chapman& Hall/CRC, London et al, 2001. 248 pp.
2. Dreiden, G.V. et al., *Techn. Phys. Letters* **21**, 415-417, (1995).
3. *Physical data*. Grigoriev, I.S., Meilikhov E.Z., eds., Moscow, Energoatomizdat, 1991, *(in Russian)*.
4. Peselnick L., Zietz, I. *Geophysics*, **24**, 285-291 (1959).
5. Chung, D.D.L., Review. Materials for vibration damping. *J Material Sci.*, **36**, 5733-5737 (2001)
6. Krasil'nikov, V.A., Krylov, V.V., *Introduction to physical acoustics*. Moscow, Nauka, 1984, *(in Russian)*.
7. Perepechko, I.I., *Introduction to polymer physics*. Moscow, Khimiya, 1978, 312 pp. *(in Russian)*.

On the role of cubic nonlinearity in localization of longitudinal strain waves

A.V. Porubov* and G.A. Maugin†

*Ioffe Physico-Technical Institute of the Russian Academy of Sciences, St. Petersburg, Russia
†Laboratoire de Modélisation en Mécanique associé au CNRS, Université Pierre et Marie Curie,
Paris, France

Abstract. New governing equations with combined quadratic and cubic nonlinearities are obtained to account for nonlinear strain waves in an elastic rod and in a plate. It is shown that strain solitary wave solutions of these equations arise as a result of balance between quadratic nonlinearity and dispersion and exists even in the absence of cubic nonlinearity. However, the amplitude, the width and the velocity of the wave are affected by the cubic nonlinearity causing, in particular, a narrowing of the longitudinal solitary wave. This allows to agree better with experiments on strain solitary wave generation in the rod and in the plate.

Keywords: Non-linear waves, elastic rod and plate, Murnaghan's materials
PACS: 62.30.+d, 03.40.kf, 43.25.+y

INTRODUCTION

The study of long nonlinear strain localized waves in elastic wave-guides is of interest for nonlinear acoustics and seismology. The most important of these are the waves that keep their shape during propagation since they transmit considerable energy over long distances. Usually these waves arise due to a balance between nonlinearity and dispersion. In particular, quadratic nonlinearity is considered to be responsible for longitudinal-wave localization. Despite numerous theoretical works, only a few experimental data are available on longitudinal strain solitary waves detection [1, 2, 3]. Moreover, the theory developed by now predicts higher width of the wave than the observed one. It points to a possible lack of correct theory. Possible improvement is in taking into account higher-order nonlinearities. Despite quadratic nonlinearity dominates other nonlinearities for the longitudinal wave description, it will be seen later that cubic nonlinearity provides important alterations in the wave behavior.

Here we briefly present recently obtained [4] new model equation for longitudinal strain waves in a rod that contains both quadratic and cubic nonlinearity. An exact solitary wave solution of this equation arises as a result of a balance between quadratic nonlinearity and dispersion, while cubic nonlinearity affects the permitted interval for the velocity of the solitary wave, its amplitude and the width.

Then 2D coupled governing equations with combined nonlinear terms are obtained for longitudinal and SH waves in a plate where additional diffraction terms appear that accounts for a transverse evolution of the wave. Transverse instability of plane solitary wave is studied, and important role of the mixed dispersive term is revealed.

CP838, *Innovations in Nonlinear Acoustics: 17th International Symposium on Nonlinear Acoustics,*
edited by A. A. Atchley, V. W. Sparrow, and R. M. Keolian
© 2006 American Institute of Physics 0-7354-0330-9/06/$23.00

SOLITARY WAVES IN A ROD

Let us consider an isotropic cylindrical elastic rod with free lateral surface. We consider cylindrical Lagrangian coordinates (x, r, φ) where x is directed along the axis of the rod, $-\infty < x < \infty$, r is the radial coordinate, $0 \leq r \leq R$, φ is a polar angle, $\varphi \in [0, 2\pi]$. Assuming that torsion can be neglected, then the displacement vector is $\vec{V} = (u, w, 0)$. The so-called nine constants Murnaghan model is used for the density of potential energy,

$$\Pi = \frac{\lambda + 2\mu}{2} I_1^2 - 2\mu I_2 + \frac{l + 2m}{3} I_1^3 - 2m I_1 I_2 + n I_3 + \\ \nu_1 I_1^4 + \nu_2 I_1^2 I_2 + \nu_3 I_1 I_3 + \nu_4 I_2^2, \tag{1}$$

where $I_k, k = 1, 2, 3$ are the invariants of the Cauchy-Green deformation tensor $\mathbf{C}$. Using zero boundary conditions for the stresses at the lateral surface one can obtain the following approximations for the displacements,

$$u = U + a_2 r^2 U_{xx}, \tag{2}$$

$$w = b_1 r U_x - b_3 r^3 U_{xxx} - B_1 r U_x^2 - B_2 r U_x^3. \tag{3}$$

where a_2, b_i, B_i depend upon elastic moduli [4]. An employment of the Hamilton variational principle allows us to obtain the governing equation for a longitudinal strain function $v = U_x$ [4],

$$v_{tt} - a v_{xx} - c_1 (v^2)_{xx} - c_2 (v^3)_{xx} + b_1 v_{xxtt} - b_2 v_{xxxx} = 0. \tag{4}$$

where

$$a = \frac{E}{\rho_0}, \ c_1 = \frac{\beta}{2\rho_0}, c_2 = \frac{\gamma}{3\rho_0}, b_1 = \frac{\nu(1-\nu)R^2}{2}, b_2 = \frac{\nu E R^2}{2\rho_0}.$$

β and γ may be of either sign [4]. This equation generalizes previously used double-dispersive equation (DDE) [3]. Since the ordinary-differential equation reduction of Eq.(4) coincides with that of the Gardner equation, one can directly use known exact travelling solitary wave solution of the Gardner equation [6]. In our case it may be written as

$$v = \frac{A}{Q \cosh(k\xi) + 1}, \tag{5}$$

where $\xi = \theta - V t$,

$$A = \frac{3(V^2 - a)}{c_1}, \ Q = \sqrt{1 + \frac{9c_2}{2c_1^2}(V^2 - a)}, \ k^2 = \frac{V^2 - a}{b_2 - b_1 V^2}. \tag{6}$$

First, we note that this solution does not result from the balance between cubic nonlinearity and dispersion. Indeed, the amplitude of the wave, $A_G = A/(Q+1)$, depends upon the value of c_2. However, when $c_2 \to 0$, then $Q \to 1$, and Eq.(5) transforms into the known solitary wave solution of the DDE [3],

$$v = A_D \cosh^{-2}(\frac{1}{2}k(x - Vt)), \qquad (7)$$

where

$$A_D = \frac{3(V^2 - a)}{2c_1}$$

A comparison of solitary wave solutions (5) and (7) allows us to conclude that the sign of the amplitude of both solutions is defined by the sign of the coefficient c_1, and tensile strain solitary waves propagate if the material of the rod provides $c_1 > 0$, while it is the case of compression waves for $c_1 < 0$. Comparing derivatives of A_G and A_D with respect to V^2 one can find that A_G growths faster with V^2 for $c_2 < 0$ and slowly for $c_2 > 0$. Hence, if we assume equal amplitude A^* for both solitary waves (5) and (7), the former wave propagates faster at positive c_2 while the latter one- at negative value of the cubic term coefficient. On the other hand, if we assume equal velocities for both solutions, the DDE solitary wave (7) will propagate with higher amplitude than the wave (5) for $c_2 > 0$ and with smaller amplitude for $c_2 < 0$. Finally, cubic nonlinearity affects the shape of the solitary wave (5). Indeed, comparing solitary waves (5) and (7) having equal amplitudes, one can see that the solution of the DDE is wider than (5) if $c_2 < 0$, while at positive c_2 the solution of our equation is wider. Formation of the solitary waves from an arbitrary input may be found in [4].

SOLITARY WAVES IN A PLATE

Let us consider an isotropic elastic plate that occupies the region $-\infty < x < \infty$, $-\infty < y < \infty, -h < z < h$, in Cartesian coordinates (x,y,z). Assume the displacement vector in the plate is $\vec{V} = (u,v,w)$. Again the Murnaghan model (1) is employed. Then we have for the displacements,

$$u = U(x,y,t) + C z^2 (U_x + V_y)_x, \qquad (8)$$

$$v = V(x,y,t) + C z^2 (U_x + V_y)_y, \qquad (9)$$

$$\begin{aligned} w = \ & -2Cz(U_x + V_y) - D z^3 \triangle(U_x + V_y) - z(q_1 [U_x^2 + V_y^2] + q_2[U_y^2 + V_x^2] + \\ & q_3 U_x V_y + q_4 U_y V_x) - z(p_1 [U_x^3 + V_y^3] + p_2[U_x + V_y][V_x^2 + U_y^2] + \\ & p_3 U_y V_x[V_y + U_x] + p_4 U_x V_y[U_x + V_y]), \end{aligned} \qquad (10)$$

where $U(x,y,t)$ and $V(x,y,t)$ are new unknown functions, $\triangle$ is the Laplace operator, the coefficients depend upon the elastic moduli. The governing equations for new unknown functions are obtained using the Hamilton principle. A simplification may be achieved assuming weak transverse variations. Then we obtain the following two coupled equations,

$$U_{tt} - a_1 U_{xx} - a_2 U_{yy} - (a_1 - a_2)V_{xy} - a_4 (U_x^2)_x - a_5 (U_x^3)_x - a_6 U_{xxxx} + a_7 U_{xxtt} = 0, \quad (11)$$

153

and

$$V_{tt} - a_1 V_{yy} - a_2 V_{xx} - (a_1 - a_2)U_{xy} = 0, \tag{12}$$

where a_i are constants depending upon elastic moduli, a_4 and a_5 may be of either sign. A plane travelling solitary wave solution of Eqs.(11), (12) is similar to the (5). In this case $\partial/\partial y = 0$, $V = 0$, and the dependent variables are functions of $\theta = x - c\,t$. Then plane longitudinal strain solitary wave solution, $\eta = U_\theta$, is

$$\eta = \frac{A}{Q\cosh(k\theta) + 1}, \tag{13}$$

The deviations in the solitary wave parameters due to the cubic nonlinearity are the same as for the rod.

Assume small transverse disturbances on the plane solitary wave in the form

$$U(\theta,t,y) = U_0 + \delta\,w_u(\theta,\tau)\exp(\kappa\,t + \imath\,s\,y), \tag{14}$$

$$V(\theta,t,y) = \delta\,w_v(\theta,t)\exp(\kappa\,t + \imath\,s\,y), \tag{15}$$

where $\delta \ll 1$, $U_{0,\theta} = \eta$. Since weak transverse variations are studied, s is small; then the solution is sought in the form

$$w_u = w_{u0} + s\,w_{u1} + s^2\,w_{u2} + ..., w_v = s\,w_{v1} + s^2\,w_{v2} + ..., \kappa = s\,\kappa_1 + s^2\,\kappa_2 + ...$$

Standard procedure allows us to obtain an equation for κ_1 that indicates an instability thanks to the mixed derivative term, at $a_7 \neq 0$. Due to instability a fully localized longitudinal wave arises. In this case cubic nonlinearity does not play a significant role.

CONCLUSIONS

Among the peculiarities of the solitary wave caused by an addition of the cubic nonlinearity one can note the decrease in the width of the wave at $c_2 > 0$ and similarly in the plate. In the experiments on propagation of strain solitary wave in a polystyrene rod and in a plate [1, 2, 3], it was found that the value of the width predicted by the theory based on the DDE equation is larger than the observed one. Also we have found a faster formation of the solitary waves for a positive cubic nonlinearity term coefficient. This is also of importance since finite-size specimens of the rod and the plate are employed.

REFERENCES

1. G.V. Dreiden, A.V. Porubov, A.M. Samsonov, I.V. Semenova, and E.V. Sokurinskaya, *Tech. Phys. Lett.*, **21**, 415–417 (1995).
2. A.M. Samsonov, G.V. Dreiden, A.V. Porubov, and I.V. Semenova, *Tech.Phys.Lett.*, **22**, 891–893 (1996).
3. A.M. Samsonov, *Strain soltons in solids and how to construct them*, Chapman & Hall/CRC, 2001.
4. A.V. Porubov, and G.A. Maugin, *Intern. J. Non-Linear Mech.* **40**, No 7, 1041–1048(2005).
5. A.V. Porubov, *Amplification of Nonlinear Strain Waves in Solids*, World Scientific, Singapore, 2003.
6. T. Kakutani, and N. Yamasaki, *J. Phys. Soc. Jpn.*, **45**, 674–679 (1978).

SECTION 3
ELASTIC-WAVE EFFECTS ON
FLUIDS IN POROUS MEDIA

Vibrate –Seismic Waves Can Change Oil-Gas Reservoir State

Victor N. Nikolaevskiy

*Schmidt Institute of Earth Physics, Russian Academy of Sciences,
Moscow 123995, RUSSIA*

Abstract. This paper describes the field data on vibration action at oil water-flooded reservoirs. It is shown that at a pore scale the main factor is the gas release form live oil and traps. The latter happens under ultrasound appeared due to solid friction at joints contacts during seismic oscillations. If gas forms small bubbles, the oil recovery is increasing. Free gas phase decrease the phase permeability. The conclusion is supported by laboratory experiments and theoretical calculations. Original results are added with short review of the available literature.

Keywords: ultrasound, seismic waves, gas bubbles, oil recovery, resonances, noise, vibration.

FIELD DATA

Dynamics of oil and water fraction in well production during the earthquake swarm had shown the effect of very strong vibrations [1]. One can see that that in these cases, the smaller phase production is increasing (Fig. 1).

The first changes of water/oil ratio (WOR) under surface vibration (20 min/hour, 15-20 hours/day, duration 37 days) on the territory of deleted reservoir Abuzy (depth - 1400 m) were found during the fieldwork done by our Institute.

The major result was that the selection of low frequency is essential for vibrations effect, the maximum oil production being at 12 Hz (Fig.2). Reached regime was stable 17 days afterwards. Some results are shown also in the table. Lot of gases was released in wells during the vibration works.

More fresh data approve again [2] that total production is not changed but oil production is increased at flooded wells and decreased at wells with low WOR. At reservoir Sutorminskoe (West Siberia), the positive effect was mentioned at pay zone inside radius R ~ 3 km. The negative effect was at small number of wells with high production of oil.

Direct action by ultrasound emitter at borehole vicinity cannot penetrate deeper into oil pay zones and are reflected that usually is interpreted as a "secondary emission" [3].

On the other hand, recently the special microseisms of low frequency (up to 20 Hz) were observed at territory of hydrocarbon reservoirs [4] and they were absent in water - saturated parts of seams.

CP838, *Innovations in Nonlinear Acoustics: 17th International Symposium on Nonlinear Acoustics,*
edited by A. A. Atchley, V. W. Sparrow, and R. M. Keolian
© 2006 American Institute of Physics 0-7354-0330-9/06/$23.00

PROCESS SCHEMATIZATION

Let assume the following scheme of the process (Fig.3):

- Seismic waves created at the reservoir surface or inside boreholes are nonlinearly transforming into high frequency noise up to ultrasound.
- The release of gases happens under action of ultrasound, generated inside a porous space.
- Only gas bubbles, adjoining to oil/water drops, can explain relative preference of smaller phase mobility in partly saturated pore space. Then the flotation regime will act during pumping of water - oil mixture towards production wells.
- If gas micro bubbles are attached at solid walls of pore channels, totally saturated with one of fluids, the channel resistance is falling down and this fluid gets the privilege to flow.

HIGH FREQUENCY NOISE GENERATION

The first problem that is met in the field is what a frequency is optimal for the vibrate action. Low frequency (~ 10 Hz) of surface vibrators and regular seismic waves (~ 100 Hz) for borehole sources were selected [5]. However, a nonlinear transformation of wave spectra is needed for getting short waves necessary to act in pore mesoscale for stimulation of oil finite recovery

Professor Garagash has suggested very simple mechanical model. It consists in the oscillating blocks hierarchy put inside each other. The calculations [6] have shown that long period of bigger block oscillation are converted into thousands times smaller periodical motions of smaller pieces. In Fig. 4 one of his unpublished results is shown.

More sophisticated approach is necessary for the case of running waves. Corresponding theory of the wave spectrum nonlinear transformations is connected with the dominant frequency effect [1] that was found in the Vilcinska data for marine sand. Let explain it by adding free oscillators to the Maxwell type rheological element (Fig. 5) as it was done in [7]. The reason was to repeat continuously the Fermi –Pasta - Ulam lattice in every material coordinate cell.

The resulting state equation, connecting stress and strain, includes higher time derivatives:

$$-p - \sum_{q=1}^{n} a_q \frac{D^q p}{D t^q} = Ke + \sum_{q=1}^{m} b_q \frac{D^q e}{D t^q}$$

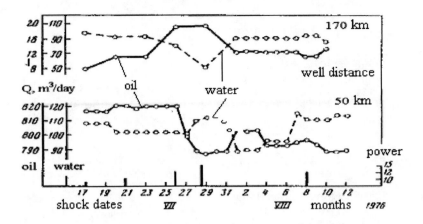

Figure 1. Hydro geologic evidence of the earthquake swarm near city of Groznyi (Oil wells were far from the epicenter zone) [1].

TABLE

Well #	Pay zone, m	WOR before vibration (%)	WOR during/after vibration (%)	Hz	Distance (m) from vibrator
33	1471-1496	96.2	93.3 – 92.0	12	1000
56	1259-1452	92.7	87.5 – 82.2	12	100

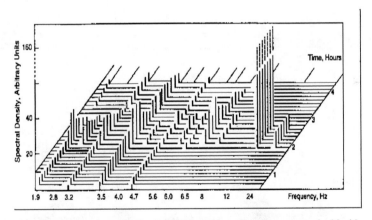

Figure 2. The appearance of 12 Hz oscillations is shown by the gage attached inside [1] the well column at the depth of 400 m. Time zero was beginning of surface vibrations.

159

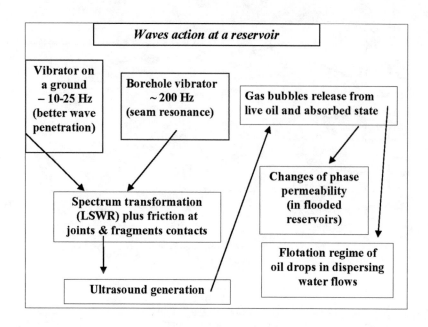

Figure 3. The stages of physical processes are developed for the process under consideration

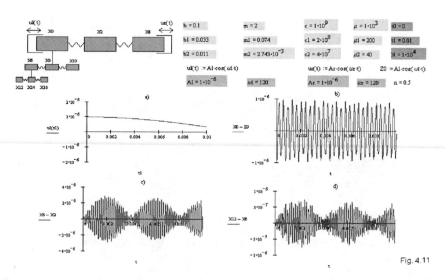

Figure 4. Blocks hierarchy can generate higher frequencies oscillations (Courtesy of I. Garagash)

Then the evolution equation is the generalization of the Burgers-Korteweg-De Vries equation:

$$\frac{\partial v}{\partial \tau} + N v \frac{\partial v}{\partial \xi} + \sum_{p=1}^{n} \Gamma_{2p+1} \frac{\partial^{2p+1} v}{\partial \xi^{2p+1}} = \sum_{p=1}^{m} \Gamma_{2p} \frac{\partial^{2p} v}{\partial \xi^{2p}}$$

Here p = pressure, e = strain, v = particle velocity and all coefficients are assumed to be positive. Moreover, their values are selected in such a manner that dispersion curve has the form, shown in Fig. 6. That is, this equation permits the interval of frequencies with negative dissipation, typical for the Brusselator [8]. The dominant frequency corresponds to the center of the interval of instability [7, 9] and the theory results in the spatial-temporary chaos phenomena [10-12] corresponding to in high frequency noise generation without a threshold.

As it was shown earlier [1], the low dominant frequency is the indicator of the Biot [13] second P-wave besides its slowness. This wave is real in porous media, saturated with gas under very low pore pressure. On the other hand, if the medium contains gas in a bubble state, at high frequency interval the P-2 wave becomes again real because here the skeleton is easy deformed with gas bubble resonance frequency [14].

One more source of ultrasound is solid friction between grains and at joint edges [1, 15] as well as due to grain rotations [1]. The appeared gas bubbles have their own dynamics in pore space that is described also by KdV equation [16]. The turbulent deviation of the Darcy law generates its own noise (~ 10 – 30 kHz according to [17]). Therefore, the real seam is full itself [18] of oscillations of different modes and the energy exchange between them is going under Long-Short-Wave Resonance conditions [1] (Fig. 7).

As it is well known, ultrasound is cleaning any objects of trapped gases but in our case its interaction with flows is extremely interesting to understand the mesa-scale picture of the process.

Thus, the seismic waves should create ultrasound noise inside an acting reservoir and, on the other hand, ultrasound is suitable tool for wave action at objects under laboratory conditions.

These arguments bring to a number of special laboratory experiments (with leading role of Professor Stepanova) with ultrasound action on oils and its flow through a standard one-dimension model of the seam. Of course, the case of live oil that exists in a reservoir had a privilege.

GAS RELEASE UNDER ULTRASOUND PULSES

The first lab experiments with water flooding of oil saturated sand samples but without gases and under outer actions with low frequencies (~ *40 Hz*) gave the oil recovery coefficient less than *0.3* at the moment of water breakthrough (*BT*) and the finite value was ~ *0.5* (private communication by V. Barabanov and A. Nikolaev, 1996).

161

But ultrasound action (*1, 3* and *5 MHz*) at the flooding had led [19] to oil recovery > *0.4* at *BT* if initially there was air in sandstone samples (*24 – 40 %*). The finite oil recovery was *0.6*.

We have performed the laboratory flooding of sand sample with live oil (that is, transform oil with propane dissolved) under pulse ultrasound action. The pressure in a saturated sand sample was increased drastically after the ultrasound séance (*~ 20 s*) for short period (Fig. 8).

The fluid pressure appraise takes place because of gas bubbles appearance in the system, although pore pressure was drop around *12 bars* and does not surpass the saturation static pressure of the oil (*~ 8 bars*).

If bubbles are small enough, they adjoin oil drops and recreate their mobility of in water flows. But if the gas becomes free phase, it is reaching the sample exit and water flooding takes the form of pulse process (Fig. 9), including very high-pressure jumps (noted by Dr. A. De Vries).

The process of water- flooding had a scale of days. After waterfront reached the sample exit, the depletion regime began. The oil and water were leaving the sample due to residual pore pressure mainly in a gas phase. Here, small surfactant additions, those are necessary for better modeling of real reservoir oil, became essential. In this case oil recovery increased up to 90% (Fig. 10).

Free gas phase is eluding from the seam (Fig. 11) and the corresponding oil part was lost in a pore space as a residual one.

EXPLANATION OF OBSERVED PHENOMENA

Oil reservoirs are full of hydrocarbon gases. They are in a free state, trapped or diffused in a pore space, in gas cap, in adsorbed state. Sound can release them and, likely, transform into bubble state. Releasing of dissolved gases has a special interest.

Let recall that in live oils there is the pressure difference of the moment of bubbles appearance and their joining into free phase. It is especially evident and exploited in the case of heavy oil [20]. In this interval of pressures the ultrasound action is changing the state of oil parameters. One can see it in Fig. 12 where the test data for oil out of porous media are shown.

If gas gets the form of micro-bubbles, they are located at interfacial surfaces and there are two possibilities. They can adjoin contacts of oil and water. This helps to oil (water) drop mobility in a flux of carrying phase and the resulting will be flotation effect. Then oil (water) phase permeability will change its form. Convexity of oil permeability can be considered as result of micro bubbles appearance in oil. The case of convex form creates more slow motion of the water/oil front and more full displacement. D. Mikhailov has showed it numerically, see [5].

If micro-bubbles increase the mobility of oil along the channel due to their adsorption at its walls, the case of phase permeability higher than unit is observed (Fig. 13). Of course, it will be so only because we use, as usual, the liquid viscosity that is separated in reality from solid walls by micro-bubbles layer.

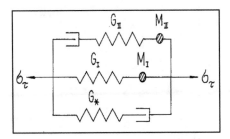

Figure 5 Generalized rheology [1] models internal oscillators in elastic-viscous media.

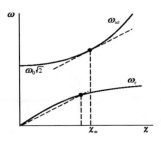

Figure 7 LSWR can transfer the seismic energy to ultrasound noise [1] due to availability of other high frequency sources.

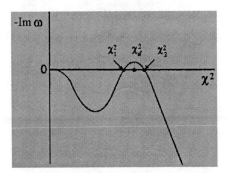

Figure 6 Effect of the dominant frequency [1] with the zero attenuation at $\chi = 0$ creates extreme spatial-temporal chaos.

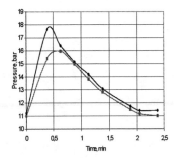

Figure 8 Changes of pore pressure after ultrasound séance at entrance and exit of the sample after recommitment of flooding process.

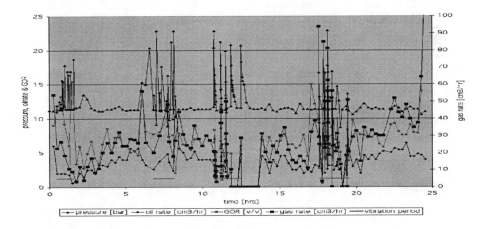

Figure 9. Pulse regime of oil production under water flooding process in a seam model

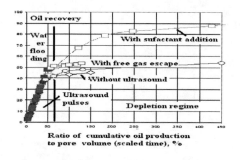

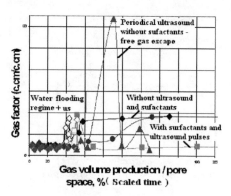

Figure 10 Finite sample states after flooding and depletion regimes of oil production in laboratory

Figure 11 Gas measurements at the exit from the sand sample

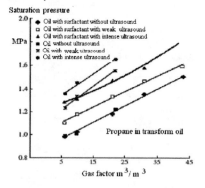

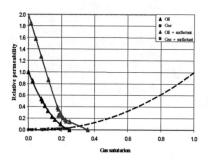

Figure 12 Oil saturation pressure is growing under ultrasound pulses

Figure 13 Changes of permeability if gas bubbles are attached to walls of pore channels (Courtesy of G. Stepanova)

If gas is converting into a free state, it fills channels and blockades the pore throats for oil drops. Oil permeability becomes zero in such channels. This corresponds to conventional understanding of gas state inside reservoirs.

In depleted reservoirs, the pressure is below saturation pressure and pore space is full of gases in bubble and a free state. In this case, as it follows from field [1, 2] and our laboratory data, vibrations can really transform them into bubble state. According to [19], the bubbles were concentrated at the water/oil displacement front as a screen and it corresponds to the loss of sound transmission to the outer gage. At the depletion stage the pulsed regime, corresponding to free gas breakout, existed [19] as in our tests.

The gases effects may explain also the data on "secondary emission" [3] by the blockage of powerful ultrasound sources as generation of bubble screen in well vicinity. The specific microseisms above the hydrocarbon reservoirs are also connected with bubbles [4]. Really, water basins do not contain any essential amount of gases and there were no such phenomenon out of reservoirs rims.

CONCLUSION

Thus, vibrate-seismic technology, patented earlier [21], acts very positively on oil recovery if regular pumps are not stopped. The generated ultrasound transfers the fluids of oil reservoirs into the state, close to the heavy oil. However, such a state can be unstable because the eluding gases try to join into free phase and leave the scene without any help to finite oil recovery. The further aim is to find cases where micro-bubbles are stable in reservoirs or to create such a situation. Such a possibility exists as it follows from spreading field experience.

The paper reflects the some result of works done with Drs. Oleg Dinariev, Stanislav Dunin, and Igor. Garagash, Dmitry Mikhailov and Galina Stepanova et al.

Other papers of our team, as well as of other authors with alternative points of view on the problem under consideration, are published elsewhere.

The author is thankful to Drs. Arnold de Vries and Alexander Mollinger for permanent attention and useful criticisms.

REFERENCES

1. V. Nikolaevskiy, *Geomechanics and Fluidodynamics,* Nedra, Moscow and Kluwer; Dordrecht, 1996.
2. G. Zibulchik, ed. *Active Seismology with Powerful Vibration Sources.* Inst. Compute. Math. & Geoph. Novosibirsk, 2004.
3. V. Drjagin et al, *Induced seismic-acoustic emission in a well,* Technology of Seismic Exploration. # 4, 2004.
4. S. Dangel et al, *Phenomenology of tremor – like signals observed over hydrocarbon reservoirs,* J. Volcanology and Geothermal Research, v. 128, 2003.
5. O. Rudenko, O. Sapozhnikov, Eds, *Nonlinear Acoustic at the Beginning of the 21-th Century,* (ISNA-16), Moscow State University, 2002.
6. I. Garagash, *Dynamic model of fragmented media with moving blocks,* Physical Mesomechanics, v. 5, # 5, 2002.
7. V. Nikolaevskiy, *Dynamics of viscoelastic media with internal oscillators.* Lecture Notes in Engineering, v. 29. Eds. C. Brebbia and S. Orszag, Springer, Berlin, 1989.
8. H. Haken, *Synergetics,* 2nd edn, Springer, Berlin, 1978.
9. I. Beresnev and V Nikolàevskiy, *A model for nonlinear seismic waves in a medium with instability,* Physics D, v. 66, 1993.
10. M. Tribelsky, *Short-wave instability in extended systems with additional symmetry,* Int. J. Bifurcation and Chaos, v. 7, N 5, 1997.
11. M. Tribelsky and M. Velarde, *Short-wavelength instability in systems with slow long-wavelength dynamics,* Phys. Rev. E, v. 54, N 5, 1996.
12. H.-W. Xi, et al. *Extensive chaos in the Nikolaevskii mode,* Phys. Rev. E, v. 62, N 1, 2000.
13. M. Biot. *Theory of elastic waves in fluid saturated porous solids,* J. Acoustic. Soc. Amer., v. 28, 168-191, 1956.
14. S. Dunin et al. *P-waves in realistic formations: gas bubbles effect,* Proc. 2nd Biot Conf. Poromechanics, Swets and Zeitlinger. Lisse. Aug. 2002.

15. T. Richard and E. Detournay, Stick-slip motion in a friction oscillator with normal and tangential mode coupling, C.R. Acad. Sci., Paris, v. 328, Ser. II b, 671-678, 2000.
16. S. Dunin and V. Nikolaevskiy, Nonlinear waves in porous media saturated with live oil. Acoustical J., v. 51, Supplement, 2005.
17. S. Nikolaev and M. Ovchinnikov, *Generation of sound by a flow through porous media,* Acoustical J., v. 38, # 1, 1992.
18. V. Nikolaevskiy et al, Residual oil reservoir recovery with seismic vibrations, Soc. Petrol. Eng., J. Production & Facilities, May 1996.
19. R. Duhon, "An investigation of the effect of ultrasound energy on the flow of fluids in porous media", PhD Dissertation, University of Oklahoma, Norman, 1964.
20. D. Joseph et al, *Modeling foamy oil flow in porous media,* Int. J. Multiphase Flow, v. 28, 1659 – 1686, 2002.
21. V. Asan-Djalalov et al., *Stimulation method for water-flooded oil reservoirs,* Authors Certificate, cl. E21 B43/00, Bull. #36, SU 1596081 A1, with priority of 1988, 1990.

Modeling and Field Results from Seismic Stimulation

E. Majer, S. Pride. W. Lo, T. Daley, Garrison Sposito[*], Seiji Nakagawa and P. Roberts[#]

Lawrence Berkeley National Labotatory, Berkeley, California
[*]University of California at Berkeley
[#]Los Alamos National Laboratory

Abstract. Modeling the effect of seismic stimulation employing Maxwell-Boltzmann theory shows that the important component of stimulation is mechanical rather than fluid pressure effects. Modeling using Biot theory (two phases) shows that the pressure effects diffuse too quickly to be of practical significance. Field data from actual stimulation will be shown to compare to theory.

INTRODUCTION

Seismic wave stimulation technology has the potential to provide a low-cost procedure for enhancing oil recovery in depleted fields and making it economically feasible to return abandoned wells to production as well as increase production in currently producing fields [1,2]. Tests of the technology indicate that seismic stimulation technology potential is greatest in fields with high water-cut and large amounts of immobile oil, making the technology exceptionally suitable for mature domestic oil fields. There is also interest in seismic stimulation for virgin lower-gravity consolidated-sand oil reservoirs where predicted recoveries are low (Alaska), and improved water injection and water flood sweep. Field tests with different seismic sources, however, have yielded promising but mixed or inconclusive results for enhancing oil production. In some cases seismic stimulation increased production rates by 50% or more, but in other cases production was unchanged or actually declined. At the present time a variety of mechanisms have been proposed for the effect, but there is not a clear understanding of the phenomenon. Needed were controlled field tests which measure the seismic energy of the sources at the depth of stimulation and tight monitoring on changes in production fluids. This work coupled with lab and theory will allow the community to make intelligent choices of stimulation procedures.

CP838, *Innovations in Nonlinear Acoustics: 17th International Symposium on Nonlinear Acoustics*,
edited by A. A. Atchley, V. W. Sparrow, and R. M. Keolian
© 2006 American Institute of Physics 0-7354-0330-9/06/$23.00

APPROACH

The major goal of this project is to obtain the comprehensive scientific and empirical knowledge needed to optimize reservoir stimulation for a wide range of field applications. LBNL and LANL have been pursuing this work for the past four years with the main emphasis on laboratory work and initial theoretical work. This effort, however is a more focused and balanced approach with increased emphasis on field validation and results. The industry partners are a mix of technology providers and users, each bringing significant expertise, field sites, field tests, and valuable data to the effort. The field work will be performed at sites of application in a variety of geologic and reservoir conditions. The overall goal is to provide a definitive assessment of the utility and applicability of stimulation.

FIELD EXPERIMENTS

The field sites for the work are providing the necessary control and background information lacking in the work that has been performed to date. The sites selected provided a wide range of geologic conditions as well as scales of measurement. Different sources were used in a variety of modes to determine the effectiveness of stimulation. Varied were the duration, duty cycle, amplitude and if possible frequency content of the sources. Both vibrational and pressure pulsing devices were used. Also considered was sources that provided elastic as well as electromagnetic energy. The goal of the field tests was to provide well controlled data sets that can be compared to determine the level of stimulation as well as provide insight on the mechanisms responsible for the stimulation. The initial tests were conducted at three different sites. The first site was the North Burbank field in Oklahoma. This is a mature sandstone reservoir which has been under production for a number of years. A vibrational source developed by Oil and Gas Consultants International was used to stimulate the reservoir. This is a very strong vibrational source with energy in the fifty to the few hundred hertz range. Initial tests at the Amoco Mounds test facility several years ago showed that energy can be recorded well over a thousand feet away (it was being tested as a cross well seismic source). Since that time it has been improved and developed for stimulation work. A new well was drilled at this site which served as the source well. Core is also available from this well. A second site was in the Central Valley of California in the Diatomite. Stimulation work has been ongoing with initial successful results, but monitoring has only been carried out at large distances with little information being obtained. New tests provided a unique opportunity to measure both the seismic as well as the pressure response in the near field (tens of meters) to the far field. The third site was also in California in a harder rock, sandstone, in the Elk Hills reservoir. Scales of measurement similar to the diatomite was also available at this site.

In the different sites multi-component geophones, accelerometers, hydrophones, and if possible pressure transducers, were deployed in wells at varying distances and depths from seismic stimulation sources as the reservoir is being stimulated. The goal being to collect the frequency and amplitude of the complete wave field and pressure response in the formation before, during and after the stimulation activities. This provided never-before-obtained data to evaluate and understand the situations under which effective stimulation occurs and the scale at which it is occurring. Particular attention was be paid to correlating the seismic results to the production results. An additional goal is to extend the results in these fields to conventional oil hard-rock primary-recovery applications. Recent work [3] has focused on the OXY Elk Hills site and possibly on an OXY site in Texas ..

THEORY AND NUMERICAL MODELING

In our past work [4], the exact equations for mass and momentum balance among all three components in an elastic porous medium containing two immiscible fluids have been derived in the linear limit. These equations generalize earlier work of Johnson, Berryman, and Spanos, reducing to Biot theory in the case of a porous medium containing a single fluid. They lead to diffusion equations for pressure/stress and for porosity (or total fluid mass) in the long-wavelength limit. Next, the long-wavelength limit was investigated in detail by developing coupled partial differential equations for wave propagation and solving them under boundary conditions appropriate for fluid pressure-pulsing experiments [4]. Special attention was devoted to establishing the frequency dependence of the porous medium permeability and its behavior under a pulsing scenario. Currently the results are being modeled numerically with data on reservoir properties and the characteristics of stimulation. The results will then be compared to the laboratory and field results. The momentum balance equations for an elastic porous medium containing two immiscible fluids also will be used to study the resonance behavior of a porous medium corresponding to a fluid pressure-pulsing experiment at the laboratory scale. To investigate resonance behavior, the natural frequencies of a cylindrical porous medium without interstitial fluids will be calculated theoretically and estimated numerically. These natural frequencies will be interpreted in respect to the experimental results in the case of interstitial fluids driven by a dynamic pore pressure to investigate whether their resonance behavior will effectively stimulate fluid flow rates.

More recent analysis [3] has focused on the related questions of how oil bubbles (ganglia) become stuck by capillary forces on pore-throat barriers and how a seismic wave may shake the material thus pushing the stuck bubbles over their bounding capillary barriers. Analytical criteria have been derived that determine whether a seismic wave of a given amplitude will be able to liberate a stuck bubble. To verify thiscriteria, lattice-Boltzmann numerical simulations of the stimulated two-phase flow have been performed in 2D. The theoretical predictions are verified by the simulations. The simulations have shown that over a significant range of reservoir conditions and for realistic ranges of the seismic amplitudes, seismic stimulation actually works.

RESULTS TO DATE

During the last year several critical results were obtained which indicate that seismic stimulation may be a viable technology in providing increased production. This was achieved through actual measurements in the field during stimulation with pressure pulsing technology and through improved modeling and theory.

Field Measurements.

Initially LBNL made measurements with a 3-C locking geophone in a well 43 feet from the stimulation well used by AERA LLC at Lost Hills at the depth of the stimulation source (ASR pressure pulse device). The source creates a pulse every stroke of the pump jack, about every 10 seconds). The signal was very strong and was recorded up to 800 feet away up the well. Analysis of the data found that the peak signal strength was in the 200 hertz range. It was suspected that the well was ringing to cause such high amplitudes. The modeling and theory (see modeling/theory task below), however, indicated that a fluid coupled wave could have such propagation characteristics. Last year we returned to Lost hills to look for such a wave with a simultaneous pressure sensor (1 sec to 10,000 hertz bandwidth) and seismic sensor (3-C geophone) in a ChevronTexaco well near the AREA LLC well which LBNL used for the close-in measurements. The well was 1000 feet away from the stimulation source (same source as used in previous experiments). Our modeling indicated that we should not see a seismic wave but if fluid coupling is occurring we should see a weak pressure wave in the stimulated formation. That is what was observed , a weak (1pa) pressure pulse which we could correlate with the stimulation pulse rate! The seismic energy coupled through the rock was not observed (what the modeling predicted). More importantly we observed the pressure signal in the 200 hertz range, the same as the input 1000 feet away. I.e. there was little attenuation in frequency of the pressure wave, i.e.,no dispersion. **This has profound implications on the mechanism of the propagation of the fluid coupled wave and its effect on increasing the effective permeability of the formation.** In 2003 and 2004 measurements were made at Oxy and Chevron again on energy from the ARA source. Signals up to 1500 feet were recorded above the noise level in the 10 to the minus 8 and 9 range. This is small but detectable. OXY reports up to 20 percent increased water cut at these levels. The question remains on which energy to introduce in the formation seismic or pressure.

North Burbank: LBNL monitored the OGCI mechanical source at the North Burbank field in Oklahoma during full scale field tests at a depth of 4000 feet and at distances up to 1200 feet away. Significant energy in the band width (up to 120 hertz) was observed at these depths and distances. The source was swept at varying frequencies to determine repeatability and controllability. It performed as designed. The test ended before correlations could be made with production but it appears the OGCI source has the energy to perform mechanical stimulation at a full field scale application.

Recent field measurements at the Elk Hills site in central California (sandstones) showed that the stimulation tool is providing energy at distances up to 2400 feet away from the stimulation well. The level of the strain signals are on the order of 10E- 9 to 10E-11 for distances of 900 feet to 1350 feet. The pressure signals are on the order of 1Pa or less. The lab results indicate that mechanical (solid matrix) strain levels on the order of 10E-6 or higher (in the 10 to 100 hertz range) were required to observe stimulation effects in porous cores. However, the equivalent pore fluid pressure oscillations induced at these strain levels were much less than 70 Pa. This is closer to the field-measured value of 1 Pa than the equivalent field strain levels, and may mean that fluid pressure pulsing is the primary mode of wave transmission that yields observable stimulation effects in the field. The theoretical/modeling results indicate that the Biot slow wave is practically unmeasurable in the observed field data range (10 hertz to 200 hertz). However, the field stimulation results show measurable production increases.

Theory and Modeling

From a theoretical formulation of deformation and wave propagation in elastic porous media containing two immiscible fluids, including the effect of inertial coupling, a dispersion equation for the dilatational waves in multi-fluid porous media was derived to describe how wave natural frequency depends on wave number. For a given natural frequency, we showed analytically that there are three dilatational body waves existing in multi-fluid porous media (P1, P2, and P3). To gain insight into the fluid-dependent nature of these three waves, numerical simulations were performed to investigate the effect of fluid saturation and natural frequency on the phase velocity and attenuation coefficients containing either water and gas or water and oil. The attenuation behavior of the wave whose velocity was greatest among the three (P1) is sensitive to the presence of different pore fluids. The saturation of pore fluids also gives an influence on the phase velocity of this wave. It was found that, at almost all degrees of water saturation, the second fastest wave (P2) has a higher velocity but is less attenuated in the gas-water mixture than in the oil-water mixture. The slowest wave (P3) has the highest attenuation coefficient, which makes this wave very difficult to observe. These results confirm the argument that this kind of wave appears solely as a result of the presence of the second fluid in the pore space. This kind of wave in the oil-water mixture is slower owing to the smaller slope of the curve of capillary pressure versus non-wetting fluid saturation. The attenuation behavior of the P1 wave is found to be sensitive to the presence of different pore fluids. The attenuation exhibits entirely the opposite trend in the two fluid mixtures. Except at the nearly full water saturation, the attenuation coefficient of the P2 wave in the air-water mixture is smaller than that in the oil-water mixture. In reference to the P3 wave, its attenuation in the air-water mixture is lower than that in the oil-water mixture.

A typical simulation result of Pride and Flekkoy [3] is given below in Figure 1. The figure gives snapshots during the various stages of a typical lattice-Boltzmann run, along with the average Darcy velocity of the oil in the simulation cell. The production runs are divided into four stages: (a). The oil is allowed to separate from the water in the pores with no applied forcing; (b). The background (steady) production gradient is turned on, with oil first flowing and then becoming stuck on capillary barriers; (c). The seismic stimulation is then applied for two wave periods which mobilizes the stuck oil and allows the bubbles to coalesce into longer streams; and (d). The seismic stimulation is turned off and the long stream of oil now flows under the influence of the background production gradient alone.

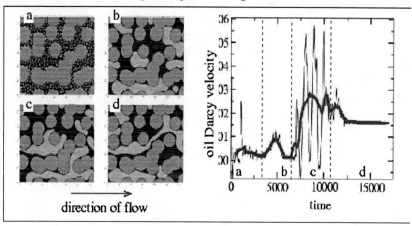

FIGURE 1. Snapshots during a lattice-Boltzmann simulation. Oil is light grey and water is dark.

Laboratory Work

To date, project laboratory experiments in the Dynamic Stress Stimulation Laboratory (DSSL) at LANL have demonstrated at least four types of stimulated flow behavior: 1.) Low-frequency (10-100 Hz) dynamic-stress cycling of sandstone cores during 2-phase fluid flow can change the oil/brine distribution in the rock without necessarily changing the bulk saturation or oil-to-brine ratio. 2.) Oil and brine flooding experiments demonstrated that changes in displaced fluid breakthrough time and residual saturation can be induced at low frequency. 3.) Steady-state oil/brine flow tests also showed that stress stimulation can affect the relative permeability to oil as a function of brine saturation. 4.) During single-phase brine flow experiments, 50 Hz stimulation is effective at restoring permeability in rocks that have been deliberately fouled with in situ clay fines. This type of formation "cleaning" has been observed before with ultrasonic energy but not at seismic frequencies. Several physical mechanisms have been identified to explain the wide range of core-scale behavior observed during this project. These include altered wettability, in-situ fines migration due to changes in colloid/surface interactions, permeability resonance (predicted by the UCB modeling), and trapped fluid ganglia mobilization. Also, physical parameters from several LANL core stimulation experiments were used as inputs to the UCB numerical modeling codes and comparisons between numerical predictions and experimental results are encouraging. Qualitative agreement has also been confirmed between the lab/theoretical parameters and those of several field stimulation tests performed under this project. In addition to the experiments above, project efforts

172

have also focused on modifications to the DSSL apparatus, refinements to the experimental procedures, and improvements to the fluid delivery and stress/strain measurement systems. These improvements allow characterization of dynamic-stress induced changes in fluid flow behavior with accuracy and precision exceeding other facilities. Finally, a new system was designed to produce direct pore-pressure stimulation in the core samples. This system was implemented in 2004 and will be used to investigate the effectiveness of pore-pressure stimulation relative to the mechanical-stress mode used in previous experiments.

REFERENCES

1. Roberts, P.M., E.L. Majer, W.-C. Lo, G. Sposito and T.M. Daley (2002). An Integrated Approach to Seismic Stimulation of Oil Reservoirs: Laboratory, Field and Theoretical Results from DOE/Industry Collaborations. Invited talk, *16th International Symposium on Nonlinear Accoustics*. Moscow, Russia, August 2002.
2. Roberts, Peter M., Esipov, Igor B., and Majer, Ernest L. (2003) Elastic Wave Stimulation of Oil Reservoirs: Promising EOR Technology? *The Leading Edge*, **22**(5), pp. 448-453,.
3. Pride, S.R. and E.G. Flekkoy (2005). Seismic stimulation for enhanced oil recovery. *J. Petroleum Science and Engineering*, (submitted April, 2005).
4. Lo, W.-C., G. Sposito, and E.L. Majer (2002). Immiscible two-phase fluid flows in deformable porous media. *Advances in Water Resources*, **25**(8-12), pp. 1105-1117, LBNL-50624.

Modeling of Oil Output Intensification in Porous Permeable Medium at Acoustical Stimulation from a Well

German A. Maksimov, Aleksei V. Radchenko

*Moscow Engineering-Physics Institute (State University), Kashirskoye shosse, 31,
Moscow, 115409, Russia
E-mail: maximov@dpt39.mephi.ru*

Abstract. Acoustical stimulation (AS) of oil production rate from a well is perspective technology for oil industry but physical mechanisms of acoustical action are not understood clear due to complex character of the phenomena. In practice the role of these mechanisms is appeared non-directly in the form of additional oil output. Thus the validity examination of any physical model has to be carried out as with account of mechanism of acoustic action by itself as well with account of previous and consequent stages dealt with fluid filtration into a well.

The advanced model of physical processes taking place at acoustical stimulation is considered in the framework of heating mechanism of acoustical action, but for two-component fluid in porous permeable media. The porous fluid is considered as consisted of light and heavy hydrocarbonaceous phases, which are in a thermodynamic equilibrium. Filtration or acoustical stimulation can change equilibrium balance between phases so the heavy phase can be precipitated on pores walls or dissolved.

The set of acoustical, heat and filtration tasks were solved numerically to describe oil output from a well - the final result of acoustical action, which can be compared with experimental data. It is shown that the suggested numerical model allows us to reproduce the basic features of fluid filtration in a well before during and after acoustical stimulation.

Keywords: Ultrasound treatment, acoustical stimulation, oil production rate simulation
PACS: 43.35.Zc, 43.30.Ky

INTRODUCTION

There are a lot of different models describing certain physical processes at AS [1-4]. But due to complexity of the problem only the complete numerical simulation of oil filtration together with processes accompanied the AS could give the answer on the principal question: which of these models is more close to reality?

The typical experimental averaged dependence of oil output on time after AS is shown in Fig.1. Plot results allow to estimate magnitude and duration of intensification process at AS. It follows from the Fig.1 the mean increase of oil output achieves 30%, and the mean duration of positive effect of AS achieves 1.5-2 months. The solid line in Fig.1 represents the averaged value of normalized (on initial value) outputs of separate wells. The dashed line is rms approximation of obtained dependence by exponential function: $Q/Q_0 = A\exp(-t/\tau)$, where the parameters have the following values $A = 1.23$, $\tau = 150$ days.

CP838, *Innovations in Nonlinear Acoustics: 17th International Symposium on Nonlinear Acoustics*,
edited by A. A. Atchley, V. W. Sparrow, and R. M. Keolian
© 2006 American Institute of Physics 0-7354-0330-9/06/$23.00

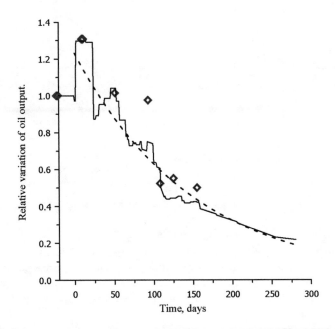

FIGURE 1. Variation of oil output (production rate) from a well after AS, normalized on its value before AS. Solid line – the data averaged by 10 wells, squares – the data of separate well, dashed line – exponential approximation.

In the report it is assumed that the porous fluid is composed from light and heavy hydrocarbonaceous phases, which are in a thermodynamic equilibrium. The external affects can vary the balance between fractions, so that the heavy fraction can be absorbed on pores walls or be dissolved. It results to change of pores diameter and hence to variation of local porosity and permeability. The variation of these parameters, in its turn, leads to change in filtration and hence to changes in distributions of pressure and velocity fields. It again leads to a shift of local thermodynamic balance etc. In the complete statement the simulation of oil output behavior at AS has to take into account the mentioned aspects of the problem dealt with behavior of fluid in porous permeable medium without and with AS.

The estimations of the basic process specific times allow us to split the complex problem on a set of quasi-independent subtasks, which are related only through their coefficients. In the framework of the given model it is possible to reproduce the principal features of oil filtration behavior (oil output rate) from a well before and after AS. Other physical mechanisms can be introduced into the developed model to understand which of them is adequate one.

SOUND ENERGY DENSITY DISTRIBUTION

The problem of acoustical energy density distribution around a well under action of ultrasound acoustical source on the well axis was considered in [5]. On the Fig. 2 the energy density distribution calculated for source frequency f =20 kHz is shown for the following parameters of porous medium: m = 0.15; k = 0.1 D.

175

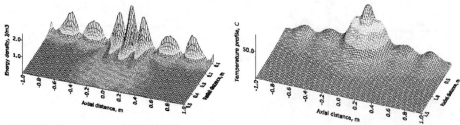

FIGURE 2. Sound energy density distribution around a well in porous permeable Biot's medium.

FIGURE 3. Space distribution of temperature field in a well vicinity at AS.

TEMPERATURE DYNAMICS DUE TO AS

With predetermined distribution of heat sources it is possible to calculate numerically the time-space variations of temperature field in the well vicinity. The dynamics of a thermal field was estimated by the finite-difference procedure [4, 6]. It was shown (see Fig.3) that at long-time AS (more than 10 hours) the temperature in the well vicinity increases on 10-14°C degrees that correspond to the data of thermometry [3].

DIFFUSION AND TRANSPORT OF HEAVY ADMIXTURE

The dynamics of a concentration field was estimated by the finite-difference procedure by analogy with temperature field. There are three basic mechanisms for variation of heavy fracture concentration. They are the transport with resolving flow, the diffusion and the absorption or dissolution on the pore walls.

The linear expansion was used for determination of pressure, temperature and energy density distribution dependence of equilibrium concentration C_*, where parameters were chosen to fit experimental data for duration of oil output decreasing. In the frameworks of the represented model the variations of local pore radius, porosity and permeability are governed by relations were represented in [6], which determine the mass balance at absorption and dissolution.

FLUID FILTRATION AND OIL OUTPUT SIMULATION

The oil filtration in porous medium is described by the filtration equation for a pressure, obtained by linearization of continuity equation, the Darcy's law and the state equation of fluid. The total fluid output is determined as integral by surface of a well perforated interval from the normal to surface component of mass fluid flow during the time period.

The resulting plots of oil output variations from a well before and after AS are shown in Fig.4 for different treating regimes of surrounding media.

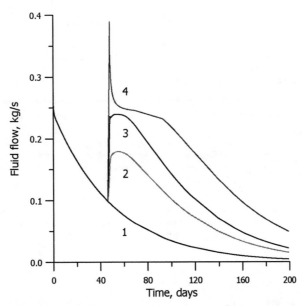

FIGURE 4. The dynamics of oil output variations from a well before and after AS for different heating regimes of surrounding medium. The lowest curve 1 corresponds to oil output dynamics without AS and it corresponds to exponential decreasing of oil output. The curve 2 corresponds to output variation dealt with heating of well surrounding in the framework of elastic model with attenuation. The curves 3 and 4 correspond to heating in the framework of Biot's model with permeability $k = 0,1$ D and opened porosity $m = 0,05$ and $m = 0,2$.

REFERENCES

1. Kuznetsov O. L., Efimova S. F. (1983), *Application of ultrasound in oil industry*, Nedra, Moscow (in Russian)
2. Gorbachev Y. I. (1998), "Physicochemical bases of ultrasonic clearing in near bottom region of oil-wells", // Geoinformatika, Vol. 3, pp. 7-12. (in Russian).
3. Pechkov A. A., Shubin A. V., (1998), "Results of operations on wells efficiency increasing by the method of ultrasonic action", Geoinformatika, Vol. 3, pp. 16-24. (in Russian)
4. Maksimov G. A., Radchenko A. V., (2001) "Role of heating at ultrasonic affecting", Geofizika, Vol. 6. (in Russian)
5. Elias S.E., Kirnos D.G., Maksimov G. A., Radchenko A. V., (2002) "Acoustical energy distribution around borehole embedded in porous permeable Biot's medium", *Proc. of Scientific session of MEPhI-2002, 21-25 January 2002*, Vol. 5, pp. 82-83. (in Russian)
6. Maksimov G. A., Radchenko A. V., (2001), "Calculations of acoustical energy density in vicinity of a borehole and oil output intensification due to acoustical stimulation", *Proc. of XI session of RAO 19 - 23 November 2001*, Vol. 2, pp. 67-71. (in Russian)

Capillary Dynamics of Elastic-Wave-Enhanced Two-Phase Flow in Porous Media

Markus Hilpert*, Chunyan Guo* and Joseph Katz†

*Johns Hopkins University, Department of Geography and Environmental Engineering, 313 Ames Hall, 3400 N. Charles St., Baltimore, Maryland, U.S.A.
†Johns Hopkins University, Department of Mechanical Engineering, 118 Latrobe Hall, 3400 N. Charles St., Baltimore, Maryland, U.S.A.

Abstract. Elastic waves may enhance two-phase flow in porous media. We investigate the role and dynamics of capillary forces during the enhancement process. We present a theory that allows us to estimate the response of trapped nonwetting phase blobs to variable frequency excitation. According to this theory capillary trapped oil blobs may exhibit resonance, depending on the properties of the fluids and the pore space. Using this theory we estimate the resonant frequencies of crude oil and gasoline blobs in sphere packings. We will also present experimental evidence showing that capillary trapped liquid blobs exhibit resonance.

Keywords: Capillarity, elastic waves, enhanced oil recovery, resonance, two-phase flow
PACS: 47.56.+r, 47.55.nb, 76.20.+q, 92.40.Kf

INTRODUCTION

One of the most challenging problems in petroleum engineering and in the remediation of groundwater contaminated by nonaqueous phase liquids (NAPL) is the formation of residual oil/NAPL [1]. The phenomenon of capillary trapping is particularly important if oil/NAPL constitutes the nonwetting phase. Whereas wetting phase can be completely recovered in the limit where it forms a film on the solid surfaces (even though these film flows can be quite slow), residual nonwetting phase can practically not be displaced or recovered by imposing a flow in the surrounding wetting phase [2].

A variety of methods have been developed to mobilize trapped nonwetting phase. These methods include vapor extraction, cosolvent flushing, surfactant flushing, steam injection, and in-situ biodegradation. A promising method for mobilizing trapped nonwetting phase consists of applying elastic waves to the subsurface. This method has originally been developed by petroleum engineers to improve the production of oil from its natural reservoirs [3, 4, 5, 6, 7]. Experimental evidence, both in the laboratory [8, 9, 10] and in the field [11], suggests that the application of elastic waves makes water-flooding of oil reservoirs more efficient, i.e., less residual oil is left behind, and oil is extracted more quickly. Also in groundwater settings, laboratory experiments have shown that NAPL removal can be improved by the application of elastic waves [10, 12, 13].

This paper contributes toward investigating the hypothesis that residual nonwetting phase can be mobilized by exploiting a phenomenon entitled *capillarity-induced resonance* [14], which implies that a trapped nonwetting phase blob represents a resonating system (like a mass-spring system) where the restoring force is due to capillarity.

CP838, *Innovations in Nonlinear Acoustics: 17th International Symposium on Nonlinear Acoustics*, edited by A. A. Atchley, V. W. Sparrow, and R. M. Keolian
© 2006 American Institute of Physics 0-7354-0330-9/06/$23.00

A MODEL FOR UNDERSTANDING CAPILLARY DYNAMICS IN POROUS MEDIA

Capillary pressure is ultimately responsible for the trapping of nonwetting phase blobs. Thus, the effects of elastic waves on the dynamics of capillary pressure across the various fluid-fluid interfaces (menisci), which a nonwetting phase blob shares with the surrounding wetting phase, are crucial to the mobilization of nonwetting phase blobs. Hilpert, in his dissertation [10], employed relatively simple concepts from fluid mechanics to understand the underlying physical mechanisms:

1. The Navier-Stokes equations in the individual fluids;
2. Young-Laplace's equation for the pressure jump across a meniscus; and
3. A model for the dynamic behavior of the three-phase contact lines between the two fluids and the solid phase (either pinned or sliding contact lines).

Hilpert et al. [14] used simple models for pore spaces to explore the phenomenon of *capillarity-induced resonance*. They assumed straight and sinusoidal capillary tubes in order to explain the resonance of nonwetting phase blobs with pinned and sliding contact lines, respectively. Figure 1 shows one of the model systems used to explore the resonance of nonwetting phase blobs with pinned contact lines. A blob of length L is trapped in a straight capillary tube of radius R due to assumed surface heterogeneities (e.g., wall roughness and/or chemical contamination).

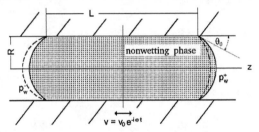

FIGURE 1. A simple model for the excitation of a trapped nonwetting phase blob in a porous medium by elastic waves consists of a blob in a straight capillary tube. Solid lines: equilibrium state. Dashed lines: nonequilibrium state due to elastic-wave action. The contact lines are assumed to be pinned.

The model describes two aspects of elastic waves that will effect the blob:

1. An oscillating pressure gradient in the wetting phase, $\Delta p_w(t) = p_w^+(t) - p_w^-(t) = \Delta p_w(\omega)e^{-i\omega t}$; and
2. An oscillatory movement of the tube wall along the tube axis with velocity $v(t) = v(\omega)e^{-i\omega t}$.

Hilpert et al. [14] developed a linear theory that relates the driving forces to the volume amplitude $\Delta V^+(t) = \Delta V^+(\omega)e^{-i\omega t}$ of the right meniscus. In the absence of gravity, the pressure difference in the wetting phase is

$$\Delta p_w(\omega) = \Delta p_c^-(\omega) - \Delta p_c^+(\omega) + \Delta p_{visc}(\omega) \qquad (1)$$

179

where $\Delta p_c^{\pm}$ are the deviations from equilibrium capillary pressures, and Δp_{visc} is the viscous pressure drop in the blob. Capillary pressure is given by

$$p_c^{\pm} = p_n^{\pm} - p_w^{\pm} = \frac{2\gamma}{R}\cos\theta^{\pm} \qquad (2)$$

where γ is the interfacial tension, and $\theta^{\pm}$ are the dynamic contact angles of the two menisci. A linear approximation for $p_c^{\pm}$ yields

$$\Delta p_c^{\pm} = \pm\frac{2\gamma}{\pi R^4}\sin\theta_0(1+\sin\theta_0)^2\Delta V^+ \qquad (3)$$

where θ_0 is the equilibrium contact angle. In Fourier space, the viscous pressure drop can be estimated from an analytical solution which assumes that the blob is infinitely long:

$$\Delta p_{visc}(\omega) = \left[\frac{\eta_n\omega}{\pi R^4}\frac{\omega/\omega_c}{h(\omega/\omega_c)}\Delta V^+(\omega) + i\rho_n\omega v(\omega)\right]L \qquad (4)$$

where

$$\omega_c = \frac{\eta_n}{\rho_n R^2} \qquad (5)$$

is the characteristic frequency, η_n is the dynamic viscosity of the nonwetting phase, ρ_n is the nonwetting phase density, and

$$h(X) = -\frac{J_2(\sqrt{iX})}{J_0(\sqrt{iX})} \qquad (6)$$

is a dimensionless function. By substituting equations (3) and (4) into Eq. (1) one obtains an expression for the blob's nondimensional frequency response χ:

$$-\frac{\Delta V^+(\omega)}{a_{eff}(\omega)}\frac{\omega_c^2}{\pi R^2} = \chi(X, X_0) = \left[X_0^2 - \frac{X^2}{h(X)}\right]^{-1} \qquad (7)$$

where

$$a_{eff}(\omega) = \frac{1}{\rho_n}\frac{\Delta p_w(\omega)}{L} - i\omega v(\omega) \qquad (8)$$

is the effective acceleration, $X = \omega/\omega_c$ is the dimensionless excitation frequency, and $X_0 = \omega_0/\omega_c$ is the dimensionless resonant frequency of the undamped system. The undamped resonant frequency, which corresponds to the limit $\eta_n \to 0$, is given by

$$\omega_0 = \sqrt{\frac{4\gamma}{\rho_n R^2 L}\sin\theta_0(1+\sin\theta_0)^2} \qquad (9)$$

A blob exhibits resonance only if

$$X_0 > \sqrt{24} \qquad (10)$$

as illustrated in Fig. 2. This mathematical condition can be interpreted with the concept of viscous penetration depth ξ. Consider a half space that is occupied by a fluid (see Fig. 3). If the solid surface undergoes a vibratory motion with amplitude $x(t) = x_0 e^{-i\omega t}$, the fluid above will oscillate too because of viscous drag. The amplitude will decay, however. The distance from the wall, where the fluid particle amplitude decreases to $1/e$ of the wall amplitude value, can be defined as the viscous penetration depth [15]

$$\xi = \sqrt{\frac{2\eta_n}{\omega \rho_n}} \tag{11}$$

This depth quantifies how far momentum propagates into the half space or up to which distance from the surface the fluid "feels" the forcing. The resonance condition can now be understood as follows. For an excitation of a blob at its undamped resonant frequency ω_0 the viscous penetration depth is $\xi = \sqrt{2\eta_n/(\omega_0 \rho_n)}$. For a capillary tube, the ratio

$$\frac{\xi}{R} = \sqrt{\frac{2\omega_c}{\omega_0}} = \sqrt{\frac{2}{X_0}} \tag{12}$$

thus quantifies the fluid region that feels the drag forces due to the tube wall. If ξ/R is small (or X_0 is large) little friction occurs. This explains the resonance condition given by Eq. (10).

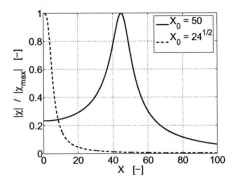

FIGURE 2. Nondimensional volume amplitude $|\chi|$ of a nonwetting phase blob in a straight capillary tube as shown in Fig. 1 versus the nondimensional excitation frequency $X = \omega/\omega_c$. The blob exhibits resonance only if the nondimensional resonant frequency $X_0 = \omega_0/\omega_c$ exceeds $\sqrt{24}$.

EVIDENCE FOR BLOB RESONANCE IN POROUS MEDIA

Hilpert and Miller [16] showed experimentally that a liquid column trapped in a capillary tube may exhibit resonance. Using microscopic imaging they measured the frequency response of a water column trapped in a vertically oriented capillary tube due to acoustic excitation. Figure 4a depicts the experimental system, while Fig. 4b compares the measured response to a theoretical model developed by Hilpert et al. [17]. The model

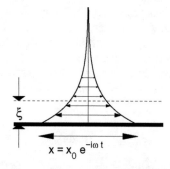

FIGURE 3. The viscous penetration depth ξ quantifies the distance at which a fluid particle feels the force exerted by a solid surface oscillating with amplitude x_0 and frequency ω. The horizontal arrows illustrate the amplitude of the fluid particles.

prediction agrees well with the measured quasi-static response and resonant frequency. The deviations in the amplitude of the dynamic response are due to the simplified treatment of the viscous pressure drop.

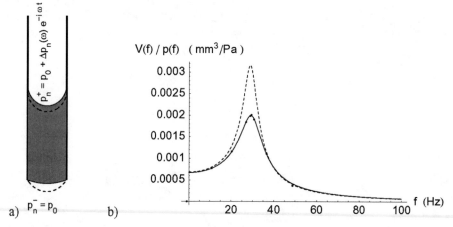

FIGURE 4. (a) A liquid column trapped in a vertical tube exhibits capillarity-induced resonance. Solid line: equilibrium state. Dashed line: nonequilibrium state due to excitation by an oscillating pressure on top of the column. (b) Measured volume amplitude of a liquid column, V, normalized by pressure amplitude p, versus sound frequency f. Solid line: measured response. Dashed line: prediction by a theoretical model [17].

Recently we have begun using Lattice-Boltzmann (LB) modeling to investigate blob resonance in porous media. Figure 5a shows a blob during excitation by an imposed oscillatory flow. Figure 5b shows how the blob amplitude depends on the excitation frequency ω. The blob clearly exhibits resonance. These numerical experiments suggest that blobs in complex pore geometries exhibit resonance too.

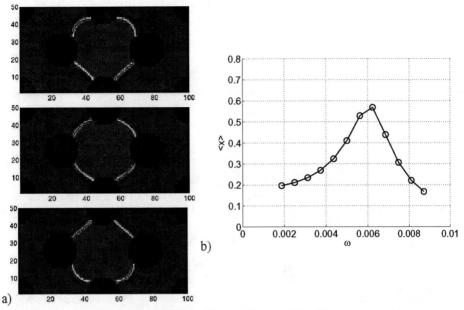

FIGURE 5. (a) Three snap-shots of a Lattice-Boltzmann simulation where a trapped nonwetting phase blob exhibits resonance. (b) Frequency response of this blob, where $< x >$ is a mean displacement amplitude in the vertical direction, and ω is the excitation frequency.

HYPOTHESIZED BEHAVIOR OF NONWETTING PHASE BLOBS IN POROUS MEDIA

Using the simple pore-scale model introduced above we can explore whether nonwetting phase blobs in porous media are able to exhibit resonance. We represent a nonwetting phase blob in a sphere packing with variable sphere diameter D by a blob in a straight capillary tube. The radius of the tube, R, is chosen to be a mean pore size R, which in turn can be estimated from the sphere size using the empirical relation $R = 0.265 \cdot D$ [10]. We assumed the blob length L to equal 10 mean pore-sizes, a contact angle of $\theta_0 = 30°$ and considered two different liquids as the nonwetting phase: (1) gasoline with $\rho_n = 730$ kg/m^3 and $\eta_n = 0.0006$ kg/(m·s), and (2) a crude oil with $\rho_n = 850$ kg/m^3 and $\eta_n = 0.005$. The wetting phase is assumed to be water.

Figure 6 shows the characteristic frequency ω_c and the resonant frequency ω_0 for the two two-liquid systems. For all sphere sizes, $\omega_0 > \omega_c$, i.e., the blobs should exhibit resonance, because frictional forces are small when the blobs are excited at their resonant frequencies.

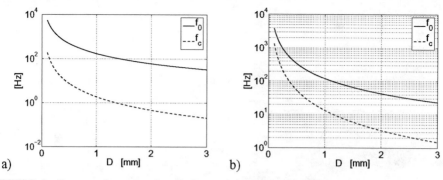

FIGURE 6. Resonant frequency f_0 and characteristic frequency f_c for blobs in a uniform sphere packing versus sphere diameter D for (a) a gasoline-water and (b) a crude oil-water system, respectively. Because $f_0 \gg f_c$, all blobs in the various sphere packings represent resonating systems according to Eq. (10).

SUMMARY AND DISCUSSION

1. We have employed relatively simple concepts in order to predict capillarity-induced resonance, namely the validity of the Navier-Stokes equation and a model for the dynamic behavior of the three-phase contact lines. Experimental investigations and Lattice-Boltzmann simulations support our modeling approach. These facts strongly suggest that *capillarity-induced resonance* exists.

2. It remains to be shown, however, whether this resonance phenomenon can be used to mobilize nonwetting phase blobs, also under the action of a superimposed external head gradient in the wetting phase. In that respect it is noteworthy that Iassonov and Beresnev [18], who appear to have generalized the model developed by Hilpert et al. [14] by accounting for such a gradient and non-Newtonian fluid behavior, did not mention the occurrence of this resonance phenomenon in their theoretical paper.

3. Our model does not account for all aspects of elastic waves in porous media saturated by two fluids, particularly not for the compression of the porous skeleton. Future models need to account for this effect that occurs for compressional waves of the first and second kind.

4. Assuming for the moment that *capillarity-induced resonance* may indeed enhance two-fluid flow, we can conclude from Eq. (8) that compressional waves of the second kind exert a greater force on nonwetting phase blobs than compressional waves of the first kind [14], because the wetting and solid phase oscillate in opposition of phase for waves of the second kind. Even though waves of the second kind are typically highly attenuated, this fact could be of considerable practical importance, because these waves may be generated at the interface between media with different acoustical impedance. Thus, subsurface heterogeneity potentially helps exploiting *capillarity-induced resonance*.

ACKNOWLEDGMENTS

This work was supported by the Petroleum Research Fund Grant PRF #38128-G9 and the National Science Foundation Grant EAR-0335766. We particularly would like to thank Igor Esipov and Peter Roberts for organizing a special session on "Elastic-Wave Effects on Fluids in Porous Media" at the 17th ISNA. This session eventually gave MH the opportunity to meet many of the key players in this exciting research area.

REFERENCES

1. M. Hilpert, J. F. McBride, and C. T. Miller, *Advances in Water Resources*, **24**, 157–177 (2001).
2. J. R. Hunt, N. Sitar, and K. S. Udell, *Water Resources Research*, **24**, 1247–1258 (1988).
3. I. A. Beresnev, and A. P. Johnson, *Geophysics*, **59**, 1000–1017 (1994).
4. P. M. Roberts, I. B. Esipov, and E. L. Majer, *The Leading Edge* (2003).
5. F. Schempf, *Hart's E & P*, **April** (2002).
6. S. A. Kostrov, W. O. Wodden, and P. M. Roberts, *Oil and Gas Journal* (2001).
7. S. Jackson, and P. Roberts, *PTTC Network News*, **7(2)** (2001).
8. A. B. Pogosyan, E. M. Simkin, E. V. Sremovskiy, M. L. Surguchev, and A. I. Shnirelman, *Transactions of the USSR Academy of Science*, **307**, 575–577 (1989).
9. H. S. Muralidhara, B. F. Jirjis, F. B. Stulen, G. B. Wickramanayake, A. Gill, and R. E. Hinchee, Development of electro-acoustic soil decontamination (ESD) process for in situ application, Tech. Rep. EPA/540/5-90/004, Environmental Protection Agency, Washington, DC (1990).
10. M. Hilpert, *Mobilisierung residualer Flüssigkeiten in porösen Medien durch Resonanz auf Schallwellen*, Ph.D. thesis, Universität Karlsruhe, Karlsruhe, Germany (1997).
11. R. M. Simkin, and M. L. Surguchev, "Advanced vibroseismic technique for water flooded reservoir stimulation, mechanism and field results," in *Proceedings of the Sixth European Symposium on Improving Oil Recovery*, 1991.
12. P. M. Roberts, A. Sharma, V. Uddameri, M. Monagle, D. E. Dale, and L. K. Steck, *Environmental Engineering Science*, **18**, 67–69 (2001).
13. W. Li, D. Vigil, I. Beresnev, P. Iassonov, and R. Ewing, *Journal of Colloid and Interface Science*, **289**, 193–199 (2005).
14. M. Hilpert, G. H. Jirka, and E. J. Plate, *Geophysics*, **65**, 874–883 (2000).
15. G. K. Batchelor, *An Introduction to Fluid Dynamics*, Cambridge University Press, Cambridge, 1988.
16. M. Hilpert, and C. T. Miller, *Journal of Colloid and Interface Science*, **219**, 62–68 (1999).
17. M. Hilpert, D. Stopper, and G. H. Jirka, *Journal of Applied Mathematics and Physics*, **48**, 424–438 (1997).
18. P. P. Iassonov, and I. A. Beresnev, *Journal of Geophysical Research*, **108**, ESE 2–1–2–9 (2003).

Microscopic Behavior Of Colloidal Particles Under The Effect Of Acoustic Stimulations In The Ultrasonic To Megasonic Range

Amr I. Abdel-Fattah, Peter M. Roberts

Los Alamos National Laboratory

Abstract. It is well known that colloid attachment and detachment at solid surfaces are influenced strongly by physico-chemical conditions controlling electric double layer (EDL) and solvation-layer effects. We present experimental observations demonstrating that, in addition, acoustic waves can produce strong effects on colloid/surface interactions that can alter the behavior of colloid and fluid transport in porous media. Microscopic colloid visualization experiments were performed with polystyrene micro-spheres suspended in water in a parallel-plate glass flow cell. When acoustic energy was applied to the cell at frequencies from 500 kHz to 5 MHz, changes in colloid attachment to and detachment from the glass cell surfaces were observed. Quantitative measurements of acoustically-induced detachment of 300-nm microspheres in 0.1M NaCl solution demonstrated that roughly 30% of the colloids that were attached to the glass cell wall during flow alone could be detached rapidly by applying acoustics at frequencies in the range of 0.7 to 1.2 MHz. The remaining attached colloids could not be detached by acoustics. This implies the existence of both "strong" and "weak" attachment sites at the cell surface. Subsequent re-attachment of colloids with acoustics turned off occurred only at new, previously unoccupied sites. Thus, acoustics appears to accelerate simultaneously both the deactivation of existing weak sites where colloids are already attached, and the activation of new weak sites where future attachments can occur. Our observations indicate that acoustics (and, in general, dynamic stress) can influence colloid-colloid and colloid-surface interactions in ways that could cause significant changes in porous-media permeability and mass transport. This would occur due to either buildup or release of colloids present in the porous matrix.

Keywords: colloid/surface interactions, acoustic stimulation, boundary layers
PACS: 82.70.Dd, 68.35.Iv, 68.35.-p

INTRODUCTION

Organic and inorganic colloids (sub-pore size particles) are found in virtually all subsurface porous mass transport systems as a mobile or immobile solid or liquid phase. Three types of colloid interactions that can affect subsurface flow by reducing the permeability of a porous medium are: a) significant deposition onto pore surfaces, b) aggregation in the bulk pore fluid, and c) bridge formation at pore throats. Colloid deposition and aggregation are consequences of different particle-surface and particle-particle interactions, respectively. These interactions are among the core subjects in colloid science [1]. Colloid bridging at pore throats is a combination of both particle-surface and particle-particle interactions that occurs under certain circumstances, such as those causing hydrodynamic jamming, or the presence of polymers that may induce

CP838, *Innovations in Nonlinear Acoustics: 17th International Symposium on Nonlinear Acoustics*,
edited by A. A. Atchley, V. W. Sparrow, and R. M. Keolian
© 2006 American Institute of Physics 0-7354-0330-9/06/$23.00

"bridging" or "depletion" flocculation. Acoustics can influence each of these interactions in ways that can cause either desirable or undesirable consequences on fluid and colloid transport. Acoustics can alter particle-surface and particle-particle interactions through the formation of "acoustically-induced" boundary layers around the particles and surfaces [2]. These boundary layers combine with those induced by the physico-chemical parameters of the colloidal dispersion. This combination of layers controls the overall interactions among particles and surrounding surfaces in the presence of acoustics. Results of microscopic colloid behavior visualization experiments are presented that demonstrate acoustic effects on colloid attachment and detachment at solid surfaces. These observed effects are likely caused by altering the various boundary layers described above.

EXPERIMENTAL APPARATUS

The main apparatus used in the experiments below is the "Automated Video Microscopic Imaging and Data Acquisition System" (AVMIDAS), which is used for real-time microscopic visualization of colloid behavior in glass flow cells [3]. The AVMIDAS apparatus (Fig. 1) combines an optical microscope with a high-resolution video monitor and recording system. The microscope utilizes dark-field light microscopy to visualize colloids as small as approximately 0.2 microns in diameter inside a parallel-plate glass flow cell with internal dimensions: 4.0 cm x 0.8 cm x 0.02 cm. (Fig. 2). Colloids appear in the video images as white dots on a black background. Real-time image data acquisition, processing and analysis are automated with a PC. In-situ measurements include surface concentration and flux of attached, detached, and deposited colloids, their mobility in the vicinity of a surface, their distribution in the bulk suspension, and their spatial and temporal distribution on the cell surface.

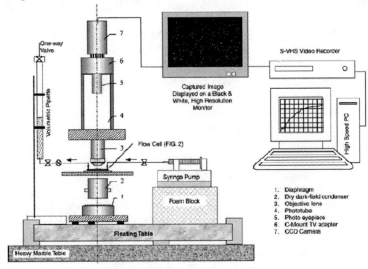

FIGURE 1. Schematic diagram of AVMIDAS.

187

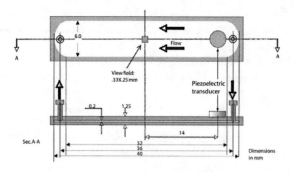

FIGURE 2. AVMIDAS glass flow cell.

To generate acoustic signals in the glass flow cell, a 1/4-inch-diameter piezoelectric transducer was attached to the top surface of the cell near the flow inlet port (Fig. 2). The transducer has a center frequency of 1 MHz, but can produce compressional acoustic energy over a frequency range of approximately 200 kHz to 10 MHz when driven by an appropriate function generator and radio-frequency (RF) amplifier. This frequency range was imposed partly by the size of the flow cell and the transducer that would fit on it, but also because colloid behavior at very high frequency must be understood first to attempt scaling to lower frequency.

QUANTITATIVE ATTACHMENT AND DETACHMENT EXPERIMENT WITH AVMIDAS

Based on previous qualitative colloid behavior experiments where acoustics was observed to induce both colloid attachment and detachment at the glass cell surface, an experiment was performed to quantify this. Dilute suspensions of 500-nm microspheres in either de-ionized water or 0.1M NaCl solution were flowed through an initially clean cell at 1.0 ml/hour. AVMIDAS was used to track and count all particles within the field of view (330 X 250 microns) as they attached or detached at the bottom cell surface. Particle attachment was observed only for the higher ionic strength (0.1M NaCl) solution and only when no acoustic energy was applied. Acoustically enhanced attachment was not observed during this particular experiment. Because the flow rate was held constant, detachment was observed only when acoustic energy was applied at various pre-determined frequencies.

For the high-ionic-strength experiment, Fig. 3 shows the attached and mobile particle counts versus elapsed time measured by AVMIDAS during repeated cycles of attachment and acoustically-induced detachment at different frequencies. Episodes when the suspension is flowing with acoustics turned off are indicated by the "OFF" arrows at the top of the figure. Acoustic stimulation episodes are indicated by the arrows labeled with the frequency used during each treatment. The numbers, (1) through (7), mark the times at which acoustics was either turned on (odd numbers) or off (even numbers). Each episode of flow alone resulted in approximately the same number of total attached particles. However, the initial rate of detachment and the final number of remaining attached particles varied with acoustic frequency. This is consistent with the measured number of mobile (detached) particles shown at the bottom of the plot.

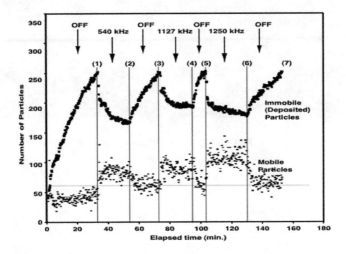

FIGURE 3. Attachment and detachment of 500-nm polystyrene microspheres suspended in 0.1M NaCl solution flowing through the AVMIDAS glass flow cell. The total number vs. elapsed time of attached (immobile) particles is plotted at the top of the figure. Detached (mobile) particles are plotted at the bottom.

Captured AVMIDAS still images of the attached particles at the times numbered (1) through (7) in Fig. 3 demonstrated two significant phenomena. First, acoustics caused detachment of particles occupying only a portion, roughly 30%, of the total sites occupied during the initial flow through the cell. This implies that both "weak" and "strong" attachment sites exist at the glass surface. The weak sites are those where acoustics caused colloid detachment and strong sites are those where colloids remained attached throughout the experiment. The second phenomenon observed was that weak sites where particles were removed by acoustics were never re-occupied. Instead, reattachment during flow alone occurred only at new, previously unoccupied sites. Apparently, acoustics accelerates both the destruction of currently "active" weak sites and the creation of new active weak sites simultaneously.

Figure 4 shows captured still AVMIDAS images at the end of the first two cycles of colloid attachment during flow alone. Initially occupied weak sites, where particles were detached during the first acoustic treatment, are marked by circles. New weak sites occupied after acoustics was turned off are marked by squares. Unmarked particles never detached during the experiment and, thus, occupy strong sites. Circled sites were never reoccupied. New sites (squares) were also never reoccupied after being vacated by acoustics. All remaining still images (not shown) captured for each cycle of flow and stimulation exhibited the same behavior. Active attachment sites are never re-occupied after their particles are acoustically detached, but new sites are continuously activated by acoustics. This phenomenon needs to be investigated.

189

Image Captured at Time (1)   **Image Captured at Time (3)**

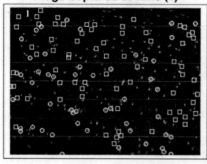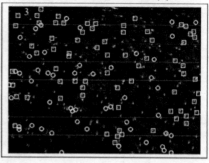

O Sites Vacated Between (1) - (2)
□ Sites Occupied Between (2) - (3)

FIGURE 4. AVMIDAS still images at time (1), left, and time (3), right, as indicated on Fig. 3.

DISCUSSION

The micro-scale experimental results presented here clearly demonstrate that high-frequency acoustic energy in the range of 500 kHz to 2 MHz can produce strong effects on colloid interactions at immobile solid surfaces. If these interactions can occur at the surfaces of a porous matrix (deposition or detachment at pore walls), this could lead to either desirable or undesirable changes in mass transport in porous media. Colloid deposition at pore surfaces, and bridging at pore throats can cause large decreases in porous media permeability, and are, thus, undesirable effects. Surface detachment and bridge breakup can increase permeability and are, thus, desirable effects. If these effects can also occur at seismic frequencies (< 500 Hz), they could alter mass transport processes in the Earth that would impact field applications such as enhanced oil recovery, colloid-facilitated transport, and groundwater remediation. The possible scaling of megasonic interactions to low frequency needs to be investigated.

ACKNOWLEDGMENTS

This work was supported by the U.S Department of Energy, Office of Basic Energy Sciences, Geosciences Division.

REFERENCES

1. Elimelech, M., J. Gregory, X. Jia, and R. A. Williams (1998), "Particle Deposition and Aggregation: Measurement, Modelling, and Simulation", Butterworth-Heinemann Ltd., Oxford (1998).
2. Dukhin, A. S., P. J. Goetz, T. H. Wines, and P. Somasundaran (2000). Acoustic and Electroacoustic Spectroscopy, *Colloids and Surfaces A*, **173**, 127-158.
3. Abdel-Fattah, A.I., M.S. El-Genk and P.W. Reimus (2002a). On Visualization of Sub-Micron Particles with Dark-Field Light Microscopy, *J. Colloid Interface Sci.*, **246**, 410-412.

Low-Frequency Dynamic-Stress Effects On Core-Scale Porous Fluid Flow Due To Coupling With Sub-Pore-Scale Particle Interactions

Peter M. Roberts, Amr I. Abdel-Fattah

Los Alamos National Laboratory

Abstract. It has been observed repeatedly that low-frequency (1-500 Hz) seismic stress waves can enhance oil production from depleted reservoirs and contaminant extraction from groundwater aquifers. The physics coupling stress waves to fluid flow behavior in porous media is still poorly understood. Numerous underlying physical mechanisms have been proposed to explain the observations. Core-scale experiments were performed to investigate one of these proposed mechanisms, the coupling of dynamic stress to sub-pore size particle (colloid) interactions with solid surfaces. This is an important mechanism because it can produce profound changes in porous matrix permeability due to either accumulation or release of natural colloids. Core-scale porous flow experiments demonstrated that both natural (in-situ) and artificial (injected) colloids can be released from the pores by applying dynamic stress to sandstone cores at frequencies below 100 Hz. Results are shown for release of in-situ particles from Fontainebleau sandstone induced by applying dynamic stress at 26 Hz.

Keywords: porous flow in rocks, dynamic-stress coupling, colloid transport, seismic stimulation
PACS: 47.55.Mh, 82.70.Dd, 83.50.-v

INTRODUCTION

Elastic (stress) waves are known to induce observable changes in fluid-flow behavior in porous media [1]. The observations vary with different combinations of dynamic-stress parameters, such as frequency, amplitude, wave mode and duration, and with the medium properties, such as permeability, elastic moduli, sub-pore-size particle (colloid) content, mineralogy, fluid saturation and ionic strength. Stress waves over a wide range of frequencies (roughly 1 Hz to 100 kHz) have been observed to influence porous fluid-flow behavior in the Earth and geomaterials over a similarly wide range of scale lengths (microns to kilometers). Published examples include oil reservoir production increases induced by seismic (1 to 500 Hz) waves [2] and permeability enhancement in sandstone by ultrasonic (10 to 100 kHz) energy [3]. Other potential applications include accelerated contaminant extraction from groundwater aquifers, and controlling colloid transport at waste facilities. Numerous physical mechanisms may control the coupling between stress waves and porous flow. Results of core-scale porous flow experiments are presented that demonstrate that sub-pore-scale particle (colloid) interaction with pore surfaces is a key mechanism by which low-frequency stress waves can produce strong effects on mass transport in porous media.

CP838, *Innovations in Nonlinear Acoustics: 17th International Symposium on Nonlinear Acoustics*,
edited by A. A. Atchley, V. W. Sparrow, and R. M. Keolian
© 2006 American Institute of Physics 0-7354-0330-9/06/$23.00

EXPERIMENTAL APPARATUS

The main component of the apparatus used is a triaxial core holder (Fig. 1) capable of applying up to 70 MPa static, axial and radial confining pressures to a cylindrical core sample. The sample is placed inside a horizontal Viton rubber sleeve designed to hold cores 2.54 cm in diameter and up to 60 cm long. Distribution plugs at each end of the sleeve accommodate fluid flow through the core. Radial confining pressure is applied to the main pressure chamber surrounding the sleeve. Static axial confinement is applied separately by a hydraulic piston attached to the inlet distribution plug (at the right end in Figure 1). During experiment runs, a constant-flow-rate pump is used to produce pulse-free flow of water through the core at flow rates ranging from 0.02 to 200 mL/min. A second pump and a two-way valve allow injection of artificial colloid suspensions into the core. The permeability along the core sample is measured using differential pressure gauges connected by tubing to taps at selected positions along the length of the rubber confining sleeve. The measured pressure drop across the core is converted to permeability using Darcy's law. Mechanical, axial dynamic stress oscillations are applied to the core during stimulation experiments using a magnetostrictive actuator, which is capable of generating dynamic force as high as ±900 N (200 lbf) peak-to-peak (P-P) with a maximum displacement of ±70 μm (0.003 in.) P-P. Depending on the rock being studied, axial dynamic strains as high as 10^{-4} can be applied with this system. Accurate measurements of applied stress/strain variations are obtained using a load cell in series with the actuator, and a Linear Variable Displacement Transducer (LVDT) coupled to the core face.

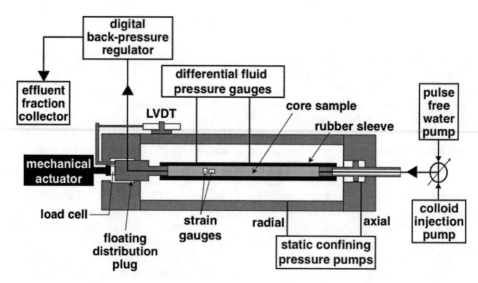

FIGURE 1. Schematic diagram of the core-flow stimulation apparatus.

IN-SITU PARTICLE RELEASE EXPERIMENT IN
FONTAINEBLEAU SANDSTONE

A Fontainebleau core, 2.54 cm in diameter by 30 cm long, was confined in the core-flow apparatus and flow of de-ionized water was initiated at 0.1 mL/min. The permeability of the sample was measured to be approximately 5 md. Before applying dynamic stress, the core was flushed for approximately four days to clean it's pore space of any untrapped particles. Effluent water samples were collected at regular time intervals and their particle concentration was characterized with a Particle Measurement System (PMS). After about 100 hours of flushing, the background particle concentration of the effluent was similar to that of the injected water, indicating that no more particles were being released from the core. Dynamic stress was then applied to the core with a frequency of 26 Hz and at three different RMS stress and displacement amplitudes. The PMS results for the effluent samples collected during stress stimulation are shown in Fig. 2. The vertical bars indicate the time and duration of each stimulation treatment. During the first treatment, 34 minutes at 170 kPa RMS stress amplitude, the effluent particle concentration increased by almost a factor 3. The concentration then gradually dropped to values below the initial background concentration over the next 17 hours. Three additional stimulations of 20 minutes were applied at higher RMS amplitudes. The effluent particle concentration increased again each time, but by lesser amounts than during the initial stimulation. These data clearly confirm the ability of low-frequency stress cycling to release trapped in-situ particles from a natural porous material at the core scale. No significant changes in permeability were observed for Fontainebleau sandstone because it's natural in-situ particle content is too low to cause pore fouling.

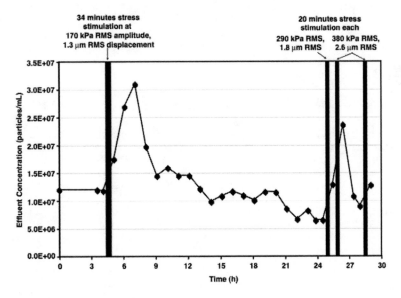

FIGURE 2. Release of in-situ particles from the Fontainebleau sample induced by 26-Hz dynamic stress stimulation.

DISCUSSION

The core-scale experimental results presented here clearly demonstrate that during fluid flow through a porous medium, low-frequency (26 Hz) dynamic stress can cause release of in-situ particles trapped in the pore space. If these particles are initially attached to the pore walls and are present at high enough concentrations, their release can cause pore-throat fouling and a subsequent decrease in permeability. However, if the ionic strength of the flowing fluid is high enough to prevent continued particle detachment, dynamic stress can also break up pore fouling caused by the initial particle release [4]. Thus, low-frequency stress waves are capable of producing both desirable and undesirable effects, depending on the physico-chemical conditions of the system. This mechanism needs to be better understood so that it can be harnessed for useful applications such as enhanced oil recovery, colloid-facilitated transport, and groundwater remediation. In particular, colloid interactions with solid surfaces occur (by definition) at microscopic, sub-pore scale ranges. Altering these interactions with acoustic energy typically requires frequencies in the megasonic range (roughly 500 kHz to 5 MHz). Thus, if low-frequency dynamic stress can affect colloid attachment or detachment at pore walls, a scaling relation may exist that needs to be investigated. Alternatively, relative motion between the solid porous matrix and the pore fluids can be induced by low frequency dynamic stress. The resulting hydrodynamic forces may be strong enough to overcome the various micro-scale forces that cause either attraction or repulsion between colloids and surfaces. Further research is required to understand the coupling of low-frequency dynamic stress to micro-scale particle behavior in porous media.

ACKNOWLEDGMENTS

This work was supported by the U.S Department of Energy, Office of Basic Energy Sciences, Geosciences Division.

REFERENCES

1. Beresnev, I.A. and P.A. Johnson (1994). Elastic-Wave Stimulation of Oil Production: A Review of Methods and Results, *Geophysics*, **59**, 1000-1017.
2. Roberts, P.M., I.B. Esipov and E.L. Majer (2003b). Elastic Wave Stimulation of Oil Reservoirs: Promising EOR Technology?, *The Leading Edge,* **22-5**, 448-453, invited paper.
3. Poesio, P., G. Ooms, M.E.H. v. Dongen and D.M.J. Smeulders (2004). Removal of Small Particles from a Porous Material by Ultrasonic Irradiation, *Transport in Porous Media*, **54-3**, 239-264.
4. Roberts, P.M. (2005). Laboratory Observations of Altered Porous Fluid-Flow Behavior in Berea Sandstone Induced by Low-Frequency Dynamic Stress Stimulation, *Acoustical Physics*, **51-4a**, 172-180, in press, Invited Paper. Simultaneously published also in the Russian edition Journal, *Akustichesky Zhurnal.*

Slow Acoustics Fluctuation
In The Granular Medium

I. Esipov, A.Vilman and K.Matveyev *⁾

N.Andreyev Acoustics institute, Moscow
e-mail: ibesipov@akin.ru
*⁾ Art Anderson Associates, Seattle, WA.

Abstract. Results of experimental research of nonlinear properties of the granular medium with fluid in the field of acoustics oscillations are considered. It is found that the acoustical respond of some separate granule of the medium is drastically changes in time under the pure tone excitation. Granular oscillation occurs essentially nonlinear rather under moderate amplitude elastic wave propagation when the granular acceleration is essentially less then gravity acceleration. Sound intensity fluctuations for the harmonic and subharmonic components are revealed in the records of up to 62 hours long. Viscous fluid suppresses fluctuations and makes the granular medium more linear for acoustical excitation. Data of the experiments are compared with the results of a simulation of sound wave propagation in granular medium with nonlinear contacts between the granules.

Keywords: nonlinear acoustics, granular media.
PACS: 43.25. Cb; 43.40. Qi.

EXPERIMENTS

Mechanical properties of granular media are essentially defined by contacts between the granules. Energy concentration of elastic deformations in the contact areas leads to anomalous high value of nonlinear acoustical parameter, which is typical for rocks [1-3]. Therefore the granular media possesses the structured acoustical nonlinearity, which differs them from typical continuous media such as pure crystals or homogeneous fluids. While nonlinearity of the homogeneous continuous media is defined by their deformation on molecular scale, the proper feature of the granular media reveals at the mesoscale level, or at the scale of separate granule dimension [4]. The main issue of this paper is the experimental research of nonlinear properties of the granular medium with fluid at the scale of separate granule in the field of acoustics oscillations.

We investigated the finite amplitude sound wave propagation in the granular medium. To receive acoustical signals in the medium following to [5], we used the gages approximately the same size as granules. Therefore we could measure the respond of separate granule on the acoustical excitation of the medium. Experiment has been done in granite gravel medium with granules dimension $0.5\text{-}2cm$. Piezoceramic plate (diameter $100mm$ and thickness $12mm$) served as an acoustical exciter. It was placed horizontally and generates an acoustical field, which propagates

CP838, *Innovations in Nonlinear Acoustics: 17ᵗʰ International Symposium on Nonlinear Acoustics*,
edited by A. A. Atchley, V. W. Sparrow, and R. M. Keolian
© 2006 American Institute of Physics 0-7354-0330-9/06/$23.00

vertically upward in a plastic container filled with the granite granules. Acoustical vibrations are registered by a pair of accelerometers with the main dimensions ø6mm × 10mm placed at 7cm above the piezoceramic plate and 10cm beneath the top medium interface. They were separated each other at 7cm symmetrically with respect to the exciter axis. They were adjusted to receive the vertical oscillations. It is founded that the transient pattern of the medium has numerous pikes which more then 10dB exceed the general trend in an experimental frequency range 2 − 14kHz with essential signal attenuation. Such pikes presentation could be explained by the interference of the signals propagated to the receiver by different ways [5].

Acoustical vibrations amplitude received by the accelerometers with respect to signal level at the exciter is shown at Fig.1. This and further experiments have been done at the frequency 5.6kHz of current radiation. One could see that the general behavior of the signals, received by both accelerometers, is similar. But their linear rise with the exciter level increase is fulfilled only in the mean and at the large interval of the signal amplitude change. Specific features in this dependence show the independent acoustical propagation from the exciter to each accelerometer. Maximum level of the received signal corresponds to piezoceramic exciter vibrations with acceleration of $0,6m/sec^2$, and amplitude of the exciter oscillation was only 5Å. That corresponds to excitation level at Fig.1 -10 dB. The level of granular vibrations measured by the accelerometer in the medium was approximately 10dB less.

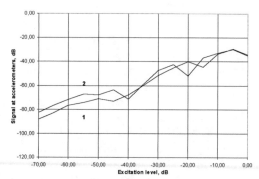

FIGURE 1. Level of dry granule vibration (for accelerometer 1 and 2) re to excitation level.

Non-monotonous dependence of the granular vibration on the acoustical excitation is related [6] to development of percolation networks between the granular, which provide the propagation of elastic vibration from the exciter to the accelerometers. Such a percolation chain is very sensitive to elastic load. Fig.2 shows the granular acoustical vibrations change in time. There are shown signal amplitude at the fundamental frequency 5.6kHz and it second harmonic and subharmonic behavior. Signal amplitude variations are similar for both accelerometers; therefore we put there the output from the only one receiver. Granule oscillation occurs essentially nonlinear rather under moderate amplitude elastic wave propagation when the acceleration under acoustical vibration is much less then gravity acceleration. Subharmonic both and harmonic components the acoustical signal are excited at the same time and their

196

amplitudes are strongly fluctuate. While the granule vibrations at the fundamental frequency 5.6*kHz* vary at 6-8*dB*, it is proved that fluctuation of the subharmonic is the most pronounced. It has giant amplitude of variations – more then 30 dB, although the fluctuations of the second harmonic are emphasized as well. That fact serves as an evidence on a stimulated process of subharmonics generation, when it excitation conditions are defined by a threshold mean [7]. It should be put attention that the fluctuations last during the whole signal record of 62 hours.

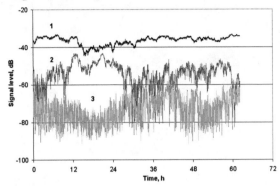

FIGURE 2. Signal level fluctuation (granules vibration) in time. 1-signal at the main frequency 3 kHz, 2 - the first harmonic at 6 kHz and 3 – subharmonic fluctuation at 1.5 kHz.

When we fill the granular medium with oil, the type of signal fluctuation essentially changed. First it noticeably increased (mainly it was related to bubble influence). After several days, when the bubbles got off, the fluctuations drastically decreased. Spectral analysis shows the regular fall in 4 decades (10^{-5}-10^{-1} Hz) of frequency band, which is usually related to fractal properties of phenomena under investigation.

COMPUTER SIMULATION

To demonstrate role of the contacts between the granules a computer simulation of sound signal propagation through the loaded chain of spherical elastics particles with Herz contacts between granules has been undertaken. Three simulations have been done. The first case corresponds to nonlinear oscillations and a random radii distribution in the chain (deviation from mean value is ±10%). The second case is done with the same nonlinear forcing, but on a regular chain with identical granules. In the third case, the nonlinear parameter of the contacts is very small, so the system response is linear; but the chain is random as in the first case. Amplitudes of the output signal component at the main frequency (relative to the input signal amplitude) are shown in fig. 3. Combined effects of nonlinearity and randomness produce fluctuations in the output amplitude, while the same forcing on the regular chain and the low-amplitude forcing on the random chain do not lead to sustained fluctuations.

These simulations demonstrate one of possible mechanisms for slow acoustic fluctuations in granular materials. In real granular objects, two- and three-dimensional effects will introduce other mechanisms. For example, contacts in a random granular matrix are unequally loaded (due to gravity) in a static state; this is different from the

197

one-dimensional chain, where all contacts carry the same load in equilibrium. Therefore, modifications occurring in the contacts between granules under unsteady forcing will significantly complicate the process. Since some contacts in a random multi-dimensional cluster are loaded very lightly, even low excitation levels may result in the appearance of fluctuations in the output amplitude, as was observed in experiments.

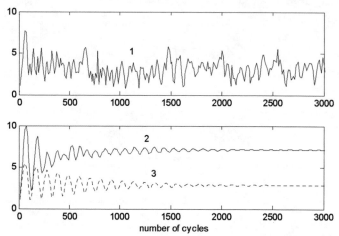

FIGURE 3. Amplitudes of the output signal at the frequency of the input signal. 1 finite amplitude, random chain, 2 finite amplitude, regular chain; 3 low amplitude, random chain.

ACKNOWLEDGMENTS

This research has been partially supported by Russian Foundation of Basic Research (project 05-02-17500).

REFERENCES

1. Problems of nonlinear seismic, Moscow, Nauka, 1987, 298 p. (in Russian)
2 Guyer R. and Jonson P., *Phys. Today,* **52**, 30 (1990)
3. Ostrovsky L.A., Johnson, P. A. *Rivista del Nuovo Cimento,* **24**(7), 1-46 (2001)
4 Belyaeva I.Yu., Zaytsev V.Yu, and Ostrovsky L.A. *Acoustical Physics,* **39**(1), 25-32 (1993)
5. Liu Chu-heng, Nagel S.R, *Phys.Rev.Lett.* **68**, 2301-2304 (1992)
6 R.C.Hidalgo, Ch.U.Grosse, F.Kun, H.W.Reinhardt, H.J.Herrmann, *Phys.Rev.Lett.* **89**, 205501-1-42002 (2002)
7. K. Naugolnykh, L Ostrovsky, *Nonlinear wave processes in Acoustics.* Cambridge University Press, 1988, 298 p

Experimental studies of Seismoelectric Effects in borehole models

Zhenya Zhu and M. Nafi Toksöz

Earth Resources Laboratory, MIT

Abstract. In a fluid-saturated porous formation, an impinging seismic wave induces fluid motion. The motion of fluid relative to the rock frame generates an electric streaming current. This current produces electric and magnetic fields, which are called seismoelectric and seismomagnetic fields, respectively. When there is a fracture or a discontinuity, a radiating electromagnetic wave is also generated, in addition to local fields. Seismoelectric and seismomagnetic fields depend on the amplitude, frequency, and mode of the seismic wave, as well as the formation porosity, permeability, pore size, and fluid conductivity. In this paper, we describe laboratory results of seismoelectric and seismomagnetic fields induced by an acoustic source in borehole models. We use a piezoelectric source for acoustic waves and a point electrode and a high-sensitivity Hall-effect transducer for measuring the localized seismoelectric and seismomagnetic fields in fluid-saturated rocks. The dependence of seismoelectric conversions on porosity, permeability and fluid conductivity are investigated. Three components of the seismomagnetic field are measured by the Hall-effect transducer. At a horizontal fracture, the acoustic wave induces a radiating electromagnetic wave. Seismoelectric and seismomagnetic well logging might be a new means to determine formation properties in a borehole.

Keywords: Electrokinetic Conversions, Seismoelectric Measurements, Borehole Models.
PACS: 43.58.+z, 91.60.Ba, 91.60.Pn, 91.90.+p.

INTRODUCTION

When a porous material is saturated with a fluid electrolyte, an electric double layer (EDL) is formed at the boundary between solid and fluid. An acoustic wave propagating in the material induces fluid motion, which generates an electric current. This current produces electric and magnetic fields, which are called seismoelectric and seismomagnetic fields, respectively. Theoretical studies [1, 2] confirm the mechanism of the conversion. Inside a homogeneous, porous medium, the seismic wave induces localized seismoelectric and seismomagnetic fields. At an interface, the acoustic wave induces a radiating electromagnetic (EM) wave. Laboratory experiments [3, 4] measured the seismoelectric fields induced by acoustic waves in scaled models. Field experiments [5, 6] measured seismoelectric signals on the ground. Seismoelectric borehole logging [7, 8] indicates a strong relationship between a seismoelectric response and a fracture.

In this paper, we demonstrate the seismoelectric phenomena with a set of laboratory experiments. The particular geometry are use is related to borehole measurements (e. g. acoustic/electric logging) in the earth. Laboratory models are scaled down using the acoustic wavelength scaling. Borehole models are made to simulate a layered earth,

CP838, *Innovations in Nonlinear Acoustics: 17th International Symposium on Nonlinear Acoustics*,
edited by A. A. Atchley, V. W. Sparrow, and R. M. Keolian
© 2006 American Institute of Physics 0-7354-0330-9/06/$23.00

and boreholes with horizontal and/or vertical fractures. An acoustic transducer, an electrode, and a Hall-effect sensor are applied to record the acoustic wave, electric and magnetic fields induced by an acoustic wave.

ELECTROKINETIC CONVERSIONS

At a solid-fluid interface of a porous medium or a fluid-saturated fracture where a fluid electrolyte comes into contact with a solid surface, anions from the electrolyte are chemically adsorbed to the solid rock leaving behind a net excess of cations distributed near the wall [9]. This region is known as the electric double layer (EDL). When an impinging seismic wave propagating in a fluid-saturated porous formation induces fluid motion, the fluid motion relative to the rock frame generates an electric streaming current. This current produces electric and magnetic fields, which are called seismoelectric and seismomagnetic fields, respectively. The fields induced inside a homogeneous medium is a localized field, which exists only in the area disturbed by the acoustic wave. At an interface between media with different properties, such as porosity, permeability, conductivity, or different lithology, the acoustic wave induces a radiating electromagnetic wave, which propagates with EM wave speed and can be received anywhere. The relationship between the acoustic wave amplitude and the amplitude of the seismoelectric field is a linear one.

Seismoelectric conversion depends on the electrolyte conductivity when the conductivity is low. When the EDL is saturated, the electric field amplitude decreases when the conductivity increases. Because the seismomagnetic field only depends on the movable charges in the fluid, the seismomagnetic amplitude increases when the conductivity increases. The seismomagnetic field is a vector field, we may measure its three components with different positions of the Hall-effect sensor. In thee reciprocal case, when an oscillating electric field is applied to a fluid-saturated porous medium, the movement of the fluid with mobile charges generates an acoustic wave. This phenomenon is referred to as electroseismic conversion.

LABORATORY EXPERIMENTS

To study the seismoelectric effects in the laboratory, three physical borehole models, a layered borehole, a borehole with horizontal fracture, and a borehole with horizontal/vertical fractures were made with natural rocks and Lucite. The layered borehole model was made of two materials (slate and Lucite) with a horizontal interface, but without a fracture between the layers. The second borehole model was made of Lucite and slate blocks with a horizontal fracture of 0.5 mm aperture. The third model has the same horizontal fracture and a vertical fracture across the borehole in the Lucite section. The diameter of the boreholes is about 10 mm.

The acoustic source is a cylinderical PZT transducer of 9 mm in diameter. A square pulse of 750 V amplitude and 10 μs width excites the source. Three kinds of receivers, acoustic transducer, point electrode, and Hall-effect device, were used to record the acoustic, electric, and magnetic fields in the borehole. The Hall-effect device used in our experiments is a magnetic sensor whose output is proportional to the magnetic flux density and the direction of the magnetic field. This device does not respond to the magnetic component of any electromagnetic wave. The models are placed in a tank with water of 65 µS/cm conductivity.

When the acoustic source is fixed in boreholes, receivers move gradually in the borehole and record the acoustic, electric, and magnetic signals in the three borehole models.

In a layered slate-sandstone borehole, we recorded the acoustic and electric signals generated by a monopole acoustic source. Figure 1 shows the borehole model (a), recorded acoustic(b) and electric(c) signals, and electric amplitude normalized by the acoustic amplitude (d), the propagation velocities of the acoustic wave and the the electric signals are the same, confirming that the acoustic wave induces the localized electric field. The amplitudes of the electric signals are directly proportional to the acoustic amplitude The electric(E)/acoustic(P) ratio depends on the porosity, permeability, conductivity, fluid mobility, and conductivity in a given rock. This is demonstrated in Fig. 1(d).

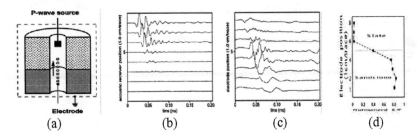

FIGURE 1. A borehole model (a) with slate and sandstone layers, acoustic waveforms (b), electric signals (c), and electric amplitude normalized by acoustic amplitude in the borehole.

Figure 2 shows the borehole model (a) with a horizontal fracture, recorded acoustic (b), electric (c), and magnetic (d) signals. The Stoneley wave in the slate section (traces 1-5 in Fig. 1c) induces electric signals, whose apparent velocity is the same as the Stoneley wave.

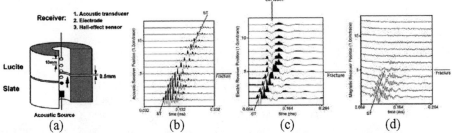

FIGURE 2. A borehole model (a) with a horizontal fracture between slate and Lucite, acoustic waveforms (b), electric signals (c), and magnetic signals (d) recived in the borehole. Lines "ST" indicate the propagation of the Stoneley waves.

At the fracture, the Stoneley wave induces an EM wave, whose velocity is that of an electromagnetic wave in the borehole (traces 6-12 in Fig. 1c). The Hall-effect device records the horizontal component of the magnetic field induced by the Stoneley wave in the slate section. The Hall-effect device does not record the magnetic component of an electromagnetic wave, as shown in Fig. 2(d)

The results confirm that acoustic waves induce stationary or localized electric and magnetic fields in a porous formation, and induce a radiating electromagnetic wave at a horizontal fracture due to its discontinuity.

Additional measurements have been made in models that contain both horizontal and vertical fractures.

CONCLUSIONS

An electric double layer is formed at the interface between rock and water. When an acoustic wave propagates in a porous medium, it induces fluid motion. The moving charges induce electric and magnetic fields. The acoustic wave in a homogeneous borehole induces localized seismoelectric and seismomagnetic fields. At a horizontal fracture, the acoustic wave induces a radiating electromagnetic wave due to the discontinuity of the borehole.

Seismoelectric conversion depends on the formation properties, such as porosity, permeability, lithology, and fluid conductivity and mobility.

Measured acoustic, electric and magnetic fields provide information about the porous medium. The seismoelectric and seismomagnetic measurements may be a new logging technique for subsurface characterization.

ACKNOWLEDGMENTS

This work was supported by the Borehole and Acoustic Logging Consortium and the Founding Members of the Earth Resources Laboratory at Massachusetts Institute of Technology.

REFERENCES

1. Haartsen, M. W., Ph. D. Thesis, MIT, 1995.
2. Pride, S. R., and Haartsen, M. W., *JASA*, **100**, 1301-1315, 1996.
3. Morgan, F. D., Williams, E. R., and Madden, T. R., *JGR*, **94**, 12449-12461, 1989.
4. Zhu, Z., Haartsen, M. W., Toksöz, M. N., *JGR*, **105**, 28005-28064, 2000.
5. Thompson, A. H., and Gist, G. A., *The Leading Edge*, **12**, 1169-1173, 1993.
6. Butler, K., Russell, R., Kepic, A., and Maxwell, M., *Geophysics*, **61**, 1769-1778, 1996.
7. Mikhailov, O. V., Queen, J., and Toksöz, M. N., *Geophysics*, **65**, 1098-1112, 2000.
8. Hunt, C. W., and Worthington, M. H., *Geophys. Res. Letters*, **27**, 1315-1318, 2000.
9. Reppert, P. M., and Morgan, F. D., *J. of Colloid and Interface Scie.*, **154**, 372-383, 2002.

Theoretical and Numerical Studies of Seismoelectric Conversions in Boreholes

Shihong Chi, M. Nafi Toksöz, and Xin Zhan

Earth Resources Laboratory, Massachusetts Institute of Technology
42 Carleton St., Cambridge, MA 02142, USA

Abstract. Elastic wave propagation in fluid-saturated porous media generates electric and magnetic fields due to electrokinetic effects. The magnitudes of converted fields strongly depend on formation properties, such as permeability, porosity, and fluid properties, such as conductivity. To explore the possibility of using seismoelectric measurement in evaluating the formation properties, we conduct theoretical and numerical studies of electrokinetic conversions in fluid-filled boreholes. First, we derive the mathematical formulations for seismoelectric responses for an acoustic source in a borehole. Then we compute the electric field in boreholes penetrating formations with different rock compressibility, permeability, and porosity. We analyze the sensitivity of the converted electric fields to formation permeability and porosity. We find that the ratio of magnitude of electromagnetic waves to that of acoustic pressure increases with increasing porosity and permeability in formations with high and low seismic velocities.

Keywords: seismoelectric, dipole, borehole, permeability, porosity, Stoneley mode
PACS: 43.20.Gp

INTRODUCTION

When a seismic wave propagates in a fluid-saturated porous medium, it causes the fluid to move relatively to the solid frame. This motion generates electromagnetic (EM) field. The electric double layer near the surface of the solid provides the fundamental mechanism of seismoelectric conversion (Pride and Morgan, 1991). Pride (1994) derived the coupled macroscopic elastic and EM equations. Zhu et al. (1999) and Mikhailov et al. (2000) extended the application of seismoelectric phenomena to borehole logging. Hu et al (2002) simulated the electric waveforms using the Pride equations. Markov and Verzhbitskiy (2004) simulated EM fields induced by acoustic multipole source in a borehole. In this paper, we model the EM fields due to acoustic sources in a borehole, and analyze the sensitivities of the electric signals to formation properties.

THEORETICAL AND NUMERICAL STUDIES

The conversion from seismic to electric fields depends strongly on formation properties, such as permeability, porosity, and fluid properties, such as conductivity.

CP838, *Innovations in Nonlinear Acoustics: 17th International Symposium on Nonlinear Acoustics*,
edited by A. A. Atchley, V. W. Sparrow, and R. M. Keolian
© 2006 American Institute of Physics 0-7354-0330-9/06/$23.00

To explore the possibility of using multipole seismoelectric measurement in evaluating the formation permeability and porosity, we conduct theoretical and numerical studies of electrokinetic conversions in fluid-filled boreholes.

Mathematical Formulation Of The Multipole Seismoelectric Field

Using to Pride's equations (1994) for seismoelectric wave propagation in porous media, the electric current density $\overline{J}$ can be written as

$$\overline{J} = \sigma\overline{E} + L\left(-\nabla p + \omega^2 \rho_f \overline{u}\right),\tag{1}$$

and the displacement of the fluid phase $\overline{w}$ can be expressed as

$$-i\omega\overline{w} = L\overline{E} + \frac{k}{\eta}\left(-\nabla p + \omega^2 \rho_f \overline{u}\right),\tag{2}$$

where $\overline{u}$ is the displacement of the solid frame, p is the pore pressure, $\overline{E}$ is the electric field strength, L is the coupling coefficient, ρ_f and η are the density and the viscosity of the pore fluid, respectively, k and σ are the dynamic permeability and conductivity of the porous medium, and ω is the angular frequency.

Hu and Liu (2002) introduced two assumptions to approximate the seismoelectric wave fields. They first showed that the converted electric field affects the elastic field negligibly and the coupling term in equation (2) can be ignored. They also assumed that the electric field is time invariant within the acoustic logging operation framework, because the EM wavelength is much longer than the tool length. Under this quasi-static condition, the electric field can be written as the gradient of an electric potential

$$\overline{E} = -\nabla\phi .\tag{3}$$

They showed that the electric potential and the potential of the gradient field of the solid displacement related as follows:

$$\nabla^2\phi = \frac{L}{\sigma}\left(-\nabla^2 p + \omega^2 \rho_f \nabla^2\varphi\right).\tag{4}$$

In wave number domain, the solution to Equation (4) is

$$\phi_n = A_{em} K_n\left(k_{em}r\right)\cos n\theta + \frac{L}{\sigma}\left(-p + \omega^2 \rho_f \varphi\right),\tag{5}$$

where A_{em} is an unknown coefficient, k_{em} is the axial wavenumber of the EM wave. The pore pressure can be written as:

$$p = \sum_{j=1}^{2}\left[\left(Q + \tilde{R}\xi_j\right)l_j^2/\phi_0\right]A_j I_n\left(k_{pf}r_0\right)K_n\left(k_{pj}r\right)\cos n\theta ,\tag{6}$$

where Q, $\tilde{R}$, and ξ_j are defined by Biot (1956a), wavenumber $l_j^2 = \dfrac{\omega^2}{\alpha_j^2}$, ϕ_0 is formation porosity, and A_j is a unknown coefficient, and r_0 is the radius of the circle of monopole point source distribution.

The potential function φ_j can be written as

$$\varphi_j = A_j I_n\left(k_{pf}r_0\right)K_n\left(k_{pj}r\right)\cos n\theta .\tag{7}$$

The potential function φ_j is a solution for a multipole source of order 2n of solid or pore fluid.

We obtain the electric field using equation (3). In formation,

$$E_z = -ik_{em}\phi_n , \qquad (8)$$

and J_r can be derived from equation (1).

In borehole fluid, we assume the electric potential to be

$$\phi_f = -ik_{em}B_{em}I_n\left(k_{em}r\right)I_n\left(k_{pj}r_0\right)\cos n\theta \qquad (9)$$

where coefficient B_{em} is to be determined. The electric current density in borehole can also be derived from equation (1).

Boundary Conditions At The Borehole Wall

Across the borehole wall, the tangential electric field and normal magnetic field are continuous. This boundary condition is equivalent to the electric potential and radial current continuity. Then we obtain a set of linear equations and can solve for unknown coefficients A_{em} and B_{em} . Finally, we can compute the electric fields in formation and borehole fluid.

Numerical Studies Of The Multipole Seismoelectric Field

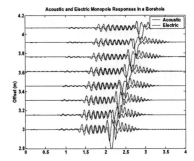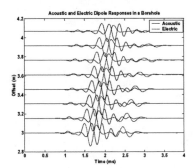

FIGURE 1. Comparison of monopole and dipole responses in borehole. For each source, the acoustic and electric signals are plotted on top of each other. They are 90 degree out of phase.

We compute the electric field in boreholes in formations with high and low rock compressibility, permeability, and porosity. Figure 1 compares the monopole and dipole wave fields. We see Stoneley wave mode has the highest amplitude in acoustic and EM wave fields. Figure 2 shows the EM wave conversion rate for the high permeability rock is about 25 times higher than that for the low permeability rock. The conversion rate increases about linearly with the logarithm of permeability. We also model porosity and compressibility effects on seismoelectric conversion.

CONCLUSION

We find that seismoelectric conversion rate increases with increasing porosity and permeability in formations with high and low seismic velocities. This observation can be used to evaluate formation properties.

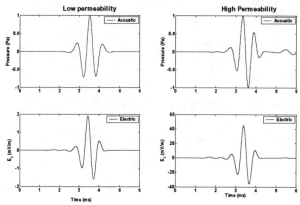

FIGURE 2. Comparison of dipole responses in low and high permeability rocks. Note the difference between the scale of the electric fields.

ACKNOWLEDGMENTS

This work is supported by the Earth Resources Laboratory Borehole and Acoustic Logging Consortium and the Founding Members of the Earth Resources Laboratory.

REFERENCES

1. Pride, S. R., 1994, Governing equations for the coupled electromagnetics and acoustics of porous media: *Physical Review* B, 50, 15678–15696.
2. Mikhailov, O., Haartsen, M. W., and Toksoz, M. N., 1997, Electroseismic investigation of the shallow subsurface: Field measurements and numerical modeling: *Geophysics*, 62, 97–105.
3. Zhu, Z., Haartsenz, M. W., and Toksoz, M. N., Experimental studies of electrokinetic conversions in fluid-saturated borehole models: *Geophysics*, 64, 1349–1356.
4. Hu, H., Wang, K., and Wang, J., 2000, Simulation of acoustically induced electromagnetic field in a borehole embedded in a porous formation: *Borehole Acoustic and Logging and Reservoir Delineation Consortia Annual Report,*, Earth Resources Laboratory, MIT.
5. Pride, S. R., and Morgan, F., 1991, Electrokinetic dissipation induced by seismic waves: *Geophysics*, 56, 914-925.
6. Hu, H., and Liu, J., 2002, Simulation of converted electric field during acoustoeletric logging: *SEG Intl. Exposition and 72nd Annual Meeting.*
7. Biot, M. A., 1956a, Theory of propagation of elastic waves in a fluid saturated porous rock. I. low frequency range: J. Acoust. Soc. Am., 28, 179-191.
8. Markov, M. G., 2004, Simulation of the electroseismic effect produced by an acoustic multipole source in a fluid-filled borehole: SPWLA 45th Annual Logging Symposium.

SECTION 4

HIGH INTENSITY FOCUSED ULTRASOUND
IN MEDICINE AND BIOLOGY

Acoustic field modeling in therapeutic ultrasound

T. Douglas Mast*, Waseem Faidi[†] and Inder Raj S. Makin[†]

*Department of Biomedical Engineering, University of Cincinnati, Cincinnati, Ohio 45267-0586
[†]Ethicon Endo-Surgery, 4545 Creek Rd., Cincinnati, Ohio 45242

Abstract. Understanding of ultrasound-tissue interaction is important for realization of clinically useful therapeutic ultrasound methods and devices. Linear acoustic propagation in homogeneous media, including diffraction and absorption effects, provides a useful first approximation but fails to accurately model many problems of interest. Depending on the therapy regime, other important effects can include finite-amplitude propagation, cavitation and other gas activity, inhomogeneous tissue structure, temperature-dependent tissue properties, and irreversible tissue modification. For bulk ablation of soft tissue using ultrasound, prediction of therapeutic effects requires accurate knowledge of space- and time-dependent heat deposition from acoustic absorption. A primary factor affecting heat deposition is local heat loss due to blood flow, both from bulk perfusion and large vessels. Gas activity due to boiling and tissue property changes due to local ablation, both of which markedly affect treatment, can be approximated by appropriate modification of the initial heat deposition pattern. Acoustically inhomogeneous tissue structure, even in nominally homogeneous organs such as the liver, can modify heating patterns enough to change treatment outcomes. These issues are illustrated by simulations of ultrasound therapy and comparison with *in vivo* and *in vitro* ultrasound ablation experiments.

Keywords: HIFU, intense ultrasound, ablation therapy, bio-heat transfer, numerical simulation.
PACS: 43.80.Sh, 43.58.Ta, 43.25.Jh

INTRODUCTION

Ablation of soft tissue using ultrasound has been investigated for a number of years [1] and is an active area of current engineering and medical research [2]. This paper reviews recent work on simulation of thermal therapy using numerical modeling of ultrasound propagation. Emphasis is given to modeling of bulk ablation using intense ultrasound beams that are unfocused or weakly focused; this approach has been developed for interstitial treatment of focal tumors [3]–[6].

Previous numerical studies of ultrasound ablation have provided insight into the dynamics of lesion formation in high-intensity focused ultrasound (HIFU) treatments, including the effects of blood flow cooling [7, 8], acoustic nonlinearity [8]–[12], cavitation [9], and thermoacoustic lensing effects [11, 12]. Modeling of interstitial ablation has also been performed, using numerical solution of the bioheat-transfer equation [13] with various methods for simulation of the ultrasound-induced heat deposition [4, 5, 14, 15]. Effects considered in these simulations have included loss of tissue perfusion and changes in ultrasound absorption due to thermal coagulation [5, 14, 15].

Here, methods for simulation of ultrasound therapy are briefly reviewed, and modeling of several propagation phenomena relevant to bulk ablation is described. For the therapy regime considered here, primary mechanisms altering ablation results include inhomogeneous cooling due to blood flow, distortion of therapy beams due to acousti-

CP838, *Innovations in Nonlinear Acoustics: 17th International Symposium on Nonlinear Acoustics*,
edited by A. A. Atchley, V. W. Sparrow, and R. M. Keolian
© 2006 American Institute of Physics 0-7354-0330-9/06/$23.00

cally inhomogeneous tissue, and alteration of heat deposition due to tissue boiling, while finite-amplitude propagation and inertial cavitation are identified as secondary effects.

INTENSE ULTRASOUND ABLATION

Background

An approach for interstitial intense ultrasound ablation has been described elsewhere [6, 16]. In this concept, an ultrasound probe capable of both B-scan imaging and intense ultrasound ablation is inserted into target tissue, using a percutaneous, laparoscopic, or open-surgery approach. The probe is then used to image the target tissue, allowing user planning of an ablation procedure, and thermal ablation of a tissue volume is performed using electronic scanning and mechanical probe rotation.

A probe configuration suggested to be suitable for such an interstitial approach is an ultrasound array operating at a frequency 3.1 MHz, capable of acoustic energy densities > 80 W/cm^2 at the the probe surface, with active dimensions of about 2.3×49 mm^2 [6]. Construction of such arrays is feasible using an acoustic stack design that provides both high power capabilities for ablation and broad bandwidth for imaging [16]. Imaging and therapy performance for arrays of this type has been described recently [6].

Numerical modeling in homogeneous media

Simulation of ultrasound ablation requires a numerical model for propagating ultrasound, which acts as a heat source due to absorption. The Pennes bio-heat transfer equation, which represents heat diffusion with an added term for perfusion losses [13], can then be solved numerically, typically by finite-difference methods [15].

Typically, cumulative thermal damage to tissue is measured using the thermal dose [17], which is defined in units of equivalent minutes at 43°C. Typically assumed tissue damage thresholds, based on the thermal dose, are on the order of EM$_{43} = 200$ min for cell death and EM$_{43} = 10^7$ min for complete protein denaturation and severe coagulative necrosis [15]. These values correspond to measurable transitions in the ultrasonic absorption of soft tissue [18].

For absorption caused by relaxation processes, the rate of heat deposition per unit volume is $Q = \alpha |p|^2/(\rho c)$ [19], where α is the acoustic absorption in nepers per unit length, $|p|$ is the pressure amplitude of a time-harmonic acoustic field, and c is the speed of sound. Acoustic fields induced by ultrasound transducers can be simulated in a number of manners suitable for efficient numerical computations, including the Fresnel approximation for the fields of rectangular elements [6, 20], exact series solutions for disk radiators in unfocused [21, 22] and focused [23] configurations, and the KZK equation for approximations of finite-amplitude propagation [24, 25]. Acoustic absorption effects can be incorporated directly into any of these models, or can be applied *post hoc* in an approximate manner. The approximate approach to absorption modeling is convenient for representing changes in heat deposition due to tissue modification without further computation of the full acoustic field [15].

FIGURE 1. Lesion from an *in vivo* interstitial, rotationally scanned exposure, showing cooling effects caused by a large blood vessel. Left: cross section approximately along the probe track (thermally coagulated tissue is indicated by its lighter shade). Right: surface rendering of reconstructed lesion.

PHENOMENA AFFECTING ULTRASOUND ABLATION

Although numerical models of ultrasound ablation employing homogeneous media provide considerable insight, their results often disagree with real therapy results. In particular, such models fail to predict the significant variability of real ablation results *in vivo*. Below, several phenomena causing such discrepancies are discussed, with attention to methods for modeling these effects.

Inhomogeneous cooling

One of the most important factor affecting ultrasound ablation is the cooling caused by blood flow. The standard Pennes bio-heat equation, which incorporates cooling by a bulk perfusion term, is sufficient to explain many differences between ablation effects in excised tissue (*in vitro*) and in living systems (*in vivo*). However, cooling effects caused by large vessels can significantly alter local ablation results, potentially causing tissue near the blood vessels to remain untreated.

An example of this effect is illustrated in Fig. 1, which illustrates a thermal lesion induced in porcine liver *in vivo* by a scanned interstitial treatment. The treatment employed a 3.1 MHz, 3 mm diameter (2.3×49 mm^2 active surface) array firing with an acoustic power of 66.7 W (source intensity of 59.5 W/cm^2) and an 80% duty cycle, for four 1.5 min treatments at adjacent angles separated by 15°. In the vicinity of a large blood vessel, tissue ablation is substantially reduced due to local cooling effects. Numerical modeling of ablation in the presence of large blood vessels can be performed using appropriate boundary conditions for heat transfer at the vessel walls [8].

Acoustic nonlinearity and inertial cavitation

For the intense ultrasound beams employed in bulk ablation, finite-amplitude propagation has a nonzero effect, due to both waveform steepening (generation of higher

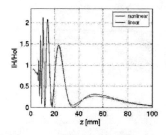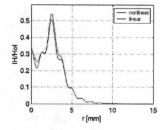

FIGURE 2. Estimated effects of acoustic nonlinearity on ultrasound heat deposition for a 12 mm, 3 MHz disk radiating at 1.5 MPa into liver tissue. The plots show the squared pressure amplitude for a linear computation together with the squared amplitude (summed over all computed harmonics) for a fully nonlinear solution of the KZK equation. Left: squared amplitude along the transducer axis. Right: squared amplitude as a function of azimuth at a distance of 30 mm from the transducer face.

harmonics) and inertial cavitation. Previous models have illustrated the effects of acoustic nonlinearity [8]–[12] and cavitation [9] on HIFU treatments.

For the ultrasound bulk ablation configuration described above, a typical acoustic pressure amplitude is about 1.5 MPa at the transducer surface, corresponding to an energy density of about 70 W/cm^2. An example computation illustrating the effects of acoustic nonlinearity on heat deposition employs a numerical solution to the KZK equation [24, 25] for a disk of diameter 12 mm (area similar to a rectangular aperture of dimensions 2.3×49 mm^2) radiating at 3 MHz with an amplitude of 1.5 MPa into a medium with acoustic properties comparable to human liver ($c = 1540$ m/s, $\rho = 1060$ kg/m^3, $\alpha = 5.75$ Np/cm/MHz$^{1.2}$, and nonlinearity coefficient $\beta = 4.7$). For these parameters, the resulting squared pressure amplitude (proportional to heat deposition) is illustrated in Fig. 2. A fully nonlinear computation is shown to cause heat deposition that varies only slightly from the linear case. Thus, in this case acoustic nonlinearity does not have a great effect on ablation results, compared to other phenomena described here.

The importance of inertial cavitation can be assessed using the mechanical index (MI) [26], defined as the maximum rarefaction pressure in MPa divided by the square root of the frequency in MHz. For an amplitude of 1.5 MPa and a frequency of 3 MHz, the MI is 0.866, and this value decreases further with depth, due to acoustic absorption. As this value is substantially less than the maximum value of 1.9 generally recognized as safe for diagnostic ultrasound, it may be assumed that inertial cavitation is a relatively minor effect in bulk ablation using unfocused intense ultrasound beams.

Inhomogeneous propagation

Most simulations of ultrasound propagation for therapy modeling have assumed tissue structure either to be uniform [8, 15] or composed of uniform layers [11]. However, the inhomogeneous structure of real tissue, which causes distortion of ultrasound beams known to cause significant aberration in ultrasound imaging [29], is equally relevant to ultrasound therapy.

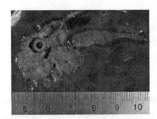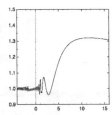

FIGURE 3. Acoustic effects of a large blood vessel on interstitial bulk ablation. Left: cross section of a thermal lesion created *in vivo* with a rotationally scanned 3 mm, 3.1 MHz array, showing anomalously large ablation distal to the vessel. Center: total squared pressure amplitude, relative to the plane wave amplitude, for a 3.1 MHz plane wave scattered by a 3 mm blood-mimicking cylinder in a liver-mimicking background, plotted on a linear gray scale. Right: squared pressure along the axis of symmetry for the same simulation.

In a common noninvasive HIFU configuration, a highly focused ultrasound beam propagates through the abdominal wall to ablate tissue in the abdominal cavity, such as a liver tumor [27, 28]. Measurements [29] and simulations [30] have indicated that propagation through the abdominal wall can cause random fluctuations on the order of 3 dB (rms) in the transmitted amplitude, due both to scattering and inhomogeneous absorption. A 3 dB change in amplitude, which corresponds to a factor of two in heat deposition, causes the time required for local tissue ablation to be correspondingly changed by a factor of two, large enough to significantly change ablation results for a given treatment plan.

In nominally homogeneous organs such as the liver, inhomogeneous tissue structure can still affect ablation results. An example from an interstitial ablation experiment performed *in vivo* in porcine liver is shown in Fig. 3. In this case, anomalously severe and deep ablation occurred distal to a large blood vessel. A plausible explanation for this result is acoustic focusing caused by the difference in sound speed between blood and liver parenchyma. To illustrate this explanation, a computation performed using the exact solution for scattering from a fluid cylinder [31] is also shown in Fig. 3. In this computation, the liver background was modeled as a fluid with sound speed 1595 m/s and the blood vessel was modeled as a cylinder with sound speed 1580 m/s. The results show substantial focusing resulting in over a 50% increase in peak heat deposition, consistent with the increased lesioning shown.

More general computations of propagation in inhomogeneous tissue for therapy modeling can be performed using three-dimensional tissue models, which can be established from image data such as cross-sectional photography or volumetric CT scans [32, 33]. For large-scale three-dimensional computations, such as those required for simulation of ultrasound therapy, a particularly useful approach is the *k*-space method [34, 35], which provides accurate results for relatively coarse spatial and temporal discretization. This method has been extended to incorporate perfectly-matched-layer absorbing boundary conditions and tissue absorption caused by relaxation processes [36]. Recent applications have included 3D simulations of propagation in breast tissue [33], simulations of focus distortion in large-scale 2D tissue models [37], and computations of propagation in a 3D prostate tissue model for simulation of ultrasonic hyperthermia [38].

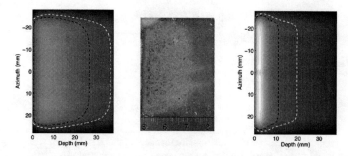

FIGURE 4. Effect of boiling and thermal dose dependent attenuation. Left: simulation without boiling and dose-dependent attenuation modeling. The simulated thermal dose is shown on a logarithmic gray scale with superimposed contours at EM_{43}=200 and 10^7 equivalent minutes. Center: *In vitro* lesion. Right: simulated thermal dose with boiling and dose-dependent attenuation modeling.

Tissue modification

Particularly severe effects on ultrasound ablation are caused by changes to tissue acoustic properties that result from thermal coagulation. These include increases in tissue absorption that have been quantified by measurements [18] and gas activity due to tissue boiling, which has been observed to alter positions and extents of HIFU lesions [39, 40]. Recent numerical models for bulk ultrasound ablation have incorporated both of these effects.

Changes in tissue absorption associated with ablation can be incorporated based on the thermal dose, in correspondence with measurements [18]. Some models have assigned a single increased absorption value to all tissue locations that have surpassed a thermal dose threshold [5]. In order to better match measurements, other models have incorporated a linear rise in absorption above a thermal dose threshold on the order of $EM_{43} = 200$ min, with a maximum set by a second thermal dose threshold on the order of $EM_{43} = 10^7$ min [14, 15].

The effects of tissue boiling, which include both strong shadowing and increased local absorption, can limit the achievable depth of ultrasound ablation. This effect has been modeled for single HIFU lesions by assuming that, after any grid point (usually near the acoustic focus) reached a temperature of 100°C, all of the thermal energy originally deposited in the distal half-space was redistributed in a 0.5 cm spherical region centered around the initial location of tissue boiling [10]. For simulation of bulk tissue ablation, in which boiling effects are not necessarily localized at a focal point, this idea has been extended to more general heat distributions [15]. In the more general approach, the thermal energy deposited within the distal region projected (shadowed) by all boiling points is equally distributed around each of the proximal grid points where the temperature has exceeded 100°C.

These tissue modification effects can substantially effect ultrasound bulk ablation, as illustrated by Fig. 4. The lesion shown was created *in vitro* in porcine liver tissue by a 3 mm, 3.1 MHz array firing for 3 min with a surface power density of 39 W/cm^2 and an 80% duty cycle. The simulated thermal dose maps shown were obtained by identical methods except for consideration of boiling and thermal dose-dependent attenuation [15]. The simulation incorporating these tissue modification effects agrees much better with the experimental result for the depth, overall shape, and rate of thermal ablation.

CONCLUSIONS

Modeling of therapeutic ultrasound requires consideration of an array of potential complicating effects. For interstitial bulk ablation, the most important effects include the thermal and acoustic effects of natural tissue inhomogeneities as well as tissue changes caused by the therapy itself. Other therapy regimes can require different considerations for accurate modeling, such as explicit treatment of finite-amplitude propagation and inertial cavitation. For any form of ultrasound therapy, it is likely that natural variations of tissue properties and structure will limit the predictive accuracy of simulations. Thus, methods for accurate monitoring and assessment of ultrasound ablation effects could increase the clinical viability of therapeutic ultrasound.

REFERENCES

1. W. J. Fry, "Biological and medical acoustics," J. Acoust. Soc. Am. **30**, 387–393 (1958).
2. J. E. Kennedy, G. R. ter Haar, and D. Cranston, "High intensity focused ultrasound: surgery of the future?" Br. J. Radiol. **76**, 590–599 (2003).
3. C. J. Diederich, W. H. Nau, and P. R. Stauffer, "Ultrasound applicators for interstitial thermal coagulation," IEEE Trans. Ultras., Ferroelect., Freq. Contr. **46**, 1218–1228 (1999).
4. R. Chopra, M. J. Bronskill, and F. S. Foster, "Feasibility of linear arrays for interstitial ultrasound thermal therapy," Med. Phys. **27**, 1281–1286 (2000).
5. C. Lafon, F. Prat, J. Y. Chapelon, F. Gorry, J. Margonari, Y. Theillere, and D. Cathignol, "Cylindrical thermal coagulation necrosis using an interstitial applicator with a plane ultrasonic transducer: *in vitro* and *in vivo* experiments versus computer simulations," Int. J. Hyperthermia **16**:508–522 (2000).
6. I. R. S. Makin, T. D. Mast, W. Faidi, M. M. Runk, P. G. Barthe, and M. H. Slayton, "Miniaturized ultrasound arrays for interstitial ablation and imaging," Ultras. Med. Biol., to appear (2005).
7. M. C. Kolios, M. D. Sherar, and J. W. Hunt, "Blood flow cooling and ultrasonic lesion formation," Med. Phys. **23**, 1287–1298 (1996).
8. F. P. Curra, P. D. Mourad, V. A. Khoklova, R. O. Cleveland, and L. A. Crum, "Numerical simulations of heating patterns and tissue temperature response due to high-intensity focused ultrasound," IEEE Trans. Ultras., Ferroelect., Freq. Contr. **47**, 1077–1089 (1999).
9. F. Chavrier, J. Y. Chapelon, A. Gelet, and D. Cathignol, "Modeling of high-intensity focused ultrasound-induced lesions in the presence of cavitation bubbles," J. Acoust. Soc. Am. **108**, 432–440 (2000).
10. P. M. Meaney, M. D. Cahill, and G. R. ter Haar, "The intensity dependence of lesion position shift during focused ultrasound surgery," Ultras. Med. Biol. **26**, 441–450 (2000).
11. I. M. Hallaj, R. O. Cleveland, and K. Hynynen, "Simulations of the thermo-acoustic lens effect during focused ultrasound surgery," J. Acoust. Soc. Am. **109**, 2245–2253 (2001).
12. C. W. Connor and K. Hynynen, "Bio-acoustic thermal lensing and nonlinear propagation in focused ultrasound surgery using large focal spots: a parametric study," Phys. Med. Biol. **47**, 1911–1928 (2002).
13. H. H. Pennes, "Analysis of tissue and arterial blood temperatures in the resting human forearm," J. Appl. Physiol. **1**, 93–122 (1948).
14. P. D. Tyreus and C. J. Diederich, "Theoretical model of internally cooled interstitial ultrasound applicators for thermal therapy," Phys. Med. Biol. **47**:1073–1089 (2002).

15. T. D. Mast, I. R. S. Makin, W. Faidi, M. M. Runk, P. G. Barthe, and M. H. Slayton, "Bulk ablation of soft tissue with intense ultrasound: modeling and experiments," J. Acoust. Soc. Am., to appear (2005).

16. P. G. Barthe and M. H. Slayton, "Efficient wideband linear arrays for imaging and therapy," 1999 IEEE Ultrasonics Symposium Proceedings, Vol. 2, pp. 1249–1252.

17. S. A. Sapareto and W. C. Dewey, "Thermal dose determination in cancer therapy," Int. J. Radiat. Oncol. Biol. Phys. 10, 787–800 (1984).

18. C. A. Damianou, N. T. Sanghvi, F. J. Fry, and R. Maass-Moreno, "Dependence of ultrasonic attenuation and absorption in dog soft tissues on temperature and thermal dose," J. Acoust. Soc. Am. 102, 628–634 (1997).

19. W. L. Nyborg, "Heat generation by ultrasound in a relaxing medium," J. Acoust. Soc. Am. 70, 310–312 (1981).

20. A. Freedman, "Sound field of a rectangular piston," J. Acoust. Soc. Am. 32, 197–209 (1960).

21. T. Hasegawa, N. Inoue, and K. Matsuzawa, "Fresnel diffraction: Some extensions of the theory," J. Acoust. Soc. Am. 75, 1048–1051 (1984).

22. T. D. Mast and F. Yu, "Simplified expansions for radiation from a baffled circular piston," J. Acoust. Soc. Am., submitted (2005).

23. T. Hasegawa, N. Inoue, and K. Matsuzawa, "A new theory for the radiation from a concave piston source," J. Acoust. Soc. Am. 82, 706–708 (1987).

24. E. A. Zabolotskaya and R. V. Khokhlov, "Quasi-plane waves in the non-linear acoustics of confined beams," Sov. Phys. Acoust. 15, 35–40 (1969).

25. V. P. Kuznetsov, "Equation of nonlinear acoustics," Sov. Phys. Acoust. 16, 467–470 (1970).

26. R. E. Apfel and C. K. Holland, "Gauging the likelihood of cavitation from short-pulse, low duty cycle diagnostic ultrasound," Ultrasound Med Biol. 17, 179–185 (1991).

27. G. R. ter Haar, D. Sinnett, and I. Rivens, "High-intensity-focused ultrasound: a surgical technique for the treatment of discrete liver tumors," Phys. Med. Biol. 34, 1743–1750 (1989).

28. F. Wu, W.-Z. Chen, J. Bai, J.-Z. Zou, Z.-L. Wang, H. Zhu, and Z.-B. Wang, "Pathological changes in human malignant carcinoma treated with high-intensity focused ultrasound," Ultras. Med. Biol. 27, 1099–1106 (2001).

29. L. M. Hinkelman, T. D. Mast, L. A. Metlay, and R. C. Waag, "The effect of abdominal wall morphology on ultrasonic pulse distortion. Part I: Measurements." J. Acoust. Soc. Am. 104:3635–3649 (1998).

30. T. D. Mast, L. M. Hinkelman, M. J. Orr, V. W. Sparrow, and R. C. Waag, "Simulation of ultrasonic propagation through the abdominal wall," J. Acoust. Soc. Am. 102, 1177–1190 (1997).

31. P. M. Morse and K. U. Ingard, *Theoretical Acoustics* (McGraw-Hill, New York, 1968), Ch. 8.

32. J. L. Aroyan, "Three-dimensional modeling of hearing in *Delphinus delphis*," J. Acoust. Soc. Am. 110, 3305–3318 (2001).

33. T. D. Mast, "Two- and three-dimensional simulations of ultrasonic propagation through human breast tissue," Acoust. Res. Lett. Online 3, 53–58 (2002).

34. N. N. Bojarski, "The *k*-space formulation of the scattering problem in the time domain: an improved single propagator formulation," J. Acoust. Soc. Am. 77, 826–831 (1985).

35. T. D. Mast, L. P. Souriau, D.-L. Liu, M. Tabei, A. I. Nachman, and R. C. Waag, "A *k*-space method for large-scale models of wave propagation in tissue," IEEE Trans. Ultrason., Ferroelectr., Freq. Contr. 48, 341–354 (2001).

36. M. Tabei, T. D. Mast, and R. C. Waag, "A *k*-space method for coupled first-order acoustic propagation equations," J. Acoust. Soc. Am. 111, 53–63 (2002).

37. M. Tabei, T. D. Mast, and R. C. Waag, "Simulation of ultrasonic focus aberration through human tissue," J. Acoust. Soc. Am. 113, 1166–1176 (2003).

38. O. M. Al-Bataineh, N. B. Smith, R. M. Keolian, V. W. Sparrow, and L. E. Harpster, "Optimized hyperthermia treatment of prostate cancer using a novel intracavitary ultrasound array," J. Acoust. Soc. Am. 114, 2347 (2003).

39. N. A. Watkin, G. R. ter Haar, and I. Rivens, "The intensity dependence of the site of maximal energy deposition in focused ultrasound surgery," Ultras. Med. Biol. 22, 483–491 (1996).

40. W.-S. Chen, C. Lafon, T. J. Matula, S. Vaezy, and L. A. Crum, "Mechanisms of lesion formation in high intensity focused ultrasound therapy," Acoust. Res. Lett. Online 4, 41–46 (2003).

216

A Theoretical Study of Gas Bubble Dynamics in Tissue

Charles C. Church and Xinmai Yang

National Center for Physical Acoustics, University of Mississippi
University, Mississippi, 38677 USA

Abstract. The behavior of cavitation bubbles in tissue driven by ultrasonic fields is an important problem in biomedical acoustics. The present solution combines the Keller-Miksis equation for nonlinear bubble dynamics with the linear Voigt model for viscoelastic media and experimental values for the model's parameters (rigidity G=0, 0.5 – 2.5 MPa and viscosity μ=0.005 or 0.015 Pa·s). Two and 3-component models are used to study the oscillations of gas bubbles in tissue and in partially digested tissue. Numerical computations are performed for a variety of cases. Inertial cavitation thresholds (P_t) are determined for various equilibrium radii, frequencies and threshold criteria. Bubble-induced tissue displacement and strain is also investigated for several representative tissue types. It is found that: 1) thresholds in tissue are up to 10 times those in liquid, 2) P_t increases nearly linearly with frequency, 3) there is an optimal relation between the microbubble and tissue radii that maximizes the displacement of the tissue adjacent to the bubble and thus the likelihood of tissue damage.

Keywords: Inertial cavitation, viscoelasticity, cavitation threshold, tissue strain
PACS: 43.80.Gx, 43.35.Wa, 43.25.Yw

INTRODUCTION

Cavitation phenomena commonly occur in water under a variety of circumstances and conditions, and as a result, bubble dynamics in aqueous media have been studied for over 80 years. Bubble dynamics models are well established for water and other simple Newtonian fluids. These models have often been applied to the case of bubble activity in tissue, even though it is understood that they are not truly appropriate. Recently, this issue has become more important due to the development of the high intensity focused ultrasound (HIFU) for therapeutic medicine. HIFU will induce cavitation in soft tissue, and these microbubbles have a huge impact on the distribution of the ultrasound energy. In these situations, the surrounding media, i.e., biological tissues, often exhibit non-Newtonian, viscoelastic behavior. More accurate modeling will certainly enhance our understanding of the behavior of cavitation bubbles *in vivo*, and it may also provide a powerful tool to improve the quality of medical ultrasound.

Previous models for viscoelastic media are based on the Rayleigh-Plesset equation, which is less appropriate when the speed of the bubble wall becomes a significant fraction of the speed of sound in the fluid. In this work, a model is employed that is capable of accounting for the potentially large-amplitude oscillations of bubbles exposed to HIFU fields. In addition, the bubble model incorporates a viscoelastic

CP838, *Innovations in Nonlinear Acoustics: 17th International Symposium on Nonlinear Acoustics*,
edited by A. A. Atchley, V. W. Sparrow, and R. M. Keolian
© 2006 American Institute of Physics 0-7354-0330-9/06/$23.00

model consistent with measured tissue properties. Although data on the viscoelastic properties of soft tissue at megahertz frequencies are very limited, the linear Voigt model has proven appropriate for the tissues studied [1,2]. The Keller-Miksis equation [3] has been shown to be suitable for large amplitude bubble oscillations [4]. The new model combines the general form of the Keller-Miksis equation with the stress-strain relation of the linear Voigt model for viscoelastic solids to study the dynamics of bubbles in soft tissue [5].

THEORY AND METHOD

The Keller-Miksis equation for the dynamics of a spherical gas bubble in an incompressible, viscous liquid of infinite extent is

$$\left(1-\frac{\dot{R}}{c}\right)R\ddot{R}+\frac{3}{2}\left(1-\frac{\dot{R}}{3c}\right)\dot{R}^2=\left(1+\frac{\dot{R}}{c}\right)\frac{p_a-p_I}{\rho}+\frac{R}{\rho c}\frac{d}{dt}[p_a-p_I], \tag{1}$$

with

$$p_a - p_I = p_g - \frac{2\sigma}{R} - p_0 + P_A g(t) + 3\int_R^\infty \frac{\tau_{rr}}{r}dr, \tag{2}$$

where p_a is the pressure at the bubble surface, p_I is the pressure at infinity, R is the bubble radius, $\dot{R}$ is the bubble wall velocity, $\ddot{R}$ is the bubble wall acceleration, p_g is the gas pressure inside the bubble, σ is the surface tension, $P_A g(t)$ is the driving pressure, p_0 is the ambient pressure, τ_{rr} is the shear stress, r is the radial direction, c is the sound speed in the surrounding medium, and ρ is the density of surrounding medium. This equation can account for the compressibility of the surrounding medium to the first order, i.e., the validity of this equation is limited to small Mach numbers [4].

To introduce the effects of the surrounding tissue, it is assumed that in the near field, the tissue behaves as an incompressible, linear Voigt viscoelastic solid. In addition to allowing a straightforward mathematical analysis, this choice also creates the potential for comparing the resulting predictions with the experimental measurements in vivo. The stress-strain relation is therefore $\tau_{rr} = 2(G\gamma_{rr} + \mu\dot{\gamma}_{rr})$, where γ_{rr} is the strain, $\dot{\gamma}_{rr}$ is the strain rate with $\dot{\gamma}_{rr} = \partial u/\partial r$, u is the velocity and G is the shear modulus (or rigidity) [6]. In the near field (near the bubble surface), $u = \dot{R} R^2/r^2$, therefore, $\gamma_{rr} = -2(R^3 - R_0^3)/3r^3$ and $\dot{\gamma}_{rr} = -2\dot{R}R^2/r^3$. For many of the results presented here, the gas inside the bubble is assumed ideal, allowing the pressure to be estimated by use of a polytropic relation, $p_g = p_{g0}(R_0/R)^{3\kappa}$, where κ is the polytropic index. Expanding $(p_a - p_I)$,

$$p_a - p_I = p_g - \frac{2\sigma}{R} - p_0 + P_A g(t) + -\left[\frac{4G}{3R^3}(R^3 - R_0^3) + \frac{4\mu\dot{R}}{R}\right], \tag{3}$$

and

218

$$\frac{d}{dt}(p_a - p_I) = \frac{d}{dt}\left(p_g - \frac{2\sigma}{R} - p_0 + P_A g(t)\right) + \frac{d}{dt}\left[3\int_R^\infty \frac{\tau_{rr}}{r}dr\right]$$

$$= \left(\frac{2\sigma}{R} - 3\kappa p_g\right)\frac{\dot{R}}{R} - p_0 + P_A\frac{dg(t)}{dt} - 4G\frac{R_o^3\dot{R}}{R^4} - 4\mu\left(-\frac{\dot{R}^2}{R^2} + \frac{\ddot{R}}{R}\right) \tag{4}$$

where $P_A g(t) = P_A \sin(\omega t)$, and ω is the radial driving frequency. Equations (1), (3) and (4) provide the necessary formulation describing the dynamics of gas bubbles in soft (*i.e.*, viscoelastic) tissue.

Notice that this formulation is simply the usual Keller-Miksis approach with additional terms to account for the elasticity of the surrounding tissue. This method accounts for the compressibility of the surrounding medium to first order, and thus it is better suited than the Rayleigh-Plesset equation to simulate large amplitude bubble oscillations. While such oscillations will be highly nonlinear, this does not necessarily imply that a nonlinear viscoelastic model is necessary to describe the motion, although the suitability of the linear Voigt model also is unclear.

When there is a layer of fluid between the bubble and the adjacent tissue, as in the case of a bubble in a blood vessel or following partial disintegration of the tissue structures immediately adjacent to a bubble exposed to one or more tone bursts of high-intensity sound, Eq. (1) must be rewritten. Assuming that the density of the tissue and liquid, indicated by subscripts T and L, respectively, differ but that the speeds of sound in the two materials are the same leads to

$$R_1\ddot{R}_1\left(1 - \frac{\dot{R}_1}{c}\right)\left[1 + \frac{\rho_T - \rho_L}{\rho_L}\frac{R_1}{R_2}\right] + \dot{R}_1^2\left(1 - \frac{\dot{R}_1}{3c}\right)\left[\frac{3}{2} + \frac{\rho_T - \rho_L}{\rho_L}\frac{4R_2^3 - R_1^3}{2R_2^3}\frac{R_1}{R_2}\right]$$

$$= \left(1 + \frac{\dot{R}_1}{c}\right)\frac{(P_a - P_I)}{\rho_L} + \frac{R_1}{\rho_L c}\frac{d(P_a - P_I)}{dt} , \tag{5}$$

where the subscripts 1 and 2 refer to the inner and outer boundaries of the fluid layer. Equation (5) is similar to a well-known expression for the dynamics of micro-bubbles used as ultrasound contrast agents [6, 7]. The expressions for p_a and p_I become

$$P_a = P_g(t) - \frac{2\sigma}{R_1} = P_{g0}\left(\frac{R_{01}}{R_1}\right)^{3\kappa} - \frac{2\sigma}{R_1} , \tag{6}$$

and

$$P_I = P_o - P_A g(t) + \frac{4\dot{R}_1}{R_1}\left(\frac{V_L\mu_L + R_1^3\mu_T}{R_2^3}\right) + \frac{4G}{3R_2^3}(R_2^3 - R_{02}^3), \tag{7}$$

where $V_L = R_{02}^3 - R_{01}^3$. Equations (5) – (7) describe the dynamics of gas bubbles in a pocket of liquid within tissue.

It is important to note that although neither the bubble-tissue model [Eqs. (1), (3) and (4)] nor the bubble-fluid-tissue model [Eqs. (5) – (7)] appears to account explicitly for any damping mechanism other than viscosity, linear analysis performed using the polytropic assumption for p_g, shows that they actually incorporate five sources of dissipation [5]. These include contributions from the three usual sources, viscous, thermal, and acoustic damping, plus two additional sources, due to the surface energy of the bubble and the rigidity of the surrounding tissue. In this case however, the magnitude of the thermal damping coefficient differs from that determined for gas bubbles in liquids [8], while the sign of that due to the surface energy is opposite to that of the other four. Due to the high value of μ_T, for linear pulsations the total damping is dominated by viscosity when ω is less than the resonance frequency and $R_{01} < 10$ μm, while the acoustic term dominates at higher frequencies and larger radii.

The predictions for the responses of air bubbles to pressure pulses were obtained using assumptions and procedures described previously for the case of an air bubble in water: a spherical bubble initially at rest (*i.e.*, $dR/dt = 0$) in an infinite medium, no exchange of gas with the surroundings, adiabatic pulsations (*i.e.*, $\kappa = 1.4$), $kR << 1$ and a fourth order Runge-Kutta technique to solve the initial value problem [9, 10]. The ambient pressure was atmospheric pressure, and the speed of sound was set to value typical of soft tissue, thus p_0=0.101 MPa, and c=1540 m/s. The material properties of blood and the soft tissues used in the simulations are given in Table 1. In the final column, a typical tissue associated with each value of G is indicated in parentheses.

Table 1. **Material properties of water, blood and soft tissue assumed for simulations.**

Parameter, symbol & unit	Water	Blood	Tissue
Gas surface tension, σ, mN/m	68.0	56.0	56.0*
Density, ρ, kg/m^3	1.00	1.05	1.10
Viscosity, μ, Pa·s	0.001	0.005	0.005 or 0.015
Rigidity, G, MPa	0.0	0.0	0.023 (liver)
			0.5 (kidney)
			1.0 – 1.5 (smooth muscle)
			2.0 (skin)
			2.5 (skeletal muscle)

* When present, the tension at the blood-tissue interface is assumed to be 0.

RESULTS AND DISCUSSION

Examples of $R(t)$ curves for 1-μm bubbles in either blood or the tissues shown in Table 1 and driven by 1-MHz pulses at driving pressures of 1 MPa and 3.2 MPa, respectively, are given in Fig. 1a) and 1b); the increase in acoustic pressure from 1 to 3.2 MPa represents a 10-fold increase in intensity, from ~29.5 to ~295.2 W/cm^2. The effect of the increasing elasticity is obvious. Comparing the results for blood and liver at 1 MPa shows that even the small increase in tissue rigidity G, combined with the increase in μ, reduces the pulsation amplitude by ~25%. Even more dramatic is the 66% lower amplitude of bubbles oscillating in kidney vs liver. As G increases further, corresponding to tissues such as muscle and skin, the pulsation amplitude continues to decrease. The nonlinearity of the oscillation also declines, and the $R(t)$ curves for tissues with $G > 1.5$ MPa essentially describe simple linear harmonic motion.

When the driving pressure is increased to 3.2 MPa, the bubble pulsation amplitudes increase in all tissues. The amplitudes of the oscillations in tissues having relatively low rigidity, e.g., liver or kidney, are still smaller than if the same bubbles were in blood, but the differences between successive pairs of the three cases are less, being

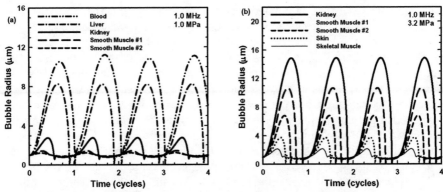

Figure 1. A comparison of radial responses for 1-μm bubbles driven by a 1-MHz pulse at a) 1 MPa and b) 3.2 MPa, for blood and the tissues shown in Table 1.

~10% and ~24%, respectively. The responses of bubbles that exhibited harmonic oscillations at 1 MPa now are clearly nonlinear, and the maximum bubble pulsation amplitude in even the most rigid tissue studied, skeletal muscle, exceeds a common criterion for the threshold for inertial cavitation, $R_{max} = 2R_0$ [11]. This indicates that the effect of rigidity will be less when the driving pressure is strong.

As would be expected from the $R(t)$ curves shown in Fig. 1, the thresholds for inertial cavitation increase greatly as the rigidity of the surrounding tissue increases. This is true regardless of the criterion used to determine the threshold, as shown in Fig. 2a for $R_{max} = 2R_0$, and in Fig. 2b) for $T_{max} = 5000K$; the threshold in blood is given for reference in each case. Several different theoretical criteria might be used to define the threshold for inertial cavitation, but because the lowest and highest thresholds for air bubbles are usually obtained for $R_{max} = 2R_o$ and $T_{max} = 5000K$, respectively [12], only these two criteria will be investigated here. Notice that for radii greater than the linear resonance radius, indicated by the open diamonds on the curves, the thresholds exhibit structure that appears to be related to the fractional-order

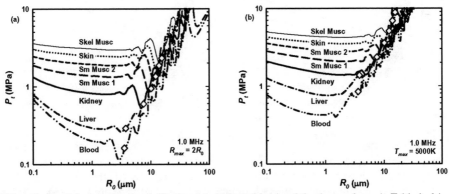

Figure 2. Inertial cavitation thresholds for air bubbles in blood and the tissues shown in Table 1, driven by 14-cycle pulses of 1-MHz ultrasound, for threshold criteria of a) $R_{max} = 2R_o$ and b) $T_{max} = 5000K$. The open diamonds show the resonance radii for each material.

(sub)harmonic resonance minima described previously [9]. The positions of the minima appear to be determined by the rigidity of the surrounding medium.

One important caveat to the use of the first threshold criterion is that simply because the condition $R_{max} = 2R_o$ is satisfied does not mean that an inertial collapse will necessarily follow. For example, the motion of the 1-μm bubble in skeletal muscle, see Fig. 1, results in $R_{max} = 2.16R_0$, but the subsequent value of $R_{min} = 0.81R_0$, while the maximum speed of the inward motion was only ~2.1 m/s, neither of which suggests that the contraction was inertia-driven. This does not mean that such bubble motion would not be harmful to the surrounding tissue, but it does mean that any injury would result from processes more subtle than the generation of free radicals or radiation of shock waves often associated with damage caused by inertial cavitation.

The 'global' threshold for inertial cavitation P_t is similar to that used by Apfel and Holland [13]. At each frequency, the minimal pressure amplitude required to satisfy the cavitation threshold criterion is determined for each equilibrium radius in the range studied, 0.1 – 100.0 μm, with the overall value of P_t chosen for the bubble requiring the least acoustic pressure to satisfy that criterion; that bubble is described as being "optimally sized". The results given in Fig. 3 show that: 1) thresholds in tissue are 4 – 10 times greater than in a viscous liquid such as blood, 2) the rate of increase in P_t with frequency is greater in tissue than in blood, and 3) the values of P_t are consistently higher for $T_{max} = 5000K$ than $R_{max} = 2R_o$. The curves for tissue in Fig. 3a) also show evidence of a transition in the optimal bubble response as the frequency increases, with the transition frequency increasing with the value of G. The nature of this change in bubble response will be the subject of future study.

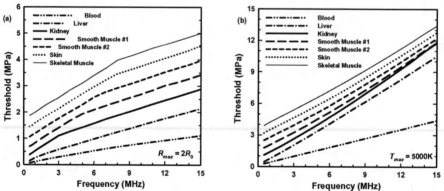

Figure 3. Thresholds for inertial cavitation of optimally sized air bubbles in blood and tissue for threshold criteria of a) $R_{max} = 2R_o$ and b) $T_{max} = 5000K$.

It is well known that violent bubble activity, such as that produced during shock-wave lithotripsy, may damage the surrounding tissue, but this is also true of short pulses of high-intensity focused ultrasound [14]. To investigate the nature of the bubble-tissue interaction, the three-component model described by Eqs. (5) – (7) was used to calculate the radial responses of bubbles either completely surrounded by tissue, or surrounded by but separated from the tissue by a spherical layer of fluid having the properties of blood as given in Table 1. This fluid layer represents either blood that has leaked from the surrounding tissue or tissue that has been physically degraded by previous bubble activity.

Examples of $R(t)$ curves for 1-μm bubbles in smooth muscle #2 ($G = 1.5$ MPa), surrounded by fluid layers and driven by 1-MHz, 1-MPa pulses are given in Fig. 4a) for fluid layers of various thicknesses, with the position of the fluid-tissue boundary, R_{02}, indicated in the figure. The amplitude of the response increases with increasing

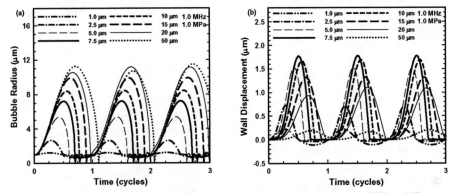

Figure 4. A comparison of (a) the radial responses of 1-μm bubbles driven by 1-MHz, 1-MPa pulses, and (b) the resulting tissue displacements for smooth muscle #2 ($G = 1.5$ MPa), see Table 1.

R_{02} because the influence to the tissue rigidity declines. The corresponding values of the displacement of the fluid-tissue boundary are given in Fig. 4b). Notice that while the radial response is greatest for the largest fluid layer, 50 μm, the displacement is maximal for a much thinner layer, 7.5 μm. This result indicates that there is an optimal value for R_{02} which will maximize tissue degradation by inertial cavitation.

Because damage to any solid, including a viscoelastic solid such as tissue, is properly quantified not by displacement but by the strain induced by the displacement, the strains corresponding to the displacements shown in Fig. 4b) were calculated at their spatial maximum, *i.e.*, at the fluid-tissue interface, $r = R_2$. The results, given in

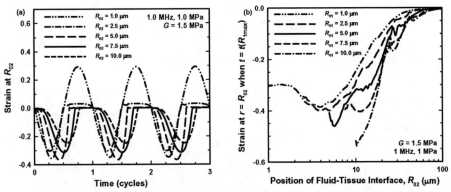

Figure 5. The strain at $r = R_{02}$ resulting from (a) the tissue displacements shown in Fig. 4b), and (b) the maximum tissue displacements for various initial bubble sizes R_{01}, as a function of fluid layer thickness.

223

Fig. 5a), show that maximum strain usually occurs at the time of the radial maximum (the sign of the strain is negative for $R_1 > R_{01}$), although considerable strain also may occur at the minimum ($R_1(t) = R_{1min}$) if the fluid layer is very thin. The value of the strain at $r = R_2$ is shown in Fig. 5b) for various equilibrium bubble radii for $R_{01} \le R_{02} \le 100$ μm. The strain generally increases with increasing bubble size, with the position of the maximum usually occurring when the fluid layer is relatively thin.

The minimum strain needed to produce permanent tissue damage, as well as the temporal characteristics of the acoustic exposure necessary to induce it, remain subjects of interest and future study.

ACKNOWLEDGMENTS

This work was supported by award number DAMD17-02-2-0014, administered by the US Army Medical Research Acquisition activity, Fort Detrick, MD. The information contained herein does not necessarily reflect the position or policy of the US government, and no official endorsement should be inferred.

REFERENCES

1. Frizzell, L. A., Carstensen, E. L., and Dyro, J. F., "Shear properties of mammalian tissues at low megahertz frequencies," *J. Acoust. Soc. Am.* **60**, 1409-1411 (1977).
2. Madsen, E. L., Sathoff, H. J., and Zagzebski, H. J., "Ultrasonic shear wave properties of soft tissues and tissuelike materials," *J. Acoust. Soc. Am.* **74**, 1346-1355 (1983).
3. Keller, J. B., and Miksis, M. J., "Bubble oscillations of large amplitude," *J. Acoust. Soc. Am.* **68**, 628-633 (1980).
4. Prosperetti, A., and Lezzi, A., "Bubble dynamics in a compressible liquid. Part 1. First order theory," *J. Fluid Mech.* **168**, 457-478 (1986).
5. Yang, X., and Church, C. C., "A model for the dynamics of gas bubbles in soft tissue," submitted to *J. Acoust. Soc. Am.*, April, 2005.
6. Church, C. C., "The effects of an elastic solid surface layer on the radial pulsations of gas bubbles," *J. Acoust. Soc. Am.* **97**, 1510-1521 (1995).
7. Hoff, L., Sontum, P. C., and Hovem, J. M., "Oscillations of polymeric microbubbles: Effect of the encapsulating shell," *J. Acoust. Soc. Am.* **107**, 2272–2280 (2000).
8. Prosperetti, A., "Thermal effects and damping mechanisms in the forced radial oscillations of gas bubbles in liquids," *J. Acoust. Soc. Am.* **61**, 17-27 (1977).
9. Church, C. C., "Prediction of rectified diffusion during nonlinear bubble pulsations at biomedical frequencies," *J. Acoust. Soc. Am.* **83**, 2210-2217 (1988).
10. Flynn, H. G., and Church, C. C., "Transient pulsations of gas bubbles in water," *J. Acoust. Soc. Am.* **84**, 985-998 (1988).
11. Flynn, H. G., "Cavitation dynamics. II. Free pulsations and models for cavitation bubbles," *J. Acoust. Soc. Am.* **58**, 1160-1170 (1975).
12. Church, C. C., "Frequency, pulse length and the mechanical index," *Acoust. Res. Lett. Online* **6**, 162-168 (2005).
13. Apfel, R. E., and Holland, C. K., "Gauging the likelihood of cavitation from short-pulse, low-duty cycle diagnostic ultrasound," Ultrasound Med. Biol. **17**, 179-185 (1991).
14. Parsons, J. E., Cain, C. A., and Fowlkes, J. B., "Characterizing Pulsed Ultrasound Therapy for Production of Cavitationally Induced Lesions" in *4th International Symposium on Therapeutic Ultrasound-2004*, edited by G. R. ter Haar and I. Rivens, AIP Conference Proceedings 754, New York: American Institute of Physics, 2005, pp. 178-180.

Nucleating Acoustic Cavitation with Optically Heated Nanoparticles

R. A. Roy, C. H. Farny, T. Wu, R. G. Holt, T. W. Murray

Department of Aerospace and Mechanical Engineering, Boston University, Boston, MA 02215

Abstract. The utilization of cavitation in high intensity focused ultrasound (HIFU) therapy requires the presence of nucleation sites; nucleation threshold pressures in tissues can exceed 4-5 MPa. We investigate the efficacy of transient vapor cavity generation from laser-illuminated gold nanoparticles as a means for nucleating cavitation. An acrylamide tissue phantom seeded with 82-nm diameter gold particle was exposed to 5 ns pulses from a 532 nm Nd:Yag laser. Acoustic emissions from inertial cavitation were detected by a 15 MHz broadband focused transducer at a laser energy of 0.10 mJ/pulse and a HIFU peak-negative focal pressure as low as 0.92 MPa. In comparison, a peak-negative focal pressure of 4.50 MPa was required to nucleate detectable cavitation without laser illumination. A simple analysis reveals that vapor cavities are formed that grow to the Blake radius, at which point they undergo rapid expansion and collapse.

Keywords: cavitation, nucleation, laser, nanoparticles.
PACS: 43.35.Ei, 43.80.Gx, 43.80.Sh

INTRODUCTION

Cavitation is implicated in a number of mechanical and thermal processes relevant to therapeutic ultrasound in general, and high-intensity focused ultrasound (HIFU) therapy in particular; see [1] for a comprehensive discussion. Bubbly media can alter the dispersive and dissipative characteristics in the propagation path, leading to defocusing effects and accelerated tissue heating. Stable cavitation generates microsteaming flows that can disrupt cells and promote mass transfer by breaking down boundary layers. Inertial cavitation promotes collapse microjets, radiated shock waves, chemical reactions, and enhanced localized tissue heating due to the rapid absorption of broadband acoustic emissions [1,2]. In order to effectively exploit these effects, practitioners must establish the cavitation field at the lowest possible acoustic pressures, however, nucleation thresholds for many tissues can be quite high, ranging from 3-4 MPa in blood to over 6 MPa in muscle tissue. Ultrasound contrast agents have been employed to promote cavitation [3,4], however, UCAs are short lived due diffusional instability and/or mechanical disruption by the HIFU acoustic field.

Below we describe an *in vitro* investigation of an alternate approach to the cavitation nucleation problem. The technique is inspired by prior work on opto-acoustic contrast enhancement using gold nanoparticles conjugated to antibodies bound to the surface of breast cancer cells implanted at 5–6-cm depths *in vitro* [5]. These particles preferentially absorb laser light, yielding transient vapor cavities and subsequent acoustic emissions upon collapse. Such cavities, when exposed to tensile

CP838, *Innovations in Nonlinear Acoustics: 17th International Symposium on Nonlinear Acoustics*,
edited by A. A. Atchley, V. W. Sparrow, and R. M. Keolian
© 2006 American Institute of Physics 0-7354-0330-9/06/$23.00

acoustic stress, will also serve as nucleation sites for acoustic cavitation. The primary advantages of this approach are (1) the moment of nucleation can be precisely timed to coincide with the peak rarefaction pressure phase and (2) the particles are durable and can be "reactivated" over an over again, thereby providing nuclei "on demand."

DESCRIPTION OF THE EXPERIMENT

We investigate this concept using 82-nm-diameter gold particles embedded in a aqueous, non-degassed, transparent, polyacrylamide gel phantom at a concentration of 10^9 particles per ml. The phantom was submerged in a tank of degassed water and exposed to individual 5 nsec pulses from a 0.8-mm diameter beam of 532-nm light generated by a frequency-doubled Nd:YAG laser. A HIFU source operating at 1.1 MHz launched 10-cycle bursts in the form of a focused beam with a focal length of approximately 6 cm. The pulse repetition frequency for the HIFU exposure was 0.1 Hz and the acoustic and optical beams were aligned collinear. To trigger the laser, we electronically delayed the HIFU drive to ensure that the laser fired at the same instant that the acoustic pulse traversed the focus. This delay could be precisely varied to explore the relationship between the onset of nucleation and the phase of the acoustic field in the cavitation sensing volume (described below). Detailed descriptions of the sample preparation, exposure geometry, and the HIU and laser generation systems are presented [6], and are not reproduced here.

To detect inertial cavitation activity, we employed a passive cavitation detector (PCD) consisting of a 15 MHz broad-band focused receiving transducer positioned confocal to the HIFU transducer. The detection axis was perpendicular to the HIFU beam and the detection beamwidth of approximately 0.6mm in the focal zone. The overlapping confocal region defined the cavitation detection zone. The received signal was linearly amplified (40 dB) and high-pass filtered above 5 MHz (passive, 4-stage Butterworth). Thus configured, the PCD was sensitive to broad-band acoustic emissions from inertial cavitation collapses. Single-shot PCD signals were digitized (8-bit, 50 Msample/sec) and stored on computer memory for processing. See [6] for a detailed description of cavitation detection instrumentation.

We applied a time gate to the PCD data to eliminate laser RF noise and then converted to the frequency domain by way of the FFT. The quantified indicator of cavitation activity is the emission level (EL) defined as follows:

$$EL = \int_{5MHz}^{25MHz} |X(f)|^2 df \qquad (1)$$

where the integrand is the square of the magnitude spectrum. This measure of the broadband noise power emanating from the cavitation detection zone serves as an approximate measure of inertial cavitation activity. An additional objective parameter indicative of activity is termed the emission gain (EG) defined by:

$$EG = 10\log\left[\frac{\langle EL \rangle}{\langle EL_{background} \rangle}\right] \qquad (2)$$

where the background emission level is measured with HIFU on and the laser beam blocked, and the averages are computed over 10 observations. The EG is an indicator of the extent to which bubble-collapse emissions exceed the background noise, and accounts for all systemic noise sources.

RESULTS AND DISCUSSION

As a first step, we measured conditions required to generate detectable cavitation activity from the laser alone and from the HIFU alone; particles were present in both cases. The minimum laser energy required to form detectable microcavitation was 4 mJ. The minimum HIFU peak negative pressure required to produce detectable microcavitation was 4.50 MPa. This is the nucleation threshold pressure for the phantom material.

Several experiments were run that demonstrated the synergistic behavior of light and sound for nucleating cavitation. We observed a suppression of nucleation threshold pressures for incident laser energies as low as 0.09 mJ, which is a factor of 45 less than the energy required to nucleate detectable vapor bubble collapses in the absence of acoustic forcing. At the maximum laser energy employed (0.11 mJ), the nucleation threshold pressure was somewhere between 0.8 and 0.9 MPa, a greater than fourfold reduction in the nucleation threshold. The suite of results is too extensive to present here; the reader is referred to [6] for a detailed report.

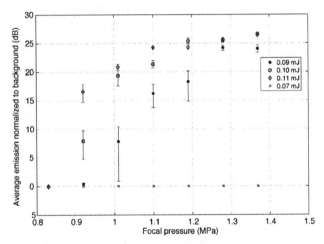

FIGURE 1. Inertial cavitation emissions relative to background as a function of laser energy and peak negative HIFU pressure. The errors bars are based on the standard error for a 10-point sample.

Figure 1 provides a summary view of the role of light and sound in suppressing nucleation thresholds in the presence of gold nanoparticles. The variable delay was set to fire the laser beam 0.15 μs after the fifth acoustic cycle arrived at the HIFU focal plane. Below a laser energy of about 0.08 mJ, no cavitation is observed; this is the illumination threshold energy for the range of acoustic pressures employed. For light exposures at and above 0.08 mJ, a clearly demarcated inertial cavitation threshold pressure exists, and this threshold is reduced as the laser energy is increased. It is important to note that both the laser energy and acoustic pressure thresholds are *significantly* lower than those measured with light and sound alone. The results summarized in Fig. 1 serve to prove the concept.

It is instructive to consider the physics that govern this process. In order to nucleate inertial cavitation, the vapor cavity produced by particle heating must grow from an initial radius of approximately 41 nm to a size equal to or greater than the Blake radius [ref 1, Sect. 4.3.1] for the acoustic pressure used. Is this, in fact, the case for the data given in Fig. 1? The energy absorbed by a particle is given approximately by:

$$Q = E_L \beta \left[\frac{R_s}{R_L} \right]^2 \qquad (3)$$

where E_L is the energy in the laser beam, β is absorption efficiency (≈ 3.3), R_S is the radius of the scatterer (41 nm) and R_L is the radius of the laser bean (0.8 mm). We assume that all this energy is absorbed as heat and forms a thin vapor shell surrounding the particle. The mass of vapor in this shell, m, is given by:

$$Q = m[C\Delta T + h] \quad ; \quad \Delta T = T_c - T_o \qquad (4)$$

where C is the heat capacity of the material and h is the heat of vaporization (both assumed to be that of water) T_o is the ambient temperature and T_c is the critical temperature (520K). In specifying the critical temperature, we account for the pressure jump at the interface due to surface tension (3.5 MPa for a 41-nm radius cavity). Once we know the mass and temperature of the vapor produced, we go to the steam tables to determine the corresponding specific volume, from which we can compute the maximum size of the cavity formed in the phantom.

To test this simple model, consider the fact that the threshold pressure for 0.01 mJ exposure was about 0.87 MPa. The resulting values for the maximum cavity size and the Blake radius at this pressure are 105 nm and 75 nm, respectively. (In this calculation, the Blake radius is corrected to account for the presence of the particle, which is assumed to be rigid). Note that the nuclei size does in, fact, exceed the Blake size, as it must. The fact that the two sizes are similar is consistent with the notion that we are measuring a threshold condition. [Work supported by the Dept. of the Army (award No. DAMD17-02-2-0014) and the Center for Subsurface Sensing and Imaging Systems (NSF ERC Award No. EEC-9986821).]

REFERENCES

1. Leighton, T. G., *The Acoustic Bubble*, Academic, London, 1997.
2. Holt, R. G., and Roy, R. A., *Ultrasound Med. Biol.* **27**, 1399–1412 (2001).
3. Tran, B. C., Seop, J., Hall, T. L., Fowlkes, J. B., and Cain, C. A., *IEEE Trans. Ultrason. Ferroelectr. Freq. Control* **50**, 1296–1304 (2003).
4. Sassaroli, E., and Hynynen, K., *J. Acoust. Soc. Am.* **115**, 3225–3243 (2004).
5. Eghtedari, M., Motamedi, M., Popov, V. L., Kotov, N. A., and Oraevsky, A. A., *Photons Plus Ultrasound: Imaging and Sensing, Proceedings of SPIE* **5320**, pp. 365–377 (2004).
6. Farny, C. H., Wu, T., Holt, R. G., Murray, T. W., and Roy, R. A., *Acoust. Res. Lett. Online* **6**, 138–143 (2005).

Bubble dynamics in constrained media

J. Cui*, M. F. Hamilton*, P. S. Wilson* and E. A. Zabolotskaya*

*Applied Research Laboratories, The University of Texas at Austin, Austin, Texas 78713-8029, USA

Abstract.
 Pulsation of an acoustically driven spherical bubble between rigid parallel plates is modeled using an augmented Rayleigh-Plesset equation that satisfies the boundary conditions. To emphasize the influence of flow constraints imposed by the plates, the only loss mechanism considered here is radiation damping. It is demonstrated that accounting for compressibility of the liquid is essential to the analysis. Expansion of the dynamical equation to quadratic order in the perturbation of the bubble radius yields an equation that is solved by successive approximations to obtain solutions at the drive frequency and its second harmonic. Amplitude responses presented as functions of drive frequency and plate separation reveal decrease in resonance frequency and increase in radiation damping as plate separation is reduced below approximately 15 bubble diameters.

Keywords: bubble dynamics, constrained media
PACS: 43.25.Yw

INTRODUCTION

Various applications of medical ultrasound involve the interaction of acoustic waves with bubbles in blood vessels and other constrained biological spaces, including the use of encapsulated microbubbles as contrast agents or vehicles for drug delivery, and cavitation generated by shock wave lithotripsy. Modeling bubble dynamics with these applications in mind requires taking into account constraints on the liquid imposed by vessel walls and other tissue interfaces. The example we use to illustrate the development of such a model is a spherical bubble in a liquid-filled channel formed by two rigid parallel plates. A linear analysis of this problem that was presented recently [1] is extended here to account for second-harmonic generation. The results are relevant to harmonic B-mode imaging employing contrast agents, which is based on detection of second-harmonic components generated and radiated by the microbubbles. [2]

THEORETICAL MODEL

The specific problem we consider is an acoustically driven, spherical gas bubble located in the midplane between two rigid parallel plates separated by distance d. For simplicity, and to focus on the influence of the plates, we ignore effects of viscosity, heat conduction, and surface tension. However, compressibility of the liquid is taken into account. In the absence of the plates, pulsation of the bubble is described by the Rayleigh-Plesset equation, augmented by a term proportional to the third time derivative of bubble volume that accounts for compressibility. In the presence of the plates, the equation is further augmented by a summation that accounts for pressure acting on the bubble due to reflec-

CP838, *Innovations in Nonlinear Acoustics: 17th International Symposium on Nonlinear Acoustics*,
edited by A. A. Atchley, V. W. Sparrow, and R. M. Keolian
© 2006 American Institute of Physics 0-7354-0330-9/06/$23.00

tions of acoustic waves radiated by the bubble. This summation is constructed using the method of images, and it satisfies the boundary condition of zero normal velocity on the surfaces of the plates. The resulting equation for the bubble dynamics takes the form

$$R\ddot{R} + \frac{3}{2}\dot{R}^2 - \frac{\ddot{V}}{4\pi c} + \frac{1}{2\pi d}\sum_{m=1}^{\infty}\frac{1}{m}\ddot{V}(t - md/c) = \frac{1}{\rho}\left[P_0(R_0/R)^{3\gamma} - P_0 - p_{\mathrm{ac}}(t)\right], \qquad (1)$$

where R is the radius and $V = \frac{4}{3}\pi R^3$ the volume of the bubble, dots over these quantities indicate time derivatives, ρ and c are the density and sound speed of the liquid, respectively, R_0 is the equilibrium bubble radius, P_0 the ambient pressure, $p_{\mathrm{ac}}(t)$ the applied acoustic pressure, and γ the ratio of specific heats.

If compressibility of the liquid is ignored, then $c \to \infty$ and the term containing $\ddot{V}$ disappears. More important, the time delays vanish and therefore $\ddot{V}$ can be removed from the summation, leaving a series in $1/m$ that diverges. This divergence corresponds to an infinite effective inertia loading the bubble wall, and it indicates that changes in bubble volume are prohibited when the liquid is incompressible. Including effects of compressibility is thus essential to modeling bubble dynamics in the channel, a conclusion that applies to any rigid channel having infinite length and constant cross section.

Equation (1) is history dependent, and direct numerical solution thus requires storage of $\ddot{V} = 4\pi(R^2\ddot{R} + 2R\dot{R}^2)$ at the delayed times $t - md/c$ for evaluation of the summation at every time step. An additional complication is that by including $\ddot{V}$ one obtains a differential equation of third order. A procedure used to solve a more general form of Eq. (1) numerically, to simulate the dynamics of bubble clusters, is presented elsewhere. [3] An alternative, for periodic excitation, is to solve Eq. (1) in the frequency domain as a set of coupled nonlinear algebraic equations. In this case it is convenient to expand Eq. (1) in terms of a perturbation of the bubble radius about its equilibrium value. This approach is the subject of another presentation. [4]

Here we solve Eq. (1) analytically by successive approximations for the case of second-harmonic generation due to a sinusoidal acoustic pressure excitation. With $R = R_0(1 + \xi)$, Eq. (1) is expanded through quadratic order in the dimensionless perturbation ξ to obtain

$$\ddot{\xi} + \omega_0^2\xi - \frac{R_0}{c}\dddot{\xi} + 2\frac{R_0}{d}\sum_{m=1}^{\infty}\frac{1}{m}\ddot{\xi}(t - md/c)$$

$$= -\xi\ddot{\xi} - \frac{3}{2}\dot{\xi}^2 + \frac{1}{2}(3\gamma + 1)\omega_0^2\xi^2 + 2\frac{R_0}{c}(3\dot{\xi}\ddot{\xi} + \xi\dddot{\xi})$$

$$- 4\frac{R_0}{d}\sum_{m=1}^{\infty}\frac{1}{m}\left[\dot{\xi}^2(t - md/c) + \xi(t - md/c)\ddot{\xi}(t - md/c)\right] - \frac{p_{\mathrm{ac}}(t)}{\rho R_0^2}, \qquad (2)$$

where linear terms are grouped on the left of the equal sign, quadratic terms and the applied acoustic pressure on the right, and $\omega_0^2 = 3\gamma P_0/\rho R_0^2$ is the square of the natural angular frequency for a bubble oscillating with infinitesimal amplitude in a free field.

Now let $p_{\mathrm{ac}}(t) = \frac{1}{2}p_0 e^{j\omega t} + \text{c.c.}$, where p_0 and ω are the pressure amplitude and angular frequency of the acoustic excitation, respectively. Then with $\xi = \xi_1 + \xi_2$, where

$\xi_n = \frac{1}{2}\Xi_n(\omega)e^{jn\omega t}$ + c.c. and $|\Xi_2| \ll |\Xi_1|$, one obtains by successive approximations the following solutions for the complex amplitudes $\Xi_n(\omega)$:

$$\Xi_1(\omega) = -\frac{p_0}{3\gamma P_0}\frac{1}{\Delta_1(\omega)}, \qquad \Xi_2(\omega) = \left(\frac{p_0}{3\gamma P_0}\right)^2 \frac{F(\omega)}{\Delta_1^2(\omega)\Delta_2(\omega)}, \tag{3}$$

where

$$\Delta_n(\omega) = 1 - n^2 \left[1 - 2\frac{R_0}{d}\ln(1 - e^{-jnkd})\right]\frac{\omega^2}{\omega_0^2} + jn^3 k_0 R_0 \frac{\omega^3}{\omega_0^3}, \tag{4}$$

$$F(\omega) = \frac{1}{4}(3\gamma + 1) + \left[\frac{5}{4} - 4\frac{R_0}{d}\ln(1 - e^{-j2kd})\right]\frac{\omega^2}{\omega_0^2} - j4k_0 R_0 \frac{\omega^3}{\omega_0^3}, \tag{5}$$

and $k = \omega/c$, $k_0 = \omega_0/c$. A benefit of solving Eq. (2) in the frequency domain is that the summations can be expressed in closed form, here in terms of the logarithms in Eqs. (4) and (5). The dimensionless quantities determining the response are d/R_0, ω/ω_0, p_0/P_0, γ, and $k_0 R_0 = (3\gamma P_0/\rho c^2)^{1/2}$. The last two are fixed once the gas-liquid combination is specified, the values of which are $\gamma = 1.4$ and $k_0 R_0 = 0.014$ for an air bubble in water under standard conditions.

RESULTS

The normalized amplitude response at the drive frequency is presented in Fig. 1(a) for several dimensionless plate separations. As plate separation decreases, the resonance frequency decreases and, because the radiation damping increases, the peak amplitude also decreases. The resonance frequency ω_r is the solution of $\mathrm{Re}\,\Delta_1(\omega) = 0$, or equivalently the solution of the transcendental relation $\omega_r = \omega_0[1 - 2(R_0/d)\ln|2\sin(\omega_r d/2c)|]^{-1/2}$. The free-field result $\omega_r \simeq \omega_0$ is obtained for $d/R_0 \gtrsim 30$, while the resonance frequency decreases rapidly for $d/R_0 \lesssim 30$ due to increased inertia caused by channeling of the liquid and thus prevention of radial flow to and from the bubble.

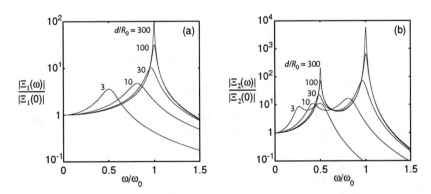

FIGURE 1. Amplitude responses (a) at the drive frequency and (b) at its second harmonic.

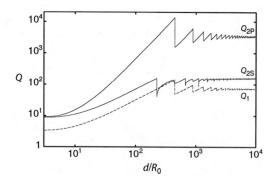

FIGURE 2. Relative amplitudes of the resonance peaks in Fig. 1.

Amplitude response curves for the second harmonic are presented in Fig. 1(b). They are seen to possess two maxima. The primary, higher-frequency resonance occurs when the drive frequency matches the bubble resonance, $\omega = \omega_r$, as in Fig. 1(a). The secondary, lower-frequency resonance at $\omega = \omega_r/2$ occurs because the second harmonic of the drive frequency, produced by the bubble, matches the bubble resonance.

Figure 2 displays the amplitudes of the resonance peaks in Figs. 1(a) and 1(b) versus plate separation using the metrics $Q_1 = |\Xi_1(\omega_r)/\Xi_1(0)|$, $Q_{2P} = |\Xi_2(\omega_r)/\Xi_2(0)|$, and $Q_{2S} = |\Xi_2(\omega_r/2)/\Xi_2(0)|$. Thus, Q_1 is the quality factor in linear theory. It is the amplitude at resonance relative to its value at zero frequency, and it is inversely proportional to damping. While the quantities Q_{2P} and Q_{2S} are defined in the same way for the second-harmonic amplitudes of the primary (P) and secondary (S) resonances, respectively, no specific connection to damping is intended. It is seen that Q_{2P} varies roughly as the square of Q_1, while apart from the oscillations for large plate separations, the dependence of Q_{2S} on Q_1 is more nearly linear.

ACKNOWLEDGMENTS

The authors are grateful to Yurii Ilinskii for many helpful discussions. This work was supported by the IR&D Program at ARL:UT, the National Institutes of Health Grant No. EB004047, and the Office of Naval Research.

REFERENCES

1. J. Cui, M. F. Hamilton, P. S. Wilson, and E. A. Zabolotskaya, "Spherical Bubble Pulsation between Parallel Plates," *J. Acoust. Soc. Am.*, **117**, 2530(A) (2005).
2. P. J. A. Frinking, A. Bouakaz, J. Kirkhorn, F. J. Ten Cate, and N. de Jong, "Ultrasound Contrast Imaging: Current and New Potential Methods," *Ultrasound Med. Biol.*, **26**, 965–975 (2000).
3. M. F. Hamilton, Yu. A. Ilinskii, G. D. Meegan, and E. A. Zabolotskaya, "Interaction of Bubbles in a Cluster near a Rigid Surface," *Acoust. Res. Lett. Online*, **6**, 207–213 (2005).
4. J. Cui, M. F. Hamilton, P. S. Wilson, and E. A. Zabolotskaya, "Periodic Nonlinear Oscillations of a Bubble between Rigid Parallel Plates," abstract to appear in *J. Acoust. Soc. Am.*, **118** (2005).

Strongly focused finite-amplitude sound beams emitted from an elliptically curved lens

Tomoo Kamakura*, Masahiko Akiyama*, and Shengyou Qian[†]

* The University of Electro-Communications, Tokyo 182-8585, Japan
† Hunan Normal University, Shangsha, 410081, China

Abstract. Strong focusing of a plane progressive ultrasound wave by a plano-concave lens with a widely opening aperture, which is made of acrylic resin and is submerged in water, is investigated theoretically and experimentally. To elliminate spherical aberration, an elliptic surface lens is introduced. The spheroidal beam equation, which is amenable to the analysis of a highly focused nonlinear beam, is used to predict the harmonic components in the beam. To make sure of the effectiveness of the present theory, experiments are done using an elliptic surface lens of eccentricity 0.544 which is attached tightly to a 1.7-MHz planar transducer with a cirular aperture of 75-mm in diameter. It is shown that the theory and experiment are in excellent agreement.

INTRODUCTION

Focusing of a sound beam is often achieved by combination of a planar transducer and an acoustic lens. The lens that is ordinarily made of plastic or solid material propagates sound at a higher speed than that at a surrounding medium such as water and body tissues, then the configuration of the lens should be concave. Although a spherically curved surface lens is widely used, spherical aberration occurs significantly, in particular for a widely opening aperture lens, being responsible for image blurring. To reduce the aberration, we introduce an elliptic surface lens whose eccentricity is equal to the ratio of sound speed in a surrounding fluid to that in the lens[1,2]. The spheroidal beam equation (SBE) whose upper limit of the applicability is at least 40° for the half-aperture angle is used for theoretical prediction of nonlinear sound beams[3,4,5].

BOUNDARY CONDITIONS APPROPRIATE TO SBE

To apply successfully the SBE model to the field analysis of a strongly focused sound beam emitted from a concave lens, it is primarily needed to specify the geometric focus. Figure 1 shows the geometry of the present lens model.

CP838, *Innovations in Nonlinear Acoustics: 17th International Symposium on Nonlinear Acoustics*,
edited by A. A. Atchley, V. W. Sparrow, and R. M. Keolian
© 2006 American Institute of Physics 0-7354-0330-9/06/$23.00

The actual lens surface S_1 is depicted in a solid line, which indicates an ellipse with semi-major length a and semi-minor length b. The origin O is the center of the ellipse. A sound beam passing through the lens attains its maximum pressure amplitude near the geometric focus O'. The distance of O' from the lens center is given $z_f = a + \sqrt{a^2 - b^2}$. The dot-dashed line in the figure is a spherical equiphase surface S_2 with curvature radius z_f, where the beam seems to be hypothetically emitted from. The idea we utilize here is based on geometrical acoustics, which projects the sound field at the

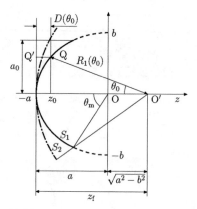

FIGURE 1. Geometry of an elliptic surface S_1 and a hypothetical bowl surface S_2.

elliptically curved surface S_1 back to the spherically curved surface S_2 on the condition that the linear dimensions of the lens are much greater than the wavelength.

This follows approximately that except for the region near the focus the spherically converging field by the elliptic lens coincides in magnitude with the spherically spreading field of a point source located at the focus O'. The magnitude of the field is expressed by A/r', where r' is the distance from O' to an arbitrarily point, and A is a constant. Hence the normal component of the particle velocity v_{2n} on S_2 is given by $(R_1/z_f)v_{1n}$ from two field equations $v_{1n} = A/R_1$ and $v_{2n} = A/z_f$, in which v_{1n} is the normal component of the velocity on S_1, and R_1 is the distance from O' to a point Q on S_2. Additionally, the wave propagaing in the lens decays inevitably due to sound absorption α in the lens. From all the above discussion, the boundary conditions for the initial pressure distribution on the surface S_2 yield

$$F(\psi) = \begin{cases} \dfrac{R_1(\theta_0)}{z_f} \cos\theta_0 \exp[-\alpha D(\theta_0)] & (0 \leq \psi \leq \psi_m), \\ 0 & (\psi > \psi_m) \end{cases} \tag{1}$$

where ψ is the angular variable in the oblate spheroidal coordinates, and is related with θ_0 by $\tan\theta_0 = \sqrt{1 + 1/\sigma^2}\tan\psi$. σ in the equation is the radial variable in the spheroidal coordinates. $D(\theta_0)$ is the distance between Q and Q' in Fig. 1 and is given by $z_0 + a$. Furthermore, ψ_m is the maximum opening half-angle of the lens.

EXPERIMENTS AND DISCUSSION

We made an elliptic surface lens of acrylic resin. A spherical lens of the same material was prepared for comparison whose half-aperture angle is 60° and curvature radius 43 mm. As is described later, the acoustic focus for the spherical lens exists

234

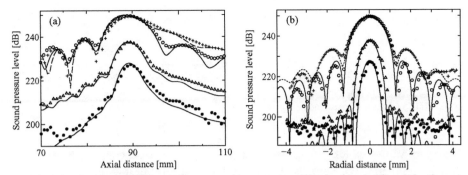

FIGURE 2. Comparison between the measured pressure amplitudes (symbols) and the theoretical prediction (solid curves) for the fundamental to third harmonics in the beam from the elliptic lens. •: fundamental, ○: second harmonic, △: third harmonic. (a): on-axis pressure amplitudes, (b): beam patterns at $z = 89.2$ mm. The measured data of the fundamental components from a spherical lens are also shown in the symbol '+' with the Rayleigh integral theory in dotted lines.

at about 89 mm from the lens, the dimensions of the elliptic lens are determined as $a = 57.6$ mm and $b = 48.3$ mm by taking into account of the speeds of longitudinal waves in the lens and water being 2670 m/s and 1480 m/s, respectively. The lens of aperture radius $a_0 = 37.5$ mm is attached tightly to an ultrasonic planar transducer of the same radius. They were immersed in degassed water. The transducer driven by a power amplifier radiates a tone-burst sinusoidal wave of 1.7-MHz with about 40 cycles duration. The sound absorption coefficient in the lens is $\alpha = 24$ Np/m.

Figure 2(a) shows on-axis sound pressure curves for a moderate intense source. Only the first three harmonics are given for the experimental data in symbols and for the theoretical prediction in solid curves. The initial pressure, which is needed for the numerical solution of the SBE model with the appropriate boundary conditions, was theoretically determined by a best fit of on-axis first harmonic pressure curve between the measured data and the theory, and was predicted to be 90 kPa. For the on-axis pressure curves, experimental data agree as a whole with the present theoretical prediction. The pressure amplitudes with less oscillation increase monotoncally during propagation before the focus, attains the maximum at the focus, and then decreases monotonically after the focus. It can be found evidently in the third harmonic that there are almost no oscillatory peaks and dips.

Figure 2(b) shows the beam patterns of the harmonics at $z = 89.2$ mm, near the focal plane. The experimentally obtained first side-lobe level relative to the main-lobe level is -25 dB for the fundamental, and is about 2 dB smaller than the theory. The beam patterns of the higher harmonics are also given in the figure. Especially, the third harmonic has almost no side-lobes within the framework of the present receiver system, whose noise floor level, including the hydrophone, is approximately 190dB. To make sure that the aberration is acutually reduced, experimental pressure data for a spherical lens with half-aperture angle 60° and

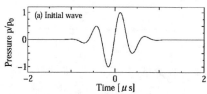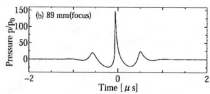

FIGURE 3. Nonlinear propagation of an intense Gaussian pulse. The initial pressure amplitude p_0 is 300 kPa. Waveforms at the source (a) and at the focus (b).

curvature radius 43 mm are given in the symbol '+' with the theory based on the Rayleigh integral (dotted lines). The pressure levels of the side-lobes by the elliptic lens are obviously 5 to 10 dB lower that those by the spherical lens.

SBE can be solved numerically in the time domain by means of an operator-splitting method, and is used to examine the nonlinear propagation of an intensely focused pulse. Figure 3 shows typical waveform changes for a Gaussian pulse whose initial pressure amplitude p_0 is 300 kPa.

CONCLUSIONS

To reduce spherical aberration for a widely opening aperture lens, an elliptic surface lens has been introduced. For describing nonlinear propagation of the beams, we used the spheroidal beam equation associated with an appropriate boundary conditions. To make sure of the effectiveness of the present theory, experiments were carried out in water using an elliptic surface lens of eccentricity 0.544 and a 1.7-MHz planar transducer. Excellent agreement between the theory and experiment supports that the present theoretical model is effective for evaluating focused sound fields by an elliptic surface lens with a widely opening aperture. [*This work was partially supported by the Ministry of Education, Science, Sports and Culture, Grant-in-Aid for Exploratory Research, 16656062.*]

REFERENCES

1. L. Schlussler, "The design and test results for an acoustic lens with elliptic surfaces," *J. Acoust. Soc. Am.* **67**, 699-701(1980).

2. A. Penttinen and M. Luukkala, "Sound pressure near the focal area of an ultrasonic lens," *J. Phys. D: Appl. Phys.* **9**, 1927-1936 (1976).

3. T. Kamakura, T. Ishiwata and K. Matsuda, "Model equation for strongly focused finite-amplitude sound beams," *J. Acoust. Soc. Am.* **107**, 3035–3046 (2000).

4. R. Xia, W. Shou, G. Cheng and M. Zhang, "The further study of the spheroidal beam equation for focused finite-amplitude sound beams," *J. Comput. Acoust.* **11**, 47–55 (2003).

5. C. Tao, J. Mu and G. Du., "The simulation of strongly focused finite amplitude ultrasound and temperature field," *Proceedings of 18th ICA*, V-3371–3374 (2004).

The Strong Effects Of On-Axis Focal Shift And Its Nonlinear Variation In Ultrasound Beams Radiated By Low Fresnel Number Transducers

Y.N. Makov[*], V. Espinosa[+], V.J. Sánchez-Morcillo[+], J. Ramis[+],
J. Cruañes[+], F. Camarena[+]

Department of Acoustics, Faculty of Physics, Moscow State University, Moscow 119992, Russia;
E-mail: Yuri@makov.phys.msu.ru
+ *Departament de Física Aplicada, Escola Politècnica de Gandia, Universitat Politècnica de València,*
Crta. Natzaret-Oliva s/n, 46730 Grau de Gandia, Spain;
E-mail: vespinos@fis.upv.es

Abstract. On the basis of theoretical concepts, an accurate and complete experimental and numerical examination of the on-axis distribution and the corresponding temporal profiles for low-Fresnel-number focused ultrasound beams under increasing transducer input voltage has been performed. For a real focusing transducer with sufficiently small Fresnel number, a strong initial (linear) shift of the main on-axis pressure maximum from geometrical focal point towards the transducer, and its following displacement towards the focal point and backward motion as the driving transducer voltage increase until highly nonlinear regimes were fixed. The simultaneous monitoring of the temporal waveform modifications determines the real roles and interplay between different nonlinear effects (refraction and attenuation) in the observed dynamics of on-axis pressure maximum. The experimental results are in good agreement with numerical solutions of KZK equation, confirming that the observed dynamic shift of the maximum pressure point is related only to the interplay between diffraction, dissipation and nonlinearity of the acoustic wave.

Keywords: sound beams, high power focused ultrasound, nonlinear absorption, Fresnel number.
PACS: 43.25.Cb, 43.25.Jh

INTRODUCTION

The study of focused sound beams has received a great attention both because its fundamental [1], and technological interest [2], due to the possibility of having a high sound intensity in a small and controlled volume. For this reason the influences of diffraction, focusing and nonlinearity on this goal have been investigated. In the linear theory of sound beams it has been shown the possibility of a shift in the position of the on-axis maximum pressure, as a result of the competition of diffraction and focusing effects. This was described, in a not complete and accurate form, with the linear gain parameter $G = L_d / R = ka^2 / 2R$, or the ratio of diffractional (Rayleigh) length, L_d, to the geometrical focusing distance, R, , where a is the aperture radius of the transducer and k the wave number. In the nonlinear regime the interest was centered in the competition between diffraction and nonlinearity [3-5]. The manifestation of nonlinear

CP838, *Innovations in Nonlinear Acoustics: 17th International Symposium on Nonlinear Acoustics*,
edited by A. A. Atchley, V. W. Sparrow, and R. M. Keolian
© 2006 American Institute of Physics 0-7354-0330-9/06/$23.00

effects has been only related to the apparition of shocks in the temporal profile of the wave. In fact, the shock formation has been blamed as responsible for the shift of the on-axis maximum pressure position, but with unclear and seemingly contradictory conclusions. In [3, 5] are related to the shift away from the transducer while in [4] to the opposite movement. It is necessary to clarify the maximum pressure shift phenomenon in focused sound beams and its dependence on the transducer geometry and driving conditions, and that constitutes the aim of this work.

THE FRESNEL NUMBER AND THE LINEAR OPERATION REGIME

It is well known from diffraction theory the concept of Fresnel zones, the annular regions on the wavefront emerging from a circular aperture, and the correspondent Fresnel number, the number of such rings bounded by the aperture from an observation point. For a point on the symmetry axis of the transducer at a distance z equal to the geometrical focusing distance the Fresnel number reads

$$N_F(z = R) \equiv N_F = a^2 / {\lambda R} = L_d / {\pi R} = G / \pi \qquad (1)$$

and therefore the Fresnel number also characterizes the relative influence of focusing and diffractive effects.

The fact of the discrepancy between the geometrical focus and main maximum pressure point for focused transducers has been pointed out in some works but without any real quantitative data and connection with the transducer parameters. For focused circular transducers with sufficiently small half-aperture (smaller than 20°) the field can be determined by the ordinary wave equation in the parabolic approximation, and in the simplest case of constant pressure p_0 along the aperture and parabolic phase profile, the on-axis pressure distribution is given by

$$\frac{p(0,\tilde{z})}{p_0} = \left| \frac{1}{1-\tilde{z}} \left(1 - e^{iG\frac{1-\tilde{z}}{\tilde{z}}} \right) \right| = \left| \frac{2}{1-\tilde{z}} \sin\left(\frac{G}{2} \frac{1-\tilde{z}}{\tilde{z}} \right) \right| \qquad (2)$$

where $\tilde{z} = z/R$ is a dimensionless coordinate along the beam axis. Looking for the largest maximum of (2), we can find to the location of the main ("focal") maximum of the on-axis pressure. The result of the numerical calculation of this root ($\tilde{z}_{max}$) from (2) depending on the Fresnel number N_F is shown in Fig. 1. and it follows that for $N_F > 6$ (corresponding to high-Fresnel-number focusing transducers) the main pressure maximum is very close to the geometrical focus, but for $N_F < 3$ (for low-Fresnel-number focusing transducers with realistic actual geometrical parameters) the difference between these points is large, and the point of the main pressure maximum is strongly shifted towards the transducer. We have studied experimentally and numerically such last case.

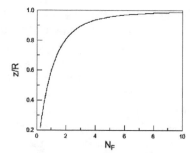

FIGURE 1. Dependence of the position of the on-axis main pressure maximum in linear regime on Fresnel number N_F.

MEASUREMENTS AND NUMERICS

We characterized the beam emitted in by a Valpey-Fisher focused transducer, based on a ceramic bowl, being $a = 1.5$ cm, R=11.7 cm which gives a Fresnel number of N_F=1.28 working in distilled water at $f = 1$ MHz. We used a membrane hydrophone NTR/Onda Corp. MH2000B, of 0.2 mm of active aperture and flat frequency response between 1 and 20 MHz. Figure 2.a) shows the on-axis maximum pressure distribution for increasing exciting voltage amplitudes, and illustrates the shift from the initial position nearer to the transducer to a place close to the geometrical focus, to come back again towards the transducer for higher values of voltage.

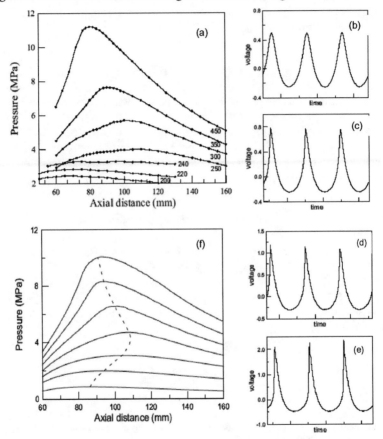

FIGURE 2. a) Measured on-axis maximum pressure amplitude for increasing voltage amplitudes (depicted). b) waveform for voltage input 200, c) 250, d) 300,and e) 450 V_{pp}, f) Numerical integration of the KZK equation for similar pressure levels and transducer parameters.

Figures 2 (b-e) show the waveform evolution with voltage. It can be noticed that the change of the motion happens when the hydrophone distortion reveals indirectly the appearance of high frequencies in the waveform peak, typically related to the beginning of shock formation. The good agreement of the result of the integration of the Kohklov-Zaboltskaya-Kutznesov (KZK) equation for the involved pressure levels and correspondent transducer parameters (Fig. 2(f)), proves that only the mechanisms of diffraction, focusing and nonlinearity are responsible of the observed phenomenon. In our opinion the nonlinearity plays two different roles: as can be seen in the KZ equation in the quasilinear approximation, the nascent upper harmonics in the slight nonlinear regime shift the on-axis maximum towards the geometrical focus, previously to the formation of shocks. When that occurs, the strong mechanism of nonlinear absorption is the responsible of movement towards the transducer because the higher exponential attenuation of higher harmonics. Measurements for bigger transducer diameters and similar curvatures, i.e., higher Fresnel numbers, show smaller initial shifts consistent with the exposed theoretical analysis and the nonlinear excursion range is severely shrunk.

CONCLUSIONS

The first cause of the large shift of the on-axis pressure maximum in initial linear regime and the large range of movement of this maximum in the nonlinear regime is the low-Fresnel-number of the focused transducer. The experimental observation of the varying on-axis pressure distribution and the temporal waveforms under increasing input voltages discloses the role of nonlinearity in the causality of one or other direction of pressure maximum shift.

ACKNOWLEDGMENTS

This work was supported by the Russian Foundation for Basic Researches (grant # 04-02-17009 and the Spanish CICYT BFM2002-04369-C04-04. The experimental results were confirmed during a research stay in the Hydroacoustics Laboratory of the Department of Physics and Technology of the University of Bergen (Norway), and Victor Espinosa wants to acknowledge greatly to Prof. Halvor Hobaek his hospitality and collaboration.

REFERENCES

1. M.F. Hamilton and D. Blackstock Eds., *Nonlinear Acoustics* , New York, Academic Press, 1997.
2. C. M. Langton, *Principles and Applications of Ultrasound*, Institute of Physics Publishing, 2004.
3. O. V. Rudenko, O. A. Sapozhnikov, *Self-action effects for wave beams containing shock fronts*, Phys.-Uspekhi 174, 2004, pp. 973-989.
4. K. Naugolnykh, and L. A. Ostrovsky, *Nonlinear Wave Processes in Acoustics*, Cambridge University press, Cambridge, 1998.
5. O. V. Rudenko, *Nonlinear sawtooth-shaped waves*, Phys.Uspekhi 38, 1995, pp.965-990.
6. Y. Li, and E. Wolf, *Focal shifts in diffracted converging spherical waves*, Optics Comm. 39(4), 1981, pp. 211-215.
7. Y-S. Lee and M. Hamilton, Time-domain modeling of pulsed finite-amplitude sound beams, J. Acoust. Soc. Am. 97, 1995, pp. 906-917.

Nonlinear and dissipative constitutive equations for coupled first-order acoustic field equations that are consistent with the generalized Westervelt equation

Martin D. Verweij* and Jacob Huijssen*

*Laboratory of Electromagnetic Research, Delft University of Technology, Mekelweg 4,
2628 CD Delft, The Netherlands

Abstract. In diagnostic medical ultrasound, it has become increasingly important to evaluate the nonlinear field of an acoustic beam that propagates in a weakly nonlinear, dissipative medium and that is steered off-axis up to very wide angles. In this case, computations cannot be based on the widely used KZK equation since it applies only to small angles. To benefit from successful computational schemes from elastodynamics and electromagnetics, we propose to use two first-order acoustic field equations, accompanied by two constitutive equations, as an alternative basis. This formulation quite naturally results in the contrast source formalism, makes a clear distinction between fundamental conservation laws and medium behavior, and allows for a straightforward inclusion of any medium inhomogenities. This paper is concerned with the derivation of relevant constitutive equations. We take a pragmatic approach and aim to find those constitutive equations that represent the same medium as implicitly described by the recognized, full wave, nonlinear equations such as the generalized Westervelt equation. We will show how this is achieved by considering the nonlinear case without attenuation, the linear case with attenuation, and the nonlinear case with attenuation. As a result we will obtain surprisingly simple constitutive equations for the full wave case.

Keywords: Nonlinear acoustics, constitutive equations, relaxation functions
PACS: 43.25.Ba, 43.25.Cb

INTRODUCTION

Many biological tissues have a noticeable nonlinear acoustic behavior. This property causes the generation of higher harmonic sound fields, in particular at those locations where an incident fundamental sound field has a large amplitude. Observation of these higher harmonics instead of the fundamental frequency enhances the focusing effect and suppresses the side lobes of a sonic beam, and thus enhances the imaging resolution of diagnostic medical ultrasound. Imaging of the second harmonic (Tissue Harmonic Imaging) is nowadays common, and a new development is to apply a combination of the third up till the fifth harmonic (SuperHarmonic Imaging) for this purpose [1]. This requires new phased array probes with an exceptionally large bandwidth. The design of these probes requires the ability to numerically simulate the generated nonlinear acoustic wavefield with sufficient accuracy for the fundamental frequency up till the fifth harmonic. For the validation, and thus for the computations, water will be used instead of tissue. Computation schemes based on the well-known KZK equation will render useless here since the underlying parabolic approximation will be invalid for the

CP838, *Innovations in Nonlinear Acoustics: 17th International Symposium on Nonlinear Acoustics*,
edited by A. A. Atchley, V. W. Sparrow, and R. M. Keolian

large steering angles envisaged. We consider the second-order generalized Westervelt equation [2] as the most promising basis for the required numerical calculations. This equation follows from combining, for a thermoviscous fluid, the continuity equation, the state equation, and the equation of motion. As a result of its derivation, it is a mixture of fundamental conservation laws and medium properties.

As an alternative, we propose to describe the nonlinear wavefield by two fundamental first-order field equations (which relate the field quantities but are independent of the behavior of the medium) and two constitutive equations (which only describe the behavior of the medium) [3]. This formulation has several benefits. First, our equations fit into a general mathematical framework for describing waves, just as Maxwell's equations and the associated constitutive equations of electromagnetics, and the first-order field equations and the associated constitutive equations of elastodynamics. This implies that succesful numerical solution methods for other fields can easily be adapted to deal with acoustic waves, and vice versa. Second, the separation between field equations and constitutive equations allows for a simple and clear way to include various kinds of medium behavior and the inclusion of medium inhomogeneities. Third, it leads in a natural way to the so-called contrast source formalism, which is very succesful for solving both forward and inverse wave problems.

After formulating the first-order equations, the derivation of the constitutive equations for the nonlinear medium is the remaining issue. Although we advocate the benefits of our first-order scheme, we also recognize the proven validity of the generalized Westervelt equation. In view of this, in the current paper we will take a pragmatic approach and obtain the constitutive equations that yield a system that is consistent with the generalized Westervelt equation.

FIRST ORDER AND SECOND ORDER EQUATIONS

The propagation of acoustic waves in a thermoviscous fluid may be described by the generalized Westervelt equation [2]

$$[1 + (D/c_0^2)\,\partial_t]\,\partial_j\partial_j p - c_0^{-2}\,\partial_t^2 p = -(\beta/\rho_0 c_0^4)\,\partial_t^2(p - p_0)^2 - c_0^{-2}\,\partial_t^2 \mathscr{L} - \partial_j\partial_j\mathscr{L}. \quad (1)$$

The term with the coefficient of nonlinearity $\beta = 1 + B/2A$ is the nonlinear term that appears in the original Westervelt equation [4], and the terms involving the Lagrangian density

$$\mathscr{L} = (\rho_0/2)\,v_j v_j - (1/2\rho_0 c_0^2)\,(p - p_0)^2 \quad (2)$$

form a nonlinear extension to account for wide-angle beams with nonplanar wavefronts [2]. The sound diffusivity D consists of the terms D^t, D^r, and D^v representing the thermal, relaxational, and viscous processes in the medium [5]. The pressure p, particle velocity v_i, density of mass ρ and wave speed c consist of a superposition of their static values, which are indicated by the subscript 0 (with $v_{i;0} = 0$), and their perturbations, which are due to the acoustic wave and will be indicated by a prime. It is easy to show that up till second order accuracy in the perturbation quantities, the generalized Westervelt equation may be obtained from the equations

$$\rho_0\,\partial_t v_k + \partial_k p - \rho_0 D^v\,\partial_j\partial_j v_k = -\partial_k\mathscr{L}, \quad (3)$$

$$\partial_t p + \rho_0 c_0^2 \partial_j v_j - (D^t + D^r) c_0^{-2} \partial_t^2 p = (\beta/\rho_0 c_0^2) \partial_t (p - p_0)^2 + \partial_t \mathscr{L}, \qquad (4)$$

which follow with the same accuracy from the equation of continuity, the equation of state, and the equation of motion.

In this paper we present an alternative to the generalized Westervelt equation through the first-order field equations [3]

$$\partial_k p + \dot{\Phi}_k = 0, \qquad \partial_j v_j - \dot{\theta} = 0, \qquad (5)$$

which are the equation of motion and the equation of volume, and the postulated constitutive equations

$$\dot{\Phi}_k = \rho^n(p, v_i) \, [\rho^d(t) *_t \mathrm{d}_t v_k], \qquad \dot{\theta} = -\kappa^n(p, v_i) \, [\kappa^d(t) *_t \mathrm{d}_t p]. \qquad (6)$$

Here $\dot{\Phi}$ is the mass flow density rate, $\dot{\theta}$ is the cubic dilatation rate, $*_t$ denotes temporal convolution, and $\mathrm{d}_t = \partial_t + v_j \partial_j$ is the material derivative. The functions ρ^n and κ^n will represent the nonlinear behaviour of the medium, and the relaxation functions ρ^d and κ^d will represent the diffusivity of the medium. Since we assume that the nonlinearities are small, we further write

$$\rho^n(p, v_i) = \rho_0 \, [1 + r(p, v_i)], \qquad \kappa^n(p, v_i) = \kappa_0 \, [1 + k(p, v_i)], \qquad (7)$$

with $\kappa_0 \rho_0 = c_0^{-2}$. In the next section we deduce the functions $r(p, v_i)$, $k(p, v_i)$, $\rho^d(t)$, and $\kappa^d(t)$ in such a way that the system in Eqs. (5) – (7) is consistent with Eq. (1) up till second order accuracy in the perturbation quantities, i.e. up till the same accuracy as used for the derivation of the generalized Westervelt equation itself.

DERIVATION OF THE CONSTITUTIVE EQUATIONS

In the *nonlinear, nondiffusive* case we have $D^{t,r,v} = 0$ and, consequently, $\rho^d(t) = \delta(t)$ and $\kappa^d(t) = \delta(t)$. Using $p = p' + p_0$ and applying straightforward mathematical manipulations, it may further be shown that up till second order accuracy the system in Eqs. (5) – (7) yields

$$\rho_0 \partial_t v_k + \partial_k p' = -\rho_0 v_j \partial_k v_j + r(p, v_i) \partial_k p', \qquad (8)$$

$$\partial_t p' + \rho_0 c_0^2 \partial_j v_j = \rho_0 v_j \partial_t v_j - k(p, v_i) \partial_t p'. \qquad (9)$$

In this case, the nondiffusive versions of Eqs. (3) and (4) are, with the same accuracy

$$\rho_0 \partial_t v_k + \partial_k p' = -\rho_0 v_j \partial_k v_j + \kappa_0 (p - p_0) \partial_k p', \qquad (10)$$

$$\partial_t p' + \rho_0 c_0^2 \partial_j v_j = \rho_0 v_j \partial_t v_j + \kappa_0 (1 + B/A) (p - p_0) \partial_t p', \qquad (11)$$

and from comparison we may conclude that

$$r(p, v_i) = \kappa_0 (p - p_0), \qquad k(p, v_i) = -\kappa_0 (1 + B/A) (p - p_0). \qquad (12)$$

243

Obviously, only the acoustic pressure plays a role in the nonlinear constitutive coefficients. In a parallel paper we show that the same results may be obtained along another line [6].

In the *linear, diffusive* case we have $\beta = 0$, $\mathscr{L} = 0$ and $d_t = \partial_t$, and consequently $r(p, v_i) = 0$ and $k(p, v_i) = 0$. The system in Eqs. (5) – (7) then yields

$$\partial_k \partial_k p - c_0^{-2} \rho^d(t) *_t \kappa^d(t) *_t \partial_t^2 p = 0. \tag{13}$$

In this case, the linear version of Eq. (1) is the diffusive equation

$$[1 + (D/c_0^2) \partial_t] \partial_j \partial_j p - c_0^{-2} \partial_t^2 p = 0. \tag{14}$$

Both equations are identical provided that the operator $\rho^d(t) *_t \kappa^d(t) *_t$ is the inverse of the operator $1 + (D/c_0^2)\partial_t$. With the aid of the Laplace transformation, it may be shown that this implies that

$$\rho^d(t) *_t \kappa^d(t) = H(t) (c_0^2/D) \exp[-(c_0^2/D)t], \tag{15}$$

in which $H(t)$ is the Heaviside step function. In the literature, relaxation effects in a medium are virtually always included in the compressibility and not in the density of mass. In view of this, we take

$$\rho^d(t) = \delta(t), \qquad \kappa^d(t) = H(t) (c_0^2/D) \exp[-(c_0^2/D)t]. \tag{16}$$

In the *nonlinear, diffusive* case we first guess that the cross products of nonlinearity and diffusivity may be neglected, so we combine the results of the previous subsections and propose

$$r(p, v_i) = \kappa_0 (p - p_0), \qquad k(p, v_i) = -\kappa_0 (1 + B/A)(p - p_0), \tag{17}$$

$$\rho^d(t) = \delta(t), \qquad \kappa^d(t) = H(t)(c_0^2/D) \exp[-(c_0^2/D)t]. \tag{18}$$

In this case, Eq. (1) applies in full, and the task is to show that our system of equations is capable to generate this equation up till second order accuracy in the perturbation quantities. Replacing $\kappa^d(t)$ by $\delta(t)$ (i.e., the limiting function for $D \downarrow 0$) in those terms that already contain two perturbation factors, and applying straightforward mathematical manipulations, it turns out that this indeed is the case. Thus we have achieved our goal.

REFERENCES

1. A. Bouakaz, and N. de Jong, *IEEE UFFC*, **50**, pp. 496-506 (2003).
2. S. I. Aanonsen, T. Barkve, J. Naze Tjotta, and S. Tjotta, *J. Acoust. Soc. Am.*, **75**, pp. 749-768 (1984).
3. A. T. de Hoop, *Handbook of radiation and scattering of waves*, Academic Press, London, 1995, pp. 20–28.
4. M. F. Hamilton, and D. T. Blackstock (eds.), *Nonlinear acoustics*, Academic Press, London, 1998, pp. 25–63.
5. M. J. Lighthill, *Waves in fluids*, Cambridge University Press, Cambridge, England, 1980, pp. 76–85.
6. J. Huijssen, and M. D. Verweij, " Nonlinear constitutive equations derived for fluids obeying an ideal gas, a Tait-Kirkwood or a B/A type equation of state," Presented at this symposium.

SECTION 5
HARMONIC IMAGING IN
DIAGNOSTIC ULTRASOUND

Mechanisms of Image Quality Improvement in Tissue Harmonic Imaging

Charles Bradley

109E₀ Plauv Preah Sihanouk, Kratie, Cambodia
and
Ultrasound Division, Siemens Medical Solutions, USA

Abstract. In this paper, the fundamental physical mechanisms by which image quality is improved in Tissue Harmonic Imaging (THI) are considered. A mathematical model that includes the effects of both nonlinearity and heterogeneity is used to compare the imaging performance of the fundamental and harmonic modes from the standpoint of image clutter. In making this comparison, the clutter reduction mechanisms of THI are revealed. Measurements that demonstrate these mechanisms are presented.

Keywords: THI, heterogeneous, harmonic, ultrasound, clutter
PACS: 43.25.Cb, 43.25.Jh, 43.80.Qf

TISSUE HETEROGENEITY, IMAGE CLUTTER, AND THI

Shortly after its introduction in the mid 1990's, Tissue Harmonic Imaging (THI), a 2nd harmonic imaging mode, replaced fundamental-mode imaging as the default imaging mode for B-mode diagnostic ultrasound scans. The reason for this paradigm shift is simple: the image haziness (commonly known as "clutter") that has plagued fundamental-mode imaging since its inception, is greatly reduced in harmonic-mode images. In spite of the fact that it has been nearly 10 years since its introduction, the fundamental physical mechanisms by which THI achieves this benefit are still poorly understood.

Image clutter has long been known to be caused by tissue heterogeneity [see. e.g., Ref (1)]. When the acoustic properties of tissue deviate from their mean values over a region that is substantial in size on the scale of an acoustic wavelength, then deformation of the acoustic beam may occur. Such deformation may include aberration (wavefront wrinkling) and reverberation (scattering, reflection), both of which give rise to clutter. Note that such *sustained* heterogeneity differs qualitatively from the very small-scale heterogeneity that gives rise to the backscattered sound fields that are used to form the image. Small-scale heterogeneity does not cause gross deformation of the acoustic beam, and as such, does not cause clutter.

It is the primary intention of this paper to uncover the fundamental physical mechanisms by which harmonic-mode imaging is less susceptible to the clutter-causing effects of tissue heterogeneity than fundamental mode imaging. This is

CP838, *Innovations in Nonlinear Acoustics: 17ᵗʰ International Symposium on Nonlinear Acoustics,*
edited by A. A. Atchley, V. W. Sparrow, and R. M. Keolian
© 2006 American Institute of Physics 0-7354-0330-9/06/$23.00

achieved by first determining the physical mechanisms by which clutter occurs in harmonic-mode images, and then by comparison of fundamental and harmonic-mode imaging in terms of the magnitude of the clutter problem.

THE IMAGE CLUTTER MODEL

The comparison of fundamental and harmonic-mode images is made between images of the same receive frequency. As receive beamformation is dynamic, it plays the dominant role in the determination of lateral resolution, and, as a consequence, like-frequency images are most directly comparable. It is also the case that when the receive frequencies are the same, (1), the backscatter of the transmitted field from the object, (2), the propagation of this backscattered field back to the transducer, and (3), the receive beamformation, are all identical. *The only difference between the two modes lies in the structure of the transmitted field.* We may therefore reduce the scope of the problem and consider the influence of (large-scale) heterogeneity on the structure of the *transmitted field alone.*

The equation used here to model the transmitted field is that derived by Kulkarni, Siegmann, and Collins[2] to model nonlinear acoustic propagation in the ocean

$$\hat{\rho}\vec{\nabla}\cdot\left(\frac{\vec{\nabla}p}{\hat{\rho}}\right)-\frac{1}{\hat{c}^2}\frac{\partial^2 p}{\partial t^2}=-\frac{\hat{\beta}}{\hat{\rho}\hat{c}^4}\frac{\partial^2 p^2}{\partial t^2}-\hat{\rho}\vec{\nabla}\cdot(\vec{u}\cdot\vec{\nabla})\vec{u}+\hat{\rho}\vec{\nabla}\cdot\left(\frac{\vec{\nabla}p^2}{2\hat{\rho}\hat{c}^2}\right)-\hat{\rho}\vec{\nabla}\cdot\left(\frac{(\vec{\nabla}\hat{\rho}\cdot\vec{\xi})\vec{\nabla}p}{\hat{\rho}^2}\right)$$

$$-\frac{1}{\hat{c}^2}\frac{\partial^2}{\partial t^2}\left(\frac{1}{2}\hat{\rho}u^2-\frac{p^2}{2\hat{\rho}\hat{c}^2}\right)+\frac{1}{\hat{\rho}\hat{c}^4}\frac{\partial}{\partial t}\left(\vec{\nabla}(\hat{\rho}\hat{c}^2)\cdot\vec{\xi}\frac{\partial p}{\partial t}\right), \qquad (1)$$

where p, $\vec{u}$, and $\vec{\xi}$ are the acoustic pressure, particle velocity, and particle displacement, and $\hat{\rho}$, $\hat{c}$, and $\hat{\beta}$ are the ambient density, small-signal sound-speed, and coefficient of nonlinearity, respectively. While it is not included in this equation, the effect of dissipation will be included in the solutions in an ad-hoc manner.

In order to solve this equation, we make a number of simplifying assumptions about the heterogeneity of the tissue. First, the heterogeneity is assumed to be spatially localized. For example, we have $\hat{\rho}(\vec{r})=\rho_0+\Delta\rho(\vec{r})$, where $\Delta\rho(\vec{r})$ is nonzero only in localized regions of heterogeneity [and the same is true for $\hat{c}(\vec{r})=c_0+\Delta c(\vec{r})$ and $\hat{\beta}(\vec{r})=\beta_0+\Delta\beta(\vec{r})$]. We further assume that these regions of heterogeneity are confined to thin layers, and that the deviations from uniformity are of order ε [e.g., $\Delta\hat{\rho}/\rho_0=O(\varepsilon)$]. It is also assumed that the source generates a band-limited signal centered on the fundamental frequency ω_0, and that the bandwidth is sufficiently narrow that the fundamental and the harmonic distortion components all fall into effectively distinct frequency bands. Under this set of assumptions, we proceed with the standard straightforward perturbation expansion in the acoustic Mach number ε [$p=p_1+p_2+p_3+O(\varepsilon^4)$, where $p_n=O(\varepsilon^n)$]:

$$O(\varepsilon): \quad \nabla^2 p_1(\omega_0) - \frac{1}{c_0^2}\frac{\partial^2 p_1(\omega_0)}{\partial t^2} = 0 \tag{2}$$

$$O(\varepsilon^2): \quad \nabla^2 p_2(\omega_0) - \frac{1}{c_0^2}\frac{\partial^2 p_2(\omega_0)}{\partial t^2} = -\frac{2}{c_0^2}\left(\frac{\Delta c}{c_0}\right)\frac{\partial^2 p_1(\omega_0)}{\partial t^2} + \vec{\nabla}\left(\frac{\Delta\rho}{\rho_0}\right)\cdot\vec{\nabla}p_1(\omega_0) \tag{3}$$

$$\nabla^2 p_2(2\omega_0) - \frac{1}{c_0^2}\frac{\partial^2 p_2(2\omega_0)}{\partial t^2} = -\frac{\beta_0}{\rho_0 c_0^4}\frac{\partial^2 p_1^2(\omega_0)}{\partial t^2} \tag{4}$$

$$O(\varepsilon^3): \quad \nabla^2 p_3^{(Ab)}(2\omega_0) - \frac{1}{c_0^2}\frac{\partial^2 p_3^{(Ab)}(2\omega_0)}{\partial t^2} = -\frac{2\beta_0}{\rho_0 c_0^4}\frac{\partial^2}{\partial t^2}[p_1(\omega_0)p_2(\omega_0)] \tag{5}$$

$$\nabla^2 p_3^{(Rev)}(2\omega_0) - \frac{1}{c_0^2}\frac{\partial^2 p_3^{(Rev)}(2\omega_0)}{\partial t^2} = -\frac{2}{c_0^2}\left(\frac{\Delta c}{c_0}\right)\frac{\partial^2 p_2(2\omega_0)}{\partial t^2} + \vec{\nabla}\left(\frac{\Delta\rho}{\rho_0}\right)\cdot\vec{\nabla}p_2(2\omega_0) \tag{6}$$

Note that the second harmonic component of the cubic equation has been divided into two parts [i.e., we have taken $p_3(2\omega_0) = p_3^{(Ab)}(2\omega_0) + p_3^{(Rev)}(2\omega_0)$], and that the fundamental and 3rd harmonic components are not reported as they are not of interest in this context. Note also that the usual terms associated with local effects have been neglected.

Consider first the fundamental field, as described by Eqs. (2) and (3). Equation (2) describes linear propagation in a homogeneous medium. The field $p_1(\omega_0)$ is therefore a fundamental frequency field that undergoes no propagation distortion due to nonlinearity and no beam deformation due to heterogeneity. As such, it represents the ideal fundamental beam; one that would yield the ideal fundamental image. Equation (3) describes linear Born scattering of this un-deformed beam. The field $p_2(\omega_0)$ therefore represents the leading-order description of the deformation of the fundamental beam by tissue heterogeneity, and therefore also represents the leading-order image clutter in fundamental-mode imaging. The ratio of the amplitudes of $p_1(\omega_0)$ and $p_2(\omega_0)$ is therefore a measure of performance of the fundamental mode from the standpoint of image clutter.

This same commentary is applicable to the 2nd harmonic fields described by Eqs. (4)-(6). Eq. (4) describes second harmonic distortion as driven by $p_1(\omega_0)$, the undeformed fundamental beam, and where the nonlinear distortion occurs in the absence of beam-deforming heterogeneity. The field $p_2(2\omega_0)$ therefore represents the ideal 2nd harmonic beam and the ideal 2nd harmonic image. Eqs. (5) and (6) describe the leading-order deformation of this ideal 2nd harmonic beam, and therefore the leading-order 2nd harmonic clutter. We may therefore take the ratio between the amplitudes of $p_2(2\omega_0)$ and $p_3(2\omega_0)$ to be a measure of performance of the 2nd harmonic imaging mode from the standpoint of clutter.

The reason for the decomposition of the harmonic beam deformation field $p_3(2\omega_0)$ into two distinct components is that these two components come about via two very different physical processes. We see in Eq. (5) that the field $p_3^{(Ab)}(2\omega_0)$ is generated by the nonlinear interaction between the un-deformed fundamental beam

$p_1(\omega_0)$ and the fundamental beam deformation field $p_2(\omega_0)$. The only component of $p_2(\omega_0)$ that overlaps with $p_1(\omega_0)$, and therefore interacts with it, is the *forward-scattered* component (the component associated with wrinkling of the fundamental wavefronts; i.e., the "*aberration*" component). As such, the field $p_3^{(Ab)}(2\omega_0)$ is labeled according to it's physical origin: aberration of the fundamental beam.

In Eq. (6), on the other hand, we see that the component labeled $p_3^{(Rev)}(2\omega_0)$ comes about due to linear scattering of whatever 2nd harmonic field has accumulated during propagation out to the heterogeneous layer. While this field generally does include forward scattered components, it is primarily composed of side and back-scattered components, which are collectively referred to as "*reverberant*" components, and are the reason $p_3^{(Rev)}(2\omega_0)$ is so-labeled.

It is of interest to note that neither of these leading-order contributors to harmonic image clutter is caused by the nonlinear interaction of the scattered fundamental field with itself (i.e. the nonlinear propagation distortion of the scattered fundamental). As $p_2(\omega_0) = O(\varepsilon^2)$, such components occur at $O(\varepsilon^4)$ and are therefore not leading-order.

ABERRATION CLUTTER

We now turn to the consideration of the "aberration" component of clutter in the 2nd harmonic image, as represented by the field $p_3^{(Ab)}(2\omega_0)$. We consider a "worst-case" aberrator: one that (1) lies at a very shallow depth, as this maximizes the range over which the clutter-generating interaction occurs, and (2), is spatially monochromatic, as this causes the beam deformation field $p_2(\omega_0)$ to remain focused, thus maximizing the strength of the interaction. For simplicity, we consider only variation in sound-speed [i.e., $\Delta\rho(\bar{r}) = 0$], and take the aberrator to be invariant in y:

$$\frac{\Delta c}{c_0} = \left(\frac{\Delta c_0}{c_0}\right) g(z)\cos(k_a x).$$

Here, $(\Delta c_0/c_0)$ represents the magnitude of the sound-speed deviation, k_a is the spatial frequency of the aberration, and the function $g(z)$, which represents the depth and thickness of the aberrator, is taken to be a compact function centered very near the face of the transducer, and is such that $\max[g(z)] = 1$ and $\int g(z)dz = L_a$ (i.e., it has a characteristic thickness of L_a).

Given such an aberrator, Eqs. (2)-(5) may be solved by Green's function methods [see, e.g., Ref. (3)] under the following assumptions. First, the source boundary condition is taken to be a focused CW Gaussian-type boundary condition given by $p|_{z=0} = p_0 e^{-(x^2+y^2)/\tilde{a}_1^2}$, where $\tilde{a}_1^2 = a^2/(1 - jG_1)$ is a complex source radius that includes the effect of focusing, a is the characteristic Gaussian source radius, $G_1 = z_1 / d$ is the focal gain, d is the focal depth, $z_1 = k_1 a^2 / 2$ is the Rayleigh distance, and $k_1 = \omega_1 / c_0$ is the wavenumber, all at the frequency $\omega = \omega_1$. The source occupies the plane at $z = 0$ and is impedance-matched to the medium (ultrasound transducers are designed to be so). The focal gain is assumed to be sufficiently moderate that the paraxial approximation is valid (typical f-numbers for cardiac and abdominal scans are in the range ~3-5), and the thickness of the aberration layer L_a is assumed to be small on all relevant scales. The resultant solution may be expressed[A]:

$$p_1(\omega_1) = \frac{p_0}{1 - jz/\tilde{z}_1} e^{-\frac{x^2+y^2}{\tilde{a}_1^2} \frac{1}{1-jz/\tilde{z}_1}} e^{-jk_1 z} e^{-\alpha_1 z} \tag{7}$$

$$p_2(\omega_1) = -\frac{k_1 L_a}{2j}\left(\frac{\Delta c_0}{c_0}\right)\frac{p_0}{1 - jz/\tilde{z}_1} e^{-\frac{[x-z\sin\varphi_1]^2+y^2}{\tilde{a}_1^2} \frac{1}{1-jz/\tilde{z}_1}} e^{-jk_1[z\cos\varphi_1 + x\sin\varphi_1]} e^{-\alpha_1 z} \tag{8}$$

$$p_2(\omega_2) = jk_1 \tilde{z}_1 \frac{\beta_0}{2} \frac{p_0^2}{\rho_0 c_0^2} \frac{\ln(1 - jz/\tilde{z}_1)}{1 - jz/\tilde{z}_1} e^{-2\frac{x^2+y^2}{\tilde{a}_1^2} \frac{1}{1-jz/\tilde{z}_1}} e^{-jk_2 z} e^{-\alpha_2 z} \tag{9}$$

$$p_3^{(Ab)}(\omega_2) = jk_1 \frac{\beta_0}{2} \frac{p_0^2}{\rho_0 c_0^2}\left(\frac{\Delta c_0}{c_0}\right)\frac{k_1 L_a}{1 - jz/\tilde{z}_1} e^{-2\frac{[x-z\sin\varphi_2]^2+y^2}{\tilde{a}_1^2}\frac{1}{1-jz/\tilde{z}_1}} e^{-jk_2[z\cos\varphi_2 + x\sin\varphi_2]} e^{-\alpha_2 z}$$
$$\int_0^z e^{-2\frac{z'^2\sin^2(\varphi_2)}{\tilde{a}_1^2}\frac{1}{1-jz'/\tilde{z}_1}} e^{-jk_2[1-\cos\varphi_2]z'} \frac{dz'}{1 - jz'/\tilde{z}_1}, \tag{10}$$

where α_n is the attenuation coefficient at frequency ω_n, $\tilde{z}_1 = k_1 \tilde{a}_1^2 / 2$ is a complex Rayleigh distance, and φ_n is an angle defined by $\sin(\varphi_n) = k_a / k_n$.

The fundamental field consists of an undeformed Gaussian beam [Eq. (7)] and another beam [Eq. (8)], also Gaussian, but of lesser amplitude and steered by an angle φ_1. The 2nd harmonic field consists of an undeformed Gaussian beam [Eq. (9)] and another beam [Eq. (10)], also Gaussian, but with some complicated range structure and steered by an angle $\varphi_2 \cong \varphi_1 / 2$. Where Eqs. (7) and (9) represent ideal images, Eqs. (8) and (10) represent clutter.

We may now introduce an SNR-like quantity that may be used as a measure of imaging performance with respect to image clutter. We take the "signal" to be the peak value of the undeformed beam (which represents the ideal image), and the "noise" to be the peak value of the associated beam deformation field (which represents the clutter). The ratio of these values is then a measure of the "Signal-to-Clutter Ratio" (SCR) of the imaging configuration.

[A] Note that whereas the cosine aberrator considered here gives rise to symmetric pairs of steered beams, the expression for one of each pair is presented here for simplicity of presentation. Note also that the real solution is given by the imaginary part of the complex phasor presented here.

251

For the fundamental mode, the fields that represent the "signal" and the "clutter" are those represented in Eqs. (7) and (8), but where we take $\omega_1 \rightarrow \omega_2$ because we are comparing fundamental and harmonic imaging of frequency ω_2. The resultant fundamental mode SCR is then given by

$$SCR_F = \left[\frac{1}{2} k_2 L_a \left(\frac{\Delta c_0}{c_0} \right) \right]^{-1}.$$

For the harmonic mode, the fields of interest are those in Eqs. (9) and (10) and the resultant SCR is given by

$$SCR_H = \frac{\int\limits_0^d \dfrac{dz'}{1-jz'/\tilde{z}_2}}{\int\limits_0^d e^{-2\frac{z'^2 \sin^2(\varphi_2)}{\tilde{a}_1^2} \frac{1}{1-jz'/\tilde{z}_1}} e^{jk_2[1-\cos(\varphi_2)]z'} \dfrac{dz'}{1-jz'/\tilde{z}_2}} \left[k_1 L_a \left(\frac{\Delta c_0}{c_0} \right) \right]^{-1} \quad (11)$$

(Note that the numerator in Eq. (11) may be expressed $j\tilde{z}_1 \ln(1-d/\tilde{z}_1)$, but is here left in integral form for ease of interpretation).

The *relative* performance of harmonic and fundamental mode imaging, from the standpoint of susceptibility to clutter, is characterized by the ratio SCR_H / SCR_F, which is simply the ratio of the integrals in Eq (11). The inverse of this ratio is shown plotted in Fig. (1a) as a function of angle (i.e., as a function of the spatial frequency of the aberrator) for a set of parameter values typical of abdominal imaging. This function represents the number of dB by which aberration-induced sidelobes are reduced in the 2nd harmonic mode as compared to the fundamental-mode. As clutter suppression of 3 or 4 dB is easily appreciable in a typical ultrasound image, clutter suppression of this degree is truly remarkable. In figure (1b) are shown two maximum-projection plots taken from ultrasound images of a point target. During imaging, a spatially monochromatic aberrating layer (a sheet of corrugated silicone) was placed on the face of the transducer. Where the fundamental image shows large aberration-induced sidelobes, the harmonic image shows none (they lie below the dynamic range floor of the image). The sidelobe suppression here is therefore *at least* 18dB.

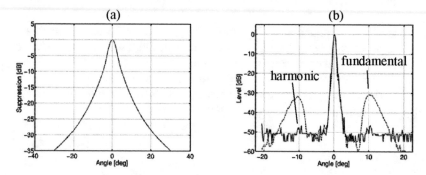

FIGURE 1. The aberration clutter (sidelobe) suppression function (a), and maximum-projection plots from 3MHz ultrasound images of a line target.

There are two distinct physical phenomena responsible for this effect, and they are represented by the two exponential terms in the denominator of Eq (11). The first represents a "non-overlap" effect: the volume and strength of the virtual source distribution that drives the 2nd harmonic clutter field [the right-hand-side of Eq. (5)] decreases monotonically as the scattering angle increases. The two interacting beams simply overlap less and less as the angle of interaction increases, and so less clutter is generated. The second exponential term represents a dispersion-like "non-resonance" effect: the wavenumber associated with the virtual source distribution along the direction of propagation matches that of the generated 2nd harmonic more and more poorly as the interaction angle increases. As a result, the 2nd harmonic level does not accumulate continuously with range, but oscillates at low level, as it does in dispersive systems [see, e.g., Ref. (4)].

REVERBERATION CLUTTER

The reverberation component of image clutter in the 2nd harmonic mode is represented by the field $p_3^{(Rev)}(2\omega_0)$, which is generated when the undeformed 2nd harmonic beam scatters from the layer of heterogeneity [see Eq. (6)]. A key simplification becomes apparent when we note that this equation is identical to Eq. (3), which describes the analogous component of image clutter for fundamental-mode imaging. In both cases, the reverberant (clutter-causing) field is caused by the linear scattering of an un-deformed, focused beam from the reverberant layer. What's more, these two un-deformed, focused beams are of the same frequency (we compare fundamental and harmonic modes operating at the same frequency). The only substantial difference between the two incident beams is that they differ in amplitude. It follows then, that the only substantial difference between the resultant reverberation fields is that they differ in amplitude. We may therefore express these scattered fields as a product of a "scattered field function" $\gamma(\vec{r})$ and the amplitude of the insonifying field. With use of Eq. (7), then, the fundamental SCR may be expressed

$$SCR_F = \frac{z_2}{d} \frac{\left|1 - jz_r / \tilde{z}_2\right|}{\left|\gamma\right|_{max}} e^{-\alpha_2(d - z_r)},$$

and with use of Eq. (9), the harmonic SCR may be expressed

$$SCR_H = \frac{\left|1 - jz_r / \tilde{z}_1\right|}{\left|\gamma\right|_{max}} \frac{\left|\ln(1 - jd / \tilde{z}_1)\right|}{\left|\ln(1 - jz_r / \tilde{z}_1)\right|} e^{-\alpha_2(d - z_r)},$$

Where z_r is the depth of the reverberant layer. The ratio of these two SCR's is then given by

$$\frac{SCR_H}{SCR_F} = \frac{d}{z_2} \frac{\left|1 - jz_r / \tilde{z}_1\right|}{\left|1 - jz_r / \tilde{z}_2\right|} \frac{\left|\ln(1 - jd / \tilde{z}_1)\right|}{\left|\ln(1 - jz_r / \tilde{z}_2)\right|} .$$

The inverse of this function, which represents the degree to which reverberant clutter is rejected in harmonic imaging as compared to fundamental imaging, is shown plotted in Fig.(2a) as a function of the depth of the reverberant layer. Particularly for shallow reverberant layers (as they tend to be), the degree of rejection is again remarkable. Also shown are 4MHz fundamental (2b) and 4MHz harmonic (2c) images of the edge

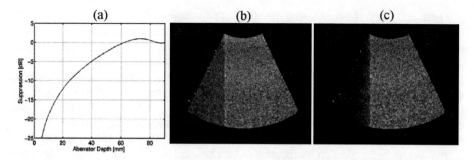

FIGURE 2. The reverb suppression function (a), a 4MHz fundamental-mode image of a phantom (b), and a 4MHz harmonic-mode image of a phantom (c).

of a tissue- mimicking phantom. The speckle field in the left of the image is a reverberant artifact caused by reflection from the phantom wall (which is *designed* to be anechoic), and is greatly reduced in the harmonic image.

Physically, this effect is due to the cumulative growth with range that is characteristic of nonlinear propagation distortion. After the 2nd harmonic field scatters from the reverberant layer, the scattered component of the field propagates linearly, but the unscattered component continues to accumulate amplitude with range. In other words, the "signal" gets boosted more and more with range, but the "noise" does not. Propagation distortion represents a channel by which energy is injected into the signal (courtesy of the fundamental field) and not into the noise.

SUMMARY

There are two distinct physical mechanisms by which clutter is reduced in the harmonic relative to the fundamental imaging mode. In the case of aberration, the clutter-causing component of the 2nd harmonic field is generated via a nonlinear interaction. Diffraction ensures that the interaction is angled, and it is a characteristic of nonlinearity that angled interactions are inefficient. As a result, the clutter-causing field is weak. In the case of reverberation clutter, the cumulative harmonic growth that is characteristic of propagation distortion ensures that the imaging field is stronger at depth (where imaging takes place) than it is shallow (where the clutter-causing reverberation field is generated).

REFERENCES

1. Hinkelman, L. M., Liu, D. L., Metlay, L. A., Waag, R. C. , *J. Acoust. Soc. Am.* **95**, 530-541 (1994).
2. Kulkarni, R. S., Siegmann, W. L., and Collins, M. D., *J. Acoust. Soc. Am.* **102**, 224-231 (1997).
3. Hamilton, M. F., "Sound Beams," in *Nonlinear Acoustics*, edited by M. F. Hamilton and D. T. Blackstock, San Diego: Academic Press, 1998, pp. 233-261.
4. Hamilton, M. F., "Dispersion," in *Nonlinear Acoustics*, edited by M. F. Hamilton and D. T. Blackstock, San Diego: Academic Press, 1998, pp. 151-175.

Investigating Mechanisms of Image Quality Enhancement Associated with Tissue Harmonic Imaging

Kirk D. Wallace, Mark R. Holland, and James G. Miller

Laboratory for Ultrasonics, Dept. of Physics, Washington University in St. Louis
Saint Louis, MO USA 63130

Abstract. Enhancements to the compactness of the nonlinearly generated second harmonic (2f) field component (narrower main-lobe and reduced side-lobe levels) and reduced impact of phase and amplitude aberration, with respect to the fundamental (1f) field component, are among the proposed factors contributing to the observed image quality improvements with harmonic imaging in diagnostic ultrasound. Despite the wide clinical use of harmonic imaging, however, the details of the mechanisms responsible for the associated image quality improvements are still not fully understood. The concept of an "effective apodization" is defined as a potentially useful tool for providing a linear approximation to the nonlinearly generated field components.

Keywords: Nonlinear acoustics, Effective Apodization, Harmonic Imaging
PACS: 43.25-x, 43.25Jh, 43.25Cb, 43.60Fg

INTRODUCTION

The concept of an "effective apodization" was introduced to describe, in terms of linear propagation, the diffraction pattern of a nonlinearly generated harmonic component of a finite amplitude ultrasonic field.[1-6] Although the effective apodization concept can be applied equally to any harmonic component of the field, because of the relevance to diagnostic imaging, we have chosen to focus on the second harmonic *(2f)*. Figure 1 illustrates one of several apodization profile pairs, a Riesz window and its corresponding effective apodization at 2f, that has been investigated.

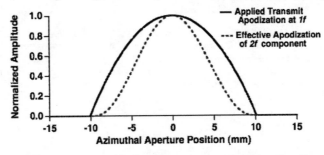

FIGURE 1. Profile of the Riesz window transmit apodization applied to the elements of the source array for the nonlinear propagation case and of the Effective Apodization at 2f for a Riesz window applied to the array elements for the linear propagation case.

CP838, *Innovations in Nonlinear Acoustics: 17th International Symposium on Nonlinear Acoustics,*
edited by A. A. Atchley, V. W. Sparrow, and R. M. Keolian
© 2006 American Institute of Physics 0-7354-0330-9/06/$23.00

Phase and amplitude aberrations resulting from tissue inhomogeneities represent a source of image degradation for both linear and nonlinear imaging modalities[7-12]; however, it is commonly believed that harmonic imaging suffers less from these effects than does fundamental imaging[1,13-17].

The effective apodization of the nonlinearly generated second harmonic portion *(2f)* of the field can be determined from measurements of the magnitude and phase of the *2f*-component in a transverse plane at one axial distance from the source.[1-6] The effective apodization is then obtained by linearly back-propagating this measured field to the source plane and extracting a profile of the magnitude across the face of this apparent source.[18-19] Figure 2 illustrates a comparison of measurements of the linear and nonlinear components of the finite amplitude fields with results of numerical simulation. As shown in Figure 3 and Figure 4, the effective apodization can be utilized to minimize the intrinsic differences in the lateral characteristics of linearly and nonlinearly generated beams.

METHODS

To conduct these experiments, the transmit characteristics of a phased array imaging system were modified to permit a Riesz-window transmit apodization at the fundamental frequency (1f) as well as an effective apodization weighting function at double this frequency (2f). These modifications to the transmit beamforming were validated with a series of hydrophone measurements both with [1] and without [2-6] the imposition of aberrating abdominal tissue speciments.

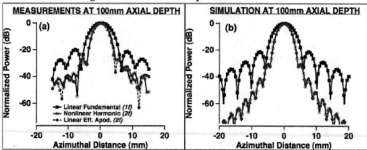

FIGURE 2. (a) Measurements of the transverse profiles of the fundamental, nonlinearly generated second harmonic, and the linearly propagated 2f transmit signal with an effective apodization for an unaberrated water path. (b) Simulated profiles of the ultrasonic fields at the same transverse slice. Agreement is observed between the nonlinear harmonic and the linear effective apodization profiles in both cases.

RESULTS

Studies have demonstrated that the shape of the effective apodization at 2f is independent of transmit frequency, F-number, focal position, and the attenuation properties of the specific propagation medium. [2, 4-6] The dependence of the effective apodization at 2f on specific choices of transmit apodization and source geometry has also been reported. [2, 4-5]

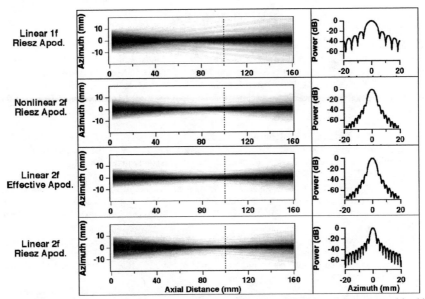

FIGURE 3. Simulated azimuthal field components from the clinical imaging array used in this study. Transverse profiles are shown normalized at each axial distance. The linearly propagated fundamental (top row, Riesz apodization) is observed to have the least compact field pattern. The nonlinearly generated harmonic (Riesz apodization) and its linear effective apodization counterpart (effective apodization of a Riesz window at 2f) are in general agreement (middle rows). Linear propagation at the same 2f frequency, but with a Riesz apodization (bottom row) has narrower main-lobe and higher side-lobes than the nonlinear and effective apodization cases. Transverse profiles are for a depth of 100 mm.

As shown in Figure 4, for an unaberrated propagation path, measurements of the linear-propagation effective apodization transmit case were in good agreement with the nonlinearly generated harmonic.[3] For the tissue aberrated paths, consistent with improved image quality, the nonlinearly generated second harmonic beam was shown to remain more compact than the corresponding linearly propagated beam patterns.[1]

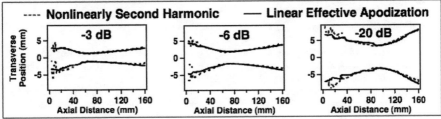

FIGURE 4. Hydrophone measurements of the nonlinearly generated second harmonic (2f) (dashed lines), and the linearly propagated field using the effective apodization at 2f (solid lines). Contour levels (-3, -6, and -20 dB) are shown at axial distances from 2 to 162 mm from the source array.

ACKNOWLEDGMENTS

This work is supported in part by NIH-R01-HL072761.

REFERENCES

1. K. D. Wallace, B. S. Robinson, M. R. Holland, M. Rielly, and J. G. Miller, "Experimental Comparisons of the Impact of Abdominal Wall Aberrators on Linear and Nonlinear Beam Patterns," presented at the 2004 IEEE Ultrasonics Symposium, (Cat. No. 04CH37553C), pp. 866-869.
2. R. J. Fedewa, K. D. Wallace, M. R. Holland, J. R. Jago, G. C. Ng, M. R. Rielly, B. S. Robinson, and J. G. Miller, "Spatial Coherence of Backscatter for the Nonlinearly Produced Second Harmonic for Specific Transmit Apodizations," IEEE Trans. Ultrason. Ferroelectr. Freq. Control 51, 576-588 (2004).
3. K. D. Wallace, R. J. Fedewa, M. R. Holland, G. C. Ng, B. S. Robinson, J. R. Jago, M. Rielly, and J. G. Miller, "Measurements Comparing the Linearly Propagated Field Using an Effective Apodization and the Nonlinearly Generated Second Harmonic Field," presented at the 2003 IEEE Ultrasonics Symposium, (Cat. No. 03CH37476C), pp. 453-456.
4. R. J. Fedewa, K. D. Wallace, M. R. Holland, J. R. Jago, G. C. Ng, M. R. Rielly, B. S. Robinson, and J. G. Miller, "Spatial Coherence of the Nonlinearly Generated Second Harmonic Portion of Backscatter for a Clinical Imaging System," IEEE Trans. Ultrason. Ferroelectr. Freq. Control 50, 1010-1022 (2003).
5. R. J. Fedewa, K. D. Wallace, M. R. Holland, J. R. Jago, G. C. Ng, M. R. Rielly, B. S. Robinson, and J. G. Miller, "Effect of Changing the Transmit Aperture on the Spatial Coherence of Backscatter for the Nonlinearly Generated Second Harmonic," presented at the 2002 IEEE International Ultrasonics Symposium, (Cat. No. 02CH37388), pp. 1624-1627.
6. R. J. Fedewa, K. D. Wallace, M. R. Holland, J. R. Jago, G. C. Ng, M. R. Rielly, B. S. Robinson, and J. G. Miller, "Statistically Significant Differences in the Spatial Coherence of Backscatter for Fundamental and Harmonic Portions of a Clinical Beam," presented at the 2001 IEEE International Ultrasonics Symposium, Atlanta, GA, (Cat. No. 01CH37263), pp. 1481-1484.
7. T. D. Mast, L. M. Hinkelman, M. J. Orr, and R. C. Waag, "The effect of abdominal wall morphology on ultrasonic pulse distortion. II. Simulations," J. Acoust. Soc. Am. 104, 3651-64 (1998).
8. L. M. Hinkelman, T. D. Mast, L. A. Metlay, and R. C. Waag, "The effect of abdominal wall morphology on ultrasonic pulse distortion. I. Measurements," J. Acoust. Soc. Am. 104, 3635-49 (1998).
9. S. E. Masoy, T. F. Johansen, and B. Angelsen, "Correction of ultrasonic wave aberration with a time delay and amplitude filter," J. Acoust. Soc. Am. 113, 2009-2020 (2003).
10. D. Raichlin, "Direct estimation of aberrating delays in pulse-echo imaging systems," J Acoust Soc Am, vol. 88, pp.191-198 (1990).
11. S. W. Flax and M. O'Donnell, "Phase aberration correction using signals from point reflectors and diffuse scatters –basic principles," IEEE Trans. Ultrason. Ferroelect., Freq. Contr., vol. 35, pp. 758-767 (1988).
12. Q. Zhu, B. D. Steinberg, and R. L. Arenson, "Wavefront amplitude distortion and image sidelobe levels: Part II – In-vivo experiments," IEEE Trans. Ultrason. Ferroelect., Freq. Contr., vol. 40, pp. 275-282 (1993).
13. T. Varslot, G. Taraldsen, T. Johansen, and B. Angelsen, "Computer simulation of forward wave propagation in non-linear, heterogeneous, absorbing tissue," IEEE 2001 Ultrasonics Symposium. Proceedings. (Cat. No. 01CH37263), 1193-1196 (2001).
14. B. J. Geimanm R. C. Gauss, G. E. Trahey, "In vivo comparison of fundamental and harmonic lateral beam shapes," presented at the 2000 IEEE Ultrasonics Symposium, (Cat. No. 00CH37121), pp. 1669-1675, (2000)
15. G. Wojcik, J. Mould, S. Ayter, and L. Carcione, "A study of second harmonic generation by focused medical transducer pulses," presented at the 1998 IEEE Ultrasonics Symposium.
16. V. F. Humphrey, "Nonlinear propagation in ultrasonic fields: measurements, modeling, and harmonic imaging," Ultrasonics 38, pp. 267-272 (2000).
17. M. A. Averkiou, "Tissue Harmonic Imaging," presented at the 2000 IEEE International Ultrasonics Symposium, (Cat. No. 00CH37121), pp. 1563-1572 (2000).
18. J. W. Goodman, Introduction to Fourier Optics (McGraw-Hill, San Francisco, 1968).
19. M. E. Schafer and P. A. Lewin, "Transducer characterization using the angular spectrum method," J. Acoust. Soc. Am. 85, 2202-14 (1989).

Statistical phase-screen model for second-harmonic beam distortion by body wall tissue in tissue harmonic imaging

X. Yan* and M. F. Hamilton*

*Department of Mechanical Engineering, The University of Texas at Austin, 1 University Station, Austin, Texas 78712–0292, USA

Abstract. In certain clinical situations, tissue harmonic imaging reduces distortion due to phase aberrations introduced by the body wall layer. A statistical model was developed to describe the effects of random inhomogeneity in the body wall on the second-harmonic beam structure. This inhomogeneity is represented by a thin random phase screen located close to the source. Phase variations across the screen are characterized statistically. An analytical solution was derived for the expected value of the intensity of the second-harmonic field for a source that radiates a focused Gaussian beam. The focal beam pattern for the second-harmonic field is compared with that of the fundamental field as a function of correlation length and variance of the phase screen, for values based on measured human abdominal wall statistics.

Keywords: inhomogeneous media, tissue harmonic imaging
PACS: 43.25.Jh, 43.80.Qf

INTRODUCTION

Tissue harmonic imaging (THI) is a technique employed in medical ultrasound that utilizes nonlinear propagation effects in ultrasound beams of finite amplitude. The basis of this imaging technique is that phase aberrations introduced by the body wall inhomogeneity affect the second-harmonic field less than the beam radiated directly by the source [1, 2]. THI can therefore reduce artifacts and clutter in the images and improve its contrast resolution [1, 3]. Here we use a random phase screen to model the effects of body wall inhomogeneity. A perturbation solution is derived for the mean intensity of the second-harmonic field generated in a beam radiated from a focused Gaussian source. The coherent and scattered components of the field are separated analytically. Results are presented for relevant statistical parameter ranges for tissue [4].

MODEL

The random phase screen in front of the source (see Fig. 1) modifies radiation at the source frequency ω as $e^{j[\omega t + \phi(\mathbf{r})]}$, where ϕ is a random function of $\mathbf{r} = (x, y)$ with zero mean, variance σ^2, and Gaussian correlation function characterized by length l:

$$\langle \phi(\mathbf{r}) \rangle = 0, \quad \langle \phi^2(\mathbf{r}) \rangle = \sigma^2, \quad C_\phi(\mathbf{r}_1, \mathbf{r}_2) = \frac{\langle \phi(\mathbf{r}_1)\phi(\mathbf{r}_2) \rangle}{\langle \phi^2(\mathbf{r}) \rangle} = \exp\left(-\frac{|\mathbf{r}_1 - \mathbf{r}_2|^2}{l^2} \right). \quad (1)$$

CP838, *Innovations in Nonlinear Acoustics: 17th International Symposium on Nonlinear Acoustics*,
edited by A. A. Atchley, V. W. Sparrow, and R. M. Keolian
© 2006 American Institute of Physics 0-7354-0330-9/06/$23.00

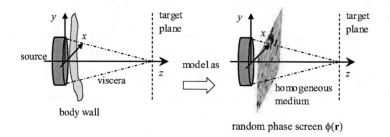

FIGURE 1. Random phase screen model of body wall.

The propagation medium (tissue) beyond the phase screen is assumed to be homogeneous. Absorption and dispersion in the tissue are not considered here, although their effect on the diffraction patterns is negligible [5].

The acoustic pressure is written $p = p_1 + p_2$ where $p_n = \frac{1}{2}q_n(\mathbf{r}, z)e^{jn\omega t} + \text{c.c.}$ and it is assumed that $|q_2| \ll |q_1|$, such that the complex pressure amplitudes are determined by the KZ equation as follows:

$$\frac{\partial q_1}{\partial z} + \frac{j}{2k}\left(\frac{\partial^2 q_1}{\partial x^2} + \frac{\partial^2 q_1}{\partial y^2}\right) = 0, \quad \frac{\partial q_2}{\partial z} + \frac{j}{4k}\left(\frac{\partial^2 q_2}{\partial x^2} + \frac{\partial^2 q_2}{\partial y^2}\right) = \frac{k\beta}{2\rho_0 c_0^2}q_1^2. \quad (2)$$

These equations are solved successively, where ρ_0, c_0, β, and $k = \omega/c_0$ are the tissue density, sound speed, coefficient of nonlinearity, and wavenumber at angular frequency ω, respectively. To obtain analytical solutions we assume radiation from a focused source with Gaussian amplitude shading: $q_1(\mathbf{r}, 0) = p_0 \exp[-(1 - jG)r^2/a^2 + j\phi(\mathbf{r})]$ and $q_2(\mathbf{r}, 0) = 0$, where $r = |\mathbf{r}|$, a is the effective source radius, $G = ka^2/2d$ is the focal gain, d is the focal length, and p_0 is the pressure amplitude in the source plane. The solutions for the primary field q_1 and second-harmonic field q_2 can be expressed as integrals over the source function [5].

We seek the ensemble averaged beam intensities $\langle I_n \rangle = \langle |q_n|^2 \rangle / 2\rho_0 c_0$. The variance of the random phase is assumed to be small, which permits the solutions for the intensities to be expanded in powers of the small parameter σ^2 and separated into terms corresponding to the coherent and scattered components of the field. Evaluated through order σ^2 and in the focal plane ($z = d$), the solution for the primary beam is

$$\langle I_1 \rangle = (1 - \sigma^2)I_1^{\text{coh}} + \sigma^2 I_1^{\text{sc}}, \quad (3)$$

$$I_1^{\text{coh}} = \frac{G^2 p_0^2}{2\rho_0 c_0}\exp(-2G^2\xi^2), \quad I_1^{\text{sc}} = \frac{G^2 p_0^2}{2\rho_0 c_0(1 + 2v^2)}\exp\left(-\frac{2G^2\xi^2}{1 + 2v^2}\right), \quad (4)$$

which was obtained previously in optics by Clarke [6], and for the second harmonic

$$\langle I_2 \rangle = (1 - 2\sigma^2)I_2^{\text{coh1}} - 2\sigma^2 I_2^{\text{coh2}} + 4\sigma^2 I_2^{\text{sc}}, \quad (5)$$

$$I_2^{\text{coh1}} = \frac{G^2 P_2^2}{2\rho_0 c_0}\exp(-4G^2\xi^2)\frac{|\ln(j/G)|^2}{1 + G^2}, \quad (6)$$

260

$$I_2^{\text{coh2}} = \frac{G^2 P_2^2}{2\rho_0 c_0} \exp\left(-4G^2\xi^2\right) \text{Re}\left\{ \frac{\ln(-jG - 2jv^2/G)\ln(j/G)}{(1 + 2v^2 - jG)(1 + jG)} \right\}, \tag{7}$$

$$I_2^{\text{sc}} = \frac{G^2 P_2^2}{2\rho_0 c_0} \exp\left(-4G^2\xi^2\right) \int_1^\infty \int_1^\infty \frac{\exp\left[4G^2 v^2 \xi^2 / M(t_1, t_2)\right] dt_1 dt_2}{M(t_1, t_2) t_1 t_2 (1 - jG + jGt_1)(1 + jG - jGt_2)}, \tag{8}$$

where $\xi = r/a$, $v = a/l$, $P_2 = \beta p_0^2 k^2 a^2 / 4\rho_0 c_0^2$, and

$$M(t_1, t_2) = \frac{v^2}{2}\left(\frac{1}{1 - jG + jGt_1} + \frac{1}{1 + jG - jGt_2}\right) + 1 + v^2. \tag{9}$$

All components of the coherent radiation (I_1^{coh}, I_2^{coh1}, and I_2^{coh2}) possess Gaussian beam profiles that are the same as when there is no phase screen. For the scattered radiation, while I_1^{sc} exhibits a Gaussian beam profile (broader than I_1^{coh}), I_2^{sc} does not. Note that the variance affects only the relative levels of the coherent and scattered radiation, whereas the correlation length also affects their field structures.

RESULTS

The statistical solution was compared with direct numerical simulations obtained using an angular spectrum approach [5]. Individual realizations of phase screens having statistics corresponding to Eqs. (1) were constructed numerically. Values selected for the parameters v and σ are based on measured statistics for human abdominal wall [4], with source dimensions and frequencies typical of those used in THI. Random phase screens having two different correlation lengths are considered. Normalized intensity level distributions calculated numerically, along the x and y axes passing through the focus, are shown as dashed lines in the first two rows of Fig. 2. The corresponding solid lines are the statistical solutions presented above for $\langle I_1 \rangle$ and $\langle I_2 \rangle$, and the dot-dash lines are beam patterns in the absence of a phase screen (equivalent to I_1^{coh} and I_2^{coh1}).

In the third row of Fig. 2, the dashed lines are the ensemble averages of the direct numerical simulations. For each distance r from the z axis, the intensity was averaged around the circle defined by that radius. The solid and dot-dash lines are repeated from the first two rows of Fig. 2. Very good agreement between the statistical theory and the simulations is achieved near the z axis. Unlike in the primary beam, the scattered energy in the second-harmonic beam is more concentrated along the z axis, and it decreases more rapidly with distance from the z axis. Note that decrease in correlation length causes broadening of the scattered radiation field.

CONCLUSION

A benefit of the statistical solution is that the coherent and scattered components of the primary and second-harmonic beams are separated analytically. The solution is based on the assumptions that the source has Gaussian amplitude distribution, inhomogeneity of the medium is confined to a thin phase screen in front of the source, the statistics of the phase screen are Gaussian, and its variance is small. Under these conditions, the

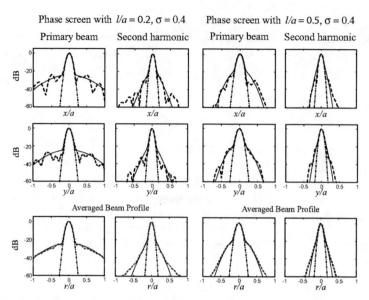

Phase screen with $l/a = 0.2$, $\sigma = 0.4$ Phase screen with $l/a = 0.5$, $\sigma = 0.4$

Primary beam   Second harmonic   Primary beam   Second harmonic

Averaged Beam Profile   Averaged Beam Profile

FIGURE 2. Intensity levels in the focal plane for phase screens with two different spatial correlation lengths. Solid lines: statistical solutions $\langle I_n \rangle$. Dot-dash lines: solutions in absence of phase screen. Dashed lines: direct numerical simulations of single realizations. Dashed lines in first two rows: realizations along x and y axes through focus. Dashed lines in third row: ensemble averages of numerical simulations.

solution is in good agreement with results obtained by direct numerical simulations. The reduction in radiation scattered off axis in the second-harmonic beam, relative to that in the primary beam, supports the observed reduction of unwanted artifacts such as clutter and haze by THI.

Financial support was provided by the Office of Naval Research and the Civilian Research and Development Foundation.

REFERENCES

1. M. A. Averkiou, "Tissue harmonic imaging," *IEEE Ultrasonics Symposium*, **2**, 1563–1572 (2000).
2. P. T. Christopher, "Finite amplitude distortion-based inhomogeneous pulse echo ultrasonic imaging," *IEEE Trans. Ultrason. Ferroelctr. Freq. Control*, **44**, 125–139 (1997).
3. B. Ward, A. C. Baker, and V. F. Humphrey, "Nonlinear propagation applied to the improvement of resolution in diagnostic medical ultrasound," *J. Acoust. Soc. Am.*, **101**, 143–154 (1997).
4. L. M. Hinkelman, D. -L. Liu, L. A. Metlay, and R. C. Waag, "Measurements of ultrasonic pulse arrival time and energy level variations produced by propagation through abdominal wall," *J. Acoust. Soc. Am.*, **95**, 530–541 (1994).
5. X. Yan, "Statistical model of beam distortion by tissue inhomogeneities in tissue harmonic imaging," Ph.D. dissertation, The University of Texas at Austin, 2004.
6. R. H. Clarke, "Analysis of laser beam propagation in a turbulent atmosphere," *ATT Technical Journal*, **64**, 1585–1601 (1985).

Comparison of mechanisms involved in image enhancement of Tissue Harmonic Imaging

Robin O. Cleveland and Yuan Jing

Department of Aerospace and Mechanical Engineering, Boston University
110 Cummington St, Boston, MA 02215

Abstract. Processes that have been suggested as responsible for the improved imaging in Tissue Harmonic Imaging (THI) include: 1) reduced sensitivity to reverberation, 2) reduced sensitivity to aberration, and 3) reduction in the amplitude of diffraction side lobes. A three-dimensional model of the forward propagation of nonlinear sound beams in media with arbitrary spatial properties (a generalized KZK equation) was developed and solved using a time-domain code. The numerical simulations were validated through experiments with tissue mimicking phantoms. The impact of aberration from tissue-like media was determined through simulations using three-dimensional maps of tissue properties derived from datasets available through the Visible Female Project. The experiments and simulations demonstrated that second harmonic imaging suffers less clutter from reverberation and side-lobes but is not immune to aberration effects. The results indicate that side lobe suppression is the most significant reason for the improvement of second harmonic imaging.

Keywords: Nonlinear acoustics, medical ultrasound, harmonic imaging, inhomogeneous media.
PACS: 43.25.Cb, 43.35.Bf, 43.80.Qf

INTRODUCTION

Muir and Carstensen [1] first proposed that diagnostic ultrasound scanners might produce large enough signals to cause nonlinear propagation distortion as the pulse travels through the body. The initial interest in nonlinear effects was because of the potential for enhanced heating which occurs because the nonlinearly generated harmonics suffer greater absorption than the fundamental [2].

Although the possibility of using nonlinearly generated harmonics to improve imaging had been recognized in the underwater acoustics community [3] the use in was delayed until the advent of wideband transducers to image microbubble contrast agents. There were three, almost simultaneous, reports of tissue harmonic imaging. Ward et al. [4] used an ultrasound transducer to image wire targets in water and found that the images from the nonlinear harmonics were better than those of the fundamental. Averkiou et al. [5] reported improved imaging on a clinical ultrasound scanner that had been programmed to form images from the backscattered second harmonic signal. Christopher [6] used numerical simulations to demonstrate that images formed from higher harmonics should have improved resolution.

It is widely accepted that imaging based on the second harmonic results in improved image quality but the reasons for the improvement are still poorly understood. Mechanisms that have been discussed in the literature are: side-lobe

CP838, *Innovations in Nonlinear Acoustics: 17ᵗʰ International Symposium on Nonlinear Acoustics*,
edited by A. A. Atchley, V. W. Sparrow, and R. M. Keolian
© 2006 American Institute of Physics 0-7354-0330-9/06/$23.00

suppression, reverberation suppression, and aberration suppression. In this paper we use numerical simulations of ultrasound pulses in inhomogeneous tissue to gain insight into the mechanisms that may be at play in THI.

METHODS

We employed a generalized form of the Khokhlov-Zabolotskaya-Kuznetsov (KZK) equation that is appropriate for the propagation of nonlinear beams in inhomogeneous media with attenuation and dispersion modeled with relaxation processes:

$$\frac{\partial p}{\partial z} = \frac{c_0}{2} \int_{-\infty}^{t'} \left(\frac{\partial^2 p}{\partial x^2} + \frac{\partial^2 p}{\partial y^2} \right) dt'' + \frac{\beta}{2\rho_0 c_0^3} \frac{\partial p^2}{\partial t'} + \frac{\delta}{2c_0^3} \frac{\partial^2 p}{\partial t'^2} + \sum_v \frac{c_v'}{c_0^2} \int_{-\infty}^{t'} \frac{\partial^2 p}{\partial t''^2} e^{-(t'-t'')/t_v} dt''$$

$$+ \frac{c'}{c_0^2} \frac{\partial p}{\partial t'} + \frac{1}{2\rho_0} \frac{\partial \rho_0}{\partial z} p - \frac{c_0}{2\rho_0} \int_{-\infty}^{t'} \frac{\partial \rho_0}{\partial x} \frac{\partial p}{\partial x} + \frac{\partial \rho_0}{\partial y} \frac{\partial p}{\partial y} dt''$$

(1)

Here p is acoustic pressure, c_0 small-signal sound speed, ρ_0 density, β coefficient of nonlinearity, δ diffusivity of sound, c_v dispersion and t_v characteristic time of the vth relaxation process, and c' is the spatial perturbation in the small-signal sound speed. In this equation all the terms, except c_0, are allowed to vary arbitrarily in space. The four terms on the right side of the upper line account for diffraction, nonlinearity, thermoviscous absorption and a finite number of relaxation processes respectively. The terms on the lower line account for the spatial variation in sound speed and density. Variation in other properties do not add new terms to the KZK equation to second order. Equation 1 was solved in the time domain using a modified version of the Texas code [7]. The numerical model was validated by measuring the pressure field of both a single-element transducer and a linear array, after propagation through a tissue mimicking phantom. Phantoms were constructed with varying thickness to simulate the presence of an aberrating layer.

For numerical simulations to capture the interaction of ultrasound pulses with tissue it is necessary to correctly model the inhomogeneities that are present in the body. A technical challenge is determining maps of acoustic properties that are appropriate for tissue. We employ a technique developed by Mast [8] whereby images provided through the Visible Human Project may be converted into approximate acoustic properties. We extended Mast's algorithm, which yielded density, sound speed, and attenuation, to also be able to create a three-dimensional map of the coefficient of nonlinearity. Figure 1(a) shows an RGB image from the thorax of the Visible Female. A fully three-dimensional data set (40 x 80 x 110 mm) of sound speed, density, attenuation and coefficient of nonlinearity was formed from 120 slices of the Visible Female. Shown in Figs. 1(b) and 1(c) are slices through the sound speed data set, one in the horizontal plane (what will be the scan plane of transducer in the simulations) corresponding to Fig 1(a) and the other in the coronal plane (elevation plane of transducer). The appropriate power law for the attenuation of tissue was modeled by fitting the coefficient for the diffusivity of sound and two relaxation processes to match an $f^{1.1}$ dependence from 300 kHz to 20 MHz.

The source for the simulations was modeled after a BK 8665 curved array driven by an AN2300 Analogic Ultrasound Engine. The transducer consists of 128 elements each 0.5 mm wide and 13 mm high with a centre-to-centre distance of 0.525 mm. The focus in the elevation plane is fixed at 70 mm and focusing in the scan plane is achieved by the 64 channel beamformer of the AN2300. For harmonic imaging a 3-cycle, 2.5-MHz center-frequency pulse was used as the source waveform and for these simulations the scan plane was focused at 70 mm (same as elevation plane).

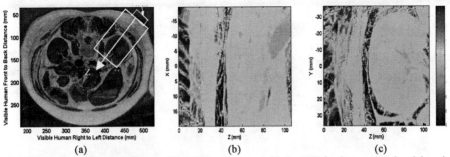

| (a) | (b) | (c) |

FIGURE 1. (a) RGB image from the Visible Female torso with the ROI for the computational domain shown by the large white rectangle (b) Slice through 3D sound speed map in transverse plane (z-axis is the propagation distance and the source is at z=0) (c) Slice through sound speed map in coronal plane. Colour scale ranges from 1450 m/s (blue) to 1600 m/s (red).

RESULTS AND CONCLUSIONS

Figure 2 shows the spatial distribution of the sound pressure levels corresponding to the fundamental (2.5 MHz) and nonlinearly generated second harmonic (5 MHz) in the tissue for both the scan and elevation planes (recall the simulation is fully three-dimensional and so these are just two-dimensional slices through the 3D field). The results show that both the fundamental and the second harmonic are strongly affected by the inhomogeneities of the soft tissue, e.g., there is a pre-focal region of high pressure from the bottom left corner in both scan plane images. The second harmonic does exhibit less near field structure and lower side-lobes. However in comparison to results for homogeneous tissue (data not shown) the change in the side-lobe levels due to the presence of inhomogeneities is at least as large (and often larger) for the second harmonic compared to the fundamental. However, the side-lobe levels of the second harmonic are still better than those of the fundamental. Simulations were carried out for a number of other locations and the results were found to be robust.

In conclusion, the simulations reported here show that the presence of aberration affects the second harmonic at least as much as fundamental. We contend that the second harmonic is **not** immune to aberration. We note that in previous work the effects of inhomogeneities were modeled as a phase screen rather than a fully 3D inhomogeneous medium and perhaps a phase-screen is not sufficient to capture the propagation of harmonic beams in tissue. Our simulations show that the second harmonic signal remains weak in the nearfield and so THI should be immune to reverberation. We also found that the side-lobes of the second harmonic, although they were affected by the inhomogeneities, remained below those of the fundamental.

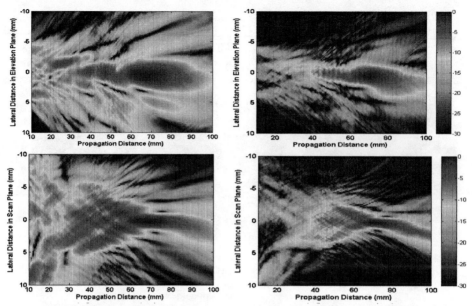

FIGURE 2. Left: Spatial distribution of the fundamental harmonic in the elevation (upper) and scan (lower) planes. Right: Spatial distribution of the second harmonic in the elevation (upper) and scan (lower) planes. The direction of propagation is left to right and the color scale is in dB re peak amplitude. It can be seen that the second harmonic has less amplitude in the near field (propagation distances less than 30 mm) and lower side lobes, but the focal region is distorted by aberration.

ACKNOWLEDGMENTS

This work was supported by CenSSIS, the Center for Subsurface Sensing and Imaging Systems, under the ERC Program of the National Science Foundation (EEC-9986821).

REFERENCES

1. T. G. Muir, and E. L. Carstensen, *Prediction of nonlinear acoustic effects at biomedical frequencies and intensities.* Ultrasound in Med.& Biol., 6:345-357 (1980).
2. E. L. Carstensen, N. D. McKay, D. Delecki, and T. G. Muir, *Absorption of finite amplitude ultrasound in tissues.* Acoustica, 51:116-123 (1982).
3. T. G. Muir, *Nonlinear acoustics, a new dimension in underwater sound,* in Science, Technology and the Modern Navy ed. E. I. Salkovitz. (1976).
4. B. Ward, A. C. Baker, and V. F. Humphrey, *Nonlinear propagation applied to the improvement of resolution in diagnostic medical ultrasound.* J. Acoust. Soc. Am., 101: 143 - 154. (1997).
5. M. A. Averkiou, D. N. Roundhill, J. E. Powers. *A new imaging technique based on the nonlinear properties of tissues.* in IEEE Ultrasonics Symposium Proceedings (Toronto, Canada 1997).
6. P. T. Christopher, *Finite amplitude distortion-based inhomogeneous pulse echo ultrasonic imaging.* IEEE Trans Ultrason Ferroelectr Freq Contr, 44: 125-139 (1997)
7. Yang, X and Cleveland, RO, "Time-domain simulation of nonlinear acoustic beams generated by rectangular pistons with application to harmonic imaging," J. Acoust. Soc. Am. 117, 113-123 (2005)
8. Mast, TD, "Two- and three-dimensional simulations of ultrasonic propagation through human breast tissue," Acoust. Res. Lett. Online 3: 53-58 (2002).

Combined Tissue Imaging Technique: Deconvolved Inverse Scattering Impedance Imaging with Account of the Second Harmonic Signal

Nikolay Kharin*, Diana Driscoll[@] and William Tobocman[@]

*Department of Biomedical Engineering, The Cleveland Clinic Foundation, Cleveland, OH 44195, USA

@ Department of Physics, Case Western Reserve University, Cleveland, OH 44106, USA

Abstract. Ultrasound tissue harmonic imaging (THI) can greatly improve the diagnostic quality of images. While a large transducer bandwidth is necessary for this technique, the axial resolution is still limited in THI. Technological limitations for THI are also more pronounced at higher frequencies because of the small size of transducer. The depth of penetration is also less due to higher attenuation. Here we describe a new combined imaging technique: Born-approximation deconvolved inverse scattering (BADIS) imaging and tissue harmonic imaging (THI). BADIS technique deconvolves the incident pulse from the reflected pulse, and uses the resulting impulse response to produce an image of the acoustic impedance distribution. These images are free of speckle and have improved axial resolution. BADIS allows the use of lower frequencies than those would be required by the pulse-echo method to achieve the same axial resolution. To provide further improvement we took into account of the nonlinear properties of tissue and implemented the second harmonic signal generated due to this nonlinearity in BADIS technique. We obtained results of soft tissue imaging that demonstrated the improvement of the image quality in comparison with the common pulse-echo imaging technique

Keywords: second harmonic imaging, ultrasound, inverse scattering, tissue harmonic imaging
PACS: 43.35.W, 43.35.Z, 43.60.+d, 43.25.+y, 03.40.K

INTRODUCTION

The conventional analysis for ultrasound imaging takes the absolute value of the reflected pulse to be the acoustic reflectivity distribution of the target. This is called the pulse-echo method. The BADIS (Born Approximation Deconvolved Inverse Scattering) method improves on the pulse-echo method by deconvolving the incident pulse from the reflected pulse. The theory and application of the BADIS method for A-scan data has been presented in earlier publications by Tobocman[1] and Tobocman et al [2]. It has been found that the BADIS method can provide vivid, high - resolution images of the layered structures. Recently, Tobocman et al [3] presented a more complete derivation of the BADIS method and applied it to B-scan data. It was shown that if the transverse variation of the impedance distribution over the lateral resolution (spot size) is small, the method provides well - resolved images of the acoustic impedance distribution of a target.

CP838, *Innovations in Nonlinear Acoustics: 17th International Symposium on Nonlinear Acoustics*,
edited by A. A. Atchley, V. W. Sparrow, and R. M. Keolian
© 2006 American Institute of Physics 0-7354-0330-9/06/$23.00

Since soft tissue structures reflect ultrasound relatively weakly, it is generally advisable to use very energetic ultrasound pulses. Thus, it is likely that nonlinear effects like second harmonic generation are taking place and it was independently suggested by number of the authors that the use of second harmonic as a possible way of enhancing the quality of medical ultrasound images. For the very short period of time the tissue harmonic imaging (THI) demonstrates its fruitfulness. Here we provide the results of our attempt to image the soft tissue using BADIS with account of the second harmonic signal reflected from the target.

EXPERIMENT

We based our experiments on the theoretical analysis provided by Kharin *et al.* [4], where the expressions for modified impulse response and impedance, taking into account the appearance of the second harmonic component, have been derived. To perform the second harmonic imaging we use a hybrid transducer by Krautkramer Branson Company. A hybrid transducer consists of a cylindrical 5 MHz transducer wrapped in an annulus-shaped 2.25 MHz transducer. Both of them are focused at a point 2 inches in front of the lens. The transducers have terminals that allow them to work individually or together. When working together, they can either transmit and receive a combined signal, or one can emit the signal and the other one can detect the reflected signal. These operations were controlled by the pulser-receiver. The transducer and soft tissue specimen were immersed in the salt water solution. An articulated mount arm holded the transducer and two laser beams, which were aimed to pinpoint the focus of the ultrasound beam. The lower frequency component of the transducer was excited by a 200-300 V pulse, produced by the pulse generator section of a pulser-receiver (Model 5052PR, Panametrics, Inc.), and the reflected ultrasonic pulses from the target were received by the higher frequency component of the same transducer. The converted voltage signals were amplified by the receiver section of the pulser-receiver and digitized with a 12 bit transient digitizer (CompuScope 8012A, Gage Applied, Inc.). The sampling rate was 80 MHz. A pulse generator served to trigger the transient digitizer for the start of data acquisition. The transducer arm motion was under control of a two-axis step motor controller (Model 6006-DB, AMSI Corp.). The arm was connected to two linear translation platforms, which were actuated by two stepping motor linear actuators, providing computer-controlled motion of the transducer arm in two directions (X and Y). The height positioning of the arm (Z-direction) was provided by the third translation platform, which was manually driven by a micrometer. The software that was utilized for the control of instruments, the data acquisition, and the data analysis had been programmed in LabView.
The porcine gastro-esophageal junction was used as a tissue specimen.

RESULTS AND DISCUSSION

The gray-scale images of a porcine gastro-esophageal junction specimen are shown in Figure 1 and Figure 2.

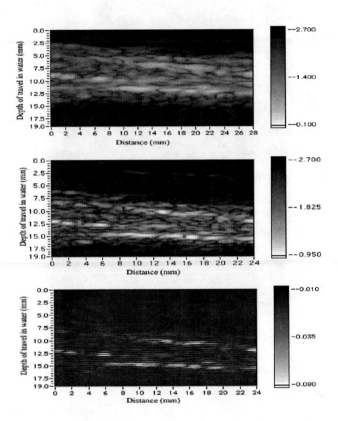

FIGURE 1. The 2.5 MHz pulse-echo (top), impulse response (middle) and acoustic impedance (bottom) images of *in vitro* porcine gastro-esophageal junction. Vertical axis – "Depth of travel in water (mm)", horizontal axis – "Distance (mm)"

The images are clearly visible, especially in the impulse response and the acoustic impedance distributions. We are interested in comparing the second harmonic image with the image generated at fundamental frequency. In other words, we need to compare Figure 1 with Figure 2. Figure 2 is the second harmonic image and it is superior to the one taken at 2.5 MHz.

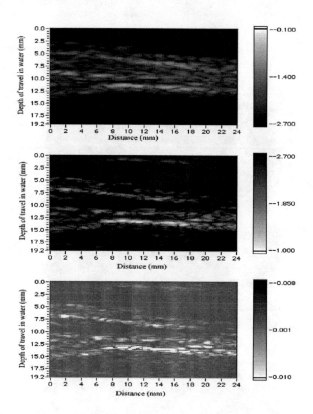

FIGURE 2. The second harmonic (pulse emitted by the 2.5 MHz and received by the 5 MHz component of the hybrid transducer) pulse-echo (top), impulse response (middle) and acoustic impedance (bottom) images of *in vitro* porcine gastro-esophageal junction.

We obtained fairly good impulse response and acoustic impedance distributions of the second harmonic signal while imaging soft tissue. The images have less speckle than their equivalent pulse-echo counterparts. We demonstrated here that we can provide an improvement of image quality using the combination of BADIS and THI imaging technique.

REFERENCES

1. Tobocman, W., *Curr.Top.Acoust* **1**, 247-265 (1994).
2. Tobocman, W., Santosh, K., Carter, J.R., and Haacke E.M., *Ultrasonics* **33**, 331-339 (1995).
3. Tobocman, W., Driscoll D., Shokrollahi N., and Izatt J., *Ultrasonics* **40**, 983-996.
4. Kharin N., Driscoll D., and Tobocman W., *Phys. Med. Biol.* **48**, 3239-3260 (2003).

Characterization and Ultrasound-Pulse Mediated Destruction of Ultrasound Contrast Microbubbles

Kausik Sarkar, Pankaj Jain, Dhiman Chatterjee*

Mechanical Engineering, University of Delaware, Newark, DE 19716, USA
*Also Mechanical Engineering, IIT Madras, Chennai 600036, India

Abstract. Intravenously injected encapsulated microbubbles improve the contrast of an ultrasound image. Their destruction is used in measuring blood flow, stimulating arteriogenesis, and drug delivery. We measure attenuation and scattering of ultrasound through solution of commercial contrast agents such as Sonazoid and Definity. We have developed a number of different interfacial rheology models for the encapsulation of such microbubbles. By matching with experimentally measured attenuation, we obtain the characteristic rheological parameters. We compare model predictions with measured subharmonic responses. We also investigate microbubble destruction under acoustic excitation by measuring time-varying attenuation data.

Keywords: Medical imaging, contrast agent, drug delivery
PACS: 43.80.Qf, 43.80.Ev, 43.25.Yw, 43.35.Wa

INTRODUCTION

A significant improvement of ultrasound image quality for procedure such as echocardiography is obtained by the use of encapsulated microbubble based contrast agents [1]. The dynamics of these encapsulated bubbles have been investigated by many researchers using in vitro ultrasound experiments[2] and models.[11,3,4] Microbubble destruction may be useful in real time blood flow velocity measurement[5], stimulating arteriogenesis[6], or targeted drug delivery.[7] In each application after microbubbles are intravenously injected, they are destroyed by an ultrasound pulse. The bubbles can be functionalized so that it preferentially attaches to specific tissues such as tumor. They can also be coated with drugs and genes that can be released to the targeted tissue upon destruction, without serious side effects. Destruction is studied by several researchers[8]. Two distinct mechanisms have been suggested for bubble destruction—slow gas diffusion and fast fragmentation.

In this paper, we offer two interfacial models of contrast microbubbles, specifically its encapsulation and apply them to contrast agent Sonazoid (GE Health Care, Oslo, Norway). Sonazoid consists of fluorocarbon gas with a surfactant membrane. We also investigate in-vitro destruction of contrast agent Definity (Bristol Myers-Squibb Imaging, North Ballerina, MA, USA). Definity is an FDA approved contrast agent containing perfluoropropane gas surrounded by a lipid encapsulation.

CP838, *Innovations in Nonlinear Acoustics: 17th International Symposium on Nonlinear Acoustics*,
edited by A. A. Atchley, V. W. Sparrow, and R. M. Keolian
© 2006 American Institute of Physics 0-7354-0330-9/06/$23.00

Encapsulation models

Previous models[11] of bubble encapsulation assumes it to be a layer of incompressible rubbery materials with homogeneous bulk material properties such as shear elasticity and viscosity, in contrast to the anisotropic nature of the shell made of only a few layers of molecules as shown by freeze-etch-TEM picture[9]. We have adopted a two-dimensional continuum fluid layer model with surface rheological parameters such as surface viscosities and elasticities[10] for the encapsulation. The model does not make the inappropriate assumption of an isotropic (in the thickness direction) continuum like existing ones[11]. The details of the models are provided elsewhere[3,4]. Briefly, the interfacial rheology determines the surface stresses in terms of rheological parameters such as surface tension γ (γ_0 for the viscoelastic model),

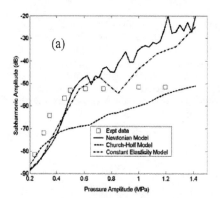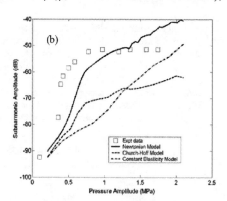

Figure 1: Predicted and measured subharmonic response for Sonazoid at various insonation frequencies: (a) 2 MHz and (b) 4.4 MHz

surface dilatational viscosity κ^s and surface dilatational elasticity E^s. Surface stresses in turn determine the jump in pressure as one passes from inside to outside of the bubble. We have developed two models—Newtonian viscous interface and viscoelastic interface. Once the equations of momentum conservation is properly integrated we obtain for the two models two Rayleigh-Plesset equations for the bubble radius $R(t)$:[4]

$$\rho\left(R\ddot{R}+\frac{3}{2}\dot{R}^2\right)=P_{G0}\left(\frac{R_0}{R}\right)^{3k}-4\mu\frac{\dot{R}}{R}-\frac{4\kappa^s\dot{R}}{R^2}-\frac{2\gamma}{R}-P_0+p_A(t)$$

$$\rho\left(R\ddot{R}+\frac{3}{2}\dot{R}^2\right)=P_{G0}\left(\frac{R_0}{R}\right)^{3k}-4\mu\frac{\dot{R}}{R}-\frac{4\kappa^s\dot{R}}{R^2}-\frac{2\gamma_0}{R}-\frac{2E^s}{R}\left[\left(\frac{R}{R_E}\right)^2-1\right]-P_0+p_A(t)$$

Where P_{G0} is the inside gas pressure initially, R_E is the unstressed equilibrium radius and μ the liquid viscosity. p_A is the excitation pressure, k the polytropic constant, and p_0 the hydrostatic pressure. We determine the parameters by matching the model predictions with measured attenuation through solution of Sonazoid[4]. The parameter values for Sonazoid are $\gamma = 0.6$ N/m, $\kappa^s = 0.01$ msP for the Newtonian, and

$\kappa^s = 0.01$ msP, $\gamma_0 = 0.0190$ N/m and $E^s = 0.51$ N/m for viscoelastic model. These models along with one from the literature[15,11] are then compared against measured subharmonic response (Figure 1). The Newtonian model performs better than others.

Ultrasound mediated destruction

We investigate in-vitro destruction of contrast agent contrast agent Definity® (Bristol Myers-Squibb Imaging, North Ballerina, MA, USA) by measuring attenuation of ultrasound as a function of time under excitation with varying pulse repetition frequency (PRF) and amplitude (Figure 2). An unfocused broadband 5 MHz transducer (-6dB bandwidth 3.1-6.55 MHz) T1 mounted on the wall of a plastic tank is excited by a function generator (33250A, Agilent, Palo Alto, CA, USA) and a power amplifier (ENI A150, Rochester, NY, USA) to produce one-cycle bursts. Another unfocused broadband transducer T2 (center frequency 3.5 MHz) mounted on the opposite wall of the tank acts as a receiver. The received signal, amplified by a Pulser-Receiver (5800PR, Panametrics, Waltham, MA, USA) was fed into an oscilloscope (TDS2012, Tektronix, OR, USA) and stored in a computer via GPIB interface. Labview® (National Instruments, USA) and Matlab® (Math-work Inc, Natick, MA, USA) were used for data acquisition and post-processing.

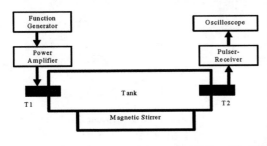

Figure 2. Attenuation setup

We have recently found that the attenuation becomes dependent on pressure for pressure values above 0.26 MPa for Definity[12] (which is much lower than the excitation levels studied here) making the linear theory of propagation and attenuation inappropriate. Therefore, rather than using attenuation as a function of frequency, we use the total attenuation of the energy of the pulse as a measure of effects of bubbles.

Figure 3 shows the variation of normalized attenuation (attenuation at any time divided by the attenuation at initial time) with time when bubbles are subjected to acoustic excitation with a PRF of 50 Hz under different pressure amplitudes. For the lowest pressure 0.78 MPa, we find that the attenuation level increases with time, indicating a transient growth in bubble radius. Presumably, the ultrasound excitation leads to structural deterioration of the encapsulation, such as appearance of small cracks, facilitating gas diffusion. Definity bubbles are made with perfluorocarbon gas which has low solubility in water. Initially, more air will diffuse in than heavy, less soluble perfluorocarbon gas diffuses out[13,14]. This causes the transient growth in bubble radius. Over longer period of time, the process reverses, and the attenuation reduces. With increased excitation levels, the attenuation decreases with time, indicating bubble destruction. We find that there is a threshold pressure, above which the measured attenuation shows a decrease with time. For all three PRFs (50,100 and 200 Hz) studied, it is around 1.2 MPa, even though one would expect that the critical pressure for destruction would reduce with increasing PRF. Increasing the pressure amplitude beyond this critical value results in faster decrease of attenuation indicating faster rate of destruction (fast fragmentation). The attenuation does not show much dependence on PRF except for intermediate excitation levels (Figure 4).

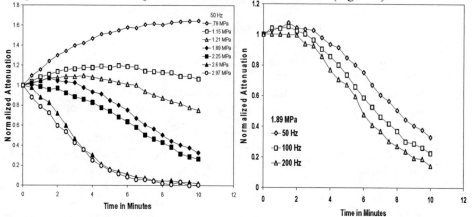

Figure 3. Variation of normalized attenuation with time for different acoustic pressure amplitudes for PRF of 50 Hz.

Figure 4. Normalized Attenuation with time under different PRF for 1.89 MPa

ACKNOWLEDGMENTS

KS acknowledges DOD contract DAMD17-03-1-0119 NSF contract CTS-0352829 and University of Delaware Research Foundation and numerous discussions with Flemming Forsberg (Thomas Jefferson University) and William Shi (Phillips Medical).

References

1. Goldberg, B.B., Raichlen, J.S. and Forsberg, F., "Ultrasound Contrast Agents: Basic Principles and Clinical Applications", edited by Martin Dunitz, London, (2001).
2. Frinking, P.J.A. and de Jong, N., *ultrasound Med. Biol.* **24(4)**, 523-533 (1998).
3. Chatterjee, D., Sarkar, K., *Ultrasound Med. Biol.* **29**, 1749-1757 (2003).
4. Sarkar, K., Shi, W.T., Chatterjee, D. and Forsberg, F., *J. Acous Soc. Am.* **118(1)**, 539-550 (2005).
5. Sonne, C, Xie, F., Lof, J., et al, *Am. Soc. Echocardiography* 16(11), 1178-1185 (2003).
6. Song, J, Ming, Q. Kaul, S. and Price, R.J., *Circulation* **106(12)**, 1550-1555 (2002).
7. Shohet, R.V., Chen, S., Zhou, Y.T. et al, *Circulation*, **101**, 2554-2556 (2000).
8. Chomas, J.E., Dayton, P.A., May, D., et al, *Appl. Phys. Lett.* **77(7)**, 1056-1058 (2000).
9. Christiansen, C., Kryvi, H., Sontum, P.C. and Skotland,T., *Biotechnol. Appl. Biochem.* **19**, 307-320 (1994).
10. Edwards, D.A., Brenner, H. and Wasan, D.T., "Interfacial transport processes and rheology" Butterworth-Heinemann, (1991).
11. Church, C.C., *J. Acoust. Soc. Am.* **97**, 1510-1521 (1995).
12. Chatterjee, D., Sarkar, K., Jain, P. and Chreppler, N. E., *Ultrasound Med. Biol*, **31(6)**, 781-786 (2005).
13. Kabalnov, A., Klein, D., Pelura, T., Schutt, E. and Weers, J.,*Ultrasound Med. Biol.*, **24(5)**, 739-749 (1998).
14. Kabalnov, A., Bradley, J., Flaim, S., Klein, D et al., Ultrasound Med. Biol., **24(5)**, 751-760 (1998).
15. Hoff, L., Sontum, P.C. and Hovem, J.M.(2000). J. Acoust. Soc. Am. **107(4)**, pp. 2272-2280.

Nonlinear behavior of ultrasound-insonified encapsulated microbubbles

Michiel Postema*, Nico de Jong[†] and Georg Schmitz*

*Institute for Medical Engineering, Department of Electrical Engineering and Information
Technology, Ruhr-Universität Bochum, 44780 Bochum, Germany
[†]Physics of Fluids, Faculty of Science and Technology, University of Twente, Enschede,
The Netherlands

Abstract. Ultrasound contrast agents consist of small encapsulated bubbles with diameters below $10 \, \mu$m. The encapsulation influences the behavior of these microbubbles when they are insonified by ultrasound. The highly nonlinear behavior of ultrasound contrast agents at relatively high acoustic amplitudes (mechanical index>0.6) has been attributed to nonlinear bubble oscillations and to bubble destruction. For microbubbles with a thin, highly elastic nanoshell, it has been demonstrated that the presence of the nanoshell becomes negligible at high insonifying amplitudes. From our simulations it follows that the Blake critical radius is not valid for microbubble fragmentation. The low maximal excursion observed and simulated for a thick, stiff-shelled microbubble is in agreement with previous acoustic analyses. The ultrasound-induced gas release from stiff-shelled bubbles has been reported. However, we also observed gas release from microbubbles with a thin, elastic shell.

Keywords: Ultrasound contrast agent, encapsulated microbubble, nanoshell
PACS: 43.25.Yw

INTRODUCTION

Ultrasound contrast agents consist of gas microbubbles encapsulated by a nanoshell. Because the resonance frequencies of these microbubbles lie in the clinical ultrasonic range, contrast agents have been used for diagnostic imaging purposes. If a microbubble is subjected to very small pressure changes with an amplitude much smaller than the static ambient pressure, its radial excursion may be considered linear [1, 2]. Contrary to tissue, however, a microbubble exhibits highly nonlinear behavior at higher acoustic amplitudes. With harmonic imaging methods, microbubbles are therefore suitable markers for perfused areas.

We investigate the influence of the nanoshell on the behavior of ultrasound-insonified encapsulated microbubbles. More specifically, we are interested in finding the conditions needed for shell rupture.

OSCILLATING MICROBUBBLES

Let us consider a microbubble with an equilibrium radius R_0 and a shell thickness $h_s \ll R_0$. In equilibrium, the gas pressure inside the bubble p_{g0} can be expressed as:

$$p_{g0} = p_0^\infty - p_v + \frac{2s}{R_0} . \qquad (1)$$

CP838, *Innovations in Nonlinear Acoustics: 17th International Symposium on Nonlinear Acoustics,*
edited by A. A. Atchley, V. W. Sparrow, and R. M. Keolian

Here, p_0^∞ is the static pressure of the liquid, p_v is the vapor pressure, and s is the surface tension. For an encapsulated gas bubble, the oscillating behavior has been described by a modified RPNNP equation, named after its developers Rayleigh, Plesset, Noltingk, Neppiras, and Poritsky [1, 2]:

$$\rho R \ddot{R} + \tfrac{3}{2}\rho \dot{R}^2 = p_{g0}\left(\frac{R_0}{R}\right)^{3\gamma} + p_v - p_0^\infty - \frac{2s}{R} - 2S_p\left(\frac{1}{R_0} - \frac{1}{R}\right) - \delta\,\omega\rho R\dot{R} - p_a(t),$$

(2)

where $p_a(t)$ is the acoustic pressure in time, R is the instantaneous microbubble radius, S_p is the shell stiffness parameter, δ is the total damping coefficient, γ is the specific heat ratio, ρ is the liquid density, and ω is the angular driving frequency. $R(t)$ is periodic with period $T = T_e + T_c$, where $_e$ stands for expansion and $_c$ stands for contraction. The excursion is defined by $a(t) = R(t) - R_0$. The shell stiffness parameter is given by [3]:

$$S_p = \frac{8\pi E h_s}{1 - v},$$

(3)

where E is Young's modulus, and v is the Poisson ratio. For albumin and lipid nanoshells, we take $0.499 < v < 0.500$. S_p can be estimated from optical observations of radius–time curves or from acoustical data using the relation [4]:

$$\omega_s^2 = \omega_r^2 + \frac{S_p}{4\pi R_0^3 \rho},$$

(4)

where ω_s is the angular resonance frequency of the nanoshelled microbubble, ω_r is the angular resonance frequency of an unencapsulated microbubble of the same size.

At high acoustic pressures (mechanical index >0.6) destructive phenomena have been observed, such as microbubble fragmentation, coalescence, and ultrasonic cracking [5]. The critical stress at which a shell ruptures σ_c, is related to Young's modulus by:

$$\sigma_c \approx E \varepsilon_c,$$

(5)

where ε_c is the critical lateral shell deformation. For most biomaterials, $\varepsilon_c < 0.5$. Here, we treat two opposite cases: I. microbubbles with a thin, very elastic shell, and II. microbubbles with a thick, fairly stiff shell.

CASE I: THIN, ELASTIC SHELL

For microbubbles with a thin, highly elastic monolayer lipid nanoshell, like SonoVue™ and other Bracco agents, it has been demonstrated that the presence of the nanoshell becomes negligible at high insonifying amplitudes [5]. Such microbubbles have been observed to expand to more than ten-fold their initial surface areas during expansion. The nanoshell behaves like an elastic membrane that ruptures under relatively small strain [6]. By the time of maximal expansion, therefore, the nanoshell has ruptured, leaving newly formed clean free interfaces. This confirms that these microbubbles may be assumed free (unencapsulated). Similar to inertial cavitation, the relatively slow

TABLE 1. Elastic properties of three contrast agents

	ω_r $[2\pi \times 10^6\,\mathrm{rad\,s^{-1}}]$	$\langle R_0 \rangle$ $[\mu\mathrm{m}]$	S_p $[\mathrm{kg\,s^{-2}}]$	E^* $[10^6\,\mathrm{Pa}]$
Albunex®	2	4.0	10	2
Quantison™	4	1.6	25	2
SonoVue™	3	1.0	1.1	2

* Estimated with $\nu \approx 0.5$

microbubble expansion is followed by a rapid collapse: $T_e > T_c$. The microbubble has a time-varying radius $R(t) > 0$. Because the expansion is virtually unlimited, however, the excursion can be asymmetric as well: $\max(a(t)) > \min(a(t))$.

We analyzed the occurrence of microbubble fragmentation with respect to the intrinsic energy of the bubble [7]. Fragmentation occurs exclusively during the collapse phase. We hypothesize that fragmentation will only occur if and only if the kinetic energy of the collapsing microbubble is greater than the instantaneous bubble surface energy. From our simulations it follows that the Blake critical radius is not a good approximation for a fragmentation threshold.

CASE II: THICK, STIFF SHELL

For microbubbles with a thick, stiff nanoshell, like Quantison™, $a(t) \ll R_0$. From high-speed optical observations, we derived that $\max(a(t)) \leq \mathscr{R}$, where $\mathscr{R} \approx 0.3\,\mu\mathrm{m}$ is the resolution of the optical system. From the difference in resonance frequency between Quantison™ and free gas microbubbles, we determined $S_p = 25\,\mathrm{kg\,s^{-2}}$ and $E = 2 \times 10^6$ Pa. The critical stress of Quantison™ is $\sigma_c \geq 80$ kPa [8], and thus $\varepsilon_c \geq 0.4$. Taking into account that $\varepsilon_c < 0.5$ and $\langle R_0 \rangle = 1.6\,\mu\mathrm{m}$, it follows that:

$$\max(a(t)) \approx 0.3\,\mu\mathrm{m} = \mathscr{R}. \tag{6}$$

Clearly, the acoustic observations are in agreement with the high-speed optical observations. The hypothesis that the rupture of the shell primarily occurs with micrububbles that have tiny flaws in the shell, has been supported by the optical observations of asynchronous cracking and cracking during a subsequent pulse.

SUMMARY OF THE RESULTS

Tabel 1 shows an overview of the shell properties of three contrast agents. SonoVue™ has a thin monolayer lipid shell, Quantison™ has a thick albumin shell, and Albunex® has a thin albumin shell.

Asymmetries with respect to the excursion axis and to the time-axis can be observed with a spherically symmetric oscillating microbubble. Although the ultrasound-induced gas release from stiff-shelled bubbles has been reported, we also observed gas release from microbubbles with a thin, elastic shell (*cf.* Fig. 1).

ACKNOWLEDGMENTS

The authors are grateful to Bracco Research SA, Genève, Switzerland, for supplying an experimental contrast agent, and Upperton Limited, Nottingham, UK, for supplying the contrast agent Quantison™.

REFERENCES

1. N. de Jong, R. Cornet, and C. T. Lancée, *Ultrasonics*, **32**, 447–453 (1994).
2. M. Postema, N. de Jong, and G. Schmitz, *Biomed. Tech.*, **accepted** (2005).
3. N. de Jong, L. Hoff, T. Skotland, and N. Bom, *Ultrasonics*, **30**, 95–103 (1992).
4. N. de Jong, A. Bouakaz, and P. Frinking, *Echocardiography*, **19**, 229–240 (2002).
5. M. Postema, A. van Wamel, C. T. Lancée, and N. de Jong, *Ultrasound Med. Biol.*, **30**, 827–840 (2004).
6. Z. Zhou, and B. Joós, *Phys. Rev. B*, **56**, 2997–3009 (1997).
7. M. Postema, and G. Schmitz, *Proc. IEEE Eng. Med. Biol. Soc.*, **accepted** (2005).
8. P. J. A. Frinking, and N. de Jong, *Ultrasound Med. Biol.*, **24**, 523–533 (1998).

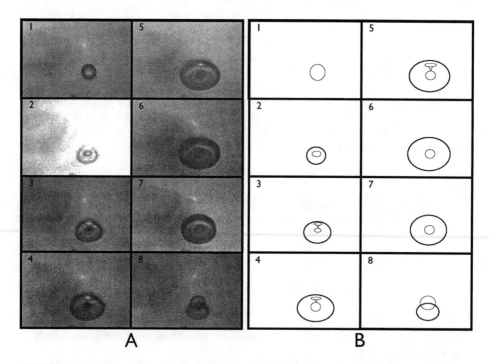

FIGURE 1. Gas release from a lipid-shelled microbubble (A) and a schematic representation thereof (B). The frames were captured at 3 million frames per second. Frame 1 has been taken prior to ultrasound arrival. Frames 2–8 cover one full ultrasonic cycle. Each frame corresponds to a $88 \times 58 \, \mu m^2$ area. The images were captured at the Department of Experimental Echocardiography, Thoraxcentre, Erasmus MC, Rotterdam, The Netherlands.

The nonlinear oscillation of encapsulated microbubbles in ultrasound contrast agents

GONG Yanjun, ZHANG Dong & GONG Xiufen

State Key Lab of Modern Acoustics, Institute of Acoustics, Nanjing University, Nanjing 210093, China

Abstract. Subharmonics or ultraharmonics provide better contrast-to-tissue ratio (CTR) in comparison with fundamental or second harmonics, which show prospective applications in medical diagnosis. In this paper, frequency response for the nonlinear oscillation of encapsulated bubbles is presented based on the Church's equation. Frequency shifting problem is explained and optimized frequency for subharmonic or ultraharmonic emissions are discussed. In addition, sound pressure dependences of the subharmonics and ultraharmonics are studied in theory as well as in measurement. Results revealed that the developments of both subharmonids and ultraharmonics have the same trend, i.e. occurrence, growth and saturation. The generation of ultraharmonic is earlier than that of subharmonic, but the component of subharmonic is larger than ultraharmonic while saturating.

Keywords: ultrasound contrast agent, subharmonic, ultraharmonic.
PACS: 43.25.Ts

INTRODUCTION

Ultrasound contrast agents (UCA) are suspensions containing gas-filled microbubbles with a diameter of 1--10 μm. At present the second harmonic imaging using UCA has been commonly used in commercial B mode scanning instruments. However, two disadvantages degrade this technique[1,2]. One is that the second harmonic signal may undergo higher attenuation than the fundamental, and the other is that the contrast-to-tissue ratio (CTR) is reduced by the second harmonic signals produced by the surrounding tissues. Nevertheless, the subharmonic and ultraharmonic signals from UCA have better CTR than the fundamental or the second harmonic signals because of the negligible subharmonic or ultraharmonic signals generated by the surrounding tissues[1,3]. Shankar et al. found that CTR of the subharmonic signal from microbubbles is better than the second harmonic signal[1]. Yu et al. reported the generations of subharmonics and ultraharmonics and their pressure dependences in Sonozoid® [4].

We use the theory of nonlinear dynamics to solve the Church's model[5] for an encapsulated microbubble, this paper tries to establish a quantitative method for estimating the dependences of the subharmonics or ultraharmonics on the sound pressure and the exciting frequency.

CP838, *Innovations in Nonlinear Acoustics: 17th International Symposium on Nonlinear Acoustics,*
edited by A. A. Atchley, V. W. Sparrow, and R. M. Keolian
© 2006 American Institute of Physics 0-7354-0330-9/06/$23.00

THEORY AND METHOD

The theoretical description of the encapsulated microbubble oscillation is based on the modified equation of Church's model[5,6], expressed as

$$\rho_L\left(R''R+\frac{3}{2}(R')^2\right)=p_0\left(\left(\frac{R_0}{R}\right)^{3\kappa}-1\right)-p_i(t)-4\mu_L\frac{R'}{R}-12\mu_S\frac{d_{Se}R_0^2}{R^3}\frac{R'}{R}-12G_S\frac{d_{Se}R_0^2}{R^3}\left(1-\frac{R_0}{R}\right) \qquad (1)$$

where R is the instantaneous radius of the bubble; R', R'' are the first and second time derivatives of the instantaneous radius; R_0 is the equilibrium radius; d_{Se} is the shell thickness at rest; p_0 is the hydrostatic pressure; ρ_L is the density of the surrounding medium; κ is the polytropic exponent; G_S is the shear modulus; μ_L is the shear viscosity in surrounding medium; μ_S is the shear viscosity of the surface; $p_i(t)=p_A\cos(\omega t)$ is the time-varying excitation acoustic pressure; ω is the angular frequency; c is the sound velocity. The linear resonance frequency f_0 of the encapsulated bubble is[6]

$$f_0=\frac{\omega_0}{2\pi}=\frac{1}{2\pi R_0}\sqrt{\frac{1}{\rho_L}\left(3\kappa p_0+12G_S\frac{d_{se}}{R_0}\right)} \qquad (2)$$

Eq. (1) is solved through numerical calculation with the initial values $R=R_0$ and $R'=0$ at $t=0$. When the sound pressure amplitude p_A is not high, a steady-state solution with a period of $1/f_R$ can be obtained. The oscillating period of the bubble $1/f_R$ is an integral multiple of the sound excitation period $1/f$, $f=mf_R, m=1,2,3,\cdots$. $1/f_R$ can also be expressed as an integral multiple of the resonance period of the bubble $1/f_0$, $f_0=nf_R, n=1,2,3\cdots$. Hence, the oscillation is found to be associated with the expression n/m, which is called the order of the resonance. When m=1, n=2, 3, ... ($n/m=2,3,...$), the resonances are corresponding to the harmonic oscillations. The case n=1, m=2, 3, ... ($n/m=1/2,1/3,...$) denote the subharmonic oscillations. The case where n= 3, m=2 is corresponding to the first ultraharmonic.

FREQUENCY RESPONSES FOR ENCAPSULATED BUBBLES

In this paper, the numerical calculation of frequency responses and its measurement are conduced for the contrast agent Levovist®. Frequency response curves for Levovist® are calculated with different sound pressure amplitudes as shown in Fig. 1. At low sound pressure, only the main resonance near $f/f_0=1$ is observed. With an increase in the sound pressure, harmonics ($f/f_0=2,3,4...$), subharmonics ($f/f_0=1/2$) and ultraharmonics ($f/f_0=3/2$) are observed. In the previous studies of the free bubble and the encapsulated microbubble[7], similar normalized curves of frequency response have been reported but the resonances lean over towards lower frequencies and jump phenomena were observed at sufficiently high sound pressure amplitudes. In this paper, the process of frequency normalization is modified, in which the resonance frequencies are determined by the mean radii instead of the initial radii. As a result, the above-mentioned problems are obviously solved (Fig. 1).

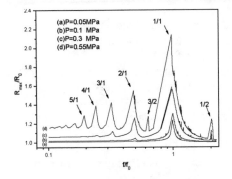

FIGURE 1. Modified frequency response curves for an encapsulated microbubble in water with a radius at rest of R0=1.5μm for different sound pressure amplitudes.

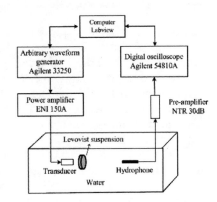

FIGURE 2. Block diagram of the experimental system.

SOUND PRESSURE DEPENDENCES OF THE SUBHARMONICS AND ULTRAHARMONICS

To investigate the characteristics of subharmonics and ultraharmonics from the encapsulated bubbles, sound pressure dependences of the subharmonics and ultraharmonics are theoretically and experimentally studied. Fig. 2 shows the block diagram of the experimental setup for measuring the subharmonic and ultraharmonic emissions from UCA. An arbitrary waveform generator was used to generate a 5 MHz burst signal containing 16 cycles at a pulse-repetition rate 100 Hz. This signal was amplified by an RF power amplifier and then sent to an planar transducer operating at 5MHz with a diameter of 10 mm. The response of the bubbles was received by a needle hydrophone located 6 cm from the surface of the transmitter. After passing through a 30 dB pre-amplifier, the received signal was captured by a digital oscilloscope with a sampling rate of 100 MHz. The Levovist® specimens used here have a concentration of 2.0mg/mL. All measurements were performed at room temperature (20.9°C).

The pressure amplitudes of the subharmonic and ultraharmonic components of the received signals as a function of the incident acoustic pressure are measured and demonstrated in Fig. 3(a). It is found that the subharmonic generation could approximately be divided into three stages, i.e. occurrence, growth and saturation. In the occurrence stage (the acoustic pressure is less than 0.35MPa), the subharmonic component is insignificant and is gradually generated. In the growth stage (the acoustic pressure ranges from 0.35 to 0.5MPa), the subharmonic component grows rapidly. As the incident pressure increases further, the growth of the subharmonic component becomes saturated due to bubble destruction. Like the subharmonic generation, the development of ultraharmonics can also be divided into three distinct stages. It grows rapidly when the acoustic pressure ranges from 0.3 to 0.4 MPa. The saturation stage of ultraharmonic component comes a little earlier than the subharmonics, and a lower acoustic sound pressure threshold is required for the ultraharmonic component. For all bubbles sizes, the ultraharmonic component is generated slightly earlier than the subharmonic component[8].

Fig. 3(b) gives the theoretically calculated acoustic pressure dependences of the subharmonic and ultraharmonic components of the received signals. The measured results are in agreement with the theoretical prediction. However, the earlier generation phenomenon is not predicted evidently by the theoretical result. The explanation has already been given above. In addition, the theoretical model is described for a single bubble, a new model involving interaction among bubbles needs to be developed. Comparing Fig. 3(a) and 3(b), one can also find the difference at low acoustic pressure (0.1MPa~0.3MPa), which is related to the noise in the measurement.

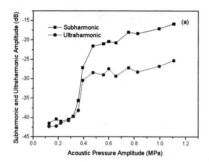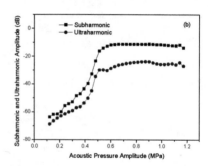

FIGURE 3. The sound pressure dependences of subharmonic and ultraharmonic components. (a) Experimental results; (b) theoretical results.

CONCLUSION

In this study, we numerically studied the nonlinear oscillation of the encapsulated microbubbles in Levovist® suspension. Furthermore, we investigated the sound pressure dependences of the subharmonics and ultraharmonics both experimentally and theoretically. Results indicated that the developments of both subharmonics and ultraharmonics exhibited similar trend, i.e. occurrence, growth and saturation, but the generation of ultraharmonic was a little earlier than that of subharmonic.

ACKNOWLEDGMENTS

This work was supported the National Natural Science Foundation of Jiangsu Province (BK2004081), SRF for ROCS, SEM, and TWAS (No. 03-390)

REFERENCES

1 Shankar P M, Krishna P D, Newhouse V L. Ultrasound in Med & Biol. 1998, 24(3): 395-399
2 Krishna P D, Shankar P M, Newhouse V L. Phys. Med. Biol. 1999, 44: 681-694
3 Basude R, Wheatley M A. Ultrasonics. 2001, 39: 437-444
4 Yu JF, Lu RR, Gong XF, et al. Chin. Phys. Lett. 2002, 19(12): 1828-1930
5 Church C C. J. Acoust. Soc. Am. 1995, 97(3): 1510-1521
6 Hoff L, Sontum P C, Hovem J M. J. Acoust. Soc. Am. 2000, 107(4): 2272-2280
7 Lauterborn W. J. Acoust. Soc. Am. 1976, 59(2): 283-293
8 Palnchon P, Bouakaz A, Klein J, et al. Ultrasound in Med & Biol. 2003, 29(3): 417-425

Phase conjugation of the second harmonic in a sound beam propagating through an immersed elastic solid with rough surfaces

M. E. Stone[*,†], O. Bou Matar[*], P. Pernod[*], V. Preobrazhensky[*] and
M. F. Hamilton[†]

[*]Joint European Laboratory LEMAC, 59652 Villeneuve d'Ascq, France : IEMN-DOAE UMR
CNRS 8520 / Wave Research Center GPI-RAS, Moscow, Russia
[†]Department of Mechanical Engineering, The University of Texas at Austin, 1 University Station
C2200, Austin, Texas 78712-0292, USA

Abstract.
    A numerical investigation is performed to explore the possibility of using wave phase conjuga-
tion to correct for phase aberrations in a nonlinear sound beam that propagates through an immersed
elastic layer with rough surfaces. Of interest here is the second harmonic generated nonlinearly in a
focused sound beam. Surface roughness at the liquid-solid interfaces is modeled by random phase
screens. The properties and dimensions of the layer are such that the wave field is progressive. Numerical simulations are based on an angular spectrum approach that accounts for both longitudinal
and transverse elastic wave interactions inside the solid, and second harmonic generation in the fluid
and solid. Phase conjugation is shown to reduce distortion of the second harmonic introduced by
the rough surfaces. Numerical results for phase screens with random distributions are shown.

Keywords: phase conjugation
PACS: 43.60.Tj

INTRODUCTION

Wave phase conjugation of acoustic waves has been shown to increase imaging resolution for medical ultrasound and NDE applications by accounting for phase aberrations
caused by inhomogeneous media. [1, 2] We present here a method of simulation for
weakly nonlinear sound propagation through an immersed elastic plate where the resulting second harmonic is conjugated and retransmitted in the reverse direction through the
solid. This method includes the effects of diffraction, and is not limited by the parabolic
approximation. Inhomogeneities are introduced in the form of phase screens at the solid
boundaries which represent the effects of rough surfaces. Transmission coefficients between the solid and fluid interfaces are utilized to account for mode conversion to and
from longitudinal and transverse waves in the solid. Results are shown for a case using
two different random phase screens at the solid surfaces.

THEORETICAL MODEL

Our numerical simulations are based upon the experimental setup currently used at
LEMAC. In our simulations, a 10 MHz circular focused transducer with radius $a =$

CP838, *Innovations in Nonlinear Acoustics: 17th International Symposium on Nonlinear Acoustics*,
edited by A. A. Atchley, V. W. Sparrow, and R. M. Keolian
© 2006 American Institute of Physics 0-7354-0330-9/06/$23.00

7.5 mm and a focal length $d = 30$ mm is situated facing a magneto-acoustic phase conjugator (MAPC) at a distance of 60 mm. [1] They are both centered on the same radial axis. A low-density polyethylene (LDPE) plate of thickness $t = 5$ mm is centered 12.5 mm from the face of the MAPC and is transverse to the propagation axis. The surrounding medium is fresh water. Both the water and LDPE plate are taken to be homogeneous. LDPE was chosen for the plate material due to its similar acoustical impedance to water. Due to this, multiple reflections inside the solid layer are ignored since the reflection coefficients will be low.

Several assumptions are made in the simulations. First, the source is assumed to be a monofrequency time-harmonic signal that has a sufficiently long pulse length as to be considered continuous. Secondly, the propagation is considered to be weakly nonlinear so that only the second harmonic generation is of significance. With these assumptions, we can calculate the complex pressure field propagating over a distance z for the fundamental wave in both the solid and fluid using the angular spectrum method [3],

$$Q_1(\mathbf{k}_\perp, z) = Q_1(\mathbf{k}_\perp, 0)e^{-jk_1 z} = Q_{1,0}(\mathbf{k}_\perp)e^{-jk_1 z}, \tag{1}$$

where $Q_{1,0} = \mathscr{F}(q_{1,0})$ is the spatial Fourier transform of the complex pressure before propagation, $\mathbf{k}_\perp = (k_x, k_y)$ is the transverse wave vector and $k_1(\mathbf{k}_\perp) = \sqrt{k^2 - |\mathbf{k}_\perp|^2}$ is the wave vector for the fundamental in the propagation direction. The resulting complex pressure is obtained from the inverse spatial Fourier transform of Q_1. Similarly, the second harmonic complex pressure is obtained from [3],

$$Q_2(\mathbf{k}_\perp, z) = \frac{j\beta k^2}{2\pi^2 \rho_0 c_0^2} e^{-jk_2 z} \int \int \frac{Q_{1,0}(\mathbf{k}_\perp - \mathbf{k}_\perp')Q_{1,0}(\mathbf{k}_\perp')[e^{-j(k_A + k_1 - k_2)z} - 1]}{(k_A + k_1 - k_2)(k_A + k_1 + k_2)} d\mathbf{k}_\perp' \tag{2}$$

where β is the coefficient of nonlinearity, $k = \omega/c_0$ is the wave number, ρ_0 and c_0 are the ambient density and sound speed, $k_2(\mathbf{k}_\perp) = \sqrt{4k^2 - |\mathbf{k}_\perp|^2}$ and $k_A = k_1(\mathbf{k}_\perp - \mathbf{k}_\perp')$. A boundary condition in the derivation of Eq. 2 requires that the second harmonic pressure be zero at the start of the propagation. Thus, any second harmonic pressure present before propagation is propagated linearly via Eq. 1 and then summed with the nonlinearly generated second harmonic from Eq. 2. [4]

The phase screens are generated from a normally distributed random noise field that is convoluted with a Gaussian correlation function. The resulting screens have a known statistical correlation length l and variance σ^2. The normalized correlation length $\nu = a/l$ strongly affects the "graininess" of the phase screen, while the variance generally affects the amplitude. A phase screen represented by $\Delta\phi_n$ is applied to the nth harmonic complex pressure by

$$q_{n,\text{out}} = q_{n,\text{in}} e^{j\Delta\phi_n}. \tag{3}$$

To account for mode conversion at the fluid-solid interface, transmission coefficients are used to split the incoming wave into its longitudinal and transverse components:

$$Q_{n,\text{in}L} = T_{n,LL} Q_{n,\text{in}}, \tag{4}$$

$$Q_{n,\text{in}T} = T_{n,LT} Q_{n,\text{in}}. \tag{5}$$

Each component is propagated individually then rejoined at the solid-fluid interface by

$$Q_{n,\text{out}} = \widehat{T}_{n,LL} Q_{n,\text{outL}} + \widehat{T}_{n,TL} Q_{n,\text{outT}}.\qquad(6)$$

The transmission coefficients are those defined by Hamilton and Landsberger. [3]

RESULTS

A case with random phase screens is presented here. The screens used have a large variance $\sigma^2 = 5$ which corresponds to a maximum phase change greater than 2π radians for the fundamental pressure. The normalized correlation length is $v = 2$. Phase screens of this size will render the fundamental wave almost completely incoherent. [4] A focused Gaussian pressure field is used as the source.

Figure 1 shows the stages of propagation for the fundamental pressure from the source to the conjugator surface. Large zero portions of the fields are trimmed for clarity. At the MAPC surface, the spreading and distortion of the wave can be clearly seen. Likewise, Fig. 2 shows the propagation of the second harmonic from the source to the MAPC. The fundamental pressure in the source plane is repeated here for reference since there is no second harmonic present at the source. At the MAPC surface, the second harmonic is distorted, but it is more coherent than the fundamental. This is mainly due to the beam radius of the second harmonic being smaller, thus the effective correlation length of the phase screens appears reduced. Also, to a lesser effect, only the coherent portion of the fundamental contributes to the second harmonic generation after passing the solid plate.

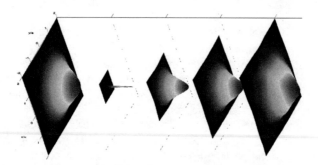

FIGURE 1. Forward propagation of the normalized fundamental pressures from the source to the MAPC surface through an LDPE plate with random phase screens on its faces. From left to right, the pressures shown are at the source plane, the focal plane, the front solid surface, the back solid surface, and the MAPC surface. Both screens have the properties of $v = 2$ and $\sigma^2 = 5$.

Finally, Fig. 3 shows the propagation of the conjugated second harmonic from the MAPC surface back to the source plane. Refocalization, and then formation of the original Gaussian source shape is clearly shown.

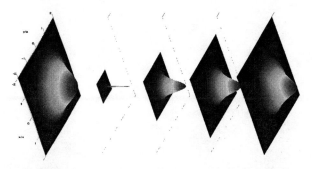

FIGURE 2. Forward propagation of normalized second harmonic pressures from the source to the MAPC surface through an LDPE plate with random phase screens on its faces. From left to right, the pressures shown are at the source plane (fundamental), the focal plane, the front solid surface, the back solid surface, and the MAPC surface. Both screens have the properties of $v = 2$ and $\sigma^2 = 5$.

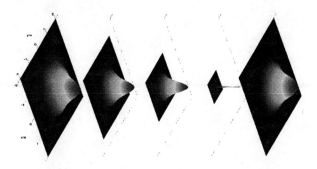

FIGURE 3. Reverse propagation of the normalized conjugated second harmonic wave for two random phase screens with $v = 2$ and $\sigma^2 = 5$. From left to right, the pressures shown are at the MAPC, the back solid surface, the front solid surface, the focal place, and the source plane.

ACKNOWLEDGMENTS

The authors wish to thank A. Brysev and L. Krutyansky for their helpful discussions and their work on the MAPC at LEMAC. This work was supported by the U.S. Civilian Research and Development Foundation.

REFERENCES

1. A. P. Brysev, L. M. Krutyanskii, P. Pernod, and V. L. Preobrazhensky, *Acoust. Phys.*, **50**, 623–640 (2004).
2. K. Yamamoto, P. Pernod, and V. Preobrazhensky, *Ultrasonics*, **42**, 1049–1052 (2004).
3. B. J. Landsberger, and M. F. Hamilton, *J. Acoust. Soc. Am.*, **109**, 488–500 (2001).
4. X. Yan, *Statistical Model of Beam Distortion by Tissue Inhomogeneities in Tissue Harmonic Imaging*, Ph.D. thesis, The University of Texas at Austin (2004).

SECTION 6
SHOCK WAVE THERAPY

25 Years of ESWL – From the Past to the Future

Bernd Forssmann

Herrsching

Abstract. It was a revolution in the treatment of urolithiasis 25 years ago, when the first extracorporeal shock wave lithotripsy (ESWL) was carried out on the prototype HM1 equipped with an electrohydraulic shock wave source. Further developments led to the HM3, the legendary bath tub that is used with high success to this day. The history of investigations to disintegrate urinary stone with one shock wave pulse by means of high power is described. Break trough for clinical application was achieved when the shock waves were applied in a sequence of pulses with low energy. In the late eighties the effectiveness of second generation lithotripters wase only judged by means of peak pressure and focal extension so that effectiveness was often misinterpreted. Despite standardization of shock wave parameters the assessment of lithotripters remains unsatisfactory. The concept of effective energy considers the whole temporal and spatial field of the shock wave and allows to determine the energy dose of stone disintegration. Thus, clinical energy dose is expected to reveal additional information to evaluate the success of shock wave lithotripsy in terms of fragmentation and side effects.

Keywords: ESWL, shock wave, effective energy
PACS:87.54.Hk

1. INTRODUCTION

In 1974 a project sponsored by the German government department for Research and Technology was initiated at Dornier in collaboration with the Urological Clinic and Institute for Surgical Research of Ludwig Maximilian University Munich to develop a lithotripter to break kidney stones with extracorporeal generated shock wave. In February 1980 the first successful extracorporeal lithotripsy was performed on a prototype lithotripter HM1 inducing a revolution in the treatment of urolithiasis [1]. Today, more than 1 million treatments of ESWL on approximately 5000 lithotripters demonstrate the acceptance of this method worldwide.

In the end of the sixties Dornier in cooperation with the Institute of Applied Physics and Electrotechnology of University of Saarland investigated the interaction of shock wave with biological tissue in a military project. The studies with light-gas-gun produced shock waves showed that shock waves can propagate within living tissue causing only limited side effects except on acoustic boundaries. During this project the idea came up to break kidney stones by means of shock waves. In 1971 at the annual spring symposium of the German Physical Society first results of shock waves generated by high speed water drops that break kidney stones in a water filled tube as wave guides were presented [2]. At Dornier the investigations on the principle feasibility of this idea were continued with a light-gas-gun from which a high speed

CP838, Innovations in Nonlinear Acoustics: 17th International Symposium on Nonlinear Acoustics,
edited by A. A. Atchley, V. W. Sparrow, and R. M. Keolian
© 2006 American Institute of Physics 0-7354-0330-9/06/$23.00

projectile up to 5 km/s hit a steel target producing the shock wave. Depending on the shape of the target a plane or focused shock wave was coupled into a free water bath where kidney stones could be exposed for fragmentation. A plane wave produced only crack network in the kidney stone, whereas a focused wave caused substantial fragmentation (Figure 1a).

2. FIRST APPROACH WITH HIGH INTENSITY SHOCK WAVES PULSES

2.1 Investigations for Producing Shock Waves

The method using light-gas guns seems to be not suitable for clinic approach; the generation of only four shock waves took an entire day.

(a)   (b)

FIGURE 1. (a) Light-gas gun with coupled water bath, and first fragmented kidney stone (1971); (b) First experimental lithotripter TM1 with electrohydraulic shock wave generator (1974).

Under the aspect that the impedance between the patient's body and coupling medium should matched as much as possible, shock wave generation was selected by means of underwater spark. To concentrate the shock wave onto the kidney stone the spark gap was located in the focal point of a semi-ellipsoid, where the kidney stone was positioned in the second focal point. The preliminary experiences with the light-gas gun suggested to destroy the stones with only one impulse. To achieve sufficient energy to fragment the stones, several shock wave generators were equipped with capacitors between 1 µF to 2.6 µF to perform the experiments. To optimize focusing ellipsoids with various aperture were investigated (Table 1).

At this time temporal and spatial field of the shock wave were measured in the surroundings of the focal point with own made piezoelectric hydrophones that were 2 mm in diameter, and 0.2 mm thick (ellipsoids EX, and TM1). To avoid resonance the crystal was fitted to a 50 cm long sliver rod; calibrating was not possible, but pressure was calculated with the piezoelectric constant of the ceramic material. Rise time of the storage oscilloscope, and hydrophone was about 30 ns. Further development of this type of hydrophone was not persecuted because the crystals only withstood several 100 shocks or lost their reproducibility.

The influence of spark gap distance, capacity, and inductance of the generator, and charging voltage were investigated to determine optimal parameters for the preclinical research program. Shock wave intensity was explored in relation to spark gap, and size of ellipsoids. Intensity increased twice, when the gap distance was changed from 1 to 5 mm. As expected the inductance of the shock wave circuit had a great influence. All efforts in the further development of the shock wave generator aimed to minimize the inductance as much as possible.

TABLE 1. Ellipsoids.

Equipement	a/b	d[mm]	Depth [mm]	P+[Mpa]	Focus [mm]
EX	1.41	156	78	350/82*)	6/17
TM1	1.69	130	89	160	17/20
HM1	1.67	150	115	(85)	n.a.
HM3	1.78	156	130	34*)	16/100
HM3 mod	1.68	172	130	41*)	7.5/40

To determine the focusing of ellipsoids the spatial pressure field was measured in the ellipsoid axis, and lateral to the ellipsoid axis around the focal point. The –6dB line of the spatial pressure field, a term that is still used today, was defined as focal extension. The smallest focal extension was achieved with the large aperture of EX ellipsoid (later used in the experimental lithotripter XL1), where pressure up to 350 MPa could be achieved in the focal point with a generator of 2.6 µF at 20 kV (82 MPa with 80nF). The maximum pressure with this ellipsoid was achieved nearly 20 mm behind the geometrical focal point indicating non-linear effects of the shock wave propagation. Pulse width (50%-line) at the maximum pressure was 280 ns for the TM1, and for the EX with 340 ns slightly higher; rise time was estimated in order of magnitude to 10ns. Focusing was also substantially improved to 10x18mm², when the ellipsoid TM1 was enlarged without changing the ratio of axis [3].

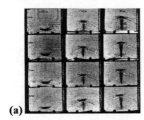 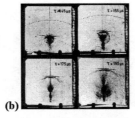

FIGURE 2. Shadowgraphs; the incident shock wave propagates from the bottom to the top of the frame: (a) Dislocation of the spark gap perpendicular to the axis a (0 mm, 1mm, 2mm ,3 mm) at 120µs, 145µs, 160µs after ignition; (b) Stone model in the focal point at 145µs, 155µs, 175µs, 190µs after ignition.

All experiments were carried out in a free water bath. To consider the influence of tissue on focusing and absorption of the shock waves the experiments were repeated with the EX ellipsoid. 4-6 cm thick layers of fat and muscle that were placed in the shock wave path in front of the focal point, did not change the size of the focal spots, the loss in the pressure amplitude was by 20% that revealed still sufficient energy to fragment the stones.

The electrode tips did not burn down symmetrically. This could cause misalignment of the shock wave propagation. Therefore, experiments with dislocated electrode tips were carried out to estimate suspected decreasing of fragmentation efficiency. The center of the tips was displaced perpendicular to the ellipsoid axis up to 3 mm, and in the ellipsoid axis 2 mm (Figure 2a). In a linear model the propagation of the shock was also calculated assuming a source in the focal point that emits a spherical wave. Between experiments and calculations a good correlation that tolerates a dislocation of the tips off the focus within 1 mm laterally was found. Dislocation on the ellipsoid axis had only minor influence on the intensity of the focused shock wave, but the location of maximum pressure moved accordingly. Later, this effect was utilized in the lithotripter MPL 9000 to provide two focal spots, one electrode with a short penetration depth for gallstone lithotripsy, and one with a larger penetration depth for kidney stone application.

2.2 Stone Fragmentation

The experiments with plane waves by means of light-gas suggested that the spalling effect would be a major contribution to the fragmentation process because the tensile strength of kidney stones of 0.6-1.1 MPa is significantly lower than the comprehensive strength of 5-8 MPa. This effect could also be observed when large stones were exposed to electrohydraulic shock waves with high intensity in the focus of the ellipsoid, but more fragmentation was achieved on the side, where the shock wave enters the stone. Systematical experiments were carried out to explore the interaction of the shock wave with glass spheres that were extremely hard model stones, using shadow photograph technique (Figure 2b). These observations suggest that stress, and shear wave play a major role in the comminution of kidney stone. Tensile waves produced by reflection on the opposite side where the shock wave exit the stone, may also contribute to disintegration, even if they have a significantly lower intensity. Cavitation was also discussed, but underrated.

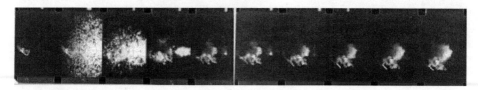

FIGURE 3. High speed photographs of stone distintegration (SYCAM-camera 1000 frames/s)

In series of tests kidney stones of different chemical and mineralogical composition were exposed to high intensity shock waves with variable pressure, that was adjusted with capacitor voltage. Chemical and mineralogical composition of the kidney sphere was determined. Chemical and mineralogical structure had only a minor influence on the fragmentation process. With increasing pressure the degree of disintegration was significantly improved when the stone were exposed to only one shock. However, several large pieces remained at pressure amplitudes up to 350 MPa that probably could not be discharged. Additional shock waves were attenuated by the surroundings fragments, and further disintegration was difficult.

An important question was how the kinetic energy is transmitted to the fragments so that they can penetrate into the surrounding tissue. In in-vitro experiments the fragmentation process was investigated with high speed photographs (SYCAM-camera, 1000 frames/s). By means of frame time and covered distance the velocity of fragments was analyzed. Fragments were weighted (Figure 3). The measurements suggested, that the estimated velocity of 0.3-0.4 cm/s, and energy of maximum 1-2 mJ will not affect the surrounding tissue [4].

Lowering the pressure amplitude to 10-50 MPa by means of generators with capacitors of 20 and 60 nF only revealed a separating of fine sand like particles at side where the shock wave hit the stone. Further shock wave exposition was also effective. With 200-300 shock wave releases the stones could be completely fragmented in fine particles. The energy dose for complete fragmentation in terms of the stored electrical energy of the shock wave generator was surprisingly lower when the pressure amplitude was in the range about the fragmentation threshold, but the number of exposed shock waves was adequately higher, than with the one-shot exposition [5].

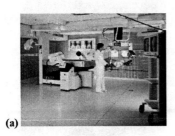 

(a)   (b)

FIGURE 4. (a) Lithotripter HM3, the Gold Standard in lithotripsy; (b) Dornier Lithotripter DLS, a modern urological work station with dry coupling shock wave system.

2.3 The Way to the HM3

In 1974 already, investigations to explore the interaction of shock waves onto living cells and tissue started with the experimental lithotripter TM1, that was equipped with water bath or a coupling membrane (Figure 1b). The results in the energy range that was used at this time, were promising. The experiments were extended to prove the safety of disintegration of kidney stones in an animal model. Shock waves could not be target to the stones because locating with ultrasound was insufficient. Poor stone fragmentation could only be achieved when volley of shocks was exposed, but no serious tissue damage was detected except when a shock wave hit the lung. Having equipped the experimental lithotripter with a three dimensional X-ray system stone localization could be used to position the stone into the focus. With the one-shot exposition several large pieces mostly remained that removed away from the focus thus complicating further fragmentation.

The break trough was achieved when the experimental lithotripter was equipped with a 60 nF shock wave generator so that stone disintegration in fine particles with repeated shock wave exposure was also possible in vivo. In extensive animal experiments the safety of the method could be proved. The lithotripter HM1 was developed using the same shock generator and a truncated ellipsoid that provided a free penetration depth of 12 cm between the ellipsoid and the focus. The first clinical results suggest to enhance the power of the shock wave generator for larger stones and the penetration depth for more flexibility in the treatment of obese patients. In 1983 these requirements could be realized with the development of HM3 that was equipped with a low inductance shock wave generation system of 150 nH and a capacitor of 80 nF that produces a focal pressure of 34 MPa (*) with PVDF-needle hydrophon). The ellipsoid was enlarged and provided a penetration depth of 13.5 cm [6]. In 1986 the shock wave generating system was modified with a 40 nF shock wave generator and an enlarged ellipsoid providing anesthesiafree lithotripsy [7].

3. ENERGY DOSE CONCEPT IN LITHOTRIPSY

3.1 Parameters of Lithotripter Field

In the eighties new shock wave sources were developed on different principles of generation and focusing, such as electromagnetic coils in combination of acoustic lens with a wide variety of aperture or piezoelectric elements attached on spherically bowl. Comparison of efficiency of different shock wave sources on the -6dB focal extension, peak pressure, pulse width, and intensity settings alone lead to more confusion than to an objective quantification of the power of a shock wave source. A great debate about the importance of these parameters to the disintegration process and the clinical relevance began.

New optical hydrophones allow to measure the complete behavior of shock wave pulse, leading to a better understanding of the disintegration process and the interaction of shock wave with biological tissue. The IEC developed standards for the measurement of shock waves, in which the relevant parameters describing a shock wave are listed (Table 2).

TABLE 2. Parameters of shock wave source.

Parameter	Unit
Peak pressure, positive (P+) and negative (P-)	MPa
Focal extension fx,fy,fz	mm
Rise time (tr)	ns
Pulse width; positive (Tp+) and negative (P-)	μs
Intensity / Energy density (PII)	mJ/mm^2
Focal energy (Ef)	mJ

In this decade of ESWL applications, the effectiveness of the shock wave source was only judged using peak pressure and focal extension. The fact that these were the only parameters to asses the effectiveness of a shock wave source was often misinterpreted as significant for the disintegrating power of a shock wave source. Peak pressure and focal size have little correlation to the stone fragmentation. Introducing energy density and focal energy by the IEC led to improved parameters to characterize a shock wave source. The energy density and the focal energy correlate to the depth of craters that are produced in materials with shock wave exposure [8]. These quantities depend heavily on the focal peak pressure, small focal extension corresponds to high focal pressure and large focal extension to low focal pressure; their significance to compare the output power of different shock wave sources is poor.

3.2 Effective Energy and Energy Dose as a Key Determinant

In order to disintegrate urinary stones energy is required. It is delivered by the shock wave pulse. It is a convention to define the energy in the region of interest of a circle of 12 mm diameter, where the shock wave interacts with the stone as effective energy Eeff(12 mm). This parameter includes the complete temporal and spatial behavior of the shock wave pulse. The energy correlates to the disintegrated stone volume and is suitable to compare different shock wave sources [9].

TABLE 3. Output of the EMSE220-types at maximum intensity level.

EMSE	220F	220F-XP	220F-XXP
Peak pressure P+ / P- [MPa]	99 / -10.4	90 / -11.7	90 / -8.0
Focal extension axial / lateral [mm]	55/2.6	55/2.6	63/5.4
Positive pulse duration Tp+ [µs]	1.4	1.8	2.2
Effective energy Eeff [mJ]	66	87	110
Energy dose [J]	7.2	5.9	4.8

To demonstrate the role of this parameter three different electromagnetic shock wave sources were compared. All three shock wave sources have the same aperture angle of 73°, and treatment depth of 150 mm, but shock wave generators are modified in power. The effective energies of these shock wave sources were determined by means of the measuring data of the temporal profiles and spatial fields of the shock wave (Table 3). Increasing the power of shock wave generators revealed only a small change in peak pressure, but pulse duration and focal extension were significantly extended. Although the shock wave pressure was not increased with the more powerful generators the effective energies changed (Figure 5 a).

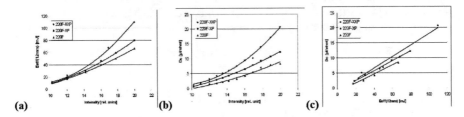

(a)　　　(b)　　　(c)

FIGURE 5. (a) Effective energy Eeff in relation to output settings; (b) Disintegration capacity Dc in relation to output settings; (c) disintegration capacity Dc in relation to effective energy.

To determine the disintegration capacity of the shock wave sources standardized artificial stones as kidney stone mimicking material were placed in a net of 2 mm mesh in the focal point and disintegrated at shock wave release frequency of 70 pulses/min. Disintegration capacity is defined as fragmented volume in µl/pulse.

The results of determining the disintegration capacity revealed a stronger increase to disintegrate the artificial stones completely than expected by the raised effective energy (Figure 5b). The relationship between disintegration capacity and effective energy is linear as postulated [10]. With the more powerful generators the disintegration capacity is improved to an equal amount of effective energy, e. g. for Eeff=50 mJ disintegration capacity is 6.1 µl/pulse, 7.4 µl/pulse, and 8.6 µl/pulse.

Energy dose plays an important part in stone fragmentation. The energy dose for complete disintegration of stones can be calculated by means of $E_{dose}=n*E_{eff}$, where n is the number of applied shock wave pulses and E_{eff} the effective energy at the used intensity setting of the shock wave generator. As listed in table 3 the energy dose decreases from 7.2 to 4.8 mJ with the more powerful generators indicating the effectiveness of the shock wave source, and nearly reaches the amount of 4.2 J using the HM3 at 30 kV [11].

Articles published so far did not included reports about clinically needed energy doses for the fragmentation of renal and ureteral stones; but only references to the number of shock waves and intensity levels, mostly defined as power index (shock wave intensity x number of shock waves). This definition contains no information about real physically applied energy. Comparison of treatment strategies and effectiveness of different lithotripters is not possible with this definition. The effective energy can be a parameter to determine the administered clinical energy dose of the shock waves in a more objective manner. The success of ESWL treatments at different intensity levels seems to remain the same when the number of shots is in the range receiving equivalent energy dose. Thus, the energy dose is expected to reveal additional information to evaluate the success of shock wave lithotripsy in terms of fragmentation and side effects.

ACKNOWLEDGMENTS

It is impossible to appreciate the contributions of all scientists on research and development of shock wave lithotripsy during a period of more than thirty years. I especially mention C. Chaussy, F. Eisenberger, and W. Hepp who belong to the core team since 1974, further C. Bohris, H. Eizenhoefer, M. Müller, and F. Ueberle working with them in the last decade of ESWL at Dornier.

REFERENCES

1. Chaussy, C., Brendel, W., Schmiedt, E., *The Lancet* **2 (8207)**, 1265-1268 (1980).
2. Haeusler, E. Kiefer, W., *Verhandl. DPG (VI)* **6**, K 36 (1971).
3. Forssmann, B., Hepp, W., Chaussy, C., Eisenberger, F. Wanner, K., *Biomed Tech* **22**, 164-168 (1977)
4. Chaussy, C. Eisenberger, F., Wanner, K. Forssmann, B., *Aktuelle Urologie* **9**, 95-101 (1978)
5. Forssmann, B., Hepp, W., Chaussy, C., Jocham, D., Schmiedt, E., Brendel, W., *Biomed Tech* **25** (Suppl.), 414-416 (1980)
6. Chaussy, C., Schmiedt, E., Jocham, D., Walther, V., Brendel, W., Forssmann, B., Hepp, W., *Extracorporeal Shock Wave Lithotripsy-New Aspects in the treatment of Kidney Stone Disease*, edited by Ch. Chaussy, Basel-Munich-Paris: Karger, 1982
7. Fischer, N., Ruebben, H., Hofsaess S., Forssmann, B., Schockenhoff, B., Giani, G., *Urologe A* **26**, 29-32 (1987)
8. Granz, B., Koehler, G., *J Stone Dis* **4**, 123-128 (1992)
9. Forssmann, B., Mueller, M., *Biomed Tech* **35** (Suppl.), 230-232 (1990)
10. Mueller, M., *Biomed Tech* **35**, 249-262 (1990)
11. Forssmann, B., Ueberle, F., Bohris, C., *J Endourol* **16** (suppl.2) 18-l 21 (2002)

Interactions of Cavitation Bubbles Observed by High-Speed Imaging in Shock Wave Lithotripsy

Yuri A. Pishchalnikov[*], Oleg A. Sapozhnikov[¶], Michael R. Bailey[†],
James A. McAteer[*], James C. Williams, Jr.[*], Andrew P. Evan[*],
Robin O. Cleveland[††], and Lawrence A. Crum[†]

[*]Department of Anatomy and Cell Biology, Indiana University School of Medicine, 635 Barnhill Drive, Indianapolis, IN 46202, USA
[¶]Department of Acoustics, Physics Faculty, Moscow State University, Vorob'evy Gory, Moscow 119992, Russia
[†]Center for Industrial and Medical Ultrasound, Applied Physics Laboratory, University of Washington, Seattle, WA 98105, USA
[††]Department of Aerospace and Mechanical Engineering, Boston University, 110 Cummington Street, Boston, MA 02215, USA

Abstract. A multi-frame high-speed photography was used to investigate the dynamics of cavitation bubbles induced by a passage of a lithotripter shock wave in a water tank. Solitary bubbles in the free field each radiated a shock wave upon collapse, and typically emitted a micro-jet on the rebound following initial collapse. For bubbles in clouds, emitted jets were directed toward neighboring bubbles and could break the spherical symmetry of the neighboring bubbles before they in turn collapsed. Bubbles at the periphery of a cluster underwent collapse before the bubbles at the center. Observations with high-speed imaging confirm previous predictions that bubbles in a cavitation cloud do not cycle independently of one another but instead interact as a dynamic bubble cluster.

INTRODUCTION

Numerical simulations applied to cavitation in shock wave lithotripsy are, for the most part, based on solutions for the growth and collapse of a solitary cavitation bubble, i.e. a single bubble in an infinite fluid [1]. Though these models have significantly improved our understanding of cavitation bubble dynamics, they do not take into account forces associated with neighboring bubbles or surfaces. High-speed camera observations suggest that cavitation bubbles can form dense clusters, and that cluster dynamics can affect bubbles in the surround [2]. Thus, theoretical studies that address collective phenomena associated with mutual interactions of cavitation bubbles can be very relevant to shock wave lithotripsy. One such report has proposed a model that takes into account an arbitrary number of bubbles interacting in a cluster [1]. The numerical results suggest that pressure waves radiated by bubbles in the cluster can accelerate, retard, or even reverse the growth and collapse of other bubbles. The present study describes high-speed camera observations that should be useful as input for numerical modeling of bubble clusters in shock wave lithotripsy.

CP838, *Innovations in Nonlinear Acoustics: 17th International Symposium on Nonlinear Acoustics,*
edited by A. A. Atchley, V. W. Sparrow, and R. M. Keolian
© 2006 American Institute of Physics 0-7354-0330-9/06/$23.00

MATERIALS AND METHODS

Experiments were conducted using a research electrohydraulic shock wave lithotripter patterned after the Dornier HM3 [3]. Typical lithotripter pulse consists of a positive spike with shock front (tens of MPa) followed by a negative-pressure tail, all with a total duration of about 10 μs [4]. An Imacon-468 (DRS Hadland, Inc., Cupertino, CA) high-speed digital camera was used to record images of cavitation bubbles. Seven frames (576x385 pixels) could be recorded with a minimum inter-frame time step of 10 ns. Sequential frames were captured to document the bubble activity generated by single shock waves administered either minutes apart (single shot regime) or at selected pulse repetition rates (PRF). The use of degassed water and single shot regime allowed us to study the dynamics of a "solitary" cavitation bubble, as bubbles generated under these conditions were dilute.

RESULTS AND DISCUSSION

Numerical simulations and high-speed photography (see, for example, [5]) show that a solitary cavitation bubble continues its inertial growth for tens or hundreds of microseconds after the passage of the lithotripter pulse, and then, driven by atmospheric pressure, starts to collapse. This first collapse is followed by subsequent rebounds. Shock waves radiated from these collapses have been documented by Schlieren imaging, as well as detected by passive cavitation detection, and fiber-optic probe hydrophone measurements. What is not usually included in numerical models is that shock wave radiation by even solitary bubbles is typically followed by a micro-jet emitted from the bubble (Fig.1).

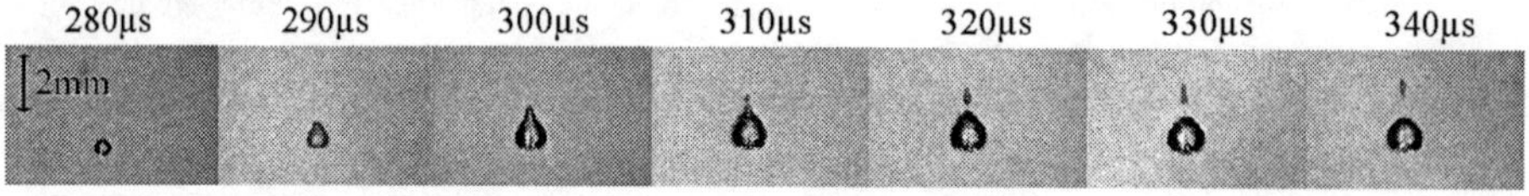

| 280μs | 290μs | 300μs | 310μs | 320μs | 330μs | 340μs |

FIGURE 1. Micro-jet emission on the rebound following initial collapse. Each "solitary" cavitation bubble was observed to collapse, and then emit a micro-jet. After emitting a jet, bubbles generally returned to a spherical shape.

Neighboring bubbles were often seen to emit micro-jets toward their mutual boundary or "gravity" center (Fig.2).

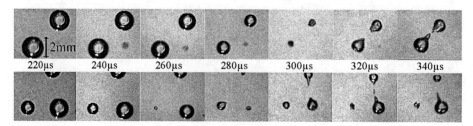

| 220μs | 240μs | 260μs | 280μs | 300μs | 320μs | 340μs |

FIGURE 2. Micro-jets emitted by the collapse of neighboring bubbles.

Perturbations in the bubble cloud (jets, hydrodynamic flows, pressure waves) may distort spherical symmetry of bubbles (Fig.3). As a result, bubbles collapse asymmetrically, such that the formation of jets is visible at the very beginning of the collapse (bubble #4 in Fig.3). The "structure" of such a jet is different from the collapses shown in Fig.1 and Fig.2. In the collapse of bubble #4 in Fig. 3, it looks like the water jet "ruptures" the opposite surface of the doughnut-shaped bubble, so that the water jet is no longer visible in the water (see bubbles 4 and 5).

FIGURE 3. HS-camera movie showing asymmetric collapse of bubbles in a cloud. Note that bubbles at the periphery of the cloud collapse earlier than bubbles in the center of the cloud: Bubbles marked #1 (see first frame) have already collapsed once and are undergoing their second collapse, and are absent in the following frames; bubble(s) #2 has just collapsed and is going to rebound; bubble #3 will collapse in the third frame; bubble #4 will collapse in the fifth frame; and bubbles #5 (two coalesced bubbles) will collapse in frame 6. As was discussed above (see Fig.2), bubbles emit jets towards their center of "gravity". That is, bubble #3 emits jet (frame 3) towards bubble #4; bubbles #4 and #5 emit jets toward the center of "gravity" for remaining bubbles. Note also that after an asymmetric collapse of a doughnut-shaped bubble, the shape of the bubble is irregular (see bubble #2 in all frames; bubbles #4 and 5 in frames 6, and 7).

Bubbles that appear to have coalesced can retain their "individuality", in that a narrow film of water is visible between them (bubbles marked #5 in Fig.3). The spherical shape that might be eventually obtained by coalesced bubbles does not necessarily prove that there is no water film that separates the coalesced bubbles [6]. Thus, "bubble coalescence" might not be a good choice of words because coalescence implies complete fusion. It has been shown that cavitation bubbles at the surface of natural and model kidney stones typically coalesce into densely packed clusters [2, 6]. Therefore, the consideration of bubble coalescence under the simplest condition, such as in Fig.3 (two coalesced bubbles in the free field), may help in understanding the bubble dynamics at the stone surface.

As was mentioned in the legend of Fig.3, cavitation clouds exhibit "concerted collapse", that is, collapse initiates at the periphery and proceeds towards the center of the cloud (Fig.4). This has been observed for ultrasound (vibratory) cavitation, as discussed in [7]. It was suggested that collapse of the outer cavities creates an additional constant pressure on the bubble cloud, thus intensifying the collapse of the other cavities. This process proceeds towards the center of the bubble cloud making the collapse of the central cavities extremely intense. It has also been observed that the collapsing bubbles in the arrays of disk-shaped cavities generate radial shock waves and jets directed towards the next row of bubbles, such that the collapse proceeds step by step in array with pressure waves from one collapsed row then collapsing the next row of cavities [8].

| 20μs | 270μs | 320μs | 370μs | 420μs | 470μs | 520μs |

FIGURE 4. Cavitation bubble cloud collapse starts at the periphery of the cloud and proceeds towards the center of the cloud. Numerous cavitation bubbles are seen in the first frame recorded 20 μs after the passage of the shock wave. 250μs later (second frame) some peripheral bubbles have collapsed. In the last frame (520μs) only bubbles at the center of the cloud (i.e. located along the axis of the lithotripter shock path) are remaining and about to collapse. The frame size is approximately 2x3 cm^2. Bubble cloud was recorded after the passage of the 10th lithotripter pulse administered at 5Hz, 20kV.

A solitary bubble collapse with a subsequent jet streaming out looks similar to jet formation observed for cavitation bubbles generated by focusing of laser beam into water [9]. Note, that the micro-jet breaks into additional bubbles (see Fig.1 and Fig.2) that could act as cavitation nuclei for subsequent shock waves, perhaps accounting for the increased number of cavitation bubbles observed at fast PRF [5].

In summary, these observations with high-speed imaging lead to several conclusions about the characteristics of the cavitation during SWL. The data show that cavitation bubble activity in the free field is influenced by the formation of bubble clouds. Cloud collapse is initiated at the periphery of the cloud and ends at the center of the cloud. Bubble coalescence does not necessarily mean the formation of a bigger bubble but instead should be probably considered as a foam-like cluster of bubbles attached to each other. High-speed camera observations suggest that such bubbles in close physical contact with one another could still collapse independently—those at the periphery collapsing before those in the center.

ACKNOWLEDGMENTS

The authors thank Philip Blomgren for his technical advice. This work was supported by grants from ONRIFO and the National Institutes of Health (DK-43881).

REFERENCES

1. M.F. Hamilton, Yu.A. Ilinskii, G.D. Meegan, and E.A. Zabolotskaya, "Interaction of bubbles in a cluster near a rigid surface", *Acoust. Research Letters Online* 6(3), 207-213 (2005).
2. Y.A. Pishchalnikov, O.A. Sapozhnikov, M.R. Bailey, *et al.*, "Cavitation bubble cluster activity in the breakage of kidney stones by lithotripter shockwaves," *J. Endourol.*, 17, 435–446 (2003).
3. R.O. Cleveland, M.R. Bailey, N. Fineberg, *et al.*, "Design and characterization of a research electrohydraulic lithotripter patterned after the Dornier HM3," *Rev.Sci.Instrum.* 71, 2514–2525 (2000).
4. Y.A. Pishchalnikov, J.A. McAteer, M.R. Bailey, *et al.*, "Acoustic shielding by cavitation bubbles in shock wave lithotripsy (SWL)," *Proceedings of ISNA 17*, (2005).
5. O.A. Sapozhnikov, V.A. Khokhlova, M.R. Bailey, *et al.*, "Effect of overpressure and pulse repetition frequency on cavitation in shock wave lithotripsy," *J.Acoust.Soc.Am.* 112, 1183–1195 (2002).
6. Yu.A. Pishchalnikov, O.A. Sapozhnikov, J.C. Williams, Jr., *et al.*, "Cavitation bubble cluster dynamics induced by lithotripter shock waves at the surface of model and natural kidney stones." *Nonlinear Acoustics at the Beginning of the 21st Century (Proc. of ISNA16)*, MSU, Moscow, Russia, Vol.1, pp.395-398 (2002).
7. I. Hansson, K.A. Morch, "The dynamics of cavity clusters in ultrasonic (vibratory) cavitation erosion," *J. Appl. Phys.* 51, 4651-4658 (1980).
8. J.P. Dear, J.E. Field, "A study of the collapse of the arrays of cavities," *J. Fluid Mech.*190, 409-425 (1988).
9. W. Lauterborn, "Liquid jets from cavitation bubble collapse," *Proc. 5th Intl. Conf. on Erosion by Solid and liquid impact (ed. J.E.Field)*, Cavendish Lab., Cambridge , UK, paper 58, (1979).

Influence of compressibility on bubble interaction

Yu. A. Ilinskii*, M. F. Hamilton*, E. A. Zabolotskaya* and G. D. Meegan*

*Applied Research Laboratories, The University of Texas at Austin, Austin, Texas 78713-8029, USA

Abstract.
Two models of collective bubble dynamics are presented, a second-order system of coupled equations based on Lagrangian formalism, and a first-order system based on Hamiltonian formalism. Consideration is given to the relative effects of bubble interaction, coalescence, and compressibity on the dynamics of 50 bubbles clustered near a rigid surface and subjected to an incident shock wave.

Keywords: bubble interaction, coalescence, compressibility
PACS: 43.25.Yw

INTRODUCTION

Clusters of cavitation bubbles produced during shock wave lithotripsy experience explosive growth, contact with one another, and violent collapse. [1] Owing to the close proximity of the bubbles and the increase in their radii by two to three orders of magnitude, the interaction of bubbles with one another and the resulting collective behavior in a cluster cannot be ignored. A theoretical model taking bubble interaction into account was developed by Doinikov [2], generalizing results obtained by others for two bubbles. His model accounts for radiation damping, but not time delays, associated with compressibility of the liquid. While instantaneous interaction is a reasonable approximation for sufficiently small bubble separation distances, time delays must be included for clusters with dimensions on the order of 1 cm, which are encountered in lithotripsy.

In the present paper we discuss a theoretical model for collective bubble dynamics in which time delay due to compressibility is taken into account. [3] Also discussed is a model for bubble coalescence. [4] The relative importance of effects due to bubble interaction, coalescence, and time delay on the collective dynamics in a cluster was touched upon briefly in earlier work. [3] One purpose of the present paper is to expand on this earlier discussion. The other purpose is to present two formulations of the dynamical equations, one based on Lagrangian formalism, and the other on Hamiltonian formalism. The latter proves to be far preferable for numerical simulations in which time delay is not taken into account.

LAGRANGIAN FORMALISM

We begin by deriving the equation of motion for dynamically coupled bubbles in an incompressible liquid. It is assumed that the bubbles remain spherical in shape, and

CP838, *Innovations in Nonlinear Acoustics: 17th International Symposium on Nonlinear Acoustics*,
edited by A. A. Atchley, V. W. Sparrow, and R. M. Keolian
© 2006 American Institute of Physics 0-7354-0330-9/06/$23.00

that the motion of the ith bubble is described by its radial velocity $\dot{R}_i$ and translational velocity $\mathbf{U}_i$, where R_i is the instantaneous radius of the bubble, $\mathbf{r}_{0i}$ the position of its center relative to a fixed origin, $\mathbf{U}_i = \dot{\mathbf{r}}_{0i}$, and dots over quantities indicate time derivatives. With R_i and $\mathbf{r}_{0i}$ taken to be generalized coordinates, Lagrange's equations describing the dynamics of the system become, for $1 \leq i \leq N$,

$$\frac{d}{dt}\left(\frac{\partial \mathscr{L}}{\partial \dot{R}_i}\right) = \frac{\partial \mathscr{L}}{\partial R_i}, \qquad \frac{d}{dt}\left(\frac{\partial \mathscr{L}}{\partial \dot{\mathbf{r}}_{0i}}\right) = \frac{\partial \mathscr{L}}{\partial \mathbf{r}_{0i}}, \tag{1}$$

where $\mathscr{L} = \mathscr{K} - \mathscr{V}$ is the Lagrangian, $\mathscr{K}$ is the kinetic energy of the system, and $\mathscr{V}$ is the potential energy. Equations (1) constitute a system of $4N$ second-order differential equations for the radius and three coordinates associated with each of the N bubbles.

The liquid is assumed to be inviscid and described by a velocity potential ϕ that satisfies Laplace's equation, $\nabla^2 \phi = 0$. For an incompressible liquid that is at rest at infinity, the total kinetic energy is [5]

$$\mathscr{K} = -\frac{\rho}{2}\sum_i \int_{S_i} \phi \frac{\partial \phi}{\partial r_i}\, dS_i, \tag{2}$$

where ρ is the density of the liquid, and r_i is the magnitude of the local coordinate vector $\mathbf{r}_i$ that defines position relative to the center of the ith bubble, $r_i = |\mathbf{r}_i|$. The summation is over all N bubbles in the cluster, and the surfaces S_i coincide with the bubble walls. Evaluation of the integral requires knowledge of the velocity potential and its normal derivative on the surface of each bubble.

The normal derivative is determined by the boundary condition on the liquid at the bubble wall: $\partial \phi / \partial r_i = \dot{R}_i + \mathbf{U}_i \cdot \mathbf{n}_i$, where $\mathbf{n}_i = \mathbf{r}_i / r_i$ is the unit vector in the direction of $\mathbf{r}_i$. For an isolated bubble, the solution of Laplace's equation that satisfies the boundary condition is $\phi_{0i} = -R_i^2 \dot{R}_i / r_i - R_i^3 \mathbf{U}_i \cdot \mathbf{n}_i / 2r_i^2$, a monopole plus a dipole. The summation $\phi_0 = \sum_i \phi_{0i}$ is taken to be the zeroth approximation for a system with more than one bubble. It satisfies Laplace's equation, but not the boundary condition.

To satisfy the boundary condition on any given bubble one must account for the flow produced by all other bubbles. The velocity potential in the vicinity of the ith bubble due to flow from the kth bubble is expressed as $\phi_{0k}(\mathbf{r}_{ki} + \mathbf{r}_i) = \phi_{0k}(\mathbf{r}_{ki}) + \mathbf{c}_{ki} \cdot \mathbf{r}_i + \cdots$, where $\mathbf{r}_{ki} = \mathbf{r}_{0i} - \mathbf{r}_{0k}$ and $\mathbf{c}_{ki} = \partial \phi_{0k} / \partial \mathbf{r}_{ki}$. Truncating the expansion after the linear term in $\mathbf{r}_i$ is sufficient for calculating terms through order R^4 / r^4 in the kinetic energy. A solution of Laplace's equation that satisfies the boundary condition through order R^2 / r^2 is obtained by adding to the zeroth approximation a solution of Laplace's equation that cancels the contribution due to $\mathbf{c}_{ki} \cdot \mathbf{r}_i$ on the boundary. The required correction is the dipole term $R_i^3 \mathbf{c}_{ki} \cdot \mathbf{n}_i / 2r_i^2$ added to the summation for $k \neq i$.

Based on the aforementioned approximations, calculation of the kinetic energy through order R^2 / r^2 yields

$$\mathscr{K} = 2\pi\rho \left[\sum_i R_i^3 \dot{R}_i^2 + \frac{1}{6}\sum_i R_i^3 U_i^2 + \sum_{i \neq k}\sum_{k \neq i} \frac{R_i^2 R_k^2 \dot{R}_i \dot{R}_k}{r_{ik}} \right.$$
$$\left. + \frac{1}{2}\sum_{i \neq k}\sum_{k \neq i} \frac{R_i^3 R_k^2 \dot{R}_k (\mathbf{U}_i \cdot \mathbf{n}_{ik}) + R_i^2 R_k^3 \dot{R}_i (\mathbf{U}_k \cdot \mathbf{n}_{ki})}{r_{ik}^2} \right]. \tag{3}$$

The potential energy of the system is expressed in terms of its differential $d\mathcal{V} = \sum_i (P_\infty - P_i)\,dV_i$, where P_∞ is the pressure far away from the bubbles, V_i is the volume of the ith bubble, and P_i is the pressure inside, given here by

$$P_i = \left(P_0 + \frac{2\sigma}{R_{i0}}\right)\left(\frac{R_{i0}}{R_i}\right)^{3\gamma} - \frac{2\sigma}{R_i} - 4\eta\frac{\dot{R}_i}{R_i}, \tag{4}$$

where P_0 is the atmospheric pressure, γ the ratio of specific heats, σ is surface tension, and η is shear viscosity. Since $V_i = \frac{4}{3}\pi R_i^3$ one can write $d\mathcal{V} = 4\pi\sum_i (P_\infty - P_i)R_i^2\,dR_i$.

From the first of Eqs. (1) one obtains

$$R_i\ddot{R}_i + \frac{3}{2}\dot{R}_i^2 = \frac{P_i - P_\infty}{\rho} + \frac{1}{4}U_i^2 - \sum_{k\neq i}\frac{R_k}{r_{ik}}\left(R_k\ddot{R}_k + 2\dot{R}_k^2\right)$$

$$+ \frac{1}{2}\sum_{k\neq i}\frac{R_k^2}{r_{ik}^2}\left[\dot{R}_k(5\mathbf{U}_k + \mathbf{U}_i)\cdot\mathbf{n}_{ik} + R_k(\dot{\mathbf{U}}_k\cdot\mathbf{n}_{ik})\right], \tag{5}$$

and from the second

$$\frac{d\mathbf{M}_i}{dt} = 4\pi\rho\sum_{k\neq i}\frac{R_i^2 R_k^2 \dot{R}_i \dot{R}_k \mathbf{n}_{ik}}{r_{ik}^2}, \quad \text{where} \quad \mathbf{M}_i = \frac{2\pi\rho}{3}R_i^3\mathbf{U}_i + 2\pi\rho\sum_{k\neq i}\frac{R_i^3 R_k^2 \dot{R}_k \mathbf{n}_{ik}}{r_{ik}^2}. \tag{6}$$

The quantity $\mathbf{M}_i$ is the generalized momentum of the ith bubble. Its first term, which may be expressed as $\frac{1}{2}\rho V_i \mathbf{U}_i$, is the momentum of the induced mass associated with translational motion of the bubble in a liquid at rest. Its second term is the contribution to the induced mass due to the flow produced by pulsations of all the other bubbles. The term on the right side of the expression for $d\mathbf{M}_i/dt$ is the secondary Bjerknes force.

Although in rather different forms, Eqs. (5) and (6) are equivalent to the system of equations derived by Doinikov [2], who also used Lagrangian formalism. Equation (5), without translation taken into account ($U_i = 0$), was used previously to model the growth and collapse of 50 bubbles adjacent to a rigid surface and subjected to a shock wave. [3, 4] Also, Eqs. (5) and (6) were used to simulate a single bubble pulsating and translating near a rigid surface, illustrating the exchange of radial and translational momentum. [4] The purpose of the present paper is to discuss more generally the properties of the model equations in connection to applications in lithotripsy.

Discussion

Our motivation for developing the theory is to model the dynamics of bubble clusters encountered in shock wave lithotripsy. In this application, the acoustic excitation is short on the time scale corresponding to growth and collapse of the bubbles. In contrast to cavitation driven by continuous, periodic ultrasound, the collective bubble dynamics considered here is principally the free response to an incident shock wave. The growth phase is explosive, with bubble radii increasing to 10^2 or 10^3 times their initial values. For initial distributions of bubble nuclei encountered in lithotripsy, the expanding bubbles inevitably come in contact with one another. Modeling issues connected with small

305

bubble separation distances and bubble coalescence constitute one topic of the following discussion. Another topic is compressibility of the liquid, which must be included for bubble clusters above a certain size. This size is determined by the time it takes sound to propagate from one side of the cluster to the other, in relation to the characteristic time for growth and collapse of the bubbles.

We first address a potential concern that Eqs. (5) and (6) may lose validity when some bubbles in the cluster are close together, since the interaction terms are obtained as an expansion in powers of the small parameter R/r. While the relative order of successive terms in the expansions is characterized by R/r for small numbers of bubbles, another ordering parameter emerges for large numbers of bubbles.

Consider the first summation in Eq. (5), the individual terms of which are seen to be of order R/r relative to the quantity on the left-hand side of the equation. We now consider the terms in this summation collectively, and evaluate the series for a bubble in the center of a spherical cluster having radius R_{cl}, volume $V_{cl} = \frac{4}{3}\pi R_{cl}^3$, and containing N bubbles with similar radii, $R_k \simeq R_i$:

$$\sum_{k \neq i} \frac{R_k}{r_{ik}} \left(R_k \ddot{R}_k + 2\dot{R}_k^2 \right) \sim R_i \left(R_i \ddot{R}_i + 2\dot{R}_i^2 \right) \frac{N}{V_{cl}} \int_{V_{cl}} \frac{dV}{r} \sim \frac{NR_i}{R_{cl}} \left(R_i \ddot{R}_i + 2\dot{R}_i^2 \right). \tag{7}$$

After summation, this term is thus of order NR/R_{cl}, rather than of order R/r, relative to the left-hand side of Eq. (5). Performing a similar calculation for the second summation in Eq. (5) reveals that it is of relative order NR^2/R_{cl}^2. For clusters containing many bubbles, the interaction terms in Eqs. (5) and (6) may be regarded as expansions not in powers of R/r, but instead in powers of the much smaller quantity R/R_{cl}. The physical explanation for this effect is that for a uniform distribution of bubbles, those farther away collectively exert greater influence because their number per unit volume increases as r^2, which more than compensates for the influence of an individual bubble, whose effect decreases as $1/r$. For N sufficiently large, Eqs. (5) and (6) therefore remain valid even when R/r is of order one for isolated pairs of bubbles.

When two bubbles come in contact during a numerical simulation, our approach is to replace them with a third bubble that satisfies the following conservation relations:

$$\text{volume:} \quad V_1 + V_2 = V_3, \tag{8}$$
$$\text{kinetic energy:} \quad \dot{V}_1 + \dot{V}_2 = \dot{V}_3, \tag{9}$$
$$\text{internal energy:} \quad P_1 V_1 + P_2 V_2 = P_3 V_3, \tag{10}$$
$$\text{mass center:} \quad \mathbf{r}_{01} V_1 + \mathbf{r}_{02} V_2 = \mathbf{r}_{03} V_3, \tag{11}$$
$$\text{momentum:} \quad \mathbf{U}_1 V_1 + \mathbf{U}_2 V_2 = \mathbf{U}_3 V_3, \tag{12}$$

where the indices 1 and 2 identify properties of the two bubbles just before contact, and the index 3 corresponds to the new bubble resulting from coalescence. In the order they appear, these relations determine the quantities R_3, $\dot{R}_3$, P_3, $\mathbf{r}_{03}$ and $\mathbf{U}_3$, respectively, for the new bubble immediately after coalescence. Equation (8) maintains continuity of radial mass flow, and Eq. (9) conserves kinetic energy of the entire system. Both relations were derived for distances far from the coalescing bubbles. Equation (10) conserves internal energy of the gas inside the bubbles. The pressure P_3 determined by Eq. (10) is used to replace the first term in Eq. (4) by $P_3(R_3^{\text{ref}}/R_3)^{3\gamma}$, where R_3^{ref} is the value of R_3 determined by Eq. (8), and which is therefore not the equilibrium radius of the new bubble. Equation (11) maintains a constant mass center for the liquid in the neighborhood of the coalescing bubbles, and Eq. (12) conserves the momentum far away. In a movie available in an open access, online publication [3], a simulation based on Eq. (5) (excluding effects of translation, but including effects of compressibility) illustrates dramatically the growth, coalescence and collapse of bubbles in a large cluster, where the coalescence algorithm is based on Eqs. (8)–(11).

For sufficiently large bubble clusters, compressibility of the liquid must be taken into account. The finite sound speed c in a compressible liquid introduces a time delay r_{ik}/c corresponding to when the motion of the kth bubble is experienced by the ith bubble. Inclusion of this delay entails evaluation of the quantities R_k, $\dot{R}_k$, $\mathbf{U}_k$, and $\dot{\mathbf{U}}_k$ appearing in the summations in Eqs. (5) and (6) at the earlier times $t - r_{ik}/c$. All other dynamical quantities, namely, those with subscript i, are evaluated at the current time t. Also associated with the finite sound speed is radiation damping of the bubble pulsations. As discussed in previous work [3, 6], radiation damping is taken into account in Eq. (5) by including the term $\dddot{V}_i/4\pi c$ on the right-hand side. Although radiation damping also results from translational motion, the translational velocities of bubbles in large clusters are sufficiently small in relation to the sound speed that these losses are negligible. Translational velocities in large clusters are typically small because Bjerknes forces on a given bubble surrounded uniformly by many others tend to cancel one another.

Results

To demonstrate the relative importance of bubble interaction, coalescence, and time delay due to compressibility, we ignore translation and present results based on numerical solutions of Eq. (5). We take the same approach as in earlier work [3], and that is to examine the total volume of gas in a cluster, $V_{\text{tot}} = \sum_i V_i$, together with the number of bubbles in the cluster, N, as functions of time. Rather than plotting V_{tot} directly, we instead plot the quantity $R_{\text{tot}} = (3V_{\text{tot}}/4\pi)^{1/3}$, the radius corresponding to a sphere of volume V_{tot}, because results for all cases considered are then easily plotted on the same scale.[1]

We consider the same geometry and initial bubble distribution as earlier. [3] A population of 50 air bubbles in water, having the same initial radius $R_{i0} = 5\ \mu$m, are randomly distributed in a cubic space of volume 1 cm^3 adjacent to a rigid surface. The method of images is used to satisfy the boundary condition on the surface. The bubbles are at rest when at $t = 0$ the shock waveform $p(t) = 2p_{\text{sh}}e^{-\alpha t}\cos(\omega t + \pi/3)$ is applied uniformly to the entire population, where p_{sh} is the pressure amplitude of the shock. With $\alpha = 0.35\ \mu\text{s}^{-1}$ and $\omega = 0.1\pi$ rad/μs, this waveform is typical of ones produced by spark sources used in lithotripsy (positive pressure of duration 1.7 μs following the shock, after which is a negative pressure of duration 10 μs, and with p_{sh} three times the magnitude

---

[1] In Fig. 3 of an earlier paper [3], the vertical axis was mistakenly labeled "V_{cl} [cm^3]". This label should be replaced by "R_{tot} [cm]".

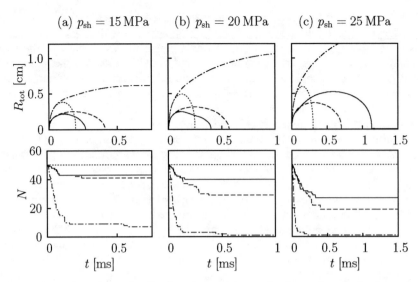

FIGURE 1. Time dependence of radius R_{tot} characterizing total volume of gas in the cluster (upper row) and number N of bubbles in the cluster (lower row) for three amplitudes of the incident shock wave. Dotted line ($\cdots$): no coalescence, no interaction. Dot-dash line ($-\cdot-$): coalescence, no interaction. Dashed line ($---$): coalescence, instantaneous interaction. Solid line (——): coalescence, delayed interaction.

of the peak negative pressure). The acoustic pressure is included in the model by letting $P_\infty = P_0 + p(t)$ in Eq. (5).

The results are shown in Fig. 1, with R_{tot} in the upper row, N in the lower row, and with shock pressure p_{sh} increasing from left to right. The four line types in each frame correspond to different combinations of effects being taken into account in the numerical simulations.

The dotted lines are simulations with neither coalescence nor interaction taken into account. That is to say, the histories of all 50 bubbles in the initial population are simulated using the Rayleigh-Plesset equation [Eq. (5) without the summations, with $U_i = 0$, and with all bubbles assumed identical] without regard to the fact that many of the predicted bubble volumes eventually overlap in space. The number N thus remains constant at 50.

Simulations represented by the dot-dash lines took coalescence into account, but not interaction. By far the largest bubbles are created in this case, as indicated by the effective radius R_{tot}, which ultimately decreases to zero well outside the figure frames. For the larger two shock amplitudes, all bubbles coalesce into one.

Both coalescence and bubble interaction were taken into account in the simulations represented by the dashed lines. Comparison with the previous two cases indicates that in the present case, interaction impedes bubble growth. This is easily understood as a consequence of neglecting time delays associated with compressibility. The initial increase in volume of any given bubble following arrival of the shock imposes a positive pressure impulse proportional to $\ddot{V}$ instantaneously on all other bubbles in the population, thus tending to offset inertial growth.

The last case, represented by the solid lines, accounts not only for both coalescence and interaction but also for the time delays associated with compressibility. This case is most complicated, as seen by comparing solid (with time delay) and dashed (without time delay) lines for the three shock amplitudes. The relative bubble sizes characterized by R_{tot} are similar for the lower two shock amplitudes, with the nominal bubble size being somewhat greater for the case without time delay. However, the relative sizes are reversed for the largest shock amplitude. This sensitivity to small changes in shock amplitude results from the critical dependence on when an acoustic wave radiated by one bubble arrives at all others in relation to the phases of their own growth-collapse cycles. Pressure waves scattered throughout the bubble population can alternately accelerate and decelerate the growth of any given bubble. A movie based on the data set from which the solid lines in Fig. 1(b) were obtained is presented in Ref. 3. There it can be seen that many of the bubbles appear to "twinkle," which is associated with jitter in bubble volume due to bombardment by the rapid succession of positive and negative pressures in the scattered waves. As the size of the cluster is reduced, time delays become less important, and the solid lines in Fig. 1 merge into the dashed lines.

HAMILTONIAN FORMALISM

A numerical difficulty is encountered for large N when solving Eqs. (5) and (6) as written, i.e., in the absence of time delay. For two or three bubbles, the system of second-order differential equations may be reduced analytically to a standard first-order form that can be solved with conventional numerical algorithms. For large N, the equations cannot realistically be recast in the desired first-order form, and a matrix inversion is required at each time step. [3] The inversions introduce substantial errors for large N. These errors may be avoided by using Hamiltonian formalism to recast the theoretical model as a set of explicit first-order equations. For large N, the model obtained from Hamilton's equations results in both increased numerical accuracy and reduced computation time.

In place of Eqs. (1) we now use Hamilton's equations as the starting point:

$$\dot{R}_i = \frac{\partial \mathcal{H}}{\partial G_i}, \quad \dot{\mathbf{r}}_{0i} = \frac{\partial \mathcal{H}}{\partial \mathbf{M}_i}, \quad \dot{G}_i = -\frac{\partial \mathcal{H}}{\partial R_i}, \quad \dot{\mathbf{M}}_i = -\frac{\partial \mathcal{H}}{\partial \mathbf{r}_{0i}}, \tag{13}$$

where $\mathbf{M}_i = \partial \mathcal{K} / \partial \mathbf{U}_i$ is the generalized momentum of translation, given by the second of Eqs. (6), and $G_i = \partial \mathcal{K} / \partial \dot{R}_i$ is the generalized momentum of expansion, given by

$$G_i = 4\pi\rho \left[R_i^3 \dot{R}_i + \sum_{k \neq i} \frac{R_i^2 R_k^2 \dot{R}_k}{r_{ik}} - \frac{1}{2} \sum_{k \neq i} \frac{R_i^2 R_k^3 (\mathbf{U}_k \cdot \mathbf{n}_{ik})}{r_{ik}^2} \right]. \tag{14}$$

Equations (13) constitute a system of $8N$ first-order differential equations. The Hamiltonian is $\mathcal{H} = \frac{1}{2} (\sum_i G_i \dot{R}_i + \sum_i \mathbf{M}_i \cdot \mathbf{U}_i) + \mathcal{V}$, and Eqs. (13) become

$$4\pi\rho \dot{R}_i = \frac{G_i}{R_i^3} - \sum_{k \neq i} \frac{G_k}{R_i R_k r_{ik}} + \sum_{j \neq i} \sum_{k \neq i} \frac{R_k G_j}{R_i R_j r_{ik} r_{kj}} + 3 \sum_{k \neq i} \frac{\mathbf{M}_k \cdot \mathbf{n}_{ik}}{R_i r_{ik}^2}, \tag{15}$$

$$4\pi\rho\dot{G}_i = \frac{3}{2}\frac{G_i^2}{R_i^4} + 9\frac{M_i^2}{R_i^4} - \sum_{k\neq i}\frac{G_iG_k}{R_i^2 R_k r_{ik}} + \sum_{j\neq i}\sum_{k\neq i}\frac{R_k G_i G_j}{R_i^2 R_j r_{ik} r_{kj}} - \frac{1}{2}\sum_{j\neq i}\sum_{k\neq i}\frac{G_i G_k}{R_k R_j r_{ki} r_{ij}}$$

$$+ 3\sum_{k\neq i}\frac{G_i(\mathbf{M}_k\cdot\mathbf{n}_{ik})}{R_i^2 r_{ik}^2} + (4\pi)^2\rho R_i^2\left(P_i - P_\infty\right), \tag{16}$$

$$4\pi\rho\dot{\mathbf{r}}_{0i} = 6\frac{\mathbf{M}_i}{R_i^3} - 3\sum_{k\neq i}\frac{G_k\mathbf{n}_{ik}}{R_k r_{ik}^2}, \tag{17}$$

$$4\pi\rho\dot{\mathbf{M}}_i = \sum_{k\neq i}\frac{G_iG_k\mathbf{n}_{ik}}{R_iR_k r_{ik}^2}. \tag{18}$$

Equations (15)–(18) are in the desired form for straightforward application of standard Runge-Kutta procedures to obtain numerical solutions for R_i, G_i, $\mathbf{r}_{0i}$, and $\mathbf{M}_i$. The translational velocity is determined by these quantities as follows:

$$\mathbf{U}_i = \frac{3}{2\pi\rho}\frac{\mathbf{M}_i}{R_i^3} - \frac{3}{4\pi\rho}\sum_{k\neq i}\frac{G_k\mathbf{n}_{ik}}{R_k r_{ik}^2}. \tag{19}$$

Equations (15)–(18) are not identical to Eqs. (5) and (6). The reason is that both systems of equations were obtained as expansions in powers of R/r, carried out through order R^2/r^2, such that differences in the two formulations are of order R^3/r^3. For cases investigated so far, numerical simulations based on the two formulations have agreed to within graphical resolution.

Time delay is taken into account in the Hamiltonian formulation as was described for the Lagrangian formulation. With time delay, there is no apparent advantage to solving Eqs. (15)–(18) rather than the first-order forms of Eqs. (5) and (6).

ACKNOWLEDGMENTS

This work was supported by IR&D funds at ARL:UT, and the Office of Naval Research.

REFERENCES

1. Yu. A. Pishchalnikov, O. A. Sapozhnikov, M. R. Bailey, J. C. Williams, Jr., R. O. Cleveland, T. Colonius, L. A. Crum, A. P. Evan, and J. A. McAteer, "Cavitation Bubble Cluster Activity in the Breakage of Kidney Stones by Lithotripter Shockwaves," *J. Endourol.*, **17**, 435–446 (2003).
2. A. A. Doinikov, "Mathematical Model for Collective Bubble Dynamics in Strong Ultrasound Fields," *J. Acoust. Soc. Am.*, **116**, 821–827 (2004).
3. M. F. Hamilton, Yu. A. Ilinskii, G. D. Meegan, and E. A. Zabolotskaya, "Interaction of Bubbles in a Cluster near a Rigid Surface," *Acoust. Res. Lett. Online*, **6**, 207–213 (2005).
4. E. A. Zabolotskaya, Yu. A. Ilinskii, G. D. Meegan, and M. F. Hamilton, "Bubble Interactions in Clouds Produced During Shock Wave Lithotripsy," Proceedings of the 2004 IEEE International UFFC Joint 50th Anniversary Conference, pp. 890–893.
5. H. Lamb, *Hydrodynamics* (Dover, New York, 1945).
6. Yu. A. Ilinskii and E. A. Zabolotskaya, "Cooperative Radiation and Scattering of Acoustic Waves by Bubbles in Liquid," *J. Acoust. Soc. Am.*, **92**, 2837–2841 (1992).

Shock Wave Interaction With Laser-Generated Single Bubbles

G.N. Sankin, W.N. Simmons, S.L. Zhu, and P. Zhong

Department of Mechanical Engineering & Materials Science, Duke University, Box 90300
Durham, NC 27708, USA

Abstract. The interaction of lithotripter shock wave (LSW) with laser-generated single vapor bubbles in water is investigated experimentally using high-speed photography and pressure measurement via a fiber optic probe hydrophone. An optimal bubble size can be determined during both the expansion and collapse phase of the bubble oscillation that leads to the maximum pressure amplification, which is produced when the collapse time matches the compressive pulse duration of the LSW.

Keywords: shock wave, laser breakdown, bubble collapse, jet, cavitation, lithotripsy.
PACS: 43.25.Cb, 47.40.Nm, 47.55.Bx

INTRODUCTION

When a shock wave impinges on a cavitation bubble, the interaction may lead to pressure amplification due to forced collapse and results in directional jet formation in consequence of the asymmetric deformation of the bubble [1-3]. With appropriate combinations of pressure amplitude and interpulse delay, this technique has been shown to improve stone comminution and macromolecular delivery [4, 5]. Despite this, the optimal shock wave profile and pulse combination have not been established. In this work, the interaction of lithotripter shock wave (LSW) with laser-generated single vapor bubbles in water is investigated.

EXPERIMENTAL

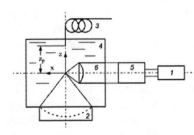

FIGURE 1. The experimental setup.

A schematic diagram of the experimental setup is shown in Fig. 1. A Q-switched Nd:YAG laser (1) with $\lambda = 1064$ nm and pulse duration = 5 ns (Tempest 10, New Wave Research) was focused into a water tank (4) to generate a single cavitation bubble via optical breakdown [6]. The laser was aligned horizontally using beam expander (5) and focusing lens (6) with its beam focus coinciding with the focal point of a piezoelectric shock wave generator (FB12, Richard Wolf) (2).

CP838, *Innovations in Nonlinear Acoustics: 17th International Symposium on Nonlinear Acoustics*,
edited by A. A. Atchley, V. W. Sparrow, and R. M. Keolian

The dynamics of laser-induced single bubble and LSW-bubble interaction were captured using a high-speed imaging system (Imacon 200, DRS Hadland), combined with a long-distance microscope (K2, Infinity) and a 5X objective lens. The camera was running at a framing rate of 2 million frames/s with up to 12 frames in each sequence. A fiber-optic coupled Xenon flash lamp (ML-1000, Dyna-Lite) was used for illumination and shadowgraph imaging. The shock waves generated by the inertial collapse of laser-induced bubble and LSW-bubble interaction were measured by using a fiber-optic probe hydrophone (FOPH-500, RP acoustics) (3) connected to a 100-MHz digital oscilloscope (TDS 2014, Tektronix). The 100-μm probe tip of the FOPH-500 was placed along the central axis of the shock wave source at a distance z_p above the focus (Fig. 1). Two digital delay generators (DG 535, Stanford Research Systems) were used to trigger the shock wave source, the laser, and the high-speed camera.

The shock wave measured is comprised of a leading compressive wave (peak pressure of $P_+ = 39$ MPa, pulse duration of $T_+ = 1$ μs), followed by a trailing tensile wave (minimum pressure of $P_- = -8$ MPa, pulse duration of $T_- \sim 2$ μs). The shock wave emitted by the collapse of the bubble measured at a distance of $z_p = 1.1$ mm has a peak pressure $P_c = 4.8 \pm 0.3$ MPa with a full width at half maximum (FWHM) pulse duration $\tau_c = 28 \pm 8$ ns (collapse time of the bubble is $T_c = 57.2 \pm 0.8$ μs). In general, the bubbles are classified into one of the three categories – E (expanding bubble), S (standing bubble near the maximum diameter), and C (collapsing bubble) [7].

RESULTS

The dynamics and consequences of shock wave-inertial bubble interaction change significantly depending on the size and phase of the oscillating bubble, which can be controlled by adjusting the interpulse delay, i.e., the time between laser-induced plasma and the arrival of the LSW at the focus. Figure 2 shows representative high-speed images of shock wave-inertial bubble interaction to produce pair of bubbles matched in size yet with opposite oscillation phases at the moment of impact. In general, the larger the bubble was at the moment of LSW impact, the longer it would be to produce the forced collapse. At the maximum bubble size, the forced collapse time (T_c') was approximately 4 μs. As shown clearly in Fig. 2, the originally spherical bubble was collapsed into a disk-shaped minimum volume perpendicular to the LSW propagation direction. Most importantly, it is observed that the forced collapse of type C bubbles occurs quickly with concomitantly larger asymmetric deformation and jetting during rebound than in their counterpart type E bubbles of the same size. Although the bubbles were nearly spherical before interaction, the asymmetric extension of the type C bubbles (with a tip velocity up to 260 m/s) along the shock wave propagation direction was significantly higher than that of the type E bubbles.

The corresponding pressure waveforms, measured simultaneously with the high-speed imaging sequences, are shown on Fig. 3a and Fig. 3b, respectively. It can be seen that the second pressure spike, corresponding to the shock wave emitted by the forced collapse of the bubble, occurs sooner and with higher pulse amplitude for the type C than for type E bubbles. Because of the presence of the bubble, the amplitude of the LSW also dropped from $P_+ = 39 \pm 1$ MPa to $P_+' = 28 \pm 1$ MPa.

a)   b)

FIGURE 2. High-speed images of the interaction between shock wave and laser-generated single bubbles at different phases of the bubble oscillation. Interframe time is 0.5 µs, exposure time for each image frame is 10 ns, frame width is 0.7 mm. Laser light comes from the right. The shock wave (not visible at this output level) propagates from the bottom to the top of the image frame and impinges on the bubble in the third frame. The interpulse delay is a) 6 and b) 50.5 µs. The corresponding bubble diameter at the moment of LSW impact is a) 0.39 and b) 0.40 mm.

Based on the pressure profile, several important parameters can be extracted. These include forced collapse time T_c', pulse amplitude P_c' and FWHM τ_c' of the second shock wave (Fig. 3b). Figure 4 shows the results obtained for various phases of the bubble oscillation normalized by the corresponding values for the laser-generated bubble (D_{max} = 0.61±0.03 mm). Overall, the data fits into two distinctive curves separating the type E from type C bubbles. The two curves appear to merge to the zero-velocity point (or type S bubble) at maximum expansion. Compared to its inertial collapse, the forced collapse of the laser-generated bubble by a LSW produces higher pressure with longer pulse duration for the resultant second shock wave emission, indicating an energy transfer from the LSW to the collapsing bubble. The maximum pressure produced by the type C bubble is 1.5 times of the value for the type E bubble, indicating a strong influence of the oscillation phase of the bubble on pressure amplification. Further, the maximum pressure amplification was produced at T_c'/T_+ = 1 for type C bubbles, compared to T_c'/T_+ = 1.3 for type E bubble.

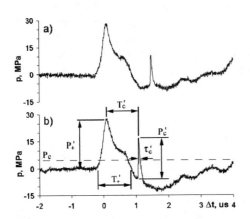

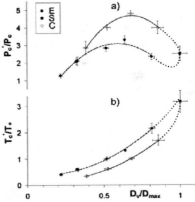

FIGURE 3. Pressure waveforms measured at z_p = 1.1 mm following shock wave-bubble interaction for a) expanding bubble and b) collapsing bubble of the same size.

FIGURE 4. Normalized a) peak pressure, and b) collapse time versus normalized bubble diameter when shock wave-bubble interaction occurs at different phase of the bubble oscillation. Quadratic fits of the experimental data are shown in dashed lines (expanding bubbles) and solid lines (collapsing bubbles). Dotted lines are hypothetical curves.

313

SUMMARY

The optimal interpulse delay is determined for shock wave driven collapse of a laser-induced bubble that leads to the strongest secondary acoustic emission, which is produced when the collapse time matches the compressive pulse duration of the shock wave. Because of the consistency in single bubble generation by a laser the interaction of shock wave with single inertial bubble with emphasis on the effect of the phase of bubble oscillation can be investigated. Using high-speed photography and pressure measurement via a fiber optic probe hydrophone it is shown that the interaction leads to non-spherical collapse of the bubble (0.6 mm in maximum diameter) with micro-jet formation along shock wave propagation direction. Under well-controlled experimental conditions an optimal bubble size is determined during both the expansion and collapse phase of the bubble oscillation. This is in contrast to the random variation of cavitation inception in water by acoustic pulses and the inherent complexity of the multi-bubble systems used in previous studies. When the interaction occurs above the optimal bubble size, the forced collapse will be slowed down by the trailing tensile component of the shock wave. Whereas below the optimal size the momentum transfer and duration of the interaction will decrease, leading to lower pressure amplification. In this work, however, for expanding bubbles the forced collapse is shown prolonged into the tensile phase of the shock wave, presumably due to the time delay it takes for the shock wave to first slow down and stop the ongoing expansion of the bubble before initiating the collapse. As a result, the pressure amplification is about half of that produced by the shock wave interaction with a collapsing bubble. There exists a direct link between maximum tip velocity and the collapse pressure, which is likely to be generated by the "water hammer" (the impact of an involuted jet on the distal wall of the bubble). Both consequences are of great importance for a diverse range of applications of ultrasound in therapeutic medicine, such as shock wave lithotripsy and macromolecule delivery into biological cells.

ACKNOWLEDGMENTS

This work was supported in part by NIH grants RO1-DK52985, RO1-EB002682, S10-RR16802 and NSF grant BES-0201921. Technical support from Richard Wolf GmbH, Germany on the use of the FB12 generator is also acknowledged.

REFERENCES

1. Tomita, Y., and Shima, A., *J. Fluid Mech.* **169**, 535-564 (1986).
2. Philipp, A., Delius, M., Scheffczyk, C., Vogel, A., and Lauterborn, W., *J. Acoust. Soc. Am.* **93**, 2496-2509 (1993).
3. Ohl, C.D. and Ikink, R., *Phys. Rev. Lett.* **90**, P214502 (2003).
4. Xi, X.F. and Zhong, P., *Ultrasound Med. Biol.* **26**, 457-467 (2000).
5. Zhong, P., Lin, H., Xi, X.F., Zhu, S.L., and Bhogte, E.S., *J. Acoust. Soc. Am.* **105**, 1997-2009 (1999).
6. Lindau, O. and Lauterborn, W., *J. Fluid Mech.* **479**, 327-348 (2003).
7. Sankin, G.N., Simmons, W.N., Zhu, S.L., and Zhong, P., *Phys. Rev. Lett.* **95**, 034501 (2005).

Modeling of Bubble Oscillations Induced by a Lithotripter Pulse

Wayne Kreider*, Michael R. Bailey* and Lawrence A. Crum*

*Center for Industrial and Medical Ultrasound, Applied Physics Laboratory
University of Washington, 1013 NE 40th Street, Seattle, WA 98105

Abstract. In therapeutic applications of biomedical ultrasound, it is important to understand the behavior of cavitation bubbles. Herein, the dynamics of a single, spherical bubble in water are modeled using the Gilmore equation closed by an energy balance on bubble contents for calculation of pressures inside the bubble. Moreover, heat and mass transfer at the bubble wall are incorporated using the Eller-Flynn zeroth-order approximation for gas diffusion, an estimation of non-equilibrium phase change based on the kinetic theory of gases, and assumed shapes for the spatial temperature distribution in the surrounding liquid. Bubble oscillations predicted by this model are investigated in response to a lithotripter shock wave. Model results indicate that vapor trapped inside the bubble during collapse plays a significant role in the afterbounce behavior and is sensitively dependent upon the ambient liquid temperature. Initial experiments have been conducted to quantify the afterbounce behavior of a single bubble as a function of ambient temperature; however, the results imply that many bubbles are present and collectively determine the collapse characteristics.

Keywords: bubble dynamics, cavitation, lithotripsy
PACS: 43.25.Yw

1. INTRODUCTION

Clinical treatments such as lithotripsy and high-intensity focused ultrasound (HIFU) utilize focused acoustic pulses for therapeutic purposes. In the course of treatment, negative acoustic pressures can activate cavitation nuclei to produce large-amplitude bubble oscillations. Because such cavitation activity can significantly impact the desired treatment, understanding the incipient motion of small bubbles exposed to high-amplitude acoustic pressures merits study. In this effort, a single bubble model developed to explore these dynamics is briefly described. In addition, trends in the predicted motion of a bubble subjected to a lithotripter pulse are presented along with some data collected from experiments designed to validate these trends.

2. BUBBLE MODEL

Large-amplitude oscillations of a single, spherical bubble in water are modeled herein by explicitly accounting for heat and mass transfer across the bubble wall. To this end, a combined model composed of various elements from the literature has been developed and implemented. A brief summary of these components is provided below.

The radial bubble dynamics are simulated with the Gilmore equation [1], thereby accounting for compressibility in the liquid. This basic formulation of the problem

CP838, *Innovations in Nonlinear Acoustics: 17th International Symposium on Nonlinear Acoustics*,
edited by A. A. Atchley, V. W. Sparrow, and R. M. Keolian
© 2006 American Institute of Physics 0-7354-0330-9/06/$23.00

is then closed by enforcing an energy balance on the bubble contents to determine the internal pressure. As reported by Prosperetti et al. [2], nonlinear bubble motions are better modeled with such an energy balance as opposed to an assumed polytropic relationship between bubble radius and pressure. In order to accurately approximate this energy balance, it is necessary to also account for the transport of heat and mass across the bubble wall. For mass transport, the diffusion of non-condensable gases is calculated from the Eller-Flynn zeroth-order solution [3], while non-equilibrium vapor transport is estimated based on the kinetic theory of ideal gases [4]. To model the heat transfer, spatial temperature profiles are assumed and the corresponding gradients are approximated in both the gaseous and liquid phases, as proposed by Yasui [5]. Overall, the model implemented herein is similar to those suggested by Matula et al. [6] and Toegel et al. [7]. However, this model additionally enables a dynamic calculation of the liquid temperature at the bubble wall. Accordingly, a framework is provided within which the interrelated effects of heat and mass transfer may be investigated for large bubble motions.

The model described above has been used to calculate bubble responses in acoustic fields characteristic of both HIFU and lithotripsy treatments. As suggested in earlier work regarding a bubble exposed to a lithotripter pulse [6], a significant amount of vapor is likely trapped inside upon collapse. Similarly, the model described above predicts that the combined effects of non-equilibrium vapor transport and temperature change in the liquid at the bubble wall during collapse can qualitatively change the bubble dynamics. While this qualitative change is perhaps most pertinent to HIFU treatments, the model does predict experimentally measurable differences in the response of a lithotripsy bubble. As illustrated in Figure 1, a bubble subjected to a shock wave is expected to respond by undergoing a prolonged growth followed by a primary collapse and multiple afterbounces. In addition, the model results displayed in Figure 1 indicate that a change in ambient temperature of $10°C$ will produce changes in the afterbounce timing of $\sim 20\%$. Moreover, this effect is essentially independent of the initial bubble size. Hence, these predictions suggest that the heat and mass transfer behavior implied by the model can be tested by altering the ambient water temperature and observing the timing of afterbounces induced by a lithotripter pulse.

3. EXPERIMENTAL OBSERVATIONS AND DISCUSSION

In order to validate model predictions regarding the relative effects of heat and vapor transport, experiments were conducted as suggested in the previous section. In these experiments, the bubble dynamics were characterized using both a high-speed camera and a pair of confocal passive cavitation detectors (PCDs). Details of the overall setup and the processing of PCD data were essentially the same as that used in earlier work by Cleveland et al. [8] and will not be discussed further here. However, it is worthwhile to note that this setup employs two aligned spherical detectors that each have a sensitive focal region of about 3.9 mm in cross section perpendicular to the transducer axis. Hence, it was expected that the dual-PCD data would efficiently identify individual bubble collapse events within the overlapping sensitive region as coincident 'spikes' on both PCD channels. In turn, the timing of these coincident 'spikes' would indicate

316

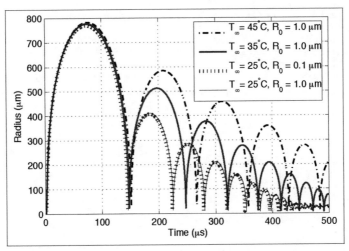

FIGURE 1. Comparison of model predictions of bubble dynamics for three ambient temperatures and two initial bubble sizes.

the timing of bubble afterbounces.

Although data were collected for ambient water temperatures of 25°C and 35°C, these data could not be interpreted to determine the afterbounce timing of individual bubbles. Instead, upon comparing PCD traces to the corresponding high-speed photographs, it became apparent that a 'spike' in the PCD signal does not necessarily correspond to a single collapse event within the sensitive region. Rather, the simultaneous collapse of a group of bubbles near the sensitive region may be recorded as a single 'spike.' This effect is demonstrated in comparison of Figures 2 and 3, wherein the time scales have been synchronized so that the shock wave arrives at the geometric focus at 184 μs. In order to improve this experiment such that the afterbounce timing of individual bubbles can be determined, additional degassing of the water (below 1.5–2 ppm dissolved oxygen) may reduce the bubble density to avoid cloud-type collapse behavior that impacts PCD data.

ACKNOWLEDGMENTS

This work was supported by the National Institutes of Health (NIH T32-EB-001650, NIH DK43881) and the National Space Biomedical Research Institute (NSBRI SMS00402).

REFERENCES

1. F. Gilmore, The growth or collapse of a spherical bubble in a viscous compressible liquid, Tech. Rep. 26-4, California Institute of Technology (1952).
2. A. Prosperetti, L. A. Crum, and K. W. Commander, *J Acoust Soc Am*, **83**, 502–514 (1988).
3. C. Church, *Journal of the Acoustical Society of America*, **86**, 215–227 (1989).

4. B. Storey, and A. Szeri, *Proceedings of the Royal Society of London Series A-Mathematical Physical and Engineering Sciences*, **456**, 1685–1709 (2000).
5. K. Yasui, *Journal of the Physical Society of Japan*, **65**, 2830–2840 (1996).
6. T. Matula, P. Hilmo, B. Storey, and A. Szeri, *Physics of Fluids*, **14**, 913–921 (2002).
7. R. Toegel, B. Gompf, R. Pecha, and D. Lohse, *Physical Review Letters*, **85**, 3165–3168 (2000).
8. R. Cleveland, O. Sapozhnikov, M. Bailey, and L. Crum, *Journal of the Acoustical Society of America*, **107**, 1745–1758 (2000).

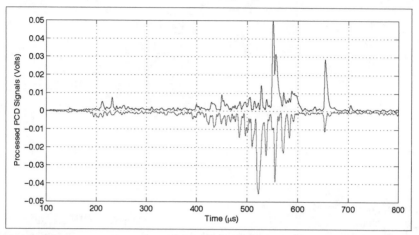

FIGURE 2. Sample PCD data acquired in the free field during a lithotripter pulse for the following conditions: 19 kV excitation, 25°C ambient temperature, 21% dissolved oxygen content.

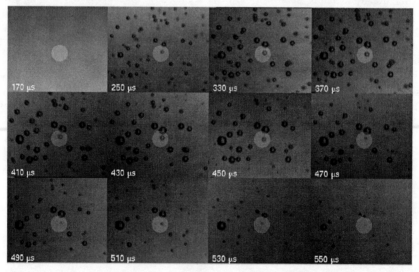

FIGURE 3. High-speed photographs corresponding to the above PCD data. The highlighted circle in each frame represents the sensitive region (3.9 mm diameter) of each PCD transducer within the focal plane of the camera.

318

Acoustic Shielding by Cavitation Bubbles in Shock Wave Lithotripsy (SWL)

Yuri A. Pishchalnikov*, James A. McAteer*, Michael R. Bailey[†], Irina V. Pishchalnikova*, James C. Williams, Jr.*, and Andrew P. Evan*¶

*Department of Anatomy and Cell Biology, Indiana University School of Medicine, 635 Barnhill Drive, Indianapolis, IN 46202, USA

[†]Center for Industrial and Medical Ultrasound, Applied Physics Laboratory, University of Washington, Seattle, WA 98105, USA

Abstract. Lithotripter pulses (~7-10 μs) initiate the growth of cavitation bubbles, which collapse hundreds of microseconds later. Since the bubble growth-collapse cycle trails passage of the pulse, and is ~1000 times shorter than the pulse interval at clinically relevant firing rates, it is not expected that cavitation will affect pulse propagation. However, pressure measurements with a fiber-optic hydrophone (FOPH-500) indicate that bubbles generated by a pulse can, indeed, shield the propagation of the negative tail. Shielding was detected within 1 μs of arrival of the negative wave, contemporaneous with the first observation of expanding bubbles by high-speed camera. Reduced negative pressure was observed at 2 Hz compared to 0.5 Hz firing rate, and in water with a higher content of dissolved gas. We propose that shielding of the negative tail can be attributed to loss of acoustic energy into the expansion of cavitation bubbles.

INTRODUCTION

It has been reported by Liebler *et al.* [1] that acoustic pulses recorded in the focus of a piezoelectric shock wave therapy head can show significant fluctuations, starting approximately at the maximum of the first tensile phase of the pulse [1]. Experimental and numerical results suggested that these fluctuations were associated with cavitation, such that with increasing bubble number density the tensile part of the acoustic pulse was shorter, and was followed by noticeable secondary oscillations. We have made similar observations for the pulses generated by shock wave lithotripters [2]. Here we demonstrate the susceptibility of the tensile phase of lithotripter shock waves to cavitation, and offer an explanation of the dynamics involved in the acoustic shielding of shock pulses.

MATERIALS AND METHODS

Studies were performed in an unmodified Dornier HM3 electrohydraulic lithotripter and a research lithotripter patterned after the HM3 (HM3-clone) [3]. To enable studies with a high-speed camera (Imacon 468, DRS Hadland, Inc., Cupertino, CA), the water tank of the HM3-clone was made of optically clear acrylic [6]. Waveforms were measured at the focus of each lithotripter using a fiber-optic probe hydrophone FOPH-

CP838, *Innovations in Nonlinear Acoustics: 17th International Symposium on Nonlinear Acoustics*,
edited by A. A. Atchley, V. W. Sparrow, and R. M. Keolian
© 2006 American Institute of Physics 0-7354-0330-9/06/$23.00

500 (Univ. of Stuttgart, Germany) [5]. Waveforms were collected in sets of 100 pulses using the Fast Frame setup of a Tektronix (TDS 5034) oscilloscope. Averaged waveforms were calculated by aligning recorded pulses to the coincidence of the half amplitude of the shock fronts by a program written in LabVIEW (National Instruments, Austin, TX) [2].

RESULTS AND DISCUSSION

Higher firing rate was used to increase the number of cavitation bubbles, which appeared to be due to persistence of cavitation nuclei (microscopic bubbles) that did not have enough time to dissolve between pulses [4]. As seen in Fig.1, these small cavitation nuclei have little effect on the leading positive-pressure phase of the pulse, but can noticeably shield propagation of its trailing negative-pressure phase. In addition to the effect of firing rate, further truncation of the negative-pressure phase was observed when water circulation in the HM3 bath was turned off, and the content of the dissolved gas was allowed to increase (compare waveforms at the same firing rates: *a* vs. *b*, and *c* vs. *d*). Thus, Fig. 1 demonstrates how readily the tensile phase of the shock pulse is affected by conditions that enhance cavitation.

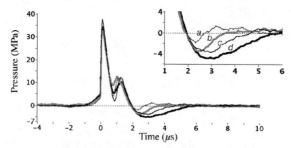

FIGURE 1. Effect of firing rate and content of dissolved gases on recorded waveforms: *a* – 2Hz at 25% oxygen saturation; *b* – 2Hz at 8% saturation, *c* – 0.5Hz at 25% saturation, and *d* – 0.5Hz at 8% saturation. The amplitude and duration of the negative-pressure phase is reduced at 2Hz compared to 0.5Hz. Increasing the gas content of the water further reduces the amplitude and duration of the negative-pressure phase. The inset is an enlargement of the negative tail. Zero time is positioned at the shock front. Each waveform is an average of 100 pulses recorded in the geometrical focus of Dornier HM3 lithotripter at 18kV. (The positive-pressure phase of the HM3 pulse has double peak structure caused by superposition of the edge waves.)

The reason why cavitation bubbles influence mostly the trailing negative tail without affecting the leading positive-pressure phase can be understood using high-speed photography. Fig.2 shows three HS-camera frames recorded with a 2µs step encompassing the arrival of the shock wave at the F2 focus of the HM3-clone lithotripter. Note, that there are no visible bubbles prior to or during propagation of the leading positive wave (above the dashed line), but the arrival of the tensile stress (below the dashed line) spawns cavitation bubbles, so that the rest of the pulse propagates through a more "bubbly" liquid. As bubbles grow, they shield the propagation of the trailing negative-pressure phase of the pulse.

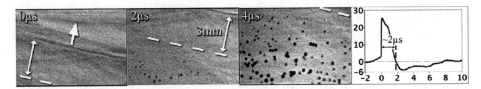

FIGURE 2. Three frames of high-speed shadowgraph photography showing the appearance of bubbles during the passage of the lithotripter shock pulse (the fifth lithotripter pulse administered at 2Hz, 20kV, and in deionized water with roughly 50% gas saturation). The diagram to the right shows the temporal profile of the HM3-clone pulse averaged over 100 shock waves administered at 18kV. Shadow of the shock front is seen at the center of the first frame (0μs) with the direction of propagation of the pulse depicted by an arrow (F2 is at the base of the single arrow). Since the positive-pressure phase lasts for ~2μs, the positive pressure occupies a 3mm strip behind the shock front, that is, an area bounded by the shadow of the shock front and the dashed line. Thus, the dashed line marks the transition from positive to negative acoustic pressure. On the second frame (2μs) the shock front is seen at the top right corner and the transition to negative pressure has moved to the center of the frame. Cavitation bubbles become visible soon after arrival of the negative-pressure phase of the pulse (below dashed line). In the third frame (4μs) the dashed line is at the right top corner, thus the positive-pressure phase has already passed, and the frame is occupied by the negative-pressure tail. Note, that the negative-pressure phase lasts for 6μs (see temporal profile of the pulse), that is, 3 times longer than the leading positive-pressure phase of the pulse (positive pressure region shown by double headed arrow). Thus, this HS-camera sequence suggests that that the tensile stress gives rise to the growth of cavitation bubbles, so that the rest of the negative tail propagates through a more "bubbly" liquid compared to the leading positive-pressure phase of the pulse.

This shielding by cavitation bubbles means that the acoustic pulse loses energy; that is, some energy delivered by the pulse is left in the medium in the form of the kinetic and potential energy of the liquid surrounding the expanding bubbles. Since the inertial growth of bubbles lasts for tens or hundreds of microseconds, this energy is not released until long after the passage of the lithotripter pulse.

Calculation of the energy needed for bubble expansion has been elegantly demonstrated by Zhong *et al.* [7]. In that report it is shown that the work done by a bubble during the expansion phase against ambient pressure is at least 2 orders of magnitude greater than the work against surface tension and viscous force. Therefore, the energy needed for a solitary cavitation bubble to grow can be estimated as a work against atmospheric pressure p_{atm} [7].

We have observed by high-speed photography that the initial radius of the bubbles is typically less than 50 μm (Fig.2), so that the energy E needed for a cavitation bubble to expand to its maximum radius R_{max} can be estimated as $E \approx \frac{4}{3} \pi R_{max}^3 p_{atm}$, and for a bubble to grow to $R_{max} = 0.5$ mm is about 0.05 mJ.

Energy of the negative-pressure phase of the pulse can be calculated as follows:

$$\int\limits_{\substack{time\ when \\ p<0}} dt \int\limits_0^\infty 2\pi\, r dr \frac{p^2(r,t)}{\rho_0 c_0}, \tag{1}$$

where p is the acoustic pressure, ρ_0 is the density of water, c_0 is sound velocity in water, r is the distance from the axis of the acoustic beam, and t is time.

The integral $\int p^2(t)\, dt$ was calculated by a LabVIEW program using measured temporal profile $p(t)$. The integral $\int p^2(r)\, rdr$ was estimated as follows:

$$\int_0^\infty p^2(r)\, rdr \approx p_{axis}^2 \int_0^{r(-6dB)} rdr, \qquad (2)$$

where p_{axis}^2 is the measured value on the axis, and $r(-6dB)$ is the radius of half maximum of the lateral distribution of the acoustic field. That is, the actual distribution perpendicular to the axis was approximated by a piston-like distribution. Taking into account that the measured lateral width for the half amplitude of the negative pressure was $r(-6dB) \approx 1cm$, the energy of the negative tail, for the pulses shown in Fig.1, was found to be a) 0.2 mJ, b) 1.6 mJ, c) 3 mJ, and d) 7.6 mJ. Thus, if the reduced negative tail in the lithotripter pulses were due entirely to loss of energy to bubble formation, one would predict the generation of about 150 more bubbles ($R_{max} = 0.5$ mm) with the pulse shown in a than with the pulse shown in d. This number of bubbles seems reasonable from the high-speed photography (Fig.2, [4]), which suggests that the shielding of the negative tail can be attributed to loss of acoustic energy into the expansion of bubbles.

In summary, this study shows that cavitation restricts the energy delivered by the negative-pressure phase of lithotripter pulses. This result underscores the value of using degassed water in lithotripters, and the need to define rates and power settings in clinical lithotripsy protocols to avoid or minimize acoustic shielding by cavitation.

ACKNOWLEDGMENTS

The authors thank Dr. Oleg Sapozhnikov for valuable discussion, Richard J. VonDerHaar and Philip Blomgren for their assistance. This work was supported by grants from ONRIFO and the National Institutes of Health (DK-43881).

REFERENCES

1. M. Liebler, T. Dreyer and R.E. Riedlinger, "Focal Pressure Variations in Shock Wave Therapies Caused by Cavitation Bubbles", *Proc. of the Joint Congress CFA/DAGA'04*, 983-984 (2004), http://www.ihe.uni-karlsruhe.de/forschung/akustik/Paper_Liebler_DAGA04.pdf
2. Y.A. Pishchalnikov, O.A. Sapozhnikov, M.R. Bailey, et al., "Cavitation selectively reduces the negative-pressure phase of lithotripter shock pulses," *Acoust. Research Letters Online*, accepted (2005).
3. R.O. Cleveland, M.R. Bailey, N. Fineberg, et al., "Design and characterization of a research electrohydraulic lithotripter patterned after the Dornier HM3," *Rev.Sci.Instrum.* 71, 2514–2525 (2000).
4. O.A. Sapozhnikov, V.A. Khokhlova, M.R. Bailey, et al. "Effect of overpressure and pulse repetition frequency on cavitation in shock wave lithotripsy," *J.Acoust.Soc.Am.* 112, 1183–1195 (2002).
5. J. Stardenraus and W. Eisenmenger, "Fiber-optic probe hydrophone for ultrasonic and shock-wave measurements in water," *Ultrasonics* 31, 267-273 (1993).
6. Y.A. Pishchalnikov, O.A. Sapozhnikov, M.R. Bailey, et al., "Cavitation bubble cluster activity in the breakage of kidney stones by lithotripter shockwaves," *J. Endourol.*, 17, 435–446 (2003).
7. P. Zhong, Y.F. Zhou, and S.L. Zhu, "Dynamics of bubble oscillation in constrained media and mechanisms of vessel rupture in SWL," *Ultrasound Med. Biol.* 27(1), 119–134 (2001).

Role of Shear and Longitudinal Waves in Stone Comminution by Lithotripter Shock Waves

Michael R. Bailey[1], Adam D. Maxwell[1], Brian MacConaghy[1],
Oleg A. Sapozhnikov[1,2], and Lawrence A. Crum[1]

[1]Center for Industrial and Medical Ultrasound, Applied Physics Laboratory, University of Washington,
1013 NE 40th St, Seattle, WA 98105 E-mail: bailey@apl.washington.edu
[2]Department of Acoustics, Physics Faculty, Moscow State University, Leninskie Gory, Moscow 119992,
Russia; E-mail: oleg@acs366.phys.msu.ru

Abstract. Mechanisms of stone fragmentation by lithotripter shock waves were studied. Numerically, an isotropic-medium, elastic-wave model was employed to isolate and assess the importance of individual mechanisms in stone comminution. Experimentally, cylindrical U-30 cement stones were treated in an HM-3-style research lithotripter. Baffles were used to block specific waves responsible for spallation, squeezing, or shear. Surface cracks were added to stones to simulate the effect of cavitation, and then tested in water and glycerol (a cavitation suppressive medium). The calculated location of maximum stress compared well with the experimental observations of where cracks naturally formed. Shear waves from the shock wave in the fluid traveling along the stone surface (a kind of dynamic squeezing) led to the largest stresses in the cylindrical stones and the fewest shock waves to fracture. Reflection of the longitudinal wave from the back of the stone - spallation - and bubble-jet impact on the proximal and distal faces of the stone produced lower stresses and required more shock waves to fracture stones, but cavitation stresses become comparable in small stone pieces. Surface cracks accelerated fragmentation when created near the location where the maximum stress was predicted.

Keywords: lithotripsy, shock wave
PACS: 43.25.Yw

INTRODUCTION

Despite over 20 years of clinical practice and fundamental research, a complete physical explanation of stone comminution by lithotripsy remains unknown. Incomplete understanding is evidenced by rising re-treatment rates despite the release of over 40 lithotripter designs over the history of shock wave lithotripsy (SWL). Four primary physical mechanisms have been proposed for fracture: stress induced by cavitation near the stone surface,[1] spallation due to summation in the distal end of the stone of the incident longitudinal wave and its inverted reflection from the acoustically soft interface at the back surface of the stone,[2] squeezing due to circumferential stresses generated by shock waves encircling the stone,[3] and shear generated by pressure gradients in the focused shock wave[4] or generated at the boundary of the stone.[5] The goal of this study was to test the role of these mechanisms of inducing the stress that leads to stone fracture. For brevity, four tests are the focus of this paper, and

CP838, *Innovations in Nonlinear Acoustics: 17th International Symposium on Nonlinear Acoustics*,
edited by A. A. Atchley, V. W. Sparrow, and R. M. Keolian
© 2006 American Institute of Physics 0-7354-0330-9/06/$23.00

cavitation is neglected.

THEORY

The linear elastic model used to simulate the stress wave propagation within a kidney stone was described in Ref. 5. The shock waveform was that defined by Church[6] and the shock front is planar on a cylindrical stone with elastic properties of a U30 model stone.[7] The location and value of the maximum tensile stress was calculated and compared with the location where the experimental stone fractured in two and the number of shock waves recorded to achieve fracture.

METHODS AND MATERIALS

Experiments were conducted in the water bath of a research lithotripter modeled after the unmodified (80 nF capacitor) Dornier HM3 electrohydraulic lithotripter (Dornier GmbH, Germany). Experiments were conducted at 1 Hz and 18 kV charging potential. Room temperature water was filtered through 10-μm pores before filling the tank and conditioned to 600 μS/cm and a 25-30% gas saturation level.

Model stones were made from Ultracal-30 gypsum (United States Gypsum, Chicago, IL).[7] Stones were cylindrical with 6.5 mm diameter and 8.5 mm length. The stones were stored under water and then were removed briefly to place them in the lithotripter bath. Stones were fixed by 3 spring-loaded plastic rods (1 mm diameter, 3 cm length) and held axisymmetrically in the acoustic field. The end point of experiments was fracture in two of the stone or 250 shock waves.

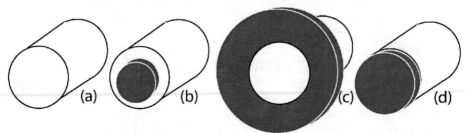

FIGURE 1. Four conditions tested: (a) the stone alone, (b) 4-mm disk on proximal face, (c) baffle on proximal face, and (d) 6.5-mm disk covering the entire proximal face.

Corprene (1 mm thick) was used to block acoustic waves in order to test the role of each mechanism. The shock wave pressure transmission through the corprene measured by hydrophone was less than 5%. Four conditions were tested: (1) the stones alone, (2) a 4-mm diameter corprene disk on the proximal face to block the longitudinal shock wave from entering the stone, (3) a corprene baffle encircling the proximal end of the stone to block squeezing around the stone, and (4) a 6.5-mm diameter corprene disk on the proximal face to block the shear wave generated at the corners of the stone. Figure 1 shows the four conditions.

RESULTS

Figure 2 contains a sequence of images showing the shock-wave-induced stress calculated within a U30 cylindrical stone. First, the longitudinal wave entered the stone from the left. Eventually this wave reached the back end of the stone and reflected and inverted from the "acoustically soft" stone-water interface. The inverted wave added to the trailing negative trough of the incident wave and created a weak local maximum tensile stress due to spallation 0.5 mm from the distal surface. Following the longitudinal wave was a wake that was generated at the surface of the stone where the wave traveled faster than the sound speed of water. Traveling at the sound speed in water along the stone surface and encircling the stone was the shock wave, and it created squeezing. In addition, mode conversion at the corners of the stone generated shear waves that focused in the distal half of the stone. Because the shear wave speed in the stone was close to the sound speed in water, the squeezing wave reinforced the shear wave and together they created the highest tensile stress within the stone. The location was 3.1 mm from the distal surface.

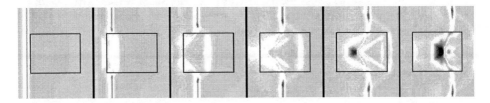

FIGURE 2. Sequence of simulations of the maximum stress induced by a lithotripter shock wave in a cylindrical stone. The wave, modeled as a plane wave, propagates to the right. The darkest regions show the highest tension. The shock wave in water ringing the stone reinforces the shear wave generated at the proximal corner of the stone and yields the highest maximum tensile stress in the stone.[5] The location of the maximum agrees well with the location where the stones fractured.

The stones alone fractured normal to the cylinder axis after 45±10 shock waves (number of stones tested N=8) at 3.6±0.2 mm from the distal surface. The agreement with the location of maximum stress calculated by the model was good. The agreement with the location of stress due to spallation was poor.

Calculations for stones with the 4-mm disk on the proximal face showed negligible change in the location and amplitude of the maximum stress compared to stones alone. Similarly, stones with the small disk broke in 54±16 shock waves (N=8) which was not significantly different from the case with stones alone. The fracture was in the same place with and without the disk. Blocking the longitudinal wave (and spallation) did not have a strong effect.

Calculations for stones with the baffle surrounding the proximal face showed a significant decrease in maximum stress compared to stones alone. The stress was nearly eliminated. Similarly in experiments, the proximal baffle required over 250 shock waves to break the stone. Removing the shock wave traveling in water along the surface of the stone had a strong effect.

Calculations for stones with the 6.5-mm disk on the proximal face also showed a significant decrease in maximum stress compared to stones alone. The decrease was due to the suppression of the shear wave generated at the edge of the proximal face of the uncovered stone. Similarly in experiments, 230±32 shock waves (N=8) were required to fracture the stones with the disk covering the entire proximal face. Five stones had not broken after 250 shock waves. The number of shock waves to fracture was significantly more than for stones without the disk. Suppressing the shear waves generated at the edge of the stone reduced the effectiveness of the shock waves in fracturing the stone.

DISCUSSION AND CONCLUSIONS

A series of experiments tested proposed mechanisms for inducing stress in kidney stones in lithotripsy. The mechanisms were spallation, squeezing, shear, and cavitation (omitted in the proceedings paper but discussed in the oral presentation). A mathematical model of linear elastic wave propagation was used with experiments on a specific cylindrical stone model to test the mechanisms. The model, supported by the experiment, described the stress mechanism as primarily shear waves generated at the corners of the stone that are reinforced by the squeezing wave running along the stone. The model created a more accurate description of how stress that leads to fracture was induced in the stone than did the definition of any one mechanism.

ACKNOWLEDGMENTS

The authors would like to thank the members of the Consortium on Shock Waves in Medicine and the Center for Industrial and Medical Ultrasound for collaboration on this study. Work supported by NIH-DK43881, NIH- DK55674, NIH-Fogarty, NSBRI-SMS00402, RFBR, and ONRIFO.

REFERENCES

1. Crum L.A., *J Urol* **140**, 1587-1590 (1988).
2. Delius M., G. Heine, and W. Brendel, *Naturwissenschaften* **75** 200-201 (1988).
3. Eisenmenger W, *Ultrasound Med. Biol.* **27** 683-93 (2001).
4. Sturtevant, B. "Shock wave physics of lithotripters" *in Smith's Textbook of Endourology*, edited by A.D. Smith et al., St. Louis, MO, Quality Medical Publishing, Inc., 1996, pp. 529-552.
5. Cleveland R.O. and Sapozhnikov O.A., *J Acoust Soc Am* **118** 2667-76 (2005).
6. Church C.C., *J Acoust Soc Am* **86** 215-227 (1989)
7. McAteer J. A., J. C. Williams, Jr., R. O. Cleveland, J. Van Cauwelaert, M. R. Bailey, D.A. Lifshitz, and A. P. Evan, *Urol Res* **33** 429–434 (2005).

SECTION 7
NONLINEAR ACOUSTICS IN
MEDICINE AND BIOLOGY

A Non - Dissipative Mechanism of Acoustic Radiation Force Generation

Lev Ostrovsky*, Alexander Sutin♣, Armen Sarvazyan♣

* Zel Technologies/NOAA ESRL, 325 Broadway, Boulder, CO 80305
♣ Artann Laboratories, 1459 Lower Ferry St., Trenton, NJ, 08618

Abstract. A general approach to the description of radiation force (RF) and shear displacement induced by an ultrasonic beam in a nonlinear rubber-like medium is developed. A new, non-dissipative mechanism of RF generation due to medium inhomogeneity is suggested, and the corresponding medium response is calculated for rubber–like media such as biological tissues.

Keywords: Radiation Force, rubber like media, shear displacement
PACS:43.25.Qp

INTRODUCTION

The acoustic radiation force (RF) is, in general, defined as a period-averaged force exerted on the medium by sound wave. Various aspects of RF action in fluids have been discussed by numerous authors, beginning from Lord Rayleigh. Most commonly, RF is presented as a unidirectional force acting on the absorbing or reflecting targets in the wave path. A classical example of such an action is sound pressure on a wall or other interface. As regards the distributed RF, such effects as acoustic streaming ("acoustic wind") [1] as well as the sound self-modulation as in the well-known parametric arrays [2] are the examples of the RF action.

In rubber-like elastic media such as biological tissues, the action of RF leads to similar effects, with an important difference that the average displacement in the medium is crucially dependent on (albeit small) shear modulus. This effect is currently being extensively explored in medical applications related to remote assessment of viscoelastic properties of internal tissues and bodily fluids. Radiation force from focused ultrasound became the basis of numerous emerging diagnostic techniques [3-6].

As shown in this presentation, there exist previously unexplored mechanisms of shear displacement generation other than the conventional mechanism related to attenuation of sound waves. Such non-dissipative mechanisms for generation of non-potential RF are related to medium inhomogeneity and/or nonlinear effects in the primary ultrasonic field. Here we present a theoretical model for a mechanism of shear stress and shear motions generation due to inhomogeneity of the acoustic parameters of a biological medium.

CP838, *Innovations in Nonlinear Acoustics: 17th International Symposium on Nonlinear Acoustics*,
edited by A. A. Atchley, V. W. Sparrow, and R. M. Keolian
© 2006 American Institute of Physics 0-7354-0330-9/06/$23.00

AVERAGE STRESS TENSOR IN AN ACOUSTIC BEAM

In the framework of the classical five-constant description, nonlinear acoustic waves in an isotropic solid can be characterized by two linear and three nonlinear elastic parameters, for which the second-order moduli, λ and μ, and the third-order Landau moduli, A, B, and C, can be taken [7]. As known, the stress–strain dependence can be derived through variations of elastic energy leading to the known Piola-Kirchhoff pseudo tensor', σ_{ik}.

$$\sigma_{ik} = \mu\left(\frac{\partial u_i}{\partial x_k} + \frac{\partial u_k}{\partial x_i}\right) + \lambda\left(\frac{\partial u_l}{\partial x_l}\right)\delta_{ik} + \left(\mu + \frac{A}{4}\right)\left(\frac{\partial u_l}{\partial x_i}\frac{\partial u_l}{\partial x_k} + \frac{\partial u_i}{\partial x_l}\frac{\partial u_k}{\partial x_l} + \frac{\partial u_i}{\partial x_l}\frac{\partial u_l}{\partial x_k}\right) + \tag{1}$$

$$\left(\frac{\lambda+B}{2}\right)\left[\left(\frac{\partial u_l}{\partial x_l}\right)^2\delta_{ik} + 2\frac{\partial u_i}{\partial x_k}\frac{\partial u_l}{\partial x_l}\right] + \frac{A}{4}\frac{\partial u_k}{\partial x_l}\frac{\partial u_l}{\partial x_i} + \frac{B}{2}\left(\frac{\partial u_l}{\partial x_m}\frac{\partial u_m}{\partial x_l}\delta_{ik} + 2\frac{\partial u_k}{\partial x_l}\frac{\partial u_l}{\partial x_i}\right) + C\left(\frac{\partial u_l}{\partial x_l}\right)^2\delta_{ik}$$

Upon representation of the acoustic field in the standard form, $\mathbf{u} = \mathbf{v} + \mathbf{U}$, where $\mathbf{v}$ is oscillating part, so that $\langle\mathbf{u}\rangle = \mathbf{U}$, where angular brackets denote period averaging, the radiation stress tensor, RS, can be defined as period-average of Eq. (1): $RS = \langle\sigma_{ik}^N\rangle$, where σ_{ik}^N is nonlinear part of stress. Here we limit ourselves by the case of a narrow-angle primary acoustic beam, for which the following inequalities are fulfilled:
$\frac{\partial}{\partial x} \propto k \gg \frac{\partial}{\partial y}, \frac{\partial}{\partial z}; v_x \gg v_y, v_z$ In this approximation, from (1) the nonlinear aort of radiation stress tensor follows :

$$<\sigma^N_{xx}> = G_1\left\langle\left(\frac{\partial v_x}{\partial x}\right)^2\right\rangle, \quad G_1 = \frac{3}{2}\lambda + 3\mu + A + 3B + C, <\sigma_{yy}> = <\sigma^N_z > G_2\left\langle\left(\frac{\partial v_x}{\partial x}\right)^2\right\rangle, \quad G_2 = \frac{\lambda}{2} + B + C,$$

$$<\sigma^N_{xy}> = <\sigma^N_{yx}> = G_3\left\langle\left(\frac{\partial v_x}{\partial x}\frac{\partial v_x}{\partial y}\right)\right\rangle, \quad <\sigma^N_{xz}> = <\sigma^N_{zx}> = G_3\left\langle\left(\frac{\partial v_x}{\partial x}\frac{\partial v_x}{\partial z}\right)\right\rangle, G_3 = \lambda + 3\mu + A + 2B \tag{2}$$

These expressions have first been obtained by Rudenko and Il'inskii (private communication). Note that
$$G_1 - G_2 = G_3 = \lambda + 3\mu + A + 2B = Q. \tag{3}$$

EQUATIONS FOR DISPLACEMENTS INDUCED BY THE RADIATION FORCE

The standard equation for waves in an elastic medium (in Lagrangian coordinates) reads
$$\rho\ddot{u}_i = \frac{\partial\sigma_{ik}}{\partial x_k} \tag{4}$$

Here $\mathbf{u}$ is displacement vector and σ_{ik} is stress tensor. The latter is represented as a sum of linear and nonlinear parts, $\sigma_{ik} = \sigma^L_{ik} + \sigma^N_{ik}$, where σ^L_{ik} and u^L_{ik} are, respectively, the linear parts of stress and strain tensor (u_{ik}) tensors, respectively, and σ^N_{ik} is a small nonlinear part. It is supposed that both second-order and third-order moduli are, in general, dependent on coordinates, and this dependence is slow as compared with the ultrasound wavelength. From (4), a modified nonlinear wave equation follows in the form

$$\rho[\ddot{\mathbf{u}} - c_t^2 \Delta \mathbf{u} + (c_l^2 - c_t^2)\nabla div\mathbf{u}] - \mathbf{S} = \mathbf{\Phi}, \tag{5}$$

where $c_l = \sqrt{(\lambda + 2\mu)/\rho}$ and $c_t = \sqrt{\mu/\rho}$ are the velocities of linear longitudinal and transverse waves, respectively, and the term S_i reflects spatial variations of the linear parameters; in what follows it is neglected.

The right-hand side of (5) is $\mathbf{\Phi} = \partial\sigma^N_{ik}/\partial x_k$, and its period-average defines the radiation force, RF, acting on the medium. In a linear, homogeneous material, equation (5) becomes the classical equation describing linear elastic wave propagation in solids [7]. In what follows we consider the case when the medium parameters depend on one coordinate x that is directed along the primary beam axis.

After averaging Eq. (5) over the period of the high-frequency wave we obtain an equation for the average displacement, $\mathbf{U}$. As usual for linear elastic waves, (e.g.,[7]), one can represent $\mathbf{U}$ as a sum of two vectors, potential one, $\mathbf{U}_1$ so that $\nabla \times \mathbf{U}_1 = 0$, and solenoidal, $\mathbf{U}_2$, for which $(\nabla \cdot \mathbf{U}_2) = 0$. $\langle\mathbf{\Phi}\rangle = \langle\mathbf{\Phi}\rangle_1 + \langle\mathbf{\Phi}\rangle_2$. After that we have for shear displacement

$$\frac{\partial^2 \mathbf{U}_2}{\partial t^2} - c_t^2 \Delta \mathbf{U}_2 - \mathbf{S}_2/\rho - \mathbf{F} = 0, \tag{6}$$

where $\mathbf{F} = \langle\mathbf{\Phi}\rangle_2/\rho$. In general, separating the potential and solenoidal parts of the RF vector needs a solution of the corresponding equations for scalar and vector potentials. However, under the simple geometries considered below it can be done in a simplified way.

EQUATIONS FOR RF INDUCED DISPLACEMENT

Let us return to the general Eq. (6) for shear displacement. First we consider a simple example of a plane (collimated) primary acoustic beam propagating along the x axis in the (x, z) plane. To find the shear motions, we determine the solenoidal part of radiation force by taking its rotation. Evidently, in the case considered, the rotation is directed along the third axis, y, so that the resulting expression is essentially scalar. To evaluate the shear forcing, $\mathbf{F}$, we use the above expressions (2) for the average nonlinear part of stress tensor. As mentioned, we suppose that primary field is a narrow beam, and that in the average field $\partial/\partial z \gg \partial/\partial x$, hence (from $divu = 0$), $u_x \gg u_z$. As a result, Eq. (6) becomes

$$U_{tt} - c_t^2 U_{zz} = \frac{Q}{\rho}\left[\frac{\partial M_a^2}{\partial x} + \frac{1}{k}\frac{\partial}{\partial z}\left(M_a^2\frac{\partial\theta}{\partial z}\right)\right] + M_a^2\frac{1}{\rho}\frac{\partial Q}{\partial x}. \tag{7}$$

Here $M_a = \partial v_x/c_l\partial t$ is dimensionless acoustic velocity in the ultrasound beam (the acoustic Mach number).

Using the fact that a linear narrow beam satisfies the classical parabolic equation for complex amplitude, for a low-angle, 2-D wave beam in a linear medium it can be shown that the term in square parentheses of Eq. (7) turns to zero. Hence, a linear,

331

non-dissipative beam in a homogeneous medium ($\partial Q / \partial x = 0$) does not produce shear force.

Here we are interested in the non-dissipative radiation force in inhomogeneous media where the radiation force (gradient of radiation stress in x - direction) is defined by the expression $RF = \dfrac{\partial Q}{\partial x}\dfrac{1}{\rho}M_a^2$.

Among the features that distinguish the non-dissipative radiation force from the classical dissipative radiation force are its practical independence of ultrasound frequency and also the dependence of its direction on the sign of the gradient of the parameter Q. Thus, the displacement produced by non-dissipative RF can be directed both outward and toward the transducer, depending on the direction of media parameters variations. For a localized inhomogeneity such as a lesion, the non-dissipative RF changes its direction along the acoustic beam.

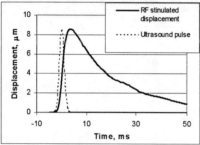

Figure 1. (a) Time dependences of displacement at the beam axis, $r = 0$ (solid line, microns) and of the primary impulse (dashed line, arbitrary units).

The estimations were conducted for a Gaussian primary beam with a characteristic focal diameter of about 4 mm and focal pressure amplitude of 2 MPa. Transverse wave velocity is taken $c_t \sim 3$ m/s, shear viscosity $v = 0.015$ m²/s, and the half-intensity pulse duration of 1 ms. We also suppose that Q varies by 20% at a scale of 0.5 cm. Time dependence of RF - produced displacement is shown in Fig. 1 together with the profile of the pump impulse (radiation force).

The authors believe that the possibility of shear displacement generation by a weakly attenuating ultrasonic beam may have both a conceptual and practical significance in nonlinear acoustics and its medical applications.

REFERENCES

1. R.T.Beyer, *Nonlinear acoustics*, American Institute of Physics, Woodbury, NY, 1997
2. P. J. Westervelt, " Parametric acoustic array", *J. Acoustical Soc. Am.* 35, 535-537 (1963).
3. A.P.Sarvazyan, O.V.Rudenko, S.D.Swanson., J.B.Fowlkes., and S.Y.Emelianov, "Shear Wave Elasticity Imaging - a new ultrasonic technology of medical diagnostics", *Ultrasound Med. Biol.*, 24, 1419-1435 (1998)
4. K.Nightingale, M.Palmeri, G.Trahey,. "Analysis of contrast in images generated with transient acoustic radiation force", *Ultrasound in Medicine and Biology*, 32, 1, 61-72 (2006)
5. J.Bercoff, M.Pernot, M.Tanter, M.Fink, "Monitoring thermally-induced lesions with supersonic shear imaging", *Ultrasonic Imaging*, **26**, 2, 71-84 (2004)
6. A.Alizad, L.E.Wold, J.F.Greenleaf, M.Fatemi., "Imaging mass lesions by vibro-acoustography: modeling and experiments", *IEEE Trans Med Imaging*, 23(9), 1087-1093 (2004)
7. L.D.Landau and E.M.Lifschitz, *Theory of Elasticity* , Pergamon, New York, 1986.

Modelling Nonlinear Ultrasound Propagation in Bone

Robin O. Cleveland[1,2], Paul A. Johnson[3] Marie Muller[2], Maryline Talmant[2], Frederic Padilla[2] and Pascal Laugier[2]

1 Dept of Aerosp. and Mech. Eng. , Boston University, 110 Cummington St., Boston, MA 02215, USA
2 Laboratoire d'Imagerie Paramétrique, Université Paris VI - UMR CNRS 7623, 15 rue de l'Ecole de Médecine , 75006 Paris, FRANCE
3 Geophysics Group, MS D443, Los Alamos National Laboratory, Los Alamos, NM 87545, USA

Abstract. Simulations have been carried out to assess the possibility for detecting the nonlinear properties of bone *in vivo*. We employed a time domain solution to the KZK equation to determine the nonlinear field generated by an unfocussed circular transducer in both cancellous and cortical bone. The results indicate that determining nonlinear properties from the generation of higher harmonics is challenging in both bone types (for propagation distances and source amplitudes appropriate in the body). In cancellous bone this is because the attenuation length scale is very short (about 5 mm) and in cortical bone because the high sound speed and density result in long nonlinear length scales (hundreds of millimeters). An alternative approach to determine the nonlinear properties was considered using self-demodulation of sound. For cancellous bone this may result in a detectable signal although the predicted amplitude of the self-demodulation signal was almost 90 dB below the source level (1 MPa). In cortical bone the self-demodulated signal was even weaker that in cancellous bone (~110 dB down) and, for a practical length signal, was not easy to separate from the components associated with the source

Keywords: Nonlinear acoustics, bone properties, harmonic generation, self-demodulation
PACS: 43.25.Jh, 43.80.Qf

INTRODUCTION

Today, the standard technique for assessment of skeletal status is based on X-ray absorptiometry techniques that measure bone mineral density (BMD). However, measuring BMD alone does not provide conclusive prediction of bone strength [1]. Several bone properties such as micro-architecture, tissue material properties (visco-elasticity or plasticity) and tissue alteration (micro fractures), which are not captured by BMD, also contribute to bone strength independently of bone mass. A promising alternative to X-ray is quantitative ultrasonic (QUS) methods which were introduced in 1984 [2]. The elastic waves used in QUS mechanically interact with the structure of the bone and therefore offer potential to probe the mechanical properties of the bone, including microarchitecture and tissue material properties [3-5]. In addition, ultrasound is non-ionizing radiation so does not have the risks associated with X-ray and ultrasound devices are significantly cheaper than the X-ray devices. However, to-date ultrasonic techniques, based on sound speed and attenuation measurements, do not appear to offer dramatic advantages over X-ray techniques. Nonlinear acoustic

CP838, *Innovations in Nonlinear Acoustics: 17th International Symposium on Nonlinear Acoustics*,
edited by A. A. Atchley, V. W. Sparrow, and R. M. Keolian
© 2006 American Institute of Physics 0-7354-0330-9/06/$23.00

techniques appear to have potential in bone interrogation. Nonlinear effects have been used in evaluate fracture in rocks and structural materials and have proven to be very powerful tool in being able to detect the presence of micro-cracks [6]. We carried out simulations to consider whether classical nonlinearity, harmonic generation and self-demodulation, might be detectable in bone.

METHOD

Simulations were carried out using a time-domain solution[7] to the KZK equation:

$$\frac{\partial p}{\partial z} = \frac{c_0}{2} \int_{-\infty}^{\tau} \nabla_\perp^2 p \, d\tau + \frac{\beta}{2\rho_0 c_0^4} \frac{\partial p^2}{\partial \tau} + \frac{\delta}{2c_0^4} \frac{\partial^3 p}{\partial \tau^3}$$

Here p is acoustic pressure, c_0 small-signal sound speed, ρ_0 density, β coefficient of nonlinearity, and δ diffusivity of sound. The source was taken to be an unfocused circular transducer with a diameter of 12.7 mm. The source pulse was taken to be 10-cycle tone burst with a centre frequency of 1 MHz and a source amplitude of 1 MPa. In these simulations the absorption was matched to that of the medium at the carrier frequency but followed a frequency squared law, which is appropriate for thermovisocus absorption but not such a good model for bone which has a near linear frequency dependence. For the simulations presented here, where harmonic generation was weak, the impact of using a frequency squared law was small.

RESULTS AND DISCUSSION

The left side of Fig. 1 shows the amplitude of the first few harmonics along the axis of the transducer for propagation in cancellous bone. The properties used to describe cancellous bone were: c_0=1500 m/s ρ_0=1200 kg/m^3 and β=10. The attenuation at 1 MHz was taken to be 200 Np/m (~20 dB/cm). For the transducer described above the Rayleigh distance was 84.5 mm, the absorption length about 5 mm, and the plane wave shock formation distance 64.5 mm. These are the characteristic lengths scales of diffraction, absorption and nonlinearity, respectively. The figure illustrates that nonlinear generation of the higher harmonics only occurs in the first 5 mm after which attenuation dominates.

In a thermoviscous fluid the relative importance of nonlinearity and attenuation can be adjusted either with the source amplitude or the source frequency. However, the attenuation of bone increases linearly with frequency and so the Gol'dberg number does not change with change in frequency. Therefore, only source amplitude can be used to enhance nonlinear effects. However, generating source pressures significantly above 1 MPa is challenging, even for high-power materials, further using higher amplitude pressures calls into question whether the ultrasound will result in bioeffects.

An alternative detection of the nonlinear properties of bone is to employ self-demodulation. Self-demodulation refers to the nonlinear generation of the envelope of a pulse. The demodulated envelope typically is at low-frequency and so will not suffer from strong attenuation in bone. The right side of Fig. 1 shows predicted axial waveforms and at 75 mm the demodulated envelope can be observed. The results

334

indicate that a detectable signal may be present for propagation through cancellous bone although the amplitude was almost 90 dB below the source level. Note that this pulse has a duration of about 10 µs which corresponds to a spatial extend of about 15 mm which is about the longest length pulse that can be practically used in the body without signals from multiple travel paths corrupting the data. Therefore it appears that it may be practical to detect the nonlinear self-demodulated signal in cancellous bone but the signals to be detected will be very weak.

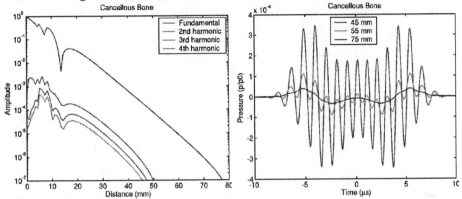

FIGURE 1. Left: Amplitude of the fundamental (blue), second harmonic (green), 3rd harmonic (red) and 4th harmonic (cyan) as a function of propagation distance. There is evidence of a small amount of harmonic generation in the near field (< 5 mm) after which attenuation dominates. Right: Predicted on-axis waveforms at 45 mm (red), 55 mm (green) and 75 mm (blue). At 45 mm the presence of the 1 MHz carrier can be observed but by 75 mm only the envelope remains.

Simulations in cortical bone were also carried out. The properties used to describe cortical bone were: c_0=3800 m/s ρ_0=1700 kg/m^3 and β=10. The attenuation at 1 MHz was taken to be 50 Np/m (~5 dB/cm) The change in the acoustic properties, particularly the sound speed, means that the length scales in cortical bone are quite different from those in cancellous bone. For the transducer above, the Rayleigh distance is 33.3 mm, the absorption length about 20 mm, and the shock formation distance 1485 mm (~1.5 m). In this case a 10-cycle pulse has a spatial extent of about 45 mm which is too large for use in the body. In addition, the absorption length is on the order of the Rayleigh distance which is not optimal for self-demodulation as geometrical spreading will result in a loss of nonlinear effects. Increasing centre frequency of the pulse improves the parameters for self-demodulation e.g., 2 MHz halves the length of the pulse, doubles the Rayleigh distance (67 mm), and halves both the shock formation distance and absorption length (740 mm and 10 mm respectively).

Simulations for various frequencies and pulse lengths were considered and the most promising results are shown in Fig. 2 for a 5-cycle pulse at 2 MHz which yields a pulse length of about 12 mm. Even in the best case the difference between the linear and nonlinear predictions is very small and the detectable region is restricted to a small band below 100 kHz. These data indicate that it is very unlikely that self-demodulation can be detected in cortical bone for propagation over length scales that are physiological relevant. The higher sound speed and density of cortical bone result in nonlinear length scales that are very long and so the self-demodulated signal is dramatically weaker than in cancellous bone and not easy to separate from the components associated with the source for a practical length signal.

In conclusion, simulations have been carried out to determine the possibility of employing classical nonlinearity as a detection modality for the nonlinear properties of bone. The results indicated that detection of higher harmonic generation is not practical due to the strong attenuation of bone. However, detection of the self-demodulation signal may be possible in cancellous bone but it very unlikely that it will be detectable in cortical bone.

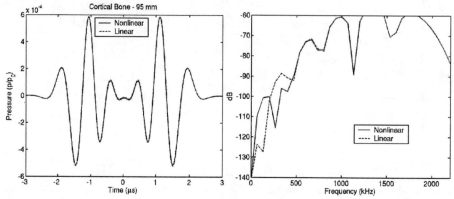

FIGURE 2. Waveforms (left) and spectra (right) at 95 mm from a source for a 5-cycle 2-MHz pulse in cortical bone for both nonlinear (solid) and linear (dashed) simulations. There was no observable difference in the time waveforms and only for frequencies below 200 kHz was there a potentially measurable (no more than 20 dB) difference in the signals.

ACKNOWLEDGMENTS

This work was supported by a grant from "Bourses de Recherche" from the Scientific Advisory Board of the Mayor of Paris and by the National Science Foundation through CenSSIS (Center for Subsurface Sensing and Imaging Systems).

REFERENCES

1. Genant, H. K., K. Engelke, et al. (1996). "Noninvasive assessment of bone mineral and structure: state of the art." J Bone Miner Res 11(6): 707-730.
2. Langton, C. M., S. B. Palmer, et al. (1984). "The measurement of broadband ultrasonic attenuation in cancellous bone." Eng Med 13(2): 89-91.
3. Laugier, P., P. Droin, et al. (1997). "In vitro assessment of the relationship between acoustic properties and bone mass density of the calcaneus by comparison of ultrasound parametric imaging and QCT." Bone 20: 157-165.
4. Padilla, F., Peyrin, F, Laugier, P. (2003). "Prediction of backscatter coefficient in trabecular bones using a numerical model of 3D microstructure". J Acoust Soc Am 113(2): 1122-1129.
5. Bossy, E., Talmant, M., Laugier, P. (2002). "Effect of bone cortical thickness on velocity measurements using ultrasonic axial transmission : a 2D simulation study. " J Acoust Soc Am 112: 297-307.
6. Van Den Abeele, K.E.-A.; Johnson, P.A.; Sutin, A. (2000), "Nonlinear elastic wave spectroscopy (NEWS) techniques to discern material damage. I. Nonlinear wave modulation spectroscopy (NWMS)," Research in Nondestructive Evaluation, 12:17-30.
7. Cleveland R.O., Hamilton M.F., and Blackstock DT (1996), "Time domain modeling of finite-amplitude sound in relaxing fluids," J. Acoust. Soc. Am. 99, 3312—3318.

High Speed Observation of Spatiotemporal Chaos in a Voice Production System

Yu Zhang and Jack J. Jiang

Department of Surgery, Division of Otolaryngology Head and Neck Surgery, University of Wisconsin Medical School, Madison, WI 53792-7375

Abstract. The spatial-temporal chaos has recently become the subject of intensive experimental and theoretical investigations. Studying vibratory dynamics of the vocal folds is important for natural voice productions. Disordered voices are usually observed in laryngeal pathologies, such as laryngeal paralysis and vocal polyps. In this study, we report spatiotemporal chaotic vibratory behavior in a biomedical vocal fold system. Spatiotemporal chaos in excised larynx vibrations is recorded using high-speed digital imaging. Spatiotemporal analyses effectively describe the spatiotemporal dynamics of the vocal fold vibrations and investigate the effects of subglottal pressure. With the increase of subglottal pressures, correlation dimension and glottal entropy are increased. Spatiotemporal chaos plays an important role in understanding irregular dynamics of disordered voice production.

Keywords: Spatial-temporal chaos, voice production, high-speed imaging.
PACS: 43.70.Aj, 05.45.Tp

INTRODUCTION

Spatially extended systems such as Taylor-Coquette flow, Rayleigh-Bernard convection, lasers, coupled-map lattices, fibrillating hearts, and the mammalian visual cortex system have extremely complicated spatiotemporal dynamics. The spatial-temporal chaos of spatially extended systems has recently become the subject of intensive experimental and theoretical investigations [1-3]. In this study, we report the spatiotemporal chaotic vibratory behavior in a biomedical vocal fold system. As well as modeling vocal tracts [4], studying vibratory dynamics of the vocal folds is important for natural voice productions. Disordered voices are associated with laryngeal pathologies such as laryngeal paralysis and vocal mass lesions. The study of temporal instability in disordered voices has received considerable interest in the field of voice science [5-8]. Temporal chaos has been observed in animal vocalizations and human voices. However, neither these chaotic models nor chaotic time series analyses have revealed the spatial complexities of the vocal fold vibrations. In this paper, we study spatiotemporal chaotic vibration of the vocal folds with high-speed digital imaging. Spatiotemporal analyses are applied to describe the spatiotemporal chaotic dynamics of vocal fold vibrations and to investigate the effect of subglottal pressure.

CP838, *Innovations in Nonlinear Acoustics: 17th International Symposium on Nonlinear Acoustics*,
edited by A. A. Atchley, V. W. Sparrow, and R. M. Keolian
© 2006 American Institute of Physics 0-7354-0330-9/06/$23.00

METHODS

Excised larynx experiments facilitate direct observation and measurement of vocal fold vibrations, and have proven to be advantageous in the study of laryngeal physiology. The parameters controlling phonation can be systematically monitored and independently controlled, which is difficult to achieve with a living human larynx. The experimental system consisted of an excised larynx setup and a high-speed camera system. A canine larynx harvested from a healthy laboratory dog was used in an experimental trial 12 to 36 hours after excision. The freshly excised canine larynx was mounted with a section of trachea on top of a pipe.

The airflow was generated using an Ingersoll-Rand (Type 30) conventional air compressor. The subglottal pressure P_s in the artificial lung was measured with an open-ended water manometer (Dwyer No. 1211). When P_s was increased above 3 cm H_2O, the larynx vibrated as a regular pattern. However, sufficiently increasing Ps (36 cm H_2O measured in this experiment), vocal fold vibrations became irregular.

A high-speed digital camera was mounted vertically above the larynx. The high-speed digital camera system acquired images at a sampling rate of 4000 frames per second with a resolution of 256 × 512 pixels. The image data was stored in digital form on a computer. During vocal fold vibrations, the vocal fold edges were extracted using image edge detection on a frame-by-frame basis.

RESULTS

Figure 1(b) shows seven successive frames of the glottal edges at the subglottal pressure P_s = 6 cm H_2O (the upper row) and P_s = 36 cm H_2O (the lower row), respectively. The glottis at the low pressure of 6 cm H_2O shows regular shapes, differing from the irregular shapes observed at the high pressure of 36 cm H_2O.

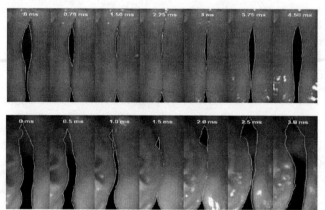

FIGURE 1. High-speed images of vocal fold vibration, where the upper and lower images correspond to subglottal pressure Ps = 6 cm H2O and 36 cm H2O, respectively.

The image of the glottis was divided into 108 horizontal segments with spatial index $j = 1, 2, \cdots, N$ (N = 108). Each segment shows the oscillation behavior $u(j,t)$ with respect to time. Thus, the spatiotemporal vibratory patterns of the vocal folds were recorded using high-speed digital imaging. Figure 2(a) shows the spatiotemporal plot of $u(j,t)$ (the upper curve) and the time series of the glottal area signal (the lower curve), $a(t) = \sum_{j=1}^{N} u(j,t)$, under the subglottal pressure $P_s = 6$ cm H_2O, respectively. Figures 2(b) shows the results of $u(j,t)$ and $a(t)$ under $P_s = 36$ cm H_2O. In comparison with the low subglottal pressure that produced regular spatiotemporal vibratory patterns and normal phonations, the high driving pressure produced irregular spatiotemporal vibratory patterns and chaotic phonation.

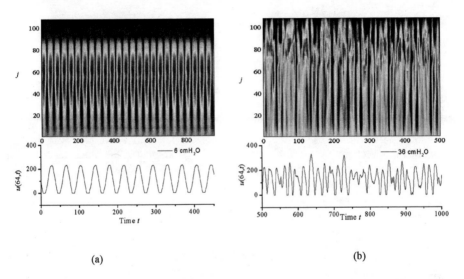

(a)   (b)

FIGURE 2. The spatiotemporal plot of $u(j,t)$ (the upper curve) and the glottal area $a(t)$ (the lower curve). (a) Ps = 6 cmH2O; (b) Ps=36 cmH2O.

In order to quantify the vibratory complexity of the vocal fold vibrations, we apply eigenmode analysis for the spatial-temporal data $u(j,t)$ and correlation dimension analysis for the time series of the glottal area $a(t)$. Figure 3 gives the effects of subglottal pressure P_s on nonlinear dynamic parameters including glottal entropy and correlation dimension. For low P_s, entropy and correlation dimension approach 0 and 1, respectively, indicating the ordered spatiotemporal behavior and periodic glottal time series. With the increase of P_s, entropy and correlation dimension both increase. At high subglottal pressure, the vocal fold vibrations show extremely irregular spatiotemporal pattern and the glottal area shows chaotic behavior. Thus, high subglottal pressure tends to increase the complexity of vocal fold vibration and induce spatiotemporal chaos in disordered voice production.

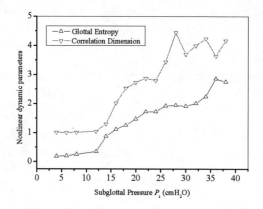

FIGURE 3. The effects of subglottal pressure on glottal entropy and correlation dimension.

CONCLUSION

In conclusion, we have observed spatiotemporal chaos in excised larynx vibrations using high-speed digital imaging. By applying spatiotemporal analysis and correlation dimension analysis, we investigated the effects of subglottal pressure on the spatiotemporal dynamics of the vocal folds. High subglottal pressure induced spatiotemporal chaotic vocal fold vibrations. Physiologically, the mechanical trauma at high subglottal pressure may result in a variety of human voice disorders, such as vocal nodules and bleeding. Spatiotemporal chaos found in this excised larynx study might be helpful to study the spatiotemporal chaos in disordered voice production and develop new clinical tools for detection of human vocal disorders.

ACKNOWLEDGMENTS

This study was supported by NIH Grant No. (1-RO1DC05522) and NIH Grant No. (1-RO1DC006019) from the National Institute of Deafness and other Communication Disorders.

REFERENCES

1. Patil, D. J., Hunt, B. R., Kalnay, E., Yorke, J. A., and Ott, E., *Phys. Rev. Lett.* **86**, 5878 (2001).
2. Ghosh, A., Ravi Kumar, V., and Kulkarni, B.D., *Phys. Rev. E* **64**, 056222 (2001)
3. Mininni, P. D., Gomez, D. O., and Mindlin, G. B., *Phys. Rev. Lett.* **89**, 061101 (2002).
4. Fee, M. S., Shraiman, B., Pesaran, B., and Mitra, P. P., *Nature* **395**, 67 (1998).
5. Jiang, J. J., Zhang, Y., and Stern, J. , *J. Acoust. Soc. Am.* **110**, 2120 (2001).
6. Tao, C., Zhang, Y., Du, G. H., and Jiang, J. J., *Phys. Rev. E* **69**, 036204 (2004)
7. Jiang, J. J. and Titze, I. R., *Laryngoscope* **103**, 872-882 (1993).
8. Jiang, J. J., Zhang, Y., and Ford, C. N., *J Acoust Soc Am.* **114**, 1 (2003).

Estimation Of Temperature Distribution In Biological Tissue By Acoustic Nonlinearity Parameter

X.F.Gong, X.Z.Liu, Y.Lu and D.Zhang

State Key Lab of Modern Acoustics, Institute of Acoustics,
Nanjing University, Nanjing, 210093,China

Abstract. In the rapid development of biomedical ultrasound, especially in the ultrasound therapy, it is desirable to monitor the temperature distribution of the tissues. In this paper the dependence of nonlinearity parameter on the temperature distribution in tissues can be estimated by measuring the nonlinearity parameter B/A. The obtained results are compared with the theoretical calculation of heat conductive model as well as the experiments by the thermocouple. Research work indicated that B/A could be used as an effective tool to monitor the temperature distribution in biological media.

Keywords: temperature, biological tissue, nonlinearity
PACS: 43.25.Ba

INTRODUCTION

In the development of the high-intensity focused ultrasound in biomedical ultrasound, much attention has been paid to the noninvasive temperature estimation in biological tissue in order to determine the region and degree of the ultrasound-induced lesions. Up to now, the thermocouple needle is used to monitor the temperature, which probably leads to the metastasis of malignant tumor. Previous studies [1-3] indicated that the ultrasound could be an effective tool to noninvasively estimate the temperature change. Acoustic nonlinearity parameter B/A is an important parameter due to its sensitivity to the changes of structures, compositions and pathological states of biological tissues. In this paper, the temperature dependence of B/A in biological tissue is studied, and compared with the theoretical estimation and the results measured by thermocouple.

THEORETICAL ANALYSIS

A heat conductive model is used to study the temperature distribution of biological sample, immersed into water at a desirable temperature (Fig.1) as follows:

$$\frac{\partial u}{\partial t} - a^2 \Delta u = 0, \left(r < R, 0 < z < d\right) \tag{1}$$

where R,d-radius and thickness of the sample, $a^2 = \lambda / \rho\mu$, λ-conductivity, ρ-density, μ-specific heat. u(r,z,t) is the temperature of the point (r,z) in the sample at time t. Δ is the Laplacian operator.

CP838, *Innovations in Nonlinear Acoustics: 17th International Symposium on Nonlinear Acoustics*,
edited by A. A. Atchley, V. W. Sparrow, and R. M. Keolian

The initial condition and boundary condition are:
$$u(r,z,0) = T_0$$
$$u\mid_{z=0} = u\mid_{z=d} = u\mid_{z=R} = T_f \qquad (2)$$

where T_0 is the temperature of the sample at t=0 and T_f is the temperature at the surface of the sample.

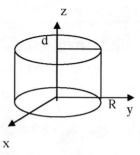

Figure 1. The sample model

Solving the above equations using the method of separation of variables, the temperature at (r,z) can be written as:

$$u(r,z,t) = T_0 + \frac{(T_0 - T_f)}{\pi} \sum_{m=1}^{\infty} \sum_{n=1,odd}^{\infty} \frac{1}{nJ_1\left(\frac{x_m^{(0)}}{R}r\right)x_m^{(0)}}$$

$$\times J_0(\frac{x_m^{(0)}}{R}r) \times \sin(\frac{n\pi}{d}z) \times \exp\{-[(\frac{x_m^{(0)}}{R})^2 + (\frac{n\pi}{d})^2]a^2 t\} \qquad (3)$$

where $x_m^{(0)}$ is the mth zero of $J_0(x)$.

As an example, the parameters for the porcine muscles are as follows:
$R = 2.75cm, d = 3cm, \mu = 3.477(KJ/kg \cdot {}^0 C), \rho = 1070kg/m^3$ and $\lambda = 0.642w/m \cdot {}^0 C$. Then, the temperature distribution in the sample at 3min after immersing in water with 60^0C is demonstrated in Fig.2. The temperatures at the boundary of the sample are larger than those at the middle of the sample.

EXPERIMENTAL METHOD AND RESULTS

Principle of measurement. The finite amplitude insert-substitution method[4] is used to measure B/A by detecting the second harmonic component p_2 in echo signals reflected from a reflective plate(with the reflection coefficient V). For biological tissue, the B/A parameter can be determined by the following equation:

$$\left(\frac{B}{A}\right)_x = \left[\frac{p_{2x}}{p_{20}} \frac{1}{D_{0x}^2 D_{x0}^2 e^{-2\alpha_2 d}} - \frac{h}{L} \frac{1 + D_{0x} D_{x0} V}{1 + V} - \frac{l}{L} D_{0x} D_{x0} \right]$$

$$\frac{L}{d} \frac{1}{D_{0x}} \frac{\rho_x c_x^3}{\rho_0 c_0^3} \times \frac{1 + V}{1 + D_{0x} D_{x0} V} [(\frac{B}{A})_0 + 2] - 2 \qquad (4)$$

where $D_{0x} = \frac{2\rho_x c_x}{\rho_0 c_0 + \rho_x c_x}$, $D_{x0} = \frac{2\rho_0 c_0}{\rho_0 c_0 + \rho_x c_x}$; c-sound velocity of the sample; the subscripts "0" and "x" correspond to water and sample, α_2 - sound attenuation coefficient of the second harmonics; L, h-the distances between the transducer T and the reflective plate, T and front interface of the sample respectively; l-distance between the back interface of the sample and the reflector; (B/A)$_0$=5.3+0.02×(T-30) for water; p_{20}/p_{2x} is the ratio of the second harmonic component measured before and after inserting the sample box in the water.

Experimental method. A compound PZT ultrasound transducer[5] is used to transmit the primary wave at 2MHz and to receive the received second harmonics at 4MHz. A 2MHz tone burst signal is generated by the pulse generator (81101A) and amplified by a broadband power amplifier (ENI A150). The received signal is recorded by a digital oscilloscope and sent to the computer for further analysis in order to detect the second harmonics. For comparison the thermocouple is used to measure the temperature of biological tissues.

Results

(1) Dependence of the nonlinearity parameter B/A, sound velocity, attenuation on the temperature in porcine muscle.

The sample was immersed into water at a measured temperature and kept for more than 15 minutes, so the temperature in sample was uniform. The experimental results of porcine muscle are shown in Fig.3 and Table 1.

TABLE 1. Relationship between acoustic parameter with the temperature.

^0C	15	22	31	41	50	60	Change*(%)
B/A	5.75	6.19	7.25	8.24	9.12	9.68	12.4
c(m/s)	1520	1529	1533	1535	1536	1538	0.2
α(dB/cm)	1.06	1.10	1.18	1.21	1.26	1.28	4.3

*This is the change rate per 10 degree in average from 20^0C to 60^0C

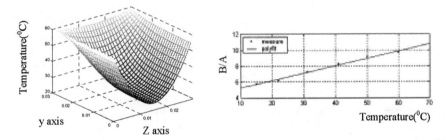

Figure 2. The temperature distribution at 3min   Figure 3. B/A changing with temperature in muscle

(2) B/A distribution inside the sample

The sample was immersed into water at a 26^0C and kept for 15 minutes, then put it into 60^0C water. Three minutes later, the B/A lateral distribution in the sample was measured as shown in Fig.4. Due to the short heat conductive time the values of B/A on both sides are larger than those in the middle part.

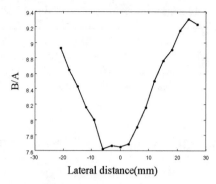

Figure 4. B/A distribution in muscle	Figure 5. Temperature distribution in muscle

(3) Temperature deduced from B/A and comparison with other results.

From Fig.4. the temperature distribution can be obtained inversely by Fig.3 and is shown in Fig.5. (Solid circles and fitted solid curve).The theoretical results of Eq.(3) are presented in Fig.5 as broken curve. Further, the temperature was measured inside the sample directly by thermocouple (as "o"). The results from these three approaches are in agreement quite well.

CONCLUSION

The results show that compared with the sound velocity and attenuation, the nonlinearity parameter B/A is the most sensitive parameter to the change of temperature. The temperature distribution in biological tissue can be deduced inversely by measuring the dependence of B/A on temperature and is coincide well with the theoretical   estimation by heat conductive model and the experimental values by thermocouple. Therefore, the nonlinearity parameter B/A is an effective tool to noninvasively measure the temperature in tissues. Further research of B/A changing with the temperature induced the ultrasound irradiation is in progress. This work was supported by the National Natural Science Foundation of China.(No.10474044).

REFERENCES

1. Roberto M.M., Christakis A.D., Noninvasive temperature estimation in tissue via ultrasound echo-shift. Part I. Analytical model. J. Acoust. Soc. Am., 1996, 100,2514-2521(1996).
2. Roberto M.M, Christakis A.D., Noninvasive temperature estimation in tissue via ultrasound echo-shift. Part II. In vitro study. J. Acoust. Soc. Am., 100,2522~2530(1996).
3. Ralf S., Nonvasive estimation of tissue temperature response to heating fields using diagnostic ultrasound. IEEE transactions on biomedical engineering, 42,828-839(1995)
4. Gong X.F., Zhu Z.M., Shi T. et al., Determination of acoustic parameter for biological tissues using FAIS and ITD methods, J. Acoust. Soc. Am., 86,1-5(1989).
5. Lu R.R., Zhang D., Gong X.F., et al., The measurement of reflected second harmonics and nonlinearity parameter of media using a transducer with complex structure. Acta Acustica, 26,515-515(2001,in Chinese).

The Mechanism of Antitumor Effects of Sonochemistry in the Low Frequency Limited Amplitude Ultrasound

Zhiyuan SHANG, Senlin GENG, Ya BAI, Hua TIAN

Institute of Applied Acoustics, Shaanxi Normal University, 710062 Xi'an Shaanxi, PRChina

Abstract: By using 1,3-diphenylisobenzofuran(DPBF) as a single oxygen trap, Designing a light trapping spot to collect sonoluminescence, the relative loss of DPBF being used to measure the $^1O_2$ productivity.the mechanism of activation of HpD by ultrasound was investigated . the results show the mechanism of single oxygen predominates on the HpD antitumor activated by ultrasonic, the main inducement for producing single oxygen is high temperature hot spot effects after ultrasonic cavitation and luminescence is not strong enough to activate HpD to produce single oxygen.

Keywords: hematoporphyrin derivative, ultrasound activation, singlet-oxygen, sonoluminescence

PACS: 43.25.+y

INTRODUCTION

The possible main role of single oxygen ($^1O_2$) in the mechanism of sonochemistry therapy of cancer, the synergistic effect of ultrasound and hematoporphyrin derivative (HpD), was investigated [1-3]. In the year 1990, Umemura, S. et al [4] experimentally have indicated the sonodynamic therapy which is based on activation of HpD by ultrasound on tumor cell killing. They have proposed that the mechanism of sonodynamic therapy involves the photoexcitation which is due to sonoluminescence produced in collapsing cavitation bubble, so led to the formation of single oxygen which is a known cytotoxic agent of violence. In this paper, we used 1, 3-diphenylisobenzofuran (DPBF) as trapping of a single oxygen, the mechanism of activation of HpD by ultrasound was investigated . Based on the relative loss of the DPBF we have found that the single oxygen is formed, when simultaneously both DPBF and HpD are present during irradiated by the low frequency limited amplitude ultrasound. Experimental results show that relative loss of DPBF increases with the sonication interval and ultrasonic intensity, which supports the argument, single oxygen is formed when HpD is activated by ultrasound. However, the sonoluminescence from the sound field of cavitation was collected

CP838, *Innovations in Nonlinear Acoustics: 17th International Symposium on Nonlinear Acoustics*,
edited by A. A. Atchley, V. W. Sparrow, and R. M. Keolian
© 2006 American Institute of Physics 0-7354-0330-9/06/$23.00

using a setup of isotropic light trap and irradiated both DPBF and HpD being present, and experimental results show that relative loss of DPBF is null almost

EXPERIMENT

1 Experiment setup

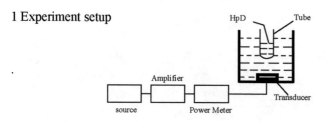

Fig.1 Ultrasound irradiated HpD experiment setup

2 Materials

HpD was purchased from Sigma Company(USA). In the experiment 100mg HpD is dissolved in 10mL NaOH solution (0.1 mol/L), neutralized by 10mL HCl solution(0.1 mol/L) till PH 7.2-7.4, fixed to 100mL with 0.9% NaCL resolution in the capacity bottle and stored under room temperature. Single oxygen trapping agent(DPBF) is the product of Fluke Company(USA). 63.5g DPBF is totally dissolved in the 100mL absolute ethanol after two days. The experimental reagent (NaOH, HCl, absolute ethanol ,$CHCl_3$) are all analytically pure.

3. Principal and method

. Weishaupt [5] provided the mechanic of photoexcitation as follow

$$HpD + h\,v \longrightarrow {}^1HpD^*$$
$${}^1HpD^* \longrightarrow {}^3HpD^*$$
$${}^3HpD^* + {}^3O_2 \longrightarrow {}^1O_2 + HpD$$

The asterisk in the equation indicates an activated state.

The above equations show that activated HpD in triplet state and oxygen in triple state interact and change into ${}^1O_2$. To detect it a trapping agent of single oxygen must be found. Then the activation efficiency of HpD can be attained by detection the trapping agent's loss. In this paper, the loss of DPBF is used to measure the ${}^1O_2$ productivity.

In the experiment DPBF+HPD solution was divided into two groups: control group and ultrasonic group. They were extracted respectively after the experiment and their absorbencies were mensurated at 400nm. The absorbency of control group was supposed to be A_0 and that of ultrasonic process group to be A_1, then the relative loss of DPBF is defined as $(A_0 - A_1)/A_0$.

RESULTS AND DISCUSSION

This study includes two aspects:

1 Mechanism of single oxygen produced by HpD activated by ultrasonic.

Relative loss of DPBF is used to measure the quantity of single oxygen. To test whether DPBF reduces after acoustic insonation 1mL DPBF($2*10^{-4}$mol/L) was put into a tube wrapped with black plastic, mixed with a 1mL 0.9% NaCl solution, insonified for 120s with 20kHz ultrasonic (electrical power are 5w ,10w.), extracted with 4mL $CHCl_3$ and the absorbency was measured. The result shows the average DPBF absorbency is less than 1%, which means that DPBF doesn't affect the insonation results(the whole experiments conducted in the darkroom, the same as the following). Then 1mL HpD is mixed with 1mL, $2*10^{-4}$mol/L DPBF solution and put into ultrasonic cavitation area in the container and sonicated with electrical power 5w,10w ultrasonic for 60,80,100,120s. The experiment shows loss of DPBF increases with time, gets to culmination at 100s and then goes steadily, which indicates under the concentration of HpD and in the ultrasonic sonicated time 100s, single oxygen probably becomes saturated. In the same time loss of DPBF increases with the electric power.

2 HpD activated whether by sonoluminescence or by the hot spot effect results from high temperature decomposing in the acoustic cavitation.

It has been tested by many experiments that HpD activated by ultrasonic can kill tumor cells. However, it has not arrive at a conclusion that whether the activation is directly caused by unbalanced high temperature spot in the ultrasonic cavitation or photochemical activation resulting from sonoluminescence. For the moment of cavitation bubble collapsing in ultrasonic cavitation there are many physical and chemical effects, such as high temperature, high pressure, radiation, shock wave and all kinds of free radical. Photo activating HpD has been studied [6]. Because HpD is a photosensitizer, to test if sonoluminescence can activate HpD, we design an experiment setup shown in the figure2.

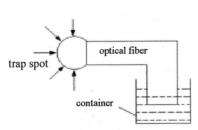

Fig. 2. Setup of light collection by a light trapping spot

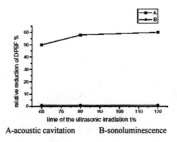

A-acoustic cavitation   B-sonoluminescence

Fig. 3. Relation between the relative loss of DPBF and radiation time by ultrasound

The sonoluminescence in the ultrasonic cavitation area is collected by isotropic light trapping spot and sent to HpD+DPBF solution through optical fiber (transmission efficiency is 85%). The experiment and detection methods are the same as the fore-going ones and the results are shown in the figure3

from which we can see HpD radiated by white light of certain intensity can produce single Oxygen for relative loss of DPBF increases with time. At 30s the loss get to the peak, i.e. 36%. However, curve B in the figure 5 is the result of sonoluminescence radiation whose max is no more than 1% although sonoluminescence can make the photoplate sensitized. the results of Curve A show most DPBF relative loss. The above results indicate that sonoluminescence can't activate HpD and produce single oxygen under the above-mentioned experiment conditions. Although the spectrum of sonoluminescence is wide its intensity is weak at 402nm(HpD's absorption peak) and can't activate HpD to produce single oxygen. There is a threshold for sonoluminescence to activate HpD which is about $1.27*10^{-6}\sim15.2*10^{-6}$w/cm^2. No optical power has been detected when optical power meter collects fluorescence at the end of optical fiber, which indicates the small fluorescence intensity of solution after ultrasonic cavitation is not enough to activate HpD to produce single oxygen.

CONCLUSION

HpD activated by Ultrasonic for antitumor has been studied widely for sound penetrates much deeper in biological tissues than light and is more important for clinical application. This paper studies the mechanism of HpD activated by ultrasonic and think the mechanism of single oxygen predominates on the HpD antitumor activated by ultrasonic. The main inducement for producing single oxygen is high temperature hot spot effects after ultrasonic cavitation and luminescence is not strong enough to activate HpD to produce single oxygen. However, how high temperature hot spot activates HpD is to be further researched.

ACKNOWLEDGMENTS

This work has been supported by the Chinese fund of natural science, grant number 19874042 from Chinese committee of natural science fund.

REFERENCE

1. liu, Q.H.,Sun,S.H., Xiao,Y.P. et al, *Sci. China, Ser.* C **2003**,46,253

2.Umemura,S., Yumita, N., Nishgaki,R., *Jpn. J. Cancer Res.* **1993**, 72, 195

3.Yumita, N., Nishgaki, R., Umemura, K. et al, *Jpn. J. Cancer Res.* **1989**, 80, 219

4. Umemura, S., Yumita, N., Nishgaki, R. et al, *Jpn. J. Cancer Res.***1990**, 81, 962

5.Weishaupt, K. R., Gomer, C. J., Gougherty, T. *J. Cancer Res.* **1976**, 36, 2326

6.Tachibana, H. *Cancer Lett.* **1994,** 74, 177

Studying the Vocal Fold Vibration Using a Nonlinear Finite-Element Model

Chao Tao, Jack. J. Jiang and Yu Zhang

Department of Surgery, Division of Otolaryngology Head and Neck Surgery, University of Wisconsin Medical School, Madison, WI 53792-7375

Abstract. The vocal fold vibration and voice production are highly complex nonlinear processes. Nonlinear relationship of glottal pressure to airflow and the nonlinearities of vocal fold collision are two important nonlinear factors of vocal fold vibration. In this paper, we will study the vocal fold vibration using a nonlinear finite-element model. In this model, the nonlinear relationship of glottal pressure to airflow, the nonlinearities of vocal fold collision, and the interaction between the airflow and vocal folds are taken into account. The impact pressure, vocal fold vibration, and glottal pressure under various lung pressures are studies. The results show that the nonlinear finite-element model is a useful tool for studying the voice production and predicting mechanical trauma leading to injurious abuse, misuse of the voice and vocal nodule.

Keywords: finite-element model, vocal fold, impact stress
PACS: 43.70.Aj, 43.70.Bk

INTRODUCTION

The studies of vocal fold biomechanics not only give an insight into voice production but also provide important information about laryngeal pathology development [1-2]. Recently, the finite-element model (FEM) has been employed to study vocal fold vibrations [3-5]. The FEM is based on continuum mechanics. As the size of the elements becomes smaller, the solution approaches the actual solution. The advantage of the FEM is the ability to deal with complex boundaries and provide more spatial information about the object than lumped mass models.

In this paper, we will present a nonlinear finite-element vocal fold model. This nonlinear finite-element model couples a FEM depicting the vocal folds and a FEM describing airflow. The nonlinear relationship of glottal pressure due to airflow, the nonlinearities of vocal fold collision and the interaction between the airflow and vocal folds are taken into account. Based on this model, we studied the vocal fold vibration and the vocal fold impact under natural phonation conditions.

METHOD

This nonlinear model contained three parts: (1) An airflow region was used to simulate the airflow in the glottis and drive the vocal fold; (2) One rigid plane was defined in the midline ($x = 0$) to simulate the nonlinear interaction between two vocal

CP838, *Innovations in Nonlinear Acoustics: 17ʰ International Symposium on Nonlinear Acoustics*,
edited by A. A. Atchley, V. W. Sparrow, and R. M. Keolian
© 2006 American Institute of Physics 0-7354-0330-9/06/$23.00

folds [5]; (3) A 3D vocal fold model was used to simulate the vibration of the vocal fold. The systematic diagram of our model is given in Fig. 1(a). When solving the oscillation of the vocal fold, the airflow model and the vocal fold model were alternately solved. In step 1, airflow was calculated, and the calculated airflow pressure was applied as the driving force on the surface of the vocal fold through a fluid-solid interaction surface. In step 2, the movement of the vocal fold under the driving force of the airflow was solved. The airflow region was then adjusted according to the calculated vocal fold displacement. By looping the above two steps, we obtained the oscillatory solution of the vocal folds.

(a)   (b)

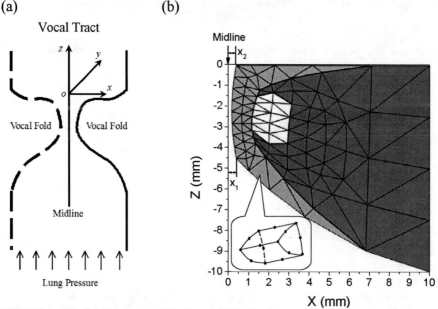

FIGURE 1. The finite-element model of vocal fold vibration system. (a) The systematic diagram of the vocal fold model. (b) One cross-section of the finite-element model of the vocal fold.

The 3D vocal-fold model is meshed by a high order prism-shaped element which is defined by 15 nodes. Figure 1 (b) gives the geometry of the prism-shaped element at the left-lower corner. One cross-section of the finite element model of the vocal fold is shown in Fig. 1(b), where the light gray region, the dark gray region and the white region represent the cover, the body and the ligament of vocal fold respectively. The vocal fold dimensions are 1.0cm in depth, 1.6cm in long, and 0.45cm in thickness. The input parameters used in the finite-element model of the vocal fold are given in Table I.

TABLE 1. The input parameters used in the finite-element model of the vocal fold

Parameters	Value
Transverse Young's modulus of the body	4 kPa
Transverse Young's modulus of the cover and ligament	2 kPa
Longitudinal shear modulus of the body	12 kPa
Longitudinal shear modulus of the cover	10 kPa

Longitudinal shear modulus of the ligament	40 kPa
Viscosity of the body, cover, & ligament	5 Poise

RESULTS AND CONCLUSION

Figure 2 presents the output of the finite-element model with lung pressure $P_L = 1.0$ kPa, where (a), (b) and (c) corresponds to impact pressure (P_{imp}), intraglottal airflow pressure (P_{air}), and intraglottal pressure (P_i) respectively. The intraglottal pressure is a combined effect of the impact pressure and the intraglottal airflow pressure. Thus intraglottal pressure can be expressed as the sum of the impact pressure and intraglottal airflow pressure, that is, $P_i = P_{imp} + P_{air}$. It is shows that there are two peaks in each circle of intraglottal pressure waveform. One peak is generated by the intraglottal airflow pressure and another is generated by vocal fold impact. This result agrees with the previous excised larynx experimental observation very well [6].

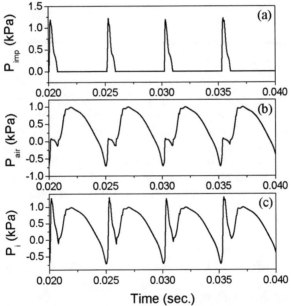

FIGURE 2. Impact pressure (a), intraglottal airflow pressure (b), and intraglottal pressure (c) as functions of time. The lung pressure is 1.0 kPa.

Lung pressure controls phonation production and vocal fold vibration; therefore, the effect of lung pressure on collision dynamics should be investigated. Figure 3 shows the impact pressure distribution in the superior-inferior glottal direction (z-direction) under different lung pressure, where (d) is the medial surface shape of the vocal fold. The surface is divided into eight elements and these elements are numbered 1 to 8 from the superior glottal position to the inferior glottal position. (a), (b), and (c) correspond to the lung pressures of 0.5kPa, 1.0kPa, and 2.5kPa, respectively. The y-axis in Fig.3 (a-c) represents the element index shown in Fig.3 (d), and the heights of columns in Fig.3 (a-c) are the impact pressure on each element.

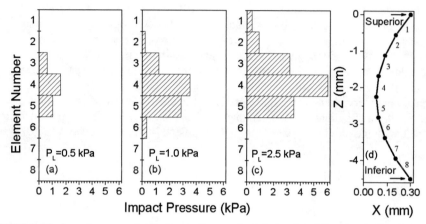

FIGURE 3. The impact pressure distribution on the superior-inferior glottal direction (z-direction) with varying lung pressure (P_L): (a) P_L = 0.5 kPa, (b) P_L = 1.0 kPa. (c) P_L = 2.5 kPa. (d) is the medial surface shape of vocal fold.

The above results show that the finite-element model can effectively study the vocal fold vibration. Compared with a lumped mass model, the nonlinear finite-element model can simulate a complex boundary condition and provide more spatial information [7]. Collision kinematics is an important output of our FEM. When collision occurs, it is thought that the contact force between two vocal folds may cause tissue damage [8]. Therefore, the study of vocal fold collision can provide valuable information for laryngeal pathologies such as vocal nodules and polyps [8].

ACKNOWLEDGMENTS

This study was supported by NSF of China (No. 30328029) and NIH Grant (No. 1-RO1DC006019 and No. 1-RO1DC05522) from the National Institute of Deafness and other Communication Disorders.

REFERENCES

1. Holmberg, E. B., Hillman, R. E., and Perkell, J. S., *J. Acoust. Soc. Am.* **84**, 511-529 (1988).
2. Titze, I. R., *J. Voice* **8**, 99-105 (1994).
3. Jiang, J. J., Diaz, C. E., and Hanson, D. G., *Ann. Otol. Rhinol. Laryngol.* **107**, 603-610 (1998).
4. Alipour, F., Berry, D. A., and Titze, I. R., *J. Acoust. Soc. Am.* **108**, 3003-3012 (2000).
5. Gunter, H. E., *J. Acoust. Soc. Am.* **113**, 994-1000 (2003).
6. Jiang, J. J. and Titze, I. R., *J. of Voice* **8**, 132-144 (1994).
7. Ishizaka, K., and Flanagan, J. L., *Bell Syst. Tech. J.* **51**, 1233-1268 (1972).
8. Stemple J. C., Glaze L. E., and Klaben B. G., *Clinical Voice Pathology Theory and Management*, San Diego: Singular, 2000.

SECTION 8
THERMOACOUSTICS

Nonlinear and edge effects in a thermoacoustic refrigerator

Philippe Blanc-Benon* and David Marx *

*LMFA, UMR CNRS 5509, Ecole Centrale de Lyon, 69134 Ecully Cedex, France

Abstract. In the present work, the full compressible Navier-Stokes equations are solved numerically, and the flow and heat transfer around a 2-D stack plate immerged in an acoustic standing wave
are computed. Distortion of the waveform temperature are found and are explained using the results
of a former nonlinear analysis. The temperature difference between the ends of the plate is investigated and compared to linear theory. The effects of the acoustic Mach number and geometrical
parameters on refrigerator performance are investigated.Results reported here may explain a part of
the difference between theoretical predictions and experimental results.

Keywords: Thermoacoustics
PACS: 43.35.Ud

In thermoacoustic refrigerators, the acoustic energy provided by an acoustic source is
used to pump heat from a cold reservoir. Such refrigerators may possibly be miniaturized, offering an application to microelectronic refrigeration. However the linear theory
frequently used to predict thermoacoustic refrigerators performance, can not account
for all the phenomena involved in thermoacoustic heat pumping. To design and predict
the performance of thermoacoustic devices, the most widely used tool is the standard
"linear theory"[1]. Although very advanced, this theory does not take into account some
phenomena such as nonlinear effects, including complex flows. Complex flows include
turbulence and motions resulting of abrupt changes of section. Vortical motions at the
extremities of the stack plates and heat exchangers have been observed both experimentally and numerically [2]. They can generate minor losses [3, 4, 5] which are detrimental
to performance. Acoustic nonlinear effects consist of acoustic streaming [6, 7, 8], and
waveform distortions .

NUMERICAL MODELS

To include most of the physical phenomena involved in thermoacoustic heat pumping (boundary layers, nonlinear acoustics, heat transfer, temperature gradient, two-
dimensional flow), a numerical solver has been developed [10]. It is based on the solution of the two-dimensional compressible and unsteady Navier-Stokes equations, coupled with mass and energy conservation equations. The governing equations are:

$$p = \rho r T \tag{1}$$

$$\frac{\partial \rho}{\partial t} + \nabla \cdot (\rho \mathbf{u}) = 0 \tag{2}$$

CP838, *Innovations in Nonlinear Acoustics: 17th International Symposium on Nonlinear Acoustics*,
edited by A. A. Atchley, V. W. Sparrow, and R. M. Keolian
© 2006 American Institute of Physics 0-7354-0330-9/06/$23.00

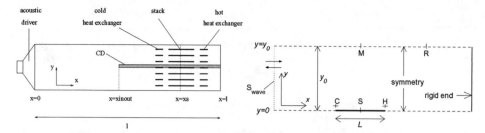

FIGURE 1. Sketches of the thermoacoustic refrigerator and the computational domain.

$$\frac{\partial(\rho\mathbf{u})}{\partial t} + \nabla\cdot(\rho\mathbf{u}\mathbf{u}) + \nabla p = \nabla\cdot\tau \tag{3}$$

$$\frac{\partial T}{\partial t} + \mathbf{u}\cdot\nabla T + (\gamma-1)T\nabla\cdot\mathbf{u}\frac{\partial T}{\partial t} + \mathbf{u}\cdot\nabla T = \frac{(\gamma-1)}{\rho r}(\Phi + \nabla\cdot(K\nabla T)), \tag{4}$$

where τ is the viscous strees tensor and Φ the viscous dissipation. In these expressions T is the temperature, p is the pressure, ρ is the density, $\mathbf{u} = [u; v]$ is the velocity vector; r is the gas constant, γ is the ratio of specific heats, μ is the shear viscosity, and K the thermal conductivity. For air: μ=1.8 10^{-5} Pa s; K=2.5 10^{-2} W K^{-1} m^{-1}; r=287 J K^{-1} kg^{-1}; γ=1.4. The temperature of the plate T_s is governed by:

$$\rho_s c_s \frac{\partial T_s}{\partial t} = \nabla\cdot(K_s\nabla T_s) + \frac{K}{l}\left(\frac{\partial T}{\partial y}\right)_{plate} \tag{5}$$

ρ_s, c_s and K_s are respectively the density, the specific heat and the thermal conductivity of the plate. l is half the thickness that an equivalent 2D plate would have. The second term on the right hand side of Eq. (5) represents the coupling with the flow. In deriving this term use has been made of the continuity of temperature and heat flux at the plate surface. On the plate surface, the boundary condition $T = T_s$ is enforced. To solve Eqs. (1)-(4) and Eq. (5), fourth order Dispersion-Relation-Preserving finite differences are used for calculating spatial derivatives, and time integration is performed using a four-step Runge-Kutta method [12]. The solver allows the simulation of the flow and heat transfer in the vicinity of a stack plate in the presence of an acoustic standing wave. The plate may be either of negligible or finite thickness, and be either heat-conducting or isothermal. The thermoacoustic refrigerator and the computational domain are represented in Fig. 1. The length of the resonator is half the wavelength, λ. The acoustic standing wave that is required in a standing wave refrigerator is created in a physical way: by superimposing two counter-propagating traveling waves.

RESULTS

The time variations of velocity and temperature were recorded at several points of the computational domain. The temperature time variation at point M of the domain is shown in Fig. 2 for different values of the drive ratio. Temperature T'_M is made

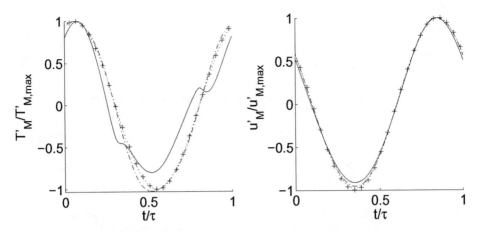

FIGURE 2. Normalized temperature and velocity time variations at point M, for different values of the drive ratio: +++ D_r=0.7%; $\cdots$ D_r=2.8%; $-\cdot-$ D_r =5.6 % ; — D_r=11.2 % .

dimensionless using its maximal value during the acoustic cycle $T'_{M,max}$. Time is made dimensionless using the acoustic period, τ. At lower drive ratios, $D_r = 0.7\%$ and $D_r = 2.8\%$, the temperature variation is sinusoidal. For a higher drive ratio, $D_r = 5.6\%$, the temperature time variation slightly departs from a sinusoidal curve. For the higher drive ratio represented, D_r=11.2%, the temperature variation clearly contains harmonics of the fundamental, showing the importance of nonlinear effects. The time variation of the velocity at point M, for different drive ratios is shown in Fig. 2. As can be seen, the velocity time variation, unlike the temperature variation, remains harmonic, even for the highest value of the drive ratio, D_r=11.2%. The normalized velocity is indeed slightly modified in its most negative values when the drive ratio increases, in such a way that the time-averaged velocity at point M over one acoustic cycle becomes slightly positive. This is consistent with observation of acoustic streaming above the plate [11] (here the non-zero mean velocity is due to "inner" vortices [8]). Nevertheless, this effect remains small.The temperature time variation was also recorded at point C of the domain, located in the fluid, just above the extremity of the plate, in the cold region (the stack is pumping heat from cold point C to hot point H). This variation is shown in Fig. 3, for different values of the drive ratio. Even for the lower drive ratios, $D_r < 5.6\%$, the temperature variation is not sinusoidall. At the highest drive ratio, D_r=11.2%, the distortion of the temperature at point C is much more pronounced than that at point M at the same drive ratio. Velocity variation at point C is shown in Fig. 3. Again, unlike the temperature variation, the velocity variation at point C remains sinusoidal for the whole range of the drive ratio. It is almost obvious from Fig. 3 that the time-averaged temperature and velocity over one acoustic cycle are not zero. The results presented above can be explained by considering the analysis made by Gusev et al.[9]. Using an inviscid but nonlinear model (the term $u \cdot \nabla T$ is fully preserved in their energy equation), these authors have shown that the nonlinear convective term $u \cdot \nabla T$ in the energy equation is responsible for temperature harmonics generation at the edges of the plate.

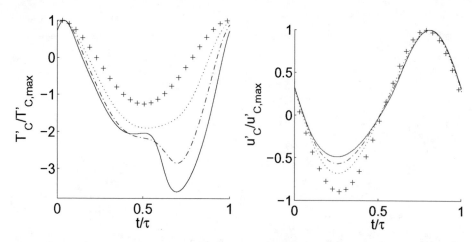

FIGURE 3. Normalized temperature and velocity time variations at point C, for different values of the drive ratio: +++ D_r=0.7%; $\cdots$ D_r=2.8%; $-\cdot-D_r$ =5.6 % ; — D_r=11.2 % .

ACKNOWLEDGMENTS

The authors acknowledge the French Ministry of Defense for its financial support.

REFERENCES

1. Swift, G. W., "Thermoacoustic engines", J. Acoust. Soc. Am. **84**, 1145–1180 (1988).
2. Blanc-Benon, Ph., Besnoin, E., and Knio, O., "Experimental and computational visualization of the flow field in a thermoacoustic stack", C. R. Mecanique **331**, 17–24 (2003).
3. Backhaus, S., and Swift, G. W., "A thermoacoustic Stirling heat engine: detailed study", J. Acoust. Soc. Am. **107**, 3148–3166(2000).
4. Wakeland, R. S., and Keolian, R. M, "Influence of velocity profile non uniformity on minor losses for flow exiting thermoacoustic heat exchangers", J. Acoust. Soc. Am. **112**, 1249–1252 (2002).
5. Morris, P. J., Boluriaan, S., and Shieh, C. M., "Numerical simulation of minor losses due to a sudden contraction and expansion in high amplitude acoustic resonators", Acta Acustica united with Acustica **90**, 393Ŭ-409 (2000).
6. Waxler, R., "Stationary velocity and pressure gradients in a thermoacoustic stack", J. Acoust. Soc. Am. **109**, 2739–2750 (2001).
7. Bailliet, H. A., Gusev, V., Raspet, R., and Hiller, R. A., "Acoustic streaming in closed thermoacoustic devices", J. Acoust. Soc. Am. **110** 1808Ŭ-1821 (2001).
8. Hamilton, M. F., Ilinskii, Y. A., and Zabolotskaya, E. A., "Acoustic streaming generated by standing waves in two-dimensional channels of arbitrary width", J. Acoust. Soc. Am. **113**, 153Ŭ-160 (2003).
9. Gusev, V., Lotton, P., Bailliet, H., Job, S., and Bruneau, M., "Thermal wave harmonics generation in the hydrodynamical heat transport in thermoacoustics", J. Acoust. Soc. Am. **109**, 84–90 (2001).
10. Marx, D., Blanc-Benon, Ph., "Numerical simulation of stack-heat exchangers coupling in a thermoacoutic refrigerator.", AIAA Journal **42**, 1338–1347(2004).
11. Marx, D., and Blanc-Benon, Ph., "Computation of the mean velocity field above a stack plate in a thermoacoustic refrigerator", C. R. Mécanique **332**, 867Ŭ-874 (2004).
12. Bogey, C. and Bailly, C., "A family of low dispersive and low dissipative explicit schemes for flow and noise computations", J. Comput. Phys., **194**, 194–214 (2004).

Thermoacoustic effects on propagation of a nonlinear pulse in a gas-filled tube with a temperature gradient

N. Sugimoto, H. Horimoto, M. Masuda and Y. Araki

Department of Mechanical Science, Graduate School of Engineering Science, University of Osaka, Toyonaka, Osaka 560-8531, Japan

Abstract. This paper examines experimentally thermoacoustic effects on propagation of a nonlinear compression pulse in a gas-filled tube with temperature gradient imposed spatially repeatedly by stacks of plates placed therein. It is shown that when the temperature gradient is present, the effects tend to increase not only in the peak pressure but also the total acoustic power flux and momentum flux.

Keywords: nonlinear acoustics, thermoacoustics
PACS: 43.35.Ud, 43.25.Cb

INTRODUCTION

The previous paper [1] has considered propagation of a nonlinear compression pulse in a gas-filled tube subjected to a spatially periodic temperature distribution to examine a possibility of amplification of the total acoustic power flux. It is shown numerically that the amplification takes place if stacks of plates are used when the ratio of the high temperature to the low one is two. The purpose of this paper is to check the theoretical findings by doing experiments.

Thermoacoustic effects appear in two ways. One appears through the local sound speed $a_e(x)$, x being the axial coordinate, to give rise to change in amplitude as well as in phase. If a temperature gradient is gentle, the peak pressure decreases as $T_e^{-1/4}$, whereas the velocity of the gas increases as $T_e^{1/4}$, where $T_e(x)$ denotes the temperature of the gas in equilibrium [1]. The other appears through the hereditary and irreversible effect due to the boundary layer on the solid surface. The velocity at the edge of the boundary layer $v_b(x,t)$, t being the time, directed normal to the solid surface into the gas is given in terms of a minus half-order derivative of the velocity $u(x,t)$ in the main-flow region (outside of the boundary layer) as

$$v_b = C\sqrt{\nu_e}\frac{\partial^{-\frac{1}{2}}}{\partial t^{-\frac{1}{2}}}\left(\frac{\partial u}{\partial x}\right) + C_T\frac{\sqrt{\nu_e}}{T_e}\frac{dT_e}{dx}\frac{\partial^{-\frac{1}{2}}u}{\partial t^{-\frac{1}{2}}}, \tag{1}$$

with $C = 1 + (\gamma-1)/\sqrt{Pr}$, $C_T = 1/2 + 1/(\sqrt{Pr}+Pr)$, and the definition

$$\frac{\partial^{-\frac{1}{2}}u}{\partial t^{-\frac{1}{2}}} \equiv \frac{1}{\sqrt{\pi}}\int_{-\infty}^{t}\frac{u(x,\tau)}{\sqrt{t-\tau}}d\tau, \tag{2}$$

CP838, *Innovations in Nonlinear Acoustics: 17th International Symposium on Nonlinear Acoustics*,
edited by A. A. Atchley, V. W. Sparrow, and R. M. Keolian
© 2006 American Institute of Physics 0-7354-0330-9/06/$23.00

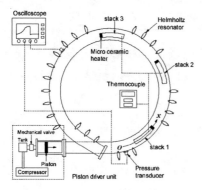

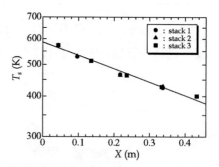

FIGURE 1. Experimental setup.

FIGURE 2. Distribution of the temperature T_s at the middle plate in each stack.

where $\nu_e(x)$, γ and Pr denote, respectively, the kinematic viscosity of the gas, the ratio of specific heats and the Prandtl number, the shear viscosity being assumed to be constant. If the temperature gradient is suitably positive so as to make the time-integral of the product of the excess pressure $p'(x,t)$ and v_b positive, the boundary layer pumps the energy into the pulse so that the total acoustic power and momentum fluxes of the pulse will be increased spatially.

EXPERIMENTS

Experiments use a part of a looped tube of inner diameter 78 mm and of loop diameter 4 m terminated with a straight tube of the same diameter on one side (see Fig. 1). The total tube length is about 13.6 m along the centerline. The tube is filled with air at the atmospheric pressure. The straight tube is connected to the piston driver unit, which launches a pulse into the tube. The other end is closed by a flat plate.

A compression pulse is generated by driving pneumatically a piston. To avoid a shock formation and its resulting nonlinear damping, an array of Helmholtz resonators is connected to the tube. The Helmholtz resonators, which have the natural frequency 240 Hz at 18.8 °C and are used in [2], are connected with axial spacing 0.2 m along the tube except for intervals where the temperature gradient is to be imposed. The axial spacing is wider than assumed in [1]. In Fig. 1, the position and the number of 48 resonators connected are not accurately drawn.

A stack of copper plates is placed inside of the tube, along which the temperature gradient is imposed by heating ceramic heaters attached to one end of the stack. The stack consists of five, thin, parallel plates of thickness 1 mm, of width 40 mm and of arc length 0.81 m, supported horizontally by the vertical plate in the same size. The spacing between the parallel plates is 10 mm wide. Three identical stacks are placed tandem in the tube with axial spacing about 1.6 m apart, so that the temperature gradient is imposed repeatedly. The heater in each stack is controlled to maintain the temperature at about 300 °C.

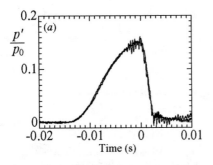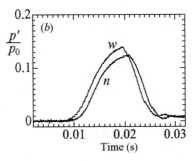

FIGURE 3. Comparison of the pressure profiles: (a) the profiles at $x = x_0$ and (b) the profiles at $x = x_1$ where the labels n and w attached to the profiles refer to the cases with no temperature gradient and with it, respectively.

Figure 2 shows the semi-logarithmic plot of the temperature in a steady state T_s measured on the surface of the middle plate in each stack against the distance X from the center of the heater along the stack. The distribution may be fitted with the straight line: $\ln T_s = -0.9465X + 6.374$, X being measured in meter. If T_e may be regarded as being equal to T_s, $T_e^{-1}\mathrm{d}T_e/\mathrm{d}x$ in (1) gives an inverse of a typical length of the temperature gradient, l^{-1}, which is constant $0.9465/\mathrm{m}$. But because the temperature at the heater to the ambient one is about two, an interval with the substantial temperature gradient is much shorter than l and also the one in the theory.

Launching the initial pulse, the pressure is measured at two positions well upstream and downstream from the three stacks, respectively, where the temperature of air is close to the ambient one. Taking the x axis counterclockwise along the centerline of the tube, the position of measurement located upstream is taken as the origin of the x axis and is designated by x_0. The cold end of the stack 1 is at about $x = 0.6$ m and the hot end at $x = 1.4$ m. The hot end of the stack 3 is at 6.2 m. The other downstream position of measurement is set at $x = x_1 \approx 7.6$ m.

RESULTS AND DISCUSSIONS

Figure 3(a) and 3(b) show the temporal pressure profiles measured at x_0 and x_1, respectively where the abscissa measures the time since the peak of each profile has just passed the position x_0, and the ordinate measures the excess pressure p' relative to the ambient pressure p_0. The labels n and w attached to the profiles in Fig. 3(b) refer to the cases with no temperature gradient and with it, respectively. In Fig. 3(a), both profiles appear to agree with each other except for some noises. The peak pressure of the pulses is set as high as 15% of p_0 ($= 0.09993$ MPa). Letting the temperature at x_0 and x_1 to be T_0 and T_1, respectively, T_0 in the case with no temperature gradient is 25.8 °C and is almost equal to T_1 ($= 25.9$ °C). In the case with the gradient, T_0 is 25.7 °C, whereas T_1 ($= 27.5°$C) is a little higher than T_0.

361

TABLE 1. Quantitative comparison of the total acoustic power fluxes and the total momentum fluxes transferred by the pulses measured at $x = x_0$ and $x = x_1$.

gradient	I_0 (kJ/m^2)	I_1 (kJ/m^2)	I_1/I_0	M_0 (N·s/m^2)	M_1 (N·s/m^2)	M_1/M_0
no	3.887	2.712	0.698	152.0	127.8	0.840
with	3.646	3.199	0.877	146.0	139.9	0.958

In Fig. 3(b), the pulse with w arrives at x_1 earlier than the one with n because the former has passed a region with higher sound speed. Both profiles appear to steepen forward a little by nonlinearity but no shocks are formed. The peak in the profile with w is seen to be the same as the initial one, whereas the one with n is decreased. This difference is obviously due to the thermoacoustic effects. The half-value width Δ of the pulse with w changes little from the initial one.

In order to compare the difference in the profiles quantitatively, the total acoustic power flux I and momentum flux M in the main-flow region are calculated from the data of p' by using

$$I = \int_{-\infty}^{\infty} p' u \, dt \approx \int_{-\infty}^{\infty} \frac{p'^2}{\rho_e a_e} dt \quad \text{and} \quad \frac{dI}{dx} = \frac{2}{R^*} \int_{-\infty}^{\infty} p' v_b \, dt, \tag{3}$$

and

$$M = \int_{-\infty}^{\infty} (p' + \rho u^2) dt \approx \int_{-\infty}^{\infty} \left(p' + \frac{p'^2}{\rho_e a_e^2} \right) dt \quad \text{and} \quad \frac{dM}{dx} = \frac{2}{R^*} \int_{-\infty}^{\infty} \rho_e u v_b \, dt, \tag{4}$$

with $p' \approx \rho_e a_e u$ where $\rho_e(x)$ is the density of the gas in equilibrium, ρ is the density, and $\rho_e a_e$ is the acoustic impedance, R^* being a local hydraulic radius of the tube. Where the stacks are placed, this radius is about one third of the tube radius. Incidentally the mass flux ρu cannot be computed in terms of p' alone up to quadratic terms.

Table 1 compares the total acoustic power fluxes and momentum fluxes. The integrals are computed numerically from the profiles in Fig. 3. Both acoustic power fluxes decrease due to the wall friction, but the decrease is small when the temperature gradient is present. This is also the case with the momentum flux. Since the temperature at the tube wall is not equal to the one at the stacks, the temperature gradient is not enough to yield the amplification predicted by the theory. But it is observed that the thermoacoustic effects tend to increase the acoustic power flux and the momentum flux.

REFERENCES

1. N. Sugimoto and D. Shimizu, "Effects of spatially periodic temperature distribution on amplification of energy flux of a nonlinear acoustic pulse propagating in a gas-filled tube" in *Nonlinear Acoustics at the Beginning of the 21st Century*, edited by O. V. Rudenko & O. A. Sapozhnikov, Moscow State University, 179–182 (2002).
2. N. Sugimoto, M. Masuda, K. Yamashita and H. Horimoto, *J. Fluid Mech.*, **504**, 271–299 (2004).

Response compensation of a thermocouple using a short acoustic pulse

Yusuke Tashiro, Tetsushi Biwa, Taichi Yazaki*

Department of Crystalline Materials Science, Nagoya University, Furo-cho Nagoya 464-8603, JAPAN
*School of Physics, Aichi University of Education, Igaya-cho, Kariya-city, 448-8542, JAPAN

Abstract. We report on the experimental technique of the measurement of temperature oscillations in acoustic waves. The temperature by a thermocouple is compared with the analytical one derived from a laminar oscillating flow theory. From the differences between them, we derive the response function of the thermocouple. We show continuously the response function in the frequency below 50 Hz using a short acoustic pulse. We can determine the temperature oscillation from the measured one by using the response function.

INTRODUCTION

Work flow I and heat flow Q constitute basic concepts in the understanding of thermoacoustic phenomena [1][2]. Recently, it became possible to measure I on the basis of its definition [3]. However, experimental derivation of Q is difficult, since it requires the direct measurement of the entropy oscillation. We focus on the total energy flow $H=I+Q$, rather than on Q itself. The energy flow H is written as $H = \rho_m C_p \langle TU \rangle_t$ for an ideal gas, where ρ_m and C_p are a mean density and an isobaric specific heat, U and T are the oscillating part of the velocity and temperature, respectively, angular brackets represent a time average. If the oscillating temperature T at the acoustic frequency can be measured without any amplitude-damping and phase delay, H is obtained from the measured T and U. Q is obviously obtained by subtracting the measured I from H.

We have been trying to establish the measurement technique of T using a thermocouple [4]. We have shown the quantitative disagreements between the measured temperature oscillation T_{ex} and the theoretical one T for both the amplitude and the phase angle using sinusoidal waves, and obtained a response function $Z_{ex} = \Gamma e^{i\theta}$ of the thermocouple, which relates T_{ex} with T as

$$T_{ex} = Z_{ex}T = \Gamma e^{i\theta}T . \tag{1}$$

Once Z_{ex} is determined for a given thermocouple, the oscillating temperature T can be accurately determined from T_{ex}. Z_{ex} would depend on the frequency of the acoustic wave. It is, therefore, of critical importance to study the frequency dependence of response function Z_{ex}.

CP838, *Innovations in Nonlinear Acoustics: 17th International Symposium on Nonlinear Acoustics*,
edited by A. A. Atchley, V. W. Sparrow, and R. M. Keolian

In this paper, we use a short pulse wave, instead of the sinusoidal one. We demonstrate that a pulse method is effective to capture the acoustic variables in a wide frequency range at a time. We show the response function Z_{ex} of the thermocouple as a continuous function of the frequency below 50 Hz using a single pulse. By using Z_{ex} obtained in this experiment, we can determine the temperature oscillation T of acoustic waves.

TEMPERATURE OSCILLATION IN A TUBE

We consider an acoustic plane wave in an ideal gas with the angular frequency $\omega (=2\pi f, f$ is the frequency). The temperature in such a wave is expressed as $T_m + T$, where T_m is a mean value and T represents oscillatory component. In a circular tube of radius r_0 without the axial temperature gradient, T is written as $T(r)/T_s = 1 - f_\alpha(r)$ from a laminar oscillating flow theory. Here T_s represents the temperature oscillation given under the adiabatic condition, which is written using a pressure oscillation P as

$$T_s = \frac{\gamma - 1}{\gamma}\frac{T_m}{P_m}P, \tag{2}$$

where P_m is a mean pressure, and γ is a ratio of isobaric to isochoric specific heats of a gas. The function $(1 - f_\alpha)$ represents a r-dependent non-dimensional temperature. At the center of the tube ($r=0$), $1 - f_\alpha(0)$ is written as

$$1 - f_\alpha(0) = 1 - \frac{1}{J_0(\sqrt{2}i^{\frac{3}{2}}r_0 / \delta_\alpha)}, \tag{3}$$

where δ_α is a thermal penetration depth given as $\delta_\alpha = \sqrt{2\alpha/\omega}$ using a thermal diffusivity α of a gas and J_0 is the complex Bessel function of the first kind. Since all the parameters in equations (2) and (3) are uniquely determined from the experimental conditions, the temperature oscillation $T(0)$ can be deduced from the measured P.

In this experiment, we evaluate the amplitude and the phase of the temperature oscillation $T(0)$ as a function of the frequency from the measured pressure pulse. These are compared with those of the temperature $T_{ex}(0)$ measured by a thermocouple to derive the response frequency Z_{ex}.

EXPERIMENTAL SETUP

The present experimental setup is schematically illustrated in Fig.1. A cylindrical glass tube of 0.6 m in length and $2r_0=21$ mm in inner diameter was used. One end of the tube was connected to an orifice valve to keep the mean pressure P_m at atmospheric pressure, and the other end was connected to a solenoid valve. An acoustic pulse was generated by opening the solenoid valve only for a short time. The axial temperature gradient along the tube was absent throughout the experiment. This is because the thermoacoustic effect [5] associated with the single acoustic pulse is very small. Helium was employed as working gas, and hence α and γ are 180×10^{-6}

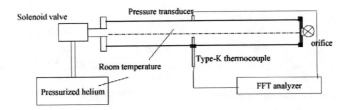

FIGURE 1. Schematic illustration of the experimental apparatus.

m²/s and 1.67, respectively.

We used a chromel-alumel (K-type) thermocouple with the wire diameter of 15 μm. The sensitivity of the thermocouple was experimentally determined as 39 μV/K. The thermocouple was covered with a support sheath made of a stainless-steel tube, but the junction was exposed to its surrounding working gas. The sheathed thermocouple was inserted into a glass tube through a narrow duct mounted on the wall, and fixed in such a way that the junction was positioned on the central axis of the tube. The pressure at the same axial position was also measured using a small pressure transducer mounted on the wall.

Electrical signals from the pressure transducer and the thermocouple were simultaneously recorded with a multi-channel 24-bit spectrum analyzer. From the power and phase spectra, we determined the amplitude ratio and the phase delay of $T_{ex}(0)$ to P. The temperature oscillation T_S for the adiabatic acoustic wave was determined by inserting values of T_m=290 K, P_m=101 kPa and measured P into equation (2). The temperature oscillation $T_{ex}(0)$ is expressed as the non-dimensional temperature $T_{ex}(0)/T_S$, which allows us to directly compare it with the theoretical one $1 - f_\alpha(0)$ given by equation (3).

EXPERIMENTAL RESULTS

The pressure P is shown in Fig. 2(a). Here the mean pressure P_m are subtracted from the measured pressure. As shown in Fig. 2(a), the pulse wave with width of 20 ms was

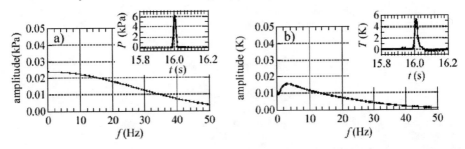

FIGURE 2. Time dependences of the pressure (a) and temperature (b) obtained in this experiment, and its power spectrum.

obtained. Consequently, the amplitude of P shown in Fig.2(a) extends up to $f = 50$ Hz. Theoretical temperature is calculated as a function of the frequency from the amplitude and the phase spectra of P. The measured temperature T_{ex} is also shown in Fig. 2 (b) as the difference from the mean temperature T_m. We determine Z_{ex} by compensating the measured and the theoretical ones.

The response of the thermocouple, $Z_{ex} = \Gamma e^{i\theta}$, is shown in Fig.3. The value of Γ, the amplitude ratio of T_{ex} to T, decreases with increasing f, while that of θ, the phase delay of T_{ex} relative to T, becomes large. As is clear from equation (1), an ideal thermocouple with a perfect response to the temperature oscillations should possess $\Gamma = 1$ and $\theta = 0$. Therefore, the decrease of Γ and θ represents that the response of the present thermocouple becomes worse with increasing f. However, it is clear that the present thermocouple is available to the measurement of T up to 50 Hz with the use of the obtained response Z_{ex}. An acoustic pulse method is effective to capture the response function Z_{ex} in a wide frequency range.

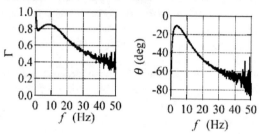

FIGURE 3. Frequency dependence of Γ and θ.

SUMMARY

We proposed the experimental technique of measuring a temperature oscillation in acoustic waves. Frequency dependence of a response function Z_{ex} of a thermocouple was derived by using an acoustic pulse. The Z_{ex} strongly depend on frequency f. With increasing f, the value of Γ decreases, while that of θ becomes large. The oscillating temperature T in a thermoacoustic device can be accurately determined by measuring T_{ex} with the thermocouple with its subsequent division by the value of Z_{ex}.

REFERENCES

1. Tominaga, A., *Cryogenics*, **35**(7), 427 (1995)
2. Swift, G. W., *J.Acoust.Soc.Am*, **84**(4), 1145 (1988); *Thermoacoustics ; Unifying perspective for some engines and refrigerators*, Acoustical Society of America Publications (2002)
3. Yazaki, T., and Tominaga, A., *Proc.R.Soc.Lond*, **454**, 2113 (1998)
4. Tashiro, Y., Biwa, T., Yazaki, T., and Mizutani, U., *Proceeding ICA2004*
5. Merkli, P., and Thoman, H., *J.Fluid Mech*, **70**(1), 161 (1975)

Experimental study on resonant frequency of the thermoacoustic cooling system

Shin-ichi SAKAMOTO, Hiroyuki HIRANO,
Takashi FUJITA and Yoshiaki WATANABE

Faculty of Engineering, Doshisha University, Kyotanabe, Kyoto 610-0321, JAPAN

Abstract. The purpose of our study is to construct a new cooling system applying the thermoacoustic effect. Stainless loop-tube is employed as our thermoacoustic cooling system and temperature decrease of 40 degrees C from the room temperature has been confirmed. In this paper, it is investigated that the relation between the viscosity boundary layer and the resonant frequency of the generated sound is investigated. Also, the sound pressure and temperature variation are observed with various total lengths of the loop-tube, with the view toward improvement in the cooling effect of the thermoacoustic cooling system. It was generally considered that the sound generated in the thermoacoustic cooling system is resonated with the tube length by 1 wavelength. However, when the total length of the loop-tube is over 2600 mm and inner pressure is 0.1 MPa, the resonant wavelength is 2. This is resulted from the influence of the viscosity boundary layer. It is found that the loop-tube decides the resonant frequency so that the thickness of the viscosity boundary layer is smaller than the stack channel radius. As a result, the resonant wavelength is 2 in a certain condition. The frequency is an important parameter for the thermoacoustic cooling system. From obtained results, one of the factors to select the frequency is found.

Keywords: thermoacoustic cooling system, viscosity boundary layer, resonant frequency
PACS: 43.25.+y, 43.35.+d, 43.90.+v

INTRODUCTION

Recent technical advance has brought about environmental destruction. Cooling system is appointed as one of the causes of environmental issues. In an existing cooling system, poisonous cooling media is generally used.

Our studies are intended to construct a new practical cooling system applying the thermoacoustic effect[1]. The thermoacoustic effect induces a mutual energy conversion of the sound energy and the heat energy[2,3,4]. By applying the thermoacoustic effect, it will be possible to construct a new epoch-making cooling system with many unique advantages: effective use of waste heat, no use of poisonous cooling media and no moving parts.

EXPERIMENTAL

Loop-tube is employed as our thermoacoustic cooling system[1]. The loop-tube consists of stainless tubes connected by 90 degrees elbows. The total length of the

CP838, *Innovations in Nonlinear Acoustics: 17th International Symposium on Nonlinear Acoustics*,
edited by A. A. Atchley, V. W. Sparrow, and R. M. Keolian
© 2006 American Institute of Physics 0-7354-0330-9/06/$23.00

loop is 1900, 2600, 3270 and 3970 mm. A block diagram of the loop-tube is shown in Figure 1. The diameter of the tube is 40 mm and the thickness of tube wall is 5 mm. Two stacks, stacks 1 and 2, are placed in the loop-tube. Each stack is sandwiched between heat exchangers. Stack 1 is employed as a prime mover and stack 2 as a heat pump. A stack is 50mm-long honeycomb ceramic. For stack 1, one 0.45mm-channel-radius stack is used and for stack 2 one 0.35mm-channel-radius stack is used. Stacks 1 and 2 are placed so that the distance between the stacks is a half of the total length of the loop-tube. For the heat exchanger A on stack 1, a whorl-shaped electric heater is used. The electric energy of 330W is supplied to the electric heater. The heat exchangers B are placed under stack 1 and on stack 2. In the heat exchangers B, there are copper fins, which are 5 mm in length and 1 mm in thickness. The fins are maintained at the reference temperature (T_R), 18 degrees C, by circulating water. As working fluid, Air is filled in the loop-tube at 0.1 MPa. Inner pressure is measured with a pressure gauge (KISTLER 601A).

The temperature at the cooling point under stack 2 is measured at the center axis of the tube with a K-type thermocouple 1.6 mm in diameter. The measurements of the temperature and the heat supply are simultaneously started. The heat supply is stopped 800 s later, and in the next 400 s the measurements are continued.

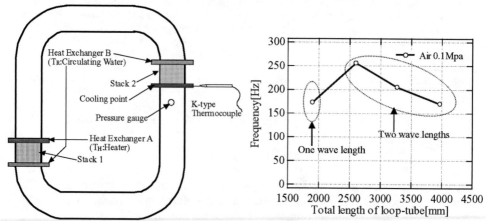

FIGURE 1. A block diagram of the loop-tube.

FIGURE 2. Frequencies of the thermoacoustic self-sustained sound as a function of total length of tube.

RESULTS AND DISCUSSIONS

40 s after the heat energy is supplied, the self-sustained sound generates in the loop-tube by the thermoacoustic effect. The sound pressure is over 160 dB. At the same time, the temperature starts to decrease. The greatest temperature decrease of 4.5 degrees C is observed, when the total loop-tube length is 3270 mm. It is found that loop-tube length has little influence upon the cooling effect in this experiment condition.

Figure 2 shows the resonant frequency of the sound generated in the loop-tube. It was generally considered that the sound generated in the thermoacoustic cooling system is resonated with the loop-tube length by 1 wavelength. As shown in Fig. 2, however, when the total length of the loop-tube is over 2600 mm, the resonant wavelength is 2 wavelengths. Under other conditions in this experiment, the wavelength is 1 and the resonant frequencies are inversely proportional to the total tube length.

It seems that the resonant frequency depends on the thickness of the viscosity boundary layer formed in stack 1 of the prime mover.

The thickness of the viscosity boundary layer is represented as follows[5]:

$$\delta_v = \sqrt{\frac{2v}{\omega}}$$ Formula (1)

$$v = \frac{\mu}{\rho}$$ Formula (2)

where δ_v is the thickness of the viscosity boundary layer, v is dynamic viscosity rate, ω is angular frequency, μ is coefficient of viscosity and ρ is gravity.

Figure 3 shows hypothetical thickness of the viscosity boundary layer, in the hypothetical condition that the working fluid is Air at 0.1 MPa and the resonant wavelength is 1 wavelength. R_{PM} in the figure is the channel radius of stack 1 of the prime mover and R_{HP} is the channel radius of stack 2 of the heat pump. δ_{vHP} is the thickness of the viscosity boundary layer formed in stack 2 and δ_{vPM} is that in stack 1.

Figure 3 indicates that the thickness of the viscosity boundary layer in stack 1 is bigger than the stack channel radius of stack 1. In this condition, an influence of the viscosity is dominant. Viscosity obstructs the energy conversion from the heat into sound, and the self sustained sound does not generated. To prevent this, the loop-tube, the system, decides the resonant frequency so that the thickness of the viscosity boundary layer is smaller than the stack channel radius.

The calculated thickness of the viscosity boundary layer from the observed resonant frequency is shown in Fig. 4. It is shown in Fig. 2 as observed frequencies that the resonant wavelength is 1 wavelength when the total tube length is 1900mm, and that the wavelength is 2 when the tube length is over 2600mm. In the both case, Fig. 4 indicates that the thickness of the viscosity boundary layer in stack 1 is smaller than the stack channel radius. This is different from Fig. 3. When the thickness of the viscosity boundary layer is smaller than the stack channel radius, an influence of the viscosity is not dominant and the sound does generate.

From Fig. 4, it is regarded that an influence of the viscosity boundary layer in stack 2 of the heat pump is little as the thickness of the viscosity boundary layer is quite smaller than the channel radius of stack 2.

CONCLUSION

In this paper, it is investigated that the relation between the viscosity boundary layer and the resonant frequency of the generated sound is investigated. It is found that the loop-tube, decides the resonant frequency by itself so that the thickness of the viscosity boundary layer is smaller than the stack channel radius. As a result, the resonant wavelength is 2 in a certain condition. The resonant frequency is an important parameter to select other parameters of the thermoacoustic cooling system such as stack channel radius, total loop-tube length, inner pressure etc. From obtained results, one of the factors to select the frequency is found. And this will induce the effective energy conversion of the thermoacoustic cooling system.

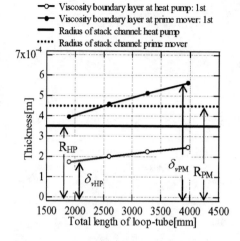

FIGURE 3. Relationship between stack channel radius and hypothetical thickness of viscosity boundary layer formed in stacks as a function of total length of tube. Resonant wavelength: 1 wavelength(1st), working fluid: Air at 0.1MPa.

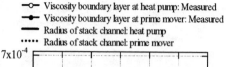
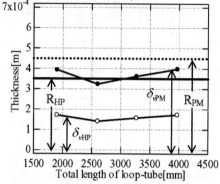

FIGURE 4. Relationship between stack channel radius and calculated thickness of viscosity boundary layer formed in stacks as a function of total length of tube. Resonant wavelength: measured wavelength, working fluid: Air at 0.1MPa.

ACKNOWLEDGEMENTS

This work was conducted in part under the project of Creation of Intelligent Cluster, Ministry of Education, Culture, Sports, Science and Technology. This study was partly supported by a Grant - in - Aid for JSPS Fellows.

REFERENCES

1. S. Sakamoto and Y. Watanabe, "The experimental studies of thermoacoustic cooler," Ultrasonics, Vol. 42, pp. 53-56, 2004
2. T. Yazaki, T. Biwa, and A. Tominaga, "A pistonless Stirling cooler," Appl. Phys. Lett. , vol. 80, No. 1, pp. 157-159, Jan. 2002
3. G. W. Swift, "Thermoacoustic Engines and Refrigerators," Physics Today, pp. 22-28, July 1995
4. P. H. Ceperley, "A pistonless Stirling engine – The traveling wave heat engine," J. Acoust. Soc. Am., vol. 66, No. 5, pp. 1508-1513, Nov. 1979
5. L. A. Wilen, "Measurements of scaling propaties for acoustic propagation in a single pore," J. Acoust. Soc. Am., vol. 101, No. 3, pp. 1388-1397, Mar. 1997

On Some Nonlinear Effects of Heat Transport in Thermal Buffer Tubes

Konstantin Matveev, Scott Backhaus, and Greg Swift

*Condensed Matter and Thermal Physics Group, Los Alamos National Laboratory,
Los Alamos, NM 87545*

Abstract. Thermal buffer tubes and pulse tubes, elements of thermoacoustic systems, should thermally isolate their two ends at substantially different temperatures while acoustic power freely flows from end to end. In practical high-power operating regimes of thermoacoustic devices, large heat leaks due to acoustic streaming can appear along a thermal buffer tube, which may degrade the overall system performance. To study this effect, a controlled experimental system is under development that will allow measurements of the heat leak due to acoustic streaming under various acoustic and thermal conditions. The phenomenon under consideration has very rich physics, and some of the accompanying effects are discussed. Gravity may significantly reduce or augment streaming and heat transport, and a simple model has been derived to account for its effect. At the ends of a thermal buffer tube, the gas moves periodically between the nearly adiabatic environment of the tube and the nearly isothermal environment of the adjacent heat exchanger. This establishes specific temperature joining conditions at the tube ends, which are important for overall heat transport by acoustic streaming.

Keywords: Thermoacoustics, Acoustic Streaming, Thermal Buffer Tube, Joining Conditions.
PACS: 43.25.Nm, 43.35.Ud

INTRODUCTION

Thermoacoustic engines and refrigerators either produce acoustic energy from, or pump thermal energy up, a mean temperature gradient [1]. They can be environmentally friendly, contain no moving mechanical parts, and have a great potential for applications where reliability is important, such as deep space travel and natural gas liquefaction on offshore platforms. At the present time, the low-amplitude regimes in these devices are well understood and can be modeled accurately, but finite-amplitude effects degrade our ability to accurately predict the performance in high-power regimes, which are of the most interest for practical applications.

One of these nonlinear effects is mass streaming, a second-order steady flow superimposed on and driven by the first-order acoustic oscillations. An arrangement typical for many thermoacoustic systems is shown in Fig. 1(a). Between the main ambient heat exchanger and the hot or cold heat exchanger lies a porous medium where the principal thermodynamic processes occur: entropy transport along the temperature gradient between heat exchangers and production or consumption of acoustic power. Many thermoacoustic systems also have a thermal buffer tube (known as a "pulse tube" in pulse tube refrigerators) and a secondary ambient heat exchanger. Acoustic power is transmitted along this tube with minimal heat leak

CP838, *Innovations in Nonlinear Acoustics: 17th International Symposium on Nonlinear Acoustics,*
edited by A. A. Atchley, V. W. Sparrow, and R. M. Keolian
© 2006 American Institute of Physics 0-7354-0330-9/06/$23.00

between the secondary ambient heat exchanger and the hot or cold heat exchanger. However, at high acoustic amplitudes, a second-order streaming flow appears that considerably increases the heat leak and results in significant degradation of the system performance. Rayleigh streaming, depicted in Fig. 1(a), is an internally circulating steady flow driven by boundary-layer interactions of thermoacoustic oscillations with the solid wall of the tube.

Significant progress in understanding acoustic streaming has been achieved in recent years. However, there are still many important open questions. Common analytical approaches usually assume that the streaming flow and its influence on the mean temperature field are small, and complicated solution procedures are described [2,3]. Under some assumptions, a practical recommendation based on analytical approaches can be made for preventing streaming, e.g., by tapering the tube [4]. Such small-streaming results are sometimes verified experimentally [5,6], but they may not be directly relevant to the high-power thermoacoustic systems where strong streaming is present [7,8]. CFD tools are promising for computing strong streaming flows [9-11], but significant problems, associated with multi-scale modeling in both space and time, remain. Also, specific boundary conditions appear at the ends of thermal buffer tubes [1,12]. The situation becomes even more difficult if other phenomena are accounted for, such as gravity, turbulence, and so on. There is a great need for a practical method to estimate streaming in thermoacoustic systems and for systematic and accurate experimental data on Rayleigh streaming in the presence of significant temperature gradients. Such a method and data series will be of great help to thermoacoustic practitioners. This paper describes our recent efforts in this direction.

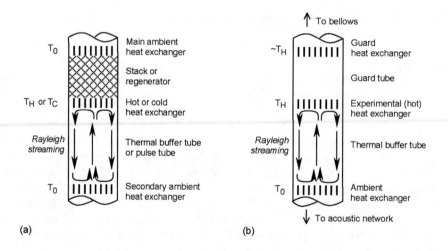

FIGURE 1. (a) System elements common to Stirling, pulse-tube, and stack-based thermoacoustic devices. Rayleigh streaming is shown inside a thermal buffer tube. (b) A part of the experimental system for measuring heat transport by Rayleigh streaming.

EXPERIMENTAL SYSTEM

A schematic of the main component of the experimental system, a vertically oriented TBT, is shown in Fig. 1(b). A linear-motor-driven bellows located above the tube produces acoustic oscillations, and an acoustic network attached to the bottom of the TBT allows us to regulate the ratio of the acoustic pressure and velocity amplitudes and the phase between them. The main goal of this system is to measure heat transfer by Rayleigh streaming from the experimental (hot) heat exchanger to the lower (ambient) heat exchanger. The heat transported by streaming in the TBT is the thermal power released at the experimental heat exchanger minus heat leaks by all non-TBT-streaming mechanisms. An additional guard tube section and a guard heat exchanger are placed above the experimental heat exchanger. The temperature of the guard heat exchanger is controlled to minimize heat transfer mechanisms in this part of the system, including streaming heat transfer in the guard tube.

The heat exchangers are copper cylinders with circular channels drilled through parallel to the direction of acoustic motion. Their temperature is set using either band heaters or aluminum blocks ported for circulating water. Flow straighteners, comprising a few layers of copper mesh, are employed at both ends of every heat exchanger to suppress jets produced by the heat exchanger channels.

At high amplitudes of the acoustic flow, significant nonlinear losses of acoustic power occur at the experimental heat exchanger, causing significant uncertainty in the thermal power released at this element. In order to minimize the heat added by acoustic dissipation, the heating element must be made as transparent as possible to the flow, though still providing effective heating. We have decided that a few sparse layers of nichrome wire weaved in a macor frame should be a good substitute for the heat exchanger currently employed, though we have not yet completed it.

To minimize heat leaks in the radial direction (to the environment), several layers of insulation cover the system, and a large-diameter copper pipe (shield) with distributed heaters and thermocouples is placed outside the insulation. By measuring temperatures on the experimental system components and comparing them with the local temperatures of the copper pipe at the same altitude, the shield is heated locally to minimize the temperature difference between the shield and system elements. Very thin walls on the TBT, just strong enough to sustain the high pressure inside the tube, minimize heat conduction along the tube walls.

From a methodological view, heat transfer by Rayleigh streaming should be measured as a function of various parameters including the difference in temperatures of the heat exchangers at the ends of the TBT. However, defining the end temperatures is not as trivial as in steady and/or incompressible flows. Gas parcels that cross the interface between the heat exchanger and the TBT during an acoustic cycle oscillate between nearly isothermal and nearly adiabatic environments, leading to complicated temperature joining conditions discussed below. To measure this temperature end effect, new parts were added to the system: a short tube section with a set of thermocouple wires stretched across the TBT, mounted between the TBT and the heat exchanger and used for determining mean gas temperatures far from the tube walls.

GRAVITY'S EFFECT ON STREAMING

Since the streaming in thermal buffer tubes occurs in the presence of a large temperature gradient, buoyancy can be important. Gravity can also play a significant role in predicting the behavior of thermoacoustic systems intended for the micro-gravity environment on spacecraft when such devices are developed and tested on Earth. We have derived a simple mathematical model to account for the influence of gravity on Rayleigh streaming and the associated convective heat transfer between two heat exchangers [13].

Assuming that the streaming is laminar and that the acoustic boundary layer thickness is small in comparison with the tube radius, the streaming flow in the TBT will have a pattern as shown in Fig. 2(a). The effective wall velocity U_w is computed from the streaming mass flux at the outer edge of the boundary layer [4]. This velocity depends on the first-order acoustic variables, temperature gradient, and gas properties. The variation of the flow velocity along the tube is assumed to be small in comparison with the variation of the flow velocity in the radial direction, the counterflowing streams are considered to have temperatures T_1 and T_2 for the inner and outer flows respectively, and the radial temperature variation within each of the two streams is ignored. With these assumptions, the appropriate momentum equation for the second-order (streaming) variables in each of the counterflowing streams is

$$-\frac{dp}{dx} + \frac{\mu}{R}\frac{d}{dR}\left(R\frac{dU}{dR}\right) - \rho g = 0 , \qquad (1)$$

where density ρ and viscosity μ are different for the two streams because they depend on temperature.

The relevant dimensionless parameters in this problem are the following:

$$G = \frac{\rho_1 g R_t^2}{\mu_1 U_w}, \qquad \alpha = \frac{\mu_1}{\mu_2}, \qquad \beta = \frac{\rho_2}{\rho_1}, \qquad (2)$$

where the first parameter characterizes the importance of the buoyancy with respect to viscous shear (index 1 indicates the inner stream), and the other parameters are the ratios of viscosities and densities in the two counterflowing streams. An implicit analytical solution for the velocity profiles has been obtained [13]. Examples of dimensionless velocity profiles across the tube are given in Fig. 2(b) for one value of G and three values of the difference in the stream temperatures. The driving wall velocity is directed upward, gravity is oriented downward, and the hot heat exchanger is above the cold heat exchanger. In the isothermal case, the profile is parabolic [dashed line in Fig. 2(b)]. As the temperature difference rises, the streaming-reducing buoyancy increases. A similar effect is produced by the magnitude of G: the higher this number, the stronger the suppression of the streaming by gravity.

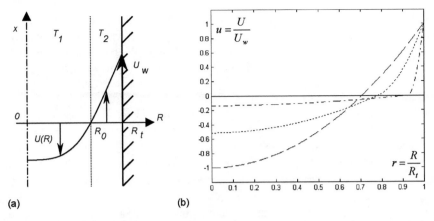

(a)   (b)

FIGURE 2. (a) A magnified view of a region inside the tube. The mean flow pattern $U(R)$ is caused by Rayleigh streaming. (b) Streaming velocity profiles at $G = 2500$. Dashed line, $\alpha = \beta = 1$; dotted line, $\alpha \approx \beta \approx 1.01$; dash-dotted line, $\alpha \approx \beta \approx 1.1$.

This approach can be extended for approximate estimations of the effect of gravity on streaming. Practical analysis can be organized in an iterative procedure as follows. First, the temperature field inside the tube can be guessed; for example, as in the absence of acoustic oscillations. Then, assuming the acoustic pressure does not vary across the tube, the acoustic wave equation can be solved for the first-order acoustic variables. The boundary-layer mass flux that drives streaming can then be found along the tube wall [4]. This mass flux determines the effective wall velocity. Using a solution of the steady Navier-Stokes equation (e.g., from a suitable CFD tool), the second-order steady streaming flow and the temperature field inside the tube can be calculated, applying appropriate boundary conditions on the tube walls and at the tube ends. The boundary-layer entropy flow and the heat conduction through the tube walls can be included. After that, one can return to the wave equation step, and carry on iterations until a converged solution for the streaming flow and mean temperature distribution is found. Then, the heat leaks by both mass streaming and the boundary-layer entropy flow can be calculated.

TEMPERATURE JOINING CONDITIONS

The thermal boundary conditions at the tube ends for the streaming problem are complicated due to oscillating motion of gas parcels through the interface between nearly isothermal and adiabatic regions. A simple explanation for this effect is given in [1] and more detailed experimental and modeling study is presented in [12].

Some understanding of this phenomenon can be obtained from the idealized temperature–position trajectories of the gas parcels in the vicinity of the heat exchanger, as shown in Fig. 3. The heat exchanger has a fixed temperature T_{HX}, and this space is nearly isothermal. The space in the TBT is nearly adiabatic. The characteristic length in this problem is a peak-to-peak acoustic displacement. Assuming one-dimensional flow, Fig. 3(b) shows the trajectories of the gas parcels on a temperature–position diagram for the case of acoustic pressure in phase with acoustic velocity, which corresponds to a traveling wave. Pressure and temperature oscillations lead the gas parcel displacement by $90°$, and the result is an elliptical trajectory for the "particular" gas parcel [shown by a bold ellipse in Fig. 3(b)] that just touches the heat exchanger at its leftmost position. Parcels to the right of the particular parcel follow similar elliptic trajectories, while parcels to the left follow truncated ellipses with a temperature of T_{HX} while inside the heat exchanger. It is interesting that these gas parcels return to the heat exchanger with temperatures different from T_{HX}. The mean temperature profile inside the peak-to-peak zone acquires a curved shape, illustrating the complexity of defining the thermal boundary conditions at the tube ends. The mean temperature profile in the gas at the tube ends depends on the acoustic pressure and displacement amplitudes, the phase between them, the temperature of the heat exchanger, the temperature gradient in the TBT beyond the peak-to-peak zone, and gas properties.

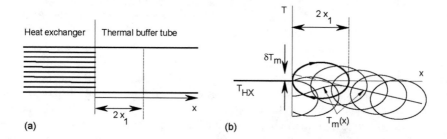

FIGURE 3. (a) An interface between an ideal heat exchanger and a thermal buffer tube. (b) Temperature-position diagram of the gas parcels with traveling-wave phasing. Adapted from [1].

A mathematical model that employs a Lagrangian-based method to track gas parcels in space and time has been developed for predicting the temperature profile. The mean temperature at a certain location is found by averaging temperatures of the gas parcels crossing this location over an acoustic cycle. Though the molecular heat conduction plays a minor role in the temperature end effect, the effective heat conduction can be augmented by vortices shed from the flow straighteners that are installed at the interface between the heat exchanger and the TBT. Modeling was carried out for both zero conduction and enhanced heat conduction.

Using the short TBT section described earlier, the temperature profile at the TBT ends was measured. An example of experimental data and model results for the mean gas temperature profiles in the TBT close to its interface with the heat exchanger is shown in Fig. 4 for traveling-wave phasing. Depending on the traveling-wave direction, the temperatures of gas parcels either drop or rise when they leave the heat exchanger, leading to the appearance of a dip or bump in mean temperature profiles, noticeable in Fig. 4. This deviation ends at the peak-to-peak displacement. The model that neglects heat conduction qualitatively predicts the temperature deviation, but the magnitude of this effect is exaggerated. The enhanced heat conduction smoothes the temperature deviation, giving results in good quantitative agreement with the test data at traveling-wave phasing.

When the phase of the acoustic wave was significantly different from $0°$ or $180°$ (e.g., by $60°$), some discrepancy developed between the theory and the experiment. Unfortunately, in the present experimental system configuration, it is not possible to closely approach the standing wave phasing, so we cannot conclude how well the model predicts the temperature joining conditions in a standing wave.

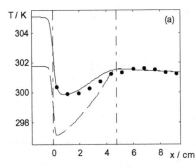 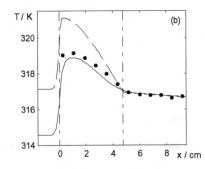

FIGURE 4. Mean temperatures of the gas (helium) at the TBT ends in the traveling wave. Points, experimental data;   solid line, calculations with enhanced heat conduction;   dashed line, calculations with zero heat conduction. Vertical dash-dotted line at $x = 0$ corresponds to the interface between the heat exchanger and the TBT;   the other vertical dash-dotted line is located at one peak-to-peak displacement from the heat exchanger. Direction of the acoustic wave propagation: (a) from heat exchanger to TBT;   (b) from TBT to heat exchanger.

CONCLUDING REMARKS

Higher performance and robustness are required from the next-generation high-power thermoacoustic devices in order to achieve their successful commercialization. However, nonlinear effects arising in high-amplitude regimes of operation of thermoacoustic systems are not yet sufficiently understood.   From a scientific standpoint, these nonlinear phenomena also represent very interesting research challenges. Mass streaming, which is induced by first-order acoustic oscillations, has significant consequences for heat transfer in thermoacoustic devices and can be either harmful or useful.   Turbulence in oscillating flows, stability and hysteresis of streaming patterns, and variable system orientation with respect to the gravity are problems waiting for future research.

ACKNOWLEDGMENTS

This work was supported by the Office of Basic Energy Sciences, Division of Materials Science, in the US Department of Energy's Office of Science.

REFERENCES

1. Swift, G.W., *Thermoacoustics: A Unifying Perspective for Some Engines and Refrigerators*, Acoustical Society of America, 2002.
2. Rott, N., "The Influence of Heat Conduction on Acoustic Streaming," *Zeitschrift für angewandte Mathematik und Physik* **25**, 417-421 (1974).
3. Bailliet, H., Gusev, V., Raspet, R., and Hiller, R.A., "Acoustic Streaming in Closed Thermoacoustic Devices," *Journal of the Acoustical Society of America* **110** (4), 1808-1821 (2001).
4. Olson, J.R. and Swift, G.W., "Acoustic Streaming in Pulse Tube Refrigerators: Tapered Pulse Tubes," *Cryogenics* **37**, 769-776 (1998).
5. Campbell, M., Cosgrove, J.A., Greated, C.A., Jack, S., and Rockliff, D., "Review of LDA and PIV Applied to the Measurement of Sound and Acoustic Streaming," *Optics and Laser Technology* **32**, 629-639 (2000).
6. Thompson, M.W. and Atchley, A.A., "Simultaneous Measurement of Acoustic and Streaming Velocities in a Standing Wave Using Laser Doppler Anemometry," *Journal of the Acoustical Society of America* **117** (4), 1828-1838 (2005).
7. Thompson, M.W., Atchley, A.A., and Maccarone, M.J., "Influences of a Temperature Gradient and Fluid Inertia on Acoustic Streaming in a Standing Wave," *Journal of the Acoustical Society of America* **117** (4), 1839-1849 (2005).
8. Menguy, L. and Gilbert, J., "Non-Linear Acoustic Streaming Accompanying a Plane Stationary Wave in a Guide," *Acta Acustica united with Acustica* **86**, 249-259 (2000).
9. Kamakura, T., Matsuda, K., Kumamoto, Y., and Breazeale, M.A., "Acoustic Streaming Induced in Focused Gaussian Beams," *Journal of the Acoustical Society of America* **97** (5), 2740-2746 (1995).
10. Yano, T., "Turbulent Acoustic Streaming Excited by Resonant Gas Oscillation with Periodic Shock Waves in a Closed Tube," *Journal of the Acoustical Society of America* **106** (1), L7-L12 (1999).
11. Boluriaan, S. and Morris, P., "Numerical Simulation of Streaming in High Amplitude Standing Wave Resonators", *Journal of the Acoustical Society of America* **113** (4), 2282 (2003).
12. Matveev, K.I., Swift, G.W., and Backhaus, S.N., "Temperatures near the Interface between an Ideal Heat Exchanger and a Thermal Buffer Tube or Pulse Tube," *International Journal of Heat and Mass Transfer*, in press.
13. Matveev, K.I., Backhaus, S.N., and Swift, G.W., "The Effect of Gravity on Heat Transport by Rayleigh Streaming," ASME International Congress & Exposition, Anaheim, CA, 2004, ASME paper IMECE 2004-59076.

Numerical Study of High Reynolds Number Acoustic Streaming in Resonators

Takeru Yano

Division of Mechanical and Space Engineering, Hokkaido University, Sapporo, 060-8628, Japan

Abstract. Acoustic streaming caused by a large amplitude resonant oscillation of an ideal gas in a two-dimensional resonator is numerically studied by solving the system of Navier–Stokes equations with a finite-difference method, without the assumption of symmetry of the flow field. The sound field including shock waves is precisely determined, and then, the streaming velocity field is evaluated in terms of a time-averaged mass flux density vector. We shall demonstrate that, in the case where the amplitude of gas oscillation is moderately large, an asymmetric quasi-steady streaming is established after more than a thousand of oscillations of sound source.

Keywords: Numerical study, Acoustic streaming, Resonance, Shock wave
PACS: 43.25.Nm, 43.25.Gf, 43.25.Cb

INTRODUCTION

Streaming motions induced by acoustic standing waves are classical topics in physics [1, 2, 3]. Today, the active control of streaming in resonators becomes an important subject in various applications, in particular in thermoacoustic devices (see, e.g., [4] and Fig. 1). Some authors have recently carried out accurate measurements for slow streaming motions [5] in a resonator. However, its behavior in the case of large Reynolds number remains unresolved.

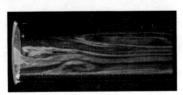

FIGURE 1. Example of asymmetric acoustic streaming in a standing-wave type thermoacoustic engine. Photograph courtesy of T. Yazaki.

Recently, the present author has numerically studied the resonant gas oscillation with a periodic shock wave in a closed tube by solving the system of compressible Navier–Stokes equations [6]. The result has suggested the occurrence of turbulent acoustic streaming when a streaming Reynolds number is sufficiently large. This is the first numerical evidence for the prediction based on the experiment [3]. Numerical studies of streaming motion with large Reynolds number have also been carried out by Alexeev and Gutfinger [7], Morris et al. [8], and Aktas and Farouk [9]. Furthermore, detailed gas motions in a vicinity of a stack plate in the thermoacoustic device have been computed by Besnoin and Knio [10] and Marx and Blanc-Benon [11]. Nevertheless, the direct

CP838, *Innovations in Nonlinear Acoustics: 17th International Symposium on Nonlinear Acoustics*,
edited by A. A. Atchley, V. W. Sparrow, and R. M. Keolian
© 2006 American Institute of Physics 0-7354-0330-9/06/$23.00

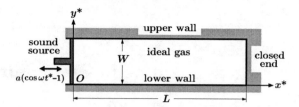

FIGURE 2. Schematic of model.

numerical simulation of viscous compressible flow is an extraordinarily hard task if one tries to resolve all phenomena from an initial state of uniform and at rest to an almost steady oscillation state throughout the entire flow field including the boundary layer. Therefore, our knowledge of streaming with large Reynolds number is still limited.

In previous papers [12, 13, 14], we have adopted a simple model based on the linear standing wave solution and a boundary layer analysis. This model employs the incompressible Navier–Stokes equations as the governing equations for the streaming velocity. As a result, we have numerically demonstrated the bifurcation and multiple existence of steady state solutions for a region of moderately large Reynolds number in a two-dimensional rectangular box. In the present paper, we shall investigate the problem of bifurcation of steady streaming, which is expected to occur before the transition to turbulent motions, not by utilizing the incompressible model, but by the direct simulation of compressible Navier–Stokes system. We treat the large Reynolds number acoustic streaming, but the Reynolds number is not so large that the turbulent streaming occurs.

PROBLEM

We shall consider the streaming motion induced by resonant gas oscillations in a two-dimensional rectangular box filled with an ideal gas (see Fig. 2). The box, whose length is L and width is W, is closed at one end by a solid plate and the other by a piston (sound source) oscillating harmonically with an amplitude a and angular frequency ω.

We assume that the sound excitation is moderately weak, the thickness of the boundary layer is sufficiently thin compared with the width of the box, and the wavelength of the excited sound is comparable with the width of the box,

$$M = \frac{a\omega}{c_0} \ll 1, \quad \epsilon = \frac{\sqrt{\nu_0 \omega}}{c_0} \ll w, \quad w = \frac{W\omega}{c_0} = O(1), \tag{1}$$

where M is the acoustic Mach number at the sound source (c_0 is the speed of sound in the initial undisturbed state), ϵ is a measure of the ratio of the thickness of acoustic boundary layer to the wavelength $\lambda = 2\pi c_0/\omega$, and w is the normalized width of the box. Furthermore, we assume the second-mode resonance, which is prescribed by

$$b = 2\pi, \tag{2}$$

where $b = L\omega/c_0$ is the normalized box length. The wave motion in the bulk of the gas can then be a resonant gas oscillation including shock waves [15].

Governing equations, initial and boundary conditions

We obtain the wave and streaming motions by solving the initial and boundary value problem for the system of compressible Navier–Stokes equations,

$$\frac{\partial \rho}{\partial t} + \frac{\partial \rho u_j}{\partial x_j} = 0, \tag{3}$$

$$\frac{\partial \rho u_i}{\partial t} + \frac{\partial p \delta_{ij} + \rho u_i u_j}{\partial x_j} = \epsilon^2 \frac{\partial \sigma_{ij}}{\partial x_j}, \quad \sigma_{ij} = \mu \left(\frac{\partial u_i}{\partial x_j} + \frac{\partial u_j}{\partial x_i} - \frac{2}{3} \frac{\partial u_k}{\partial x_k} \delta_{ij} \right), \tag{4}$$

$$\frac{\partial E}{\partial t} + \frac{\partial (E+p) u_j}{\partial x_j} = \epsilon^2 \left(\frac{\partial \sigma_{ij} u_i}{\partial x_j} + \frac{\partial q_j}{\partial x_j} \right), \quad E = \frac{1}{2} \rho u_i^2 + \frac{p}{\gamma - 1}, \quad q_j = \frac{\mu}{(\gamma - 1) Pr} \frac{\partial T}{\partial x_j}, \tag{5}$$

where $x_i = x_i^* \omega / c_0$ and $t = \omega t^*$ are the nondimensional space coordinates and time; $\rho = \rho^* / \rho_0$ and $p = p^* / (\rho_0 c_0^2)$ are the nondimensionalized gas density and pressure; $u_i = u_i^* / c_0$ is the nondimensional gas velocity; E is the nondimensional total gas energy per unit volume; σ_{ij} and q_i are the nondimensional viscous tensor and heat flux vector; $\mu = \mu^* / \mu_0$ is the nondimensional viscosity coefficient; γ is the ratio of specific heats; Pr is the Prandtl number. The equation of state for ideal gas $\gamma p = \rho T$ is used to close the system.

The initial condition is given as

$$u_i = 0, \quad p = \gamma^{-1}, \quad \rho = 1, \quad T = 1. \tag{6}$$

The boundary conditions on the oscillating piston face are

$$u = -M \sin t, \quad v = 0, \quad T = 1 \quad \text{at} \quad x = M(\cos t - 1) \text{ and } 0 \leq y \leq w, \tag{7}$$

and the boundary condition except for the piston face are

$$u = v = 0, \quad T = 1, \tag{8}$$

where $x = x_1$, $y = x_2$, $u = u_1$, $v = u_2$, and the isothermal condition is imposed on the gas temperature on the wall.

The viscosity coefficient is assumed to obey the Sutherland's law and the Prandtl number Pr to be a constant (0.7).

The results presented in the following are the cases where the acoustic Mach number at the sound source $M = 0.01$ and 0.001, the normalized box width $w = 2\pi/5$ and $\pi/5$, and the ratio of the boundary layer thickness to the wavelength $\epsilon = 0.00316$, which corresponds to $\omega/(2\pi) = 12.5$ kHz in the air in the standard state.

NUMERICAL METHOD

The initial and boundary value problem (3)–(8) is solved with the high-resolution up-wind finite-difference TVD scheme [16]. The method has been used for various non-linear acoustics problems by the present author (e.g., see [6, 17]), and the details are omitted here.

We shall remark that in the present computation we don't assume any symmetry of flow pattern of acoustic streaming, and therefore we solve the entire field in the box $0 \leq x \leq b$ and $0 \leq y \leq w$. The entire field is subdivided into a boundary-fitted 700×300 nonuniform mesh, where the mech points are clustered near the boundary. The minimum grid size is 0.0002 in the vicinity of the wall and this is so small compared with $\epsilon = 0.00316$ that we can resolve the acoustic boundary layer. The resolution of boundary layer is crucially important because the acoustic streaming in the bulk of the gas is mainly induced by the streaming motion in the boundary layer or the so-called limiting velocity of the inner streaming.

The time step is $2\pi/50000$, and the CFL number is about 0.5. The CPU time for computation of one oscillation cycle of piston is about 4.4 hours on a state-of-the-art PC (dual-cpu machine).

The streaming velocity u_S is evaluated by the time-averaged mass flux vector,

$$u_S = \begin{pmatrix} u_S \\ v_S \end{pmatrix} = \int_t^{t+2\pi} \begin{pmatrix} \rho u \\ \rho v \end{pmatrix} dt. \tag{9}$$

We shall further remark that we don't give any artificial seed of asymmetry in the computation. The numerical code is written symmetrically in the algebraic sense. The asymmetry inherent in the numerical operations of finite figures spontaneously grows due to the instability of the system concerned.

RESULTS

Evolution of resonant gas oscillation

At the initial instant $t = 0$, the gas in the box is uniform and at rest. After the beginning of oscillation of the piston, the wave amplitude grows in proportion to Mt. At $t = O(1/\sqrt{M})$, the wave amplitude reaches the maximum value of $O(\sqrt{M})$, where two shock waves are formed since the excitation at the sound source is the second mode. Figure 3 shows the temporal evolution of pressure amplitude at the closed end. At almost $t/(2\pi) = 15$, the wave amplitude reaches its maximum value, and thereafter a quasi-steady oscillation state continues.

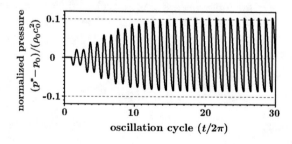

FIGURE 3. Initial evolution of pressure amplitude at the closed end.

382

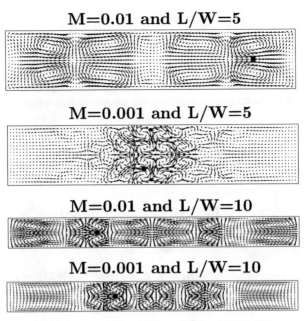

M=0.01 and L/W=5

M=0.001 and L/W=5

M=0.01 and L/W=10

M=0.001 and L/W=10

FIGURE 4. Streaming velocity field at $t/(2\pi) = 300$.

Development of acoustic streaming

Figure 4 shows the streaming velocity fields at $t/(2\pi) = 300$. Except for the case of $M = 0.01$ and $L/W = 5$, the streaming patterns are considerably different from that of the classical Rayleigh streaming, which consists of the regular arrangement of four vortex pairs and the flow pattern is symmetric with respect to $x = b/2$ and $y = w/2$. Although the flow pattern for $M = 0.01$ and $L/W = 5$ is apparently similar to that of Rayleigh streaming, the flow directions of its major vortexes are opposite to those of Rayleigh streaming. This has been reported by Alexeev and Gutfinger [7], although their computation assumes the symmetry of flow field with respect to $y^* = W/2$ and they have truncated the computations at one hundred cycles.

The streaming velocity fields at $t/(2\pi) = 500$ are shown in Fig. 5. An origin of asymmetry appears near $x = 0$ and $y = w/2$ in the case of $M = 0.01$ and $L/W = 5$. The flow patterns in the other cases constantly change from those shown in Fig. 4. This clearly means that computations for a few hundreds of cycles are insufficient for the analysis of high Reynolds number acoustic streaming.

We therefore continue the computations for the two cases of $L/W = 5$ over a thousand of cycles. Figure 6 shows the development of asymmetric streaming pattern for the case of $M = 0.01$ and $L/W = 5$, and Fig. 7 shows the case of $M = 0.001$ and $L/W = 5$. The time from the beginning of oscillation and the maximum of streaming velocity $U_{\max}$ are shown in each plot in Fig. 6. Note that, as can be seen from Figs. 6 and 7, the maximum velocity $U_{\max}$ is of the order of M, because the maximum wave amplitude is of $O(\sqrt{M})$

M=0.01 and L/W=5

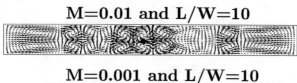

M=0.001 and L/W=5

M=0.01 and L/W=10

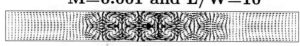

M=0.001 and L/W=10

FIGURE 5. Streaming velocity field at $t/(2\pi) = 500$.

and the streaming motion is a second-order nonlinear phenomenon.

The top figure in Fig. 6 shows that the streaming velocity field is fully asymmetric at 860th cycle. From the comparison of the streaming pattern at 1160th cycle and that at 1260th cycle, we may conclude that an asymmetric streaming almost reaches a quasi-steady state. On the other hand, the symmetry of streaming velocity field for the case of $M = 0.001$ and $L/W = 5$ is hardly destroyed up to 1020th cycle, as shown in the top figure in Fig. 7. Nevertheless, an origin of asymmetry appears at around the center of the resonator at 1120th cycle, and then the asymmetry grows and prevails in the entire field at 1320th cycle. At this stage, however, we cannot conclude that the streaming velocity field for the case of $M = 0.001$ and $L/W = 5$ reaches a quasi-steady state.

Here, we shall comment on the streaming Reynolds number. The streaming Reynolds number Rs may be defined by

$$\text{Rs} = \frac{U_\text{S} L_\text{S}}{\nu_0}. \tag{10}$$

From the bottom figure in Fig. 6, we take the characteristic speed of streaming $U_\text{S} = 0.007c_0$. The characteristic length of streaming L_S may be taken as $W = (2\pi c_0)/(5\omega)$ (width of box). Consequently, we have Rs = 880 for the case of $M = 0.01$ and $L/W = 5$.

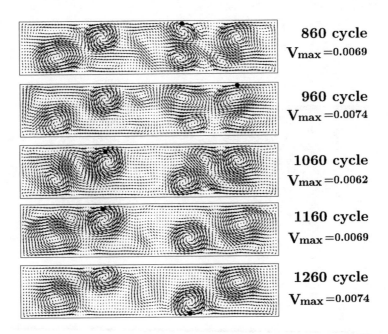

FIGURE 6. Development of asymmetric streaming pattern for $M = 0.01$ and $L/W = 5$.

CONCLUSIONS

We have demonstrated that acoustic streaming in a resonator develops into an asymmetric quasi-steady flow. The previous authors [7, 9] assumed the symmetry with respect to the centerline of resonator in their computations, and therefore, they couldn't find the asymmetric solutions. Furthermore, they truncated the computation at 100 or 200 cycles from the beginning of sound excitation, which is clearly insufficient for the analysis of streaming motion of moderately large Reynolds number, unless the strong nonlinearity rapidly excites a turbulent streaming motion as shown in [6].

The existence of the steady asymmetric flow regime prior to the transition to turbulent motions is important for understanding of acoustic streaming with large Reynolds number in resonators.

REFERENCES

1. Lord Rayleigh, *The Theory of Sound* Dover, New York, 1945.
2. E. N. da C. Andrade, On the circulations caused by the vibration of air in a tube, *Proc. R. Soc. A*, **134**, 445–470 (1931).
3. P. Merkli and H. Thomann, Transition to turbulence in oscillating pipe flow, *J. Fluid Mech.*, **68**, 567–575 (1975).
4. G. W. Swift, Thermoacoustic engines and refrigerators, *Phys. Today.* **48**, American Institute of Physics, New York, 1995, pp.22–28.

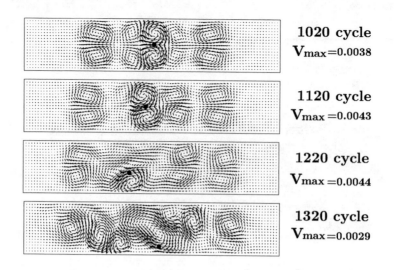

1020 cycle
$V_{max}=0.0038$

1120 cycle
$V_{max}=0.0043$

1220 cycle
$V_{max}=0.0044$

1320 cycle
$V_{max}=0.0029$

FIGURE 7. Development of asymmetric streaming pattern for $M = 0.001$ and $L/W = 5$.

5. M. W. Thompson and A. A. Atchley, Simultaneous measurement of acoustic and streaming velocities in a standing wave using laser Doppler anemometry, *J. Acoust. Soc. Am.*, **117**, 1828–1838 (2005).
6. T. Yano, Turbulent acoustic streaming excited by resonant gas oscillation with periodic shock waves in a closed tube, *J. Acoust. Soc. Am.*, **106**, L7–L12 (1999).
7. A. Alexeev and C. Gutfinger, Resonance gas oscillations in closed tubes: Numerical study and experiments, *Phys. Fluids*, **15**, 3397–3408 (2003).
8. P. J. Morris, S. Boluriaan, and C. M. Shieh, Numerical simulation of minor losses due to a sudden contraction and expansion in high amplitude acoustic resonators, *Acta Acust. United Ac.*, **90**, 393–409 (2004).
9. M. K. Aktas and B. Farouk, Numerical simulation of acoustic streaming generated by finite-amplitude resonant oscillations in an enclosure, *J. Acoust. Soc. Am.*, **116**, 2822–2831 (2004).
10. E. Besnoin and O. M. Knio, Numerical study of thermoacoustic heat exchangers in the thin plate limit, *Numer. Heat Transf. A-Appl.*, **40**, 445–471 (2001).
11. D. Marx and P. Blanc-Benon, Computation of the mean velocity field above a stack plate in a thermoacoustic refrigerator, *C. R. Mec.*, **332**, 867–874 (2004).
12. T. Yano, S. Fujikawa, and H. Muranaka, Numerical Study of Rayleigh Type Acoustic Streaming with Large Reynolds Number, *Proceedings of 17th International Congress on Acoustics*, **1**, 2001, pp.56–57.
13. T. Yano, S. Fujikawa, and M. Mizuno, Bifurcation of acoustic streaming induced by a standing wave in a two-dimensional rectangular box, *Nonlinear Acoustics at the Beginning of the 21st Century*, edited by O. V. Rudenko and O.A. Sapozhnikov, Moscow State University, Moscow, 2002, pp.227–230.
14. T. Yano, Numerical study of acoustic streaming in a resonator with large Reynolds number, *Proceedings of the World Congress of Ultrasonics 2003*, 2003, pp.661–664.
15. W. Chester, Resonant oscillations in closed tubes," *J. Fluid Mech.*, **18**, 44–65 (1965).
16. S. R. Chakravarthy, Development of upwind schemes for the Euler equations, *NASA contractor report*, **4043** (1987).
17. T. Yano and Y. Inoue, Strongly nonlinear waves and streaming in the near field of a circular piston, *J. Acoust. Soc. Am.*, **99**, 3353–3372 (1996).

Oscillating Flow in Adverse Pressure Gradients

Barton L. Smith, Adam J. Dean, Zachary E. Humes, Kristen V. Mortensen
and Spencer Wendel

Mechanical and Aerospace Engineering, Utah State University

Abstract.
Results from a series of experimental and numerical studies of oscillating flow are presented. Particle Image Velocimetry (PIV) measurements reveal that the displacement amplitude in a typical thermoacoustic demonstration engine is insufficient to generate jetting. We show numerically that enthalpy flux from a similar engine can be enhanced by placing a hollow cone at the end of the tube. In an experimental study of oscillatory flow in a 2-D wide-angle diffuser, we measure full-field velocity and pressure simultaneously. The minor losses due to this flow are shown to be small for small amplitudes and approach steady flow values for large amplitudes.

INTRODUCTION

In virtually any geometry or flow regime, the behavior of oscillatory flow is significantly different from steady flow. Attempts to treat oscillatory flow in a quasi-steady manner, assuming that at any instant the flow has no memory and behaves as a steady flow would at the instantaneous Reynolds number, have met with only limited success [1]. The temporal accelerations/decelerations present in oscillatory flow, but absent in steady flow, have interesting effects on separated shear layers which can lead to reattachment of separated boundary layers.

Boundary layer separation is always dependent on the boundary layer state (laminar or turbulent), with turbulent flow being less likely to separate. The complex nature of the transition to turbulence in oscillatory flow strongly impacts flow behavior in cases for which separation is likely, such as adverse pressure gradient flows. Meanwhile, reattachment is strongly influenced by the rollup of the separated shear layer, especially during the accelerating portion of the cycle.

Such flows are important to most thermoacoustic devices. While these devices typically operate at high pressure, much can be learned from a study of oscillating flow in atmospheric air and at incompressible conditions. A series of such studies was begun at Los Alamos National Lab in 2000 [2, 3] and has been continued at Utah State University (USU) [4]. The present paper presents an overview of some of the most recent results. In some cases, more detail can be found in earlier presentations [5, 6].

We begin with a demonstration of the application of state-of-the-art velocity measurements techniques (Particle Image Velocimetry, or PIV) to an atmospheric-air standing-wave heat engine. This is followed by a demonstration of the efficacy of using a hollow cone placed in an oscillating flow to circulate rejected heat out of a thermoacoustic engine. Finally, we describe a fundamental study of oscillating flow in an adverse pressure gradient (APG) geometry.

CP838, *Innovations in Nonlinear Acoustics: 17th International Symposium on Nonlinear Acoustics*,
edited by A. A. Atchley, V. W. Sparrow, and R. M. Keolian
© 2006 American Institute of Physics 0-7354-0330-9/06/$23.00

MEASUREMENTS AND NUMERICAL MODEL

Most of the measurements are made in the Utah State University Oscillatory Flow Facility. A description of the facility can be found in [5, 6]. All velocity measurements are made using a 2-D digital PIV system. The instantaneous velocity field is measured phase-locked to the driving signal. For measurements in the test-tube engine, phase-locked acquisition is performed by triggering the PIV system with a microphone placed near the exit. For the APG study, one hundred samples were acquired at 18 points in phase. Pressure was measured simultaneously at 8 locations, (we will refer these as set A, see Fig. 4) with piezo-resistive pressure sensors built into the wall. For each PIV sample, 2 periods are measured (100 samples each) at each sensor. Subsequent to velocity measurements, the sensors are moved to new locations (set B), and the cases are repeated. In order to avoid errors caused by the nonlinearity of these transducers, the full voltage waveforms from the transducers are digitized and a nonlinear calibration curve is used to convert voltages to pressures. These techniques have been found to provide accurate time-averaged pressure in the presence of large amplitude fluctuations [2]. In the study of steady circulation generated by a cone, Fifty samples were acquired at 18 points in phase.

In addition to the geometrical parameters, these problems are described by two flow parameters: the dimensionless displacement amplitude (or stroke length) based on the small flow dimension, and a Reynolds number. The displacement amplitude can be computed from the velocity data: $L_0 = \int_0^{T/2} u_0(t)\, dt$ where T is the driving period and u_0 is the cross-stream average velocity. For sinusoidal volume flow rate, $L_0 = u_{\max}/\pi f$, where f is the driving frequency and $u_{\max}$ is the amplitude of u_0. The Reynolds is based on the viscous penetration depth, $\mathrm{Re}_\delta = u_{\max}\delta/\nu$, where $\delta = \sqrt{2\nu/\omega}$, ν is the kinematic viscosity and $\omega = 2\pi f$.

The numerical RANS model is an axisymmetric representation of the thermoacoustic engine from the hot end of the stack to beyond the outlet. The simulations were performed with the commercial URANS solver FLUENT. Turbulence is modeled with the standard $k - \epsilon$ viscous model with enhanced wall treatment. This model has been shown to be effective for simulating the global effects of oscillating flow [4]. The boundary conditions employed and other details of the model are described in [6]

DEMONSTRATION ENGINE

PIV measurements were performed near the end of a test-tube-based 1/4-wavelength standing-wave thermoacoustic engine. The test tube is 2.3 mm in diameter and 20 mm long. In order to facilitate optical measurements near the end of the engine, the open end of the pyrex tube was ground flat. Note that it was necessary to allow the engine to run for several minutes to stabilize its operating frequency. During this time, seeds used for PIV collected on the edges of the exit rendering near wall data impossible, especially at the exit. Examples of the vector field are shown in Fig. 1 for two instances in time corresponding to the peak flow in either direction. It is clear from the data that the displacement amplitude of this engine is insufficient to form a synthetic jet [4], which

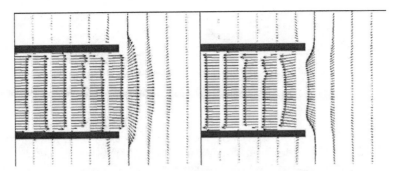

FIGURE 1. Velocity vectors acquired near the end of a standing wave demonstration engine at peak blowing (left) and peak suction (right).

would greatly improve heat transfer from the cold end of the engine. This motivates the use of a cone to promote circulation in the Chip Cooler study below. Based on this data, the engine operated with a peak average velocity of 3.3 m/s, a frequency of 465 Hz, and thus Re = 20 and $L_0/D \approx 0.1$. The outgoing flow has a velocity profile consistent with laminar oscillatory flow [7] (i.e. peaks near the edges), while in incoming flow separates as it enters the tube and generates a recirculation region. Because of errors in the PIV correlation due to distortion of the optical field, it is not possible to resolve the velocity field sufficiently to determine if time-averaged streaming removes heat from this engine.

THERMOACOUSTIC CHIP COOLER

To meet the need for increased cooling reliability, we are developing a passive chip-cooling system that has similar cooling performance to an integrated fan/heat sink but with reduced cost and no moving parts. This device, based on a thermoacoustic engine, moves heat away from a chip via forced air convection. The air motion is generated by converting acoustic power generated in the stack to steady flow. These devices require no external power. In fact, if an increase in power dissipation is experienced by the chip, the engine's output will also increase making the system inherently stable. A schematic of the proposed engine is shown in Fig. 2. The engine is based on a circular tube closed at one end to create a 1/4 wavelength resonator. The cone converts the oscillatory motions generated by the engine to time-averaged circulation with the ambient.

Conversion of oscillatory flow to steady flow has been recently proposed as a method to bring heat into thermoacoustic engines [1]. Conversion is accomplished by passing the oscillatory flow through a flow element with larger losses in one direction than the other. An attractive strategy to improve the efficiency and cost of construction of thermoacoustic machines is the replacement of a heat exchanger with enthalpy flux from a steady flow. New thermoacoustic designs call for the replacement of the hot heat exchanger with a loop including acoustic diodes to convert acoustic power to steady flow [8].

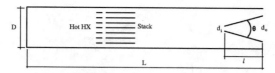

FIGURE 2. Proposed thermoacoustic engine. The engine length is one-quarter the acoustic wavelength.

We numerically study oscillatory zero-mean flow in an axisymmetric geometry consisting of a $D = 5$-cm tube 7.9 cm long. An exit plane is affixed to the end of the tube. A hollow cone of length $l = D$ and having large diameter d_o and a small diameter d_i is suspended in the end of the tube as shown in Fig. 2.

The heat flux $\dot{q}$ out of the engine is shown in Fig. 3a as a function of L_0/D for various Reynolds numbers and cone angles. The heat flux is normalized to remove the influence of our choice of inlet temperature: $\dot{Q} = \dot{q}/\rho U_{\max} C_p (T_h - T_c)$, where $\dot{q}$ is the heat flux, $U_{\max}$ is the volume flow rate amplitude, C_p is the constant pressure specific heat, and T_h and T_c are the hot and ambient temperatures, respectively. It is evident that:

1. The heat flux increases with L_0/D
2. Adding of a cone results in an order of magnitude improvement in the heat flux
3. The heat flux increases somewhat with Reynolds number for large cone angles
4. At small cone angles, the heat flux is Reynolds number independent
5. The heat flux is maximum near $\theta = 24°$

This heat removal comes at the cost of the acoustic power that must be supplied to the cone. Acoustic power is also non-dimensionalized: $\dot{W} = \dot{E}/(\rho U_{\max}^3/A^2)$, where A is the area of the tube and ρ is the fluid density. Fig. 3b shows that the required power $\dot{W}$ is largely a function of cone angle. Larger cone angles remove more heat from the engine, but require more $\dot{W}$.

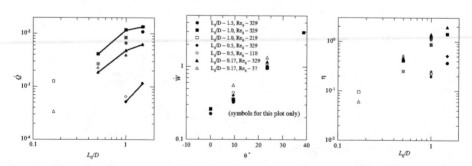

FIGURE 3. (a) Dimensionless heat flux $\dot{Q}$ from the tube as a function of L_0/D. The symbols denote the cone angle: (•) $\theta = 39°$, (■) $\theta = 24°$, (▲) $\theta = 9.5°$, (♦) $\theta = 0°$ (no cone). The shades represent Reynolds number values: (Black) Re = 329, (Dark grey) Re = 219, (Light Grey) Re = 110, (Open) Re = 36.7. (b) The non-dimensional acoustic power $\dot{W}$ as a function of θ. (c) Efficiency (η) or $\dot{Q}/\dot{W}$ as a function of displacement amplitude L_0/D. Symbols as in (a).

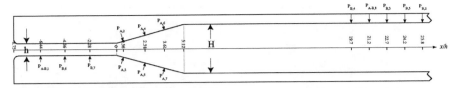

FIGURE 4. Schematic of the geometric parameters of the APG study and pressure tap locations.

This motivates the definition of an efficiency parameter $\eta = \dot{Q}/\dot{W}$. Fig. 3c demonstrates how η increases with L_0/D. Using a cone angle of $\theta = 9.5°$ increases the efficiency over no cone by nearly an order of magnitude. This data also suggests an optimum cone angle and that using large angles ($\theta = 39°$) has no benefit.

ADVERSE PRESSURE GRADIENT STUDY

We study zero-mean oscillatory flow in a nominally 2-D expansion from a width $h = 1.68$ cm to $H = 6.67$ cm with a total angle of $30°$ (Fig. 4. Additional details of the geometry are contained in [2]). Two Reynolds numbers (400 and 800) at displacement amplitudes in the range $10 < L_0/h < 36$ are studied. The resultant values are determined from phase-averaged velocity data sets by computing the cross-stream average velocity at several points near the inlet to the diffuser. Since the flow rate waveform is not a pure sinusoid, the flow rate data are least-squares fit to a waveform containing the fundamental and the first two harmonics. The Reynolds number and displacement amplitudes are calculated based on this fitted function. An example of the fit is shown in Fig. 6a. The value of dimensionless time t^*/T relative to the driving signal at which the flow rate crosses zero, moving in the positive x direction, is chosen as $t/T = 0$.

By assuming a distinct minor loss in each direction that is assumed valid for the entire half cycle [9], we can relate time-averaged pressure to these loss factors. During the blowing part of the cycle (flow from left to right), we call the minor loss K_B. The suction loss is denoted K_S. By writing down the one-dimensional energy equation with minor losses for each flow direction and time-averaging, we obtain

$$P = \frac{\rho u_{\max}^2}{8}(K_B - K_S - 2 - 2\eta^2),$$

(1)

where P is the time-averaged pressure difference, $u_{\max}$ is the maximum cross-stream average velocity at the narrow end, and $\eta = h/H$. Details of this derivation are in [5].

Depending on the displacement amplitude, the expanding flow may separate from the diffuser walls resulting in large losses–similar to steady flow. The dependence of this phenomena on the displacement amplitude is demonstrated using streamfunction contour maps for two values of L_0/h at the same time in the cycle, slightly after the peak in the blowing (Fig. 5). Note that the volume flow rate between any two streamlines is equal [10]. For $L_0/h = 36$, separation occurred much earlier, and the separated flow has rolled into a vortex pair. The phase-averaged flow downstream of the vortex pair appears

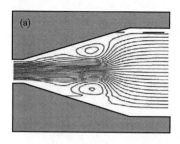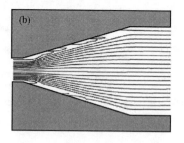

FIGURE 5. Streamline maps for $L_0/h = 36$ (a) and 10 (b) at $t/T = 0.28$.

reattached. This may be an artifact of the phase-averaging process. For $L_0/h = 10$, the flow has just begun to separate. In fact, the separation for this case does not become significant before the flow reverses, and, as shown below, this results in much smaller losses. Animations of the streamlines for these two cases are available at the USU Experimental Fluid Dynamics Laboratory website [11].

The set A pressure sensors are distributed on both sides of the diffuser and allow investigation of variations between the pressure on either side of the diffuser as the flow flaps from side to side.. Repeatable features in the pressure signals are easily detected when the pressure data are phase averaged. The phase-averaged pressures for several sensors are shown in Fig. 6a along with the fit to the volume flow rate. There are three features to note:

1. The flow is symmetric left to right
2. The pressure amplitude during the blowing portion of the cycle $(0 < t/T < 0.5)$ changes appreciably between the first and second row of sensors, but very little between the second and third indicating that the flow has separated in between these rows
3. Prior to the convergence of the second and third row pressure values, there is a sudden loss of pressure recovery (i.e. an upward jump in the pressure) experienced at sensors 2 and 3. This jump corresponds to the passing of the separation point through this location. A similar but less pronounced jump in the pressure is experienced at sensors 4 and 5 somewhat later.

The time average of the pressure P measured with both sets of pressure sensors is shown in Fig. 6b. Smith and Swift showed that oscillatory flow through an entrance such as the one at the top of the facility will generate time-averaged pressure below the ambient [12], which is also the case for the present data at $x/h = 25$. As one moves from the top of the facility downward through the diffuser, P becomes increasingly negative, and reaches a constant value in the small channel. Fugal *et al.* [4] demonstrated numerically that in entrance regions, significant cross-stream variations in pressure can exist. Entrance effects from the diffuser and from the end of the facility are detectable in the data in Fig. 6b at $x/h = 22$. While our channels are long enough to ensure that changing their length would not impact the results, one must still take care in choosing a location to assess the time-averaged pressure, and that this location is sufficiently

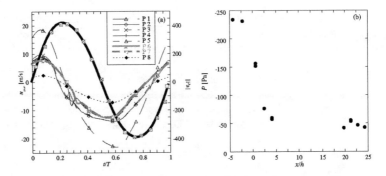

FIGURE 6. (a) Phase-Averaged Pressure for taken with the set A sensors for Re= 800 and $L_0/h = 36$. The thick black line is the fit to the cross-stream average velocity. The original 18 data points are also shown. (b) Time-Averaged Pressure for Re= 800 and $L_0/h = 22$. The start of the expansion is $x = 0$.

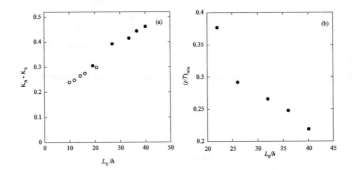

FIGURE 7. (a) The difference in the minor losses computed from the velocity waveform and the time-averaged pressure using Eq. 1. (b) The time of arrival of the separation point at the same location for increasing displacement amplitudes.

removed from both the exit of the facility and the diffuser.

The time-averaged pressure results and velocity information can be related to the loss factors in Eq. 1. The results for the full set of data are shown in Fig. 7a. Three conclusions are drawn from the data:

1. The results are Reynolds number independent despite the fact that Re= 400 likely represents laminar flow while Re= 800 is turbulent
2. The loss grows with L_0/h
3. The loss tends to zero at small displacement amplitudes and to the steady flow value for large displacement amplitudes.

The arrival of the separation as a function of L_0/h is evident in the phase-averaged pressure traces (not shown, see [6]). As the displacement amplitude becomes longer, the flow separates earlier. The separation arrival time relative to the period is plotted against L_0/h in Fig. 7b. Separation occurs earlier for longer displacement amplitudes. For $L_0/h < 22$, separation is not detectable at the location of this pressure sensor. There is a clear correlation between the separation time and the losses, with earlier times leading to larger losses.

It is interesting to compare the present results to the steady flow literature. According to [13], a diffuser of the geometry reported here will have a loss factor for flow moving from left to right in the range $0.42 < K < 0.57$ (depending on the velocity profile shape) and 0.044 for flow moving from right to left. These values are consistent with the present results for large amplitudes.

CONCLUSIONS

A series of experiments and simulations for separated oscillatory flow are presented. A numerical parameter study of oscillatory flow over a hollow cone placed in the end of a tube demonstrates that this scheme is effective for removing waste heat from the tube. Heat removal was found to increase with cone angle (up to a maximum) and displacement amplitudes, as did the acoustic power required to drive the cone. The efficiency of the cone for heat removal also increased with displacement amplitude.

Our study of oscillatory flow in a gradual expansion shows that oscillatory flow minor losses are smaller than for those steady flow. The losses grow with increasing displacement amplitude in the range $10 < L_0/h < 40$. Losses are independent of Re in the range $400 < \mathrm{Re} < 800$. Finally, the extent and duration of boundary separation increases with L_0/h.

REFERENCES

1. Swift, G. W., and Backhaus, S., *J. Acoustical Soc. Am.*, **116**, 2923–2938 (2004).
2. Smith, B. L., and Swift, G. W., *J. Acoust. Soc. Am.*, **110**, 717–723 (2001).
3. Smith, B. L., and Swift, G. W., *J. Acoust. Soc. Am.*, **113**, 2455–2463 (2003).
4. Fugal, S. R., Smith, B. L., and Spall, R. E., *Phys. Fluids*, **17** (2005).
5. Smith, B. L., Mortensen, K. V., and Wendel, S., "Oscillating Flow in Adverse Pressure Gradients," in *Proceedings of FEDSM2005 ASME Fluids Engineering Summer Conference*, 2005, paper number 2005-77458.
6. Dean, A. J., Smith, B. L., and Humes, Z., "Steady circulation generated from oscillating flow through a hollow cone at the exit of a pipe," in *Proceedings of FEDSM2005 ASME Fluids Engineering Summer Conference*, 2005, paper number 2005-77454.
7. Schlichting, H., *Boundary-Layer Theory*, McGraw Hill, 1968.
8. Swift, G. W., *Thermoacoustics: A unifying perspective for some engines and refrigerators*, Acoustical Society of America, 2002.
9. Backhaus, S., and Swift, G. W., *J. Acoust. Soc. Am.*, **107**, 3148–3166 (2000).
10. White, F. M., *Fluid Mechanics*, McGraw Hill, 1986.
11. Smith, B. L., Experimental Fluid Dynamics Laboratory website (2005), http://www.mae.usu.edu/faculty/bsmith/EFDL/EFDL.htm.
12. Smith, B. L., and Swift, G. W., *Exp. Fluids*, **34**, 467–472 (2003).
13. Fried, E., and Idelchik, I. E., *Flow Resistance: a design guide for engineers*, Hemisphere, 1989.

Visualization of Acoustic Streaming in a Looped Tube Thermoacoustic Engine with a Jet Pump

Tetsushi Biwa, M. Ishigaki*, Y. Tashiro, Y. Ueda**, T. Yazaki***

Department of Crystalline Materials Science, Nagoya University, Chikusa-ku, Nagoya, 464-8603, JAPAN
*Department of Computational Science and Engineering, Nagoya University, Chikusa-ku, Nagoya, 464-8603, JAPAN
**Institute of Industrial Science, University of Tokyo, Komaba 4-6-1, Meguro-ku, Tokyo,153-8505, Japan.
***School of Physics, Aichi University of Education, Igaya-cho, Kariya-city, JAPAN

Abstract. The steady flow induced in a looped tube thermoacoustic engine having a jet pump is visualized using a sheet-like laser light and small tracer particles. It is shown that the acoustic streaming velocity and its direction depend on the configuration of the jet pump. The performance of the looped tube cooler is also shown to be significantly affected by the acoustic streaming induced in the loop.

INTRODUCTION

Use of a looped tube as a waveguide of a thermoacoustic engine has enabled the execution of a Stirling thermodynamic cycle by a traveling wave [1]. Since the Stirling thermodynamic cycle can ideally have the efficiency of Carnot's cycle, thermoacoustic Stirling engines [2] can have the efficiency comparable to the conventional heat engines. However, it is pointed out that a nonzero mass flow called acoustic streaming carries heat away from the hot heat exchanger and generates unwanted heat losses in the thermoacoustic prime-movers and coolers[3-6]. Hence, a deeper understanding of the acoustic streaming is important to increase the efficiency of the thermoacoustic engine having a looped tube.

In this work, we built a looped tube thermoacoustic engine having an asymmetric constriction called a jet pump [3] and observed the acoustic streaming using a sheet-like plane Ar-ion laser light and tracer particles. It is found that the direction and the magnitude of the streaming velocity [7] depends on the orientation of the jet pump. On the basis of this observation, we constructed the looped tube thermoacoustic cooler to see the effect of the streaming on the performance of the cooler.

EXPERIMENTAL APPARATUS AND PROCEDURE

The loop, with the average length 2.1 m, is made of glass and stainless-steel tubes of 40 mm in inner diameter and three 90° elbows. The fourth elbow was replaced with

CP838, *Innovations in Nonlinear Acoustics: 17th International Symposium on Nonlinear Acoustics*,
edited by A. A. Atchley, V. W. Sparrow, and R. M. Keolian
© 2006 American Institute of Physics 0-7354-0330-9/06/$23.00

a tee closed by a thin glass plate, which was used as a view port through which a sheet-like plane light of an Ar-ion laser (500 mW) was introduced in the glass tube.

The jet pump is made of a narrow tube (2.4 cm in inner diameter) having conical waveguide at both ends. One end of the narrow tube is extended into the inside of the cone, but the other end is connected to the cone smoothly without an abrupt change in the cross section. The largest inner diameter of the cone is 40 mm, which is the same as that of the straight tubes. The jet pump is installed at the velocity antinode, which is shown by the region enclosed by dashed lines (see Fig. 1). Two different orientations are tested; in one orientation (A) the acoustic intensity flows out of the narrow tube from the extended end, and in the other (B) it flows into the narrow tube from that end.

A prime-mover regenerator consisting of a ceramic honeycomb with many square channels of the cross section 0.9×0.9 mm^2 was sandwiched by hot and cold heat exchangers [8]. The hot heat exchanger was heated by an electrical heater and its temperature T_H was monitored by a thermocouple. Also, the temperature T_G of the gas, on the central axis of the tube and 9-cm away from the hot heat exchanger, was monitored, in order to see the extent of the streaming qualitatively. The cold heat exchanger was kept at room temperature T_R by cooling water. The cooling water was also used to keep the temperature of the stainless-steel tube near the hot heat exchanger at T_R. A cooler-regenerator also consisting of a ceramic honeycomb was installed in the loop to make a pistonless Stirling cooler.

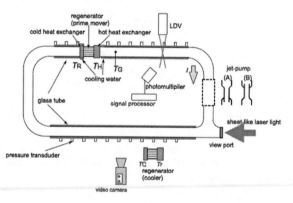

FIGURE 1. Schematic illustration of the present looped tube engine. Glass tube regions are drawn by thick gray lines. A long glass tube region was visualized ay a sheet-like light.

The present engine was filled with air at atmospheric pressure. When the heat Q_H supplied from the electrical heater exceeded 50W, the gas column in the loop began to oscillate at the fundamental mode (~146Hz), regardless of the orientation of the jet pump. While keeping $Q_H = 100$ W, the tracer particles of 20 μm in typical diameter was introduced within the glass tube, to visualize the acoustic streaming [5]. The pressure and velocity fields were also determined when $Q_H = 100$ W, using a series of small pressure transducers and a laser Doppler velocimeter, respectively.

RESULTS

Figure 2 shows T_H and T_G as a function of Q_H. In the region above Q_H = 50 W, T_H with the jet-pump (A) decreases its slope in response to the excitation of the spontaneous gas oscillation. As a result, T_H with the jet-pump (A) becomes lower than that with (B) when the same amount of Q_H is supplied to the hot heat exchanger. In contrast to T_H, the temperature T_G with the jet-pump (A) becomes higher than T_G with (B). The lower T_H and higher T_G imply that the mass flow runs clockwise in the loop in the case of (A), and that the mass flow carries the heat from the hot heat exchanger. Also it is suggested that when the jet pump (B) is used, the streaming running clockwise in the loop is suppressed and its direction is reversed.

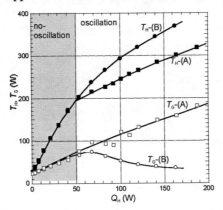

FIGURE 2. Q_H-dependence of the hot heat exchanger temperature T_H and the gas temperature T_G at the position 9-cm away from the hot heat exchanger.

In order to directly observe the acoustic streaming, we visualized and recorded the streaming when Q_H=100 W. Figure 4 shows a typical photograph taken from the digital video. The traces of the particles are seen as horizontal straight lines. The length of the line corresponds to the peak-to-peak displacement of the tracer particle. It is found that the lines moves along the tube axis with finite velocity. This represents the mass flow velocity called Lagrangian average of the velocity [7]. The value and the direction were read off from the videos when the jet-pumps (A) and (B) were used. It was found that the mass flow velocity V_0 near the centerline of the tube was about 44 mm/s in the *clockwise* direction when the jet-pump (A) was used. On the other hand, V_0 was about 32 mm/s in the *anticlockwise* direction with the jet-pump (B). This is the direct demonstration that the asymmetric waveguide can control the velocity and the direction of the mass flow. For the rough estimation of the heat ΔQ carried by the flowing mass, we assumed that the mass flow velocity V_0 is uniform over the cross section of the tube, and deduced ΔQ as

$$\Delta Q = A\rho_m C_P V_0 (T_H - T_R) \qquad (1)$$

where A is the cross-sectional area of the tube, and ρ_m, C_P represent a mean density, and isobaric specific heat of the gas, respectively. The heat ΔQ was estimated to be 16 W in the case of the jet-pump (A), and 13 W in (B). These losses are not negligible compared to the supplied heat Q_H =100W, and are much larger than the acoustic output power (~0.4 W) of the present prime-mover regenerator, which was determined from the simultaneous measurements of pressure and velocity.

To see the influence of the mass flow on the cooler, we modified the looped tube engine and constructed the cooler having an additional cooler-regenerator in the loop. We observed that the cooling temperature of the looped tube cooler with the jet-pump (A) showed a minimum as a function of the acoustic pressure. As a result, the cooling temperature with the jet pump (B) was consistently lower than that with (A). The temperature rise observed in the case of the jet pump (A) can be attributed to the presence of the mass flow running clockwise in the loop.

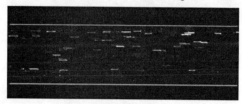

Figure 3. Visualization of the acoustic streaming. Thick horizontal lines at the top and the bottom represents the glass tube wall.

SUMMARY

Acoustic streaming in the looped tube thermoacoustic Stirling engine with an asymmetric constriction called a jet pump was visualized. It was found that the acoustic streaming velocity and its direction were dependent on the orientation of the jet pump. The performance of the looped tube cooler was significantly influenced by the acoustic streaming induced in the loop.

REFERENCES

1. Yazaki, T., Iwata, A., Maekawa, T., and Tominaga, A., *Phys. Rev. Lett.*, **81**, 3128-3132.
2. Ceperley, P. H., *J. Acoust. Soc. Am.* **66**, 1508-1513 (1979).
3. Backhaus, S. and Swift, G. W., *Nature* (London), **399**, 335-338 (1999); *J. Acoust. Soc. Am.*, **107**, 3148-3166 (2000).
4. Yazaki, T., Biwa, T. and Tominaga, A., *Appl. Phys. Lett.* **80**, 157-159 (2002).
5. Ueda, Y., Biwa, T., Mizutani, U. and Yazaki, T., *J. Acoust. Soc. Am.*, **115**, 1134-1141 (2004).
6. Job, S., Gusev, V., Lotton, P. and Bruneau, M., *J. Acoust. Soc. Am.*, **113**, 1892-1899 (2003).
7. Nyborg, W. L., "Acoustic Streaming" in *Nonlinear Acoustics*, edited by M. F. Hamilton and D. T. Blackstock, Academic, San Diego, 1998, pp. 207-231.
8. Ueda, Y., Biwa, T., Mizutani, U. and Yazaki, T., *J. Acoust. Soc. Am.*, **117**, 3369-3372 (2005).

Gas Diodes for Thermoacoustic Self-circulating Heat Exchangers

Greg Swift and Scott Backhaus

*Condensed Matter and Thermal Physics Group, Los Alamos National Laboratory,
Los Alamos, NM 87545*

Abstract. An asymmetrical constriction in a pipe functions as an imperfect gas diode for acoustic oscillations in the gas in the pipe. One or more gas diodes in a resonant loop of pipe create substantial steady flow, which can carry substantial heat between a remote heat exchanger and a thermoacoustic or Stirling engine or refrigerator; the flow is driven directly by the oscillations in the engine or refrigerator itself. This invention gives Stirling and thermoacoustic devices unprecedented flexibility, and may lead to Stirling engines of unprecedented power. We have built two of these resonant self-circulating heat exchangers, one as a fundamental test bed and the other as a demonstration of practical levels of heat transfer. Measurements of flow and heat transfer are in factor-of-two agreement with either of two simple calculation methods. One calculation method treats the oscillating and steady flows as independent and simply superimposed, except in the gas diodes. The other method accounts for the interaction between the oscillating and steady flow with the quasi-steady approximation. The mutual influence of superimposed turbulent oscillating and steady flows is a theoretical challenge.

Keywords: Thermoacoustics, acoustic streaming, heat transfer.
PACS: 43.35.Ud, 43.25.Nm

INTRODUCTION

A gas diode is an asymmetrical structure that tends to favor one direction of flow over the other. Figures 1(a) and (b) illustrate a simple and common example, a pipe constriction that is abrupt on one side and gently tapered on the other. In Fig. 1(a), the flow passes through the constriction with difficulty, because separation occurs in the abruptly diverging flow and essentially all of the kinetic energy gained in the gradually converging flow is dissipated by turbulence. In Fig. 1(b), the flow passes through the constriction more easily, because separation is avoided. Gas diodes include the vortex diodes described by Mitchell [1], the valvular conduit described by Tesla [2], and the tapered structures sometimes called jet pumps [3,4]. Gas diodes are much less perfect than electronic diodes, often with the ratio of backward and forward flow impedances less than a factor of ten. Nevertheless they can partially convert an oscillating flow into a significant and useful steady flow.

Figures 1(c)-(f) illustrate gas diodes used for external heat transfer in thermoacoustic and Stirling engines and refrigerators [5-7]. Figure 1(c) illustrates the traditional approach, in which geometrically intricate heat exchangers are used to interweave the external fluid (e.g., cooling water or hot combustion products) and the working gas, bringing them into very good thermal contact with no mass exchange. A

CP838, *Innovations in Nonlinear Acoustics: 17th International Symposium on Nonlinear Acoustics*,
edited by A. A. Atchley, V. W. Sparrow, and R. M. Keolian
© 2006 American Institute of Physics 0-7354-0330-9/06/$23.00

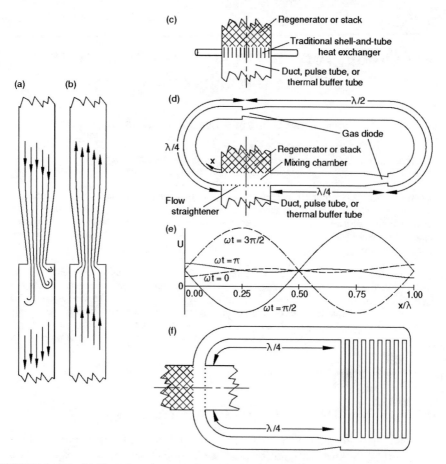

FIGURE 1. (a), (b) Flow through a gas diode. (a) High-impedance direction. (b) Low-impedance direction. (c) Part of a thermoacoustic or Stirling engine or refrigerator, with traditional heat exchanger. (d) Replacement of traditional heat exchanger with two gas diodes and external pipe one wavelength (λ) long. (e) For case (d), volumetric flow U as a function of position x at four different times t, with ω the angular frequency. (f) Alternative geometry with one gas diode, two quarter-wavelength pipes, and one heat exchanger with large surface area.

shell-and-tube heat exchanger or finned-tube heat exchanger is typical. In Fig. 1(c), the thermodynamic working gas oscillates vertically through one set of streamlined passages, while the external fluid flows horizontally through another set of passages.

The nature of such oscillating-gas engines and refrigerators imposes constraints on their traditional heat exchangers as they are scaled up to higher power. Higher power demands more heat-transfer surface area. However, passage lengths cannot be increased in the direction of oscillation of the thermodynamic working gas, because doing so increases viscous dissipation while providing no increase in the effective heat transfer area when the length exceeds the "stroke" of the oscillating gas flow. Hence, the number of passages is usually increased in proportion to the power, keeping the size of each passage constant. Such heat exchangers can have thousands of passages, causing expense and unreliability.

Thermally induced stress poses an additional challenge to reliability when such a geometrically complex heat exchanger is at an extreme temperature, i.e., a red-hot temperature for an engine or a cryogenic temperature for a refrigerator. Yet another shortcoming of the traditional heat exchangers in these engines and refrigerators is that they must be located close to one another, simply because each heat exchanger must be adjacent to one end of the nearest regenerator or stack, and these components themselves are typically short. Hence, intermediate heat-transfer loops must typically be used, adding complexity and expense.

Figures 1(d) and 1(f) show two variations of our alternative approach [8] to heat exchange for thermoacoustic and Stirling engines and refrigerators. In Fig. 1(d), the traditional heat exchanger has been replaced by a mixing chamber and one long loop of pipe. Oscillations of the gas in the loop are caused by those in the mixing chamber. Gas diodes in the loop create nonzero steady flow, so the motion of the working gas in the loop is a superposition of oscillating and steady flows. Most of the extensive outside surface area of the loop is available for thermal contact with the external fluid, which can flow either parallel or perpendicular to the pipes. The steady flow through the loop carries heat between this surface area and the mixing chamber.

The fact that the loop in Fig. 1(d) is one wavelength long leads to beneficial features, shown in Fig. 1(e). The gas diodes are at oscillating-volume-flow maxima, so they can create a large steady flow. Meanwhile, the ends of the loop are at oscillating-volume-flow minima, so the loop perturbs the oscillations in and near the mixing chamber minimally. Figure 1(e) illustrates such minimal perturbation with the pipe ends presenting a real impedance to the mixing chamber, but a slightly shorter pipe would add a positive imaginary part to that impedance, which could cancel unwanted gas compliance in and near the mixing chamber.

Figure 1(f) illustrates a half-wave option with increased heat-transfer surface area and reduced dissipation of acoustic power. The resonant enhancement of oscillating volume flow at the gas diode is still present. However, beyond the gas diode, the pipe is subdivided into many passages in parallel, with a large increase in surface and cross-sectional areas. The higher cross-sectional area causes reduced acoustic velocity and reduced oscillating pressure, decreasing viscous and thermal-hysteresis dissipation of acoustic power despite the increased surface area for heat transfer.

Relative to either Fig. 1(d) or Fig. 1(f), dissipation of acoustic power in the pipes between the mixing chamber and the gas diodes can be reduced further by proper variation of the cross-sectional area of those pipes with x, with smaller area near the gas diodes and larger area near the mixing chamber. The principles of dissipation reduction by this method are described by Hofler [9] for laminar oscillations.

Another variant of the basic idea of a resonant self-circulating heat exchanger is the use of several gas diodes in parallel where one diode is shown in Fig. 1. The use of several diodes in parallel can shorten the diode assembly while keeping the taper angle of the diodes gentle enough to avoid flow separation during the times of gradually diverging flow in the tapered parts.

In this paper, we briefly review two calculational approaches [8] and two experimental investigations [8,10] of this concept. Much more detail is in those references.

CALCULATIONAL APPROACHES

Swift and Backhaus [8] described two approaches for calculating the flows in a self-circulating loop. The "independent-flows" approach treats the oscillating and steady flows as independent and simply superimposed, except at the gas diodes where a quasi-steady approximation treats the two flows jointly. The "coupled-flows" approach applies the quasi-steady approximation to the entire loop.

In the quasi-steady approximation, steady-flow analysis is applied at each instant of time and the results are time-averaged over a full cycle of the sound wave. The quasi-steady approximation is valid if the flow at any instant of time depends only on circumstances of that instant, without memory of recent past history.

In both approaches, the quasi-steady approximation is used to couple the steady and oscillating flows at the small ends of the gas diodes. The irreversible part of the turbulent-flow pressure difference δp across any lumped-element flow impedance can be expressed using the minor-loss coefficient K via $\delta p = K \rho u^2/2$, where ρ is the gas density and u is its velocity at a reference location in the lumped element. Hence, for superimposed steady and oscillating flows, the irreversible part of the time-averaged pressure difference developed by the minor loss at the small end of a gas diode can be estimated using

$$\frac{\omega}{2\pi} \oint_0^{2\pi/\omega} \delta p(t)\, dt = \frac{\omega}{2\pi S^2} \left[\int_{t_0}^{\pi/\omega - t_0} K_+ \frac{1}{2}\rho \left(|U_1| \sin \omega t + \dot{M}/\rho \right)^2 dt \right.$$
$$\left. - \int_{\pi/\omega - t_0}^{2\pi/\omega + t_0} K_- \frac{1}{2}\rho \left(|U_1| \sin \omega t + \dot{M}/\rho \right)^2 dt \right], \qquad (1)$$

where K_+ and K_- are the minor-loss coefficients for the two directions of flow through the diode, S is the area on which the K's are based (conventionally, the smallest cross-sectional area of the component), $\dot{M}$ is the steady mass flow, $|U_1|$ is the amplitude of the oscillating volume flow, t is time, and t_0 is the time at which the flow crosses zero. Steady-flow values of the K's are tabulated in Ref. 11.

The independent-flows approach for pressure gradients elsewhere in the loop is simple. Standard acoustics expressions are used to calculate acoustic pressure gradients in the loop without consideration of the steady flow. Independently, the time-averaged pressure gradient in the loop due to the steady flow is estimated without consideration of U_1, using standard equations of fluid mechanics.

The coupled-flows approach is much more complex. It includes the Doppler shift of the acoustic phenomena by the steady flow and the influence of the oscillating flow and steady flow on each other via turbulence, which is due to the nonlinear nature of turbulent flow resistance. Where flow resistance is described by $\delta p(t) = R[U(t)]^n$ with $n \neq 1$, the extra pressure difference caused by an increment $+\varepsilon$ of mass flow above average is not canceled when an equal decrement $-\varepsilon$ below average occurs half a cycle later, so the time-averaged pressure gradient depends on both the oscillating and steady flows. Similarly, steady flow shifts the oscillating flow into a regime of higher flow impedance, increasing the acoustic pressure gradient.

Reference 8 gives closed-form results for both calculational approaches.

EXPERIMENTS

We have experimented with two self-circulating loops similar to Figs. 1(d) and 1(f).

The full-wave loop shown in Fig. 2(a) used 2.4 MPa argon near and slightly above ambient temperature and was piston driven near 50 Hz. The loop had a total length of 6.33 m and was made mostly of stainless-steel pipes with an inside diameter of 2.21 cm. Two gas diodes were located as shown in Fig. 1(d). Heat was applied to the loop with electric heater tapes wrapped around the four straight parts of the pipes between elbows and diodes. In operation, the gas circulating around the loop absorbed heat from the electric heaters, causing the temperature of the gas to rise from one end of the loop to the other, and delivered the heat to the mixing chamber. Further details are given in Ref. 8.

Figure 3(a), adapted from Ref. 8, shows temperature as a function of position around the full-wavelength loop of Fig. 2(a), with a wave amplitude of 190 kPa in the mixing chamber, for four different rates of heating. Filled symbols are measurements on the tube wall, and open symbols are measurements in the gas in the mixing chamber. The four heated regions are indicated by the four sloping straight-line segments in Fig. 4, which show the calculated gas temperature as a function of x for one rate of heating. The deviation of the erect triangles from the line segments is due to velocity-dependent heat transfer.

Figure 2(b) shows a thermoacoustic-Stirling hybrid engine [3] that was modified by replacing its traditional hot heat exchanger with a half-wave loop similar to that of Fig. 1(f). The loop consisted of two 2.2-cm diameter, 6.5-meter long (~ ¼ wavelength) pipes connected by a 10-liter tank to mimic the volume of the extended heat exchanger shown in Fig. 1(f). Approximately 2/3 of the way from the mixing chamber to the tank, each pipe turned 180° [at the right edge of Fig. 2(b)], so the tank was between the pipes. The straight sections of the pipe were heated by electric tube furnaces, and all was surrounded by the white insulation visible in Fig. 2(b). Detailed drawings of the loop and its interface to the engine can be found in Ref. 10. The engine and loop were filled with 3.1-MPa helium gas and oscillated spontaneously at 80 Hz.

In steady operation, the pressure oscillations in the engine drove the oscillations in the loop, so the gas diode created a steady flow around the loop to carry the heat from the tube furnaces to the mixing chamber near the hot end of the engine's regenerator. The oscillating flow through the regenerator drew the heated steady flow into the regenerator, so the engine could maintain the pressure oscillations.

Figure 3(b), adapted from Ref. 10, shows the temperature of the steady flow entering and leaving the mixing chamber vs the regenerator's hot-face temperature. The temperature of the steady flow leaving the mixing chamber matched the regenerator hot-face temperature to within 10°C (perfect equality is indicated by the straight line) while the heat delivered by the steady flow varied over a factor of four and the difference between the inlet and outlet temperatures ranged from 45°C to 100°C. A simple explanation of the interaction between the steady cross flow from the loop and the oscillating flow through the regenerator provides qualitative understanding of this plot. The steady flow is too small to completely flush out the mixing chamber volume in each acoustic cycle, so all of the incoming steady flow is drawn into the regenerator at least once before it leaves again. Thus, the excellent heat transfer in the regenerator assures that the flow bound for its next trip around the loop begins at the regenerator hot-face temperature.

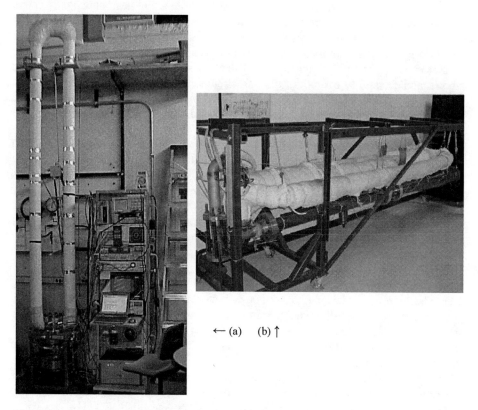

← (a)   (b) ↑

FIGURE 2. (a) Full-wavelength and (b) half-wavelength apparatus. White insulation covers both loops.

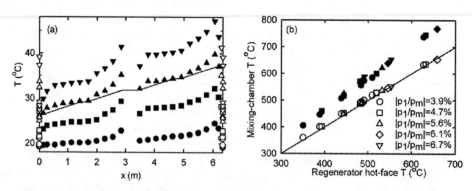

FIGURE 3. (a) Temperature T as a function of position x around the loop of Fig. 2(a), for four different heater powers. (b) Temperature of the steady flow entering (filled symbols) and leaving (open symbols) the mixing chamber of Fig. 2(b), as a function of the regenerator hot-face temperature.

COMPARING CALCULATIONS AND MEASUREMENTS

Figure 4, adapted from Ref. 8, shows some measurements with the full-wave loop of Fig. 2(a), and corresponding calculations.

In Fig. 4(a), the symbols represent measurements of the oscillating pressure at nine locations around the loop ($x = 0$ and $x = 6.33$ m being the same location—the mixing chamber). The solid curves in Fig. 4(a) represent calculations with the independent-flows approach, and the dashed curves represent calculations with the coupled-flows approach. The Doppler effect is largely responsible for the measured values of $\text{Im}[p_1]$, and the coupled-flows approach accounts for it well. Figure 4(b) shows that the two calculational approaches give very nearly the same result for $|U_1|$ at the gas diodes, so the Doppler effect in the loop has little effect on the pumping strength of the diodes.

Figure 4(c) shows the steady mass flow around the loop as a function of pressure amplitude in the mixing chamber. The open and filled symbols represent steady flows inferred from data such as those in Fig. 3(a), based on two slightly different ways of using the data to obtain the net temperature gain as the flow goes around the loop. The circles represent measurements with the loop average temperature in the range 43–47 °C and the temperature rise around the loop in the range 13–17 °C. The triangles represent measurements with higher heats and temperatures, with the loop average temperature in the range 57–60 °C and the temperature rise around the loop in the range 20–24 °C. The highest measured mass flow is 50 gm/s. The total mass of argon in the loop is only 95 gm, so the steady flow clears the entire loop every two seconds at the highest acoustic amplitude.

The solid line in Fig. 4(c) is based on the independent-flows approach, and the dashed line is based on the coupled-flows approach. The independent-flow calculation overestimates the steady flow, so it must overestimate the strength of the diodes or underestimate the steady-flow resistance of the rest of the loop. The coupled-flows approach underestimates the steady flow, so it is likely that it overestimates the effect of the oscillations on the resistance to steady flow.

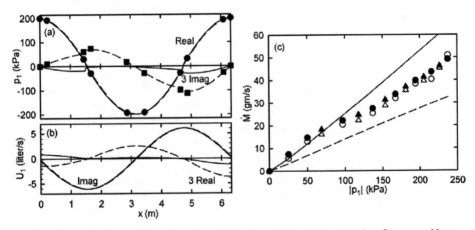

FIGURE 4. (a) Pressure and (b) volume flow in the one-wavelength loop. (c) Mass flow around loop.

In the diodes, for the conditions of Fig. 4, the maximum Reynolds number is 10^6, but the gas displacement amplitude is only about 10 times the gas-diodes' small diameter, perhaps not large enough to justify confidence in the quasi-steady approximation.

CONCLUSIONS

Gas diodes in a resonant loop of piping can cause a substantial steady flow, which in turn can carry substantial heat. They offer unprecedented flexibility in the design of Stirling and thermoacoustic engines and refrigerators.

The steady flow created by gas diodes in a loop can be measured by its effect on the acoustic pressure wave and by the heat that it carries. Two calculational approaches— "independent flows" and "coupled flows" —yield results in *rough* agreement with measurements, but neither approach is in accurate agreement with measurements of steady mass flow. Whether the disagreement is due to shortcomings in the under-standing of the gas diodes, the flow in the pipe, or both is not known.

The mutual influence of turbulent steady and oscillating flows in pipes is a significant theoretical challenge; there, the quasi-steady approximation is clearly inadequate for calculating the time-averaged pressure gradient and the heat-transfer coefficient.

ACKNOWLEDGMENTS

This work was supported by the Office of Basic Energy Sciences, Division of Materials Science, in the US Department of Energy's Office of Science, and by NASA's George C. Marshall Space Flight Center's Director's Discretionary Fund.

REFERENCES

1. M. P. Mitchell, "Pulse tube refrigerator," US Patent No. 5,966,942; continuation in part 6,109,041 (1999).
2. N. Tesla, "Valvular conduit," US Patent No. 1,329,559 (1920).
3. S. Backhaus and G. W. Swift, "A thermoacoustic-Stirling heat engine: Detailed study," *J. Acoust. Soc. Am.* **107**, 3148-3166 (2000).
4. A. Petculescu and L. A. Wilen, "Oscillatory flow in jet pumps: Nonlinear effects and minor losses," *J. Acoust. Soc. Am.* **113**, 1282-1292 (2003).
5. G. W. Swift, *Thermoacoustics: A Unifying Perspective for some Engines and Refrigerators* (Acousti-cal Society of America Publications, Sewickley, Pennsylvania, 2002).
6. G. Walker, *Stirling Engines* (Clarendon, Oxford 1960).
7. R. Radebaugh, "A review of pulse tube refrigeration." *Adv. Cryogenic Eng.* **35**, 1191-1205 (1990).
8. G. W. Swift and S. Backhaus, "A resonant, self-pumped, circulating thermoacoustic heat exchanger," *J. Acoust. Soc. Am.* **116**, 2923-2938 (2004).
9. T. J. Hofler, "Thermoacoustic refrigerator design and performance." Ph.D. Thesis, University of California, San Diego, 1986.
10. S. Backhaus, G. W. Swift, and R. S. Reid, "A self-circulating thermoacoustic heat exchanger," to be published in *Applied Physics Letters*, July 2005.
11. E. Idelchik, *Handbook of Hydraulic Resistance* (3rd edition, Begell House, New York, 1994).

Eliminating Nonlinear Acoustical Effects From Thermoacoustic Refrigeration Systems

Steven L. Garrett, Robert W. M. Smith, and Matthew E. Poese

Penn State Graduate Program in Acoustics and Applied Research Laboratory
State College, PA 16804-0030

Abstract. Nonlinear acoustical effects dissipate energy that degrades thermoacoustic refrigerator performance. The largest of these effects occur in acoustic resonators and include shock formation; turbulence and boundary layer disruption; and entry/exit (minor) losses induced by changes in resonator cross-sectional area. Effects such as these also make the creation of accurate performance models more complicated. Suppression of shock formation by intentional introduction of resonator anharmonicity has been common practice for the past two decades. Recent attempts to increase cooling power density by increasing pressure amplitudes has required reduction of turbulence and minor loss by using an new acousto-mechanical resonator topology. The hybrid resonator still stores potential energy in the compressibility of the gaseous working fluid, but stores kinetic energy in the moving (solid) mass of the motor and piston. This talk will first present nonlinear acoustical loss measurements obtained in a "conventional" double-Helmholtz resonator geometry (TRITON) that dissipated four kilowatts of acoustic power. We will then describe the performance of the new "bellows bounce" resonator configuration and "vibromechanical multiplier" used in the first successful implementation of this approach that created an ice cream freezer produced at Penn State for Ben & Jerry's.

Keywords: thermoacoustics, shock suppression, minor loss, turbulence, bellows bounce resonator, vibromechanical multiplier, ice cream.

PACS: 43.35.Ud, 43.25.Gf

INTRODUCTION

Since the invention of the first thermoacoustic refrigerator [1], researchers have attempted to increase the efficiency of the thermoacoustic heat pumping process while also increasing the cooling power density. The useful heat pumping power, $\dot{Q}_{net}$, of a thermoacoustic chiller exhibits a quadratic dependence upon the linear acoustic field variables such as the oscillatory pressure amplitude, p_1, and the average acoustic fluid velocity, $<u_1>$ [2],

$$\dot{Q}_{net} \simeq \frac{p_1}{p_m} \frac{\langle u_1 \rangle}{a} p_m a A .$$ (1)

The mean gas pressure is p_m, a is the speed of sound, and A is the cross-sectional area of the second thermodynamic medium (*i.e.,* stack or regenerator [3]) used to temporarily store the heat that is being pumped thermoacoustically from a lower temperature to a higher temperature.

CP838, *Innovations in Nonlinear Acoustics: 17th International Symposium on Nonlinear Acoustics,*
edited by A. A. Atchley, V. W. Sparrow, and R. M. Keolian
© 2006 American Institute of Physics 0-7354-0330-9/06/$23.00

To reach commercially competitive cooling power densities, thermoacoustic refrigerators have had to increase peak-to-mean pressure ratios, p_1/p_m. As this dimensionless amplitude increases, nonlinear acoustical effects become more significant. This paper will present some measurements of those effects in complete thermoacoustic refrigeration systems and will describe some remedies that have been used to overcome those limitations while allowing p_1/p_m to increase.

RESONATOR ANHARMONICITY

Variation in resonator cross-section has been used to suppress harmonic generation in thermoacoustic resonators. In the first well-instrumented thermoacoustic refrigerator, built at Los Alamos National Laboratory, Hofler showed that the linear thermoviscous losses could be reduced by cross-sectional variation [4]. Shock formation within the resonator was also suppressed by that variation, although Hofler made no mention of this in the discussion of his design.

The suppression of shock formation in resonators has been studied extensively [5,6,7] and its use in thermoacoustic refrigerators is well-established [8]. MacroSonix Corporation [9] has developed the details of the interaction of the anharmonic components of the sound field within shaped resonators to such a high level that they were able to create "custom" waveforms with high pressures peaks and broad pressure minima [6].

RESONATOR TURBULENCE AND "MINOR LOSS"

In studies of resonator loss directed at development of acoustic compressors, Ilinskii, et al. [10] found that harmonic generation can account for only 10-20% of the losses that exceed the linear thermoviscous boundary layer predictions. Although Hofler did not seem to be concerned about shock formation, he did make resonator loss measurements above $p_1/p_m = 3\%$ showing that there was resonator losses that exceeded the predicted thermoviscous losses. Hofler attributed this excess resonator loss to the onset of boundary layer turbulence observed by Merkli and Thomann [11]. The Merkli-Thomann criterion for viscous boundary-layer stability is expressed in terms of an acoustic Reynolds number, $R_{ac} = \delta_\nu <u_1>/\nu, \leq 300$, where $\delta_\nu = (2\nu/\omega)^{1/2}$ is the viscous penetration depth [2], $\nu = \mu/\rho$ is the kinematic viscosity, and $\omega = 2\pi f$ is the radian frequency of the acoustic fluid oscillation.

Although Hofler was able to achieve pressure ratios in excess of $p_1/p_m = 8\%$ in his empty resonators, his loudspeaker [12] could only produce $p_1/p_m \leq 3\%$ when the resonator contained the thermoacoustic core (i.e., stack and heat exchangers). Years later, similar behavior was observed by Poese and Garrett [13] in a standing-wave, stack-based refrigerator. As can be seen in Figure 1, the measured cooling power agrees with the linear theory, as implemented in the DELTAE software [14], also up to

408

$p_1/p_m = 3\%$. At higher amplitudes (up to $p_1/p_m \cong 6\%$), which again corresponds to $R_{ac} \geq 300$, the heat pumping power is seen to be diminished from that predicted by the "linear" theory.

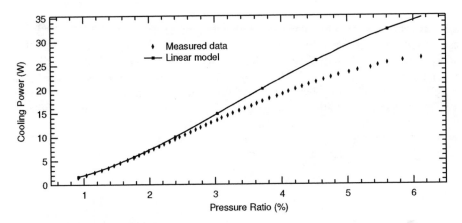

FIGURE 1. Comparison between measured heat pumping power and the predictions of the linear theory as implemented by the DELTAE software [14]. (Figure taken from Ref. 13)

In addition to turbulence, another major source of nonlinear dissipation in thermoacoustic resonators operated at high pressure ratios is entry and exit losses (minor loss) at changes in resonator cross-section. This loss is created by the asymmetry in the convergent and divergent flows at such discontinuities. The power budget for the TRITON thermoacoustic refrigerator is shown in Figure 2 at one operating point.

The acoustic power delivered by the driver (3,844 W) was calculated from the product of acoustic pressure at the driver piston and the volume velocity of the piston. The power flow through the center of the neck (2,957 W) was measured using the two-microphone technique of Fusco, *et al.* [15]. The two-microphone technique requires measurement of the pressures at the two microphone locations as well as the relative phase difference between the pressures at those two locations. The amplitude and phase measurements were made using two lock-in amplifiers. Those lock-ins were also capable of measurement of the sensor output at twice the drive (synchronization) frequency. By measuring the $2f$-component of the microphone output voltage, it was also possible to measure the oscillatory component of the Bernoulli pressure, $p_B = (1/2)\rho v^2$, and thus deduce the mean velocity of the gas oscillating in the resonator's neck.

As can be seen in Figure 2, 887 W of acoustical power were consumed by turbulence and minor loss (only a small fraction of that loss was due to linear thermoviscous losses) between the driver and the center of the resonator neck. Because the power loss due to turbulence and to minor loss are both proportional to $|U_1|^3$ [2], it was not possible to distinguish the relative contribution of those two loss mechanisms without further experimentation (*e.g.,* changing the cone angle or length of the neck) [16]. If it is assumed that an equal loss occurred between the center of the neck and the cold heat exchanger, then 46% of the power delivered by the driver was consumed by resonator loss at $p_1/p_m = 5.6\%$.

It may be worthwhile pointing out that the nonlinear losses in the TRITON resonator were exacerbated by changes to the resonator volume that were required by modest modifications to the driver and heat exchangers that were not anticipated in the original design. Earlier tests at $p_1/p_m \cong 4\%$ had shown a total power dissipation for the entire resonator of only approximately 200W; far less than the nearly 1.8 kW dissipated under the conditions represented in Fig. 2. Because the gas displacement is substantially smaller than the lengths of the transitions, it is not possible to accurately predict this nonlinear dissipation using the "minor loss" approach given by Swift [2], since the conditions underlying the Iguchi Hypothesis are not satisfied.

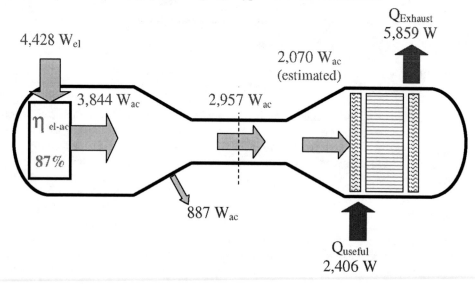

FIGURE 2. Schematic diagram of the TRITON refrigeration system operating at $T_{cold} = 49$ °F (9.3 °C) and $T_{exhaust} = 103$ °F (39.2 °C) and $p_1/p_m = 5.6\%$ showing measured acoustic and thermal power flows [16].

BELLOWS BOUNCE RESONATOR

There were several ways in which the nonlinear losses in a TRITON-like resonator could be reduced but they all would involve increases in resonator length that would make such a design even more commercially unacceptable. An early attempt by Hofler and Grant [17] to replace the hydrodynamic (gas) mass in a standing-wave, stack-based thermoacoustic refrigerator with a mechanical mass was unsuccessful due to a nonlinear oscillatory instability in the mass-stiffness combination. Fortunately, during the development of TRITON two breakthroughs showed how a new generation of thermoacoustic refrigerators could be both more efficient and more compact.

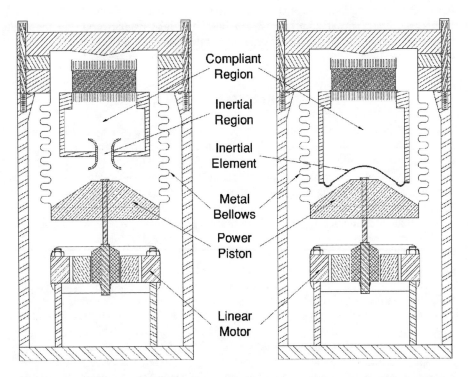

FIGURE 3. Cross-sectional drawing of two "bellows bounce" resonators of the type used in the thermoacoustic ice cream freezer developed for Ben & Jerry's [18]. The moving mass of the linear motor and power piston is resonated against the gas stiffness provided by the gas trapped within the metal bellows (and between the bellows and pressure vessel). The elastic stiffness of the bellows does not contribute to the stiffness since the bellows is designed to be one-quarter of a wavelength based on the bellows' compressional wave speed [19]. The drawing on the left shows an acoustical phasing network (Helmholtz resonator) of the Swift [20] or de Blok [21] style that employs a gas inertance. The drawing on the right shows the same system employing a loudspeaker cone as a physical mass in a hybrid "vibromechanical multiplier" [22].

The efficiency breakthrough came with the development of the thermoacoustic-Stirling engine by Backhaus and Swift [23]. That prototype demonstrated that a regenerator and acoustical phasing network could combine the high acoustical impedance of a standing wave and the traveling wave phasing required between oscillating pressure and flow velocity required for heat pumping in a regenerator [24]. The elimination of nonlinear hydrodynamic resonator losses and the opportunity to increase power density evolved from the process of optimizing flexure seals (bellows) for the TRITON refrigerator [19]. It is possible to eliminate the standing-wave resonators and instead resonate the mass of the piston attached to the bellows and the moving mass of the linear motor that displaced the piston against the stiffness of the gas (working fluid) inside the bellows. Since the kinetic energy of the oscillator was provided by solid mass, instead of moving fluid, there would be neither minor loss nor turbulence loss to dissipate acoustic power in such a "bellows bounce" resonator [25]. Figure 3 shows cross-sectional drawings of two such bellows bounce resonators. In both cases, the thermal core (*i.e.*, regenerator and heat exchangers) and the entire acoustical phasing network (*i.e.*, inertia and compliance) could be contained entirely within the bellows.

VIBROMECHANICAL MULTIPLIER

Figure 4 provides a plot of the losses in the inertance tube of the acoustical phasing network shown on the left-hand side of Figure 3. In such a phasing network, the inertance and its associated compliant region form a Helmholtz resonator. That resonator is used to increase the pressure inside the compliant region slightly above that outside the compliant region. As the pressure fluctuates in the main cavity, the magnitude of the pressure fluctuation within the compliant region is about 3-10% greater than the fluctuating pressure in the main volume. When the instantaneous pressure is above p_m in the main cavity, the pressure is 3-10% higher in the compliant region. This pressure difference across the regenerator causes a small flow of the gaseous working fluid (helium) from the compliant region to the main cavity. During the next half cycle, as the motor lengthens the main cavity, the pressure drops below p_m in the main cavity and the pressure in the compliant region becomes even lower. This pressure difference (equal in magnitude, but opposite in sign) reverses the flow of gas through the regenerator so that gas flows from the main cylinder back into the compliant region. This oscillating flow of gas is (nominally) in-phase with the global acoustic pressure fluctuation, since the Helmholtz resonator is driven well below its resonance frequency, as required to execute the acoustic Stirling cycle [24] and move heat from the side of the regenerator that faces the main cylinder to the side of the regenerator that faces into the compliant region.

In the thermoacoustic ice cream freezer we built for Ben & Jerry's [27], the motor is operated at 100 Hz and the pressure amplification provided by the Helmholtz resonator is 8%. For that case, the resonance frequency of the Helmholtz resonator needs to be about three time that of the drive frequency. As can be seen from Figure 4, the losses associated with an inertance tube exhibit a minimum of about 13 watts for a tube of that is about 45 cm in length and 4 cm in diameter. Aside from the unacceptably large dissipation within the inertance tube alone, a tube with those dimensions would not fit within the available space.

Once again, our approach was to eliminate hydrodynamic mass in favor of physical mass and replaced the inertance tube of the Helmholtz resonator with a loudspeaker cone having the required mass to create another hybrid acousto-mechanical resonator that we call the "vibromechanical multiplier"[22]. The loudspeaker cone was joined to the compliant region by an ordinary elastomeric (Santoprene™) surround. Since the

412

compliant region and the section of the bellows near the piston are held at ambient temperature, an elastomeric flexure seal was possible. The loudspeaker cone had the additional advantage of also suppressing Gedeon streaming [26].

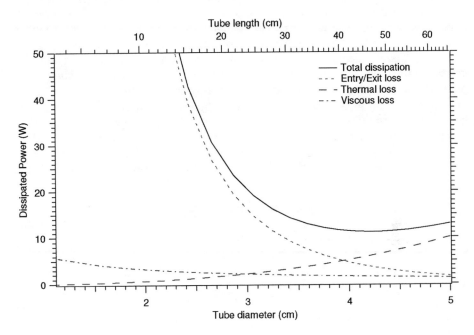

FIGURE 4. The power dissipation in an inertance tube of the type shown in the left-hand drawing of Figure 3. The dissipation is plotted as a function of tube diameter (lower axis) or tube length (upper axis) so that the inertance of the tube [2] is held fixed at $L = 500$ kg/m^4. When the inertance tube is short and narrow, the loss is dominated by minor losses associated with the high velocities and large change in cross-section. When the tube is long and wide, the losses are due to linear thermoviscous effects on the large surface area of the tube.

Although we were not able to directly measure the dissipation in the vibromechanical multiplier directly, we were able to infer that loss from the DELTAE model of the overall system performance that was in very good agreement with the performance parameters we could measure directly (*e.g.,* cooling power, exhaust power, acoustic power, and regenerator and fluid temperatures) [27]. Based on that model, the mechanical losses in the resistance of the suspension and on the thermal relaxation loss on the surfaces of the loudspeaker cone were 2.4 W and 0.9 W respectively. The thermal relaxation loss on the surface of the compliance region was 1.9 W, so the total loss was only 5.2 W. That total is only 1.6% of the total acoustic power (326 W) circulating through the thermoacoustic-Stirling refrigerator.

CONCLUSIONS

Ever since the development of the first thermoacoustic refrigerators, the increase in pressure amplitudes necessary to achieve commercially competitive efficiencies and cooling power densities has been made difficult due to losses produced by nonlinear acoustic phenomena such turbulence and minor loss. Both of those mechanisms are consequences of high Reynolds number flows. By elimination of fluid inertia and replacement by physical mass in the acoustic resonators required to produce the high pressure ratios and acoustic phasing networks required to produce Stirling cycle phasing in regenerators, those losses have been eliminated. At the pressure ratios now in common use for thermoacoustic refrigeration applications ($p_1/p_m \le 10\%$), we have been able to create flexure seals for those moving masses that appear to be capable of "infinite lifetime" operation. Hopefully, as our understanding of flexure seal fatigue improves, we will be able to use moving masses instead of fluid inertia to reach even higher pressure ratios and thus still higher cooling power densities at high efficiency.

ACKNOWLEDGMENTS

The Office of Naval Research, the Penn State Applied Research Laboratory, Ben & Jerry's Homemade, Inc., and the ThermoAcoustics Corporation have supported this research reported. Chris Davies, John Heake, and Nathan Naluai, Tony Shearer, Thomas Gabrielson, and Robert Keolian all contributed to the results reported for the TRITON Project.

REFERENCES

[1] J. C. Wheatley, G. W. Swift and A. Migliori, "Acoustical heat pumping engine," US Pat. No. 4,398,398 (1983).

[2] G. W. Swift, *Thermoacoustics: A unifying perspective for some engines and refrigerators*, (Acoustical Society of America, New York, 2002); ISBN 0-7354-0065-2.

[3] S. L. Garrett, "Resource Letter TA-1: Thermoacoustic engines and refrigerators," Am. J. Phys. **72**(1), 11-17 (2004).

[4] T. J. Hofler, "Thermoacoustic Refrigerator Design and Performance," Ph.D. thesis, Physics department, University of California, San Diego (1986).

[5] D. F. Gaitan and A. A. Atchley, "Finite amplitude standing waves in harmonic and anharmonic tubes," J. Acoust. Soc. Am. **93**(5) 2489-2495 (1993).

[6] Y. Ilinskii, B. Lipkens, T. Lucas, T. W. Van Doren and E. Zabolotskaya, "Nonlinear standing waves in an acoustical resonator," J. Acoust. Soc. Am. **104**(5), 2664-2674 (1998).

[7] M. F. Hamilton, Y. A. Ilinskii and E. A. Zabolotskaya, "Linear and nonlinear frequency shifts in acoustical resonators with varying cross sections," J. Acoust. Soc. Am. **110**(1), 109-119 (2001).

[8] S. L. Garrett, "High power thermoacoustic refrigerator," U. S. Pat. No.5,647,216 (July 15, 1997); South African Letters Patent No. 96/6512 (31 July 1996).

[9] MacroSonix Corp., 1570 East Parham Road, Richmond, VA 23228.

[10] Y. A. Ilinskii, B. Lipkens and E. A. Zabolotskaya, "Energy loss in an acoustical resonator," J. Acoust. Soc. Am. **109**(5), Pt. 1, 1859-1870 (2001).

[11] P. Merkli and H. Thomann, J. Fluid Mech. **68**, 567-576 (1975).

[12] T. Hofler, "Accurate acoustic power measurements with a high-intensity driver," J. Acoust. Soc. Am. **83**(2), 777-786 (1988).

[13] M. E. Poese and S. L. Garrett, "Performance measurements on a thermoacoustic refrigerator at high amplitudes," J. Acoust. Soc. Am. **107**(5), Pt. 1, 2480-2486 (2000).

[14] The DELTAE software is described at the Los Alamos National Laboratory thermoacoustics web site: http://www.lanl.gov/thermoacoustics/doc-options.html, where the entire software manual is available for download   The fully-functional software for IBM-PC or Mac is included with Swift's textbook (Ref. 2).

[15] A. M. Fusco, W. C. Ward and G. W. Swift, "Two-sensor power measurements in lossy ducts," J. Acoust. Soc. Am. **91**(4), Pt. 1, 2229-2235 (1992).

[16] C. L. Davies, "Loss and Heat Exchanger Evaluation for a Large Thermoacoustic Refrigerator," Masters Thesis in Acoustics, December 2002.

[17] L. A. Grant, "An Investigation of the Physical Characteristics of a Mass Element Resonator," Masters Thesis, Naval Postgraduate School, Monterey, CA (1992).

[18] K. Buscemi, "Sound Power," Appliance Design **53**(4), 42-45 (2005).

[19] R. W. M. Smith, "High Efficiency Two Kilowatt Acoustic Source for a Thermoacoustic Refrigerator," Masters Thesis in Engineering Mechanics, Penn State (December 2000).

[20] G. W. Swift, S. N. Backhaus and D. L. Gardner, "Traveling-wave device with mass flux suppression," US Pat. No. 6,032,464 (2000).

[21] C. M. De Blok and N. A. H. J. Van Rijt, "Thermo-acoustic system," US Pat. No. 6,314,740 (2001).

[22] R. W. M. Smith, M. E. Poese, S. L. Garrett and R. S. Wakeland, "Thermoacoustic device," US Pat. No. 6,725,670 (Apr. 27, 2004).

[23] S. Backhaus and G. W. Swift, "A thermoacoustic Stirling heat engine," Nature **399**, 335-338 (1999); "A thermoacoustic Stirling heat engine: Detailed study," J. Acoust. Soc. Am. **107**(6), 3148-3166 (2000).

[24] P. H. Ceperley, "A pistonless Stirling engine – The traveling wave heat engine," J. Acoust. Soc. Am. **66**(5), 1508-1513 (1979).

[25] M. E. Poese, R. W. M. Smith, R. S. Wakeland and S. L. Garrett, "Bellows bounce thermoacoustic device," US Pat. No. 6,792,764 (Sept. 21, 2004).

[26] D. Gedeon, "DC gas flows in Stirling and pulse-tube cryocoolers," in *Cryocoolers 9*, edited by G. Ross (Plenum Press, New York, 1997), pp. 385-392.

[27] M. E. Poese, R. W. M. Smith, S. L. Garrett, R. van Gerwen and P. Gosselin, "Thermoacoustic refrigeration for ice cream sales," Int. Inst. Refrigeration 6th Gustav Lorentzen Natural Working Fluids Conference, Glasgow, UK, 1 Sept 2004.

AEROACOUSTIC THERMOACOUSTICS

Jos Zeegers[*] and William V. Slaton[¶]

* Faculty of Applied Physics, Eindhoven University of Technology, 2 Den Dolech, 5612 AZ, Eindhoven, The Netherlands

¶ Department of Physics and Astronomy, The University of Central Arkansas, 201 Donaghey Avenue, Conway, Arkansas 72035-0001, United States

Abstract. Mean flow of gas in a pipe with side-branches is used to generate high amplitude aeroacoustic sound. This sound is used to generate a temperature difference over a thermoacoustic stack. Several experiments were made to study the thermoacoustic performance of different pore sizes and length of the stacks. Furthermore a study was made to dampen the aeroacoustic sound to estimate the power that is available in the aeroacoustic sound source. An outlook is given for a new research project.

INTRODUCTION

In the period 2002-2004 research was performed in the Low Temperature group of TUE on the combination of aeroacoustic sound generation and thermoacoustics. This research item came from an idea out of the oil industry where downhole equipment and instrumentation has to be powered a few miles underground in case of gas wells. A technique was needed that yielded electric power without any moving parts downhole, solely using the hydrodynamic energy of the flow. Shell proposed to convert part of the hydrodynamic power into acoustics, using natural vortex shedding at side-branches of a gas pipeline to generate acoustic resonance. By inserting a thermoacoustic stack in one of the side-branches, thermal power can be generated. The temperature difference that is established is finally converted into electricity using a thermoelectrical element. In this paper our investigations are summarized and an outlook is given for future work. A more detailed description can be found in [1, 2].

EXPERIMENTAL SET-UP.

In order to generate high intensity aeroacoustic sound, the high-pressure flow set-up of the Fluid Dynamics group of the Applied Physics Faculty of TUE has been used. A schematic description of this set-up is shown in figure 1.

The flow set-up consists of supply tanks that store 18000 liters compressed air up to 200 bars that is reduced to the pressure needed for the experiments, which is basically in the range from 1 to 15 bars. Downstream from the regulator valve the air passes a flow meter and enters a 3 m long and 30 mm diameter steel pipe. At the end of the pipe a cross-junction with two side-branches is mounted. These side-branches have an internal diameter of 25 mm. The radius of curvature of the connection from main pipe to side-branch is about 5 mm. Downstream of the cross-piece there is an extra section of about 1 m main pipe and a muffler to dampen the higher harmonics of the acoustic waves generated by the cross-piece. Downstream of the muffler a control valve is inserted by which the flow can be controlled. Apart from the flow meter there are pressure transducers installed and the temperature is measured.

CP838, *Innovations in Nonlinear Acoustics: 17th International Symposium on Nonlinear Acoustics*, edited by A. A. Atchley, V. W. Sparrow, and R. M. Keolian

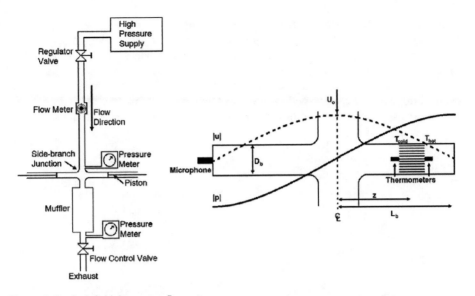

Figure 1. On the left, high pressure flow set-up
Figure 2. On the right, detailed outline of the side branch situation.

The side branches are equipped with pistons and their length L_b can be set from 15 cm up to about 90 cm each. In figure 2 the side-branches are shown in more detail. The main flow speed U_0 can be varied up to speeds of around 50 m/s at even 10 bar average pressure.

In one of the side-branches a microphone is installed to measure the pressure, and the other side-branch contains a thermoacoustic stack that is instrumented with small chip Platinum thermometers. This stack can be slid through the tube at any position, but in most tests was about 5 to 10 mm from the piston. This as here the largest temperature differences will occur. In the figure the acoustic pressure and velocity fields are indicated with a solid and dashed line respectively. This indicates the first harmonic standing acoustical wave that fits into the side-branch system.

Under steady state flow conditions, vortices are generated at the edges where the side-branches are connected to the main pipe. This has been studied extensively in the past [3]. At some speed the characteristic generation frequency of the vortices coincides with the first acoustic harmonic of the side-branches. In that case lock-in occurs and the microphone will measure a large sound pressure. Relative pressure levels p_{ac}/P_0 of 25% or at 10 bar 200 dB can be reached.

STACK MEASUREMENTS.

In figure 3 the pressure amplitude for an empty side-branch (no stack inserted in the tube) is plotted versus the Strouhal number defined as $St = f \cdot D_b/U_0$. Here f is the characteristic frequency, D_b the diameter of the side-branch and U_0 the mean speed in the main pipe. These results are obtained at a fixed side-branch length and frequency of 350 Hz by changing the flow through the system. At low speeds (at St=0.46) the resonance onset occurs. The amplitude increases when increasing the speed and decreasing the Strouhal number. At St=0.15 the resonance shuts down as lock-in cannot occur anymore due to a too large difference in frequency between the natural vortex generation and resonance frequency. If speed is increased more, a second or even third harmonic can be excited (not shown). Remark that indeed these dimensionless pressure intensities are huge and there is a dependency on pressure P_0.

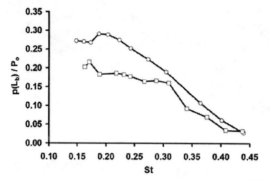

Figure 3. Dimensionless acoustic amplitude as a function of Strouhal number at 350 Hz no stack is inserted in the side-branch. Circles P_0 = 10 bar, Squares P_0 = 2 bar.

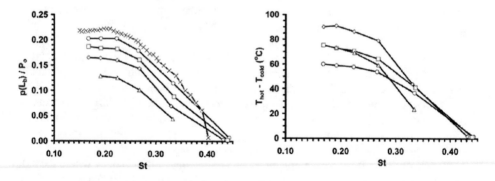

Figure 4. Left, pressure amplitude as a function of the Strouhal number for 900 cpsi stacks at 10 bars with various lengths: crosses, no stack; circles, 2 cm stack; squares, 4 cm stack; diamonds, 6 cm stack; triangles, 8 cm stack. **Right**, temperature difference versus Strouhal number, symbols same meaning as in left. All data were taken at 300 Hz.

In figure 4 left, the situation is shown when ceramic 900 cpsi (cells per square inch) stacks were inserted in the side-branch. The behavior is similar, but there is a definite trend that the acoustic pressure drops when the stacks become longer. This can be understood as at increasing length the viscous damping increases, loading the acoustic sound source, hence decreasing the pressure amplitude. When a stack with a length of 10 cm was inserted in the side-branch the resonance did not occur. In figure 4 on the right the temperature differences are shown. It is clearly shown that at these large pressure amplitudes a temperature difference of nearly 100 K can be attained. This is sufficient for an application where a thermo-electrical element has to be sourced with a temperature difference to yield electrical power output. The 6 cm stack yields the best results.

In order to investigate a possible influence of the cell density, different cell density ceramic stacks were inserted in the side-branch. This as the efficiency of the thermoacoustic pumping is strongly related to the thermal penetration depth. In figure 5 this aspect is shown in more detail for stacks with a length of 4 cm.

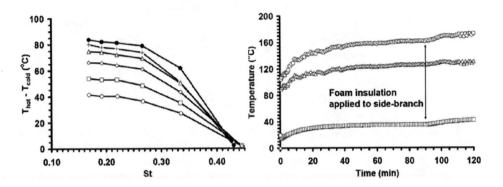

Figure 5. Left Temperature difference for six different 4 cm long stacks as a function of the Strouhal number at 10 bar and 300 Hz. Open circles, 300 cpsi; squares, 400 cpsi; diamonds, 600 cpsi; triangles, 900 cpsi; pluses, 1200 cpsi; closed circles, 1500 cpsi.
Figure 6.Right Time elapse at 10 bar and with a 6 cm length 1500 cpsi stack inserted in the side-branch. Circles: T_{hot}, Squares: T_{cold}; Triangles: ΔT.

At 300 Hz and 10 bars the thermal penetration length in air is about 50 micrometer, whereas the smallest cell spacing is still about 500 microns for the 1500 cpsi density. This makes clear that there is a steady rise of the temperature difference at increasing cell density up to 1500 cpsi.

In order to obtain the largest temperature difference with our system a 6 cm length 1500 cpsi stack was inserted in the side branch and the temperature difference recorded over time. In figure 6 the time elapse is shown. The graphs make clear that temperature differences of even 140 K can be obtained. We estimate that in our experiment the effective thermal power must be of the order of 25-50 Watts taking into account the heating that occurs of the whole side-branch of the flow set-up.

CONCLUSION: the stack measurements give us a first qualitative picture of an aeroacoustic sound source when combined with a thermoacoustic stack. Estimates are that about 25-50 Watts of thermal power can be produced using thermoacoustic heat pumping. In applying such thermal power to drive a thermoelectric element, electric outputs of 0.5 W can be expected. In commercial applications these loads are a factor 100 larger as diameters downwell are about a factor 3-4 larger than in our experimental set-up.

SOUND LOADING BY A DAMPER: SET-UP.

In order to obtain more quantitative data on the power of the aeroacoustic sound source a second set-up was developed to load the sound source by a known and controllable damping method. The same set-up has been used as shown in figure 1, however, the side-branch section is different. This part is now replaced by the one shown in figure 7.

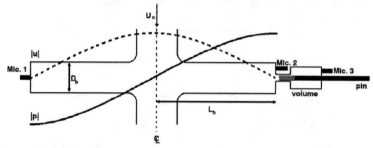

Figure 7. Junction piece with a controllable damper and volume in the right side-branch.

The side-branch on the right hand side contains a viscous resistance that can be tuned by means of a stack shaped set of plan parallel slots that can be slid through a cylindrical housing using a controllable pin. Furthermore the volume behind the resistance can be tuned. From Fusco's work [4] formulas were obtained to calculate the load via a measurement of the pressure amplitudes and phases of microphones 2 and 3. Furthermore it is necessary to calculate the normal viscous damping that occurs in the side-branch as well as estimate the radiation of higher even order acoustic modes out of the main pipe. For more details one is referred to Slaton [1, 2].

DAMPER MEASUREMENTS.

A typical result for the acoustic speed is shown in figure 8 left. Here the acoustic speed is plotted versus the Strouhal number for two pressures, 2 and 10 bars respectively. For each pressure the acoustic speed was determined for three different damping settings. These are a so called 0 mm setting representing no damping and a 25 mm and 28 mm setting, where the resistance valve was partially open with a fixed damping. The curves show a clear optimum as would be expected from the stack investigations. When the aeroacoustic sound source is loaded, the acoustic speed drops with 10-25% with respect to the no-load situation. It is not always possible to decrease the amount stepwise to 50% lower values, as then a threshold occurs and the sound source turns off. Remark here that acoustic pressure amplitude and speed are computed via: $p_{ac} = \rho c u_{ac}$. From the measurements the total source power is computed and shown in figure 8 right. This graph corresponds to the same situation as figure 8 left. The source power is made dimensionless using the main speed, acoustic speed, density and cross section area. If a typical value of 40 m/s for u_{ac} and U_0 at a Strouhal number of 0.3 is used, one finds that roughly 30 W of acoustical power is available as a sound source, being in good agreement with earlier estimates we made from the stack measurements.

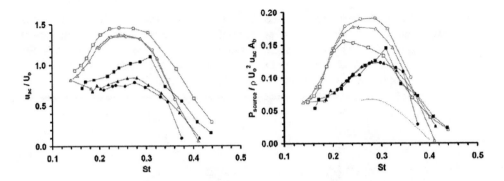

Figure 8. Left The dimensionless acoustic speed at two pressures as a function of the Reynolds number. Open symbols 10 bar, filled symbols 2 bar. Squares, 0 mm pin setting; triangles, 25 mm pin setting; circles, 28 mm pin setting. **Right** The computed acoustic source power for the data of figure 8 Left. Same symbols nomenclature.

The measurements as discussed above have been made by setting the damper at a fixed value, while changing the speed in the main flow. In this way the Strouhal number is changed. Some measurements were made by setting a fixed main speed of the flow set-up and thus fixing the Strouhal number and then varying the damping factor by step-by-step changing the resistance. The result of such a measurement is shown in figure 9. Here at a constant frequency of 350 Hz and at 10 bars pressure the load power is plotted as a function of the dimensionless acoustic speed. Measurements have been performed at three different Strouhal numbers of 0.18, 0.26 and 0.4.

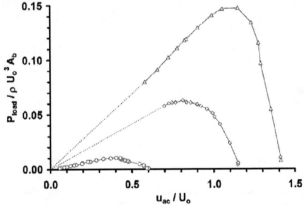

Figure 9. Acoustic (damping) load power versus the dimensionless acoustic speed for three different Strouhal numbers. Diamonds, St=0.18; triangles, St=0.26; circles, St=0.4.

At the high speed end the valve is closed, i.e. there is no connection with the volume. When slowly increasing the resistive damping from 0 the load rapidly increases at slowly decreasing acoustic velocity reaching a maximum that depends on the Strouhal

421

number. When further increasing the load resistance, however, the load power decreases again, and finally the sound source will stop. The reason that the load power finally decreases although the resistance and damping are increased has to do with the fact that at increasing load there is a moment that the increase in load has resulted in a too large drop of the available source power and acoustic speed.

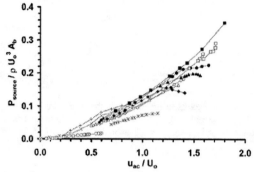

Figure 10. Compilation of all data measured in the aeroacoustic flow set-up for the acoustical source power.

Finally in figure 10 a review graph is given for the amount of source power that is available in the aeroacoustic set-up. With these data the source power can be determined and estimates be made for practical applications.

CONCLUSION: By loading the aeroacoustic source with a damper, estimates can be made to determine the load power that can be extracted. In principle for Shell it opens up this novel technology to generate from the thermoacoustically converted thermal power, electrical power downhole to source and operate continuously instrumentation. The efficiency from hydraulic flow energy to electrical output is of the order 0.05%, so every kW delivers then 0.5 W of electrical power, which is sufficient under practical circumstances.

NEW RESEARCH WORK.

In our opinion new research work should focus on the **high amplitude** oscillatory gas flow and interaction with solid boundaries. This is a new research 2-PhD project that recently was granted to our group. Questions that can be addressed are: what happens when acoustic waves with 25% or more acoustic amplitude interact with solid boundaries? What happens when acoustic turbulence occurs? How far can one-dimensional linear models be extracted to predict phenomena in thermoacoustic machines? Etc.

If for instance our stack measurements are scaled to Reynolds number and dimensionless hydraulic radius as listed in Swift's book [5], figure 11 can be made up. This figure shows that the measurements lie in the region that Swift defines as weakly turbulent. It is much of interest to investigate in more detail the phenomena that occur in and around the stacks or heat exchangers at these high amplitudes.

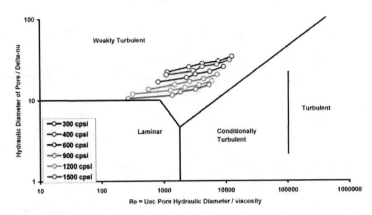

Figure 11. The dimensionless ratio of hydraulic diameter and thermal penetration depth as a function of the Reynolds number for the 4 cm stacks at maximum amplitudes.

In the new project the intentions are to study in more detail the fundamental phenomena by mathematical analysis, CFD techniques and experimentally using PIV visualization techniques. Final outcome of the work should be a better understanding where in current thermoacoustic heat pumps losses occur and what can be done to improve the efficiency. Furthermore an improved model can be developed.

REFERENCES

1. W.V. Slaton and J.C.H. Zeegers. *An aeroacoustically driven thermoacoustic heat pump.* J. of the Acoust. Soc. of Am. **117**, 3628-3635 (2005).
2. W.V. Slaton and J.C.H. Zeegers. *Acoustic power measurements of a damped aeroacoutically driven resonator.* J. of the Acoust. Soc. of Am. **118**, 83-91 (2005).
3. J.C. Bruggeman, A. Hirschberg, M.E.H. van Dongen, A.P.J. Wijnands and J. Gorter. *Self sustained aeroacoustic pulsations in gas transport systems.* J. of Sound and Vib. **150**, 371-393 (1991).
4. A.M. Fusco, W.C. Ward and G.W. Swift. *Two-sensor power measurements in lossy ducts.* J. of the Acoust. Soc. of Am. **91**, 2229-2235 (1992).
5. G.W. Swift. *Thermoacoustics* (2002) AIP Press and Acoustical Society of America. ISBN 0-7354-0065-2

Regenerator/Open Duct Interface Losses In A Traveling Wave Thermoacoustic Engine With Mean Flow

Nathan T. Weiland

Sound Thermochemical Solutions
106 Ascot Place, Pittsburgh, PA, 15237

Abstract. In an open cycle traveling wave thermoacoustic engine, a mean flow of hot gas passing through the regenerator can replace the hot heat exchanger as the source of heat input to the engine. Previous work has shown that a substantial difference must exist between the temperature of the solid material at the hot face of the regenerator and the mean temperature of the hot gas flowing into the regenerator at this interface. Heat transfer across this finite temperature difference has the potential to be a large source of irreversibility in this type of engine. This work analyzes the entropy generation due to the joining conditions between the regenerator and the open duct, and relates it to a loss in the ability to generate acoustic energy in the engine. This analysis allows us to draw some pertinent analogies between the mean flow case and the more traditional case, where a hot heat exchanger provides the heat input to the engine. In addition, this study also investigates the gain or loss of acoustic energy flux across the regenerator/open duct interface, and reveals it to be a small effect.

Keywords: Thermoacoustics, Interface Losses, Joining Conditions, Mean Flow
PACS: 43.35.Ud, 43.25.Ed, 43.28.Py

INTRODUCTION

In an open cycle traveling wave thermoacoustic engine, a slow mean flow of hot gas flows into the hot side of the regenerator, replacing the hot heat exchanger as the means for transferring heat into the engine. A previous investigation of this engine reveals that a potentially large mean temperature difference must exist between the hot gas flowing into the regenerator and the regenerator's solid material, due to isentropic temperature oscillations that occur in the open space adjacent to the regenerator and the traveling wave nature of the acoustic gas motions at this location [1]. Heat transfer from the incoming gas to the regenerator solid across this temperature difference is expected to be a significant source of irreversibility in the proposed open cycle engine.

The interface loss mechanisms examined here are similar to those between a heat exchanger and an adjoining open duct in a Stirling engine [2], pulse tube refrigerator [3], or other type of thermoacoustic device [4]. In addition, the mean flow is shown to create a source of irreversibility that can be likened to the irreversible heat transfer across a finite temperature difference between the solid and gas within a heat exchanger [5]. The entropy generation analysis of these loss mechanisms is then linked to the acoustic power loss across the interface, as understanding both the

CP838, *Innovations in Nonlinear Acoustics: 17th International Symposium on Nonlinear Acoustics*,
edited by A. A. Atchley, V. W. Sparrow, and R. M. Keolian
© 2006 American Institute of Physics 0-7354-0330-9/06/$23.00

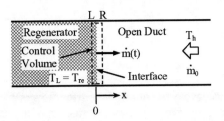

FIGURE 1. Schematic of the Regenerator/Open Duct Interface Model

thermodynamic and acoustic loss mechanisms at this interface is of critical importance in designing useful, efficient open cycle traveling wave thermoacoustic engines.

INTERFACE MODEL SUMMARY

The following investigation is an extension of the paper by Weiland and Zinn [1]. For completeness, the pertinent modeling details of this paper are reproduced below.

The regenerator/open duct interface model is depicted in Figure 1, where a mean flow of hot gas with mean temperature T_h flows into a regenerator with a mean solid temperature T_{re}. The left and right surfaces of the interface are denoted by L and R subscripts, respectively, and enclose a control volume of negligible internal mass. As a result, the mass fluxes $\dot{m}_L$ and $\dot{m}_R$ can both be expressed as:

$$\hat{m}(\hat{t}) = \frac{\dot{m}(t)}{|\dot{m}_1|} = \hat{m}_0 + \sin(\hat{t} + \phi),\qquad(1)$$

where mass fluxes have been non-dimensionalized by dividing by the magnitude of the oscillating mass flux, $|\dot{m}_1|$, the dimensionless mean mass flux is $\hat{m}_0$, $\hat{t} \equiv \omega t$ is the dimensionless time, and ϕ is chosen to give $\dot{m} = 0$ at $t = 0$, yielding $\phi = \sin^{-1}(-\hat{m}_0)$. Note that $\hat{m}_0$ is negative according to the sign convention of Figure 1, and that the mean mass flux is generally much smaller than the acoustic mass flux $(-1 << \hat{m}_0 < 0)$.

The acoustic cycle can be divided into three time periods, one for each type of mass flux at the interface. Gas flows from left to right out of the regenerator during the first time period, $0 \le \hat{t}_a \le \hat{t}_{mid}$, where $\hat{t}_{mid} = \pi - 2\phi$ and the subscript a is assigned to all variables in this time period. Likewise, the subscript b denotes all variables in the second time period, during which the gas that began the acoustic cycle within the regenerator returns to the regenerator. The time at which a particular gas parcel leaves the regenerator, $\hat{t}_a$, is linked to the time at which it returns, $\hat{t}_b$, by noting that:

$$\int_{\hat{t}_a}^{\hat{t}_b} \hat{m} d\hat{t} = \hat{m}_0(\hat{t}_b - \hat{t}_a) - \cos(\hat{t}_b + \phi) + \cos(\hat{t}_a + \phi) = 0 .\qquad(2)$$

Similarly, the "discontinuity" time, $\hat{t}_d$, marks the end of the second time period, $\hat{t}_{mid} \le \hat{t}_b \le \hat{t}_d$, when all of the gas originating in the regenerator has returned to it, i.e.:

$$\int_0^{\hat{t}_d} \hat{m}d\hat{t} = \hat{m}_0\hat{t}_d - \cos(\hat{t}_d + \phi) = 0. \tag{3}$$

The discontinuity time also marks the start of the third time period, $\hat{t}_d < \hat{t}_c \le 2\pi$, assigned the subscript c, during which fresh gas enters the regenerator for the first time at the end of the acoustic cycle due to the slow mean flow of gas into the regenerator.

The pressure across the interface is assumed to be continuous, and is expressed as:

$$\hat{p}(\hat{t}) = 1 + \hat{p}_1 \sin(\hat{t} + \phi + \theta), \tag{4}$$

where $\hat{p} \equiv p/p_0$, $\hat{p}_1 \equiv p_1/p_0$, and p_1 leads $\hat{m}_1$ by the phase angle θ.

Assuming perfect thermal contact in the regenerator, the temperature of the gas at the left side of the interface is equal to T_{re} at all times, i.e. $T_L = T_{re}$. Likewise, the gas temperature at the right side of the interface is equal to T_{re} as it flows out of the regenerator during the first time period. This gas undergoes isentropic temperature oscillations in conjunction with acoustic pressure oscillations in the open duct before returning to the regenerator during the second time period. The gas that begins the acoustic cycle just outside the regenerator also experiences isentropic temperature oscillations about its elevated mean temperature, T_h, before it enters the regenerator during the third time period. Using isentropic temperature/pressure relations yields:

$$\hat{T}_R(\hat{t}) = \begin{cases} 1 & 0 \le \hat{t}_a \le \hat{t}_{mid} \\ \left[\dfrac{1 + \hat{p}_1 \sin(\hat{t}_b + \phi + \theta)}{1 + \hat{p}_1 \sin(\hat{t}_a + \phi + \theta)} \right]^{(\gamma-1)/\gamma} & \hat{t}_{mid} < \hat{t}_b \le \hat{t}_d, \\ \hat{T}_h [1 + \hat{p}_1 \sin(\hat{t}_c + \phi + \theta)]^{(\gamma-1)/\gamma} & \hat{t}_d < \hat{t}_c \le 2\pi \end{cases} \tag{5}$$

where dimensionless temperatures are defined as $\hat{T} \equiv T/T_{re}$, γ is the ratio of specific heats, and each time $\hat{t}_a$ has a corresponding time $\hat{t}_b$ which satisfies Eq. (2).

If we restrict our attention to $-\pi/2 < \theta < \pi/2$, which yields the desired acoustic energy flux out of the regenerator, we find that $T_{R,b}$ is always smaller than T_{re}, thus heat is transferred from the regenerator's solid material to the gas as it travels back into the regenerator during the second time period. In steady state operation, this heat transfer must be balanced by heat transfer from the gas to the regenerator solid during the third time period in order to maintain a constant regenerator interface temperature, giving rise to the cited temperature difference between T_{re} and T_h. To a good approximation, this temperature difference can be expressed as:

$$\hat{T}_h \equiv \frac{T_h}{T_{re}} \approx 1 - \frac{\gamma-1}{\gamma} \frac{\hat{p}_1 \cos\theta}{2\hat{m}_0} + \frac{\hat{Q}_k}{\hat{m}_0}, \tag{6}$$

where $\hat{Q}_k \equiv Q_k / |\hat{m}_1| c_p T_{re}$ is the dimensionless conductive heat flux through the solid material of the regenerator, and c_p is the constant pressure specific heat of the gas.

ENTROPY GENERATION MODELING

The second law of thermodynamics for an open system can be expressed as [4]:

$$\frac{dS_{CV}}{dt} = \frac{1}{T_{in}}\frac{dQ_{in}}{dt} + s_L\frac{dm_L}{dt} - s_R\frac{dm_R}{dt} + \frac{dS_{gen}}{dt},$$ (7)

where s is the entropy per unit mass of the fluid, S_{gen} is the total entropy generation, and S_{CV} is the control volume's total entropy, assumed here to be negligible since the control volume is negligibly small. The rate of heat transfer into the control volume can be written as $dQ_{in}/dt = \dot{m}c_p(T_R - T_L)$, and occurs at the temperature $T_{in} = T_{re}$ [1]. Noting that $\dot{m}_L = \dot{m}_R$, integration of Eq. (7) over one acoustic cycle yields:

$$S_{gen} = \int_0^{2\pi/\omega} \dot{m}\left[s_R - s_L - \frac{c_p(T_R - T_L)}{T_{re}} \right] dt.$$ (8)

In the ideal gas approximation: $s_R - s_L = c_p \ln(T_R/T_L) - R \ln(p_R/p_L)$. Recalling that pressure is continuous across the control volume and noting that $T_L = T_{re}$ at all times, the dimensionless entropy generation per acoustic cycle, $\hat{S}_{gen} \equiv S_{gen}\omega/c_p|\dot{m}_1|$, is:

$$\hat{S}_{gen} = \int_{\hat{t}_{mid}}^{\hat{t}_d} \hat{m}(\hat{t}_b)\left[\ln(\hat{T}_R(\hat{t}_b)) - \hat{T}_R(\hat{t}_b) + 1\right]d\hat{t}_b + \int_{\hat{t}_d}^{2\pi} \hat{m}(\hat{t}_c)\left[\ln(\hat{T}_R(\hat{t}_c)) - \hat{T}_R(\hat{t}_c) + 1\right]d\hat{t}_c,$$ (9)

where the acoustic cycle has been broken into its time periods and the integral over the first time period has been neglected because $\hat{T}_R(\hat{t}_a) = 1$.

The second term on the right side of Eq. (9) describes the entropy generated in the third time period, $\hat{S}_{gen,c}$, due to heat transfer across the finite temperature difference from the hot incoming gas to the regenerator solid, while the first term on the right side of Eq. (9) is the second time period's entropy generation, $\hat{S}_{gen,b}$, due to heat transfer across the finite temperature difference from the regenerator solid to the colder gas that returns to the regenerator from the open duct. This term is essentially the same loss mechanism first described by Smith and Romm for heat exchanger/open duct interface losses in Stirling engines [2], with the addition of mean flow effects which alter the limits of integration for this term in Eq. (9).

The results of Smith and Romm can be obtained by considering Eq. (9) in the limiting case where there is no mean flow, i.e.:

$$\hat{S}_{gen,noflow} = \int_\pi^{2\pi} \hat{m}(\hat{t}_b)\left[\ln(\hat{T}_R(\hat{t}_b)) - \hat{T}_R(\hat{t}_b) + 1\right]d\hat{t}_b,$$ (10)

where the second time period now covers exactly half of an acoustic cycle, and the third time period does not exist.

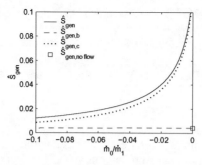

FIGURE 2. Entropy generation components versus mean mass flux for $\hat{p}_1 = 0.1$, $\gamma = 5/3$, and $\theta = 0$.

The effects of mean mass flux on $\hat{S}_{gen,b}$, $\hat{S}_{gen,c}$, and the total entropy generated during an acoustic cycle are shown in Figure 2. The model predicts that most of the entropy generation occurs during the third time period, for smaller magnitudes of mean mass flux. This is not surprising, since the interface temperature difference, $\hat{T}_h$, increases as the magnitude of $\hat{m}_0$ decreases according to Eq. (6), resulting in higher entropy generation rates. As the magnitude of $\hat{m}_0$ decreases, the entropy generated during the second time period increases slightly in Figure 2, because the length of the second time period increases while the temperature difference between the gas and the solid does not change significantly during this time period. The square on the right ordinate of Figure 2 describes the entropy generation in the absence of mean flow as expressed in Eq. (10). This result matches the prediction of Smith and Romm [2] for the values of $\hat{p}_1$, θ and γ used in Figure 2, and shows that mechanism of Smith and Romm is responsible for $\hat{S}_{gen,b}$ in the open cycle engine.

Although it may appear from Figure 2 that adding mean flow to a thermoacoustic engine only serves to increase its losses, note that the no mean flow case requires the use of a hot heat exchanger to bring heat into the engine, which has additional sources of irreversibility that are not accounted for in Figure 2. Most notably, the short times and distances over which large amounts of heat must be transferred in the hot heat exchanger often necessitate large temperature differences between the gas and the solid material of the heat exchanger, which is a significant source of irreversibility [5]. To a lesser extent, heat transfer across a temperature difference between the hot heat exchanger and the regenerator can also be a source of irreversibility, though it is often negligible in traveling wave thermoacoustic devices [6]. Since these two heat exchanger irreversibilities are associated with supplying heat to the regenerator, they can be viewed as analogous to, and are on the same order of magnitude as, the entropy generated during the third time period, $\hat{S}_{gen,c}$, which is when the heat from the mean flow is supplied to the regenerator in an open cycle engine.

Figure 3 plots the general effects of $\hat{p}_1$ and θ on the total entropy generation per acoustic cycle, $\hat{S}_{gen}$. The highest values of entropy generation occur near $\theta = 0$, where traveling wave acoustic phasing maximizes the difference in a gas parcel's regenerator exit and re-entry temperatures. This temperature difference decreases to zero for $\theta =$

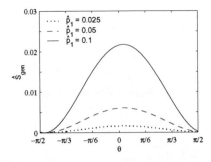

FIGURE 3. Entropy generation versus $\hat{p}_1$ and θ for $\hat{m}_0 = -0.05$ and $\gamma = 5/3$.

$\pm \pi/2$, where gas parcels return to the regenerator with the same temperature as when they left. Also note that the dependence of $\hat{S}_{gen}$ on $\hat{p}_1$ is not linear, with higher pressures resulting in much higher generation of entropy per acoustic cycle, similar to the trends observed by Smith & Romm [2]. This nonlinearity is introduced in Eq. (9) in the natural logarithm of the temperature, which a function of pressure in Eq. (5).

Although the dependence of entropy generation upon γ is not shown, higher values of γ result in heat transfer across larger temperature differences according to Eq. (6), which generates more entropy than smaller values of γ. Likewise, conductive heat transfer in the regenerator solid, $\hat{Q}_k$ in Eq. (6), also serves to increase the interface temperature difference, hence increasing the resulting entropy generation.

ACOUSTIC ENERGY DISSIPATION EFFECTS

In an open cycle thermoacoustic system, entropy generation at the regenerator/open duct interface results in a loss of acoustic energy flux across the interface, as well as a loss in the ability to convert the mean flow's thermal energy to acoustic energy in the regenerator. This "conversion" loss can be inferred from the difference between the total entropy generation and the interface's acoustic power loss, as shown below.

To begin, the volumetric velocity is written as: $U(t) = \dot{m}(t)/\rho(t)$, where ρ is the gas density. On the left side of the control volume and on the right side of the control volume during the first time period, the ideal gas law yields $\rho(t) = p(t)/RT_{re}$, as gas temperatures are equal to T_{re} at these locations and times. In the open duct, ρ evolves according to the isentropic relation: $\rho/\rho_{ref} = [p/p_{ref}]^{1/\gamma}$, where reference conditions are those at the time the gas exits the regenerator for the first and second time periods, and the mean properties of the upstream gas for the third time period. Using these expressions for ρ, the dimensionless volumetric velocity, $\hat{U} \equiv Up_0/|\dot{m}_1|c_p T_{re}$, on the left and right sides of the control volume can be expressed, respectively, as:

$$\hat{U}_L(\hat{t}) = \frac{\gamma - 1}{\gamma} \frac{\hat{m}(\hat{t})}{\hat{p}(\hat{t})} \qquad 0 \le \hat{t} \le 2\pi, \qquad (11)$$

$$\hat{U}_R(\hat{t}) = \begin{cases} \dfrac{\gamma-1}{\gamma}\dfrac{\hat{m}(\hat{t}_a)}{\hat{p}(\hat{t}_a)} & 0 \le \hat{t}_a \le \hat{t}_{mid} \\[2mm] \dfrac{\gamma-1}{\gamma}\dfrac{\hat{m}(\hat{t}_b)}{\hat{p}(\hat{t}_a)}\left[\dfrac{\hat{p}(\hat{t}_a)}{\hat{p}(\hat{t}_b)}\right]^{1/\gamma} & \hat{t}_{mid} \le \hat{t}_b \le \hat{t}_d \,, \\[2mm] \dfrac{\gamma-1}{\gamma}\hat{T}_H\dfrac{\hat{m}(\hat{t}_c)}{[\hat{p}(\hat{t}_c)]^{1/\gamma}} & \hat{t}_d \le \hat{t}_c \le 2\pi \end{cases} \tag{12}$$

where $\hat{m}(\hat{t})$ is given by Eq. (1) and $\hat{p}(\hat{t})$ is given by Eq. (4).

Using Eqs. (11) and (12), the left and right first order acoustic volumetric velocities are computed numerically by finding the fundamental Fourier component, i.e.:

$$\hat{U}_1(\hat{t}) = a_1\cos(\hat{t}+\phi) + b_1\sin(\hat{t}+\phi), \tag{13}$$

$$a_1 = \frac{1}{\pi}\int_0^{2\pi}\hat{U}(\hat{t})\cos(\hat{t}+\phi)d\hat{t}; \qquad b_1 = \frac{1}{\pi}\int_0^{2\pi}\hat{U}(\hat{t})\sin(\hat{t}+\phi)d\hat{t}. \tag{14}$$

The acoustic energy flux on either side of the control volume can be shown to be:

$$\hat{E} \equiv \frac{\dot{E}}{|\dot{m}_1|c_p T_{re}} = \frac{1}{2\pi}\int_0^{2\pi}\hat{p}_1\tilde{\hat{U}}_1 d\hat{t} = \frac{1}{2}|\hat{p}_1|[a_1\sin\theta + b_1\cos\theta]. \tag{15}$$

Using Eq. (16), the total dimensionless dissipation of acoustic energy flux over one cycle, $\delta\hat{E}$, is given by: $\delta\hat{E} = 2\pi\left(\hat{E}_L - \hat{E}_R\right)$. The entropy generation at the interface is linked to $\delta\hat{E}$ by noting that [4]: $W_{lost} = T_0 S_{gen}$, where T_0 is the temperature at which the entropy is generated, equal to T_{re} in this case. Nondimensionalizing and noting that $\hat{T}_{re} = 1$, we find that $\hat{S}_{gen,dE} = \delta\hat{E}/\hat{T}_{re} = \delta\hat{E}$, thus the entropy generation associated with the loss in ability to convert the mean flow's thermal energy into acoustic energy in the regenerator can be expressed as: $\hat{S}_{gen,conv} = \hat{S}_{gen} - \delta\hat{E}$.

Figure 4a describes the effects of the mean mass flux on $\hat{S}_{gen}$, $\hat{S}_{gen,dE}$, and $\hat{S}_{gen,conv}$. Comparison to Figure 2 suggests that $\hat{S}_{gen,conv}$ can generally be associated with the third time period, thus it is analogous to the entropy generated in transferring heat across a finite temperature difference in a hot heat exchanger, while $\hat{S}_{gen,dE}$ is primarily generated in the second time period, and is directly related to the Smith and Romm-type of irreversibility in a closed cycle engine. In fact, Figure 4a shows that $\delta\hat{E}$ across the interface is smaller than that of the no mean flow case from Eq. (10), which matches the results of Smith and Romm [2] and Swift [4]. Thus, the interface acoustic energy dissipation is roughly a third order effect as described in these studies.

In Figure 4b, the dependence of $\hat{S}_{gen}$, $\hat{S}_{gen,dE}$, and $\hat{S}_{gen,conv}$ upon the phase lag θ reveals some very interesting trends. Most notably, $\hat{S}_{gen,dE}$ is asymmetric with respect to the phase lag, where more acoustic dissipation occurs for $\theta > 0$ than for $\theta < 0$, and

430

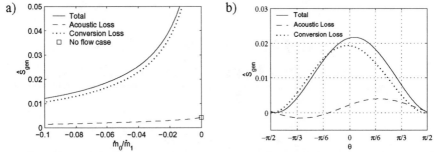

FIGURE 4. Entropy generation components versus a) $\hat{m}_0$, and b) θ, for $\hat{p}_1 = 0.1$ and $\gamma = 5/3$.

there appear to be acoustic *gains* predicted (i.e., $\delta\hat{E} < 0$) for some phase lags below θ = 0. This occurs because lower phase lags shift the acoustic pressure in time to more closely align the discontinuity in the volumetric velocity with the acoustic pressure minimum, thereby increasing the acoustic energy flux at the right side of the interface.

CONCLUSIONS

The entropy generated at the regenerator/open duct interface in the open cycle engine primarily occurs when the hot mean flow cools down to the solid temperature upon first entering the regenerator. This irreversibility is similar in size and function to heat transfer from the solid to the gas across a finite temperature difference in a hot heat exchanger. The acoustic energy loss across the interface is shown to be small, and is roughly equal to the acoustic losses seen across heat exchanger/open duct interfaces in other closed cycle thermoacoustic devices.

ACKNOWLEDGMENTS

This work was supported by Ben T. Zinn's Lewis Chair account at Georgia Tech's School of Aerospace Engineering.

REFERENCES

1. Weiland, N. T. and Zinn, B. T., "Open cycle traveling wave thermoacoustics: Mean temperature difference at the regenerator interface," *J. Acoust. Soc. Am.* **114**, 2791-2798 (2003).
2. Smith, J. L. and Romm, M., "Thermodynamic Loss at Component Interfaces in Stirling Cycles," *Proceedings of the 27th Intersociety Energy Conversion Engineering Conference*, San Diego: Society of Automotive Engineers, 1992, pp. 5.529-5.532.
3. Kittel, P., "The temperature profile within pulse tubes," *Adv. Cryogenic Eng.* **43**, 1927-1932 (1998).
4. Swift, G. W., *Thermoacoustics: A unifying perspective for some engines and refrigerators*, Melville, NY: Acoustical Society of America, 2002.
5. Swift, G. W., "Analysis and performance of a large thermoacoustic engine," *J. Acoust. Soc. Am.* **92**, 1551-1563 (1992).
6. Brewster, J. R., Raspet, R., and Bass, H. E., "Temperature discontinuities between elements of thermoacoustic devices," *J. Acoust. Soc. Am.* **102**, 3355-3360 (1997).

431

Recovery of Thermal Energy Losses in a Traveling Wave Thermoacoustic Engine with Mean Flow

Nathan T. Weiland

Sound Thermochemical Solutions
106 Ascot Place, Pittsburgh, PA, 15237

Abstract. The use of mean flow in a traveling wave thermoacoustic engine can be used to eliminate the hot heat exchanger in traditional traveling wave engines, which can lead to improved efficiency and reductions in cost, size, complexity, and thermal expansion stresses. In one form of this engine, a mean flow of hot gas provides heat input to the engine by flowing from hot to cold through the regenerator. Previous work has shown that an inherent loss mechanism in this type of engine converts much of the thermal energy input into conductive and entropy flux losses in the regenerator. The present work describes a means to recover much of this lost thermal energy by preheating the mean flow before it undergoes its primary heat addition process. This reduces the heat input required to reach a desired hot temperature, or equivalently, allows higher temperatures to be attained in the engine, thus increasing the engine's efficiency. Appropriate modeling of the preheat process is shown to yield analytic solutions for the regenerator temperature profile and the mean flow preheat temperature. The relevant parameters and design considerations influencing the effectiveness of the preheating process are also explored.

Keywords: Thermoacoustics, Mean Flow, Thermal Losses
PACS: 43.35.Ud, 43.28.Py

INTRODUCTION

In an open cycle traveling wave thermoacoustic engine, a mean flow of hot gas provides the thermal energy input to the engine by flowing into and through the regenerator, thereby replacing the engine's hot heat exchanger, as well as its irreversibilities and thermal expansion stresses [1]. Previous studies on this type of engine show that much of the thermal energy entering the regenerator from the mean flow is essentially wasted. To understand how this occurs, consider the total energy flux passing through the regenerator, $\dot{H}_{re}$, in an open cycle engine [2]:

$$\dot{H}_{re} = \dot{M}c_p\left(T_m(x) - T_0\right) - \left(\psi + Ak + A_s k_s\right)\frac{dT_m(x)}{dx}, \qquad (1)$$

where $\dot{M}$ is the mean mass flux, c_p is the constant pressure specific heat of the gas, T_m is the regenerator and gas mean temperature (assuming an approximately isothermal regenerator), T_0 is the ambient temperature, ψ describes entropy flux losses due to any

CP838, *Innovations in Nonlinear Acoustics: 17th International Symposium on Nonlinear Acoustics*,
edited by A. A. Atchley, V. W. Sparrow, and R. M. Keolian
© 2006 American Institute of Physics 0-7354-0330-9/06/$23.00

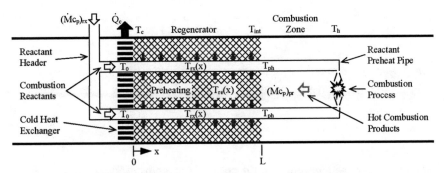

FIGURE 1. Schematic of the reactant preheating process.

non-isothermality in the regenerator, and Ak and $A_s k_s$ are the cross-sectional area and thermal conductivity of the gas and solid, respectively. In a radially insulated regenerator, $\dot{H}_{re}$ is constant along the regenerator's axial length, thus the temperature gradient increases as the mean flow passes from hot to cold through the regenerator according to Eq. (1). Consequently, its convective thermal energy is converted into thermal conduction and entropy flux losses, which then exit the engine out the cold heat exchanger. Using the strategy described below, most of these thermal losses are recovered, thus increasing the open cycle thermoacoustic engine's efficiency [3].

REACTANT PREHEATING MODEL

A schematic of the reactant preheating process used to recover thermal losses is shown in Figure 1. A mean mass flux of combustion reactants, $\dot{M}_{rx}$, enters a reactant header and is evenly distributed between n reactant preheat pipes with inside diameter D_{ph}. The reactants enter the regenerator at ambient temperature, T_0, exit at a preheated temperature, T_{ph}, and combust to yield hot combustion products at a temperature, T_h. The combustion products flow towards the regenerator with a mean mass flux $\dot{M}_{re}$, cool down to the regenerator interface temperature, T_{int}, upon entering [4], then flow through to the regenerator's cold side at temperature T_c. The mean temperature at any axial location is denoted by T_{rx} for the reactants, and by T_{re} for the combustion products flowing through the approximately isothermal regenerator.

The heat transfer rate to the combustion reactants, per unit length, is written as [5]:

$$\left(\dot{M}c_p\right)_{rx}\frac{dT_{rx}}{dx} = n\pi D_{ph}h\left(T_{re} - T_{rx}\right), \tag{2}$$

where $n\pi D_{ph}$ is the total circumference of all of the preheat pipes across which heat is transferred. The convective heat transfer coefficient, h, is an empirical coefficient correlated to experimental data though the use of the Nusselt number, $Nu = hD_{ph}/k$, where $Nu \approx 0.022Re^{4/5}\sigma^{1/2}$ is a common correlation for turbulent, fully developed gaseous flow in a pipe [6], $Re = 4\dot{M}/\pi\mu nD_{ph}$ is its Reynolds number, μ is its dynamic viscosity, and $\sigma = \mu c_p/k$ is its Prandtl number. Rearranging Eq. (2) yields:

$$\frac{dT_{rx}}{dx} = N(T_{re} - T_{rx}), \tag{3}$$

where N is assumed to be a constant given by $N \equiv n\pi k Nu/(\dot{M}c_p)_{rx}$.

The heat transfer to the reactants affects the total energy flux in the regenerator by:

$$\frac{d\dot{H}_{re}}{dx} = -n\pi D_{ph}h(T_{re} - T_{rx}). \tag{4}$$

Assuming that the regenerator solid, combustion products, and reactant preheat pipe walls are equal to T_{re} at any axial location, applying Eq. (1) and rearranging yields:

$$\frac{d^2 T_{re}}{dx^2} - \Xi\frac{dT_{re}}{dx} = MN\Xi(T_{re} - T_{rx}), \tag{5}$$

where we have defined: $\Xi \equiv (\dot{M}c_p)_{re}/(\psi + Ak + A_s k_s)$ and $M \equiv (\dot{M}c_p)_{rx}/(\dot{M}c_p)_{re}$.

If Eq. (5) is broken into two first order differential equations, noting that M, N, and Ξ are approximately independent of position and temperature within the regenerator, then Eqs. (3) and (5) can be solved as a system of coupled first order differential equations for T_{rx}, T_{re} and dT_{re}/dx [3]. Applying the boundary conditions of Figure 1, $T_{rx}(x=0) = T_0$, $T_{re}(x=0) = T_c$, and $T_{re}(x=L) = T_{int}$, yields the following solutions:

$$T_{re}(x) = c_1 + c_2 \tfrac{1}{2}(\Xi + N + R)e^{\frac{1}{2}(\Xi - N + R)x} + c_3 \tfrac{1}{2}(\Xi + N - R)e^{\frac{1}{2}(\Xi - N - R)x}, \tag{6}$$

$$T_{rx}(x) = c_1 + c_2 N e^{\frac{1}{2}(\Xi - N + R)x} + c_3 N e^{\frac{1}{2}(\Xi - N - R)x}, \tag{7}$$

$$
\begin{aligned}
c_1 &= T_0 + \frac{2R(T_{int} - T_0) + (T_c - T_0)[(\Xi + N - R)e^{\lambda_3 L} - (\Xi + N + R)e^{\lambda_2 L}]}{R(2 - e^{\lambda_2 L} - e^{\lambda_3 L}) + (\Xi + N + 2M\Xi)(e^{\lambda_3 L} - e^{\lambda_2 L})} \\
c_2 &= \frac{(T_{int} - T_0)(\Xi - N - R) + (T_c - T_0)[2N - (\Xi + N - R)e^{\lambda_3 L}]}{NR(2 - e^{\lambda_2 L} - e^{\lambda_3 L}) + N(\Xi + N + 2M\Xi)(e^{\lambda_3 L} - e^{\lambda_2 L})} \\
c_3 &= \frac{-(T_{int} - T_0)(\Xi - N + R) - (T_c - T_0)[2N - (\Xi + N + R)e^{\lambda_2 L}]}{NR(2 - e^{\lambda_2 L} - e^{\lambda_3 L}) + N(\Xi + N + 2M\Xi)(e^{\lambda_3 L} - e^{\lambda_2 L})}
\end{aligned}
\tag{8}
$$

with $R \equiv [(\Xi + N)^2 + 4MN\Xi]^{1/2}$, $\lambda_1 = 0$, $\lambda_2 = [\Xi - N + R]/2$, and $\lambda_3 = [\Xi - N - R]/2$.

MODEL RESULTS

Figure 2a plots regenerator and reactant temperature profiles for the preheating case and the non-preheating case, where the mean flow creates an exponential temperature profile [2]: $T_{nph}(x) = T_c + (T_{int} - T_c)(1 - e^{\Xi x})/(1 - e^{\Xi L})$. Note that preheating the combustion reactants has created a more linear regenerator temperature profile, which

434

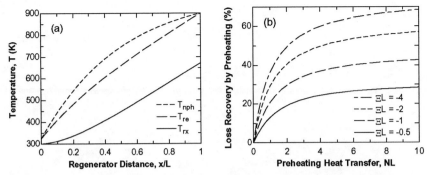

FIGURE 2. For $M = -0.9$, $T_0 = 300$ K, $T_c = 325$ K, and $T_{int} = 900$ K, (a) regenerator and reactant temperature profiles for $\Xi L = -2$ and $NL = 2$, and (b) percent of thermal losses recovered by preheating.

reduces the average gas temperature, viscosity, and hence viscous acoustic losses in the regenerator, in addition to reducing thermal conduction and entropy flux losses exiting the cold side of the regenerator. In this example, 41% of these thermal losses are recovered by preheating the reactants from 300 K to 672 K in the regenerator, effectively increasing the engine's efficiency by reducing the amount of fuel required to reach the desired combustion temperature. Figure 2b shows the percent of thermal energy loss recovery by reactant preheating versus the dimensionless groups NL and ΞL. Note that increasing the regenerator length, L, drastically improves thermal loss recovery, though this benefit must be balanced against increased acoustic viscous losses in the design of a real engine. The preheating heat transfer is most effectively increased by reducing the reactant preheat pipe diameter ($N \propto D_{ph}^{-4/5}$), and, to a lesser extent, by increasing the number of preheat pipes ($N \propto n^{1/5}$), though their effects on the preheat piping system's pressure drop and fabrication must also be considered.

ACKNOWLEDGMENTS

This work was supported by Ben T. Zinn's Lewis Chair account at Georgia Tech's School of Aerospace Engineering.

REFERENCES

1. Weiland, N. T., Zinn, B. T., and Swift, G. W., "Traveling-wave thermoacoustic engines with internal combustion," U. S. Patent No. 6,732,515 B1, (2004).
2. Weiland, N. T. and Zinn, B. T., "Open cycle traveling wave thermoacoustics: Energy fluxes and thermodynamics," *J. Acoust. Soc. Am.* **116**, 1507-1517 (2004).
3. Weiland, N. T., "Feasibility Analysis of an Open Cycle Thermoacoustic Engine with Internal Pulse Combustion," Ph.D. Thesis, Georgia Institute of Technology (2004).
4. Weiland, N. T. and Zinn, B. T., "Open cycle traveling wave thermoacoustics: Mean temperature difference at the regenerator interface," *J. Acoust. Soc. Am.* **114**, 2791-2798 (2003).
5. Incropera, F. P. and DeWitt, D. P., *Fundamentals of Heat and Mass Transfer*, 4th edition, New York: Wiley, 1996.
6. Kays, W. M., and Crawford, M. E., *Convective Heat and Mass Transfer*, 3rd edition, New York: McGraw-Hill, 1993.

SECTION 9
SOUND BEAMS, RESONATORS,
AND STREAMING

Higher Order Combination Tones Applied To Sonar Waveform Design And Underwater Digital Communications

Stephen L. Fogg

Kildare Corporation
New London, CT

Abstract. Nonlinear 'parametric' sonar is distinguished by highly predictable in-water formations of identifiable von Helmholtz spectral energies produced directly as a result of two or more preselected primaries simultaneously contained in a transmit waveform. In the nearly half-century of scientific endeavors within the field of parametric sonar, the methodical investigation into formulation techniques and practical applications using higher-order combination tones has been noticeably lagging the attention received by their more commonly recognized kin of second-order sum and difference frequencies. Generalized mathematical and graphical viewing techniques are presented for elucidating the abundance of cross-band complexities and facilitating preliminary design efforts specifically employing any of these higher-order parametric frequency components on operational systems. Recent sonar experiments implementing pulsed parametric transmit waveforms intended to fully exploit their intrinsic broadband nonlinear energy have demonstrated the potential for improved underwater target detection and classification in acoustically harsh environments. However, research efforts could benefit from more efficient and universal tools for predetermining all of the desired in-water spectral-temporal characteristics. New developments utilizing this methodology have led to unique approaches for designing stepped CW, LFM and hyperbolic FM detection waveforms incorporating enhanced signal processing qualities and constructing coding schemes for reliable underwater acoustic digital communications.

Keywords: parametric sonar, nonlinear, underwater acoustics, digital communications, combination tones, intermodulation, broadband, spread spectrum, FSK, waveform
PACS: 43.25.Lj, 43.30.Lz, 43.58.Ry, 43.60.Dh, 43.60.Wy

One major difficulty with operating active sonar in shallow-water is due to the self-induced, unwanted reverberation from surface, bottom, and volume backscatterers that can blanket the returning coherent target echo with high levels of incoherent background interference. Incidental target-like clutter from daytime fish schools, rocky outcroppings, or 'Lloyd Mirror-like' coherent bottom backscatter [1] is also a prevalent sonar display problem that can cause increased false alarm rates. In the interest of improving detection performance in this less than ideal acoustical environment, nonlinear spectrum-spreading mechanisms can be exploited. Spin-off underwater digital communication coding applications are envisioned from the technology gained from spectrum spreading echo-ranging techniques described below.

From transmission of the original excitation waveform to reception of the target echo, the finite-amplitude acoustic pulses of active sonar struggle to maintain fidelity as the signal passes through nonlinear mechanisms (amplifier, transducer, perhaps

CP838, *Innovations in Nonlinear Acoustics: 17th International Symposium on Nonlinear Acoustics*,
edited by A. A. Atchley, V. W. Sparrow, and R. M. Keolian
© 2006 American Institute of Physics 0-7354-0330-9/06/$23.00

cavitation, target, channel media, and receiver). Whether the effect is weak or strong, measurable levels of harmonic distortion will accumulate throughout the process. This unavoidable nonlinear phenomenon of spectrum spreading the energy is not necessarily a drawback for an object-sensing device designed on maximizing the utility of its spectral resources. The time-spreaded, multipath phase delays of a propagating wave in shallow water destructively interfere so often that amplitude fading is a severe reliability problem that precludes using amplitude-based acoustic sensors. The components of a propagating wave in the frequency domain, however, behave more predictably and can be expected to remain fairly spectrally stable.

Nonlinear Wave Modulation Spectroscopy has seen success using two-tone intermodulation distortion effects for sensing the existence of irregularities and discontinuities within objects. From a sonar application standpoint, intermodulation distortion can also be leveraged for a processing advantage. Beamformed arrays can insonify targets with a predetermined spread spectrum structure containing fundamental, harmonic, and intermodulation frequencies (FHIF). To generate predetermined intermodulation distortion requires a mixing device that acts on two or more chosen fundamental (primary) frequencies that spawn new spectra. The spectral identities are attributable to all possible permutations of the summations and differences of the given primary frequency and its harmonics with any other given primary and its respective harmonics. The total predicted number of these Helmholtz combination tones is limited only by the number of primaries and the number of harmonics for each primary that are present. Helmholtz was notable for demonstrating Tartini's tone was not a specific example of linear beating subjectively 'heard' by the listener but was a member of a prolific set of nonlinear sum and difference combination tones produced by two loud musical pitches. These intermodulations existed independent of any receiver and as in the case of the low difference tone, it could be objectively heard after the procreating frequencies had dissipated. It is this 'difference-frequency' that can populate the spectrum below the primary band. As with harmonic distortion, intermodulation can originate in any number of nonlinear mechanisms within a process. Higher order intermodulations though by definition must first rely on the presence of higher harmonics of the primaries.

Qualitatively, there are four general classes of FHIF systems. A single transmitted tone creating harmonics can passively intermodulate with themselves. A system that initially creates only odd harmonics can begin to produce the evens as well. Dual frequency signals are the classic 'parametrics'. Multi-tone systems contain three or more discrete primaries. Finally, any of the previous three interacting with broadband noise [2] - or voice [3] - makes for a fourth category. To unify the management and prediction of FHIF spectra, a single approach to identification nomenclature was chosen to combine harmonics with sum and difference intermodulation frequencies. A generalized system consisting of p distinct primaries of fundamental frequencies, F_i, where $i = 1, 2, \ldots p$, produces an FHIF set of identifiable spectra, T, given by:

$$T = k_1 F_1 + k_2 F_2 + \ldots k_p F_p, \quad k_1 \in 0, 1, 2, \ldots \infty \text{ and } k_{(2 \text{ to } p)} \in \pm 0, 1, 2, \ldots \infty \quad (1)$$

where k are the harmonic coefficients. Using the expression,

440

$$\sum_{i=1}^{p} |k_i| = d, \qquad d \in 1,2,3, \dots \infty \qquad (2)$$

then subsets of all possible members of T can be defined as Orders of T where all the permutations of (1) that satisfy $d = 1$ are spectra belonging to the First Order. There are just p First Order permutations. Each represents a fundamental primary frequency. It must be noted that in general there are 2^{p-1} mathematically possible frequencies for *each* permutation but whenever the condition $k_i = 0$ exists for a particular permutation, the sum and difference results on F_i are redundant. In the case of First Order, 2^{p-1} reduces to just 2^0 leaving a single unique primary frequency for each permutation.

Permutations that satisfy $d = 2$ belong to the Second Order and consist of all the second harmonics and the sum and difference frequencies between pairs of fundamental primaries. Tones created which satisfy $d = 3$ are Third Order and so on. Third Order encompasses all third harmonics and sum and differences (SADs) between a fundamental and any Second Order tone. N^{th} Order ($d = n$) contains n^{th} harmonics, SADs between a fundamental and ($n^{th} - 1$) orders, SADs between 2nd harmonics and ($n^{th} - 2$) orders, etc. In general, the total *number* of permutations for *each* order is a function of both the number of primaries, p, and the given order number, d. **Figure 1** shows subset element sizes of T as a function of p and d. Grid entries indicate the number of permutations (white cells) and the 'reduced' total number of unique tones (shaded cells) resulting from those permutations. For example, the 7th order of 4 primaries has 120 possible permutations resulting in 476 unique tones. In reality the 'reduced' number is decreased even further due to degeneracy. Interestingly, the matrix of white cells is symmetric about the diagonal dashed line but the matrix of shaded cells is not.

Even though F_i are arbitrary and time independent, eventually degeneracy occurs. This is when one or more tones from higher orders coincide or nearly coincide with those from lower orders and cause either amplification or interference. It is important to predict instances of degeneracy. This is illustrated by examining the 'difference' band for the case of two CW primaries. Depending on the actual primary frequencies, the first indication of degeneracy can happen as soon as second order tones (highest degree of degeneracy) or not appear until the 20th or even higher orders. This property is directly related to *minimum inter-order separation* (MIOS) which realistically

p \ d	1	2	3	4	5	6	7	8	9
2	2	3	4	5	6	7	8	9	10
	2	4	6	8	10	12	14	16	18
3	3	6	10	15	21	28	36	45	55
	3	9	19	33	51	73	99	129	163
4	4	10	20	35	56	84	120	165	220
	4	16	44	96	180	304	476	704	996
5	5	15	35	70	126	210	330	495	715
	5	25	85	225	501	985	1765	2945	4645
6	6	21	56	126	252	462	792	1287	2002
	6	36	146	456	1182	2668	5418	10128	17718
7	7	28	84	210	462	924	1716	3003	5005
	7	49	231	833	2471	6321	14407	29953	57799
8	8	36	120	330	792	1716	3432	6435	11440
	8	64	344	1408	4712	13504	34232	78592	166344
9	9	45	165	495	1287	3003	6435	12870	24310
	9	81	489	2241	8361	26677	74313	187137	432073
10	10	55	220	715	2002	5005	11440	24310	48620
	10	100	670	3400	14002	48940	149830	411280	1030490

FIGURE 1. Order matrix of T for FHIF systems. Number of permutations (white) and distinct tones (shaded) are given for each order, d, as a function of the number of primaries, p.

defines the largest predicted bandwidth that is free of any member of *T*. **Figure 2** depicts two different sets of CW primaries with both upper primary (P_U) frequencies normalized to unity. A trace of the calculated combination tone frequencies in the difference-frequency band vs. the order number, *d*, shows characteristic patterns signifying the degree of degeneracy. In (a) the pattern (for $P_L = 0.625$) repeats after the 6th order (d6) and in (b) it repeats after d27 (for $P_L = 0.719$) for a MIOS of 0.125 and 0.029 respectively. Letting one F_i be a time dependent LFM allows for constructing a simple graphical viewing technique to resolve which pairs of CW primaries produce a desired MIOS. **Figure 3** plots all predicted spectrolines (frequency vs. time curves) belonging to *T* up to d9 for two primaries over a unit pulselength time, *t*. P_U is a CW normalized to 1000 Hz and P_L (the lower primary) sweeps through the entire difference band for the entire pulselength. Points of intersecting spectrolines indicate instantaneous degeneracy. If a MIOS of 125 Hz were required, it is seen that this occurs for the four pairs of CW primaries exactly at *t* = .125, .375, .625 and .875. Realistically, a tolerable band of neighboring values of *t* representing near-perfect degeneracy would suffice. In **Fig.** 4 the predicted MIOS for two CW primaries at *t* = .3749 will eventually start to drift with ever increasing values of order number, *d*. Although as *d* increases, very high order intermodulations have decreasing levels of spectral energy and ultimately just become part of the background noise.

Although the goal of more efficient use of frequency band vs. time (signal space) is currently addressed with ongoing improvements in active sonar and underwater digital communications (UDC), it typically involves operating frequencies governed by a transducer's transmitting voltage response (TVR). This conventional approach to employing projector hardware puts physical limitations to the in-water frequencies available. The discussion of FHIF systems above suggests ways to overcome this barrier and enable operations in bands far above and below the peak of the TVR curve. Such approaches reserve the transducer response band for the primaries and manage the spectral distributions of harmonics and intermodulations by modeling a specific set of tones in *T* up to any desired order, *d*. Each tone has its own particular beam pattern radiation characteristics that vary in relative strength, beam width and side-lobes depending on its frequency, order number and the distortion mechanism.

Apart from the issue of degeneracy, it is critical in predicting the higher orders beyond the usual d2 sum and difference when dealing with time-varying frequency content. Instead of instantaneous degeneracy, the issue in this case is that over the duration of a pulse there can exist bands of higher orders that spectrally overlap bands

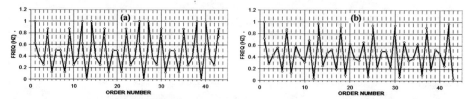

FIGURE 2. Degeneracy patterns for two primary pairs illustrate how tones coincide as a function of *d*. Pattern in (a) is more degenerate than (b) as tones effectively repeat after d6 and d27 respectively.

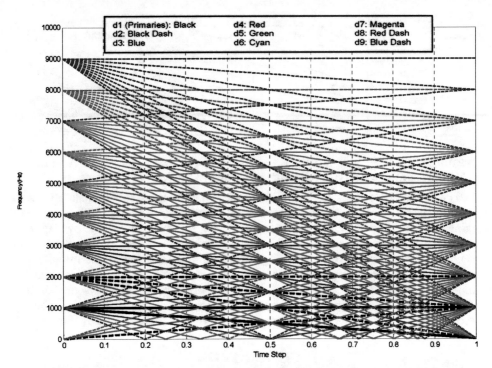

FIGURE 3. Normalized spectrolines plot for all possible primary pairs up to ninth order where intersections indicate degeneracy. Vertical gap size between points of degeneracy represents predicted bandwidth (MIOS$_N$) absent of any higher order tones.

from lower orders producing a higher buildup of reverberation that could reduce the Signal-to-Background ratio.

In addition to the avoidance of cross-band overlapping, **Fig. 3** provides a capability for predicting suitable primary combinations for obtaining certain MIOS values. This is of practical importance when developing methods to implement banks of band pass filters in FSK (frequency shift keying) decision algorithms for a UDC application. The intrinsic wide-band coverage of *T* allows for more than one set of decision algorithms to operate in parallel for the same set of primaries. This redundancy mitigates against the total loss of data frames from intermittent channel-induced phase

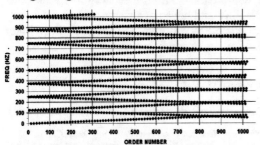

FIGURE 4. Predicted divergent drift of tones with increasing values of order number, *d*, for two primaries with near-perfect degeneracy (*t* = 0.3749).

cancellations or noise. Narrow band jamming or unauthorized monitoring would also become a lesser threat against FHIF systems. Narrow band transducers could virtually become wide-band and implement FSK protocols. A narrow band source might improve its potential for reliable telemetry if data packets became less susceptible to the effects of irrecoverable propagation fading or blockages.

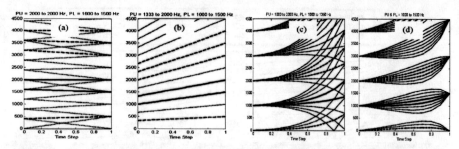

FIGURE 5. Spectroline plots: (a) CW-LFM primaries: zero-gap, zero-overlap up to d8; (b) constant LFM degeneracy; (c) and (d) hyperbolic chirps. (solid=d1, dashed=d2, dotted= higher orders).

The linear FM sweeps in **Fig. 5(a)** manage higher orders to eliminate all gaps between bands. Overlaps do not appear until eighth order. In **Fig. 5(b)** two LFM primaries are chosen to develop alternate ways for managing band overlap by maintaining perfect degeneracy of all orders throughout the sweep of the pulse. The echo-ranging hyperbolic chirps in **Fig. 5(c) and 5(d)** develop target insonifying signals with not only sweeping impulses but also sweeping pulsations by using higher orders that intentionally avoid perfect degeneracy for a more efficient signal space. The pulsation rate, adjusted by modifying the primary waveform hyperbolic eccentricity, could induce target responses to amplify excitations of structural modes.

CONCLUSION

Designers of 2-way echo-ranging and 1-way data telemetry systems can exploit the predictable spectral identities from nonlinear harmonic and intermodulation distortion spectrum-spreading effects that are produced when two or more preselected primaries are transmitted. Conventional projectors deliver limited spectral resources governed by the transducer TVR curve. The intentional generation of intermodulations through management of the deterministic signal space characteristics affords greater creativity in utilizing previously unobtainable bandwidth as described by the FHIF set T. The resulting redistribution of spectral quantities – especially the difference-frequencies below the primary band – is desirable in the interest of enhancing performance in the poorest of operating environments. The increased bandwidth diversity allows for addressing performance and security related issues such as band fading and band privacy. Without prohibitive sacrifices, non-broadband transducers can be transformed for increased gains in data transfer reliability and advantageous control of frequency dependent factors. The performance of some FHIF system designs in shallow water channels will depend on obtaining a practical understanding of degeneracy effects.

REFERENCES

1. B. Cole et al, (2004). Coherent bottom reverberation: Modeling and comparisons with at-sea measurements. *J. Acoust. Soc. Am.* **116**, pp. 1985-1994.
2. Hamilton, M. and Blackstock, D., eds. (1998). *Nonlinear Acoustics* (Academic Press, San Diego, CA) p. 393.
3. E. C. Gannon, W. L. Konrad, L. F. Carlton, and J. L. Nelson, "Parametric Voice and Standard AN/UQC Comparison Tests in Long Island Sound; Spring 1972," NUSC Technical Memorandum No. TD1X-13-72, Naval Underwater Systems Center, New London, CT, 11 September 1972.

Theoretical and experimental study of the topological charge of linear and nonlinear acoustical vortices

R. Marchiano* and J.-L. Thomas†

*Laboratoire de Modélisation en Mécanique (UMR CNRS 7607), Université Pierre et Marie Curie, 4, place Jussieu, 75252 Paris Cedex 05, France
†Institut des NanoSciences de Paris (UMR CNRS 7588), Université Pierre et Marie Curie, 140 rue de Lourmel, 75015 Paris, France

Abstract. An acoustical or optical vortex is a beam whose phase is winding along a line of phase singularity. The intensity associated to such a beam, also called screw dislocation, has a doughnught shape. The number of 2π jumps of the phase around the singularity line is called the topological charge. Using the concept of pseudo-momentum, a general relation linking the pseudo-energy, the angular pseudo-momentum and the topological charge of an acoustical vortex is derived. In an inviscid and isotropic medium, this relation can be interpreted as a conservation law of the topological charge for linear vortices, while it implies a linear increase of the topological charge with the order of harmonics for nonlinear vortices. These behaviors of the topological charge have been experimentally confirmed for linear and nonlinear acoustical vortices. Experimental results obtained by a new and versatile experimental set-up will be presented.

Keywords: Phase singularity, screw dislocation, acoustical vortex
PACS: 43.25.+y, 42.25.-p

INTRODUCTION

An acoustical vortex, also called acoustical screw dislocation, is a beam whose plane of equiphase has an helical shape and is winding around a line along which it is not determined. At a specified range along that line, the number of jumps of 2π achieved by the phase on a close contour is called the topological charge denoted l and the sense of the winding gives the sign of the topological charge: positive for anticlockwise and negative for clockwise [1] (for an illustration, see Fig.2 b) for which $l = -1$). On that line, the amplitude of the wave field is null and it forms a dark core (Fig.2 a)) because of the destructive interferences: the field onto two points symetric in relation to the center of the beam are phase shifted by π. This kind of phase singularity has been widely studied in optics where they have been called optical vortices because of many analogies between these structures and hydrodynamical vortices. Not only are the studies, in optics, interesting from a theoretical point of view but there also exist many applications such as optical spanners, atom guiding or information transmission (see [2] and [3] for reviews). However, there exist only few studies in acoustics. Hefner and Marston [4] proposed to use acoustical vortices for underwater alignement. Gspan et al. [5] showed that these structures could be produced by opto-acoustic generation. In this paper, we summarize results obtained in acoustics for linear and nonlinear vortices [6] and [7]. First of all, we recall the theoretical law about the behavior of the topological

CP838, Innovations in Nonlinear Acoustics: 17th International Symposium on Nonlinear Acoustics, edited by A. A. Atchley, V. W. Sparrow, and R. M. Keolian
© 2006 American Institute of Physics 0-7354-0330-9/06/$23.00

charge for linear or nonlinear acoustical vortices. Then, the experimental procedure, which has been used to validate the theoretical study, is presented.

THEORY

For sake of simplicity, we restrict the study to Gauss Laguerre (GL) beams [7]. Indeed, for these beams, an analytical expression exists and they are easily made in acoustics or optics. It has been established in optics, that these beams carry an angular momentum [8]. Nevertheless, this result is valid only for electromagnetic waves in vaccuum. But in acoustics, we have to deal with the medium of propagation. This difference is important and leads to distinguish the total momentum and the pseudo-momentum of the wave. Introducing the concept of pseudo-momentum, Thomas *et al.* [6] derived a general relation linking the pseudo angular momentum (M_z), the pseudo energy ($\langle E \rangle$), the angular frequency (ω) and the topological charge:

$$M_z = \frac{l}{\omega} \langle E \rangle .$$ (1)

Unlike the total momentum, the conservation laws for the pseudo-momentum and the pseudo-energy are related to the symmetry of the medium of propagation and not to the symmetry of the space. Consequently, if the medium is isotropic and inviscid, the pseudo angular momentum and the pseudo energy are both conserved and the ratio $\frac{l}{\omega}$ has to be constant. This means the topological charge is constant during the propagation of the beam. This result is an extension of the work of Allen *et al.* [8]. It is valid both in acoustics and in optics for electromagnetic waves circulary polarized in dielectric media. Moreover, this result can be extended to the weakly nonlinear regime. It is well known, that if the amplitude of a wave is high enough harmonics are naturally generated. In this case formula 1 becomes:

$$\sum_i M_{iz} = \sum_i \frac{l_i}{\omega_i} \langle E_i \rangle ,$$ (2)

where the integer i indexes the number of harmonics. That last equation shows that the pseudo angular momentum and the pseudo-energy are both conserved only if the ratio between the topological charge and the angular frequency is constant. If the medium is isotropic and inviscid, these two quantities are supposed to be conserved, consequently, the ratio between the topological charge and the angular frequency have to be constant. Considering the phenomenon of harmonics generation, the topological charge has to increase proportionnaly to the frequency. This point has been experimentally checked [6], the results are presented in the next section.

EXPERIMENTS

The experimental set-up is depicted in Figure 1. The acoustical vortices are produced in a water tank by an hexagonal array of 61 circular piezoelectric transducers (diameter: 11mm). It is noticeable that water is an isotopic and inviscid medium of propagation. The central frequency of the transducers is 1MHz (the wavelength is $\lambda =1.5$mm). Electronic

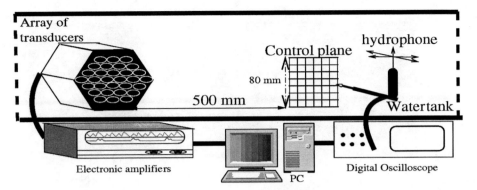

FIGURE 1. Experimental setup

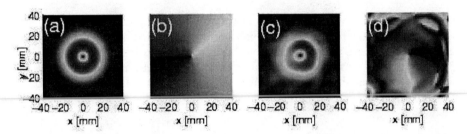

FIGURE 2. RMS amplitude and instantaneous phase of a theoretical ((a) and (b)) and an experimental ((c) and (d)) acoustical vortex of charge -1

amplifiers allow us to control the amplitude, the phase and the shape of each signal emitted by each transducer. The pressure field is recorded by a hydrophone moved in the three directions of space by a step by step motor device. The acoustical vortices are synthesized using the inverse filter technique. A control plane is defined 500mm away from the array of transducers. It is sampled with 2916 points regularly set on a square grid of 80×80mm with a spatial step of 1.5mm. The propagation operator is recorded between these points and the array of transducers. The inverse filter technique allows to calculate the signals to emit by the transducers to synthesize chosen patterns in the control plane. To synthesize a -1 single vortex, the field depicted in figures 2 a) and b) has been prescribed in the control plane. Then, the pressure field has been measured in the control plane for initial low amplitude waves.

Figure 2 c) shows the RMS amplitude. The "doughnut" shape, well known in optics, is recovered: the center of the beam exhibits a zero amplitude surrounded by a ring of constant amplitude. The size of the center at half-depth is about 4.5λ. For the phase (2 d)), the line of discontinuity is clearly visible (transition between white and black): the jump of 2π being particularly sharp. The helical structure is well recovered and points out the singularity at the extremity of the discontinuity line.

The present experimental setup permits to produce high amplitude waves. Thus, a

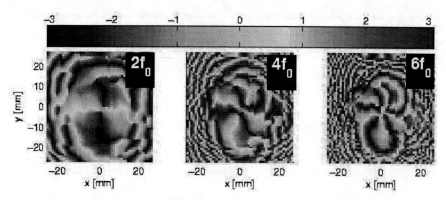

FIGURE 3. Phases of the second, fourth and sixth harmonic of a −1 nonlinear acoustical vortex

nonlinear acoustical vortex can be generated by sending the same signals than for the −1 vortex presented before, but with high amplitude (10 times higher than previously). The measured phases on the second, fourth and sixth harmonics are presented in Figure 3. These three pictures show that the number of 2π jumps, the toplogical charge, increase linearly with the number of the harmonics. This observation is in perfect agreement with the theoretical law derived for the nonlinear regime since the ratio l_i/ω_i is constant as predicted.

CONCLUSIONS AND OUTLOOKS

The present results are an important generalisation of previous works. But they also outline some original properties of beams with screw dislocation, especially in nonlinear acoustics. From this study, it is now envisageable to use the properties of the topological charge of nonlinear vortices to explore properties of different media. Another application of the nonlinear vortices could be the manipulation of small objects using the radiation pressure at the centre of the vortices. A last possibility is to create acoustical dark solitons, or solitary waves, using balance effect between diffraction and nonlinearity on the spatial distribution of the nonlinear vortices.

REFERENCES

1. M. Mansuripur, and E. M. Wright, *Optics and Photonics News*, **1**, 40–43 (1999).
2. M. S. Soskin, and M. V. Vasnetsov, *Progress in Optics*, **42**, 221–276 (2001).
3. A. S. Desyatnikov, L. Torner, and Y. S. Kivshar, *Progress in optics*, **47** (2005).
4. B. T. Hefner, and P. L. Marston, *J. Acoust. Soc. Am.*, **106**, 3313–3316 (1999).
5. S. Gspan, S. Bernet, and M. Ritsch-Marte, *J. Acoust. Soc. Am.*, **115**, 1142–1146 (2004).
6. J.-L. Thomas, and R. Marchiano, *Phys. Rev. Lett.*, **91**, 244302–(1–4) (2003).
7. R. Marchiano, and J.-L. Thomas, *Phys. Rev. E*, **in press** (2005).
8. L. Allen, M. W. Beijersbergen, R. J. C. Spreeuw, and J. P. Woerdman, *Phys. Rev. A*, **45**, 8185–8189 (1992).

Numerical Study of Nonlinear Ultrasonic Resonators

Christian Vanhille and Cleofé Campos-Pozuelo

Universidad Rey Juan Carlos, ESCET, Tulipán, s/n., 28933 Móstoles, Madrid, Spain
Instituto de Acústica, CSIC, Serrano 144, 28006 Madrid, Spain

Abstract. In the framework of high-power ultrasonic applications, this communication deals with the numerical simulation of nonlinear resonators. The objective is the understanding of the behavior of nonlinear acoustic waves in resonant cavities. For this purpose several models are developed in the time domain. They are all based on conservation laws written in Lagrangian coordinates and isentropic state equation. A system of differential equations is deduced and numerically solved in the one, two and three dimensional (axisymmetric configuration) cases. These numerical tools predict the behavior of strongly nonlinear acoustic waves in realistic cavities. Some results are presented and show how the models allow us to simulate in the time domain the behavior of complicated nonlinear modes. The quasi-standing character of the waves is especially pointed out.

Keywords: Nonlinear acoustics, Resonators, Quasi-standing waves, Numerical simulation.
PACS: 43.25.Gf

MODELS

In the framework of nonlinear acoustics [1], the equations governing the behavior of the nonlinear standing acoustic waves in a resonator are deduced from the continuum mechanics laws. They are presented in the time domain and in the most general case: the three-dimensional configuration. Afterward the problem is solved by means of three different numerical models, each one corresponding, respectively, to the one, two, and axisymmetric (three-dimensional) problems.

All the variables are considered in Lagrangian coordinates. The rigid-walled resonator (domain $\Omega \subset \mathbb{R}^3$) is full-filled by an irrotational and thermoviscous Newtonian fluid. The fluid is assumed to be initially at rest and excited by means of a time-dependant mechanical piston placed anywhere at the boundary of the cavity. The evolution of the wave is simulated in the time domain $T \subset \mathbb{R}^+$, i.e., the displacement vector $u(x,y,z,t)$ is obtained, from which the pressure field $p(x,y,z,t)$ is deduced.

The conservation laws (mass and momentum) [2], together with the assumption of Newtonian and irrotational fluid ($\nabla \times u = 0$), and state equation (by considering an ideal gas and isentropic process) lead to the following set of equations, written in Lagrangian coordinates

$$\frac{\rho_0 - \rho}{\rho} = \nabla \cdot u \tag{1}$$

CP838, *Innovations in Nonlinear Acoustics: 17th International Symposium on Nonlinear Acoustics*,
edited by A. A. Atchley, V. W. Sparrow, and R. M. Keolian
© 2006 American Institute of Physics 0-7354-0330-9/06/$23.00

$$\rho_0 \frac{\partial^2 \boldsymbol{u}}{\partial t^2} = -\nabla p + \mu \overline{\nabla}^2 \frac{\partial \boldsymbol{u}}{\partial t} + \left(\mu_B + \frac{1}{3}\mu \right) \left(\nabla \cdot \left(\nabla \cdot \frac{\partial \boldsymbol{u}}{\partial t} \right) \right) \qquad (2)$$

$$\frac{p}{p_0} = \left(\frac{\rho}{\rho_0} \right)^\gamma \qquad (3)$$

where ρ_0 is the ambient density, ρ is the density, p is the acoustic pressure, μ is the shear viscosity, μ_B is the bulk viscosity, $p_0 = \rho_0 c_0^2 / \gamma$ is the ambient pressure, c_0 is the small amplitude value of the sound speed, and γ is the specific heat ratio. The combination of these equations leads to the vectorial wave equation proposed here:

$$\left\{ \begin{array}{c} \rho_0 \dfrac{\partial^2 \boldsymbol{u}}{\partial t^2} = \gamma p_0 \dfrac{\nabla(\nabla \cdot \boldsymbol{u})}{(1+\nabla \cdot \boldsymbol{u})^{\gamma+1}} + \mu \overline{\nabla}^2 \dfrac{\partial \boldsymbol{u}}{\partial t} + \left(\mu_B + \dfrac{1}{3}\mu \right) \left(\nabla \cdot \left(\nabla \cdot \dfrac{\partial \boldsymbol{u}}{\partial t} \right) \right) \\[2mm] \nabla \times \boldsymbol{u} = \boldsymbol{0} \end{array} \right\} \quad (x,y,z,t) \in \Omega \cup T$$

$$p(x,y,z,t) = p_0 \left(1 + \nabla \cdot \boldsymbol{u}(x,y,z,t) \right)^{-\gamma} - p_0$$

$$(4)$$

The appropriate boundary, traducing the null normal displacement and driven piston, as well as the initial conditions, corresponding to the rest of the fluid at the outset (displacement and velocity null), are also considered.

Four dimensionless independent variables are created and introduced in Eq. (4) as well as in the auxiliary conditions. The new space and time dimensionless domain is discretised by means of a 4-D finite-differences grid. A second-order implicit finite-differences scheme is applied to the equations and the value of $\boldsymbol{u}$ is evaluated at each point of this grid. Pressure values on the grid are calculated by applying a first-order explicit finite-differences scheme to the last equation of Eq. (4).

RESULTS

In this section the models are applied to some representative cases and the corresponding results are outlined. The resonators are assumed to be full-filled by air: $c_0 = 340$ m/s, $\gamma = 1.4$, $\rho_0 = 1.29$ kg/m^3.

In the one-dimensional case we consider the resonator described in Fig. 1.a. A strongly nonlinear plane wave is generated in the cavity by considering the driven harmonic frequency $f = 20$ kHz and the amplitude displacement at the piston equal to $250\ \mu$m. The attenuation parameter value used is $\alpha = 2.06$ m^{-1} [3]. The pressure waveform at the reflector (see Fig. 1.a) during one period of the steady state is plotted in Fig. 2.a. An asymmetric sawtooth is observed.

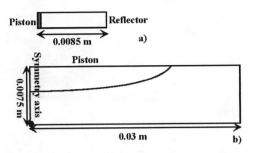

FIGURE 1. Configuration of the resonator in the one (a) and axisymmetric (b) cases.

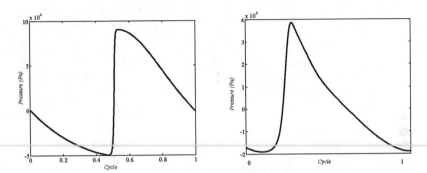

FIGURE 2. (a) Pressure waveform in the one-dimensional case at the reflector; (b) Pressure waveform in the axisymmetric case at the centre of the reflector ($\bullet$).

In the three-dimensional case we consider the resonator with axial symmetry (cylinder) presented in Fig. 1.b. The complicated axisymmetric $(1,1)$ mode is excited at the constant driven frequency $f = 23600$ Hz. The displacement amplitude at the centre of the piston is 30 μm, which generates an extremely nonlinear field. Fig. 2.b presents the pressure waveform at the centre of the reflector, ($\bullet$) in Fig. 1.b, at the steady state. A strong harmonic distortion (asymmetric sawtooth) is also observed.

The quasi-standing character of the wave is defined in the one-dimensional configuration [3]. The rms pressure in the one-dimensional resonator is displayed in Fig. 3.a: harmonic distortion and nonlinear attenuation imply that the harmonic components interfere and transform the pressure field into a quasi-standing nonlinear wave. Two time scales exist:

- the large time scale (rms pressure) exhibits the standing behaviour of the wave, the pressure profiles are smoother for high than for low amplitudes: the pressure is more homogeneously redistributed and no node exists.
- the small time scale shows the propagation effect of the wave, the pressure profiles become much sharper (shock) when nonlinear effects are considered.

In the axisymmetric resonator, the quasi-standing behaviour of the wave only exists in the longitudinal direction [4]. Fig. 3.b displays the evolution of the pressure during a cycle along the symmetry axis. Pressure nulls at many different points of the axis during a cycle, and forms a propagating shock.

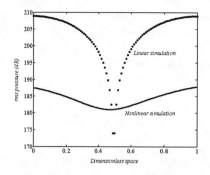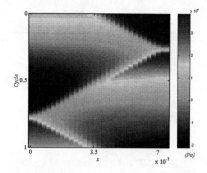

FIGURE 3. (a) rms pressure distribution in the one-dimensional case, compared to the linear simulation; (b) Evolution of the pressure distribution on the symmetry axis during one cycle in the axisymmetric case.

The two-dimensional case was illustrated by means of the simulation of the $(1,1)$ mode of the resonator, much more complicated than the plane wave mode [5]. The constant driven frequency was $f = 20$ kHz, the displacement amplitude at the piston was $140\ \mu\mathrm{m}$, and $\alpha = 2.5\ \mathrm{m}^{-1}$. Pressure waveform and rms pressure distributions were calculated and the quasi-standing nature of the nonlinear waves was verified as well for this complicated mode.

CONCLUSIONS

The establishment of a quasi-standing nonlinear wave inside a resonator is a very important feature to understand the high-power acoustic behaviour in applications. The models presented in this paper predict this character and can be used to simulate the behaviour of complicated quasi-standing nonlinear waves and understand the phenomenon in industrial processes.

ACKNOWLEDGMENTS

This work has been developed within the frame of the following projects: MCYT DPI2002-03409 and URJC PPR-2003-46.

REFERENCES

1. Enflo, B. O., and Hedberg, C. M., *Theory of Nonlinear Acoustics in Fluids*, Dordrecht: Kluwer Academic, 2002.
2. Temkin, S., *Elements of Acoustics*, New-York: John Wiley & Sons, 1981.
3. Vanhille, C., and Campos-Pozuelo, C., *Journal of the Acoustical Society of America* **109**, 2660-2667 (2001).
4. Vanhille, C., and Campos-Pozuelo, C., *Ultrasonics* **43**, 652-660 (2005).
5. Vanhille, C., and Campos-Pozuelo, C., *Journal of the Acoustical Society of America* **116**, 194-200 (2004).

452

Localization of Ultrasound in Acoustic Interferometers

I. Pérez-Arjona, V.J. Sánchez-Morcillo and V. Espinosa

Departamento de Física Aplicada, Universidad Politécnica de Valencia,
Ctra. Natzaret-Oliva S/N 46730 Grao de Gandia, (Spain)

Abstract. We theoretically consider the spatio-temporal dynamics of large aperture acoustic interferometers where the parametric genration of ultrasound takes place. Under given conditions the parametric field shows bistability (hysteresis) between different homogeneous solutions. This bistability is the basis for the generation of two types of localized structures.

Keywords: Acoustic interferometers, parametric generation, localized structures
PACS: 43.25.Ts, 43.25.Rq.

INTRODUCTION

One of the most outstanding characteristics of extended nonlinear systems is the existence of localised structures (also known as solitary waves or spatial solitons), which consist in a region of space where the field takes a different value form the rest, where it remains uniform. This possibility has been largely explored in optics, specially in a resonator configuration. In this work we consider the existence of similar solutions in a particular acoustic system. The considered model describes the parametric interaction of acoustic waves in a driven large aspect ratio resonator, that is, when the transverse dimension of resonator is large compared to its longitudinal dimension, and so considering the role played by diffraction.

THE MODEL

The system under study consists in an acoustic interferometer of length L, composed by two parallel plane walls with high reflectivities, containing a fluid medium inside. The resonator is pumped by an external acoustic field of amplitude p_{in} and frequency ω_0. As follows from previous experimental studies on this problem[1], the study of parametric sound generation can be described in terms of the interaction of three field modes, the driving or fundamental (ω_0) and subharmonic frequencies (ω_1,ω_2), satisfying $\omega_0 = \omega_1 + \omega_2$, assuming that any of them is close to a resonator eigenfrequency. We will consider here the particular degenerate case, where $\omega_1 = \omega_2$.

Under this condition, the dimensionless pressure p has the general form of the superposition of standing waves

CP838, *Innovations in Nonlinear Acoustics: 17th International Symposium on Nonlinear Acoustics*,
edited by A. A. Atchley, V. W. Sparrow, and R. M. Keolian
© 2006 American Institute of Physics 0-7354-0330-9/06/$23.00

$$p(x,y,z,t) = A_0(x,y,t)\cos[\omega_0 t - \phi_0(t)]\cos[k_0 z] + A_1(x,y,t)\cos[\omega_1 t - \phi_1(t)]\cos[k_1 z] \quad (1)$$

where z is the coordinate along the longitudinal direction, (x,y) define the plane transverse to the resonator axis, $(\phi_j; A_j)$ are the phases and amplitudes of fundamental $(j = 0)$ and subharmonic $(j =1)$ fields and $k_j = \omega_j/c_j$ corresponds to the wave number of a cavity mode. Assuming slowly varying envelopes of the mode amplitudes and phases, and in the limit of sufficiently small losses, that allows to consider an effective loss parameter γ_j for each mode, the pressure evolution is given by the equations[2]

$$\frac{\partial}{\partial t} p_0 = -(\gamma_0 + i\delta_0)p_0 - i\frac{\sigma\omega_0}{\rho_0 c_0^2}p_1^2 + i\frac{c_0^2}{2\omega_0}\nabla_\perp^2 p_0 + \frac{c_0}{L}p_{in},$$

$$\frac{\partial}{\partial t} p_1 = -(\gamma_1 + i\delta_1)p_1 - i\frac{\sigma\omega_1}{\rho_0 c_0^2}p_1^* p_0 + i\frac{c_0^2}{2\omega_1}\nabla_\perp^2 p_1. \quad (2)$$

together with their complex-conjugates. In Eqs. (2), the fields $p_j = A_j \, exp \, (i \, \phi_j)$ are the deviations respect to equilibrium pressure values, $\delta_j = \omega_j^c - \omega_j$ is the detuning between the frequency of the field ω_j and the closest frequency of the cavity ω_j^c, and the parameter σ is related with the nonlinearity parameter B/A by $\sigma = (1 + B/(2A))/4$.

Equations (2) can be written in a simpler way introducing the normalizations

$$p_0 = i\frac{2\rho_0 c_0^2 \gamma_1}{\sigma\omega_0}P_0; \; p_1 = \frac{\rho_0 c_0^2 \sqrt{2\gamma_0\gamma_1}}{\sigma\omega_0}P_1; \; p_{in} = i\frac{2L\rho_0 c_0 \gamma_0 \gamma_1}{\sigma\omega_0}\varepsilon \quad (3)$$

and a dimensionless detuning parameter $\Delta_j = \delta_j/\gamma_j$. This leads to the final form of the model given by

$$\frac{1}{\gamma_0}\frac{\partial}{\partial t}P_0 = -(1 + i\Delta_0)P_0 - P_1^2 + i\frac{c_0^2}{2\gamma_0\omega_0}\nabla_\perp^2 P_0 + \varepsilon,$$

$$\frac{1}{\gamma_1}\frac{\partial}{\partial t}P_1 = -(1 + i\Delta_1)P_1 + P_1^* P_0 + i\frac{c_0^2}{2\gamma_1\omega_1}\nabla_\perp^2 P_1. \quad (4)$$

BISTABILITY AND LOCALIZED STRUCTURES

The necessary condition for the existence of localized structures is the bistability between different solutions of the system. If in a spatially extended system exist different stable solutions, these solutions can coexist in different regions, connected in a localized spatial region. These types of structures have been studied in optical cavities for a time, and our aim is to show that they can be also present in acoustical resonator.

Two stationary homogeneous solutions are sustained by Eqs. (4): the trivial one, where $P_0^{tr} = \varepsilon/(1 + i\Delta_0)$ and $P_1^{tr} = 0$, and the nontrivial solution, where the subharmonic field is switched on, given by

$$\left| P_0^{nt} \right|^2 = 1 + i\Delta_1^2,$$

$$\left| P_1^{nt} \right|^2 = -1 + \Delta_0 \Delta_1 \pm \sqrt{\left| \varepsilon \right|^2 - \left(\Delta_0 + \Delta_1 \right)^2}. \tag{5}$$

and that exists above a given threshold pump value, ε_{th}. At this pump value the trivial solution loses its stability and bifurcates into the nontrivial one.

When this bifurcation is subcritical, that is, subharmonic field intensity growing up in the sense of pump decreasing, both trivial and nontrivial solutions are stable in a region of the system parameters [Fig.1(a)]. Under this condition, a localized structure can exist connecting them, as will be shown in next section.

Another type of bistability can be also present in this system. This instability is of topological nature. We note that Eqs. (4) present a phase indetermination of π, which means that above ε_{th}, two nontrivial branches of opposite phase are equally stable [Fig.1(b)]. These branches are bistable, and so they can be connected by a localized structure, now of the type called "wall".

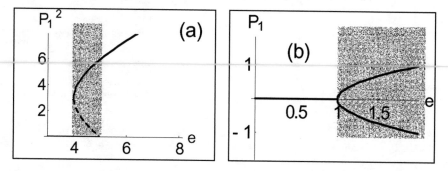

FIGURE 1. (a) Subcritical bifurcation of trivial solution. Continuous (dashed) lines represent stable (unstable) branches. Parameters are $\Delta_0 = \Delta_1 = 2.0$ and $\varepsilon_{th} = 5.0$. (b) Bistability between two equivalent solutions with opposite phases. Shaded regions are the zones of bistability between solutions, where localized solutions can exist.

NUMERICS

In order to test the analytical predictions, and to check whether this type of solutions could really exist in an acoustical interferometer, we have numerically integrated the Eqs. (4). In Fig.2 the two different localized structures discussed above are shown, both in one (left pictures) and two (right pictures) transverse-dimensional resonators.

In the first case [Fig. 2(a)] a number of localized structures of the first type is found (bright spots in the figure). They exist in the hysteretic region, and corresponds to local excitations of the subharmonic field on a background of zero amplitude. In the second case [Fig. 2(b)] the structures of the second type are obtained, and have the from of a wall (in one dimension) or a ring (in two dimensions) of zero amplitude immersed on a background of constant (nonzero) amplitude.

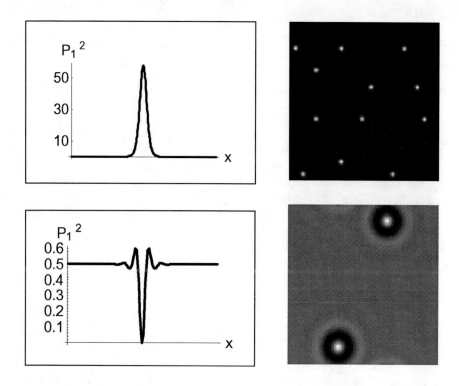

FIGURE 2. (a) Bright localized srtuctures for $\Delta_0 = 6.0$, $\Delta_1 = 3.0$ and $\varepsilon_{th} = 14.6$. (b) Dark localized structures for $\Delta_0 = \Delta_1 = 0.0$ and $\varepsilon_{th} = 1.5$.

The solutions reported here are robust and stable against perturbations. They are obtained only in the degenerate case (subharmonic generation at half frequency of the driving). In the nondegenerate case, the wall structure is however not stable, and instead localized structures with topological character (acoustic vortices) arise.

ACKNOWLEDGMENTS

This work supported by project BFM2002-04369-C04-04 of the CICYT of the Spanish Government.

REFERENCES

1 Pérez-Arjona I. and Sánchez-Morcillo V.J., http://arxiv.org/abs/nlin.PS/0505016
2. N.Yen, J. Acoust. Soc. Am. 57, 1357-1362 (1975)

Standing Waves In Quadratic And Cubic Nonlinear Resonators: Q-Factor And Frequency Response

Bengt O.Enflo*, Claes M.Hedberg**, and Oleg V.Rudenko**

*Department of Mechanics, Kungl Tekniska Högskolan, S-1044, Sweden
**Blekinge Institute of Technology, S-371 79 Karlskrona, Sweden

Abstract. High-Q acoustic resonators are used to significantly increase the wave energy volume density. For high-intensity waves the magnitude of Q-factor is determined not only by the design of the resonator, but by the strength of internal field as well. Methods to calculate the Q-factor and both the spatio-temporal and spectral structure of this field are described. Results are given for quadratic and cubic nonlinear resonators demonstrating quite different physical properties.

Keywords: Quadratic and cubic nonlinear media, nonlinear acoustic resonator, nonlinear Q-factor.
PACS: 43.25.Gf, 43.25.Dc

INTRODUCTION

Different definitions of Q-factor are known. If one resonator wall ($x = 0$) is immovable and the other wall ($x = L$) vibrates according to $u(x = L, t) = A\sin(\omega t)$, where $u(x, t)$ is the particle velocity, one can define Q as the ratio of "internal" amplitude of $u(x, t)$ and "external" amplitude A of the driving force. The inequality $Q \gg 1$ means that inside the resonator cavity the medium vibrates much more than the boundary - the "quality" of resonator is high. As distinct from a linear system, a nonlinear generates a series of harmonics. Therefore, it is necessary to specify the meaning of "internal" amplitude of vibration. It can be the amplitude of fundamental mode, for example. However, the mean square velocity $\langle u(x,t)^2 \rangle$ averaged over the period and over the volume of the resonant cavity seems more sophisticated. Another possible definition exploits the maximum positive peak of disturbance u_+. The Q-factor can be also the normalized maximum of the corresponding frequency response, which is the dependence of the 1st harmonic amplitude, $\langle u(x,t)^2 \rangle$, or u_+ on the frequency shift (discrepancy) between driving frequency ω and a resonant frequency.

METHOD FOR ANALYTICAL SOLUTION

The approach based on "nonlinear superposition" of time-oscillating counter-propagating waves is used. This approach is described in detail in [1] and used by

CP838, *Innovations in Nonlinear Acoustics: 17th International Symposium on Nonlinear Acoustics*,
edited by A. A. Atchley, V. W. Sparrow, and R. M. Keolian
© 2006 American Institute of Physics 0-7354-0330-9/06/$23.00

several authors before. For quadratic nonlinearity and harmonic vibration of one wall the functional equation is derived [1]:

$$F\left(\omega t - kL + \frac{\varepsilon}{c}kLF\right) - F\left(\omega t + kL - \frac{\varepsilon}{c}kLF\right) = A\sin(\omega t). \qquad (1)$$

The unknown function F describes the velocity profile of both counter-propagating waves, $k = \omega/c$ is wave number, c is sound velocity and ε is nonlinearity coefficient. Equation (1) can describe not only steady-state vibrations, but transient processes as well [1], including unlimited linear amplitude growth at resonance (with $\varepsilon = 0$). If the following requirements are fulfilled - namely, the length L of resonator is small in comparison with the shock formation distance, frequency ω of driving force differs slightly from the resonant frequency and the Q-factor is large, Q>>1 - the general equation (1) can be reduced [1] to the inhomogeneous Burgers equation:

$$\frac{\partial U}{\partial T} + \Delta\frac{\partial U}{\partial \xi} - \pi\varepsilon\frac{\partial U}{\partial \xi} - D\frac{\partial^2 U}{\partial \xi^2} = \frac{M}{2}\sin(\omega t). \qquad (2)$$

The following dimensionless variables and coefficients are used here:

$$U = \frac{F}{c}, \ M = \frac{A}{c}, \ \xi = \omega t + \pi, \ T = \omega t/\pi, \ D = \frac{b\omega^2 L}{2c^3\rho} << 1. \quad (3)$$

The first temporal variable ξ describes "fast" oscillations, whereas the second one T is responsible for "slow" evolution due to nonsteady-state build-up and decay processes, as well as for nonlinear and dissipative distortion of the wave profile. Parameter D is the small ratio of the length of resonator to the characteristic absorption distance. The absorption coefficient b and corresponding dissipative terms are not presented, for simplicity, in the initial functional equation (1), but this modifying can be easily performed in the differential equation (2). Exact analytical solutions to equation (2) for travelling waves were obtained in Refs. [2, 3]. These solutions can be adopted for calculation of standing nonlinear waves.

The scheme of solution consists of 2 stages. At first stage, the wave profile is calculated on the base of exact solutions to equation (2). This procedure and corresponding results are described in [4] not only for harmonic excitation, but for arbitrary periodic right-hand-side of equations (1) and (2). At the second stage, the wave field $u(x,t)$ is used for calculation of Q-factor and frequency response

RESPONSE AND Q FOR QUADRATIC RESONATOR

The nonlinear frequency response defined as rms velocity is shown in Fig.1. Curves are constructed for different amplitudes of vibration of the wall, corresponding to values $(M/\pi\varepsilon) = (1, 4, 9, 16, 25)\cdot10^{-2}$. Along horizontal axis the dimensional discrepancy is shown, $\Delta/\pi\varepsilon$; $\Delta = \pi(\omega - \omega_0)/\omega_0$, where ω_0 is the resonant frequency. It is necessary to emphasize, that frequency response curves shown in Fig.1 are symmetric, their form is the same at positive and negative discrepancies.
Another definition as positive peak velocity of the wave profile is illustrated in Fig.2. The amplitude of vibration of the wall equals here to $(M/\pi\varepsilon) = 0.09, 0.25$ for two constructed curves. As distinct from Fig.1, these curves are not symmetric, because

positive and negative half-periods of the wave profile are different; at positive discrepancies the positive peak of profile is higher than the negative one.

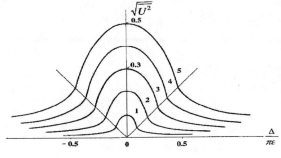

FIGURE 1. Quadratic resonator root-mean-square velocity as function of discrepancy parameter.

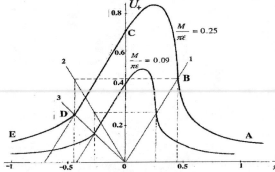

FIGURE 2. Quadratic resonator positive peak velocity as function of discrepancy parameter.

If linear dissipation caused by viscosity and thermal conductivity is taken into account, the exact expression for Q-factor is

$$Q = \frac{1}{\sqrt{\pi\varepsilon M}} \left[-\frac{D^2}{\pi\varepsilon M} a_0 \left(\frac{\pi\varepsilon M}{2D^2} \right) \right]^{1/2}, \qquad (4)$$

where $a_0(q)$ is the eigenvalue of the Mathieu function $ce_0\left(\frac{\xi}{2}, \frac{\pi\varepsilon M}{2D^2}\right)$. Numerical evaluation of nonlinear Q-factor for gaseous media at the wall vibration amplitudes $A \sim 10 cm/s$ gives for Q the order of several tenths, whereas the corresponding linear Q-factor for frequencies of order of kHz is 2-3 orders higher.

CUBIC RESONATOR

In acoustics cubically nonlinear systems are studied much less than quadratic. It is interesting to analyse the nonlinear behaviour of standing waves in cubic resonators because of two reasons. First of all, the physical evolution is radically different from the evolution of quadratic nonlinear waves. Secondly, cubic media are of great interest in connection with new applied problems. Such problems are connected with strong shear wave excitation for medical purposes [5-7], and with diverse geo-physical applications [8]. The approach described above can be successively applied to standing waves in cubic resonators, but now the idea of complete independence of counter-propagating waves breaks down. This method must be radically modified. The functional equation corresponding to the equation (1) now has the following structure

$$F\left[\omega t + kL - \frac{\varepsilon}{c^2}kL(I + F^2)\right] - F\left[\omega t - kL + \frac{\varepsilon}{c^2}kL(I + F^2)\right] = A\sin(\omega t) \tag{5}$$

where I is the unknown integral of the squared wave profile. Consequently, the equation (5) is integro-functional. Nevertheless, it can be studied analytically. In some special cases it is reducible to cubic Burgers-type equation studied in Ref. [9] for plane waves and in Ref.[5, 10] for spatially-limited beams. Study of it is in process, however, some important results are already derived, and one is shown in Fig.3.

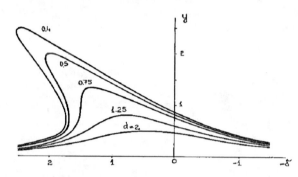

FIGURE 3. The cubic resonator frequency response as a function of discrepancy parameter.

That is the frequency response $y(\delta)$ for cubic resonator constructed for the following magnitudes of parameter $d = 2, 1.25, 0.75, 0.5, 0.4$ (curves 1-5).

$$\delta = \Delta\left(\frac{16}{3\pi\varepsilon M^2}\right)^{1/3}, \; d = D\left(\frac{16}{3\pi\varepsilon M^2}\right)^{1/3}, \; y = \frac{I}{c^2}\left(\frac{3\sqrt{2}\pi\varepsilon}{M}\right)^{2/3} \tag{6}$$

With absorption decrease the frequency response distorts. At weak absorption $d < \frac{\sqrt{3}}{2}$ the curve becomes non-single-valued at some positive values of discrepancy δ.

REFERENCES

1. O.V.Rudenko, C.M.Hedberg and B.O.Enflo, *Acoustical Physics* **47**, 452-460 (2001).
2. O.V.Rudenko, *JETP letters* **20**, 203-204 (1974).
3. A.A.Karabutov, E.A.Lapshin and O.V.Rudenko, *Sov.Phys.JETP* **44**, 58-63 (1976).
4. B.O.Enflo, C.M.Hedberg and O.V.Rudenko, *J.Acoust.Soc.Am.* **117**(2), 601-612 (2004).
5. O.V.Rudenko and O.A.Sapozhnikov, Sov.Phys.*JETP* **79**(2),.220-229 (1994).
6. A.P.Sarvazyan, O.V.Rudenko, S.D.Swanson, J.B.Fowlkes and S.Y.Emelianov. *Ultrasound in Medicine and Biology* .**24**,1419-1436 (1998).
7. S.Catheline, J.-L.Gennisson and M.Fink. *J.Acoust.Soc.Am.* **114**, 3087-3091 (2003).
8. L.A.Ostrovsky and P.A.Johnson. *La Rivista del Nuovo Cimento* **24** (7), 1-47 (2001).
9. I.P.Lee-Bapty and D.G.Crighton, *Philos.Trans.Royal Soc.London* A **323**, 173-209 (1987).
10. O.V.Rudenko and O.A.Sapozhnikov, *Physics-Uspekhi* **174**(9), 973-997, (2004).

Nonlinear Resonant Oscillations Generated between Two Coaxial or Eccentric Cylinders

Eru Kurihara* and Takeru Yano*

*Division of Mechanical and Space Engineering, Graduate School of Engineering, Hokkaido University, Sapporo 060-8628, Japan

Abstract. A resonant gas oscillation generated between two coaxial or eccentric cylinders is studied. In the resonance of plane waves, it is known that shock waves can be generated in an acoustic resonator even if the acoustic Mach number M is sufficiently small. In contrast to the plane wave, however, the resonance of the cylindrical wave does not always show such a discontinuity. In this case, a periodical amplitude modulation of the resonant oscillation of the gas is observed. In the resonant oscillation generated between two eccentric cylinders, depending on the frequency of the sound source (the outer cylinder), the oscillation or pulsation mode can be generated even if the sound source pulsates symmetrically about the axis.

Keywords: Resonance, Cylindrical wave, Eccentric cylinders, Coaxial cylinders
PACS: 43.25.Gf, 46.40.Ff

INTRODUCTION

For a resonance of a cylindrical wave, in contrast to that of a plane wave, a shock free oscillation can be generated when an amplitude of a sound source is not so large because of dispersive effect due to the geometry of the sound field[1].

The present authors have studied the gas oscillation between coaxial cylinders theoretically and numerically[2]. The result shows that a parameter of the geometry of the cylinders (the ratio of the radius of the inner cylinder to the outer one) plays a very important role in the behavior of the resonant gas oscillation and the condition of shock formation.

In order to investigate the effect of the geometry of the sound field in detail, in this paper, we analyze the resonant gas oscillation between eccentric cylinders.

FORMULATION OF THE PROBLEM

Figure 1 shows a schematic illustration of the eccentric cylinders. The inner cylinder of radius r_a is rigid, whereas the outer one has a mean radius $\bar{r}_b$ and the radius of the outer cylinder oscillates harmonically with the amplitude a and the angular frequency ω. The axis of the inner cylinder is separated from that of the outer one by a distance δ. The annular region between the two cylinders is filled with an inviscid ideal gas which is initially at rest.

The behavior of the gas oscillation generated between the cylinders is governed by the following nondimensional system of equations representing the conservation law of mass, momentum, and energy:

CP838, *Innovations in Nonlinear Acoustics: 17th International Symposium on Nonlinear Acoustics*,
edited by A. A. Atchley, V. W. Sparrow, and R. M. Keolian
© 2006 American Institute of Physics 0-7354-0330-9/06/$23.00

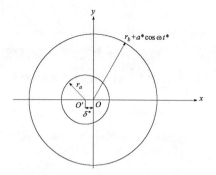

FIGURE 1. Schematic illustration of the eccentric cylinders.

$$\frac{\partial Q}{\partial t} + \frac{\partial E}{\partial x} + \frac{\partial F}{\partial y} = 0,$$ (1)

$$Q = \begin{bmatrix} \rho \\ \rho u \\ \rho v \\ E_t \end{bmatrix}, \quad E = \begin{bmatrix} \rho u \\ p + \rho u^2 \\ \rho uv \\ (E_t + p)u \end{bmatrix}, \quad F = \begin{bmatrix} \rho v \\ \rho uv \\ p + \rho v^2 \\ (E_t + p)v \end{bmatrix}.$$ (2)

Here, u and v are the x and y componentsof the velocity normalized by the speed of sound c_0, respectively, ρ is the mass per unit volume normalized by ρ_0, that in the undisturbed gas, p is the pressure normalized by $\rho_0 c_0^2$, and $E_t = \frac{1}{2}\rho(u^2 + v^2) + \frac{p}{\gamma - 1}$ is the normalized total energy per unit volume of the gas. The boundary conditions are written as

at the outer cylinder$(r = r_s) : u_r = -M\sin t,$ at the inner cylinder $: u_r = 0,$ (3)

where r_s is a nondimensional radius of the outer cylinder and u_r is the radial component of the velocity. The boundary of the inner cylinder is represented by the following equation,

$$(r\cos\theta - \delta)^2 + (r\sin\theta)^2 = (\alpha r_s)^2.$$ (4)

In Eqs. (3) and (4), there are two parameters which play important roles in the phenomenon under consideration. The first parameter is the acoustic Mach number $M = a\omega/c_0$ at the outer cylinder. The second parameter α is the ratio of the radius of the inner cylinder to the mean radius of the outer cylinder, defined by $\alpha \equiv r_a/\bar{r}_b$. The geometry of the system is characterized by this parameter.

If the amplitude of the gas oscillation is adequately small, so that the nonlinearity is quite small, the governing equation (1) can be reduced to the linear wave equation for the velocity potential Φ as,

$$\Delta\Phi - \frac{\partial^2 \Phi}{\partial t^2} = 0.$$ (5)

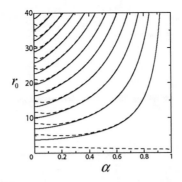

FIGURE 2. Dependence of resonant radius r_0 on α. Solid lines denote r_0 for the pulsation modes and dashed lines for the oscillation modes.

LINEAR SOLUTION

In order to investigate a resonant condition and the behavior of the resonant gas oscillation by small excitations, we now analyze the wave equation (5). When the acoustic Mach number M at the sound source and eccentricity of the inner cylinder δ are quite small, boundary conditions (3) can be written as follows,

$$\frac{\partial \Phi}{\partial r} = M \sin t = \varepsilon a^* \sin t \quad \text{at } r = r_s, \qquad \frac{\partial \Phi}{\partial r} = 0 \quad \text{at } r = \alpha r_s + \varepsilon \Delta \cos \theta, \quad (6)$$

where ε is a small parameter and a^* and Δ are the normalized amplitude at the sound source and eccentricity, respectively. Expanding Φ and the boundary condition as

$$\Phi(r, \theta, t) = \varepsilon \Phi_1(r, \theta, t) + \varepsilon^2 \Phi_2(r, \theta, t) + O(\varepsilon^3), \qquad (7)$$

$$\frac{\partial \Phi}{\partial r}\bigg|_{r = \alpha r_s + \varepsilon \Delta \cos \theta} = \frac{\partial \Phi}{\partial r}\bigg|_{\alpha r_s} + \varepsilon \Delta \cos \theta \frac{\partial^2 \Phi}{\partial r^2}\bigg|_{\alpha r_s} + O(\varepsilon^3) = 0, \qquad (8)$$

and substituting Eq. (7) into Eq. (5), then we can obtain a solution satisfying the condition (6) and (8) as follows,

$$\Phi(r, \theta, t) = \varepsilon a^* \sin t \left[\frac{J_0(r) N_1(\alpha r_s) - J_1(\alpha r_s) N_0(r)}{Z_1(r_s, \alpha r_s)} \right.$$

$$\left. + \varepsilon \Delta \cos \theta \frac{J_1'(r_s) N_1(r) - J_1(r) N_1'(r_s)}{Z_1(r_s, \alpha r_s)} \frac{J_1'(r) N_1(\alpha r_s) - J_1(\alpha r_s) N_1'(r)}{Z_2(r_s, \alpha r_s)} \right], \quad (9)$$

$$Z_1(r_s, \alpha r_s) = J_1(r_s) N_1(\alpha r_s) - J_1(\alpha r_s) N_1(r_s), \qquad (10)$$

$$Z_2(r_s, \alpha r_s) = J_1'(r_s) N_1'(\alpha r_s) - J_1'(\alpha r_s) N_1'(r_s). \qquad (11)$$

In this study, we call the r_s satisfying $Z_1 = 0$ or $Z_2 = 0$ the resonance radius for a given α, and represent such a value of r_s as r_0. When $Z_1 = 0$, the gas oscillation of radially symmetrical mode (puslation mode) is generated, while for the case of $Z_2 = 0$, the oscillation mode is generated [3].

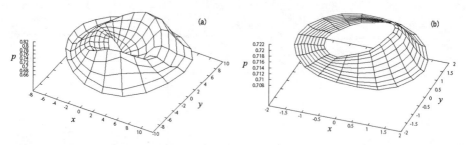

FIGURE 3. Pressure distribution in the cylinder. (a): $\alpha = 0.1$, $\delta = 0.1$, $M = 1 \times 10^{-2}$, and $r_0 = 3.940$ (pulsation mode). (b): $\alpha = 0.5$, $\delta = 0.1$, $M = 1 \times 10^{-4}$, and $r_0 = 6.393$ (pulsation mode).

The resonance radius r_0 satisfying $Z_1 = 0$ or $Z_2 = 0$ for given α is shown in Fig. 2. Solid lines denote resonance radiuses for pulsation modes, and dashed lines for oscillation modes.

NUMERICAL RESULTS FOR NONLINEAR PROBLEM

If the acoustic Mach number M or eccentricity δ is large, the linear solution (9) cannot represent the phenomenon correctly. Therefore, we investigate the nonlinear resonant problem by the numerical calculation for the governing equations. (1) and (2). [4, 5] Numerical calculations have been carried out for two cases. The one is the fundamental pulsation mode for $\alpha = 0.1$ ($r_0 = 3.940$) and the other is the fundamental pulsation mode for $\alpha = 0.5$ ($r_0 = 6.393$).

The pressure distribution for $\alpha = 0.1$, $\delta = 0.1$, and $M = 1 \times 10^{-2}$ is illustrated in Fig. 3(a). It is found that there are the waveform distortion due to the nonlinearity and asymmetry of wave reflection at the inner cylinder due to the large eccentricity. Figure 3(b) shows the pressure distribution for $\alpha = 0.5$, $\delta = 0.1$, and $M = 1 \times 10^{-4}$. Although the radius of the outer cylinder is the resonant radius of the pulsation mode, in this case, the behavior of the gas oscillation is similar to that of the oscillation mode. The reason is that a difference between the resonant radius of the pulsation mode and that of the oscillation mode is small in the region of large α and then the oscillation mode becomes dominant [see Fig. 2 and Eq. (9)].

REFERENCES

1. W. Ellermeier, *Acta Mech.*, **121**, 97–113 (1997).
2. E. Kurihara, Y. Inoue, and T. Yano, *Fluid Dyn. Res.*, **36**, 45–60 (2005).
3. S. Temkin, *Elements of Acoustics*, John Wiley and Sons, 1981.
4. S. R. Chakravarthy and S. Osher, *AIAA. J*, **21**, 1241–1248 (1983).
5. S. R. Chakravarthy and S. Osher, *Large-Scale Computation in Fluid Mechanics*, edited by B. Engquist, S. Osher, and R. C. J. Somerville, American Mathematical Society, Providence, RI, 1985, pp. 57–86.

Rayleigh streaming simulation in a cylindrical tube using the vorticity transport equation

Debbie Sastrapradja* and Victor W. Sparrow*

*Graduate Program in Acoustics, The Pennsylvania State University, 316B Leonhard Building,
Univeristy Park, PA 16802, dsastrapradja@vocollect.com/vws1@psu.edu

Abstract. Current numerical calculation of acoustic streaming can involve major computing time and resources. To develop a quicker model, the vorticity transport equation (VTE) is used. In this paper Rayleigh streaming in a cylindrical tube is simulated using the VTE. The goal of using the VTE is to obtain a relatively fast solution with minimal computational resources, which in this case is a single PC. A clustered grid is utilized to capture the boundary layer effect on the acoustic streaming. The governing equations used are the VTE, Poisson's equation, and an equation that relates the stream function with the velocity. It is demonstrated that the VTE method to calculate Rayleigh streaming works well.

Keywords: streaming, vorticity, thermoacoustics
PACS: 43.25.Nm

INTRODUCTION

The numerical calculation of acoustic streaming usually involves the solution of the continuity equation, the Navier Stokes equation, and the energy equation. When a direct numerical simulation (DNS) is used to solve those equations, it requires considerable computing power and time [1]. In this paper, instead of using the three mentioned equations, the vorticity transport equation (VTE) is utilized to calculate the streaming behavior. This formulation allows the computation to be performed on a single PC and give relatively fast results. Although the predicted streaming behavior is not as precise in detail as the DNS method, the speed can be beneficial when a fast calculation of streaming is needed as a part of the design of a thermoacoustic device.

VORTICITY STREAM FUNCTION METHOD

Existing numerical calculations of acoustic streaming are made by solving the full Navier Stokes equations using direct numerical simulation (DNS), without making any assumption of the fluid characteristics. The complexity of the numerical model may be lessened by making a set of assumptions about the flow. The current approach makes use of several assumptions about the flow and the use of the vorticity transport equation as the primary equation to calculate the acoustic streaming velocity.

CP838, *Innovations in Nonlinear Acoustics: 17th International Symposium on Nonlinear Acoustics*,
edited by A. A. Atchley, V. W. Sparrow, and R. M. Keolian
© 2006 American Institute of Physics 0-7354-0330-9/06/$23.00

Vorticity transport equation

The **time-averaged** VTE written in a 2-D vector form is:

$$-\nabla \times \mathbf{F}_2 = \nu \nabla^2 \omega_2 \tag{1}$$

where $\mathbf{F}_2 = F_{2x}\hat{i} + F_{2y}\hat{j}$, $\omega_2 = (\partial u_{2y}/\partial x) - (\partial u_{2x}/\partial y)$, $-F_{2x} = \frac{\partial}{\partial x}(u_{1x}u_{1x}) + \frac{\partial}{\partial y}(u_{1x}u_{1y})$, $-F_{2y} = \frac{\partial}{\partial x}(u_{1x}u_{1y}) + \frac{\partial}{\partial y}(u_{1y}u_{1y})$, and ν is the kinematic viscosity. x denotes the longitudinal direction and y denotes the radial direction. It should be noted that since this analysis is in a planar (x,y) coordinate system, we interpret the vorticity ω_2 as the rotating fluid in the (x,y) plane, the z component of the vector vorticity $\boldsymbol{\omega}_2$.

Poisson's equation

In a two dimensional incompressible flow, Poisson's equation can be written as:

$$-\omega_2 = \frac{\partial^2 \psi_2}{\partial x^2} + \frac{\partial^2 \psi_2}{\partial y^2} = \nabla^2 \psi_2 \tag{2}$$

Therefore, one can find the streaming velocity with these steps:

1. Calculate the forcing function $\mathbf{F}_2$ from the first order acoustic velocities u_{1x} and u_{1y}.
2. Knowing $\mathbf{F}_2$, solve for the vorticity ω_2 by solving Eq. 1.
3. Knowing ω_2, solve for the stream function ψ_2 by solving Eq. 2.
4. Solve for the acoustic axial and transverse streaming velocities u_{2x} and u_{2y}:

$$u_{2x} = \frac{\partial \psi}{\partial y}, \quad u_{2y} = -\frac{\partial \psi}{\partial x} \tag{3}$$

DIRECT METHOD OF SOLUTION

One way to solve the VTE and Poisson's equation is through an *iterative method* and another way is through a *direct method*. In this research we use the direct method because here the VTE is a time-averaged equation and the method itself is unconditionally stable. The direct method involves solving a set of linear algebraic equations. To do this, we first need to form a matrix of the coefficients of the vorticity (in the case of VTE) or the stream function (in the case of Poisson's equation). We then have to decompose each matrix into its lower and upper matrices, taking advantage of the fact that the solution to a triangular set of equations is quite trivial. The solution is then obtained through *backsubstitution* [4]. Details on the solution method including the matrix formulation are available [3].

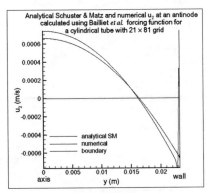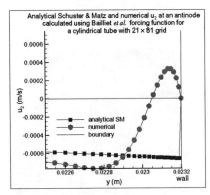

FIGURE 1. Schuster & Matz analytical streaming velocity vs. calculated result using Bailliet *et al.* forcing function at an antinode along y direction with 21×81 grid points. $l = \lambda/2 = 0.553$ m, $r = 0.0232$ m. The left plot shows the upper half of the tube, and the right plot is near the tube's top wall.

ANALYTICAL VS. NUMERICAL STREAMING

Due to the lack of a closed form solution for streaming in a cylindrical tube, the VTE numerical streaming is compared to the analytical streaming in a cylindrical tube derived by Schuster and Matz [5].

The two plots on Fig. 1 show the VTE numerical streaming velocities calculated using the Bailliet *et al.* forcing function for a cylindrical tube and the analytical streaming velocity obtained by Schuster and Matz. On the axis, the numerical streaming velocity is 7.9×10^{-5} m/s (12%) higher than the analytical streaming velocity (based on the 21×81 grid). Near the boundary, Schuster and Matz's calculation does not show the inner streaming unlike the calculated numerical streaming. Schuster and Matz's negative peak value near the boundary is 1×10^{-4} m/s (17%) lower than Bailliet *et al.*'s value (based on the 21×81 grid).

The two plots on Fig. 2 show the acoustic streaming velocity vector plots. Note that the tube dimensions are not to scale. The plot on the left shows the upper half of the tube for the entire length (one half of a wavelength), and the plot on the right shows the region near the top wall where inner streaming occurs in the boundary layer.

CONCLUSIONS

Typical acoustic streaming calculations using a DNS method require considerable computing time and resources. Although the results are very detailed, the long computing time can sometimes be a problem if results are needed fairly quickly (such as in the early design stage). The current method requires a relatively short computing time and it can be done on a single PC. The finest grid used in this research is 21×81 grid points, and it required about 1 hour and 15 minutes of computing time on a 2 GHz Intel Pentium 4, 512 RAM PC using an Absoft compiler. This reduced the amount of computing time significantly as DNS sometimes requires a few days to finish the calculation on multi-

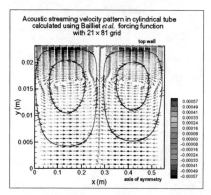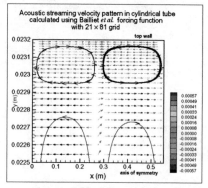

FIGURE 2. Streaming cells in a cylindrical tube. $l = \lambda / 2 = 0.553$ m, $r = 0.0232$ m. The Bailliet *et al.* forcing function was used for the calculation. The left plot shows the upper half of the tube, and the right plot is near the tube's top wall.

ple processors. The resulting numerical solutions agreed fairly well with the analytical solutions. This is proof that this method is a valid method to be used as a tool in thermoacoustic design. As long as one has a knowledge of the first order acoustic velocities inside a thermoacoustic device, then this method can be used to predict the Rayleigh streaming in that device.

Besides the forcing function, the other crucial aspect of the calculation is establishing the correct boundary conditions. Different wall conditions and different types of streaming may require different types of boundary conditions.

ACKNOWLEDGMENTS

This work was supported in part by Office of Naval Research Grant N00014-99-1-0921. We also gratefully acknowledge discussions with Philip Morris, Said Boluriaan, Anthony Atchley, and Cengiz Camci.

REFERENCES

1. M. K. Aktas, and B. Farouk, *J. Acoust. Soc. Am.*, **116**, 2822–2832 (2004).
2. H. Bailliet, V. Gusev, R. Raspet, and R. A. Hiller, *J. Acoust. Soc. Am.*, **110**, 1808–1821 (2001).
3. D. Sastrapradja, *Rayleigh Streaming Simulation Using The Vorticity Transport Equation*, Ph.D. thesis, The Pennsylvania State University (2004).
4. W. H. Press, S. A. Teukolsky, W. T. Vetterling, and B. P. Flannery, *Numerical Recipes In Fortran 77 The Art Of Scientific Computing*, Press Syndicate of the University of Cambridge, Cambridge, 1992, chap. 2, pp. 22–89.
5. V. K. Schuster, and W. Matz, *Akust. Zeitschrift*, **5**, 349–352 (1940).

MRI Detection Of Acoustic Streaming In Gases

Igor Mastikhin, Ben Newling, Scott Culligan

mast@unb.ca, MRI Centre, Department of Physics, University of New Brunswick, PO Box 4400,
Fredericton, NB E3B 5A3, Canada
Keywords: Acoustic Streaming, Magnetic Resonance Imaging
PACS: 43.25.Nm, 76.60.Pc

INTRODUCTION

Acoustic streaming (AS) is the time-independent fluid motion generated by a sound field. This motion is caused by the transfer of acoustic momentum, through attenuation or absorption of a sound beam. Raleigh acoustic streaming, first described by Lord Raleigh [1], is circular flow from node to antinode that occurs when a standing wave is set up inside an enclosure.

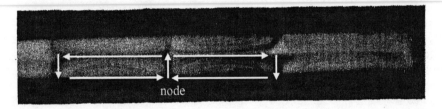

node

FIGURE 1. AS in a standing wave: theoretical arrows [1] and experimental [2] results with smoke particles.

Prevalent methods of AS detection are particle image velocimetry (PIV) techniques, based on injection of small particles (smoke _etc._) into the studied gas volume [2]. It is difficult to perform PIV measurements without disturbing the AS gas dynamics in a vessel. Magnetic Resonance Imaging (MRI) is potentially capable of non-invasively obtaining 3-dimensional information on velocity and diffusion of gas. However, gas densities are 1000 times lower and NMR signal relaxation times are often 2-3 orders of magnitude shorter than those of liquids, requiring unconventional MRI methods.

We tested several MRI methods for detection of acoustic streaming in gases. The first method, aimed at measuring low velocities of gas (mm/s), follows a preparation-readout scheme, with a Stimulated Echo (STE) sequence or a single-point imaging method (SPRITE) as a readout [3]. The second method is a modification of a displacement-sensitized SPRITE sequence for higher gas velocities (m/s) [4]. It is also possible to encode motion sensitivity in the magnitude of an MR image, by the use of

CP838, _Innovations in Nonlinear Acoustics: 17th International Symposium on Nonlinear Acoustics_,
edited by A. A. Atchley, V. W. Sparrow, and R. M. Keolian
© 2006 American Institute of Physics 0-7354-0330-9/06/$23.00

tags [5]. Tags are areas of nulled magnetisation, which appear as dark stripes in an MR image. The distortion of the tags with time gives a measure of the flow field.

EXPERIMENTAL

We used propane as a gas for imaging on a Nalorac (Martinez, USA) 2.35 T, horizontal-bore superconducting magnet with quadrature birdcage radiofrequency (RF) probe. A 30 W speaker was situated coaxially inside a 10-cm-i.d. 122 cm-long propane-filled cylindrical tube. At 790 Hz, a standing wave formed in the tube which was positioned so that one

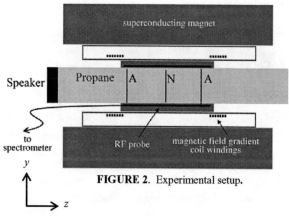

FIGURE 2. Experimental setup.

of nodes was located at the RF probe centre.
MRI measurements were performed with motion encoding separately in the y-direction (across the tube) and z-direction (along the tube), with parameters as follows: $\tau = 50$ ms, $G_{max} = 0.9$ G/cm, 64 scans, with 8 min acquisition time. In Fig. 3, we show the resulting images. Along the horizontal axis are profiles of gas along the tube (resolution 2 mm). Along the vertical axis are velocity spectra (velocity resolution 1.2 cm/s). The intensity of each pixel is proportional to the volume of fluid at the particular position along the tube, which has the velocity indicated by the position along the vertical axis.

RESULTS

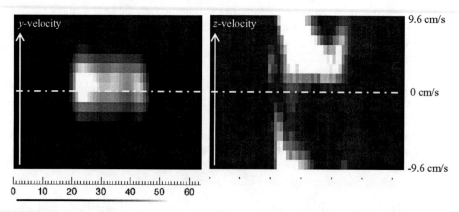

FIGURE 3. Spatially resolved velocity spectra of y-velocity across the tube (left) and z-velocity along the tube (right).

MRI measurements (Fig. 3) show a developed streaming in the tube. The measurements are profiles along the tube, summed along other dimensions. While gas y-velocities across the tube are symmetrical (left), z-velocities along the tube are not symmetrical at all. To explain this discrepancy, we performed DANTE-tagging measurements with a SPRITE MRI readout. The gas volume was "tagged" with a DANTE-tagging sequence and a 100 ms evolution time was allowed between the tagging and the readout, during which tag position evolved with the flow field. A shift in the tagged gas, indicated a developed circulation in the tube along the z-direction. The circulation was caused by convection of gas due to a temperature difference between the ends and the centre of the tube.

CONCLUSIONS

Detection of slow gas motion caused by sound pressure is possible by means of MRI within minutes of acquisition time. The resulting spatially-resolved velocity spectra show the expected velocity distributions for the case of developed Raleigh streaming (Fig 3 compared to Fig. 1). In Fig.3 (left), the maxima of y-velocity occur near the nodes and antinodes of the standing acoustic wave, corresponding to the ends of the Raleigh cells. In Fig.3 (right), maxima of z-velocity occur between the nodes and antinodes. A marked asymmetry in the z-velocity spectra may result from a combination of convection and streaming or from unsteadiness (on the order of minutes) of counterflow, leading to loss of the signal from the velocity maps.

ACKNOWLEDGMENTS

The authors thank the Natural Science and Engineering Research Council of Canada for Discovery Grants (I.M. & B.N.) and an Undergraduate Student Research Award (S.C.). We also thank Murray Olive and Brian Titus for fabricating parts of the experimental setup. UNB MRI Centre is supported by Major Facilities Award.

REFERENCES

1. Lord Rayleigh, "On the circulation of air observed in Kundt's tubes," *Philos. Trans. R. Soc. London*, Ser. A **175**, 1–219(1884).
2 Andrade, E. N. Da C., "On the Circulations Caused by the Vibration of Air in a Tube," *Proc. R. Soc. Lond.* **A134**, 445-470 (1934).
3. I.V. Mastikhin *et al*, *J.Magn.Reson.* **136**, 159-168, (1999).
4. B. Newling *et al.*, *Phys. Rev. Lett.* **93**, 154503 (2004).
5. Mosher & Smith, *Magn. Reson. Med.* **15**, 334-339 (1990).

Inverse Acoustic Wave Field Synthesis

Georgios N Lilis*, Srinivas Telukunta† and Sergio D Servetto**

*School of Electrical and Computer Engineering, Cornell University
†Sibley school of Mechanical and Aerospace Engineering, Cornell University
**School of Electrical and Computer Engineering, Cornell University[1]

Abstract. Acoustic wave field synthesis is a method of sound reproduction of a primary sound source with the use of secondary sources, implemented as arrays of micro-speakers. This method is based on wave field theory. We have developed a finite element based scheme to solve this problem posed in an inverse setting. Huygens principle has been used to validate the method for a few simple problems.

Huygen's principle suffers from the drawback that incorporation of secondary sources at desired locations is rather difficult. We present a method for inverse acoustic wave field synthesis which overcomes the above limitation in the context of Two-Dimensional problems and show that incorporation of any geometric or material non linearities is relatively straight forward. This has significant implications for problems in geophysics or biological medium where material inhomogeneities are quite prevalent. Numerical results are presented for sample problems in noise cancellation and wave field synthesis for inhomogeneous media. The method is capable of being extended to three dimensional problems involving dissipation and anisotropic medium and is a topic of continuing research.

Keywords: finite element method, inverse problem, wave field synthesis, noise cancellation, non linear, inhomogeneous media.
PACS: 43.25.Ba

INTRODUCTION

Acoustic wave field analysis or simply wave field analysis (WFA) refers to the recording of sound fields in enclosures with arrays of microphones and to the processing of the recorded data. Acoustic wave field synthesis or simply wave field synthesis (WFS) refers to the generation of sound fields with desired or prescribed temporal and spatial properties. The idea of wave field synthesis has been in existence for many years, and is often credited to have been first introduced by Berhkout in 1988. A good introductory review of this literature can be found in [1, 2].

In traditional applications of sound enhancement or reproduction, individual (or groups) of loudspeakers are used to generate a replica of the desired sound pattern. Use of high-quality systems in appropriate manner will help in generation of required temporal properties of sound, however the spatial properties are determined by the interference patterns and often the spatial signal is correct only within a very limited listener area. As an example, consider the use of two loud speakers to enhance the signal of a primary point source behind them. Most listeners perceive the signal of the loudspeakers earlier than the primary signal which leads to mislocalization, since the first arriving sound wave determines the direction from which the sound is heard. Another drawback

[1] Work supported by the National Science Foundation, under awards CCR-0238271(CAREER), CCR-0330059 and ANR-0325556.

CP838, *Innovations in Nonlinear Acoustics: 17th International Symposium on Nonlinear Acoustics*,
edited by A. A. Atchley, V. W. Sparrow, and R. M. Keolian
© 2006 American Institute of Physics 0-7354-0330-9/06/$23.00

of the traditional approach of sound reproduction, is that the sound field is measured first, at a few chosen "representative" positions, assuming that the acoustic parameters (sound speed, attenuation) are valid for some (usually) large region around these points. This however is not a reliable approach as it does not include the spatial and temporal inhomogeneities of the acoustic medium.

In the current state of art, these problems are overcome with the use of array technology involving the use of arrays of microphones and micro-speakers. These are placed at suitable positions either on the boundary or within an enclosed volume. This is based on the use of techniques from wave field synthesis (WFS) and wave field analysis (WFA). This in general involves a significant amount of experimentation and hence considerable cost. Here we describe a numerical technique to carry out this procedure in a smart way and to solve the wave synthesis and analysis as a solution to an inverse problem.

WAVE FIELD THEORY

One essential aspect of wave field theory in the context of acoustics is the use of both wave field analysis and wave field synthesis to record sound fields and produce a desired sound pattern with the use of arrays of microphones and micro-speakers. A short description of various applications of these ideas are provided next.

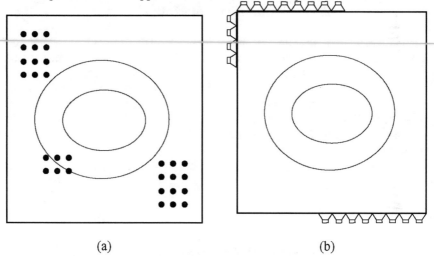

(a)   (b)

FIGURE 1.   (a) Schematic of microphone Arrays for WFA (b) Schematic of microspeaker Arrays for WFS

Wave Field Analysis: Wave field analysis [Fig.1(a)] is based on the measurement of air pressure variation along closely spaced microphone arrays. This generates detailed spatial coherence of neighboring responses, and hence a better insight of the wave fields than the use of individual far spaced measurements.

Wave Field Synthesis: Wave field synthesis [Fig.1(b)] generates the desired spatial and time dependent sound pattern over a given volume of region. This is accomplished with an array of microspeakers and is based on the Huygens principle.

Applications: The applications of this technology find use in direct sound enhancement of cinemas, theaters, various simulators, auditoriums, teleconference systems etc.

NUMERICAL APPROACH TO WAVE FIELD THEORY

Wave field theory has tremendous implications in acoustics, and typically these are accomplished with the use of arrays of microphones and micro-speakers. However, this is often very expensive and time consuming to be based only on extensive experimentation. We propose a numerical approach to this problem. Specifically, we propose a finite difference scheme to numerically generate desirable time evolving wave fields (with numerical probes to simulate the microphone arrays). This information is used in the finite element method to obtain the signal distribution of microspeaker arrays in order to generate the sound of required quality. We show that this is a well conditioned problem and provide sample results for few simple cases. The goal is to extend these ideas for the case when a few sample measurements are made with actual microphones, then use this data to calibrate the region of interest and deduce the distribution of microspeaker arrays to generate appropriate field distribution.

Finite Element Method

Finite Element Method (FEM) is a numerical method that solves the equations of motion in discrete, continuous increments of time and space variables. For an introductory treatise on FEM refer [3, 4].

For our analysis, a simple model that describes the wave propagation inside a 2 dimensional isotropic and homogeneous medium S and time T. Therefore the whole space-time topological space can be expressed as: $\Omega = S \times T$ with boundary $\partial\Omega$. Inside such a space, a scalar wave field Φ satisfies the 2 dimensional wave equation shown below (1).

$$k\nabla^2\Phi - \rho\frac{\partial^2\Phi}{\partial t^2} = -F \qquad k[\Phi_{xx} + \Phi_{yy}] - \rho\Phi_{tt} = -F \qquad (1)$$

Constrained Inverse Wave Field Problem

An inverse wave field problem is the estimation of signal distribution, which is often ill-conditioned and may not have a unique solution. To overcome this problem we propose a constrained inverse problem illustrated below (2).

$$[K^{(G)}]\{\Phi\} = -\{F\} \qquad [K^{(G)}]\{\widetilde{\Phi}\} = -\{\widetilde{F}\} \qquad ||\widetilde{\Phi} - \Phi|| < \varepsilon \qquad (2)$$

Here ε represents the error between the field Φ and the field $\widetilde{\Phi}$ created by the source $\widetilde{F}$, that is constrained in space and time.

Results

We present two sample results in this paper, one for homogenous medium (fig.2(a)) and the other, a non-homogenous case (fig.2(b)). A central step function kind of a signal is applied at the center of a room ($3m \times 3m$ in dimension). The field evolution is solved via finite difference scheme. This field is used as input for our scheme and the central signal is captured as shown below for various cases.

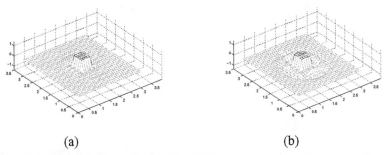

(a)   (b)

FIGURE 2. (a) Signal distribution estimated from FEM (Homogenous medium 3×3 units, signal applied in central 0.3×0.3 region.) (b) Signal distribution estimated from FEM (Non-homogenous medium 3×3 units, signal applied in central 0.3×0.3 region.)

Conclusions

Wave field synthesis has broad applications and involves extensive experimentation and cost. We propose a numerical approach based on inverse formulation with FEM to alleviate some of these concerns. The original signal corresponding to a given acoustic wave field distribution has been generated quite accurately with this scheme for both homogenous and non-homogenous 2-D media. The method is straightforward to be extended to 3-D media with dissipation [6]. The mesh size needed to capture the source distribution with a reasonable accuracy does indeed depend on the frequency and propagation velocity of the medium [6]. Finally, its essential to use various sparse algorithms for the solution of the resultant linear system of equations [5].

ACKNOWLEDGMENTS

Authors would like to acknowledge the financial support from National Science Foundation to Mr. Lilis, during the period of this research. Authors would also like to express their sincere thanks to Prof. Subrata Mukherjee, Theoretical and Applied Mechanics Department, Cornell University for helpful discussions on the subject.

REFERENCES

1. A.J. Berkhout, *Applied Seismic Wave Theory*, Elsevier, 1987.
2. L.E. Kinsler, A.R. Frey, A.B. Coppens and J.V. Sanders, *Fundamentals of Acoustics*, Wiley, 1999.
3. T.J.R Hughes, *Finite Element Method - Linear Static and Dynamic Finite Element Analysis*, Prentice Hall, Englewood Cliffs, 1987.
4. O.C. Zienkiewicz, R.L. Taylor and J.Z. Zhu, *The Finite Element Method: Its Basis and Fundamentals*, Butterworth-Heinemann, 2005.
5. T.F. Coleman and Y. Li, *Large-Scale Numerical Optimization*, Society for Industrial and Applied Mathematics, 1990.
6. G.N. Lilis, A. Halder, S. Telukunta and S. Servetto, "Hybrid numerical scheme for inverse time evolving wave fields," *In Preparation for IJNME*.

SECTION 10

BUBBLES, PARTICLES, AND FLOWS

Variation of Periodicity In Non-spherical Bubble Vibration

K Yoshida, Y Watanabe

Faculty of Engineering, Doshisha University in Japan

Abstract. The periodicity of non-spherical vibration of a microbubble driven by ultrasound are discussed. To examine the periodicity of non-spherical vibration, the observation with both the high-speed video camera and the streak camera are carried out. Two kinds of non-spherical periodical vibrations are observed. One is synchronous with the fundamental period of the driving ultrasound and the other is synchronous with the double period. It is shown that the periodicity of non-spherical vibration largely depends on the driving condition as the ambient radius and the sound pressure. These vibration periodicities change repeatedly and are clearly divided for ambient radius.

Keywords: Microbubble, Ultrasound, Non-spherical vibration, Periodicity, Streak camera, High-speed video camera.
PACS: 43.25Yw.

INTRODUCTION

Under ultrasonic irradiation, a microbubble becomes a secondary sound source since the bubble vibrates. This feature of bubble is applied to ultrasound imaging as contrast agent in medical field. The analysis of bubble vibration, in particular the one of vibration periodicity which decides the frequency of the sound emitted by the bubble, is an important matter. For the spherical bubble, the theoretical analysis of vibration periodicity is carried out by present study.[1],[2] For non-spherical bubble, many theoretical and experimental analyses of bubble collapse are reported, however, few study paying attention to the vibration periodicity. In this report, the periodicity of the non-spherical vibration is discussed from the optical observation with both the high-speed video camera and the streak camera.

OPTICAL OBSERVATION SYSTEM

The optical observation system of bubble vibration with the streak camera and the high-speed video camera is shown in Fig.1, where a Langevin transducer employed as sound source to form a standing wave of frequency 27kHz in a cylindrical cell of diameter 60mm and height 60mm. A bubble is trapped at the anti-node of standing wave. The observation of bubble vibration is carried out with shadowgraph method. The incident light of xenon lamp from a side of the cell is focused near the bubble with a lens. Under this condition, the instantaneous shadow image of bubble is

CP838, *Innovations in Nonlinear Acoustics: 17th International Symposium on Nonlinear Acoustics*,
edited by A. A. Atchley, V. W. Sparrow, and R. M. Keolian
© 2006 American Institute of Physics 0-7354-0330-9/06/$23.00

observed with the streak camera (PHOTRON) or the high-speed video camera (HAMAMATSU PHOTNICS). The slit is set behind the long distance microscope only in the observation with the streak camera. The bubble radius is calculated from the shadow area of the bubble in the observed image.

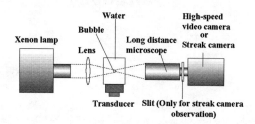

FIGURE 1. The optical observation system of bubble vibration

OBSERVATION WITH THE STREAK CAMERA

Various behaviors such as spherical or non-spherical vibration and break up of the bubble are observed with the high-speed video camera. It is shown that the bubble behavior changes in order of spherical vibration, non-spherical vibration and break up of the bubble as the sound pressure rise.

To examine the periodicity of non-spherical vibration, the observation with the streak camera was carried out. With the streak camera, the bubble vibration can be observed in real time. However, due to the slit, the whole bubble image cannot be observed. The problem of observation with the streak camera is to judge whether the bubble vibrates spherically or non-spherically. This problem is solved by the light transmitted through the bubble. When the bubble is a sphere, the light from the xenon lamp is transmitted only in the center of bubble, because the area in the center of bubble is perpendicular against the bubble wall. When the bubble is extremely distorted, the area of transmitted light becomes complex and different from the one of spherical bubble. If the bubble vibrates non-spherically, the transmitted light changes temporary because the area perpendicular against the bubble wall changes. In the image observed with the streak camera, it is judged by the temporal variant of light transmitted through the bubble whether the bubble vibrates spherically or non-spherically.

Figure.2 shows the bubble vibration in increasing driving sound pressure from the spherical vibration region to the break up region and the temporal variant of effective bubble radius R_e measured from observed streak images is shown. In Fig.2, it is shown that the bubble vibrates non-spherically since the transmitted light changes and the time variant of effective bubble radius repeats same vibration with about 72 μ s. The bubble vibrates non-spherically and synchronously with the double period of the driving ultrasound.

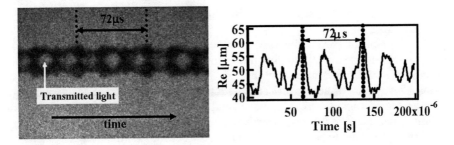

FIGURE 2. The streak image of non-spherical vibration and temporal variant of effective bubble radius.

OBSERVATION WITH THE HIGH-SPEED VIDEO CAMERA

In the observation with the streak camera, the ambient radius cannot be measured. To measure the relativity between the periodicity of non-spherical vibration and the driving conditions which are ambient radius R_0 and sound pressure P_a, the observation with the high-speed video camera, with which the whole bubble image can be observed, is carried out in various R_0 and P_a. However, the maximum recording rate of the high-speed video camera in this observation system, which is 40500frame/sec, is still an insufficient speed to observe the bubble vibration in 27 kHz. However, the bubble vibration can be reproduced if the bubble keeps vibrating periodically for a much longer time than the fundamental period of the driving ultrasound.

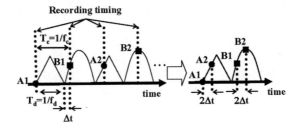

FIGURE 3. The principle of reproduction for bubble vibration.

The principle of reproduction for bubble vibration is shown in Fig.3. f_c, f_d represent the recording rate of the high-speed video camera and driving frequency respectively. f_c and f_d are set to differ slightly. The image is recorded at $t=0$, $t=T_d+\Delta t$, $t=2T_d+2\Delta t,...(T_d=1/f_d, \Delta t=1/f_c-1/f_d)$. If the bubble vibrates synchronously with the double period of the driving ultrasound, the two different vibration mode is recorded alternatively every frame. In this case, the reproduction of vibration can be carried out by reproducing the temporal axis of different two vibration mode separately. In this way, Even if the bubble vibrates synchronously with any times period of driving ultrasound, the bubble vibration can be reproduced. In this observation, the recording

481

rate of the high-speed video camera is 27000 frame/sec and driving frequency is 27.3kHz respectively.

The variation of periodicity for non-spherical vibration is shown in Fig.4. The circles represent generation condition of non-spherical vibration. The domain above the circles is the break up region and the lower domain is the spherical vibration region. Two kinds of non-spherical vibration are observed. In Fig.4, the open circle represents synchronous with the fundamental period of the driving ultrasound and the solid one represents synchronous with the double one respectively. It is shown that the periodicity of non-spherical vibration largely depends on the driving condition of both the ambient radius and the sound pressure. These vibration periodicities change repeatedly and are clearly divided into four regions for the ambient radius from 20μm to 100μm.

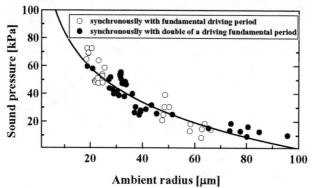

FIGURE 4. Variation of periodicity for non-spherical vibration

CONCLUSION

From the aspect of the vibration periodicity, non-spherical vibration is observed with both the streak camera and the high-speed video camera. Two kinds of non-spherical periodical vibrations are observed. One is synchronous with the fundamental period of the driving ultrasound and the other one is synchronous with the double period. It is shown that the periodicity of non-spherical vibration largely depends on the driving condition for the ambient radius and the sound pressure, and these vibration periodicities change repeatedly and are clearly divided for the ambient radius.

REFERENCES

1.   A. Eller and H. G. Flynn, *J.Acoust.Soc.Am* **46**, pp 722-727(1969)
2.   D. Koyama, A. Sakai and K. Watanabe, *IEICE Trans.A* **84**, pp.1500-1507(2001)

Radial and Translational Motion of Gas Bubbles in the Field of a Plane Piston and in a Progressive Flat Wave

Emmanuil M. Agrest[*] and Gennady Nikolayevich Kuznetsov[†]

[*]Johnson & Wales University, 801 West Trade Street, Charlotte, NC 28202
[†]The Russian Academy of Sciences Wave Researches Centre of Institute
of General Physics, 123007, Russia, Moscow, 5th Magistralnaya, 11

Abstract. The radial and translational motion of small gas bubbles in the near field region of a pulsating piston and in a plane propagation sound wave was studied experimentally and by numerical simulations. It was established that the sub-resonance size gas bubbles move toward the axis of symmetry of the piston-zone of the pressure antinodes. As a result of convective diffusion and coagulation, those small bubbles, the nucleus of cavitation, grow in size. The special role of the rigid surface of the piston in keeping the bubbles near the surface is noted. It is established that while moving toward the node of pressure, the amplitude of the bubble pulsation increases, and nonlinearity results in typical cavitation noise on the sub-harmonics of the main frequency. A series of cavitation impulses is noted. The fact that gas bubbles eject at a high speed from the zone of the antinodes of pressure away of the piston surface is established both by numerical simulations and experiments. Growth in size and the fact that the resonance radius depends on the amplitude both contribute to this ejection. Some generalization about the behavior of trans-resonant bubbles in different inhomogeneous sound fields is made.

Keywords: Cavitation, bubble, nonlinear, oscillation.
PACS: 47.55.Bx, 47.55.Dz

INTRODUCTION

It is a well-known fact that small gas bubbles -- nucleus of cavitation -- while pulsating in size, have a tendency to move toward the maximum of the amplitude of an inhomogeneous ultrasonic field (antinode) [1-4]. Contrarily, bubbles with a radius greater than resonance size move in the opposite direction (toward the node). For higher intensity, the resonant radius depends on the pulsating radius amplitude. Long-term bubble spatial oscillations are possible [5] for bubbles close to linear resonant size in the field of the standing sound wave.

EXPERIMENTS

Experiments were performed in a water pool with rod type emitters. A metallic cover plate of rectangular form of 20 cm per side is attached to the cylindrical piezoceramic rods. Films are taken under water with the simultaneous recording of

CP838, *Innovations in Nonlinear Acoustics: 17th International Symposium on Nonlinear Acoustics*,
edited by A. A. Atchley, V. W. Sparrow, and R. M. Keolian
© 2006 American Institute of Physics 0-7354-0330-9/06/$23.00

audible signals. Saturated water was used. The experimental study of cavitation was conducted in the following three regimes: continuous harmonic emission with different but constant amplitudes; pulse emission with constant amplitude, and different intervals and pulse duration; and continuous emission with a linear growth and then decrease of pressure amplitude. The underwater filming frequency was 25 frames per second. Maximum pressure was formed near the center of the plate with about 6 to 10 dB decrease toward the edges of the plate. A frequency of 3 kHz and the duration of 0.5 sec were used for the pulse regime. The amplitude of sound pressure along the axis at a distance of 30 cm from the surface was about 0.7 bar. The static pressure was 1 bar. For the continuous emission, the variable pressure amplitude varied from 0 to 1.5 bar and then again to zero.

It follows from this experiment that the cavitation appears as a result of micro bubbles' growth. These micro bubbles – embryos of cavitation – were initially located in the micro-cracks of the metallic cover plate as well as in the water itself. This is confirmed by the fact that the place of origin of cavitation during repeated start and disconnection of sound changes insignificantly. The positions of the zones of origination and development of cavitation practically do not change with a continuous increase in the pressure. The gas diffusion and the coagulation of micro-embryos contribute to an increase in these gas embryos. While their sizes are still smaller than the resonance size for the master frequency of sound, bubbles formed this way move toward the antinode as a cumulative result of the alternative pressure gradient. As a result of interaction with the rigid surface of the emitter, the pulsating gas bubbles practically roll along the surface, coagulating with other gas bubbles on their way. It is also observed that some bubbles, which are not exactly near the surface, move toward the surface of the emitter in the vicinity of the antinode around the area of about 5 cm. Small bubbles, as they approach the antinode, increase their size and as they become trans-resonant (with radius greater than the resonance radius for the emitted frequency), they lose dynamic stability and eject from the zone of antinode in the direction normal to the surface of the emitter. Those ejections occur practically in equal time intervals while other small bubbles keep moving toward the center of the transducer, replacing those that were already ejected. A space bubble train is formed. In their motion, some bubbles attract each other and coagulate, others divide into two or more smaller bubbles, forming splitting trajectories of translating bubble motion.

With further development of cavitation, especially in the case of an increase in the sound pressure, a chaotically moving cloud of discrete bubbles is formed in the zone of the antinode, far from the surface of emitter. Bubbles pulsate, but mostly they do not coagulate. The reason they fail to coagulate any more is that the pressure gradient vanishes because of the dissipation of energy in this two-phase medium; this may also decrease the amplitude of bubble pulsations. The pulsating energy may sometimes be insufficient for overcoming the forces of surface tension. Another reason for not coagulating is that the trajectories of individual bubbles are rather dispersing in the area far of the surface of the emitter, while close to the surface they are radial convergent. On the outer side of the cloud, that is located at a distance of several centimeters from the surface of the emitter, a compact zone of small bubbles is formed. These small bubbles coagulate and continuously generate bubbles of trans-resonant sizes. These trans-resonant bubbles are ejected from the antinode at a high speed toward the periphery. Streams of micro-bubbles continuously supplement the zone of the antinode with gas phase, and trans-resonant bubbles mentioned above are rapidly ejected from this zone, dragging neighboring liquid particles along and forming a liquid micro flow. These liquid micro flows themselves carry along other, much smaller, bubbles. So, in the film we can see trans-resonant bubbles ejected along with a cloud of tiny bubbles.

It is possible to observe the coagulation of large bubbles, and small bubbles following them in both the areas near the emitter and farther from the surface. Simultaneously, with a change in the structure of cavitation zone and a change in the dynamics of individual cavities, the spectrum of cavitation noise also changes dramatically. A continuous part of the spectrum appears and this tells us about the formation of stochastic process.

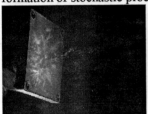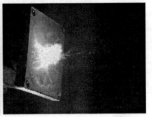

FIGURE 1. Stages of cavitation.

NUMERICAL CALCULATIONS

The numerical calculations were made to show that even a simple model of differential equations that describes both radial and translational motion of a bubble might be used to illustrate the above-mentioned phenomenon, and what one can see in the experimental results shown on the film.

The system of equations of radial and translational motion is taken in the form [4].

$$R\ddot{R} + \frac{3}{2}\dot{R}^2 + \frac{2\sigma}{R} - \frac{1}{\rho}\left[\left(P_0 + \frac{2\sigma}{R_0}\right)\left(\frac{R_0}{R}\right)^3 + P_H\right] - \frac{1}{4}u^2 = -\frac{1}{\rho}(P_0 + P_m\varphi\sin\omega t) \quad (1)$$

$$\frac{d}{dt}\left[\rho'V\vec{u} + \frac{1}{2}\rho V(\vec{u} - \vec{v})\right] = -D(\vec{u} - \vec{v}) - V\nabla P \quad (2)$$

Where R and R_0 are the current and initial radii of a gas bubble, $\varphi(r) = \varphi(Z/L)$ is a function, that characterizes the heterogeneity of acoustic field of the scale L; ω is the angular frequency, σ is the coefficient of surface tension, ρ', ρ are the density of gas and liquid, V is the volume of the bubble, $\vec{u}$ is the velocity vector of spatial motion of the bubble, $\vec{v}$ is the velocity vector of liquid, ∇P is the gradient of sound pressure. $D = 6\pi\mu R\left[1 + 0.065\,\text{Re}^{2/3}\right]^{1/3}$ is the drag coefficient, which characterizes liquid resistance to translational motion of the bubble, $\text{Re} = \frac{2\rho R|\vec{u} - \vec{v}|}{\mu}$ is the Reynolds number, μ is the viscosity of liquid. In the area close to the axis of symmetry and close to the surface of the emitter, the function $\varphi(r)$ may be approximated with a cosine function and this makes some similarities with the results obtained on spherical and cylindrical concentrators.

Numerical calculations show that in addition to spatial bubble migration toward the center of the transducer plate in some cases (see fig. 2), large scale time spatial oscillations take place. This phenomenon is due to the fact that while moving in space, a small under-resonant bubble moves toward the antinode and reaches the area where the amplitude of its radial pulsation is big enough for the bubble to become sub-resonant (see fig. 3). From this point, the bubble changes the direction of its spatial movement to the opposite. This process repeats periodically. Generally speaking, translational oscillations will take place for bubbles of a certain range of sizes in any sound field if the pressure amplitude changes with the space coordinate. This kind of oscillation does not take place in the field of a plane progressive wave. All bubbles are dragged by the sound field toward the direction of the wave propagation, regardless of their size (see fig. 4).

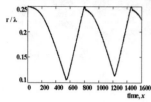 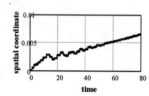

FIG. 2. Large scale time spatial oscillations

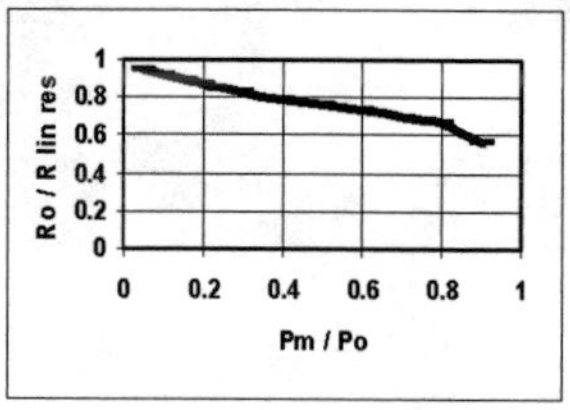

FIG. 3. Critical radius versa amplitude

FIG. 4. Spatial bubble motion in a progressive wave

CONCLUSIONS

1. Radial pulsations and spatial movement of gas bubbles obey general peculiar features regardless of the specific type of transducer and the structure of the inhomogeneous sound field.
2. Micro-cracks of the piston surface, together with bubbles interaction, play a significant role in the origination of cavitation.
3. The closer the bubble's size is to the resonance size, the faster the ejection of the bubble away from the surface of the emitter in the zone of the progressive wave will be.
4. Large, over-resonant bubbles are not formed in the central zone of cavitation – they are either split into smaller ones or ejected.

REFERENCES

1. Crum L.A., and Eller A.I., Motion of Bubbles in a Stationary Sound Field, J. Acoust. Soc. America, 48, 181-189 (1970).
2. Agrest, E. M., and Kuznetsov, G. N., *Sov. Phis. Acoust.* 18, 143-147 (1972)
3. Agrest, E. M., and Kuznetsov, G. N., *Sov. Phis. Acoust.* 19, 212-215 (1973)
4. Agrest, E. M., and Kuznetsov, G. N., *Sov. Phis. Acoust.* 20, 1-5 (1974)
5. Agrest, E. M., and Korets, V. L., *Sov. Phis. Acoust.* 24, 1-5 (1978)

Spatial and Temporal Measurements Of The Impulsive Pressure Generated By Cavitation Bubble Collapse Near A Solid Boundary

Yi-Chun Wang, Ching-Hung Huang, and Ho-Hsun Tsai

Department of Mechanical Engineering, National Cheng Kung University, Tainan 701, TAIWAN

Abstract. The impulsive pressure generated by the collapse of a spark-produced cavitation bubble near a solid boundary is studied experimentally. The spatial distribution and the temporal variation of the transient event are directly recorded using a custom-made PVDF piezoelectric array transducer. The features and the possible mechanisms of the impulsive pressure are discussed. The high sensitivity, low cross-talk, and low cost of the piezoelectric array transducer indicate its applicability in high amplitude impulsive field measurements.

Keywords: Cavitation, Bubble, Impulsive Pressure, PVDF, Transducer Array
PACS: 47.55.Bx, 47.55.Dz, 07.07.Mp

INTRODUCTION

As a cavitation bubble appears in a liquid and expands to its maximum size, the ambient pressure in the surrounding liquid much exceeds the pressure inside the bubble. The bubble will begin to collapse. The collapse is so violent that the bubble can reach a minimum volume several orders smaller than its original value and then rebounds. This process can proceed a number of cycles. When the growth and collapse cycle of the bubble occurs near a solid boundary, the symmetry of the process is destroyed. A liquid jet is formed, threads the bubble, and impacts on the boundary. Very large pressure pulses can be produced in these processes and are responsible for cavitation erosion of the material. Experimental evidences have shown that the amplitude and the duration of these pulses are mainly affected by the stand-off parameter $\gamma \equiv L/R_{max}$, where L is the initial distance of the bubble from the boundary and R_{max} is the maximum bubble radius [1]. The process of the bubble collapse happens so quickly that extremely high-speed photography is necessary to offer enough temporal resolution. Nevertheless, the impact events are covered by the bubble and are very difficult to visualize.

Since one of the most important consequences of cavitation is its erosion on nearby boundaries, the lack of direct measurement of the impulsive pressure generated by the bubble collapse with enough temporal and spatial resolution is considered to be a shortcoming of the previous work [2-5]. To explore the composition of the impulsive pressure generated by cavitation bubble collapse, we construct a transducer array on a 25 μm thick PVDF (polyvinylidene fluoride) piezoelectric film using a laser micro-fabrication technique.

CP838, *Innovations in Nonlinear Acoustics: 17th International Symposium on Nonlinear Acoustics*,
edited by A. A. Atchley, V. W. Sparrow, and R. M. Keolian
© 2006 American Institute of Physics 0-7354-0330-9/06/$23.00

FABRICATION AND CALIBRATION OF ARRAY TRANSDUCER

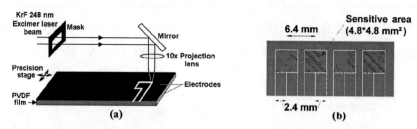

(a)

(b)

FIGURE 1. (a) Electrode Patterning of PVDF film using Excimer laser micro-machining technique. (b) Layout of the PVDF array transducer. Dashed lines indicate the patterning on the reverse side.

The transducer array is fabricated directly on a 25-μm-thick aluminum-metalized polarized PVDF film (FV301926, Goodfellow Cambridge Ltd., U.K.) using an excimer laser micro-machining system (PS-2000, Excitech Ltd., Oxford, U.K.). As illustrated in Fig. 1(a), the laser system emits KrF ultraviolet light pulses with energies up to 350 mJ and pulse duration of 30 ns. After passing a mask, the laser beam is modulated in shape and size and is focused onto the aluminum electrode of the PVDF film using a 10x projection optics. Instant ablation of the metal material occurs due to localized absorption of the high density laser energy. The position and path of the laser beam relative to the target surface are controlled using a PC-based programmable x-y-z precision stage (with resolution of 1 μm). Plotted in Fig. 1(b) is the layout of the PVDF transducer array. The sensitive area is formed by overlapping isolated electrodes from both sides. As subjected to dynamic pressure, electrical potentials only between overlapping electrodes can be extracted. The size of the sensing elements is 4.8x4.8 mm^2, and the spacing between centers of these elements is 6.4 mm. To eliminate the electrical loading effect, a buffer circuit for each of the sensing elements is designed using an operational amplifier (AD843JR, Analog Devices, Inc.).

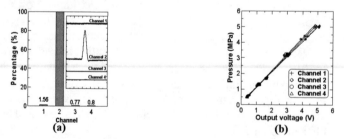

(a)

(b)

FIGURE 2. (a) Coupling between the sensing elements. Impact signals from ball dropping (on channel 2) are also shown in the figure. (b) Calibration results of the PVDF array transducer.

Coupling between the sensing elements is estimated by dropping a steel ball of 8 mm in diameter on one of the elements and monitoring the output signals from all channels. Results are illustrated in Fig. 2(a). Comparison of the signal amplitudes shows an insignificant crosstalk level less than 2%. Dynamic calibration of the array transducer is accomplished using a gas shock tube (PCB Model 901A10, PCB Piezotronics Inc.). A shock wave is generated and is reflected at the end plate of the tube on which the array transducer is mounted. Shock speed is measured and is used to calculate the amplitude of the reflected shock (with accuracy approximately of ±1.5%). Calibration results are demonstrated in Fig. 2(b), showing a good uniformity of the sensing elements.

MEASUREMENTS OF CAVITATION IMPULSE PRESSURE

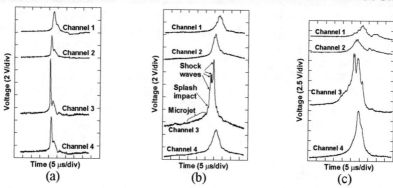

FIGURE 3. Impulsive signals generated by bubble collapse for (a) $\gamma = 2.8$ (b) $\gamma = 0.9$ (c) $\gamma = 0.35$.

Cavitation bubble is generated in de-ionized water by discharging a capacitor (1.2 μF) across a pair of tungsten electrodes (1 mm in diameter and 0.3 mm in gap distance). The maximum radius of the cavitation bubble, R_{max}, is measured directly using a high-speed CCD camera (Kodak SR-Ultra, maximum framing rate of 10000 frames per second). By careful control of the electrode gap distance and the discharge voltage, cavitation bubbles of a quite stable size ($R_{max} = 9.82$ mm) can be realized. Various values of the stand-off parameter, $\gamma = L/R_{max}$, can then be set by adjusting the distance L (using a micrometer).

Figure 3 shows the impulsive signals induced by the first collapse of the bubble for $\gamma = 2.8$, 0.9 and 0.36 respectively. For the relative large value of $\gamma = 2.8$ (Fig. 3a), the pressure pulses measured by all of the four channels show a fast rising and short duration shock wave characteristics. The shock first reaches the channel 3 (which registers the earliest and the largest pulse of 136 ns rise time and 595 ns duration) and propagates outwards. The speed of the shock can be estimated based on the distance of the sensing elements and the time difference between the signals and is found to vary from 8889 to 4571 m/s as it propagates from channel 3 to channel 1. About 1.72 μs later, a second spike is observed, although it is much smaller than the main pulse. To explain the pressure loading in Fig. 3(a), we refer to the recent work by Lindau and Lauterborn [5]. By employing a shadowgraph illumination and an extra high-speed camera of maximum framing rate of 100 million frames per second, they visualized shock waves originating from different flow dynamic mechanisms for the case of $\gamma = 2.6$. The strongest shock, so called the *tip bubble shock wave*, was emitted (at a speed of about 7000 m/s) toward the solid boundary due to the convergence of the *jet torus shock wave* which was generated as the liquid jet impacted on the lower bubble wall. Behind the tip bubble shock wave, two *compression shock waves* appeared; one was only seen to propagate away from the boundary and the other propagated in all direction. It should be mentioned that the bubble in the present work is about 6.6 times larger than that of [5] and, therefore, can induce much stronger shocks.

For $\gamma = 0.9$ (Fig. 3b), several (at least five) overlapping but distinguishable fast rising spikes appear first on channel 3. However, compared to Fig. 3(a), a much longer lasting and smaller amplitude jet-impact pressure is recorded before these spikes. It is well known that, for this value of γ, the liquid jet penetrates the bubble before it is compressed to minimum volume [1-6]. The first spike right follows the jet signal has a slower rising time (520 ns) compared to others and is probably caused by the impact between the projecting splash (which is generated after the liquid jet collides on the boundary and flows radially outwards) and the inwards moving liquid induced by the bubble collapse [6]. The train of four spikes immediately following the splash impact corresponds to the loading of multiple shock waves. These shocks overtake each other, coalesce into a single wave form, and propagate outwards. The other three channels receive the shock loading at different time. The shock speed calculated from the output signals of channel 1 and channel 2 is approximately 2319 m/s. The source of these shocks was highlighted recently by Brujan et al. [4]. They focused on the bubble stand-off parameter $\gamma \approx 1$ and pictured emission of multiple shock waves after the liquid jet penetrated the bubble, splashed away from the solid boundary, and led to the formation of two toroidal cavities split from the main bubble; the torus farther from the boundary disintegrated into several parts, collapsed separately, and emitted several different shock waves.

Figure 3(c) shows the transducer output for the small value of $\gamma = 0.36$. The signal of channel 3 (which is placed right under the bubble) is similar to that in Fig. 3(b); a liquid jet impacts on the surface and then several shock waves are recorded. However, since the liquid film between the bubble and the boundary is now very thin, the jet impact is more directly and pronounced. Another interesting difference between Fig. 3(c) and 3(b) is the relative timing of the signals. In Fig. 3(c) channel 2, 3 and 4 receive impulses almost simultaneously. Previous study [3, 5] has demonstrated that for small value of γ the liquid jet setups an unstable torus bubble right on the surface. Multiple simultaneous collapses happen along the major axis of the torus which, we believe, covers the area of the sensing elements corresponding to the above three channels.

ACKNOWLEDGMENTS

This work was sponsored by National Science Council, Taiwan under contract No. NSC 91-2212-E-006-154.

REFERENCES

1. Tomita, Y. and Shima, A., *J. Fluid Mech.*, **169**, 535-564 (1986).
2. Vogel, A., and Lauterborn, W., and Timm, R., *J. Fluid Mech.*, **206**, 299-338 (1989).
3. Philipp, A. and Lauterborn, W., *J. Fluid Mech.*, **361**, 75-116 (1998).
4. Brujan, E. A., Keen, G. S., Vogel, A., and Blake, J. R., *Phys. Fluids*, **14**(1), 85-92 (2002).
5. Lindau, O. and Lauterborn, W., *J. Fluid Mech.*, **479**, 327-348 (2003).
6. Shaw, S. J., Schiffers, W. P., Gentry, T. P., and Emmony, D. C., *J. Acoustic Soc. of Am.*, **107**, 3065-3072 (2000).

Magnetic Resonance Imaging of Acoustic Streaming in Cavitating Fluid

Igor Mastikhin, Ben Newling

MRI Centre, Physics Department, University of New Brunswick, 8 Bailey Dr.,
Fredericton, E3B 5A3, Canada. Email: mast@unb.ca

Abstract. Acoustic streaming (AS) is a bulk flow caused by attenuation of an acoustic wave propagating in the medium [1]. When cavitating bubbles are present in the fluid, they actively absorb acoustic waves, generating acoustic streaming. Therefore, measurements of acoustic streaming can provide information on the cavitation field [2]. In this work, Magnetic Resonance Imaging was applied to studies of cavitating fluid in a standing acoustic wave at an acoustic frequency of 31kHz. A spin echo Pulsed Field Gradient sequence was employed to sensitize the measurement to motion. Velocity spectra, kinetic energy maps and maps of the hydrodynamic dispersion coefficient were obtained for air-saturated water, water with surfactant ([SDS] = 1 mM) and water with SDS/NaCl ([NaCl] = 0.1 M). Cavitation bubbles cause an increase in dispersion coefficient and acoustic streaming. These effects are not observed in degassed samples. Streaming was most developed in samples with surfactants, which also demonstrate a pronounced anisotropy of the dispersion coefficient. Stabilization of the bubble surface and reduction of bubble coalescence by the surfactant can explain the observed differences.

Keywords: Cavitation, Magnetic Resonance Imaging, Acoustic Streaming

PACS: 76.60.Pc, 47.55.Bx, 43.25.Nm

INTRODUCTION

Acoustic cavitation is the generation, oscillation and collapse of gaseous bubbles in fluids under the action of high-power ultrasound. The prevalent methods used in studies of cavitation are optical and acoustical. By their nature, these techniques are sensitive to changes in optical or acoustical transparency respectively. In this work, we apply magnetic resonance imaging (MRI) methods to the study of a cavitating fluid. Linear gradients of magnetic field allow spatially resolved information on fluid dynamics to be obtained, non-invasively and from an optoacoustically opaque medium. We show the applicability of MRI to studies of the dynamics of cavitating fluid, with spatially resolved measurements of velocity spectra and hydrodynamic dispersion coefficient.

The measurements were performed during the initial stage of degassing cavitation for air-saturated water, degassed water, water containing a surfactant

CP838, *Innovations in Nonlinear Acoustics: 17th International Symposium on Nonlinear Acoustics*,
edited by A. A. Atchley, V. W. Sparrow, and R. M. Keolian
© 2006 American Institute of Physics 0-7354-0330-9/06/$23.00

(sulphur dodecyl sulphate, SDS), and water with SDS and NaCl. It is known that surfactants modify the dynamics of cavitating bubbles, affecting sonochemistry and sonoluminescence in the cavitation field [3,4]. We have investigated the surfactant modification of the dynamics of cavitating fluid on the macroscale.

EXPERIMENTAL

All measurements were performed on a *Nalorac* (Martinez, USA) 2.35 T, horizontal-bore superconducting magnet with *Tecmag* (Houston, USA) Apollo console. A water-cooled 20-cm-id. *Nalorac* magnetic field gradient set was driven by *Techron* (Elkhart, USA) 8710 amplifiers.

A quadrature birdcage radiofrequency (RF) coil (*Morris*, Canada) was driven by a 2-kW *AMT* (Brea, USA) 3445 RF amplifier.

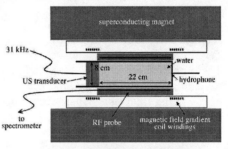

The 31-kHz Langevin-type ultrasonic transducer (*SensorTech*, Canada), with a parallel beam, was situated coaxially inside a 5-cm-i.d. 22 cm-long cylindrical vessel at the centre of the magnet (Fig.1). The vessel was aligned with the direction of the main magnetic field.

FIGURE 1. Apparatus

We employed a displacement-sensitive pulsed field gradient (PFG) sequence [5] with slice selection. Series of 32 PFG measurements were performed, requiring 10.6 min of the total experiment time. The ultrasonic transducer was activated after the 4th measurement in a series and turned off after the 25th. The series were acquired for degassed water, air-saturated water, water with sodium dodecyl sulfate (SDS, 1mM), and water with SDS and NaCl (0.1M). Velocity spectra, kinetic energy maps and maps of dispersion coefficient were obtained for all water samples.

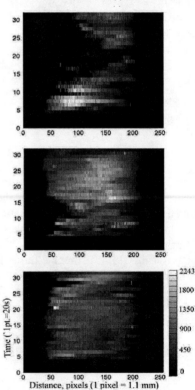

RESULTS & DISCUSSION

No streaming was observed in degassed water. In non-degassed water samples, streaming appeared after the ultrasound was switched on (#5): a coherent flow along the vessel with circulation was detected in all samples.

FIGURE 2. Kinetic energy maps for a) water, b) water with SDS, c) water with SDS/NaCl. US transducer is at the left. Each row is a profile along the tube of water; the interval between rows is 20 s.

492

Clearly, the presence of gaseous bubbles, which are excellent acoustic energy absorbers, is critical for the observed streaming. When the experiments were performed on the same water sample several times in a row at 5 min intervals, the streaming became weaker with every subsequent experiment, indicating a partial degassing of the sample.

The observed streaming must be produced by cavitation bubbles, as energy absorbers (Eckart streaming) or fluid motion activators (quasi-streaming). Most probably, both types of streaming are present. Quasi-acoustic streaming is characterised by a decelerating flow, while Eckart streaming is characterized by a flow accelerating away from the transducer. Immediately after the start of sonication, all the velocity profiles were decelerating. In the second half of the experiments (from the 15th PFG measurement), there was an accelerating flow in SDS-containing samples.

In relative kinetic energy maps (Fig.2), it is shown that in water without surfactants (Fig.2a), the coherent flow was destroyed within first minute of sonication while in water with surfactants (b-SDS, c-SDS/NaCl), it stayed during the whole sonication interval (#25). Note that in water with SDS and salt, the streaming is reduced in its intensity.

The dynamics of both types of acoustic streaming should reflect the behaviour of cavitating bubbles in the acoustic field. Indeed, our measurements demonstrate that for three water samples (air-saturated distilled water, water with SDS, and water with SDS/NaCl), the acoustic streaming patterns are different. An observed enhancement of streaming by SDS can be explained by the stabilization of the bubbles and reduction of their coalescence.

SDS is known for its effects on the cavitation field, especially on bubble interactions [3,4]. A working hypothesis [4] is that, as the anionic surfactant accumulates at the bubble interface, the bubbles acquire a negative charge, and the charged bubbles then tend to repel each other. The maximum effect of surfactant addition on sonochemistry was found experimentally to be at [SDS] of 1-3 mM [4]. An addition of NaCl reduces the SDS effects, because NaCl screens the electrostatic field.

Hydrodynamic dispersion maps showed a strong increase in dispersion soon after the sonication start (#5). Profiles in Fig.3 are dispersion maps summed from #5 till #15, and for two directions: dispersion along the

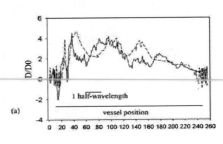

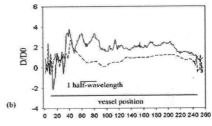

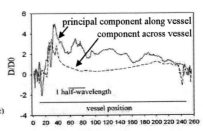

FIGURE 3. Averaged relative dispersion coefficient maps for a) water, b) water with SDS, c) water with SDS/NaCl. US transducer is at the left. D0 is the free self-diffusion coefficient of water at 20 C. These profiles along the tube are the sum over 10 measurements (20 s each).

vessel, and dispersion across the vessel. A very unusual feature of the dispersion increase is its anisotropy in water with SDS: the principal component of dispersion coefficient along the vessel increases (solid line), while the component in the perpendicular (broken line) direction does not change significantly.

At least two factors can increase the dispersion in cavitating fluid: microstreaming and bubble migration. Microstreaming is caused by cavitating bubbles and is thought to be responsible for an active mixing of the fluid in cavitation fields. The second source of dispersion in the standing wave is the bubble migration between the nodes and antinodes of pressure. Origins of such migration are rectified diffusion that causes bubble growth, and the primary Bjerknes force that attracts bubbles smaller than the resonant size towards, and repels larger bubbles away from the nodes.

Bubble migration is expected to affect the dispersion along the direction of the sound wave propagation, i.e. along the principal axis of the vessel in our case. Migration will generate an anisotropic dispersion. Microstreaming should not have a preferred direction in the bulk of the liquid. Only near the rigid boundaries will the microflows be directed towards the boundaries [9]; this boundary effect can be neglected in our experiments, because the most of the detected signal comes from the bulk. Microstreaming will generate an isotropic dispersion in the fluid.

Therefore, the observed anisotropy of the dispersion for water with SDS and SDS/NaCl can be explained by changes in the relative importance of these two factors in solutions with surfactants, for example, a reduction of microstreaming. While the dispersion profiles of the air-saturated water show approximately the same locations of zones with high dispersion coefficient, both for displacements along and across the vessel, the dispersion profiles of water with surfactants show no high dispersion zones for the displacements across the vessel. The stabilization of bubbles with surfactants reduces bubble fragmentation, decreasing the microstreaming and reducing the isotropic component of dispersion.

ACKNOWLEDGMENTS

The authors thank Prof. B. J. Balcom for useful discussions and the Natural Sciences & Engineering Research Council of Canada for financial support. The UNB MRI Centre is supported by an NSERC Major Facilities Access Award.

REFERENCES

1. W.L. Nyborg. Acoustic streaming. In *Physical Acoustics*, edited by W.P.Mason, Vol.2, Part B (Academic Press, New York 1965).
2. C.Campos-Pozuelo, C.Granger, C. Vanhille, A.Moussatov, B.Dubus. *Ultrasonics Sonochemistry* 12, 79 (2005).
3. N.Segebarth, O.Eulaerts, J.Reisse, L.A.Crum, and T.J.Matula. *J.Phys. Chem B* 106, 9181 (2002).
4. M. Ashokkumar, R.Hall, P.Mulvaney, and F.Grieser. *J.Phys. Chem. B* 101,10845 (1997).
5. O.Stejskal and J.E.Tanner. *J.Chem.Phys.* 42, 288 (1965).

Acoustic Radiation Force On Elliptical Cylinders And Spheroidal Objects In Low Frequency Standing Waves

Philip L. Marston, Wei Wei, and David B. Thiessen

Physics and Astronomy Dept., Washington State University, Pullman, WA 99164-2814

Abstract. The acoustic radiation force on symmetric objects in standing waves is expressed using partial-wave coefficients for the more elementary problem of scattering of traveling waves. From known low frequency scattering properties of elliptical cylinders, ellipsoids, and disks, deformation into flatter profiles should increase the radiation force on dense objects in air.

Keywords: Radiation Pressure.
PACS: 43.25.Qp

INTRODUCTION AND REVIEW OF APPLICATIONS

The acoustic radiation force of standing waves is used to trap, manipulate, or separate small objects such as drops, bubbles, and particles. By modulating the acoustic radiation pressure it is possible to excite low frequency capillary or elastic modes of objects [1,2]. While small fluid objects such as drops or bubbles may ordinarily be a spherical shape when capillary forces are dominant, even in the absence of modulation these objects deform in a standing wave [2,3]. Similarly, liquid cylinders that would ordinarily be circular take on an elliptical shape [4]. We outline here a simple way to describe how the radiation force on small objects changes as the shape of the object changes. The approximations presented here concern situations where the deformation is weak and where radiation forces are well approximated by our analysis that assumes an inviscid outer fluid. If R denotes the radius of the object and k the acoustic wavenumber in the outer fluid, this condition can be met even if kR << 1, provided the kinematic viscosity of the outer fluid is small. The evaluation of the effects of shape on the radiation force can be complicated in the traditional nearfield approach of King [5] because the relevant surface projection of the radiation force is evaluated and integrated over the surface of the object. Instead we evaluate the radiation force from the farfield scattering. Initially we review this approach for cylinders [6]. Shifting the relevant stress integration to a surface having a large radius, the farfield form of the scattered wave may be used in the radiation force without introducing any approximation. For cylinders, scattering contributions that fall off faster than $1/\sqrt{r}$ give a vanishing contribution to cylindrical surface integrals for the radiation force when r goes to infinity. In 3-dimensions, scattering contributions for spherical surface integrals that fall off faster than $1/r$ do not affect the radiation force

CP838, *Innovations in Nonlinear Acoustics: 17th International Symposium on Nonlinear Acoustics*, edited by A. A. Atchley, V. W. Sparrow, and R. M. Keolian
© 2006 American Institute of Physics 0-7354-0330-9/06/$23.00

when r diverges. In both cases the scattering by an incident traveling wave is expressed by a partial wave series having partial wave coefficients $a_n = \alpha_n + i\,\beta_n$ as explained below. These coefficients are determined by solving the scattering problem with the appropriate shape, size, and material properties of the scatterer. Let the spatial part of the complex amplitude of the incident traveling wave be denoted as $P_{i0} = P_a \exp(ikz)$ and the time dependence as $\exp(-i\omega t)$ where $k = \omega/c$ where c is the speed of sound. The coefficients are then used to express the radiation force on symmetric objects in standing waves. The z axis is also taken to be the wavevector axis of the standing wave. In our approach it is necessary for the symmetry of the object to be an even function of the scattering angle θ. An example is shown in Fig. 1. The radiation force depends on the location of the object in the standing wave. This is expressed using the distance of the origin used in the expansion of the outgoing radiation from the location of the pressure antinode that would be present in the standing wave if the object were not present. Denote this quantity by h where $z = -h$ is the location of the pressure antinode nearest to the origin of the coordinate system used in the specification of the a_n.

FORCE ON CIRCULAR AND ELLIPTICAL CYLINDERS

The cylinder's axis is perpendicular to the z axis. Using a dimensionless function Y_{st} the radiation force-per-length on a cylinder in a standing wave becomes [6]:

$$F_z/L = (R/4)\,P_s^{\,2}\,\kappa_o \sin(2kh)\,Y_{st} \qquad (1)$$

where κ_o is the compressibility of the surrounding fluid and P_s is the standing wave pressure amplitude. In the case of a circular cylinder R is the radius of the circular cylinder. For other shapes, R is a reference radius and may be selected as discussed below. Equation (19) of [6] expresses Y_{st} exactly from the coefficients a_n. Subsequently, Mitri [7] gave what may appear to be a different series for Y_{st} in the circular cylindrical case, however, with an appropriate grouping of terms [8], the series in [6] may be converted to Mitri's result. For a circular cylinder of radius R when $kR \ll 1$ only the monopole ($n=0$) and the dipole ($n=1$) values of a_n are needed. For a fixed (or massive) rigid circular cylinder with $kR \ll 1$, $Y_{st} \approx 3\pi kR$. Consider now the fixed (or massive) rigid elliptical case shown in Fig. 1 where $ka \ll 1$ and $kb \ll 1$ and let $\varepsilon = [(b/a) - 1]$ denote the deviation from a circular shape where $|\varepsilon| \ll 1$. The coefficients a_0 and a_1 may be approximated from results of Rayleigh [9] and Twersky [10]. Let $R = \sqrt{(ab)}$ in Eq. (1) denote the radius of the circle having the same area as the elliptical cylinder. In Fig. 1, $\varepsilon > 0$ and the radiation force function on a small ellipsoid is found in [6] to have the following approximate expansion: $Y_{st} \approx 3\pi kR\,[1 + (\varepsilon/3)]$. The attraction of the cylinder to the velocity antinode increases when ε increases. Liquid bridges in air deform in standing waves so as to make $\varepsilon > 0$.

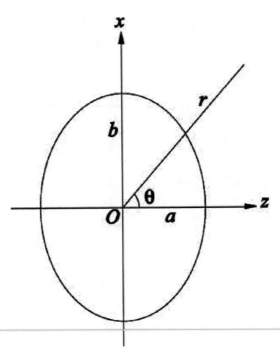

FIGURE 1. Sectional view of an ellipsoidal cylinder or an oblate spheroidal drop. In the case of an acoustically levitated oblate drop, the axis z of rotational symmetry is vertical. In the case of an ellipsoidal cylinder, the cylinder's axis is the y axis. The traveling and standing waves discussed are along the z axis.

FORCE ON SPHERICAL, OBLATE, AND PROLATE DROPS

When a traveling wave is incident along the symmetry axis of an oblate or prolate spheroid, the scattering for the region outside the smallest sphere that bounds the spheroid may be written as

$$P_{ss} = P_a \sum_{n=0}^{\infty} i^n (2n+1) a_n h_n^{(1)}(kr) P_n(\cos\theta) \tag{2}$$

where $h_n^{(1)}$ is a spherical Hankel function, P_n is a Lengendre polynomial of $\cos\theta$, and $a_n = \alpha_n + i\beta_n$ is the partial wave coefficient. The radiation force depends on the a_n through the same relationship that is found for the special case of a spherical object. For the special case of a standing wave, the radiation force is found from the appropriate special case of Hasegawa [11]. It is convenient to define a new dimensionless radiation force function for standing waves Q_{st} which is related to the radiation force by:

$$F_z = \kappa_o P_s^2 \pi k^{-2} Q_{st} \sin(2kh) \tag{3}$$

where κ_o is the compressibility of the surrounding fluid and P_s is the standing wave pressure amplitude. Inspection of Eq. (27) of [11] gives:

$$Q_{st} = \sum_{n=0}^{\infty} (n+1)(-1)^{n+1} [\beta_n (1+2\alpha_{n+1}) - \beta_{n+1}(1+2\alpha_n)]. \tag{4}$$

For a sphere of radius R the usual dimensionless force function [11] is $Y_{st} = 8Q_{st}/(kR)^2$. At sufficiently low frequencies only the monopole (n=0) and dipole (n=1) terms are important [9] and to leading order, in the absence of dissipation, $Q_{st} \approx -\beta_0 + 3\beta_1$. The following example is relevant to the levitation of small drops in air. At low frequencies, for the case of a small fixed (or massive) rigid spheroid the relevant monopole and dipole coefficients may be written [12,13]:

$$\beta_0 = -k^3 V / (4\pi), \qquad \beta_1 = k^3 M_{zz} / (12\pi) \tag{5}$$

where V is the volume of the spheroid and M_{zz} is a component of the *magnetic-polarizability tensor* for the spheroid with a symmetry axis z. (This tensor is based on a magnetic analogy for an object of the same shape [12].) The radiation force function becomes.

$$Q_{st} \approx k^3 V [1+ (M_{zz}/V)] / (4\pi). \tag{6}$$

In the case of a sphere $M_{zz}/V = 3/2$ and the force F_z reduces to King's result [5,14]. Consider now the case of a spheroid where as in Fig. 1, $2a$ denotes the length along the symmetry axis and $2b$ denotes the width. The volume is $V = (4/3)\pi ab^2$. Let $\varepsilon = [(b/a) - 1]$ denote the dimensionless deviation from a spherical shape. For an *oblate* spheroid $\varepsilon > 0$. The function M_{zz}/V may be evaluated as described by Senior [12] and originally by Rayleigh [9] to give $Q_{st}(\varepsilon)$. We find the following expansion for the force ratio relative to the case of a sphere having the same volume:

$$Q_{st}(\varepsilon) / Q_{st}(0) \approx 1+ (6/25)\varepsilon + (9/875)\varepsilon^2 + \dots \tag{7}$$

Since $Q_{st}(0) = 5k^3V/(8\pi) > 0$, it follows that the attraction to the velocity antinode increases for drops levitated in air the more oblate the drop. Prolate spheroids have $\varepsilon < 0$ and are found to have the same expansion for this ratio. For the unusual case of a prolate drop, the radiation force decreases the more prolate the drop.

FORCE ON SMALL THIN DISKS IN STANDING WAVES

The rotational symmetry axis of the disk is the z-axis so that Eq. (6) applies for a small disk. Let b denote the disk radius and $2a$ the thickness. It follows that $V =$

498

$2\pi ab^2$ and from Pierce [13]: $M_{zz} = (8/3)b^3$. The low-frequency radiation force on a dense rigid disk from Eqs. (3) and (6) agrees with King [15]. Introducing a shape parameter of $\gamma = b/a$ used by Xie [3] (in terms of our parameters), the ratio of the radiation force on a disk to that on a sphere of the same volume is predicted to be $f_0 = [1 + (M_{zz}/V)]/(5/3) = 0.4 + 0.1697\,\gamma$. From numerical solutions based on an integral equation, Xie fits computational results to $f_0 = 0.758 + 0.177\,\gamma$. The agreement with our model is satisfactory since when γ is small our thin-disk assumption breaks down.

ACKNOWLEDGMENTS

This research was supported by NASA. Aspects of the scattering research were supported by ONR.

REFERENCES

1. Marston, P. L., and Apfel, R. E., *J. Colloid Interface. Sci.*. **68**, 280-286 (1979).
2. Marston, P. L., and Thiessen, D. B., *Annals of the New York Academy of Sciences* **1027**, 414-434 (2004). [In Eq. (24) for the drop shape replace ") (" by ") / (".]
3. Xie, W. J., and Wei, B., Phys. Rev. E 70, 046611 (2004).
4. Marr-Lyon, M. J., Thiessen, D. B., and Marston, P. L., *Phys. Rev. Lett.* **86**, 2293-2296 (2001). [An erratum that K-V should read as V-K has been published in *Phys. Rev. Lett.* **87**, 209901 (2001).]
5. King, L. V., *Proc. Roy. Soc.* **A147**, 212-240 (1934).
6. Wei, W., Thiessen, D. B., and Marston, P. L., *J. Acoust. Soc. Am.* **116**, 201-208 (2004); **118**, 551 (E) (2005).
7. Mitri, F. G., *Eur. Phys. J. B*, **44**, 71-78 (2005).
8. Wei W., and Marston, P. L., submitted to *J. Acoust. Soc. Am.*
9. Lord Rayleigh, *Philos. Mag.* **44**, 28-52 (1897).
10. Twersky, V., *J. Acoust. Soc. Am.* **36**, 1314-1329 (1964).
11. Hasegawa. T., *J. Acoust. Soc. Am.* **65**, 32-40 (1979).
12. Senior, T. B. A., *J. Acoust. Soc. Am.* **153**, 742-747 (1973).
13. Pierce, A. D., *Acoustics: An Introduction to Its Physical Principles and Applications*, Woodbury, NY: Acoustical Society of America, 1989, pp. 425-428.
14. Wang, T. G., and Lee, C. P., "Radiation Pressure and Acoustic Levitation," in *Nonlinear Acoustics*, edited by M. F. Hamilton and D. T. Blackstock, San Diego, California: Academic Press, 1998, pp. 177-205.
15. King, L. V., *Proc. Roy. Soc.* **A153**, 1-16 (1935).

Profile of Liquid Droplet Agitated by Counter-Propagating Rayleigh Waves: Standing-Wave Soliton or Ultrasonic Fountain?

Boris A. Korshak, Vladimir G. Mozhaev and Anna V. Zyryanova

Acoustics Department, Faculty of Physics, Moscow State University, 119992 GSP-2, Russia

Abstract. An experimental study of the interaction between droplets of various liquids and 15 MHz-surface acoustic waves (SAWs) of Rayleigh type on lithium niobate substrate has been undertaken. The results are recorded as short movies. In the traveling-wave mode of operation, the known phenomena of acoustic droplet transport and vortex acoustic streaming inside the droplets are observed. In the standing-wave mode, a new phenomenon of quasi-stationary change of droplet profile under the action of counter-propagating SAWs is observed. The dynamic profile consists of a pedestal of uniform thickness above which a smooth liquid column is formed. Two alternative interpretations of this phenomenon namely standing-wave soliton and ultrasonic fountain are suggested. The calculation of radiation force generated by two counter-propagating leaky Rayleigh waves at the boundary between solid and liquid half-spaces is presented to support the last mentioned interpretation of the found phenomenon.

Keywords: surface acoustic waves, biochips, droplet shape, acoustic radiation pressure
PACS: 43.25.Nm, 43.25.Qp, 43.35.Pt, 62.35.+v, 68.08.De

INTRODUCTION

Planar microfluidic systems based on the technology of surface-acoustic-waves (SAWs) in piezoelectric substrates offer new and exciting opportunities to develop programmable biochips [1-6]. These new devices are of great importance and offer much promise for future applications in medicine and molecular biology, as well as micro- and nanochemistry. Prototypes of these devices are already available and are described on the website www.advalytix.de. Nevertheless, some important aspects of the physics of the interaction between SAWs and liquid droplets are still not understood in detail, although they are of interest for developing high-performance programmable biochips. In particular, this concerns the shape of droplets.

Surface tension forces dominate in liquid behavior at the microscale. Namely these forces are responsible for the formation and shape of liquid droplets. On the other hand, droplet oscillations under the action of surface acoustic waves give rise to additional forces which can greatly change the droplet shape. The experimental observation of such changes is the primary focus of the present study. The calculation of radiation force in a liquid half-space generated by counter-propagating leaky Rayleigh waves is also performed to assist in interpreting experimental results.

CP838, *Innovations in Nonlinear Acoustics: 17th International Symposium on Nonlinear Acoustics*,
edited by A. A. Atchley, V. W. Sparrow, and R. M. Keolian
© 2006 American Institute of Physics 0-7354-0330-9/06/$23.00

EXPERIMENTAL OBSERVATIONS

The experiments have been performed using droplets of different liquids, including water, white spirit solvent, acetone and ethyl alcohol, placed on YZ-lithium niobate 15-MHz SAW delay line. Interdigital transducers deposited on the surface at opposite ends of the substrate allows us to generate one traveling SAW or two counter-propagating SAWs. The results are recorded as short movies. In the traveling-wave mode of operation, the known phenomena of acoustic droplet transport and vortex acoustic streaming inside the droplets are observed. The velocity of acoustic transport of droplets of white spirit solvent reaches 15 mm/s for SAW power about 0.1 Wt. The visualization of acoustic streaming inside the droplets is achieved by the use of potassium permanganate crystals as coloring agent. The turning velocity of acoustic vortices of 100-300 degrees per second is detected in droplets of ethyl alcohol. In the standing-wave mode, a new phenomenon of quasi-stationary change of droplet profile under the action of counter-propagating SAWs is observed. A snapshot from the movie demonstrating the peak appearance on the surface of acetone droplet is shown in Fig. 1.

FIGURE 1. Quasi-stationary change of droplet profile under the action of counter-propagating SAWs.

FIGURE 2. Profile of liquid droplet: (a) equilibrium state, (b) traveling-wave interaction, (c) counter-propagating wave interaction.

The sketches of droplet profiles and peaks observed in the experiment are given in Fig. 2. The dynamic profile formed by counter-propagating waves consists of a pedestal of uniform thickness above which a smooth liquid column is formed (Fig.

2c). Two alternative interpretations of this phenomenon such as standing-wave soliton [7] and ultrasonic fountain [8] are considered. The first interpretation looks less convincing since we did not observe a change of frequency for stroboscopic flashing effect in movies for droplets of different size.

RADIATION FORCE

The calculation of radiation force generated by two counter-propagating leaky Rayleigh waves at the boundary between solid and liquid half-spaces is presented to support the last of the suggested interpretations of the found phenomenon. The geometry of this problem is shown in Fig. 3.

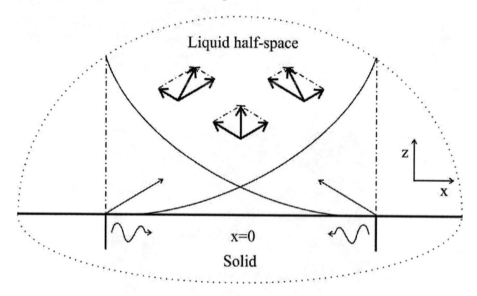

FIGURE 3. Geometry of the problem.

The potentials φ of the particle velocities v in liquid ($v = grad\varphi$) for counter-propagating leaky waves are assumed to be of the form:

$$\varphi_+(z \geq 0) = \varphi_0 \exp(ik_x x + ik_z z - \alpha x + \beta z), \qquad (1)$$

$$\varphi_-(z \geq 0) = \varphi_0 \exp(-ik_x x + ik_z z + \alpha x + \beta z). \qquad (2)$$

These expressions imply that the counter-propagating leaky waves have the same amplitudes at x = 0. The result of calculation of the amplitude of radiation force component, given by $f_i = \rho_0 (\partial/\partial x_k)\overline{(v_i v_k)}$, is the following

$$|f|^2 = 4\varphi_0^2 \varphi_0^{*2} \rho_0 \exp(4\beta z)\left[A^2 \sinh^2(2\alpha x) + C^2 \cosh^2(2\alpha x)\right], \qquad (3)$$

where $A = -(k_x^2 + \alpha^2)\alpha + (k_x k_z - \alpha\beta)\beta$, $C = (k_z^2 + \beta^2)\beta - (k_x k_z - \alpha\beta)\alpha$. Eq. (3) has been used to calculate the profile of the surface where the amplitude of the radiation force is constant (Fig. 4). It should be noted that the direction of the force along this profile is not constant. Far from the plane x = 0 this force is predominantly determined by one of the waves and so it is directed upward and very close to the side lines of the curve shown in Fig. 4. On the other hand, this force becomes pure vertical and it is directed upward at x = 0. Similar behavior of the radiation force might be expected in the center of drops. This supports the second suggested interpretation of the found phenomenon.

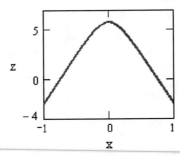

FIGURE 4. Profile of the surface where the amplitude of radiation force is constant.

ACKNOWLEDGMENTS

A.V. Zyryanova acknowledges the support of her study by DAAD.

REFERENCES

1. C. J. Strobl, A. Rathgeber, A. Wixforth, C. Gauer, and J. Scriba, "Planar Microfluidic Processors" in *2002 IEEE Ultrasonics Symposium Proceedings*, IEEE, 2002, vol. 1, pp. 255-258.
2. A. Wixforth, C. Gauer, J. Scriba, M. Wassermeier, and R. Kirchner, "Flat Fluidics: A New Route Toward Programmable Biochips" in *Proceedings SPIE*, 2003, vol. 4982 Microfluidics, BioMEMS, and Medical Microsystems, pp. 235-242.
3. A. Wixforth, *Superlattices and Microstructures* **33**, 389-396 (2003).
4. Z. Guttenberg, A. Rathgeber, S. Keller, J. O. Raedler, A. Wixforth, M. Kostur, M. Schindler, and P. Talkner, *Phys. Rev.* **E 70**, 056311 (2004).
5. C. J. Strobl, Z. von Guttenberg, and A. Wixforth, IEEE Trans. **UFFC-51**, 1432-1436 (2004).
6. S. Alzuaga, S. Ballandras, F. Bastein, W. Daniau, B. Gauthier-Manuel, J. F. Manceau, B. Cretin, P. Vairac, V. Laude, A. Khelif, and R. Duhamel, "A Large Scale X-Y Positioning and Localisation System of Liquid Droplet Using SAW on LiNbO3" in 2003 IEEE Ultrasonics Symposium Proceedings, IEEE, 2003, pp. 1790-1793.
7. R. Wei, B. Wang, Y. Mao, X. Zheng, and G. Miao, *J. Acoust. Soc. Am.* **88**, 469-472 (1990).
8. C. Cinbis, N. N. Mansour, B. T. Khuri-Yakub, *J. Acoust. Soc. Am.* **94**, 2365-2372 (1993).

Transient high frequency boosting of single bubble sonoluminescence

Jean-Louis Thomas and Nicole Bras

*Institut des NanoSciences de Paris, UMR CNRS 7588, Université Pierre et Marie Curie,
140, rue de Lourmel, 75015 Paris, France*

Abstract. Boosting of single bubble sonoluminescence is achieved by sending on the bubble a high frequency pulse of ultrasound. This pressure pulse is generated by eight piezo-electric transducers focused adaptively on the bubble. By changing the location of the pulse different regimes are available and the corresponding experimental results are compared with numerical simulations.

Keywords: Sonoluminescence, transient boosting, ultrasound.
PACS: 78.60.Mq, 47.55.Dz, 43.35.+d

Single bubble sonoluminescence is characterized by a fast compression of the gas contained in a collapsing bubble leading to the generation of UV and visible photons with ultrasound of centimetric wavelength[1,2]. The estimate of the gas temperature at the end of the bubble collapse has varied of four orders of magnitude according to the hypothetical existence of an acoustic shock wave inside the bubble. Recent developments show that water vapour plays a very important role in damping the temperature increase but it may also help to generate the shock wave by lowering the sound velocity[3,4]. This work presents experimental results of transient boosting of sonoluminescence for which both the dynamic and the water vapour content is affected[5]. The boosting is get by focusing the pulse generated by eight high frequency transducers. A complete description of the experimental procedure has been published previously[5]. Different kinds of transient boosting of sonoluminescence can be achieved by changing the shape of the waveform and its timing. The experimental results are compared with a model taking into account water vapour condensation-evaporation with a diffusion limited behavior[6], chemical reactions limited by the effect of the exluded volume[7,8], and radiation from a weak ionised plasma[9]. Note that this model assumes that the gaz pressure remains uniform or at least that the non uniformity has a negligible effect on the photon emission. Recently, first experimental measurements of temperature of up to 15000K has been achieved for single bubble sonoluminescence in H_2SO_4 solutions[10].

In order to measure the bubble radius dynamic, the Mie scattering technique is used[2]. This measurement is fitted with a RP equation taking into account water vapour condensation-evaporation and chemical reactions. One finds the best agreement for $R_0=4.6$ µm and $P_0=1.45$ atm. The experimental conditions were : the frequency of the sinusoidal acoustical signal of 27536 Hz, a temperature of 290 K, a degasing at 400 mbars resulting in an O_2 concentration of 3.4 mg/l.

CP838, *Innovations in Nonlinear Acoustics: 17th International Symposium on Nonlinear Acoustics,*
edited by A. A. Atchley, V. W. Sparrow, and R. M. Keolian
© 2006 American Institute of Physics 0-7354-0330-9/06/$23.00

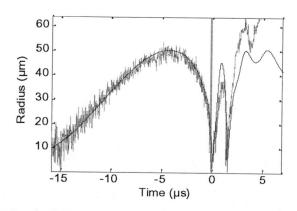

FIGURE 1. Time evolution of the bubble measured with Mie scattering (gray curve) and the fit by a Rayleigh-Plesset equation (black solid line). The pulse of pressure impinges on the bubble at t=-0.6 µs

The first case investigated is a pulse of pressure arriving around the end of the collapse. An example of bubble dynamics around the time of arrival of the pulse is displayed in Fig. 1. Note that the pulse is about 2 cycles[5]. Therefore after the first positive half cycle that accelerates the collapse, there is a negative one that induces a fast expansion followed by a very violent collapse, around t=1.5 µs. However even if this second collapse is much more violent than the first one, no photon are detected. Experimentally a maximum brightness gain of 300% is measured when the pulse of pressure arrives around 0.6µs before the end of the collapse, circle of Fig. 2. These results are compared with the numerical simulations in three cases. In the first case, solid line, water vapor is not allowed to evaporate in the bubble during the expansion stage. In these conditions the maximum temperature is 30820K (rising to a maximum of 46500K with the pressure pulse and 77830 K on the second collapse discussed above) and the number of photon radiated is $2.1 \ 10^7$. The brightness gain is larger than the experimental results. In the second case, solid line with cross marker, the water vapor evaporation-condensation process is taken into account but chemical reactions are not allowed. In this case strong disagreement with measurement is observed. The maximum temperature is now only 15490K and the corresponding number of radiated photons is $4 \ 10^4$ (rising to a modest maximum of 16110K with the pressure pulse). The amount of water vapor trapped in the bubble is 21% rising to 24% with the pressure pulse. In the last case, solid line with square, the eight main chemical reactions are taken into account. The maximum temperature is only 10570K. In this last case the agreement with measurements is good even if only 210 photons are emitted (rising to a maximum of 11280K with the pressure pulse). However the amount of water vapor trapped inside the bubble at the end of the collapse is strongly dependent on the diffusive boundary layer that is determined by a dimensional analysis[8]. This parameter possesses is strongly uncertain and could be adjusted to decrease the amount of water vapor trapped inside the bubble in order to recover a number of photons comparable to the measurements. Reducing it by a factor of 0.65 provides a good fit with the experimental data, Fig. 2 dotted line, a maximum temperature of 12340K and 1730 photons, 13540K with the pressure pulse.

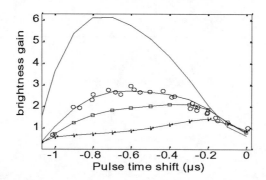

FIGURE 2. Brightness gain when the pressure pulse arrives on the bubble from 0 μs to -1 μs before the end of the collapse stage. Circle : measurements; Solid line : no water vapor evaporation; Solid line with cross : with water vapor evaporation, without chemistry; Solid line with square : with water evaporation, with chemistry, Dotted line : with water evaporation, with chemistry but boundary layer thickness multiply by 0.65.

As a conclusion good agreement is observed taking into account water vapor evaporation-condensation and chemistry and a thinner boundary layer. The pressure pulse significantly accelerates the collapse although the temperature increase is only important when no vapor enter the bubble. However the number of radiated photons is too small compared to experimental value reported in the literature [2]. Moreover a brighter second flash is expected to occur at the second collapse for which higher temperature and collapsing speed of about 3000 m.s^{-1} are calculated. Experimentally, in most cases no flash is observed. This discrepancy may be due to instability of the bubble surface limiting the efficiency of the collapse. To address this point, the pulse can be time shifted in the rest period of the bubble at t=3.3 μs. In these conditions a flash of about the same amplitude is observed, Fig. 3. In the following the numerical and experimental parameters are the same but R_0=4.21 μm and P_0=1.38 atm. The maximum temperature are 12390K at t=0 μs and 15850K at t=5μs with the pressure pulse. The percentage of water vapor plus chemical products is 66% of the whole content. The calculated brightness gain is 14 wheras experimentally it is less than 1. The calculated collapsing speed reached 5500 m.s^{-1} !

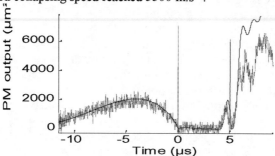

FIGURE 3. Brightness gain when the pressure pulse arrives on the bubble at +3.3 μs after the end of the collapse stage. Gray curve photomultiplier output, solid line : computed curve with water evaporation, chemistry but boundary layer thickness multiplied by 0.65.

506

Another case interesting case is when the pressure pulse is applied during the beginning of the expansion stage of the bubble, at t=18.3μs. The pressure pulse is 5 times weaker than in the previous case and of opposite sign. This pressure pulse induces an increase of the maximum radius reached by the bubble and a corresponding brightness gain of 400%, Fig. 4. Again, the amount of water vapor trapped inside the bubble is increased and contributes to dampen the effect of the boosting. The brightness gain is computed to 500% with water vapor, chemistry and the factor of 0.65 on the boundary layer thickness. The maximum temperature are 12390K and 14450K with the pressure pulse. The collapsing speed reaches 3000 m.s^{-1}.

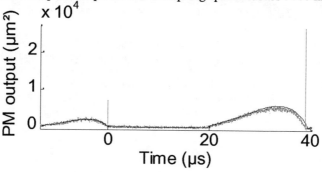

FIGURE 4. Brightness gain when the pressure pulse arrives on the bubble at +18.3 μs after the end of the collapse stage. Gray curve photomultiplier output, solid line : computed curve with water evaporation, chemistry but boundary layer thickness multiplied by 0.65.

New kind of transient boosting of single bubble sonoluminecence have been presented. Very violent bubble collapse are achieved for which the flash of sonoluminescence can be significantly increased or on the contrary decreased. This behavior remains unclear. The collapsing speed are probably outside the range of validity of the Rayleigh-Plesset equation. Moreover the non uniformity of the gas may play an important rôle. For more moderate boosting, a relatively simple model of single bubble sonoluminescence gives good agreement with experimental parametric results by adjusting the boundary layer thickness. In this case, the temperature increase is small due to the damping resulting from the increase heat capacity. This kind of boosting should on the contrary be very efficient for liquid with very low vapor pressure like concentrate aqueous sulphuric acid solutions used in [10]

REFERENCES

1. Gaitan, D. F., et al, *J. Acoust. Soc. Am.* **91**, 3166-3183 (1992)
2. Barber, B. P. et al, *phys. Rep.* **281**, 65 (1997)
3. Moss, W. C. et al, *phys. Rev. E* **59**, 2986-2993 (1999)
4. Storey, B. D. and Szeri A. J., *Proc. R. Soc. Lond. A* **456**, 1685-1709 (2000)
5. Thomas, J.-L., *Phys. Rev. E* **70**, 016305 1-5 (2004)
6. Toegel, R. et al, *Phys. Rev. Lett.* **85**, 3165-3168 (2000)
7. Toegel, R., Higenfeldt S., and Lohse D., *Phys. Rev. Lett.* **88**, 034301 1-4 (2002)
8. Lu, X. et al, *Phys. Rev. E.* **67**, 056310 1-9 (2003)
9. Hilgenfeldt, S., Grossmann, and Lohse D., *Nature (London)* **398** 402 (1999)
10. Flanningan D. J. and Suslick K. S., *Nature (London)* **434** 52-55 (2005)

Nonlinear Acoustic Wave Profile Transformation in Dissipative Gas-Liquid Media

D.C. Kim

Siberian State University of Telecommunications and Informatics
86 Kirov Str, 630102 Novosibirsk, Russia, Email: dck@osmf.sscc.ru

Abstract. We predict two new types of soliton-forming mechanisms induced only by nonlinear bubble dynamics for the firs time. First, we found a new type of soliton-forming mechanism induced only by adiabatic bubble dynamics. Second, to explore the soliton in dissipative medium, we developed the fourth-order compact finite-difference schemes for nonlinear parabolic and hyperbolic PDEs that arise for nonlinear wave problems in bubbly liquids. As results, we have discovered a transformation of acoustic wave into a stationary solitary wave in dissipation environment.

Keywords: Bubbly flow, heat transfer, compact finite-difference schemes, KdV equation.
PACS: 43.25.+y; 44.30.+c; 47.55.-t

INTRODUCTION

The heat exchange between phases has not been adequately investigated in the past because of the simulations of two-phase gas-liquid flows are a challenging problem in science and technology. Our previous research works [1] calculated the bubble oscillations under the assumption that gases inside a bubble changed adiabatically or isothermally. The heat transfer effect was neglected in those cases. This is fine in some cases (e.g. for shock pulse) but not all.

In resent years has been hard interest in designing numerical methods for nonlinear thermal bubble dynamics. In work [2] a second-order finite difference technique and in [3] a several of spectral methods were used to solve the heat conduction equation of single bubble. Motivated by soliton-forming mechanisms study we abandoned those methods in favour of much accuracy and stable and more transparent method based on an implicit compact difference scheme of solving nonlinear parabolic differential equation. In addition, the parallel implementation of the heat diffusion process of single bubble into the continuum model for bubbly flows is done. For that the fourth-order compact finite-difference scheme for the solution of nonlinear hyperbolic wave equation of Lighthill-type is also developed [4].

In present paper the mathematical model describing the thermal dynamic behaviour of bubbly mixture is presented. Profound changes are found between wave patterns with the adiabatic and heat-conducting gas bubbles. Numerical results demonstrated that the thermal gradient at the moving bubble wall has a significant impact on the wave evolution process. It is discovered the stationary solitary wave in a dissipative systems.

CP838, *Innovations in Nonlinear Acoustics: 17th International Symposium on Nonlinear Acoustics*,
edited by A. A. Atchley, V. W. Sparrow, and R. M. Keolian
© 2006 American Institute of Physics 0-7354-0330-9/06/$23.00

THE MODEL EQUATIONS

To describe self-consistency the unsteady two-phase bubbly flow one has to use the unsteady energy equation inside a spherical bubble, the momentum equation, the Rayleigh-Lamb bubble dynamic equation (the Keller formulation), the generalized inhomogeneous wave equation of Lighthill-type [5, 1], and equation of state

$$\frac{\gamma-1}{\gamma}\frac{P_g}{T}\left(\frac{\partial T}{\partial t}+u\frac{\partial T}{\partial r}\right)-\dot{P}_g = \nabla(\kappa\nabla T),\tag{1}$$

$$\frac{dP_g}{dt} = \frac{3}{R}\left((\gamma-1)\kappa\frac{\partial T}{\partial r}\bigg|_{r=R}-\gamma P_g\dot{R}\right),\tag{2}$$

$$\left(1-\frac{\dot{R}}{c}\right)R\ddot{R}+\frac{3}{2}\dot{R}^2\left(1-\frac{\dot{R}}{3c}\right) = \left(1+\frac{\dot{R}}{c}\right)\frac{1}{\rho}\left[P_B(t)-P_S(t+\frac{R}{c})-P_0\right]+\frac{R}{c}\frac{dP_B(t)}{dt},\tag{3}$$

$$\frac{\partial^2\rho}{\partial t^2}-c^2\frac{\partial^2\rho}{\partial x^2} = \frac{\partial}{\partial t}\left(\rho\frac{\partial}{\partial t}\ln\left(1+\frac{\varphi_0}{1-\varphi_0}\bar{\rho}R^3\right)\right),\tag{4}$$

$$P-P_0 = c_L^2(\rho-\rho_0),\tag{5}$$

in which

$$u = \frac{1}{\gamma P}\left((\gamma-1)\kappa\frac{\partial T}{\partial r}-\frac{1}{3}r\dot{P}_g\right),\quad P_B = P_g(t)-\frac{2\sigma}{R}-\frac{4\mu\dot{R}}{R}.\tag{6}$$

In the above T is temperature, κ - thermal conductivity, u – the radial velocity field of gas. Other notations are as in [1]. To integrate the system of coupled equations (1)-(5), a set of initial and boundary conditions is stated. The heat equation (1) inside a bubble subject to the initial condition and von Neumann and Dirichlet boundary conditions

$$T(r,0) = T_L,\ 0 < r < R;\quad \frac{\partial T(0,t)}{\partial r} = 0,\ T(R,t) = T_L.\tag{7}$$

The corresponding boundary and initial conditions for Eqs (3)-(4) have been presented earlier. We emphasize that the efficiency of high accuracy schemes can be reached only if they are combined with the adequate numerical boundary conditions. The wave propagation is calculated with the given sinusoidal function on the left-hand boundary and the non-reflecting Sommerfeld conditions for the right-hand boundary.

FINITE DIFFERENCE APPROXIMATION

For numerical analysis, we shall rewrite Eq. (1) in more conventional form introducing thermal potential Ψ and normalized radius y. Then, for numerical approximation we shall introduce new function $\vartheta = y\Psi$. The scheme stencil consists of nine points: three points on the bottom-time layer, three points on the middle-time layer and three points on the upper layer. The implicit fourth-order accuracy compact scheme for ϑ one can obtain using asymptotic expansion

$$\vartheta_i^n = \zeta^n\Lambda_0^y\frac{\vartheta^{n+1}+\vartheta^{n-1}}{2}-\frac{h^2}{12}\Lambda_0^y\frac{\vartheta_i^n}{\zeta^n}-\phi^n-\frac{h^2}{12}\Lambda_0^y\frac{\phi^n}{\zeta^n},\tag{8}$$

in which

$$(\phi)_i^n = y\left[\frac{\gamma-1}{\gamma P_g R^2}\left(\frac{\partial\Psi}{\partial y} - \frac{\partial\Psi}{\partial y}\bigg|_{y=1}\right)y\frac{\partial\Psi}{\partial y} - D\dot{P}_g + \frac{y}{R}\dot{R}\frac{\partial\Psi}{\partial y}\right],$$ (9)

$$\vartheta_t^n = \left(\frac{\vartheta^{n+1} - \vartheta^{n-1}}{2\tau}\right); \quad \zeta^n = \frac{D(P_g^n, T^n)}{R^2},$$ (10)

where Λ_0^y - central second order difference operator associated with y coordinate direction, ϕ^n is the sours terms of Eq. (1) after appropriate transformation.

NONLINEAR WAVES IN GAS-LIQUID MIXTURES

We investigate the acoustic wave propagation in water with distributed air bubbles. It is known that for propagation of sound in bubbly mixture could, if the bubble vibration is assumed linear, be derived the KdV equation [6]. However bubble oscillations are intrinsically nonlinear. It may be added here that the nonlinearity of bubbles is larger by two orders of magnitude then one of convective term [7]. First, we revealed a new type of soliton-forming mechanism induced only by adiabatic bubble dynamics. We have shown that acoustic wave dissociate into a train of solitons [1]. These phenomena were observed earlier in the numerical solution of the KdV equation [8]. Contrary to Wijngaarden's mechanism [6] the soliton arises when the bubble nonlinearity balances by the bubble dispersivity. For this case we have obtained another KdV equation in low-frequency approximation

$$R_\tau + \varepsilon\frac{3c_1}{2R_0}(\gamma+1)(c_1^2/c_L^2-1)RR_\eta + \varepsilon\frac{1}{2\omega_0^2}(c_L/c_1)^3(1-c_1^2/c_L^2)R_{\eta\eta\eta} = 0.$$ (11)

where c_1 is the speed of sound in the low frequency approximation, ε - small parameter, $\tau = \varepsilon^{1/2}t$ and $\eta = \varepsilon^{1/2}(x-c_1 t)$ - slow variables.

Second, it is clear that the adiabatic bubbles are not enough to fully understand soliton-forming mechanisms in bubbly liquids because the realistic bubble behavior is always dissipative. Dissipative systems are more complicated than Hamiltonian ones in the sense that, in addition to nonlinearity and dispersion, they included energy exchange with the environment or external sources. Energy exchange cannot be explained by replacing the dissipation with an effective viscosity in Rayleigh-Lamb equation since this will result in absolute damping any waves. Neither can be fully explained by introducing effective Nusselt numbers that depend only on the magnitude of the average temperature of gas and liquid, since the observed behavior also depend on the instantaneous thermal gradient at the moving surface boundary. The heat transfer effect on thermal and dynamics interactions of gas bubbles with the liquid induced by sound wave must be solved from solution the heat conduction equation. Numerical simulations revealed a transformation of acoustic wave into a stationary solitons as shown in Fig. 1. The sound propagates, steepens, and decays strongly over the first several periods due to the thermal damping effect. However, the form of first loop is deformed as shown in curve 1 and 2 at instant of time $t = 50.6$ ms (solid) and $t = 68.5$ ms (dash-dotted) and it evolves to a solitary-like wave and then propagates with zero-decrement as shown in Fig. 1(b) at instant of time $t = 212.4$ ms (thin solid).

510

This phenomenon can be explained by the reverse effect of mowing boundary wall on the thermal process inside a bubble that is revealed in our simulations. Indeed, as is known, the heat flux q at the interface is expressed as follows: $q = -\kappa(T)(\partial T / \partial r)\big|_{r=R}$. In Fig. 1(b) the dashed line corresponds to the normalized thermal gradient, and the solid line (bold) to the normalized bubble wall velocity, and thin solid line to the pressure profile (solitary wave) at time $t = 212.4\,\mathrm{ms}$. This figure shows that the correlation between the thermal gradient and the bubble wall velocity is evident. If the bubble expands the gas layer near a wall will be cool down because of the timescale of thermal diffusion is much bigger than the time of expansion. As a result, the thermal gradient is positive at the bubble wall, the heat transport from relatively warm liquid into the colder gas. And vice versa, during compression the thermal gradient is negative at the bubble wall, the heat transport from relatively warm gas into the colder liquid. This means that the bubble can able be in thermal equilibrium with the surrounding liquid, i.e. that the bubble oscillates without loss of energy at each collapsing process. So, the soliton appears as a result of two balance conditions in a dissipative medium. First condition is the balance between nonlinear waveform distortion incurred by bubble nonlinearity and bubble induced dispersivity. Second condition is the balance between penetrating into and leaking out heat flows of oscillating bubbles.

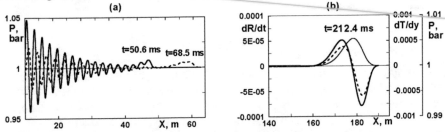

FIGURE 1. Acoustic wave profile transformation in dissipative gas-liquid medium. Here $\varphi_0 = 10^{-4}$, $R_0 = 0.5$ mm, the initial amplitude of sound $P_A = 0.05$ bar. The frequency of sound is $\omega = \omega_0 / 20$.

As conclusion, I wish to thank the corresponding member of the RAS, Prof. G.A. Mikhailov for his support and valuable assistance. I am grateful to Dr. Y.N. Morokov of the ICT SB RAS for helpful discussions about compact schemes.

REFERENCES

1. Kim, D.C., Computer Physics Communications, **147**, pp. 526-529 (2002).
2. Prosperetti, A., Crum, L.A., Comander, K.W., J. Acoust. Soc. Am., **83** (2), pp. 502-514 (1988).
3. Kamach, V., and Prosperetty, A., J. Acoust. Soc. Am., **85** (4), pp. 1538-1548 (1989).
4. Kim, D.C., Proceedings of the 8th Korea-Russia International Symposium on Science and Technology, Russia, Tomsk, June 26- July 3, 2004, **2**, pp. 225-229. ISBN: 0-7803-8383-4; LC USA Card Number: 2004101655.
5. Crighton, D.G., and Ffowcs Williams, J.E., J. Fluid Mech., **36**, pp. 585-603 (1965).
6. van Wijngaarden, L., J. Fluid Mech., **33**, pp. 465-474 (1968).
7. Zabolotskaya, E.A., Akusticheskiy Zhurnal, **21**, pp. 334-337 (1975).
8. Zabusky, N.J., Kruskal, M.D., Phys. Rev. Let., **15**, pp. 240-243 (1965).

Dynamics of a Tethered Bubble

Alexey O. Maksimov [1], Timothy G. Leighton[2], Peter R. Birkin [3]

[1]*Pacific Oceanological Institute, Far Eastern Branch of the Russian Academy of Sciences, 43 Baltic Street, Vladivostok 690041, Russia*
[2]*Institute of Sound and Vibration Research, University of Southampton, Highfield, Southampton SO17 1BJ, UK*
[3]*School of Chemistry, University of Southampton, Highfield, Southampton SO17 1BJ, UK*

Abstract. Small gas bubbles adhering to solids occur in a range of manufacturing processes, including printing, casting, coating and electroplating. The behavior of a gas bubbles tethered to a rigid plane boundary in an oscillatory pressure field is investigated by use conformal symmetry of the problem. The dynamics of the tethered bubble differ from those of the free bubble. The inertial (or added) mass depends on the contact angle and this variation is not monotonic. As a result, the natural frequency depends on the contact angle. Viscous damping of the tethered bubble is increased by more than two orders of magnitude, firstly, owing to the greater dissipation near rigid wall in comparison with free air/water bubble interface and secondly, because of a contact line dissipation effect.

INTRODUCTION

Small gas bubbles adhering to solids occur in a range of manufacturing processes, including printing, casting, coating and electroplating. These so-called 'tethered' bubbles generate a boundary between the solid and the liquid, preventing complete coverage. This is usually disadvantageous to production. In the pottery industry, the occurrence of bubbles in the liquid 'casting slip' which is injected into moulds results in loss of production time and raw materials, as the expansion of these bubbles in the kiln when the pottery is 'fired' generates unwanted holes and pitting in the finished product [1]. Similar problems can occur in the casting of metals. The importance of the tethered bubbles has also been apparent in the many studies of the tensile strengths of liquids. Micron-sized bubbles of contaminant gas form important weaknesses in the liquid. These bubbles persist for long periods, stabilized against dissolution or buoyant rise through their adhesion to crevices within the solid boundary. Unlike free microbubbles, only static thresholds of acoustic cavitation have been theoretically derived for the crevice model [2]. Sonoluminescence (SL) from an isolated bubble on a solid surface [3] differs from single-bubble SL. The forced oscillation of a gas bubble can have important biological effects. Rooney [4] showed that hemolysis of an erythrocryte was possible using an acoustically driven tethered gas bubble.

Clearly the forced oscillation of a tethered gas bubble is of significant technological importance for a thorough understanding of the problem to be necessary. Hence we present here a theoretical approach to this problem.

CP838, *Innovations in Nonlinear Acoustics: 17th International Symposium on Nonlinear Acoustics*,
edited by A. A. Atchley, V. W. Sparrow, and R. M. Keolian

BUBBLE DYNAMICS IN TOROIDAL COORDINATES

Consider an air bubble of radius curvature R_0 tethered to the rigid wall (see Figure 1) and driven by an acoustical wave of amplitude P_m and angular frequency ω. The basic bubble equilibrium shapes show a segment of the spherical profile with contact angle ϑ_c. The diameter of the ring of contact is denoted by L ($L = 2R_0 \sin \vartheta_c$). As the radius of the bubble is often much smaller than the acoustic wavelength λ, the pressure within the bubble is constant, when $R_0 / \lambda \ll 1$ (homobaric bubble) and hence the bubble wall is an equipotential surface, if surface tension and inertial nonlinearities are neglected.

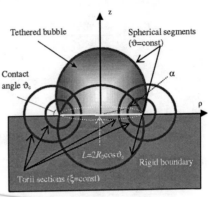

Fig. 1. Toroidal coordinates – natural coordinates for the tethered bubble.

The inversion method was utilized to obtain an exact solution for a gas bubble tethered to a rigid plane boundary in an oscillatory pressure field [5]. This method is based on the invariance of the Laplace equation to conformal transformations. It was also shown [6], by using the symmetry of the problem, that the toroidal coordinates are most suitable for analysis of the oscillations of the tethered bubble. Thus the dynamics of the tethered bubble in toroidal coordinates can be analyzed by using an analytical approach and by analogy to the dynamics of a free spherical bubble.

The solution can be expressed in terms of the Green's function of the Laplace's equation with a homogeneous (vanishing) condition on the boundary (bubble wall):

$$\nabla^2 G(\mathbf{r} \mid \mathbf{r'}) = -4\pi\delta(\mathbf{r} - \mathbf{r'}), \tag{1}$$

The Green's function can be expressed in toroidal coordinates $z = (L/2)\sin\vartheta \times [\cosh\xi - \cos\vartheta]^{-1}$, $x = (L/2)\sinh\xi\cos\alpha[\cosh\xi - \cos\vartheta]^{-1}$, $y = (L/2)\sinh\xi\sin\alpha \times [\cosh\xi - \cos\vartheta]^{-1}$, where ξ goes from 0 to ∞, ϑ goes from $-\vartheta_c$ to ϑ_c and α (the azimuthal angle) goes from 0 to 2π. The surface $\xi = const$ is a torus (which is a cyclide) and the surface $\vartheta = const$ is a spherical bowl. It then takes the following form

$$G(\xi,\vartheta,\alpha \mid \xi',\vartheta',\alpha') = \sqrt{\cosh\xi - \cos\vartheta}\sqrt{\cosh\xi' - \cos\vartheta'}\frac{1}{\sqrt{2}\vartheta_c}\left(\frac{2}{L}\right)\int_s^\infty \frac{\sinh\left(\pi t/2\vartheta_c\right)dt}{(\cosh t - \cosh s)^{1/2}} \times$$

$$\times \left\{\cos\left[\frac{\pi}{2\vartheta_c}(\vartheta - \vartheta')\right]\left\{\cosh\left(\frac{\pi}{\vartheta_c}\right) - \cos\left[\frac{\pi(\vartheta - \vartheta')}{\vartheta_c}\right]\right\}^{-1} + \cos\left[\frac{\pi}{2\vartheta_c}(\vartheta + \vartheta')\right]\left\{\cosh\left(\frac{\pi}{\vartheta_c}\right) - \cos\left[\frac{\pi(\vartheta + \vartheta')}{\vartheta_c}\right]\right\}^{-1}\right\}, \tag{2}$$

where $\cosh s \equiv \cosh\xi\cosh\xi' - \sinh\xi\sinh\xi'\cos(\alpha - \alpha')$. This is the main result of this section.

VOLUME OSCILLATIONS OF A TETHERED BUBBLE

As an application of the explicit form of the Green's function (2) we describe the volume oscillations of a tethered bubble. The modified Rayleigh equation for the volume pulsation $V = V_0 + \Delta V$ has the form:

$$\left[2R_0 \sin \vartheta_c C(\vartheta_c)\right]^{-1} d^2V/dt^2 + \left(\gamma P_0/\rho_0 V_0\right)\Delta V = -\left(P_m/\rho_0\right)\sin(\omega t), \qquad (3)$$

$$C(\vartheta_c) = \frac{\pi^2}{2\sqrt{2}\vartheta_c^2} \int_{\vartheta_c}^{\pi} \frac{\sin \vartheta_c \sin \kappa d\kappa}{\left(\cos \vartheta_c - \cos \kappa\right)^{3/2}} \int_{arch\left(\frac{1-\cos\vartheta_c \cos\kappa}{\cos\vartheta_c-\cos\kappa}\right)}^{\infty} d\xi \left(ch\xi - \frac{1-\cos\vartheta_c \cos\kappa}{\cos\vartheta_c - \cos\kappa}\right)^{-1/2} \frac{sh\left(\pi\xi/2\vartheta_c\right)}{ch^2\left(\pi\xi/2\vartheta_c\right)},$$

where ρ_0 and P_0 are the equilibrium density and pressure, γ is the polytropic exponent. It follows directly from this equation that the fundamental frequency for the tethered bubble depends on the contact angle and this dependence is not monotonic

$$\Omega_0^2(R_0,\vartheta_c) = \Omega_*^2(R_0)\sin\vartheta_c\left(C(\vartheta_c)/2\pi\right)\left[1-\left(1-\cos\vartheta_c\right)^2\left(2+\cos\vartheta_c\right)/4\right] \qquad (4)$$

here $\Omega_*^2(R_0) = \sqrt{3\gamma P_0/\rho_0}R_0^{-1}$ is the fundamental frequency of a free bubble. The expression for the natural frequency is given for a fixed radius of curvature R_0 of the spherical segment. However it is more interesting to trace the manner in which the natural frequency of a tethered bubble of fixed volume V_0 varies as the contact angle is changed. The dependence of $\Omega_0(V_0,\vartheta_c)/\Omega_*(V_0)$ on the contact angle ϑ_c is shown in Fig. 2 by solid line.

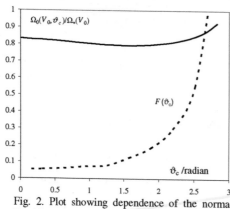

Fig. 2. Plot showing dependence of the normalized frequency and damping on the contact angle.

We have so far assumed the flow near the bubble and the rigid wall to be inviscid and irrotational. The viscous damping factor (for the kinematic viscosity ν) of the volume oscillations of the tethered bubble can be obtained by using the approach based on the total energy conservation in incompressible liquid when the time derivative of energy is balanced by a dissipative function – Longuet-Higgins [7] used this approach in finding the viscous damping of surface modes on the bubble wall. The relative decay of amplitude per cycle (known as the 'damping factor') is:

$$\gamma_0 = \left(2\nu/R_0^2\right)\sqrt{\omega R_0^2/2\nu}\,F(\vartheta_c), \qquad (5)$$

where we have separated out the co-factor $(2\nu / R_0^2)$ which describes damping of a free non-tethered bubble. The computed angular dependence of the normalized damping $F(\vartheta_c) = \gamma_0\left(2\nu/R_0^2\right)^{-1}\left(\omega R_0^2/2\nu\right)^{-1/2}$ is shown in Fig. 2 by the dotted line.

DISCUSSION AND CONCLUSIONS

This section discusses two examples for the application of this theory. The first follows from the development of a new electrochemical sensing technique [8, 9] in which millimetre sized bubbles were suspended under a glass rod. As this has proved to be the most precise technique for determining the resonance of a single bubble [1], it follows a comparison of the pressure threshold measurements for the onset of surface oscillations, and examination of the variation in this threshold due to the shift of the natural frequency of the tethered bubble (4) is thus possible.

The second example follows from the experiment of MacDonald *et al.* [10], where oscillations of a bubble at the tip of an optical fibers in liquid nitrogen have been studied. When a pulse of laser is delivered to the fiber tip, a bubble forms (with a radius of just a few micrometers), initially as a shallow meniscus, on the fiber core. It then grows toward the more spherical shape, and freely oscillates. The oscillations are fastest immediately after the pump pulse, when they have a frequency of ~ 17 MHz. The oscillation frequency then drop rapidly, settles at ~ 1.35 MHz within a few microseconds, and remains almost constant for 25-30 cycles.

The time history of these oscillations can be approximated by the Rayleigh equation (3) with varying inertial mass. Assuming adiabatic variation of this mass we can use the expressions (4) for natural frequency of the bubble with fixed volume and varying contact angle that is varying inertial mass. The initial state can be approximated by the spherical segment with contact angle $\vartheta_c \sim \pi$. The drop of oscillation frequency from 17 till 1.35 MHz has a natural explanation and can be described on the order of magnitude by the expressions derived for the natural frequency of the tethered bubble.

The dynamics of the tethered bubble differ from those of the free bubble mainly by the value and dependence of the inertial (or added) mass on the contact angle. This provides variation of the natural frequency with the contact angle and this variation is not monotonic. Viscous damping of the tethered bubble is increased by more than two orders of magnitude, firstly, owing to the greater dissipation near rigid wall in comparison with free air/water bubble interface and secondly, because of the contact line dissipation effect.

REFERENCES

1. T. G. Leighton, *International Journal of Modern Physics B*, **18**, 3267-3314 (2004).
2. A. Atchley, A. Prosperetti, *J. Acoust. Soc. Am.*, **86**, 1065-1084 (1989).
3. K. R. Weninger et al., *Phys. Rev. E.*, **56**, 6746-6749 (1997).
4. J. A. Rooney, *Science*, **169**, 869-871 (1970).
5. A. O. Maksimov, *J. Sound & Vibr.*, **283**, 915-926 (2005)
6. A. O. Maksimov, *Communications in Nonlinear Science and Numerical Simulation*, **9**, 83-92 (2004).
7. M. S. Longuet-Higgins, *J. Fluid Mech.* **201**, 543-565 (1989).
8. P. R. Birkin, Y. E. Watson, T. G. Leighton, K. L. Smith, *Langmuir Surfaces and Colloids.* **18**, 2135-2140 (2002).
9. Y. E. Watson, P.R. Birkin, T. G. Leighton, *Ultrasonics Sonochemistry*, **10**, 65–69 (2003).
10. K. F. MacDonald *et al.*, *Phys. Rev. E*, **68**, 027301 (2003).

Hamiltonian structure, symmetries and conservation laws in bubble dynamics

Alexey O. Maksimov

Pacific Oceanological Institute, Far Eastern Branch of the Russian Academy of Sciences, 43 Baltic Street, Vladivostok 690041, Russia

Abstract. Lie group analysis is applied to the nonlinear bubble dynamics accounting both distortion and breathing modes according to the perfect-liquid model. The Hamiltonian formulation of the problem has been presented and the closed form of the Hamiltonian as a functional of canonical variables has been obtained. The equations of the prolongation theory have been derived and particular solutions corresponding subgroups of the full symmetry group have been determined. The availability of certain continuous group of symmetry generally offers a possibility of reducing the order of the corresponding differential equation. It appears possible to describe analytically the nonlinear dynamics of a bubble driven by an external perturbation in the conditions ensuring the scale invariance. The efficiency of bubble extension under the action of a periodically prolonged scale-invariant external acoustic field has been studied.

INTRODUCTION

The inertia-dominated dynamics of a bubble in weakly viscous fluid subjected to a standing acoustic wave has been the subject of intense investigations for many years. Much early work on the motion of gas bubbles was aimed at understanding the bubble as a source of sound. For this problem, it is the time-dependent changes in bubble volume that are of dominant interest. Because shape oscillations are a weaker source of sound in linear theory, and because the spherical bubble problem is easier, most theoretical studies of bubble dynamics have focused on this problem. Although gas bubbles are very frequently non-spherical, or become so as a consequence of instabilities of the oscillating spherical bubble, the dynamics of a non-spherical bubble are much more difficult to study.

The key to bringing notations of Hamiltonian mechanics usefully to bear on the non-spherical bubble dynamics is a remarkably concise interpretation of the bubble-surface nonlinear boundary conditions in terms of functional derivatives of the energy integral. This simplifying formalism appears to have been noticed first by Zakharov [1] for water waves problem. Representation of the nonlinear boundary conditions by functional derivatives of energy and momentum integrals were also considered by Benjamin and Olver [2] and we shall closely follow this approach in the current study.

CP838, *Innovations in Nonlinear Acoustics: 17th International Symposium on Nonlinear Acoustics,*
edited by A. A. Atchley, V. W. Sparrow, and R. M. Keolian
© 2006 American Institute of Physics 0-7354-0330-9/06/$23.00

HUMILTONIAN FORMALIZM FOR NON-SPHERICAL BUBBLE

Consider an air bubble of radius curvature R_0 driven by an acoustical pressure $P_1(t)$. The water compressibility can be ignored as the radius of the bubble R_0 is assumed to be much smaller than the acoustic wavelength λ and the velocity potential, denoted by φ, is governed by the Laplace's equation $\Delta\varphi = 0$, $\mathbf{v} = \nabla\varphi$, where $\mathbf{v}$ is the velocity. We shall use the spherical coordinates (r,θ,α) and with the equation of the bubble surface as $r = R_0 + \xi(\theta,\alpha,t)$, the kinematic boundary condition takes the form $\left[\partial/\partial t + (\mathbf{v},\nabla)\right](r - R)_{r=R_0+\xi} = 0$ or

$$\frac{\partial\xi}{\partial t} = \left(\frac{\partial\varphi}{\partial r}\right)_{r=R_0+\xi} - \frac{1}{(R_0+\xi)^2}\frac{\partial\xi}{\partial\theta}\left(\frac{\partial\varphi}{\partial\theta}\right)_{r=R_0+\xi} - \frac{1}{(R_0+\xi)^2\sin^2\theta}\frac{\partial\xi}{\partial\alpha}\left(\frac{\partial\varphi}{\partial\alpha}\right)_{r=R_0+\xi}. \quad (1)$$

The dynamical boundary condition is that the pressure on the two sides of the surface differ only because of surface tension i.e. if P_l and P_g denote the pressure in the water and in the bubble correspondingly, than $P_l = P_g - \sigma(\nabla,\mathbf{n})$, $P_g = P_0(V_0/V)^\gamma$, where $\mathbf{n}$ is the unit vector normal to the surface $r = R_0 + \xi(\theta,\alpha,t)$, σ is the coefficient of the surface tension. We adopt a polytropic law for the gas bubble and V, V_0 are the instantaneous and equilibrium bubble volume, γ is the polytropic exponent, P_0 is the equilibrium pressure in the bubble and P_l is governed by the Bernoulli equation

$$P_l = -\rho_0\left[\partial\varphi/\partial t + (\nabla\varphi)^2/2\right] + P_1(t), \quad (2)$$

where ρ_0 are the equilibrium density and $P_1(t)$ is the external pressure.

This problem admits a Hamiltonian formulation with the Hamiltonian $H = T + U + A$, where $T = \rho_0\int(\mathbf{v}^2/2)d\mathbf{r}$ is the kinetic energy, U is the potential energy, which is the sum of the internal energy $V_{in} = -\int P_g dV = P_0V_0(\gamma-1)^{-1}(V_0/V)^{\gamma-1}$ and the surface energy $V_s = \sigma\int\sqrt{1 + (R_0+\xi)^{-2}(\partial\xi/\partial\theta)^2 + (R_0+\xi)^{-2}\sin^{-2}\theta(\partial\xi/\partial\alpha)^2} \times$ $\times(R_0+\xi)^2\sin\theta d\vartheta d\alpha$, the last term A is the work done by the external force $A = -(P_\infty + P_{ext})(V - V_0)$.

The resulting canonical equations of motion are

$$\frac{\partial\xi}{\partial t} = \frac{\delta H}{\rho_0(R_0+\xi)^2\delta\varphi^s(\theta,\alpha,t)}, \quad \frac{\partial\left(\rho_0(R_0+\xi)^2\varphi^s(\theta,\alpha,t)\right)}{\partial t} = -\frac{\delta H}{\delta\xi}, \quad (3)$$

which coincide with the kinematic and dynamic boundary conditions. The canonically conjugate variables are the surface deflection $\xi(\theta,\alpha,t)$ and velocity potential on the surface $\varphi^s(\theta,\alpha,t) \equiv \varphi(r = R_0 + \xi,\theta,\alpha,t)$ multiplied by the square of the bubble radius $(R_0+\xi)^2$. This result is generalization of [3] where it was shown that for the purely radial bubble oscillations the Rayleigh equation can be written as a Hamiltonian system.

SYMMETRIES OF THE NON-SPHERICAL BUBBLE DYNAMICS

Our object here is to establish symmetry groups for the problem just specified. The methods used to find symmetries are developed from the theory of Lie group and prolongation theory. The procedure of determining the group of symmetry is given in [2, 4] and consists of the constructions of the prolongations to the space of derivatives of the dependent variables and boundary prolongation.

Algebra of symmetry of the Laplace equation is ten-dimensional and has the basis of the following form: $P_j = \partial_j$ – translations; $J_{ij} = x_i\partial_j - x_j\partial_i$ – rotations; $K_i = (2x_i x_j - |x|^2 \delta_{ij})\partial_i - x_i\varphi\partial_\varphi$ – conformal transformations; $D = x_i\partial_i - (1/2)\varphi\partial_\varphi$ – scaling [4], here $\partial_i \equiv \partial/\partial x_i$, $\partial_\varphi \equiv \partial/\partial\varphi$.

The symmetry group for the non-spherical bubble problem is generated by the following one-parameter subgroups:

Translations $G_1 : (x+\varepsilon, y, z, t; \varphi)$,

$$P_x = (\sin\vartheta\cos\alpha)\partial_r + (r^{-1}\cos\vartheta\cos\alpha)\partial_\vartheta - (r^{-1}\sin^{-1}\vartheta\sin\alpha)\partial_\alpha;$$

$$G_2 : (x, y+\varepsilon, z, t; \varphi), \quad P_y = (\sin\vartheta\sin\alpha)\partial_r + (r^{-1}\cos\vartheta\sin\alpha)\partial_\vartheta - (r^{-1}\sin^{-1}\vartheta\cos\alpha)\partial_\alpha;$$

$$G_3 : (x, y, z+\varepsilon, t; \varphi), \quad P_z = \cos\vartheta\partial_r + (r^{-1}\sin\vartheta)\partial_\vartheta.$$

Rotations $G_4 : (x, z\sin\varepsilon + y\cos\varepsilon, z\cos\varepsilon - y\sin\varepsilon, t; \varphi)$, $J_x = \sin\alpha\partial_\vartheta + (\cos\alpha\cos\vartheta\sin^{-1}\vartheta)\partial_\alpha$;

$$G_5 : (x\cos\varepsilon + z\cos\varepsilon, y, z\cos\varepsilon - x\sin\varepsilon, t; \varphi), \quad J_y = -\cos\alpha\partial_\vartheta + (\sin\alpha\cos\vartheta\sin^{-1}\vartheta)\partial_\alpha;$$

$$G_6 : (x\cos\varepsilon - \sin\varepsilon y, x\sin\varepsilon + y\cos\varepsilon, z, t; \varphi), \quad J_z = \partial_\alpha.$$

Time translation $G_7 : (x, y, z, t+\varepsilon; \varphi)$, $T = \partial_t$ (if $\partial P_1(t)/\partial t = 0$).

Variation of base-level for potential $G_8 : (x, y, z, t; \varphi+\varepsilon)$, $\Phi = \partial_\varphi$.

Galilean boost $G_9 : (x+\varepsilon t, y, z, t; \varphi+\varepsilon x + (1/2)t^2)$,

$$B_x = t\left[(\sin\vartheta\cos\alpha)\partial_r + (r^{-1}\cos\vartheta\cos\alpha)\partial_\vartheta + (r^{-1}\sin^{-1}\vartheta\sin\alpha)\partial_\alpha\right] + (r\sin\vartheta\cos\alpha)\partial_\varphi;$$

$$G_{10} : (x, y+\varepsilon t, z, t; \varphi+\varepsilon y + (1/2)t^2),$$

$$B_y = t\left[(\sin\vartheta\sin\alpha)\partial_r + (r^{-1}\cos\vartheta\sin\alpha)\partial_\vartheta - (r^{-1}\sin^{-1}\vartheta\cos\alpha)\partial_\alpha\right] + (r\sin\vartheta\sin\alpha)\partial_\varphi;$$

$$G_{11} : (x, y, z+\varepsilon t, t; \varphi+\varepsilon z + (1/2)t^2), \quad B_z = t\left[\cos\vartheta\partial_r + (r^{-1}\sin\vartheta)\partial_\vartheta\right] + (r\cos\vartheta)\partial_\varphi.$$

Scaling $G_{12} : (\lambda^{2/(2+3\gamma)}x, \lambda^{2/(2+3\gamma)}y, \lambda^{2/(2+3\gamma)}z, \lambda t; \lambda^{(2-3\gamma)/(2+3\gamma)}\varphi)$,

$$S = \left[2/(2+3\gamma)\right]r\partial_r + t\partial_t + \left[(2-3\gamma)/(2+3\gamma)\right]\varphi\partial_\varphi, \quad \text{if } P_1(t) = P_1(0)\left(t_0/t+t_0\right)^{6/2+3\gamma},$$

$\sigma = 0$. In these expressions ε denotes each additive group parameter and $\lambda > 0$ denote each multiplicative parameter that can be considered as e^ε. Note that one can not state that all possible symmetries of the kind under consideration have been found.

The physical interpretations attaching to the various symmetries groups are reasonably clear. The first seven may seem trivial, being recognizable with little thoughts. However, it is necessary to list them all since they underline significant conservations laws. The Galilean boots G_9, G_{10}, G_{11} represent the effect observed from frames of reference moving uniformly. Finally, the scaling group G_{12} is easily understood, being evident from dimensional consideration. The present result is generalization of [6] where it was shown that scaling is one of two subgroups existing for the Rayleigh equation.

DISCUSSION AND CONCLUSIONS

Noether's famous theorem states that every one-parameter group of symmetries for a variational problem determines a conservation law satisfied by solutions of the corresponding Euler-Lagrange equations. This general principal can now be applied to find conservation laws for the bubble dynamics and to obtain analytical solutions. We use the fact that the availability of certain continuous group of symmetry generally offers a possibility of reducing the order of the corresponding differential equation.

The complete symmetry group is found for the Rayleigh equation [5, 6]. It appears possible to describe analytically the nonlinear dynamics of a bubble driven by an external perturbation in the conditions ensuring the scale invariance. The efficiency of bubble extension under the action of a periodically prolonged scale-invariant external acoustic field has been studied [7]. It was demonstrated that this driving, having the form of a train of shock waves, does not lead to a significant decompression.

Whilst nonlinear phenomena are associated with the excitation of many modes simultaneously on a bubble, we have dealt with the two that occur at the lowest driving amplitudes, the breathing mode and Faraday wave wall motions [8]. It is possible to justify synchronous fashion with which excited distortion modes oscillate relative to the phase of the breathing mode by the presence of the phase symmetry.

In the present work the study of Lie point symmetries in bubble dynamics and its applications to acoustics has been reviewed. As the first step, a Hamiltonian formulation for non-spherical bubble dynamics is presented. Then, the symmetry groups for this problem are identified systematically by use of infinitesimal-transformation and prolongation theory. It appears possible to analytically describe the nonlinear bubble dynamics using the symmetry properties of the problem.

ACKNOWLEDGEMENT

This study was supported by RFBR, project no 04-02-16412.

REFERENCES

1. V. E. Zakharov, *J. Appl. Mech. Tech. Phys.* **2**, 190-202 (1968).
2. T. B. Benjamin, P. J. Olver, *J. Fluid. Mech.,* **125**, 137-185 (1982).
3. P. Smereka, B. Binir, S. Banerjee, *Phys. Fluids,* **30**, 3342-3350 (1987).
4. N. H. Ibragimov, *Transformation groups applied to mathematical physics*, Reidel, Dodrechtl, 1985.
5. A. O. Maksimov, *Communications in Nonlinear Science and Numerical Simulation,* **9**, 83-92 (2004).
6. A. O. Maksimov, *Acoust. Phys.*, **48**, 805-812 (2002).
7. A. O. Maksimov, *Tech. Phys. Lett.,* **31**, 270-273 (2005).
8. A. O. Maksimov, T. G. Leighton, *ACUSTICA - Acta Acustica* **87**, 322-332 (2001).

SECTION 11
INFRASOUND, PROPAGATION, SHOCKS, AND NOISE

A Theory of Atmospheric Microbaroms

Roger Waxler and Kenneth E. Gilbert

National Center for Physical Acoustics, University of Mississippi, University, Mississippi 38677

Abstract. A theory of the radiation of microbaroms by ocean waves is presented.

Keywords: microbaroms, infrasound

The Microbarom Signal.   It is well known that energetic ocean wave systems radiate infrasound, both into the atmosphere and ocean, in a narrow band centered at about 0.2 Hz [1]. It is found [2, 3, 4] that that the infrasound radiation is a non-linear effect, arising in the second harmonic of the standing wave component of the ocean surface wave field. The oceanic microbaroms interact with the sea floor, producing a seismic signal known as microseisms. There are many examples of measured microbarom spectra, both atmospheric and oceanic, published in the literature [5, 6, 7, 8, 9]. Typical atmospheric spectra peak at tenths of Pascals per root Hz [9] while oceanic spectra peak at hundreds of Pascals per root Hz [5]. Note that the oceanic microbarom signals are three orders of magnitude larger than the atmospheric signals.

The Sea State.   The displacement ξ of the air/water interface is typically treated statistically[10, 11] in the sense that ξ is being taken to be a Gaussian process. One writes

$$\xi(\mathbf{x}_H, t) = \mathrm{Re} \int \hat{\xi}(\mathbf{k}) e^{i\left(\mathbf{k}\cdot\mathbf{x}_H - \omega(\mathbf{k})t\right)} d^2k \tag{1}$$

with sea state average $\langle\ \rangle_S$ given by

$$\langle \hat{\xi}(\mathbf{k})\hat{\xi}(\mathbf{q})^* \rangle_S = \mathscr{F}(\mathbf{k})\delta(\mathbf{k} - \mathbf{q}).$$

$\mathscr{F}(\mathbf{k})$ is called the wave number spectrum. It is related to the spectral density function $F(f, \theta)$ through a dispersion relation $2\pi f = \omega(\mathbf{k})$ by the change of variables $\mathscr{F}(\mathbf{k}) d^2k = F(f, \theta) df d\theta$. When needed here, the deep water dispersion relation $\omega(\mathbf{k}) = \sqrt{g|\mathbf{k}|}$, where g is the acceleration due to gravity, will be used.

For highly excited sea states the wave spectrum is sharply peaked at around 0.1 Hz. Spectra tend to have sharp low frequency cutoffs and long high frequency tails. Some model forms for the frequency spectrum $\int_0^{2\pi} F(f, \theta) d\theta$ for a variety of peak frequencies and significant wave heights $H_s = \frac{1}{4}\sqrt{\int_0^\infty \int_0^{2\pi} F(f, \theta) d\theta df}$ are shown in Fig. 1. The detailed θ dependence of $F(f, \theta)$ is not well understood [13, 14, 11]. Generally, far from land $F(f, \theta)$ is expected to be peaked in the direction of the prevailing winds. To produce the standing wave field required for microbarom radiation, however, there must be cross or upwind wave trains on the ocean surface in addition.

CP838, Innovations in Nonlinear Acoustics: 17th International Symposium on Nonlinear Acoustics,
edited by A. A. Atchley, V. W. Sparrow, and R. M. Keolian

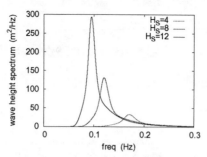

FIGURE 1. JONSWAP[11, 12] model spectra for $\int_0^{2\pi} F(f, \theta)\, d\theta$.

The Second Order Two-fluid Model. As in Brekhovskikh *et. al.* [4] we consider a two fluid model consisting of a dense fluid, the ocean, under a rare fluid, the atmosphere. The position of the interface between the fluids is assumed to be given by the displacement ξ. The equations of fluid mechanics in the air and water and the conditions of continuity of the pressure and normal component of the velocity across the air/water interface are expanded to second order in Mach number $\frac{f_0 H_S}{c_\sigma}$ where f_0 is the peak sea state frequency and c_σ is the small signal sound speed in air, for $\sigma = a$, and water, for $\sigma = w$. In addition to Mach number there is the small parameter $\frac{\omega_0}{|k|c_\sigma}$, the ratio of the sea state wavelength to the acoustic wavelength.

The first order (linear) velocity field is given by[15]

$$\mathbf{v}_1(\mathbf{x}, t) = \mathrm{Re} \int \omega(\mathbf{k}) \left(-(-1)^\sigma \frac{\mathbf{k}}{\sqrt{|\mathbf{k}|^2 - \frac{\omega^2}{c_\sigma^2}}} - i\hat{\mathbf{z}} \right) \hat{\xi}_1(\mathbf{k}) e^{i\left(\mathbf{k} \cdot \mathbf{x}_H - \omega(\mathbf{k})t\right) - \sqrt{|\mathbf{k}|^2 - \frac{\omega^2}{c_\sigma^2}}|z|} d^2 k$$

where $(-1)^\sigma$ is 1 is the air and -1 in the water. To leading order in $\frac{\omega_0}{|k|c_\sigma}$ one has

$$\mathbf{v}_1(\mathbf{x}, t) \approx \mathrm{Re} \int \omega(\mathbf{k}) \left(-(-1)^\sigma \frac{\mathbf{k}}{|\mathbf{k}|} - i\hat{\mathbf{z}} \right) \hat{\xi}_1(\mathbf{k}) e^{i\left(\mathbf{k} \cdot \mathbf{x}_H - \omega(\mathbf{k})t\right) - |\mathbf{k}||z|} d^2 k \qquad (2)$$

so that the motions of the air and water are approximately the same, other than relative phase. Note that in the linear approximation there is no acoustic radiation: the velocity field is vertically evanescent.

To leading order in $\frac{\omega_0}{|k|c_\sigma}$ the second order wave equation for the second order velocity potential ϕ_2 is

$$\left(-\frac{\partial^2}{\partial t^2} + c_0^2 \nabla^2 \right) \phi_2 = \frac{\partial}{\partial t} \|\mathbf{v}_1\|^2.$$

Similarly, the interface continuity conditions are

$$\rho_\sigma \left(\frac{\partial \phi_2}{\partial t} + g\xi_2 \right) \Big|_{z=-0^+}^{z=0^+} = -\rho_\sigma \left(\frac{\partial v_{1z}}{\partial t} \xi_1 + \frac{1}{2} \mathbf{v}_1 \cdot \mathbf{v}_1 \right) \Big|_{z=-0^+}^{z=0^+}.$$

for the pressure and

$$\frac{\partial \phi_2}{\partial z}\Big|_{z=-0^+}^{0^+} = \left(-\xi_1 \nabla \cdot \mathbf{v}_1 + \nabla_H \cdot (\xi_1 \mathbf{v}_1)\right)\Big|_{z=-0^+}^{0^+}$$

for the velocity.

Given $\xi_1 \approx \xi$ it is in principle straightforward to obtain the outwardly propagating solution ϕ_2 of the above system. For arbitrary sea states ξ one may expand in a Fourier transform as in (1) and solves the above system for a general plane wave. The resulting solution, given as a complicated superposition of the acoustic £elds which accompany colliding plane waves on the ocean surface, is quite complicated.

The solution simpli£es signi£cantly if instead of arbitrary sea states one treats the sea state statistically and obtains statistical properties of the received signal rather than attempting to produce the signal itself. As usual [16] one equates the cross correlation

$$\lim_{T \to \infty} \frac{1}{T} \int_0^T P_\sigma(\mathbf{x}_H, z, t) P_\sigma(\mathbf{x}'_H, z', t + \tau)^* \, dt$$

between the received pressure signals P_σ at different points and times with an ensemble average

$$\langle P_\sigma(\mathbf{x}_H, z, 0) P_\sigma(\mathbf{x}'_H, z', \tau)^* \rangle_T.$$

Here the total ensemble average $\langle \cdot \rangle_T = \langle \cdot \rangle_P \langle \cdot \rangle_S$ is the average over both propagation medium $\langle \cdot \rangle_P$ and sea state $\langle \cdot \rangle_S$. To £nd this cross correlation it turns out that one only need consider the collisions of ocean surface waves of equal frequency and opposite propagation direction. One obtains

$$\langle P_\sigma(\mathbf{x}_H, z, 0) P_\sigma(\mathbf{x}'_H, z', \tau)^* \rangle_T = \int_S \int_0^\infty \mathcal{Q}_\sigma(\mathbf{x}_H, \mathbf{x}'_H, z, z', \mathbf{y}_H, f) \mathcal{D}_\sigma(f) e^{-i 2\pi f \tau} \, df \, d^2 \mathbf{y}_H$$

(3)

where $\mathcal{Q}_\sigma$ is the cross spectral density[16],

$$\mathcal{Q}_\sigma(\mathbf{x}_H, \mathbf{x}'_H, z, z', \mathbf{y}_H, f) = \langle \hat{G}_\sigma(\mathbf{x}_H, z, \mathbf{y}_H, 0, 2\pi f) \hat{G}_\sigma(\mathbf{x}'_H, z', \mathbf{y}_H, 0, 2\pi f)^* \rangle_P,$$

(4)

for propagation from a point source at $\mathbf{y}_H \in S$ to receivers at $(\mathbf{x}_H, z)$ and $(\mathbf{x}'_H, z')$; here the $\hat{G}_\sigma$ are the frequency domain Greens functions; and $\mathcal{D}_\sigma$ is the source strength spectrum squared

$$\mathcal{D}_a(f) = \frac{4\rho_a^2 g^2 \pi^4 f^3}{c_a^2} \left(\frac{9g^2}{4\pi^2 c_a^2 f^2} + \frac{c_a^2}{c_w^2}\right) \int_0^{2\pi} F(\frac{f}{2}, \theta) F(\frac{f}{2}, \theta + \pi) \, d\theta$$

(5)

$$\mathcal{D}_w(f) = \frac{4\rho_w^2 g^2 \pi^4 f^3}{c_w^2} \int_0^{2\pi} F(\frac{f}{2}, \theta) F(\frac{f}{2}, \theta + \pi) \, d\theta.$$

(6)

Note that the ratio of the atmospheric to oceanic source spectra is independent of the sea state,

$$\sqrt{\frac{\mathcal{D}_a(f)}{\mathcal{D}_w(f)}} = \frac{\rho_a}{\rho_w} \sqrt{1 + \frac{9g^2 c_w^2}{4\pi^2 c_a^4 f^2}}.$$

(7)

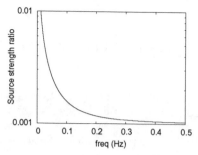

FIGURE 2. The ratio of atmospheric to oceanic source spectra, $\sqrt{\frac{\mathcal{D}_a}{\mathcal{D}_w}}$.

This ratio is plotted in Fig. 7. In the frequency band of interest, ~ 0.2 Hz, the oceanic source strength is three orders of magnitude larger than the atmospheric source strength. This discrepancy arises from Equ. (2) and the density difference between air and water. Equ. (2) shows that the accelerations of the air and water near the ocean surface are approximately the same. However, because $\frac{\rho_a}{\rho_w} \sim 0.001$ the pressure gradients needed to support these accelerations are a thousand times greater in the water.

Finally, the factor $\int_0^{2\pi} F(\frac{f}{2},\theta)F(\frac{f}{2},\theta+\pi)\,d\theta$, common to both oceanic and atmospheric source spectra, is a measure of the spectrum of ocean surface plane waves of equal frequency and opposite propagation direction (see also [3] and [4]). This factor is poorly understood, but is presumably small[5, 3] .

ACKNOWLEDGMENTS

We would like to thank Konstantine Naugol'nykh for bringing Ref. [4] to our attention.

REFERENCES

1. W. L. Donn, and B. Naini, *J. Geophys. Res.*, **78**, pp. 4482 to 4488 (1973).
2. M. S. Longuet-Higgins, *Phil. Trans. R. Soc. Lond.*, **243**, pp. 1 to 35 (1950).
3. K. Hasselmann, *Reviews of Geophysics*, **1**, §3 pp. 177 to 210 (1963).
4. I. M. Brekhovskikh, V. V. Goncharov, V. M. Kurtepov, and K. A. Naugol'nykh, *Izv. Atmospheric and Oceanic Physics*, **9**, pp. 899 to 907 (1973).
5. S. C. Webb, and C. S. Cox, *J. Geophys. Res.*, **91**, pp. 7343 to 7358 (1986).
6. B. T. R. Lewis, and L. M. Dorman, *http://faculty.washington.edu/blewis/papers/noise/paper.html* (1998).
7. J. V. Olsen, and C. A. L. Szuberla, *J. Acoust. Soc. Am.*, **117**, pp. 1032 to 1037 (2005).
8. M. Garces, M. Willis, C. Hetzer, A. L. Pinchon, and D. Drob, *Geophys. Res. Lett.*, **31**, L018614 (2004).
9. J. R. Bowman, G. E. Baker, and M. Bahavar, *Geophys. Res. Lett.*, **32**, L09803 (2005).
10. B. Kinsman, *Wind Waves*, Prentiss Hall, Englewood Cliffs, NJ, 1965.
11. G. J. Komen, L. Cavaleri, M. Dronelan, K. Hasselmann, S. Hasselmann, and P. A. E. M. Janssen, *Dynamics and Modeling of Ocean Waves*, Cambridge, Cambridge, 1996.
12. K. Hasselmann, T. P. Barnett, E. Bouws, H. Carlson, D. E. Cartwright, K. Enke, J. A. Ewing, H. Gienapp, D. E. Hasselmann, P.Kruseman, A. Meerburg, P. Müller, D. J. Olbers, K. Richter, W. Sell,

and H. Walden, *Dtsch. Hydrogr. Z. Suppl. A*, **8**, 95 (1973).

13. M. A. Donelan, J. Hamilton, and W. H. Hui, *Philos. Trans. R. Soc. London, Ser. A*, **315**, pp. 509 to 562 (1985).

14. S. C. Webb, *J. Acoust. Soc. Am*, **92**, pp. 2141 to 2158 (1992).

15. H. Lamb, *Hydrodynamics*, Dover, NY, 1945.

16. J. S. Bendat, and A. G. Piersol, *Random Data, Analysis and Measurement Procedures*, Wiley, New York, 2000, 3 edn.

Infrasound Radiation of Cyclone

K. Naugolnykh, and S. Rybak[†]

*University of Colorado/Zel Technologies, LLC and NOAA/ Environmental Technology Laboratory,
325 Broadway, Boulder, CO 80305, USA
[†] N. Andreev Acoustics Institute, 4 Shvernik ST., 117036, Moscow, Russia*

Abstract. Tropical cyclones produce strong perturbations of atmosphere and the ocean surface accompanying by an acoustical radiation. Infrasonic signals in the 0.1-0.5 frequency band can be observed at a distances of thousands of miles from the cyclone. There are specific features of the mechanism of infrasound radiation by cyclone. The radiated sound can be trapped by atmospheric wave guide. The presence of oversaturated water vapor layer in the upper domain of the cyclone leads to the development of specific nonlinear effects of radiation instability that provides amplification of infrasound radiation.

Keywords:Infrasound.
PACS: 43. 28. Dm

INFRASOUND GENERATION BY THE SURFACE WAVES

Infrasound generation

Storm regions in the ocean are sources of atmospheric infrasound, which can propagate over long distances [1]. It appears that surface wave interaction can make a substantial contribution to infrasound generation. The nonlinear interaction of counter-propagating sea-surface waves generates sound waves traveling away from the surface [2,3]. The radiation of sound into the water and into the atmosphere is not symmetrical. The ocean surface is acoustically absolutely compliant for the radiation of sound into the water. As a result, dipole sound sources appear in this case. For the radiation of sound into the atmosphere the ocean surface is a surface of volume-velocity sources (monopoles) because of its low acoustic compliance [4]. Sound waves are faster than surface gravity waves meaning that the interaction between them is weak. However two counter-propagating surface waves almost equal frequencies can produce a resonant interaction with the sound wave. The three wave resonant interaction [5] can be described by the equation of the potential flow

$$\Delta\varphi = 0$$

and the boundary conditions at $z = \varsigma$

$$\frac{\partial \varsigma}{\partial t} + \frac{\partial \varsigma}{\partial x_i}\frac{\partial \varphi}{\partial x_i} = \frac{\partial \varphi}{\partial z}, i = 1,2,$$

φ is the velocity potential, ζ is the sea surface displacement,

(1)

CP838, *Innovations in Nonlinear Acoustics: 17th International Symposium on Nonlinear Acoustics*,
edited by A. A. Atchley, V. W. Sparrow, and R. M. Keolian
© 2006 American Institute of Physics 0-7354-0330-9/06/$23.00

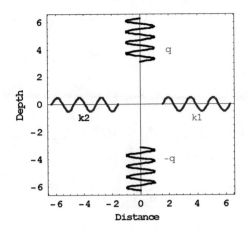

FIGURE 1. Two surface gravity waves of wave-numbers k_1 and k_2 propagating in almost opposite directions result in radiation of sound wave of wave number q and -q, propagating upward and downward.

$$(2)$$

$$\frac{\partial \varphi}{\partial t} + \frac{1}{2}(\nabla \varphi)^2 + gz = 0,$$

g is the gravity acceleration. Transfer the boundary conditions to the plane z=0 leads to the nonlinear equation [6]. The solution of this equations for the two counter-propagating surface gravity waves (axes z is directed upward)

$$\varphi_i = a_i e^{kz} \cos[k_i x - \omega_i t], \, i = 1,2 \tag{3}$$

interacting with the sound wave

$$\varphi_3 = a_3 \sin[qz - \omega_3 t] \tag{4}$$

in the approximation of the slow varying amplitudes [7] leads to the equation of the sound wave amplitude

$$a_3 = -\frac{k_1 k_2 a_1 a_2}{\omega_3}. \tag{5}$$

The sound wave frequency is sum of the gravity waves frequencies

$$\omega_3 = \omega_1 + \omega_2, \tag{6}$$

and sound wave wave number is

$$q = \frac{\omega_3}{c_a}, \tag{7}$$

c_a is the sound velocity in the air.

529

Horizontal component of the sound wave wave number is determined by the equation

$$q_{x3} = k_1 - k_2. \tag{8}$$

The pressure amplitude of the sound

$$p_a = \rho_a \upsilon_1 \upsilon_2,$$

$$\tag{9}$$

ρ_a is the density of the air. $p_a / p_{wa} = \dfrac{\rho_a}{\rho_w}$, ρ_w is the water density.

Sound Propagation in the Supersaturated Water Vapor

Due to expansion of the air in the upper layer of the cyclone its temperature decreasing and the water vapor becomes supersaturated. The sound propagation in such a medium stimulates the relaxation process of returning to the equilibrium state governed by the equation

$$\xi' = -\frac{1}{\tau}(\xi - \xi_0), \tag{10}$$

ξ is a physical characteristic of the system, ξ_0 corresponds to equilibrium state, and τ is the relaxation time of the process. In the case of harmonic perturbations produced by sound of frequency ω the presence of non-equilibrium processes leads to appearance of the complex wave number k of the sound wave

$$k = \omega \sqrt{\frac{1 - i\omega\tau}{c^2{}_0 - c^2{}_\infty i\omega\tau}}, \tag{11}$$

c_0, c_∞ are the sound velocities at the low and high frequencies.

If the frequency of sound ω is small, $\omega\tau \ll 1$, this equation can be expanded with respect to $\omega\tau$:

$$k = \frac{\omega}{c} + ic_0\tau \left(\frac{\omega}{c_0}\right)^2 - 2.5(c\tau)^2 \left(\frac{\omega}{c_0}\right)^3 - 6.5(c\tau)^3 \left(\frac{\omega}{c_0}\right)^4. \tag{12}$$

The partial differential equation (PDE) that corresponds approximately to the given dispersion equation is

$$\frac{\partial u}{\partial t} + u\frac{\partial u}{\partial x} - c^2{}_0\tau \frac{\partial^2 u}{\partial x^2} + 2.5c_0(c_0\tau)^2 \frac{\partial^3 u}{\partial x^3} + 6.5c_0(c_0\tau)^3 \frac{\partial^4 u}{\partial x^4} = 0. \tag{13}$$

530

Here u(t,x) is the sound wave particle velocity, and x is coordinate in the accompanying system of coordinate. This is the equation of long wave instability, it has the solutions $u \sim \exp(\alpha t - i\omega t)$ that indicate the possibility of the sound amplification [8], α is the increment of the perturbation increasing.

1. Shuleykin, V. V., The Voice of the Sea. Dokl. Akad. Nauk SSSR, 3, 259, 1935.
2. M. S. Longuet-Higgins, A theory of the origin of microseisms,
                          1950, Phil. Trans., Ser. A, 243, pp.1-35.
3. L. M. Brekhovskikh, "Underwater sound waves generated by surafce waves in the ocean,"Izv., Atmospheric and Oceanic Physics, 2, 9 ,(1966), pp. 970-980..
4. L. M. Brekhovskikh,V. V. Goncharov, V. M. Kurtepov, and K. A. Naugolnykh, The radiation of infrasound into the atmosphere by surface waves in the ocean,1973 Izvestia AN, Atmospheric and Oceanic Physics, 9 9,pp. 899-907.
5. Phillips, O. M. (1977). The dynamics of the upper ocean (Cambridge University Press).
6. A.C. Kibblewhite and C.Y. Wu, `The generation of microseisms and infrasonic ambient noise by nonlinear interactions of ocean surface waves`, J. Acoust. Soc. Am., 1989, 85, pp.1935-1945.
7. Naugolnykh K. A, Rybak S. A., "Sound generation due to the interaction of surface waves, Acoustical Physics 49 (1), 2003, pp100-103.
8. N. J. Balmforth, G. R. Ierley, and R. Worthing, "Pulse Dynamics in an unstable medium", SIAM J. APPL., MATH.,1997, 57, 205-251.

Nonlinear propagation of high-frequency energy from blast waves

Alexandra Loubeau* and Victor W. Sparrow*

*Graduate Program in Acoustics, The Pennsylvania State University, 202 Applied Science
Building, University Park, PA 16802, aloubeau@psu.edu/vws1@psu.edu

Abstract. High-frequency energy is generated from nonlinear propagation of finite-amplitude shock waves created by explosions. Energy at these high frequencies may be harmful to bats because their auditory systems use high-frequency information. In April 2005, measurements of blast waves were performed by the U.S. Army. Spectrographic analysis of the waveforms confirms that the high-frequency energy is concentrated near the shocks. The data from these recent experiments serve as a benchmark for analysis of computational model predictions. Using a hybrid time-frequency domain Anderson-type algorithm, the extended generalized Burgers equation is solved for the propagation of a blast wave. Computer code predictions are compared to the benchmark experimental data.

Keywords: blast noise, nonlinear propagation, shock wave, high frequency, bats, Chiroptera
PACS: 43.28.Mw, 43.25.Cb

INTRODUCTION

High-frequency energy is generated from nonlinear propagation of finite-amplitude shock waves created by explosions. Energy at these high frequencies may be harmful to bats because their auditory systems use high-frequency information for flight navigation, communication, and hunting. The rise time associated with a blast wave's initial shock jump is a parameter used to assess the presence of high-frequency energy; it is defined as the time it takes for the sound pressure to rise from 10% to 90% of its maximum value. Measurements of rise times from blast wave experiments are compared to computer code predictions.

EXPERIMENTAL RESULTS

In April 2005, measurements of blast waves were performed by the U.S. Army at the Edgewood site of Aberdeen Proving Grounds in MD. The source was 0.57 kg (1.25 lb) of C4 explosive. The aim of the experiments was to accurately measure the rise portion of the blast wave. Special equipment was used to sample at 1 MHz to get good resolution at the shock. Three 3.175-mm (1/8-in.) microphones with a large dynamic range and a bandwidth up to 140 kHz were also used. The blast waves were measured at distances of 25 and 50 m from the explosive.

Spectrographic Analysis

Spectrographic analysis of the waveforms confirms that the high-frequency energy is concentrated near the shocks. Spectrograms are found by sliding a narrow time window across the waveform and taking an FFT at each point. The spectrogram in Fig. 1 is

CP838, *Innovations in Nonlinear Acoustics: 17th International Symposium on Nonlinear Acoustics*,
edited by A. A. Atchley, V. W. Sparrow, and R. M. Keolian

representative of a blast wave measured at 25 m. Notice that the peaks in high-frequency content occur at the shocks, with substantial energy at the initial shock. It is these shocks, therefore, that contain the high frequencies that may be harmful to bat hearing.

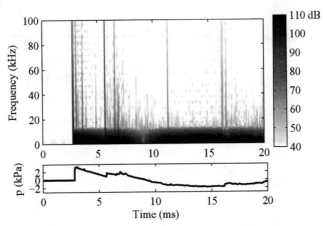

FIGURE 1. Spectrogram of blast wave measured at 25 m.

Microphone Baffles

Microphone baffles were used to reduce diffraction artifacts of the microphone housing on the rise portion of the blast waves. The baffles used in the experiments were 20.32-cm (8-in.) squares, and the microphones fit flush with the front of the baffle. Assuming a plane wave, a rigid circular baffle, and measurement at a point, there is pressure doubling at the microphone due to the superposition of the incident and reflected waves. The diffracted wave is delayed according to the radius, out of phase, and half the amplitude [1]. In reality, the experiments involve a spherical wave, a square baffle, and a microphone element of finite area. However, using these assumptions, a synthesized diffracted wave can be calculated from the unbaffled measurements and summed with the unbaffled wave. The result, shown in Fig. 2, is very similar to the measured wave with the baffle. Even though the blast waveform is distorted due to the finite size of the baffle, the rise portion is measured accurately.

Figure 3 is a comparison of the rise portions of the measured blast waves. The rise time measured at 25 m by the unbaffled microphone is 6.4 μs. By contrast, the baffled response at the same distance of 25 m shows a smoother rise with a rise time of 32.8 μs. As the blast wave propagates, nonlinear effects result in waveform steepening, and the rise time measured with a baffled microphone at 50 m decreases to 8.8 μs.

The rise portion of the waveform measured with the baffled microphone is compared to an analytical result for a stationary wave in a monorelaxing fluid [2]. The relaxation process chosen is molecular vibration of oxygen molecules, which is the dominant relaxation mechanism in air for the pressure amplitudes being considered [3]. Figure 4 compares the rise portion of the measured wave at 50 m from a baffled microphone to the analytical result. The maximum amplitude is chosen to be the same, and the calculated

533

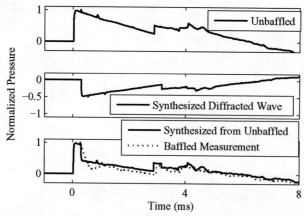

FIGURE 2. Synthesized baffled waveform comparison. The upper waveform (unbaffled) is summed with the middle waveform (-1/2 of unbaffled and delayed), yielding the dotted line in the lower plot.

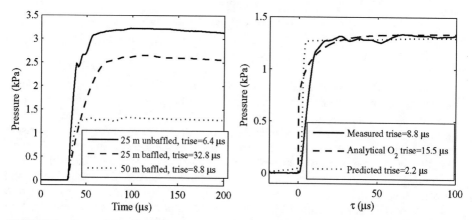

FIGURE 3. Comparison of baffled and unbaffled rise portions for 25 and 50 m.

FIGURE 4. Comparison of measured blast wave at 50 m, analytical shock wave in a monorelaxing fluid, and predicted blast wave at 50 m.

rise times are similar. The ratio of relaxation effects to nonlinear effects can be described by the parameter $D = m\rho_0 c_0^2 / 2\beta p_0$ [2]. In this case, $D = 0.0517$, which means that there are strong nonlinear effects.

COMPUTATIONAL MODEL PREDICTIONS

The evolution of a finite-amplitude wave can be described by the extended generalized Burgers equation, given in Equation 1.

534

$$\frac{\partial p}{\partial r} = \frac{\beta p}{\rho_0 c_0^3} \frac{\partial p}{\partial \tau} + \frac{\delta}{2c_0^3} \frac{\partial^2 p}{\partial \tau^2} + \sum_v R_v \left(\frac{\partial^2 p}{\partial \tau^2} \right) - \frac{m}{r} p \qquad (1)$$

This equation accounts for the physical processes of nonlinearity, thermoviscous absorption, absorption and dispersion due to molecular relaxation, and geometrical spreading. For spherical spreading, $m = 1$. This equation is solved using the method of fractional steps [4]. The code, based on work by Pestorius and Anderson, is a hybrid time-frequency domain algorithm where nonlinearity is applied in the time domain and spreading and absorption are applied in the frequency domain [4].

The rise portion of the baffled waveform is combined with the unbaffled waveform at 25 m for use as an input to the code. Predicted waveforms are then found for the distances 50, 100, and 200 m. The prediction for 50 m, shown in Fig. 4, indicates that the wave steepens up at first, and the rise time drops from $32.8\,\mu s$ at 25 m to $2.2\,\mu s$. At 100 m, the rise time increases slightly to $2.8\,\mu s$, and at 200 m the rise time is increased to $9.6\,\mu s$. The measured rise time at 50 m is longer than what is predicted, and this may be attributed to the finite bandwidth of the measurement system [5].

CONCLUSIONS

The model predictions presented in this paper show that nonlinear effects act to steepen a blast wave as it propagates up to a distance of approximately 50 m from the source. This results in a decrease in rise time and therefore an increase in high-frequency energy. This trend is in agreement with experimental results. However, it is difficult to obtain accurate measurements of the rise portion of blast waves, and future work will focus on using a system with a higher bandwidth to capture the short rise times.

ACKNOWLEDGMENTS

This work was supported by the U.S. Army ERDC-CERL under contract DACA42-003-C-0052. Blast experiments were directed by Dr. Larry Pater of the U.S. Army. The authors gratefully acknowledge T. B. Gabrielson and T. M. Marston for advice on and construction of microphone baffles.

REFERENCES

1. D. T. Blackstock, *Fundamentals of Physical Acoustics*, John Wiley & Sons, Inc., New York, 2000.
2. M. F. Hamilton, Y. A. Il'inskii, and E. A. Zabolotskaya, "Dispersion," Chap. 5 in *Nonlinear Acoustics*, edited by M. F. Hamilton and D. T. Blackstock, Academic Press, San Diego, CA, 1998.
3. A. D. Pierce, "Rise profiles of initial shocks in sonic boom waveforms," *J. Acoust. Soc. Am.*, **104**(3), 1829 (1998).
4. J. H. Ginsberg and M. F. Hamilton, "Computational Methods," Chap. 11 in *Nonlinear Acoustics*, edited by M. F. Hamilton and D. T. Blackstock, Academic Press, San Diego, CA, 1998.
5. T. B. Gabrielson, T. M. Marston, and A. A. Atchley, "Nonlinear propagation modeling: Guidelines for supporting measurements," in *Proc. NOISE-CON 2005*, 2005.

Irregular Reflection of Acoustical Shock Waves and von Neumann Paradox

S. Baskar*, F. Coulouvrat* and R. Marchiano*

*Laboratoire de Modélisation en Mécanique, Université Pierre et Marie Curie & CNRS (UMR 7607), 4 place Jussieu, 75252 Paris cedex 05, France

Abstract. We investigate the reflection of weak acoustical shock waves grazing over a rigid surface. We define a critical parameter and examine the different types of reflection structure depending on this parameter. The study of the step shock is then extended to both N-waves and periodic saw-tooth waves, which are more realistic from an acoustical point of view. The numerical simulations reveal new reflection structures for these two waves which are not observed for step shocks. The results of the model are finally compared for periodic saw-tooth waves to ultrasonic experiments.

Keywords: weak shock reflection, von Neumann paradox, nonlinear acoustics
PACS: 43.25.C, 43.25.J

INTRODUCTION

When a shock wave impinges on a rigid surface, it gives rise to a reflected wave. For a weak incident shock or a large oblique angle, the reflected and the incident shocks intersect on the rigid surface, which is called *regular reflection*. In the linear limit, it reduces to the well known Snell's law. When the oblique angle decreases or the strength of the shock increases, the regular reflection ceases to exist. Then, the incident and reflected shocks intersect above the rigid surface and a new shock, known as the *Mach shock* or *Mach stem* emerges from the point of intersection, known as the *triple point*. This type of reflection is called *irregular reflection*. It was first observed experimentally by Ernst Mach in 1878 for a strong shock and is named after him as *Mach reflection*. Von Neumann's theory (1943) on the oblique shock reflection is well suited for strong shocks, but ceases to give physically realistic results for weak shocks in contrast with experimental results (Colella and Henderson (1990)): this is known as the *von Neumann paradox*. Many reasons for the paradox have been given by several authors (Colella and Henderson, 1990, Brio and Hunter, 2000, Ben-Dor *et al.*, 2002 and Kobayashi *et al.*, 2004). Although the weak shock reflection is still not fully understood, the rigorous efforts in the experimental and theoretical investigations for several decades have given a good understanding on the reflection phenomena for step shocks. However, this problem has not been studied from an acoustical point of view, where shocks are extremely weak (Mach number of order 1.001) and are rarely step shocks but generally have more complex profiles such as N-waves or saw-tooth waves. This is the objective of the present work to investigate theoretically, numerically and experimentally irregular reflection of acoustical shock waves.

CP838, *Innovations in Nonlinear Acoustics: 17th International Symposium on Nonlinear Acoustics*,
edited by A. A. Atchley, V. W. Sparrow, and R. M. Keolian
© 2006 American Institute of Physics 0-7354-0330-9/06/$23.00

KZ EQUATION AND BOUNDARY CONDITIONS

We consider a 2D nonlinear plane wave grazing obliquely with small angle $\theta \ll 1$ over a rigid surface with acoustical Mach number $M_a \ll 1$, assuming $O(\theta){=}O(\sqrt{M_a})$. We assume a homogeneous and inviscid fluid of ambient density ρ_0 and sound speed c_0. The longitudinal and the transverse variables are respectively x and y, and the physical time is t. The dimensionless delayed time is $\tau = \omega(t-x/c_0)$, where ω is a some reference frequency. The spatial variables are scaled by $Y = y/L$, $X = x/D$, where $L = 1/(k\sqrt{2\beta M_a})$ is the transverse length scale and $D = 1/(\beta k M_a)$ is the shock formation distance with $k = \omega/c_0$, $\beta = 1+B/2A$ and B/A the nonlinearity parameter. With the above approximations the well-known KZ equation (Zabolotskaya and Khokhlov, 1969)

$$P_{X\tau} - P_{YY} = \left(P^2/2\right)_{\tau\tau} \tag{1}$$

is appropriate. Here $P = p_a/(\rho_0 c_0 U_0)$ is the dimensionless acoustic pressure with p_a the acoustical pressure and U_0 the velocity amplitude of the incident field.

We define the critical parameter $a = \sin\theta/\sqrt{2\beta M_a} = O(1)$ and specify the incoming pressure field at $X = 0$ as

$$P(X,\tau,Y) = \begin{cases} 0, & \text{if } \tau+aY+a^2X \leq -X/2 \\ 1, & \text{if } \tau+aY+a^2X > -X/2 \end{cases} \tag{2}$$

$$P(X,\tau,Y) = \begin{cases} -(\tau+aY+a^2X)/(X+1) & \text{, if } |\tau+aY+a^2X| \leq \sqrt{(1+X)/2} \\ 0 & \text{, otherwise} \end{cases} \tag{3}$$

$$P(X,\tau,Y) = \begin{cases} -(\tau+aY+a^2X+\lambda)/(X+1) & \text{, if } -\lambda < \tau+aY+a^2X \leq 0 \\ -(\tau+aY+a^2X-\lambda)/(X+1) & \text{, if } 0 < \tau+aY+a^2X \leq \lambda \end{cases} \tag{4}$$

respectively, for a step shock Eq.(2), an N-wave Eq.(3) and a periodic saw-tooth wave Eq.(4).

NUMERICAL SIMULATIONS

Following Coulouvrat and Marchiano (2003) we can show that the reflection solution is self-similar for step-shocks, which is given by $P(X,\tau,Y) = Q(\xi = \tau/X, \eta = Y/X)$. Fig. 1 shows the reflection solution for $a = 2.0$, 0.7, 0.4 and 0.1 respectively. In fig. 1a we observe a moderately strong regular reflection, where the slope of the reflected shock is greater than the incident one, which shows that the nonlinear solution still deviates from Snell's law, the one recovered only for values of a larger than about 4.0. The type of reflection depicted in fig. 1b ($a = 1$) is almost a Mach reflection with three shocks, and a discontinuity in the slope between the incident and the Mach shocks. However, because of the curvature of the Mach and reflected shocks, and non-constant reflected pressure, this is not a classical Mach reflection (see Ben-Dor, 1992 for the definition) and therefore we call it *generalized Mach reflection*. Fig. 1c ($a = 0.4$) shows the von Neumann reflection, where the reflected shock connects the incident one through a smooth compression wave. This type of reflection was observed experimentally and computed from Euler equations by Colella and Henderson (1990). As we reduce the value of a ($ = 0.1$) further, we observe on fig. 1d a new type of reflection, where the

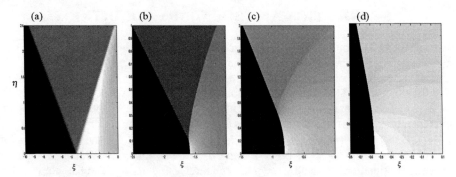

FIGURE 1: Reflection of a step shock for $a = 2$ (a), 1 (b), 0.4 (c), 0.1 (d) at $X = 1$

reflected wave is a compression wave instead of a shock, which we call *generalized von Neumann reflection*. The regular to irregular reflection occurs at $a = \sqrt{2}$ (Hunter, 1991). We further observe numerically the transition from generalized Mach reflection to von Neumann reflection around $a = 0.6$ and from von Neumann to generalized von Neumann reflection for $a < 0.3$.

The reflection pattern of the leading shock of the N-wave or of the periodic saw-tooth wave are more complex, as the reflected shock interacts with the waves behind the incident shock. Also, the incident shock amplitude decreases with propagation. As a result, the triple point first moves away from and then goes back to the rigid boundary (see fig. 2a for $a = 0.5$). Consequently, there is no self-similar solution in these cases. Fig. 2b shows the solution of the periodic saw-tooth wave for $a = 0.5$ and $X = 1$, where the triple point is approximately at its maximum $Y \approx 0.25$. We observe numerically the decrease of the maximum position of the triple point when the parameter a increases, which results in an almost regular reflection for $a > 0.9$ (saw-tooth waves) and for $a > 0.8$ (leading shock of the N-waves). Fig. 2c shows for an N-wave the triple point associated to the Mach reflection of the head shock. For the tail shock, there exists the so-called transitioned regular reflection, with an inverted triple-shock pattern emanating from the termination of the irregular reflection (Ben-Dor, 1992).

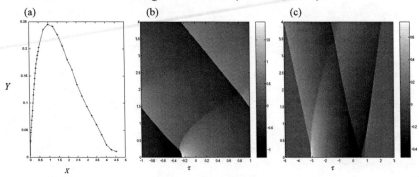

FIGURE 2: (a) Triple point trajectory for $a = 0.5$ (saw-tooth wave). (b) Reflection of saw-tooth wave for $a = 0.5$ at $X = 1$. (c) N-wave reflection for $a = 0.5$ at $X = 5$

538

EXPERIMENTAL RESULTS

The reflection phenomena described above have been reproduced experimentally for periodic ultrasonic (1 MHz) shock waves produced in water by an array of 128 piezo-electric transducers controlled electronically (Marchiano *et al.*, 2003). The inverse filter technique (Tanter *et al.*, 2000) allowed, by a control of the pressure field in the linear regime, to emit at 61 cm and over a 10 cm width, a semi-plane wave incident on the symmetry axis with angles = 1°, 3° and 5°. The symmetry condition was used as an equivalent way to impose the rigid boundary condition, by avoiding coupling between the metal reflector and water. Then the array was used in nonlinear regime, producing shock waves with amplitude about 10 bars. The pressure field measured at 90 cm with a 40 MHz bandwidth membrane hydrophone is displayed on fig. 3 for the 3 angles. It appears clearly the 3 different reflection regimes predicted theoretically: Nonlinear regular reflection at 5° (fig. 3a), generalized Mach reflection at 3° (fig. 3b) and generalized von Neumann reflection at 1° (fig. 3c). This experiment demonstrates the complex reflection pattern already observed in aerodynamics and predicted theoretically also occurs for weak acoustical shock waves. [Experiments were performed at INSP, UMR-CNRS 7588, Université Pierre et Marie Curie, with the help of Dr. Jean-Louis Thomas. S. Baskar benefits from a post-doc fellowship from the French Ministery of Research].

(a) (b) (c)

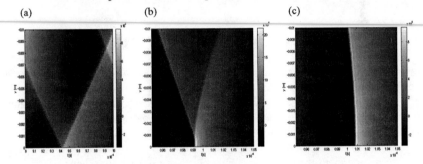

FIGURE 3:(a) Experimental results for angle 5° (a), 3° (b) and 1° (c)

REFERENCES

1. J. von Neumann, *ed. A. H. Taub, John von Neumann collected work, vol 6, (Pergamon, New York, 1963),* pp. 238 – 299 (1943).
2. P. Colella, and L. F. Henderson, *J. Fluid Mech.*, **213**, 71 – 94 (1990).
3. J. K. Hunter, and M. Brio, *J. Fluid Mech.*, **410**, 235 – 261 (2000).
4. G. Ben-Dor, M. Ivanov, E. I. Vasilev, and T. Elperin, *Progress in Aerospace Sciences*, **38**, 347 – 387 (2002).
5. S. Kobayashi, T. Adachi, and T. Suzuki, *Fluid Dyn. Res.*, **35**, 275–286 (2004).
6. E. A. Zabolotskaya, and R. V. Khokhlov, *Sov. Phys. Acoust.*, **15**, 35–40 (1969).
7. F. Coulouvrat, and R. Marchiano, *J. Acoust. Soc. Am.*, **114**, 1749 – 1757 (2003).
8. J. K. Hunter, *In Multidimensional hyperbolic problems and computations. IMA Volumes in Mathematics and its Applications Vol. 29 (ed. A. J. Majda and J. Glimm), Springer,* pp. 179 – 197 (1991).
9. R. Marchiano, J.-L. Thomas, and F. Coulouvrat, *J. Acoust. Soc. Am.*, **114**, 1758–1771 (2003).
10. M. Tanter, J.-L. Thomas, and M. Fink, *J. Acoust. Soc. Am.*, **108**, 223–234 (2000).

Nonlinear Acoustic Propagation into the Seafloor

B. Edward McDonald

Naval Research Laboratory, Washington DC 20375

Abstract. Explosions near the seafloor result in shock waves entering a much more complicated medium than water or air. Nonlinearities may be increased by two processes inherent to granular media: (1) a poroelastic nonlinearity comparable to the addition of bubbles to water, and (2) the Hertz force resulting from elastic deformation of grains, proportional to the Youngs modulus of the grains times the strain rate to the power 3/2. These two types of nonlinearity for shock propagation into the seafloor are investigated using a variant of the NPE model. The traditional Taylor series expansion of the equation of state (pressure as a function of density) is not appropriate to the Hertz force in the limit of small strain. We present a simple nonlinear wave equation model for compressional waves in marine sediments that retains the Hertz force explicitly with overdensity to the power 3/2. Numerical results for shock propagation are compared with similarity solutions for quadratic nonlinearity and for the fractional nonlinearity of the Hertz force.

Keywords: Shock Propagation, Granular Media, Sediment
PACS: 43.30.Lz, 43.25.Cb

THEORETICAL PRELIMINARIES

Explosive shock waves propagating in the seafloor are primarily compressional waves. A nonlinear theory appropriate to marine sediments including shear is rather complicated[1,2], and it is not clear that it leads to a tractable shock propagation model. A much simpler theory emerges when only the compressional wave is considered.

Elastic grains under compression contribute a term to the stress-strain relation which is proportional to the 3/2 power of the strain rate, known as the Hertz nonlinearity[3]. In order to accomodate the fractional order Hertz nonlinearity we have proposed[4] a variant of the nonlinear progressive wave equation (NPE)[5] which is cast in a wave-following frame translating in the x direction at a nominal unperturbed sound speed c_0 with respect to the medium:

$$\partial_t \rho' = -\frac{1}{2c_0}\partial_x(p(\rho') + c_0^2(\rho'^2/\rho_0 - \rho')) - \frac{c_0}{2}\int_\infty^x \nabla_\perp^2 \rho' dx$$
$$+ K\alpha c_0 \int_0^\infty \frac{\partial_x \rho'(x+\xi,y,z)}{\delta x + \xi}d\xi, \tag{1}$$

where $\rho' = \rho - \rho_0$ is the density perturbation and p is pressure, subscript zero designates the unperturbed medium, and $\nabla_\perp^2 = \partial_y^2 + \partial_z^2$. Time domain frequency-linear attenuation has been added to (1) as in [6] with α the attenuation in dB per wavelength, δx a minimum scale length, and $K = \ln 10/(20\pi^2) = 0.012$.

CP838, *Innovations in Nonlinear Acoustics: 17th International Symposium on Nonlinear Acoustics*,
edited by A. A. Atchley, V. W. Sparrow, and R. M. Keolian
2006 American Institute of Physics 0-7354-0330-9/06/$23.00

The coefficient of quadratic nonlinearity normally used in nonlinear acoustics,

$$\beta \equiv 1 + \frac{1}{2}\frac{\rho_0}{c_0^2}\frac{\partial^2 p}{\partial \rho'^2} \qquad (2)$$

diverges at small strain for a nonlinearity of order 3/2. For that reason we do not take the usual Taylor expansion of p, but leave (1) as is.

For porous media with low shear modulus and empty pores, Ostrovsky[7] has shown that the nonlinearity coefficient peaks at a high value for low porosity, analogous to the increase in nonlinearity of water when a small volume fraction of bubbles is present. When the pores are filled with water, however, particle velocities near the pores are much lower. This would prevent dramatic increases in nonlinearity in saturated sediments, especially at porosity values near 0.4 typical of the first few meters of the seafloor. When bubbles are in the sediment, however, the quadratic nonlinearity coefficient may increase significantly.

SIMILARITY SOLUTIONS AND NUMERICAL RESULTS

In the absence of attenuation, eq. (1) admits similarity solutions $\rho'(\mathbf{r}/t)$ in three dimensions. Simple analytic profiles emerge for nonlinear plane wave shocks, which are stable against three dimensional perturbation[8]. When the dominant nonlinearity in (1) is proportional to ρ'^n, and $n > 1$ is a constant, shock profiles are derived from a one dimensional version of (1) with dependent variable u proportional to ρ' :

$$\partial_t u = -\partial_x (u^n), \qquad (3)$$

which has similarity solution (allowing for change of proportionality constant across $x = 0$ to keep u real)

$$u(x,t) = \text{sign}(x)\,|x/nt|^{1/(n-1)}. \qquad (4)$$

The Rankine Hugoniot shock jump conditions (conservation of mass across the shock) for the similarity solution yields a shock location $x_s \propto t^{1/n}$ relative to the zero crossing which propagates through the medium at speed c_0. For quadratic nonlinearity $n = 2$, and (4) describes the classic N wave shock profile whose width increases as $t^{1/2}$. For the Hertz nonlinearity $n = 3/2$, and (4) predicts a parabolic shock profile whose width increases as $t^{2/3}$.

Numerical solutions illustrating the Hertz nonlinearity are obtained by nondimensionalizing x, t, and u, and integrating the equation

$$\partial_t u = -\partial_x \left(u^2 + A \cdot \max(0,u)^{3/2}\right) \qquad (5)$$

where now $u = \beta_0 \rho'/\rho_0$, β_0 is the quadratic nonlinearity coefficient of the medium excluding Hertz forces, and A is a constant determining the strength of the granular stresses. We chose A to make the quadratic and granular nonlinearities comparable. The *max* function in (5) reflects that grains transmit stress only under compression. Initial

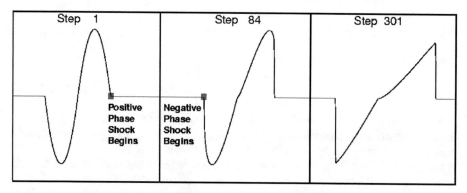

FIGURE 1. Numerical integration of eq. (5) illustrating the behavior of a nonlinear compressional wave propagating in a saturated granular medium.

conditions are taken as a single cycle of a sine wave with maximum amplitude 0.02. Results are shown in Figure 1.

Numerical methods used are second order upwinding with flux correction as described in [9] on a grid of 202 points in x. As the wave propagates the positive phase becomes concave upward and develops a shock quickly since the nonlinearity of order 3/2 is stronger at low amplitude than the quadratic nonlinearity. The divergence of the nonlinearity coefficient (2) at low amplitude for the Hertz nonlinearity has only moderate effects on the shape and width of the positive phase by the end of the simulation. The shape of the positive shock phase is in qualitative agreement with the parabolic profile predicted by (4). In the negative shock phase only the quadratic nonlinearity is active, and the profile evolves toward the linear ramp predicted by (4).

SUMMARY

The NPE has been reformulated[4] in eq. (1) for compressional waves in marine sediments with general equation of state $p(\rho')$ containing fluid and granular contributions. The granular contribution considered here was the Hertz force with constant number of grain contacts. In numerical simulations smooth sine wave profiles develop shocks and evolve toward power law profiles predicted by similarity solutions (4). In more realistic situations[4] (not shown here) taking into account results from consolidation tests[10], the contact number increases with strain rate, leading to a nonlinearity of order 5/2 or higher. Simulations with nonlinearity order higher than 2 develop late time profiles which are concave downward[4] as predicted by (4).

ACKNOWLEDGMENTS

Work supported by the Office of Naval Research.

REFERENCES

1. M. A. Biot, "Nonlinear and Semilinear Rheology of Porous Solids," *J. Geophys. Res.* **78**, 4924-4937 (1973).

2. D. M. Donskoy, K. Khashanah, and T. G. McKee, "Nonlinear Acoustic Waves in Porous Media in the Context of Biot's Theory," *J. Acoust. Soc. Am.*, **102**, 2521-2528, (1997).

3. Landau and Lifschitz, *Theory of Elasticity*, Addison- Wesley (1959).

4. B. E. McDonald, "Modeling Nonlinear Compressional Waves in Marine Sediments," *J. Acoust. Soc. Am.*, submitted (2005).

5. B. E. McDonald and W. A. Kuperman, "Time Domain Formulation for Pulse Propagation Including Nonlinear Behavior at a Caustic," *J. Acoust. Soc. Am.* **81**, 1406 - 1417 (1987).

6. J. E. White, *Seismic Waves: Radiation, Transmission, and Attenuation*, McGraw-Hill, (1965).

7. L. A. Ostrovsky, "Wave Processes in Media with Strong Acoustic Nonlinearity," *J. Acoust. Soc. Am.*, **90**, 3332- 3337 (1991).

8. B. E. McDonald, "Stability of Self Similar Plane Waves in Nonlinear Acoustics," *J. Acoust. Soc. Am.* submitted (2005).

9. B. E. McDonald and J. Ambrosiano, "High Order Upwind Flux Correction Methods for Hyperbolic Conservation Laws," *J. Comp. Phys.* **56**, 448 - 460, (1984).

10. W. J. Winters, "Stress History and Geotechnical Properties of Sediment from the Cape Fear Diapir, Blake Ridge Diapir, and Blake Ridge," *Proc. Ocean Drilling Prog.* **164**, 421 - 429, (2000).

Application of Weighted Essentially Non-Oscillatory Schemes to the Propagation of Shock Waves in the Atmosphere

Mark S. Wochner and Anthony A. Atchley

The Pennsylvania State University Graduate Program in Acoustics
University Park, PA, 16802

Abstract. Using a weighted essentially non-oscillatory scheme, a Navier-Stokes system of equations is solved in two dimensions. The model is able to account for the effects of molecular relaxation of nitrogen and oxygen and accounts for absorption due to shear viscosity, bulk viscosity, and thermal conductivity. The numerical scheme is described and the general utility of the current model is discussed. Applications are shown that demonstrate the model's ability to stably propagate multidimensional shocks.

Keywords: Shock propagation, molecular relaxation, weighted essentially non-oscillatory schemes.

PACS: 43.28.Js, 43.25.Cb, 43.25.Vt

INTRODUCTION

The goal of this research is to create an accurate computational model that will allow for a wide variety of atmospheric propagation scenarios. For this purpose, a Navier-Stokes system of equations is composed that contains terms to account for bulk viscosity, shear viscosity, and thermal conductivity, as well as two additional equations that allow the model to account for the absorption and dispersion of nitrogen and oxygen relaxation. Due to computational hardware limitations, the model is constrained to two dimensions, although the equation set could easily be extended to three dimensions.

The accurate and stable propagation of shocks is usually the most difficult aspect of wave propagation of this sort. To address the problem of accurate, stable propagation of any number of shocks in any direction, the weighted essentially non-oscillatory (WENO) method of solving systems of equations is used. As will be shown below, this method uses "smoothness indicators" to determine a waveform's level of discontinuity which it then uses to specially treat these waveforms so that the shock structure is maintained stably. This method and its applications to acoustical propagation is the focus of the current work.

CP838, *Innovations in Nonlinear Acoustics: 17th International Symposium on Nonlinear Acoustics*,
edited by A. A. Atchley, V. W. Sparrow, and R. M. Keolian
© 2006 American Institute of Physics 0-7354-0330-9/06/$23.00

MODEL EQUATIONS

The model equations used in this investigation include the continuity and the momentum equations, which will not be shown, as well as the entropy equation and a relaxation equation associated with each relaxation mechanism [1]:

$$\rho \frac{Ds_{fr}}{Dt} + \sum_v \frac{\rho}{T_v} c_{vv} \frac{DT_v}{Dt} - \nabla \cdot \left(\frac{\kappa}{T} \nabla T \right) = \sigma, \qquad (1)$$

$$\frac{DT_v}{Dt} = \frac{1}{\tau_v} (T - T_v), \qquad (2)$$

$$\sigma = \frac{\mu_B}{T} (\nabla \cdot \vec{v})^2 + \frac{\mu}{2T} \sum_{ij} \phi_{ij}^2 + \frac{\kappa}{T^2} (\nabla T)^2 + \frac{\rho}{T} \sum_v A_v \frac{DT_v}{Dt}. \qquad (3)$$

Here, σ is a variable used to represent the source terms of the entropy equation, ρ is density, t the time, $\vec{v}$ the velocity vector, μ_B bulk viscosity, μ shear viscosity, φ_{ij} the rate of shear tensor, s_{fr} the frozen entropy, v a summation variable representing the molecules whose relaxation will be taken account of, κ thermal conductivity, and T_v, c_{vv}, τ_v, and A_v the apparent vibration temperature, specific heat at constant volume, relaxation time, and affinity associated with the v-type molecule, respectively.

This equation set is then recast into conservative form and applied to a two dimensional domain with two relaxation molecules, nitrogen and oxygen. To close the system of equations, an appropriate entropy equation and a van der Waals equation of state is added to the equation set [2]. The result is six constitutive equations, all readily solvable by a time-domain numerical method and two state equations to complete the system.

NUMERICAL METHOD

A thorough discussion of the WENO scheme is provided in Reference 3. The method's ability to propagate shocks stably arises from two qualities that make it unique to time-domain methods: the use of multiple stencils and smoothness indicators specific to each stencil. The WENO method uses three three-point stencils in order to determine derivatives. For each of the three stencils there is a smoothness indicator which is simply an algorithm for determining the amount of discontinuity that is contained within the stencil. An extreme example is given in Fig. 1. Three three-point stencils are used to determine the derivative at point $i+1/2$, but there is a shock at point $i+1/2$. Stencils 1 & 2 contain this shock, so for this case, the smoothness indicators for stencil 1 & 2 will cause the WENO scheme to weigh the results from stencil 3 much higher than the others since it is the only one that does not contain the shock. This method of finding shocks and more heavily weighting stencils that do not contain them is executed for the three stencils over all space. The use of multiple stencils is the essence of how WENO allows the stable propagation of shock structures.

545

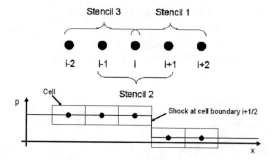

FIGURE 1. The three stencils used in the calculation. At point i+1/2 a shock has formed which is contained in stencils 1 & 2. In this case, the WENO scheme weights stencil 3 heavily in the calculation.

RESULTS

To demonstrate the WENO scheme and its ability to propagate multidimensional shocks, the phenomena of mach stem formation is simulated. In this situation, a high amplitude pulse is incident upon a rigid boundary. As the wave travels along the rigid boundary, the reflected and incident wave merge into a single elongated wave front. The amplification factor, given as the ratio of the peak pressure at the wall to the pressure of the incident wave front, then begins to increase as the propagation continues.

The calculation's initial conditions include a Gaussian spatially-distributed pulse with initial amplitude of 6.09×10^4 Pa, Gaussian half width of 1.37×10^{-2} m, spatial steps of 6.86×10^{-4} m and temporal steps of 6.86×10^{-7} s. The code is then allowed to propagate the initial condition and the results are given in Fig. 2.

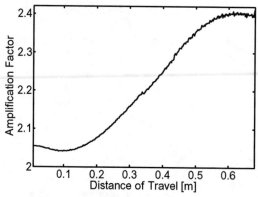

FIGURE 2 Amplification factor, given as the ratio of the peak pressure at the wall to the pressure of the incident wave front, as a function of the distance the wave traveled along the rigid boundary.

It can be seen that for near normal incidence, where the distance of travel is equal to zero in Fig. 2, the amplification factor is at a local maximum. It has been determined that the amplification factor matches that predicted by the Pfriem solution

[4] within 0.0264%. It is also apparent from Fig. 3 that as the wave travels along the rigid boundary, the incident and reflected waves begin to combine into an elongated and unified wave front, known as a Mach stem.

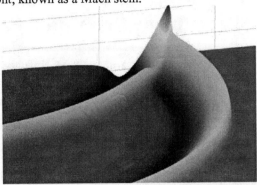

FIGURE 3 Mach stem formation waveform after propagating 0.38m, corresponding to 1500 computational iterations. On the left is the reflected wave and on the right is the incident wave. Along the wall the wave fronts have merged and elongated into a Mach stem.

CONCLUSIONS

In this investigation the WENO scheme has been introduced and shown to stably propagate multidimensional shocks. The results obtained match those given by analytical solutions in the region where they are available, and are qualitatively consistent with high-amplitude propagation effects in the regions where analytic results do not exist.

ACKNOWLEDGMENTS

This work is supported by the Office of Naval Research and the Exploratory and Fundamental Program of the Applied Research Laboratory at The Pennsylvania State University. The authors would also thank Dr. Chi-Wang Shu for his assistance.

REFERENCES

1. Pierce, A. D., *Acoustics: An Introduction to Its Physical Principles and Applications*, Acoustical Society of America, New York, 1989.
2. Sears, F. W. and Salinger, G. L., Thermodynamics, kinetic theory, and statistical thermodynamics, Addison-Wesley, Reading, 1975.
3. Shu, C.-W., "Essentially non-oscillatory schemes for hyperbolic conservation laws," in Advanced Numerical Approximation of Nonlinear Hyperbolic Equations, B. Cockburn, C. Johnson, C.-W. Shu, and E. Tadmor (Edited by A. Quarteroni), Lecture Notes in Mathematics (Springer-Verlag, Berlin/New York, 1998), Vol. 1697, pp. 325-432.
4. Blackstock, D. T., "Nonlinear acoustics (theoretical)," in *American Institute of Physics Handbook 3rd ed.* by D. E. Gray, McGraw Hill, New York, 3-183--3-205, 1972.

547

A Nonlinear Evolution of Wide Spectrum Acoustical Disturbances in Nonequilibrium Media with one Relaxation Process

N.E. Molevich[**] and V.G. Makaryan[*]

[*]Korolev's Samara state aerospace university, Moskovskoe sh. 34, 443086 Samara, Russia;
[**]Samara Branch, P.N. Lebedev Physics Institute, Novo-Sadovaya 221, 443011 Samara, Russia;

Abstract. It is investigated the solutions of a general acoustical equation, describing in the second order perturbation theory a nonlinear evolution of wide spectrum acoustical disturbances in nonequilibrium media with one relaxation process. Its low- and high- frequency limits correspond to Kuramoto-Sivashinsky equation and the Burgers equation with a source, respectively. Stationary structures of general equation, the conditions of their establishment and all their parameters are found analytically and numerically. It is obtained the condition of instability of a disturbance that has a step-like initial form. In acoustically active media it is predicted the existence of the stationary periodical roll waves and the solitary pulse with the shock front and the exponential tail. These periodical waves and solitary pulse are autowaves. Their parameters depend only on nonequilibrium medium properties.

Keywords: relaxation, nonequilibrium, solitary pulse, roll waves
PACS: 43.25.Cb, 43.25.Rq, 43.35.Ae

INTRODUCTION

Acoustics of nonequilibrium media is different significantly as compared with acoustics of equilibrium media [1,2]. It is known that in nonequilibrium media an inversion of the second viscosity and the dispersion is possible. Nonisothermal plasmas, chemical active mixtures, vibrationally excited gases are examples of such media. The media with a negative viscosity are acoustically active. In present work the nonlinear acoustical structures in such media are investigated.

GENERAL ACOUSTICAL EQUATION AND ITS LOW- AND HIGH-FREQUENCY LIMITS

In [3], it was obtained general acoustical equation

$$C_{V\infty}\tau_0(\rho_{tt} - u_\infty^2\rho_{xx} - \frac{u_\infty^2\Psi_\infty}{\rho_0}\rho_{xx}^2 - \frac{\mu_\infty}{\rho_0}\rho_{xxt})_t +$$
$$+ C_{V0}(\rho_{tt} - u_0^2\rho_{xx} - \frac{u_0^2\Psi_0}{\rho_0}\rho_{xx}^2 - \frac{\mu_0}{\rho_0}\rho_{xxt}) = 0, \tag{1}$$

CP838, *Innovations in Nonlinear Acoustics: 17th International Symposium on Nonlinear Acoustics,*
edited by A. A. Atchley, V. W. Sparrow, and R. M. Keolian
© 2006 American Institute of Physics 0-7354-0330-9/06/$23.00

describing the small amplitude disturbances in media with one relaxation process

$$\frac{dE}{dt} = \frac{E_e - E}{\tau} + Q, \tag{2}$$

where for vibrationally excited gases E is the vibrational energy, E_e is its equilibrium meaning, τ is the time of relaxation; Q is the energy source maintaining vibrational nonequilibrium; $u_\infty = \sqrt{\gamma_\infty T_0 / M}$, $u_0 = \sqrt{\gamma_0 T_0 / M}$ are the speeds of the high frequency and the low frequency sounds; $\gamma_\infty = C_{P\infty}/C_{V\infty}$, $\gamma_0 = C_{P0}/C_{V0}$; $C_{V0} = C_{V\infty} + C_K + S\tau_T$, $C_{P0} = C_{P\infty} + C_K + S(\tau_T + 1)$ are the low frequency specific heats in vibrationally excited gases at constant volume or pressure; T_0, ρ_0, τ_0 are the stationary values; M is the molecular mass; $S = Q\tau_0/T_0$ is the nonequilibrium degree; $\tau_0 = \tau(T_0, \rho_0)$; $\qquad C_K = (dE_e/dT)_{T=T_0}$; $\qquad\qquad \tau_T = \partial \ln \tau_0 / \partial \ln T_0$; $\mu_\infty = 4\eta/3 + \chi m (1/C_{V\infty} - 1/C_{P\infty})$, $\mu_0 = 4\eta/3 + \chi m (1/C_{V0} - 1/C_{P0})$ are the high frequency and the low frequency shear viscosity – heat-capacity coefficients; $\Psi_\infty = (\gamma_\infty + 1)/2$ is the high frequency nonlinear coefficient; Ψ_0 is the low frequency nonlinear coefficient. It is important that the coefficient Ψ_0 depends on the nonequilibrium degree S and can be even negative. Eq. (1) is valid for the weak dispersion $\widetilde{m} = (u_0^2 - u_\infty^2)/u_\infty^2 \sim \theta \ll 1$.

For waves travelling in one direction $(\widetilde{\rho} = \rho/\rho_0, \varsigma = (x - u_\infty t)/u_\infty \tau_0, y = \theta t/\tau_0)$, Eq. (1) reduces to

$$\left(\widetilde{\rho}_y + \frac{\Psi_\infty}{2}\widetilde{\rho}_\varsigma^2 - \widetilde{\mu}_\infty \widetilde{\rho}_{\varsigma\varsigma}\right)_\varsigma - \nu\left(\widetilde{\rho}_y + \frac{m}{2}\widetilde{\rho}_\varsigma + \frac{\widetilde{\Psi}_0}{2}\widetilde{\rho}_\varsigma^2 - \widetilde{\mu}_0 \widetilde{\rho}_{\varsigma\varsigma}\right) = 0, \tag{3}$$

where $\widetilde{\mu} = \mu/2\tau u_s^2 \rho_0$, $\widetilde{\Psi}_0 = \gamma_0 \Psi_0 / \gamma_\infty$, $\nu = C_{V0}/C_{V\infty}$.

In the low frequency approximation $(\partial \widetilde{\rho}/\partial y \sim \theta \widetilde{\rho})$, Eq. (3) reduces with an accuracy to $\sim \theta^3$ to the modified Kuramoto-Siwaszynski equation

$$\widetilde{\rho}_y + \widetilde{\Psi}_0 \widetilde{\rho} \widetilde{\rho}_\varsigma = \mu_\Sigma \widetilde{\rho}_{\varsigma\varsigma} + \widetilde{\beta} \widetilde{\rho}_{\varsigma\varsigma\varsigma} + \widetilde{\kappa} \widetilde{\rho}_{\varsigma\varsigma\varsigma\varsigma}. \tag{4}$$

In Eq. (4) $\mu_\Sigma = \widetilde{\mu}_0 + \widetilde{\xi}$ is total viscosity, $\widetilde{\xi} = \xi_0 / 2\rho_0 \tau_0 u_0^2$, where

$$\xi_0 = \frac{\rho_0 \tau_0 C_{V\infty}(u_\infty^2 - u_0^2)}{C_{V0}} = \frac{P_0 \tau_0 [(\tau_T - C_{V\infty})S + C_K]}{C_{V0}^2} \tag{5}$$

is the second viscosity coefficient, $\widetilde{\kappa} = C_{V0}\widetilde{\beta}/C_{V\infty} = C_{V0}^2 \widetilde{\xi}/C_{V\infty}^2$ (with neglect of $\sim \widetilde{\mu}_0^2, \widetilde{\mu}_0 \widetilde{\xi}$). For $C_{V0} > 0$, all these coefficients are negative if the second viscosity coefficient is negative, that is, for $(\tau_T - C_{V\infty})S + C_K < 0$.

In the high frequency approximation ($\partial \tilde{\rho}/\partial y \sim \theta^{-1}\tilde{\rho}$), Eq. (3) reduces (with an accuracy to $\sim\theta^2$) to the Burgers equation with a source and integral dispersion

$$\tilde{\rho}_y + \Psi_\infty \tilde{\rho}\tilde{\rho}_\varsigma = \tilde{\mu}_\infty \tilde{\rho}_{\varsigma\varsigma} - \tilde{\alpha}_\infty \tilde{\rho} - \tilde{\beta} \int \tilde{\rho}\,d\varsigma, \qquad (6)$$

where $\tilde{\alpha}_\infty = \xi_0 C_{V0}^2 / C_{V\infty}^2 \rho_0 \tau_0 u_\infty^2$ is the dimensionless gain (at $\xi_0 < 0$) of the high frequency sound, $\tilde{\beta} \approx C_{V0}\alpha_\infty / C_{V\infty}$ is the dispersion coefficient. The solutions of Eq. (4) and (6) are well known. The shortcoming of both equations is their disability to describe a nonstationary evolution of disturbances with a wide spectrum. Moreover, a spectrum of their stationary structures is wider than their application region.

The evolution of a disturbance with an arbitrary spectrum must be investigated on the basis of the complete equation (3), as in this study.

STATIONARY STRUCTURES

For $\Psi_0 > 0, C_{V0} > 0$ and the negative total viscosity, Eq. (3) describes three stationary structures that are shown in Fig. 1 [4].

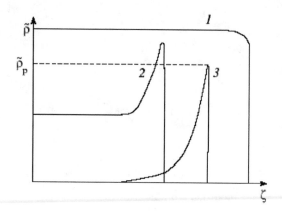

Figure 1. Stationary structures of Eq. (3).

The most interesting structure is strongly asymmetric solitary pulse (curve3, Fig. 1) with the shock front width $\sim \tilde{\mu}_\infty / \mu_\Sigma \ll 1$ and exponential trail $\tilde{\rho}_p \exp \nu \tilde{\Psi}_0\varsigma / 2\Psi_\infty$, where $\tilde{\rho}_p = -4\nu\mu_\Sigma / (2\Psi_\infty - \tilde{\Psi}_0)$ [5].

NUMERICAL SIMULATION OF EQUATION (3)

The initial step-like disturbance with amplitude $\tilde{\rho} > \tilde{\rho}_{cr} = -2\nu\mu_\Sigma / (\Psi_\infty - \tilde{\Psi}_0)$ transformed to the first stationary structure (curve 1, Fig. 1). The second structure (curve2, Fig. 1) was obtained for $\widetilde{\rho}_{cr} > \widetilde{\rho} > \widetilde{\rho}_{cr1} = -2\nu\mu_\Sigma /(2\Psi_\infty - \widetilde{\Psi}_0)$. The steps with amplitudes $\widetilde{\rho} < \widetilde{\rho}_{cr1}$ were unstable and broke down into a periodic sequence of stationary pulses (Fig.2). Each pulse had previous form and amplitude $\widetilde{\rho}_p$ (curve3, Fig. 1). Thus, such pulse is autowave (self-wave), whose form and amplitude depend on parameters of the nonequilibrium medium only.

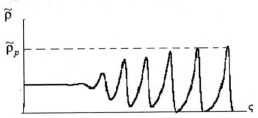

Figure 2. Nonstationary disintegration of a small-amplitude step into pulse autowaves

The bell-like disturbance transformed into the same pulses and roll waves in trail. These roll waves were autowave too. Their amplitude and period did not depend on amplitude and square of the initial bell-like disturbance.

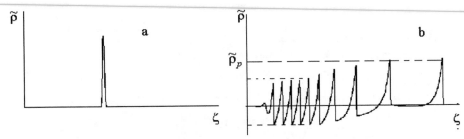

Figure 3. Nonstationary disintegration of a bell (a) into pulse and roll autowaves (b)

The obtained stationary structures have a wide spectrum and can't be described by the known equations of the low- or high-frequency approximations.

ACKNOWLEDGMENTS

This work was supported in part by the Russian-American Program "Basic Research and Higher Education" and the Administration of Samara region.

REFERENCES

1. Molevich, N.E., *AIAA Paper* 1020-2004
2. Molevich, N.E., Klimov, A.I., and Makaryan, V.G., *Int. J. Aeroacoustics* **4**, 345-355 (2005)
3 Molevich N.E., *Sib. Fisiko-Tehnich. Zh.* (In Russian), №1,.133-136 (1991).
4 Makaryan, V.G., and Molevich, N.E., *Technical Phys.* **50**, 685-691 (2005).
5 Makaryan, V.G., and Molevich, N.E., *Fluid Dynamics.* **39**,. 836-845 (2004).

Nonlinear constitutive equations derived for fluids obeying an ideal gas, a Tait-Kirkwood or a B/A type equation of state

Jacob Huijssen* and Martin D. Verweij*

*Laboratory of Electromagnetic Research, Delft University of Technology, Mekelweg 4,
2628 CD Delft, The Netherlands

Abstract. A generalized theoretical framework for acoustic, electromagnetic and elastodynamic waves would give fruitful insights into equivalent phenomena in these physical domains and would form a basis to draw up general analytical or numerical solution methods. In this contribution we adopt the structure of Maxwell's equations for electromagnetic fields, encompassing the formulation of two first-order field equations and two first-order constitutive equations, and we apply it to the area of nonlinear acoustics. We derive the constitutive equations of a fluid directly from its thermodynamic equation of state (EOS). In the constitutive equations, the nonlinear medium behaviour of the fluid is described by a pressure-dependent density and compressibility. The resulting equations are general, making them valid for phenomena occuring in applications with finite amplitude waves of any magnitude, like waveform distortion or radiation pressure. This paper concerns with obtaining constitutive equations for fluids obeying an ideal gas law, a Tait-Kirkwood EOS or a 2-term Taylor approximation of the EOS employing the B/A nonlinearity parameter. The latter EOS is used in many of the classical model equations of nonlinear acoustics. We show that all three types result in simple expressions for the density and compressibility.

Keywords: Nonlinear acoustics, constitutive equations, equation of state, Tait-Kirkwood, B/A parameter
PACS: 43.25.Ba, 43.25.Cb

INTRODUCTION

The basic equations of continuum mechanics describe the mathematical-physical fundament of the acoustic wave phenomena that can be observed in fluids. Using several simplifications and approximations, we can obtain from them the nonlinear wave equations like the Burgers equation, the Westervelt equation, the KZK equation and the Kuznetsov equation [1, 2]. These have the form of second order partial differential equations (PDEs).

In a similar way as the basic acoustic equations for fluids, the Maxwell equations describe electromagnetic waves, and equivalently the basic equations of elastodynamics describe elastic waves in solids. For all three physical domains, we can draw up equations with an identical mathematical structure, encompassing two conservation laws and two constitutive equations, all having the form of first order PDEs [3]. The conservation laws describe the interaction between the wavefield quantities, and the constitutive equations lay down the behaviour of the medium under consideration.

One advantage of having such a general mathematical form for all three domains of physics is that it facilitates the analysis of comparable wave propagation problems

CP838, *Innovations in Nonlinear Acoustics: 17th International Symposium on Nonlinear Acoustics*,
edited by A. A. Atchley, V. W. Sparrow, and R. M. Keolian
© 2006 American Institute of Physics 0-7354-0330-9/06/$23.00

and the design of general numerical solution methods that are valid for problems in all respective domains. Moreover, there are several advantages of using a set of first order PDEs and associated constitutive equations as a starting point for (numerical) analysis instead of a single second order PDE like the nonlinear wave equations described above. Firstly, the constitutive equations provide a versatile low level mathematical description of the medium. We can directly include various kinds of medium behaviour, like relaxation, nonlinearity, inhomogeneity, or anisotropy in the constitutive equations. Secondly, many techniques that exist for numerically evaluating first order PDEs, like absorbing boundary conditions, perfectly matched layers and staggered grids can be applied. At present there is a vast number of existing codes based on first order PDEs for acoustics [4, 5], electromagnetics and elastodynamics.

The classical formulations for linear and nonlinear acoustics give the conservation laws of mass, momentum and energy, together with an equation of state (EOS) for the medium under consideration [1, 6]. The medium properties show up in all four equations. Instead, in our formulation we use the conservation laws of momentum and volume. In this paper, we show how the gap between the two different formulations can be bridged by deriving the two nonlinear constitutive equations from the EOS with the help of the conservation laws of energy and mass. We present expressions for three different equations of state.

This paper is limited to nondissipative media. In another paper we show how dissipation due to medium viscosity and thermal conduction can be included in the constitutive equations [7].

CONSERVATION LAWS OF MOMENTUM AND VOLUME AND GENERAL FORM OF THE CONSTITUTIVE EQUATIONS FOR FLUIDS

For a non-viscous fluid the conservation laws of momentum and volume are [3, p. 24-25]

$$\partial_k p + \dot{\Phi}_k = f_k, \tag{1}$$
$$\partial_r v_r - \dot{\theta} = q, \tag{2}$$

where in the first equation p is the mechanical pressure, $\dot{\Phi}_k$ is the time rate of change of the mass flow density and f_k is the volume density of any external forces. In the second equation v_r is the velocity, $\dot{\theta}$ is the induced cubic dilatation rate and q is the injected cubic dilatation rate. The quantities f_k and q respectively represent external force and volume sources.

For a linear, lossless, homogeneous, time invariant, isotropic, instantaneously reacting and locally reacting fluid we write down the constitutive equations in a general form as [3, p. 26]

$$\dot{\Phi}_k = \rho d_t v_k, \tag{3}$$
$$\dot{\theta} = -\kappa d_t p, \tag{4}$$

where ρ is the mass density and κ is the compressibility of the medium, and where d_t denotes the material derivative $\partial_t + v_r \partial_r$.

Using Eqs. (3) and (4) in Eqs. (1) and (2), and taking the first order linearized forms around a rest state with $p = p_0$, $v = v_0 = 0$, $\rho = \rho_0$, and $\kappa = \kappa_0$ gives

$$\partial_k p' + \rho_0 \partial_t v_k = f_k, \qquad (5)$$
$$\partial_r v_r + \kappa_0 \partial_t p' = q, \qquad (6)$$

where $p' = p - p_0$ is the acoustic pressure. By subtracting the time derivative of Eq. (6), multiplied by ρ_0, from the spatial derivative of Eq. (5), a second order wave equation in p' is obtained as

$$\partial_k^2 p' - \rho_0 \kappa_0 \partial_t^2 p' = S, \qquad (7)$$

where $S = \partial_k f_k - \rho_0 \partial_t q$ is the source term.

CONSTITUTIVE EQUATIONS FOR AN IDEAL GAS LAW, A TAIT-KIRKWOOD EOS AND A TWO-TERM TAYLOR APPROXIMATION OF THE EOS

In order to come to a set of nonlinear model equations that is consistent with the classical formulation, we derive the constitutive equations from the thermodynamic equation of state for a specific medium. To do this, we use Eqs. (3) and (4) and obtain expressions for the nonlinear medium parameters ρ and κ. The general procedure is as follows[1]. Firstly, we formulate the EOS in terms of $\rho = \rho(p,s)$ with the help of the conservation law of energy. When we assume isentropic wave propagation – which for non-dissipative fluids is equal to adiabatic wave propagation – we then directly have an expression for $\rho = \rho(p)$. Secondly, to obtain the expression for κ, we take the material derivative d_t of $\rho(p)$ to obtain

$$d_t \rho = \frac{\partial \rho}{\partial p} d_t p. \qquad (8)$$

Next, we use the conservation law of mass, which can be written as [3, 6]

$$\partial_t \rho + \partial_k \Phi_k = \partial_t \rho + \partial_k (\rho v_k) = d_t \rho + \rho \partial_k v_k = 0. \qquad (9)$$

When we use the source free version of Eq. (2) in Eq. (9) and insert the result in Eq. (4), we can directly see the expression for $\kappa = \kappa(p)$ by comparison with Eq. (8) as

$$\kappa = \frac{1}{\rho} \frac{\partial \rho}{\partial p}. \qquad (10)$$

The described procedure has been applied to three equations of state. The first EOS is the adiabatical gas law, which can be derived from the ideal gas law using the

---

[1] In this procedure we assume identity between the mechanical pressure and the thermodynamic pressure. This is true for non-viscous fluids, and in practice for most other cases the distinction is negligible. See for a further discussion on this e.g. [8].

TABLE 1. Expressions for ρ and κ for the three equations of state. The expressions for the Taylor expansion are first order approximates.

	ρ	κ
Adiabatical gas law	$\rho_0(p/p_0)^{1/\gamma}$	$(\gamma p)^{-1}$
Tait-Kirkwood EOS	$\rho_0[(p+p_i)/(p_0+p_i)]^{1/n}$	$[n(p+p_i)]^{-1}$
Two-term Taylor approximation	$\rho_0[1+\kappa_0(p-p_0)]$	$\kappa_0[1-\kappa_0(1+B/A)(p-p_0)]$

conservation law of energy as

$$\frac{p}{p_0} = \left(\frac{\rho}{\rho_0}\right)^\gamma, \tag{11}$$

where p_0 and ρ_0 are the rest state quantities and γ is the ratio of specific heats. The second EOS is the Tait-Kirkwood EOS, which is an empirical equation used for describing isentropic flow in compressible liquids [1, 9], given by

$$\frac{p+p_i}{p_0+p_i} = \left(\frac{\rho}{\rho_0}\right)^n, \tag{12}$$

where p_i and n are constants along a particular isentrope. The third, very often used form is a two-term Taylor approximation of the EOS, written here in isentropic form expanded around a rest state $p_0 = p(\rho_0, s_0)$ as

$$p(\rho) = p_0 + c_0^2(\rho - \rho_0) + \frac{c_0^2}{2\rho_0}\frac{B}{A}(\rho - \rho_0)^2, \tag{13}$$

where $c_0 = 1/\sqrt{\rho_0\kappa_0}$ is the small-signal sound speed and B/A is a measure of nonlinearity. In Table 1 the results are listed for all three equations of state. The expressions given for the B/A type EOS are first order Taylor approximations of the resulting expressions. These expressions are identical to the ones obtained in [7].

REFERENCES

1. M. F. Hamilton and D. T. Blackstock, eds., *Nonlinear acoustics*, Academic Press, San Diego, 1998.
2. B. O. Enflo and C. M. Hedberg, *Theory of nonlinear acoustics in fluids*, Kluwer Academic Publishers, Dordrecht, 2002.
3. A. T. De Hoop, *Handbook of radiation and scattering of waves*, Academic Press, San Diego, 1995.
4. M. Tabei, T. D. Mast, and R. C. Waag, J. Acoust. Soc. Am. **111**(1), 53–63 (2002).
5. S. Ginter, M. Liebler, E. Steiger, T. Dreyer and R. E. Riedlinger, J. Acoust. Soc. Am. **111**(5), 2049–2059 (2002).
6. A. D. Pierce, *Acoustics*, McGraw-Hill, New York, 1981.
7. M. D. Verweij and J. Huijssen, "Nonlinear and dissipative constitutive equations that are consistent with the generalized Westervelt equation", presented at this symposium.
8. F. M. White, *Viscous fluid flow*, 2nd ed., Mc Graw-Hill, New York, 1991.
9. J. O. Hirschfelder, C. F. Curtiss, and R. B. Bird, *Molecular theory of gases and liquids*, fourth press, John Wiley & Sons, New York, 1967.

Molecular Relaxation Simulations in Nonlinear Acoustics using Direct Simulation Monte Carlo

Amanda Danforth-Hanford[*], Patrick D. O'Connor[†], Lyle N. Long[¶], and James B. Anderson[†]

Graduate Program in Acoustics, The Pennsylvania State University, University Park, PA 16802
†Department of Chemistry, The Pennsylvania State University, University Park, PA 16802
¶Department of Aerospace Engineering, The Pennsylvania State University, University Park, PA 16802

Abstract. The direct simulation Monte Carlo (DSMC) method describes gas dynamics through direct physical modeling of particle motions and collisions. DSMC is based on the kinetic theory of gas dynamics, where representative particles are followed as they move and collide with other particles. DSMC provides a useful tool for capturing all physical properties of interest for nonlinear acoustic problems, such as dispersion, attenuation, absorption, harmonic generation and nonequilibrium effects. The validity of DSMC for the entire range of Knudsen numbers (Kn), where Kn is defined as the mean free path divided by the wavelength, allows for the exploration of sound propagation at low Kn (low frequency, atmospheric conditions) as well as sound propagation at high Kn (high frequency, dilute gases, or in microdevices). For low Kn, nonlinear effects play an important role in waveform evolution. For high Kn, nonequilibrium effects are strong and increased absorption cancels out nonlinearity effects.

Keywords: DSMC, Monte Carlo, Absorption, Dispersion, Relaxation.
PACS: 43.25.Cb

INTRODUCTION

It is widely known that viscous, thermal and molecular relaxation losses play an important role in sound propagation in the atmosphere. Traditionally, acoustics is concerned with the treatment of the fluid as a continuum using macroscopic quantities such as density, velocity and pressure as dependent variables. However, the continuum model has its limitations and the model breaks down for Knudsen numbers (Kn) greater than 0.1, where Kn is defined as the mean free path divided by the wavelength. On the other hand, particle methods are necessary for, but not limited to, problems with Kn > 0.1. Since absorption mechanisms are inherently molecular properties, it is a natural progression to use a particle method for analysis.

Direct simulation Monte Carlo (DSMC) [1] is the particle method used in this study. DSMC is a stochastic method used to computationally model gas flows with thousands or millions of representative molecules through the direct physical modeling of particle motions and intermolecular collisions. Using DSMC to study acoustic waves allows us to explore real gas effects for all values of Kn with a molecular model that traditional continuum methods cannot offer.

The Knudsen number is large for sound propagation in very dilute gases or at high frequencies requiring a particle-method solution. DSMC results for acoustic wave

CP838, *Innovations in Nonlinear Acoustics: 17th International Symposium on Nonlinear Acoustics*,
edited by A. A. Atchley, V. W. Sparrow, and R. M. Keolian
© 2006 American Institute of Physics 0-7354-0330-9/06/$23.00

propagation in monatomic gases have shown that sound absorption depends heavily on Kn [2]. Successful application of the DSMC method to nonlinear acoustic waves has already been completed in monatomic and diatomic gases [2,3]. For this paper DSMC was used to study the interaction between acoustic absorption and nonlinearity through bimolecular collisions in mixtures of air containing water vapor for varying Kn.

DIRECT SIMULATION MONTE CARLO

Direct simulation Monte Carlo is a very versatile simulation tool, and has been successfully used in various systems including but not limited to: hypersonic flows, rarefied gas dynamics [1], chemical reactions [4], and acoustics [2,3]. The current model contains a phenomenological approach to simulate diatomic and polyatomic molecules in systems containing several species. This approach, developed by Larsen and Borgnakke [5], handles the internal energy exchange between the molecules during collisions. The theory assumes that a fraction of the collisions are regarded as inelastic, redistributing post-collision energy in a stochastic manner to produce physically realistic behavior at the macroscopic level. In this study, the molecules themselves will be considered in a classical sense as hard sphere molecules adhering to the Larsen-Borgnakke phenomenological approach.

SIMULATION APPROACH

Under standard atmospheric conditions, roughly 99% of dry air is composed of nitrogen, oxygen and trace amounts of carbon dioxide, argon, neon, methane, etc. The introduction of water vapor to a dry air mixture has been shown to be an important factor in absorption due to molecular relaxation [6,7]. The simulations generated in this research therefore focus entirely on nitrogen, oxygen and water vapor mixtures. The acoustic frequency of maximum absorption is called the relaxation frequency, and is a strong function of temperature.

Acoustic waves were generated in the simulation domain by creating a sinusoidal density distribution of specified frequency. In all cases, the acoustic density amplitude of the sinusoidal source is 20% of the ambient density, with an initial acoustic flow velocity equal to zero. Results for a single frequency at varying humidities were computed in a gas mixture, as well as a single humidity for varying Kn to investigate the results different absorption effects.

A parallel, object-oriented DSMC solver was developed for this problem. The code was written in C++ and was run on massively parallel computers. Despite excellent parallel efficiency, CPU time and memory requirements were quite large, taking approximately 60 hours on 32 processors for each run.

ABSORPTION AT LOW KN WITH VARYING HUMIDITY

Due to computational limitations and the strong dependence of the relaxation frequency to temperature, the DSMC simulations in a mixture of oxygen, nitrogen and water vapor were run at standard atmospheric pressure (ambient pressure = 101000

Pa) and a temperature of 373 K. The relaxation frequency at this temperature is approx 2 MHz, and is the frequency used in this study. The absorption coefficient for varying humidities is plotted with theory [6] in figure 1.

Due to the high amplitudes of the simulation at low Kn, we expect nonlinear effects to become important in the evolution of the sound wave. The shock formation distance at this frequency is beyond the simulation domain. However, considerable wave steepening was observed and is shown in figure 2.

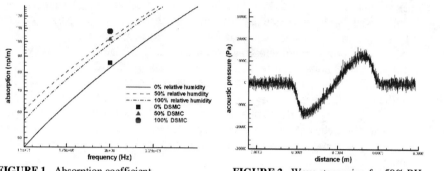

FIGURE 1. Absorption coefficient   **FIGURE 2.** Wave steepening for 50% RH wave

ABSORPTION AT HIGH KN

The scaled absorption coefficient, α/k_0 where α is the absorption coefficient and k_0 is the acoustic wave number, for Kn ranging from 0.04 to 35 is shown in figure 3. DSMC simulation results are compared to a classical Navier-Stokes prediction, experimental work by Greenspan, and previous results for monatomic gas [2]. The variation in Kn was obtained by maintaining the cells per wavelength constant at 100, but varying the cell size from 1/2 of a mean free path to 3/1000 of a mean free path.

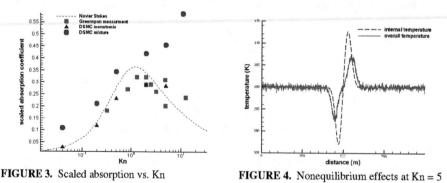

FIGURE 3. Scaled absorption vs. Kn   **FIGURE 4.** Nonequilibrium effects at Kn = 5

Figure 3 shows an increase in absorption due to molecular relaxation for gas mixtures compared to monatomic cases. When a gas is in thermal equilibrium the internal, translational, and overall temperatures are equal. However, nonequilibrium effects become more important as Kn increases. The strong nonequilibrium effects are shown in figure 4, where internal and overall temperature are plotted with distance at a snapshot in time, as the acoustic wave propagates at Kn = 5. The time delay and magnitude difference between the two curves is evident, indicating molecular relaxation. These strong nonequilibrium effects play an important role in the increased absorption at high Kn.

Despite the fact that the acoustic density amplitude is 20% of the ambient density in these high Kn cases, the absorption is so great that nonlinear effects are very small. There is no evidence of wave steepening or shock wave formation.

CONCLUSIONS

Given the statistical nature of DSMC, there is an intrinsic degree of scatter in the simulation results. This universal drawback of DSMC makes it difficult in certain cases to produce adequate resolution, and is a source of limitation for this work.

The inclusion of internal energy in diatomic and polyatomic species has allowed us to investigate absorption and nonlinearity in gas mixtures. Increasing the relaxation collision number of a species affects the amount of nonequilibrium in the relaxation process. In this study it was determined that at low Kn there are nonequilibrium effects in mixtures with high concentrations of water, but the overall effect is small compared to the classical absorption coefficient. For mixtures with high Kn the nonequilibrium effects are very prominent and result in a much larger absorption coefficient.

Future work will include more detailed investigations into the internal energy configurations of diatomic and polyatomic molecules, the temperature dependence of relaxation collision numbers, and modeling characteristic temperatures of molecules

ACKNOWLEDGMENTS

We would like to thank the National Science Foundation for funding the Consortium for Education in Many-Body Applications, grant no. NSF-DGE-9987589.

REFERENCES

1. Bird, G. A., *Molecular Gas Dynamics and the Direct Simulation of Gas Flows*, Oxford University Press, 1994.
2. Danforth, A. and Long, L, "Nonlinear acoustic simulations using direct simulation Monte Carlo" in *Journal of the Acoustical Society of America*, Vol. 116 (4) , 2004, pp. 1948-1955.
3. Danforth, A. and Long, L., "Nonlinear Acoustics in Diatomic Gases Using Direct Simulation Monte Carlo" in *24th International Symposium on Rarefied Gas Dynamics*, AIP Conference Proceedings 762, American Institute of Physics, Melville, NY, 2005, pp. 553-558.
4. O'Connor, P. Long L. and Anderson J., "Direct simulation of ultrafast detonations in mixtures" in *24th International Symposium on Rarefied Gas Dynamics*, AIP Conference Proceedings 792, American Institute of Physics, Melville, NY, 2005.
5. Borgnakke, C. and Larsen, P. S., "Statistical collision model for Monte Carlo simulation of polyatomic gas mixture" in *Journal of Computational Physics*, Vol. 18, pp. 405-420.
6. Bass, H. E., Sutherland, L. C., Piercy, J. and Evans, L., "Absorption of Sound by the Atmosphere" in *Physical Acoustics: Principals and Methods*, edited by W. P. Mason, Orlando: Academic Press, 1984, pp. 145-232.
7. Bass, H. E., Sutherland, L. C., Zuckerwar, A. J., Blackstock, D. T. and Hester, D. M., "Atmospheric absorption of sound: Further developments" in *Journal of the Acoustical Society of America*, Vol. 97, 1995, pp. 680-683.

Metrics That Characterize Nonlinearity in Jet Noise

Sally McInerny[+], Micah Downing[*], Chris Hobbs[*], Michael James[*], and Michael Hannon[+]

[*]Wyle Laboratories, Inc., 2001 Jefferson Davis Highway, Suite 701, Arlington, VA, 22202 USA
[+]Dept. of Aerospace Engineering and Mechanics, The University of Alabama, P.O. Box 870280, Tuscaloosa, AL, 35487-0280 USA

Abstract. Analyses of acoustic data recorded during a series of high thrust military jet flyovers indicate that the skewness of the pressure is not well correlated with the sound pressure level, but the skewness of the gradient is. The influence of propagation distance and engine exhaust velocity on the results requires further examination. The results presented here emphasize the need for high bandwidth, high signal to noise ratio measurements. [Work supported by the Strategic Environmental Research and Development Program.]
Keywords: jet noise, nonlinear metrics.
PACS: 43.25.Zx, 43.28.Hr

INTRODUCTION

Acoustic data were recorded during a series of high thrust military jet runs. These runs included level flight at 1000 ft at steady engine power conditions ranging from minimum and military power. A few runs were flown at 250 ft. For afterburner power, climbs were performed to reduce the forward acceleration; data were recorded with military power climbs for comparison.

Sound pressures were measured at multiple heights above the ground using both 1/2" and 1/4" microphones. A 24-bit PC based system recorded the data at 96k samples per second. The effects of microphone diameter and measurement height on data characteristics are not examined here. Tracking information (needed for source to receiver distance and directivity angle determination) is not, yet, available.

Each recording was analyzed to determine the time period during which the 0.5s averaged L_P was within 6 dB of the maximum, referred to as the 6 dB down time period. Data during the 6 dB down time period were then analyzed to determine: 1) the probability density distribution of the acoustic pressure and its skewness coefficient; 2) the probability density distribution of the gradient and its skewness coefficient; and 3) the variation of maximum gradient with pressure rise, for the pressure rise portions of the waveforms. 3rd octave band L_P distribution and the Howell-Morfey indicator will be assessed in later analysis. A discussion of the relevance of these analyses can be found in [1-4]. This paper presents a limited subset of the calculated results.

CP838, *Innovations in Nonlinear Acoustics: 17th International Symposium on Nonlinear Acoustics*,
edited by A. A. Atchley, V. W. Sparrow, and R. M. Keolian
© 2006 American Institute of Physics 0-7354-0330-9/06/$23.00

RESULTS

The L_{eq} over the 6 dB down time period ranged from 87 dB (minimum power run) to 113.5 dB (military flyover at 1000 ft). A straight line provided an excellent fit to the relationship between the L_{eq} and the maximum 0.5s averaged L_P during the 6 dB down time period, with a minor caveat for the minimum power runs where the 6 dB down time period was longer. Given the strong correlation between the maximum L_P and the L_{eq}, either one could be used to examine the variation of other metrics with sound pressure level.

Figures 1 and 2 are plots of the maximum 0.5s averaged L_P versus skewness of the pressure and versus skewness of the gradient of the pressure, respectively. (Note that the gradient was calculated using a forward difference, for the reasons cited in reference [3].) It can be seen that the skewness of the pressure is not well correlated with the maximum L_P. There is a stronger correlation between the maximum L_P and the skewness of the gradient.

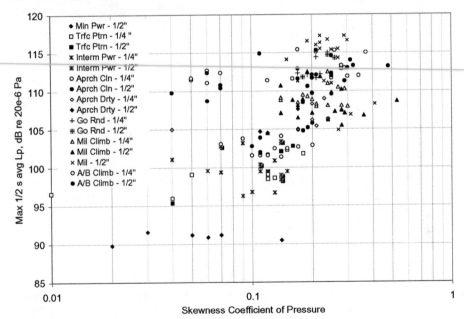

FIGURE 1. Skewness Coefficient of the Pressure versus the Maximum 0.5 s Averaged L_P over the 6 dB Down Period.

The maximum gradients for the jet noise data sets can be compared to those for the sonic boom measurements in Figure 3. The maximum instantaneous pressure, as opposed to maximum 0.5 s averaged L_P, is used in Fig. 3, because it's considered a more meaningful metric for a sonic boom waveform. The maximum gradient of sonic boom waveforms is smaller than that of the jet noise for the same maximum instantaneous pressure. That is, for the same peak pressure, the rise time of the sonic boom waveforms is larger than that of the measured jet noise data sets.

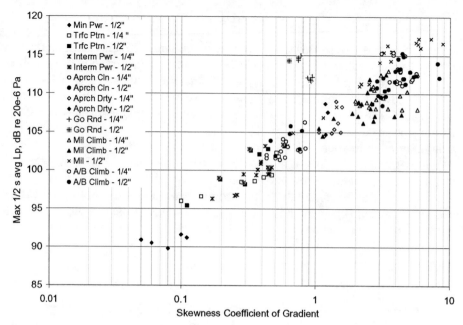

FIGURE 2. Skewness Coefficient of the Gradient of the Pressure versus the Maximum 0.5 s Averaged L_P over the 6 dB Down Period.

DISCUSSION

That the skewness and maximum value of the gradient are more closely related to the maximum L_P than the skewness of the pressure itself is unfortunate. Accurate calculation of the gradient from measured data requires a high bandwidth recording system, so that the rapid rise time of the waveforms is accurately captured. Furthermore, a high signal to noise ratio is essential for the gradient calculation – otherwise, the gradient is dominated by ambient and system noise. Given that crest factors calculated from the recorded data were as large as 21.4 dB, the requirement for a high signal to noise ratio necessitates a high bit data recording system. Data that were recorded during this series of flyovers using a 16-bit high bandwidth system with range settings set to accommodate the highest expected levels were virtually useless.

Figs. 1-3 present aggregate dependences of various metrics on the magnitude of the sound pressure without regard to exhaust velocity or propagation distance, nor with respect to microphone diameter or height above the ground. The degree to which weak shocks are present in 'source' waveforms, loosely defined as the waveforms measured at the limit of the geometric far-field of the exhaust, depends on the ratio of the exhaust velocity to the speed of sound in the atmosphere. As one moves from supersonic jets to afterburning jets to rockets, Mach wave radiation goes from being present to dominating the source radiation characteristics.[5,6] Furthermore, the exhaust velocity dictates the sound power per unit exhaust area. The larger this is the more non-linear waveform steepening can be expected to play a role. Lastly, absorption - determined by temperature and relative humidity and propagation distance - and atmospheric turbulence act to increase the rise time of weak shocks.[7,8] These shocks account for the high values of the gradient and its skewness.[9,3]

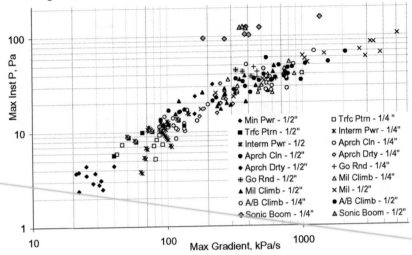

Figure 3. Maximum Gradient versus Maximum Instantaneous Pressure, Including Sonic Booms.

REFERENCES

1. C. L. Morfey, and G. P. Howell, "Nonlinear propagation of aircraft noise in the atmosphere," AIAA Journal, vol. **19**, 986-992 (1981).
2. Gee, K.L., Gabrielson, T.B., Atchley, A.A., and Sparrow, V.W., "Preliminary Analysis of Nonlinearity in Military Jet Aircraft Noise Propagation," <u>AIAA Journal</u>, vol. 43, no. 6, pp. 1398-1401.
3. McInerny, S.A, and Olcmen, S. M., "High Intensity Rocket Noise: Atmospheric Absorption and Characterization," <u>J. Acoust. Soc. Of America</u>, vol. 117, no. 2, pp. 578-591.
4. McInerny, S. A., "Launch Vehicle Acoustics, Part II - Statistics of The Time Domain Data," <u>J Aircraft</u>, vol. 33, no. 3, May-June 96, pp. 511-517.
5. **Sound and Sources of Sound**, A.P. Dowling and J.E. Ffowcs-Williams, Halstead Press: a division of John Wiley & Sons, NY, 1983.
6. McInerny, S. A., Lu, G., and Olcmen, S., "Rocket and Jet Mixing Noise, Background and Procedures," Final Report under contract #UM 0308-013 for the National Center for Physical Acoustics, Univ. of MS, Aug. 2004.
7. J. Kang, "Nonlinear acoustic propagation of shock waves through the atmosphere with molecular relaxation," Ph.D. Thesis, Department of Mechanical Engineering, Pennsylvania State University (1991)..
8. B. Lipkens, and D. T. Blackstock, "Model experiment to study sonic boom propagation through turbulence. Part I: Model experiment and general results," J. Acoust. Soc. Am., vol. **103**, 148-158 (1998).
9. H. E. Bass, R. Raspet, J. P. Chambers, and M. Kelly, "Modification of sonic boom wave forms during propagation from the source to the ground," J. Acoust. Soc. Am., vol. **111**, 481-486 (2002).

Asymptotic Behavior in the Numerical Propagation of Finite-Amplitude Jet Noise

Kent L. Gee and Victor W. Sparrow

Graduate Program in Acoustics, The Pennsylvania State University, 213 Applied Science Bldg.,
University Park, PA 16802, kentgee@psu.edu, vws1@psu.edu

Abstract. One issue of interest pertaining to the development of a numerical model applicable to the nonlinear propagation of jet noise is the behavior of spectral predictions at large distances. In this study, a recorded noise waveform from a military jet aircraft is numerically propagated via a hybrid time-frequency domain solution to the generalized Burgers equation that incorporates spherical spreading and atmospheric absorption and dispersion. Numerical results show that the spatial rate of change of the difference between the nonlinearly- and linearly-predicted power spectra appears to approach constant nonzero behavior at high frequencies. This asymptotic relationship is analogous to that predicted by analytical theory for initially-sinusoidal plane and spherical waves.

Keywords: jet noise, nonlinear, finite-amplitude, propagation, long-range
PACS: 43.25.Cb

BACKGROUND

As part of the development of a numerical model to predict the noise propagation from high-performance military jet aircraft [1], long-range propagation has been considered to determine the extent of nonlinearity and see if the nonlinear portion of the model can simply turned off at some distance. Previous studies that have treated the asymptotic behavior of nonlinear propagation analytically are now reviewed.

Blackstock [2] studied the long-range decay of planar, initially-sinusoidal waves in a thermoviscous medium with the Fay Fourier-series solution. He found that the nth harmonic of the fundamental decays as $e^{-n\alpha_0 x}$, where α_0 is the absorption coefficient of the fundamental and x is distance. This represents a slower decay than the expected small-signal solution, $e^{-n^2\alpha_0 x}$, for all of the harmonics except the fundamental ($n = 1$), signifying that "old-age" is not synonymous with "linear." Webster and Blackstock [3] extended the exact plane-wave solution by the use of a fifth-order accurate perturbation solution and showed that similar behavior holds for spherically-spreading, initially-sinusoidal waves.

The old-age behavior of initially-Gaussian, finite-amplitude noise propagation through thermoviscous media has been examined by Gurbatov and colleagues in a number of publications (see Refs. [4,5] and additional references therein). They have found that a weak-shock (f^{-2}) spectral slope at high frequencies eventually gives way to an exponential law roll-off at sufficiently large distances, due to the absorptive

CP838, *Innovations in Nonlinear Acoustics: 17th International Symposium on Nonlinear Acoustics*,
edited by A. A. Atchley, V. W. Sparrow, and R. M. Keolian
© 2006 American Institute of Physics 0-7354-0330-9/06/$23.00

properties of the medium. Extension of these thermoviscous-medium results to atmospheric propagation is only qualitative; however, similar behavior is observed in the numerical atmospheric propagation of finite-amplitude jet noise waveforms. Limited results are now shown and discussed.

RESULTS

A waveform recorded at a distance of 61 m from a tied-down F/A-22 Raptor with one engine at afterburner has been numerically propagated according to a generalized Burgers equation that accounts for nonlinearity, spherical spreading, and atmospheric absorption and dispersion. The numerical solution technique, a hybrid time-frequency domain algorithm similar to those of Pestorius [6] and Anderson [7], is documented in Ref. [1]. Nonlinearly- and linearly-predicted third-octave spectra between 61-3048 m (200-10,000 ft) are shown in Fig. 1. Energy transfer from the peak-frequency region (100-300 Hz) to higher frequencies is readily seen. Comparison of the difference between the linear and nonlinear predictions for a given third-octave band demonstrates that this difference apparently continues to grow, particularly at high frequencies, out to the extent of the propagation range.

This behavior is more readily examined in the context of the nonlinear gain, *NG*, which is the difference in decibels between the nonlinearly- and linearly-predicted sound pressure levels (SPL_n and SPL_l). *NG* is written as

$$NG(r, f) = SPL_n(r, f) - SPL_l(r, f) \qquad (1)$$

and its spatial partial derivative may be estimated over a range step Δr as

$$\partial NG(r, f)/\partial r \approx [NG(r + \Delta r, f) - NG(r, f)]/\Delta r. \qquad (2)$$

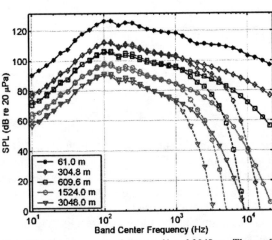

FIGURE 1. Predicted third-octave spectra between 61 and 3048 m. The nonlinear predictions are shown with solid (—) lines and the linear predictions with dashed (---) lines.

Examination of $\partial NG(r,f)/\partial r$ as a function of range permits study of the evolution of nonlinear interactions on the spectral level. Note that $\partial NG(r,f)/\partial r = 0$ at ranges and frequencies for which the propagation is linear.

Displayed in Figs. 2 and 3 are $\partial NG(r,f)/\partial r$ calculations between 61-3048 m for the third-octave bands between 63-125 Hz and 6.3-12.5 kHz, respectively. A comparison of these two figures demonstrates very different nonlinear behavior for these different frequency regions. Figure 2 reveals that, in the vicinity of the peak-frequency region of the spectrum, the rate of change between nonlinear and linear theory is very slow and eventually $\partial NG(r,f)/\partial r \to 0$. On the other hand, Fig. 3 demonstrates a very different behavior in that $\partial NG(r,f)/\partial r$ appears to approach non-zero constant behavior by several hundred meters, especially for the 10- and 12.5-kHz third-octave bands. The increase in $\partial NG(r,f)/\partial r$ for the 12.5-kHz band at about 2500 m is caused by the noise floor of the nonlinear algorithm, which, once reached, causes the high-frequency energy to decay more slowly than it otherwise would [1].

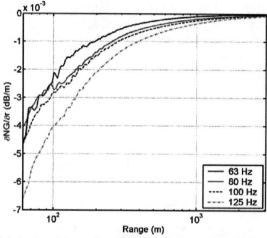

FIGURE 2. Spatial derivative of the nonlinear gain (*NG*) between 61 and 3048 m for four third-octave bands in the vicinity of the peak-frequency region of the spectrum

CONCLUSIONS

These results for the long-range numerical propagation of an F/A-22 afterburner noise waveform suggest asymptotic behavior that is analogous to that discussed by Blackstock [2] and Webster and Blackstock [3] for initially-sinusoidal waveforms. The propagation in the peak-frequency region of the spectrum eventually appears linear, but the high-frequency energy appears to always decay nonlinearly. Additional work is needed, however, to understand these how results may impact the perception of noise at large distances [1].

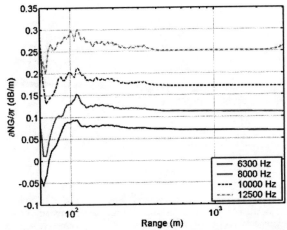

FIGURE 3. Spatial derivative of the nonlinear gain (*NG*) between 61 and 3048 m for the four third-octave bands two decades above those shown in Fig. 2. Note that the anomalous behavior at about 2500 m for the 12.5-kHz curve is attributed to the nonlinear algorithm noise floor.

ACKNOWLEDGMENTS

This work was supported by the Strategic Environmental Research and Development program, through a subcontract with Wyle Laboratories. The cooperation of the F/A-22 program office is also gratefully acknowledged. The authors are also grateful to D. T. Blackstock for helpful discussions in interpreting the numerical results.

REFERENCES

1. Gee, K. L. "Prediction of nonlinear jet noise propagation," Ph.D. Thesis, Graduate Program in Acoustics, The Pennsylvania State University, University Park, PA, 2005.
2. Blackstock, D. T., "Thermoviscous attenuation of plane, periodic, finite-amplitude sound waves," *J. Acoust. Soc. Am.* **36**(3), 534-542 (1964).
3. Webster, D. A., and Blackstock, D. T., "Experimental investigation of outdoor propagation of finite-amplitude noise," NASA Contractor Report 2992, Applied Research Laboratories., The University of Texas at Austin, 1978.
4. Gurbatov, S. N., Enflo, B. O., and Cherepennikov, V. V., "Old-age stage evolution of intense cylindrical and spherical acoustic noise," *Nonlinear Acoustics at the Beginning of the 21st Century* Vol. 1, edited by O. V. Rudenko and O. A. Sapozhnikov, Moscow, Russia: Faculty of Physics, Moscow State Univ., 2002, pp. 341-348.
5. Gurbatov, S. N. and Rudenko, O. V., "Statistical phenomena," chap. 13 in *Nonlinear Acoustics*, edited by M. F. Hamilton and D. T. Blackstock , San Diego, Academic, 1998, pp. 377-398.
6. Pestorius, F. M. "Propagation of plane acoustic noise of finite amplitude," Ph.D. Dissertation, The University of Texas at Austin, 1973.
7. Anderson, M. O., "The propagation of a spherical N wave in an absorbing medium and its diffraction by a circular aperture," Technical Report ARL-TR-74-25, Applied Research Laboratories, The University of Texas at Austin, AD 787878, 1974.

Nonlinear effects in propagation of broadband jet noise

Pablo L. Rendón

CCADET, Universidad Nacional Autónoma de México
Ciudad Universitaria, México D.F. 04510, México
Email: rendon@aleph.cinstrum.unam.mx

Abstract.
   It is normally assumed that linear acoustics mechanisms are sufficient to give a comprehensive description of the propagation of broadband jet noise. However, there are indications that the inverse square law of linear acoustics, coupled with small-signal attenuation along rays, does not adequately decsribe the evolution with distance of the high-frequency components of the jet noise spectrum. Experimental data, though limited, strongly suggests that jet noise signals are sufficiently intense for nonlinear effects to be taken into account in their propagation to the distant field. We propose a weakly nonlinear perturbation approach in the form of a Taylor series expansion for the power spectrum in order to characterize the effects of nonlinearity on the propagation of broadband jet noise, while also taking into account the effects of linear attenuation and directivity patterns.

Keywords: jet noise, broadband noise propagation, nonlinear acoustics
PACS: nonlinear acoustics

INTRODUCTION

The importance of statistical phenomena related to the nonlinear propagation of waves in many areas of physics and technology has fuelled the study of problems concerning the interaction of intense noise disturbances in non-dispersive media; see Rudenko [1]. Although it is still not clear to what extent nonlinear effects affect the propagation of jet noise, it has become increasingly clear that classical linear acoustics theory is unable to adequately portray this type of propagation. Experimental data provided by Howell and Morfey [2], though limited, strongly suggests that jet noise signals are sufficiently intense for nonlinear effects to be taken into accountin their propagation to the distant field. They found, in fact, that linear acoustic theory can only predict the propagation of noise fields accurately at low engine levels; at high power conditions, over distances ranging from 500 to 1000 metres, the high frequency attenuation (between 5 and 10 kHz) was observed to be 10 dB lower than expected [3]. More recently, Gee, Gabrielson, Atchley and Sparrow have used F/A-18E static engine run-up measurements to argue that noise propagation is nonlinear in that case for military thrust conditions [4].There are a variety of linear mechanisms which could be at least partly responsible for this underprediction, such as temperature and humidity variations in the atmosphere, or any effects associated with thrust-dependent directivity patterns. Our assumption, however, following Pernet and Payne [5], is that the anomalously high levels of the high-frequency components of the spectrum are to be accounted for by an energy transfer from the intense low-frequency components achieved through cumulative wave-form distortion.

CP838, *Innovations in Nonlinear Acoustics: 17th International Symposium on Nonlinear Acoustics,*
edited by A. A. Atchley, V. W. Sparrow, and R. M. Keolian
© 2006 American Institute of Physics 0-7354-0330-9/06/$23.00

Gurbatov [6] has singled out two processes which seem to characterize the propagation of finite-amplitude noise: waveform steepening and an increase in the time-scale of oscillations. The steepening leads to shock formation, and consequently to energy transfer to high frequencies. The second process has to do with relative velocities of different shock fronts, and in particular with their coalescence. In this way, energy is transferred to the lower end of the spectrum, and the spectrum broadens as a result of the combination of these two processes. Two limiting cases concerning the evolution of high-intensity broadband noise arise: at small distances, where the influence of the (at that moment) few shock fronts can largely be ignored, and at very large distances, where the shocks determine the statistical properties. Thus, for small distances, a Taylor series expansion for the power spectrum seems a reasonable way to implement a predictive scheme. This is the approach followed by Howell and Morfey [2], and the one we will follow as well in the next section, incorporating the effects of linear attenuation and directivity patterns.

A WEAKLY NONLINEAR PERTURBATION SCHEME

Since linear theory is almost universally used to describe the propagation of broadband noise signals, we assume nonlinear propagation effectscan be studied through a weakly nonlinear perturbation approach; nonlinear effects are dealt with as perturbations which in some sense are small.

We start with the model equation for the pressure $p(r, \tau)$ in an outgoing, spherically symmetric sound field with weak nonlinearity and dispersion and diffusion both negligible:

$$\frac{\partial P}{\partial r} + \frac{P}{r} = \beta P \frac{\partial P}{\partial \tau}, \tag{1}$$

where the variables are all dimensional; r is range, $\tau = t - (r - r_0)/a_0$ is the retarded time, a_0 is the small-signal sound speed, and $\beta = (1 + \gamma)/2\rho_0 a_0^3$ is a parameter of nonlinearity, with γ the adiabatic exponent and ρ_0 the mean density. The boundary condition, for an initial radius r_0, is given as $P(r_0, \tau) = q(\tau)$.

The dependent variables are related locally in the normal linear acoustics manner, so that $P = \rho_0 a_0 u$, where u is the radial velocity. We also will assume that, for all frequencies of interest and for all $r > r_0$, $\omega r/a_0 \gg 1$, so that by writing $p = (r/r_0)P$ and $x = r_0 \ln(r/r_0)$ we may reduce equation (1) to its plane wave form,

$$\frac{\partial p}{\partial x} = \beta p \frac{\partial p}{\partial \tau}, \tag{2}$$

with the boundary condition now $p(0, \tau) = q(\tau)$.

The exact solution to this problem can be given in characteristic form, $p(x, \tau) = q(\tau + \beta x p(x, \tau))$, which we will now write as a Taylor series, of which we keep only the first three terms,

$$p(x, \tau) = q(\tau) + \frac{\beta x}{2!} \frac{\partial q^2}{\partial \tau} + \frac{(\beta x)^2}{3!} \frac{\partial^2 q^3}{\partial \tau^2} + O(\omega^3 \beta^3 x^3 q^4), \tag{3}$$

where ω^3 stands for the magnitude of $\partial^3/\partial \tau^3$.

Now, to shift the analysis to the frequency domain, we must first assume that the Fourier transform of $q(\tau)$, denoted $\tilde{q}(\omega)$, exists and is convergent; then, taking the Fourier transform of (3) we obtain

$$
\begin{aligned}
p(x,\omega) &= \tilde{q}(\omega) + \frac{1}{2!}(i\omega\beta x)\int_{-\infty}^{\infty}\tilde{q}(\omega_1)\tilde{q}(\omega-\omega_1)d\omega_1 \\
&+ \frac{1}{3!}(i\omega\beta x)^2\int_{-\infty}^{\infty}\int_{-\infty}^{\infty}\tilde{q}(\omega_2)\tilde{q}(\omega_1-\omega_2)\tilde{q}(\omega-\omega_1)d\omega_1 d\omega_2 + O(\omega^3\beta^3 x^3 q^4)
\end{aligned}
$$

(4)

where we have used the convolution theorem a couple of times. The predictions of nondissipative linear acoustics are given by the first terms of both (3) and (4); the convolution terms in (4) represent the sum and difference frequencies generated, at the lowest order, from the pairs of frequencies present in $\tilde{q}(\omega)$ by the quadratic nonlinearity.

In dealing with the nonlinear evolution of broadband jet noise, we want our theory to require only those simple statistical measures of the initial pressure $q(\tau)$ that are needed in order to predict the values of the same quantities for the evolved pressure, $p(x,\tau)$. Suppose then that $q(\tau)$ is a stationary random function, with zero mean. The only statistical quantities normally available from measurement are the mean square, $q_0^2 = \langle q^2(\tau)\rangle$, and the power spectral density of the initial signal, $S_{qq}(\omega)$. For this information in itself to be sufficient to describe the nonlinear evolution of the given initial signal, without any further specification of higher moments, we assume that the joint probability distribution of $q(\tau)$ is Gaussian, for any finite collection of points $\tau_1, \tau_2, \ldots, \tau_n$. All the relevant statistical quantities and moments are then determined by the initial spectrum (or, equivalently, the initial autocovariance, $R_{qq}(\tau')$). It is then possible to rewrite the expansion given by (4) as

$$
\begin{aligned}
S_{pp}(x,\omega) &= (1-\beta^2\omega^2 q_0^2 x^2)S_{qq}(\omega) \\
&+ \frac{1}{2}\beta^2\omega^2 x^2\int_{-\infty}^{\infty}S_{qq}(\omega_1)S_{qq}(\omega-\omega_1)d\omega_1 + O(x^4),
\end{aligned}
$$

(5)

where $O(x^4)$ is clearly not quoted with the correct dimensions, but serves to indicate where the series is truncated. Notice especially that due to the initial Gaussian distribution there are no terms proportional to odd powers of x in the expansion.

It is easily shown that this spectral expansion is energy conserving, to second order. We can therefore interpret the term $-\beta^2\omega^2 q_0^2 x^2 S_{qq}(\omega)$ to represent the draining of energy from the region of the spectrum where most of it is concentrated in order to produce the cascade to the higher frequencies, represented by the convolution term in (4). This convolution term, however, is affected by difference frequencies as well as sum frequencies, so that the direction of the energy cascade is also determined, at least in part, by the initial shape of the low frequency spectrum.

The importance of nonlinearities on the evolution of broadband noise over sufficiently large distances could still be undermined by either deviations from spherical symmetry, which certainly do occur in the propagation of jet noise, or linear attenuation mechanisms, which can be significant over large distances, even at low frequencies.

It is not difficult to show that f the initial pressure field is directional, and once again assuming that $\omega r/a_0 \gg 1$, equation (1) still holds along any ray specified by a unit vector $\hat{\mathbf{n}} = \hat{\mathbf{n}}(\theta, \phi)$, where θ and ϕ are polar angles; the directional dependence appears only parametrically. Thus,

$$
\begin{aligned}
S_{pp}(r, \omega; \hat{\mathbf{n}}) =\ & \frac{1}{2} \left(\frac{r_0}{r}\right)^2 \beta^2 \omega^2 r_0^2 \ln^2 \left(\frac{r}{r_0}\right) int_{-\infty}^{\infty} S_{qq}(\omega_1; \hat{\mathbf{n}}) S_{qq}(\omega - \omega_1; \hat{\mathbf{n}}) d\omega_1 \\
& + \left[1 - \beta^2 \omega^2 q_0^2(\hat{\mathbf{n}}) r_0^2 \ln^2 \left(\frac{r}{r_0}\right)\right] \left(\frac{r_0}{r}\right)^2 S_{qq}(\omega; \hat{\mathbf{n}}) + O(x^4)
\end{aligned}
\tag{6}
$$

so that spectral distortion will be accompanied by changes in the directivity pattern at each frequency. In order to deal with atmospheric attenuation, we introduce $\alpha = \alpha(\omega)$, the attenuation coefficient of linear acoustics, which is determined from measurements. Since we ignore any effects associated with dispersion (we may do so here, but not in any discussion where skewness of the pressure field is considered), this coefficient is real and an even function of ω. The model equation we use here to replace (1) is a composite approximate equation suggested by Crighton [7],

$$
\frac{\partial \tilde{p}}{\partial r} + \frac{\tilde{p}}{r} + \alpha(\omega)\tilde{p} = \beta p \frac{\tilde{\partial} p}{\partial \tau},
\tag{7}
$$

and it is the simplest one possible which includes all the effects required, with the correct limiting forms. The associated boundary condition is analogous to the one for equation (1), $\tilde{p}(r_0, \omega) = \tilde{q}(\omega)$.

After some manipulation we observe, perhaps surprisingly, that equation (6) continues to hold, to $O(x^2)$, when the left hand side of the equation refers to the spectrum of $\varphi = (r/r_0)e^{\alpha(r-r_0)} \tilde{p}(r, \omega)$. We can therefore conclude that, to $O(x^2)$, atmospheric attenuation affects linear and nonlinear terms in the same way, and that the scale of the nonlinear effects involved during propagation is not determined in any manner by the attenuation.

ACKNOWLEDGEMENTS

The present research is being carried out with the support of project PAPIIT IN116205, awarded by the DGAPA, Universidad Nacional Autónoma de México.

REFERENCES

1. O. V. Rudenko, *Soviet Physics: Uspekhi* **149** No. 3, 413-447 (1986).
2. G.P. Howell, C.L. Morfey, *Journal of Sound and Vibration* **114**(2), 189-201 (1987).
3. G.P. Howell, C.L. Morfey, *AIAA Journal* **19**, 986-992 (1981).
4. K.L. Gee, T.B. Gabrielson, A.A. Atchley, V.W. Sparrow, *AIAA Journal* **43** No. 6, 1398-1401 (2005).
5. D.F. Pernet, R.C. Payne, *Journal of Sound and Vibration* **17**, 383-396 (1971).
6. S.N. Gurbatov, A.N. Malakhov, A. I. Saichev, *Nonlinear Random Waves and Turbulence in Nondispersive Media: Waves, Rays and Particles,* Manchester University Press, Manchester, 1991.
7. D. G. Crighton, "Nonlinear acoustic propagation of broadband noise" in *Recent Advances in Aeroacoustics,* edited by A. Krothapalli, C.A. Smith, Springer Verlag, New York, 1983, pp. 411 - 454.

Investigation of a Single-Point Nonlinearity Indicator in One-Dimensional Propagation

Lauren E. Falco, Anthony A. Atchley, Kent L. Gee, Victor W. Sparrow

Graduate Program in Acoustics
The Pennsylvania State University
University Park, PA 16802

Abstract. The influence of nonlinear effects in jet noise propagation is typically characterized by examining changes in the power spectral density (PSD) of the noise as a function of propagation distance. The rate of change of the PSD is an indicator of the importance of nonlinearity. Morfey and Howell [AIAA J. 19, 986-992 (1981)] introduced an analysis technique that has the potential to extract this information from a measurement at a single location. They develop an ensemble-averaged Burgers equation that relates the rate of change of the PSD with distance to the quantity $Q_{p^2 p}$, which is the imaginary part of the cross-spectral density of the pressure and the square of the pressure. With the proper normalization, spreading and attenuation effects can be removed, and the normalized quantity represents only spectral changes which are due to nonlinearity. Despite its potential applicability to jet noise analysis, the physical significance and utility of $Q_{p^2 p}$ have not been thoroughly studied. This work examines a normalization of $Q_{p^2 p}$ and its dependence on distance for the propagation of plane waves in a shock tube. The use of such a controlled environment allows for better understanding of the significance of $Q_{p^2 p}$.

Keywords: nonlinearity indicators, jet noise
PACS: 43.25.Ba, 43.25.Cb

INTRODUCTION

The study of the propagation of jet noise often uses the power spectral density (PSD) as the quantity of interest in assessing impact on the surrounding community. Many factors can influence the evolution of the PSD as the noise propagates, not least of which is nonlinearity. In any jet noise prediction scheme, it is important to determine whether nonlinearity affects the propagation and, if so, to account for it.

It has been shown that a linear prediction model does not accurately represent the propagation of noise from full-scale [1] and model-scale [2] jets. For these cases, the presence of nonlinearity was established by examining changes in the PSD relative to a linear prediction, especially for higher frequencies and larger propagation distances. This requires measurements at multiple locations. There can be difficulties and limitations with such a method. For full-scale measurements, factors such as ground reflections, wind and temperature gradients, and the complex nature and directivity of the source can affect the evolution of the PSD. It can be difficult to separate the influence of these effects from that of nonlinearity. For model-scale measurements, the ability to capture spectral changes due to nonlinearity is often limited by the

CP838, *Innovations in Nonlinear Acoustics: 17th International Symposium on Nonlinear Acoustics*,
edited by A. A. Atchley, V. W. Sparrow, and R. M. Keolian
© 2006 American Institute of Physics 0-7354-0330-9/06/$23.00

frequency bandwidth of the measurement system and the maximum propagation distances dictated by the measurement space. In light of these limitations, it would be beneficial to be able to determine the presence or importance of nonlinearity with a measurement at a single location.

Morfey and Howell [3] derive an expression containing a quantity that has the potential to serve as a single-point nonlinearity indicator. A normalization of this quantity, often referred to as "Q/S" or "the Morfey-Howell nonlinearity indicator", has recently been used by several researchers [1, 2, 4] in the analysis of high-amplitude nose. However, its physical meaning is not well understood and has not yet been thoroughly investigated.

THEORY

The present analysis follows that of Morfey and Howell [3]. Its basis is the Burgers equation,

$$\frac{\partial p}{\partial x} - \frac{\beta}{\rho_o c_o^3} p \frac{\partial p}{\partial \tau} = \frac{\delta}{2} \frac{\partial^2 p}{\partial \tau^2}, \tag{1}$$

where p is acoustic pressure, x the distance from the source, β the coefficient of nonlinearity, ρ_o the ambient density, c_o the equilibrium sound speed, τ the retarded time, and δ the diffusivity of sound [5]. After some manipulation, including transformation to the frequency domain and ensemble-averaging, they obtain an equation similar to

$$\frac{d}{dx}\left(e^{2\alpha x} S_p\right) = -\omega \frac{\beta}{\rho_o c_o^3} e^{2\alpha x} Q_{p^2 p}, \tag{2}$$

where S_p is the power spectral density and $Q_{p^2 p}$ the imaginary part of the cross spectral density of p^2 and p. The term in parentheses on the left-hand side of Eq. (2) is the absorption-corrected PSD. According to linear theory, the spatial derivative of this quantity is zero. Thus, $Q_{p^2 p}$ is a measure of the rate of nonlinear distortion. For a given frequency and propagation distance, a positive value for the right-hand side of Eq. (2) indicates a gain of energy, and a negative value indicates a loss of energy.

A physical interpretation of this equation can be drawn from the normalized harmonic amplitudes of an initially sinusoidal wave undergoing nonlinear propagation. If these amplitudes are plotted as functions of distance, as in Fig. 4 of Blackstock [6], the slope of a curve at any given point is qualitatively represented by the spatial derivative on the left-hand side of Eq. (2). With this interpretation, theory suggests that the fundamental and second harmonic will be of most importance near the source, and the higher harmonics will gain importance as propagation distance increases. The data presented in this paper are for a narrowband noise source. The theory for such a source, developed by Gurbatov & Rudenko [7], predicts that the spectral evolution of a noise source is qualitatively similar to that of a sinusoidal source but happens more quickly.

EXPERIMENTS

The data presented in this paper were obtained in a plane wave tube constructed of PVC pipe with an inner diameter of 5.21 cm and a total usable length of 9.68 m. It is fitted with two JBL 2402H drivers at one end, four B&K 6.35 mm type 4938 microphones spaced equally along its length, and a fiberglass anechoic termination at the other end. The first cross mode of the tube occurs at approximately 3.8 kHz. For a more complete description of the apparatus and a validation of the sound propagation in the tube, see Refs. [8] and [9]. All data shown here have a source frequency band centered at 2.9 kHz with half-power points at approximately 2.8 and 3.0 kHz.

RESULTS

The dependence of Q_{p^2p} on source amplitude and propagation distance was investigated by plotting the right-hand side of Eq. (2) for various experimental conditions. For Figs. 1(a) and 1(b), measurements from the first microphone (0.10 m propagation distance) were used to calculate Q_{p^2p} for four different source amplitudes.

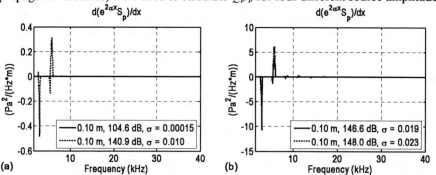

FIGURE 1. Right-hand side of Eq. (2) at a fixed distance of 0.1 m for four different source amplitudes.

In all cases in Fig. 1, energy is being lost in the fundamental frequency region and gained in the second harmonic frequency region. This is consistent with the theory discussed above. It should be noted that although the values for the lowest amplitude case (104.6 dB re 20 μPa in Fig. 1(a)) appear to be zero, they exhibit the same trend as in the other plots but at a much smaller amplitude.

Figures 2(a) and 2(b) depict the right-hand side of Eq. (2) at all four microphone locations for the 148 dB source condition. The value for the fundamental frequency region is always negative, and the (nonzero) values for the higher harmonic regions are always positive, indicating that energy is continually transferred upward in the spectrum. The higher harmonic regions do not show significant nonlinear growth until larger propagation distances. These observations are similar to those made of an initially sinusoidal signal [8].

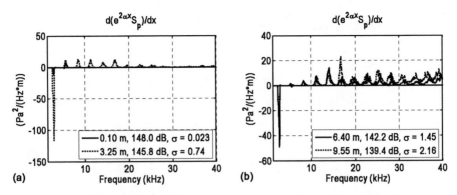

FIGURE 2. Right-hand side of Eq. (2) at four different distances for the same 148 dB source condition.

CONCLUSIONS

The quantity Q_{p^2p} has been identified as being related to the nonlinear distortion rate of narrowband noise. The quantity has been shown to behave qualitatively as would be expected given the theory of the evolution of an initially narrowband noise signal. Its dependence on amplitude and on propagation distance have been demonstrated.

ACKNOWLEDGMENTS

The authors would like to thank the National Science Foundation, the Office of Naval Research, and the Strategic Environmental Research and Development Program for supporting this work.

REFERENCES

1. Gee, K.L. et al., "Preliminary analysis of nonlinearity in military jet aircraft noise propagation," AIAA J. 43(6), 1398-1401 (2005).
2. Petitjean, B.P. et al., "Acoustic pressure waveforms measured in high speed jet noise experiencing nonlinear propagation," 43rd AIAA Aerospace Sciences Meeting & Exhibit, 2005, AIAA 2005-209.
3. Morfey, C.L. and G.P. Howell, "Nonlinear propagation of aircraft noise in the atmosphere," AIAA J. 19(8), 986-992 (1981).
4. McInerny, S.A., and Ölçmen, S.M., "High-intensity rocket noise: nonlinear propagation, atmospheric absorption, and characterization," J. Acoust. Soc. Am. 117, 578-591 (2005).
5. Lighthill, M.J., "Viscosity effects in sound waves of finite amplitude," from Surveys in Mechanics, G.K. Batchelor and R.M. Davies, eds., Cambridge University Press: Cambridge, 250-351 (1956).
6. Blackstock, D.T., "Connection between the Fay and Fubini solutions for plane sound waves of finite amplitude", J. Acoust. Soc. Am. 39, 1019-1026 (1966).
7. Gurbatov, S.N. and O.V. Rudenko, "Statistical phenomena," in Nonlinear Acoustics, edited by M.F. Hamilton and D.T. Blackstock, San Diego: Academic Press, 1998, pp. 382-388.
8. L. E. Falco, et.al, "Investigation of a single-point indicator of nonlinearity in one-dimensional propagation," Forum Acusticum Paper # 703, Budapest, Hungary, Aug. 2005.
9. Gee, K.L. et al., "Nonlinear modeling of F/A-18E noise propagation," 11th AIAA/CEAS Aeroacoustics Conference, 2005, AIAA 2005-3089.

SECTION 12
SONIC BOOM

Meteorologically Induced Variability of Sonic Boom of a Supersonic Aircraft in Cruising or Acceleration Phase

François Coulouvrat [a], Reinhard Blumrich [b] and Dietrich Heimann [b]

[a] Laboratoire de Modélisation en Mécanique, UMR CNRS 7607, Université Pierre et Marie Curie, 4 place Jussieu, 75252 Paris cedex 05, France
[b] Deutsches Zentrum für Luft- und Raumfahrt (DLR), Institut für Physik der Atmosphäre Oberpfaffenhofen, 82234 Weßling, Germany

Abstract. The influence of the meteorological variability on the characteristics of the sonic-boom from a projected commercial SST aircraft is investigated. The sonic boom is calculated by means of an advanced algorithm, taking into account nonlinear distortion, absorption, refraction including shadow zones, and focusing effects. Real meteorological situations are considered based on a full ten-year data set in 12 and/or 24 hour resolution. Three different climate regions are studied: a mid-latitude coastal sea region, a tropical coastal sea area, and a sub-polar land region. For the acceleration phase, the study is limited to a one year data set above the English Channel for flights from Paris to New York. Frequency distributions of sonic boom characteristics such as wave amplitude, rise time, carpet width, location, strength and geometrical extension of caustics are shown for each area, all seasons and opposing flight directions. While variability turns out to be low for cruising-flight boom at the ground track, it is high laterally and for focused boom (carpet width or caustic length, waveboom amplitude).

Keywords: Sonic boom, atmosphere, meteorology, shock waves
PACS: 43.25, 43.28

INTRODUCTION

Sonic boom is the distant acoustical trace of the aerodynamical flow, including shock waves, around a supersonic body. Most concern is about sonic booms near the earth surface (ground or sea surface), where it can be heard by living beings. Sonic booms produced by meteors, cracking whips or gun shells are likely to have been perceived for centuries but were not recognized as such before the 20th century. Thunder can also be considered as the natural and frequent sonic boom produced by lightning. However, it is not before the 1950's that the sonic boom phenomenon has been brought into prominence with the first supersonic flights. In the 60's, public concerns with military aircraft operations and the development of the French-British Concorde, the short-lived soviet Tupolev 144 and their cancelled US competitor Boeing B2707, paved the way for a concerted research effort on sonic boom. Since, the entry into commercial service of Concorde (1976-2003), the environmental impact of the Space Shuttle and of military training areas, and renewed interest for supersonic transport with different research programmes have maintained the interest for sonic

CP838, *Innovations in Nonlinear Acoustics: 17th International Symposium on Nonlinear Acoustics*,
edited by A. A. Atchley, V. W. Sparrow, and R. M. Keolian
© 2006 American Institute of Physics 0-7354-0330-9/06/$23.00

boom. At the present time, several projects of SuperSonic Business Jets (SSBJ) have been unveiled. They are comforted by the 2003 flight of the sonic boom demonstrator (F5-E) with atailored fuselage showing a sonic boom reduction from about 60 to 40 Pa (Quiet Supersonic Platform, 2003). However, the sonic boom annoyance remains a major barrier to be overcome for a supersonic overland flight, the key for an economically profitable civil supersonic transport.

The characteristics of sonic boom at the ground level (position of ground impact point, peak overpressure, rise time, sound level) depend on both the source and the local atmospherical state. While the source is controlled by the aircraft shape and the flight parameters, the propagation is strongly sensitive to atmospherical parameters (temperature, density, wind and humidity) along the propagation path. The key role of meteorology on sonic boom refraction was investigated as early as World War I for sonic booms produced by gun shells (Esclangon, 1925). Thus, sonic boom characteristics are expected to vary widely because atmospheric parameters vary in time and space. Moreover, the local long-term statistics of sonic boom characteristics depend on the climate, i.e. the frequency distribution of the relevant meteorological parameters and their vertical gradients. For an environmentally acceptable supersonic air transport it is necessary to estimate the size of the area affected by sonic booms and their strength beneath potential flight tracks as a function of weather conditions. As a benefit flight tracks can be defined such that unwanted impact on sensitive ground areas can be avoided depending on the actual state of the atmosphere. No long-term monitoring of primary booms from regularly occurring supersonic cruise operations has ever been performed, as this would have required a permanent offshore measuring system beneath Concorde flight path. Therefore, long-term variability has to be estimated by numerical simulations of sonic boom propagation. This is the purpose of the present study to investigate by statistical analysis the meteorologically induced variability of the primary sonic-booms emanating from high-flying aircraft in cruising or maneuvering flight conditions. Because meteorological parameters are subject to daily and seasonal variability, a long time-series with high temporal resolution is required to obtain reliable statistics of sonic-boom behavior at a particular location.

To reach that objective, first the theoretical and numerical model will be briefly presented (section 1). Then meteorological-induced variability will be examined, first in case of cruising conditions for various climates (section 2). Then the "superboom" induced by acceleration is investigated (section 3). This last study is restricted to the area of the English Channel, for the Paris to New York route operated according to procedures similar to those applied by Air France for the former Concorde flights. Simulations are performed for an Airbus mock-up of the planned European Supersonic Commercial Transport (ESCT) vehicle, with a conventional supersonic design aimed at carrying 250 passengers at a Mach 2 cruise. Detailed presentation of the study can be found in Blumrich, Coulouvrat and Heimann, 2005 (a&b).

SONIC BOOM MODEL

The proper matching between near-field aerodynamics and far-field acoustical propagation was formulated by Whitham (1952) for a body of revolution and extended by Walkden (1958) for a non-axisymmetric lifted body. Though it is often superseded by direct CFD simulations (e.g. Plotkin and Page, 2002), for the purpose of the present study focused on the influence of meteorology rather than on aircraft design, it is sufficient to model the source term with Whitham's F-functions. These ones were computed for the chosen configuration for 5 Mach numbers (between 1.2 and 2.0 with 0.2 step) and for azimuth angles of emission normal to the Mach cone between 0° and 70° in steps of 5°. Two angles of attack were considered (4° for a heavy aircraft near take-off, 2.5° for a lighter aircraft in cruise or near landing).

Like other sonic-boom codes (Hayes et al., 1969; Thomas, 1972; Cleveland et al., 1996; Plotkin, 2002) the propagation model is based on full ray tracing in a stratified, moving atmosphere Along each ray, the acoustical pressure p satisfies a nonlinear, generalized Burgers' equation. Non-linear effects are deemed essential in the long-range propagation of finite-amplitude waves such as sonic booms (Whitham, 1956). They are responsible for the slow evolution of the waveform until the typical "N" shape is achieved which is frequently recorded at ground level. The acoustical pressure is determined only for rays that touch the ground, thus forming at the ground level the so-called geometrical carpet. Because of atmospheric refraction, not all emitted rays do touch the ground. A special procedure has been developed to determine the limiting rays, e.g. the two (one starboard, one port side) extreme rays delineating the carpet. They can be of two types whether the atmosphere near the ground is upward or downward refracting. In the upward case, the limiting ray grazes over the ground and separates the geometrical carpet from the shadow zone. In the second case, the limiting ray remains horizontal above the ground so that the carpet is not confined. It is to be noticed that this situation is very frequent, occuring at least on one side of the carpet in about 50% of all investigated cases.

For an upward refracting atmosphere, geometrical acoustics does not predict any signal inside the shadow zone. However, diffraction theory allows extending the signal there (Coulouvrat, 2002) with a proper matching to geometrical acoustics. Inside the shadow zone, the complex analytical expression simplifies into a creeping-wave series expansion, with boom amplitude decreasing exponentially and rise time increasing almost linearly. Near the cut-off, the grazing boom is sensitive to the ground nature (surface roughness or porosity) with a significant increase in the rise time compared to rigid ground. Since the present study focuses on meteorological effects, a perfectly flat and rigid surface is assumed for simplicity.

Atmospheric absorption, mostly due to molecular relaxation of diatomic nitrogen and oxygen relaxation, is a key factor for sonic boom rise times. A full integration of the lossy generalized Burgers' equation along each ray (Cleveland *et al.,* 1996) would have been too demanding from a computational point of view. Therefore, an additional steady-state approximation (Hodgson, 1973; Pierce and Kang, 1990; Coulouvrat and Auger; 1996) was made, assuming that the boom wave propagates over a sufficiently long distance to reach at the ground level an almost steady waveform. Then the rise time emerges simply as a local balance between non-linear effects and absorption This approximation is the more accurate the longer the signal, the longer the propagation and the more humid the air. In this study the approximation is well justified for a rather long aircraft (compared to a SSBJ), flying at high altitudes, and, in most considered cases, over the seas. Finally, it is to be noted that inside the shadow zone near the carpet edge, the finite rise time is due mostly not to absorption but to diffraction effects, fully taken into account by the model.

Atmospherical turbulence within the planetary boundary layer is also known to produce large fluctuations in the characteristics of sonic booms recorded at the ground level (Pierce and Maglieri, 1972). Despite some recent progresses in the understanding of shoch wave propagation in a random medium (Kelly *et al.*, 2000, Blanc-Benon *et al.*, 2002), implementation of the models within a predictive sonic boom software remains a challenge. Therefore, the present model does not take into account the influence of atmospheric turbulence, and the variability of sonic boom characteristics which is investigated here, is exclusively due to the large-scale variability of the atmosphere caused by planetary waves and transient synoptic pressure systems.

Geometrical acoustics breaks down around caustics, which are surfaces of sound focusing. The theoretical model for sonic boom focusing was established by Guiraud (1965). It includes two physical effects (diffraction and non-linearity) combined in the so-called Nonlinear Tricomi Equation (NTE). One peculiarity of the NTE is its mixed-type character, depending on the conisdered side ("illuminated" or "shadow zone") of the caustic. Auger (2001) provided a recent derivation of Guiraud's theory for the general case of a 3D, heterogeneous atmosphere with slow winds. The numerical modeling of sonic boom focusing is carried out in four successive steps: ray tracing, determination of the ground position of the geometrical caustics (the envelope of rays where the ray tube algebraic area vanishes), determination of the caustic local geometry, and numerical resolution of the Nonlinear Tricomi Equation. The numerical solver for the NTE is described in Marchiano, Coulouvrat and Grenon (2003).

For the purpose of statistical analysis, we selected as relevant acoustical parameters describing sonic boom ground impact : 1) the maximum sound pressure and 2) the rise time (especially either ground track for 0° emitted angle, or laterally for limiting rays), and 3) the starboard and port-side carpet widths corresponding to sonic boom A-weighted Sound Exposure Level (*A-SEL*) above either 90 or 70 dB. This choice for metric is justified as *A-SEL* metric is the second "best" metric for correlation to annoyance (after Stevens Mark VII Perceived Level), both for *indoor* and *outdoor* booms (McCurdy *et al.*, 2004).

METEOROLOGICAL VARIABILITY FOR CRUISING BOOM

Three target areas have been selected, representing mid-latitude (TA1), tropical (TA2) and sub-polar (TA3) way points along flight routes between major population agglomerations. The location of the target areas are shown in Fig. 1. For all target areas, two opposing flight directions, corresponding to the great circle between relevant destinations, are also indicated. Cruising altitudes (between 17.5 and 19.8 km) and angles of attack (2.5° or 4°) were considered accordingly. The aircraft speed was assumed to be Mach 2 for all target areas and flight directions. Meteorological data were extracted from the ERA15 re-analysis database of the European Centre for Medium range Weather Forecast (Gibson et al., 1997).

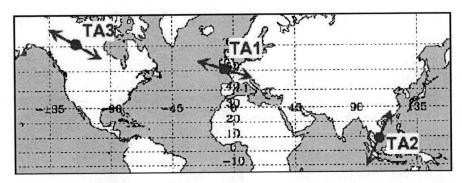

FIGURE 1. Selected target areas and corresponding headings.

The data set covers a 10-year period (1984-1993) in time intervals of 24 hours (00 UTC; TA1 & 2) or 12-hours intervals (00 and 12 UTC; TA3) with a 30-level vertical resolution varying between approx. 100 m near the ground and 4000 m near 31 km.

One example of sonic boom statistics for target area TA1 is shown Fig.2. For a given target area and flight direction, the ground track maximum pressure varies only by about 5 to 15 Pa. However, different target areas and flight directions show significantly different distributions because of various flight altitudes and weights. The maximum pressures at the outer edges of the sonic-boom carpet exhibit much broader frequency distributions with a variability range from 20 to 150 Pa. This higher variability is caused by the higher sensitivity of the laterally, grazing boom to refraction effects due to wind gradients. Though the mean maximum sound pressure is lower than at the ground track, larger values frequently occur due to ray 'pre-focusing' which cannot be handled correctly by the present model. Note the ICAO standard atmosphere results in rather small overpressures compared to the simulations based on real meteorological data, as the absence of strong gradient near the ground smoothes lateral propagation. The 70 dB A-SEL total width of the sonic boom carpet varies widely between 60 and 200 km, with smaller variability range for TA2 (60-120 km). As expected, at TA1 and TA3 where westerly winds are most frequent, the width of the carpet is smaller for westbound flights than for eastbound ones. In the majority of the cases, the port side and starboard widths of the carpet are smaller than for the standard atmosphere values. Seasonal differences exist (mostly spring at TA1 or TA3) but are not well pronounced.

METEOROLOGICAL VARIABILITY FOR SUPERBOOM

Based on Paris to New York mean ground track, corresponding meteorological data were extracted from the ERA15 ECMWF database at the vertical of points along that track. Actual tracks were then computed accorded to the estimated performances of the mock-up, including Concorde acceleration procedures preventing overland boom. For computational cost reasons, the study was restricted to the year 1993 (00 UTC).

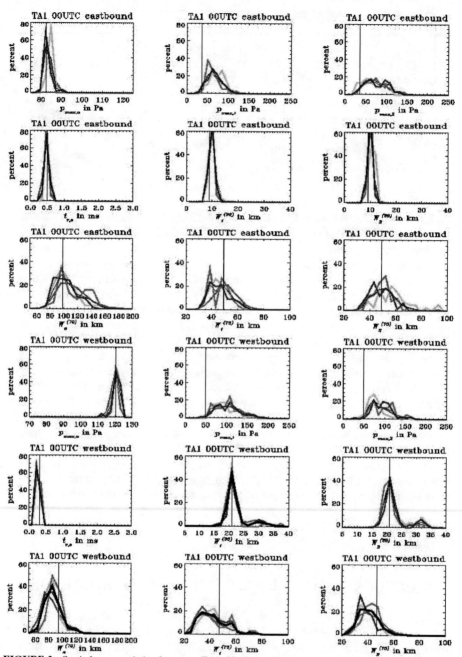

FIGURE 2. Sonic boom statistics for target TA1. Up (lines 1 to 3) : eastbound flight – Down (lines 4 to 6) : westbound flight. Lines 1 & 4 : max. overpressures at ground track, port side and starboard carpet edges. Lines 2 & 5 : ground track rise time, port side and starboard 90dB A-SEL carpet widths. Lines 3 & 6 : total, port side and starboard 70dB A-SEL carpet widths. Line colours / season correspondance : green / spring – orange / summer – red / autumn – blue / winter – black / full year.

The year 1993 was selected for its near-average behavior regarding the height of the thermal tropopause and the North Atlantic Oscillation Index. It resulted into 365 meteorology-dependant trajectories and associated sonic boom footprints. For each case, the full geometrical extent of the acceleration caustic and the focused superboom emitted at 0° angle (and impacting ground at the so-called "focus" localized by green cross on Fig.3) were computed. Fig. 3 shows 4 simulations associated to two extreme cases (extremely strong head – 01/11/93 – or cross – 10/22/93 – winds) and two near-average situations (03/28/93 and 10/19/93). These four examples, confirmed by one-year statistics, show the strong variability of the focused boom. Maximum overpressures vary mainly between 0.2 and 0.7 kPa around a 0.41 kPa average (4 times the cruise value at Mach 2) and a 0.14 kPa standard deviation, but in a few cases peak values of up to 1.5 kPa were calculated. Maximum A- and C-weighted Sound Exposure Levels spread about 10 dB around mean values (91.4 dB for A-SEL, 120.5 dB for C-SEL). Mean rise time is 4.3 ms, more than ten times the mean value at Mach 2 cruise. The ground caustics are in average 84.7 km long, with a large standard deviation of 31.8 km and strong asymmetry again due to dominant west winds.

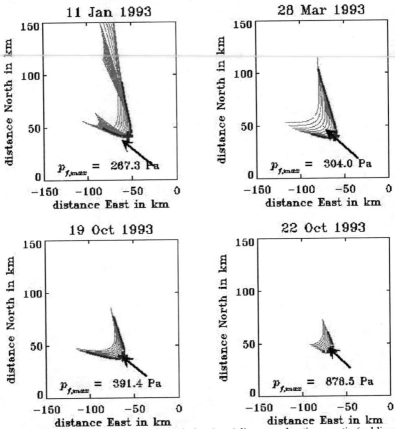

FIGURE 3. Ground track (blue arrow), isoemission (gray) lines, acceleration caustic (red lines), focal point (green cross) and associated maximum superboom overpressure (in Pa) at four different days.

In total, an area of 7,750 km² is affected by focused sonic booms which occur mostly over water, though 7.5 % the of simulations touch land (near Cherbourg). This land impact can be avoided by shifting the trajectory of about 30 km to the north-east.

The study shows that boom focusing remains a major environmental barrier to overcome. For supersonic flights restricted over the seas, future routes will have to be carefully optimized to avoid overland focusing near the coasts. For aircraft intended for unrestricted overland supersonic flight, the situation is more critical. Given high amplification and strong variability, it remains unlikely that, even for low-boom designed jets, superbooms can be kept below a (yet undefined) "acceptable" level.

ACKNOWLEDGMENTS

The study was carried out as part of the project 'SOBER' co-funded by the European Commission under contract G4RD-CT-2000-00398. The authors are obliged to J. Carla and R. Etchevest (Airbus France), Dr. K.-P. Hoinka (DLR), S. Illa (Transiciel Technologies), Dr. R. Marchiano (Univ. Paris 6), Dr. G. Zängl (Univ. Munich) and B. Vandenbroucke (Airbus France) for various assistances.

REFERENCES

1. Esclangon, E., *L'acoustique des canons et des projectiles*, Imprimerie Nationale, 1925
2. Blumrich, R., Coulouvrat, F., and Heimann, D., *J. Acoust. Soc. Am.,* in press (2005a)
3. Blumrich, R., Coulouvrat, F., and Heimann, D., *J. Acoust. Soc. Am.,* in press (2005b)
4. Whitham, G. B., *Comm. Pure Appl. Math.*, **5**, 301-348 (1952)
5. Walkden, F., *Aeron. Q.*, **IX**, 164-194 (1958)
6. Plotkin, K. J., and Page, J. A., *AIAA Paper* 2002-0922 (2002)
7. Hayes, W. D., Haefeli, R. C. and Kulsrud, H. E., *NASA CR-1299* (1969)
8. Thomas, C. L., *NASA TN D-6832* (1972)
9. Cleveland, R. O., Chambers, J. P., Bass, H. E., Raspet, R., Blackstock, D. T., and Hamilton, M. F, *J. Acoust. Soc. Am.*, **100**, 3017-3027 (1996)
10. Plotkin, K. J., *J. Acoust. Soc. Am.*, **111**, 530-536 (2002)
11. Whitham, G. B., *J. Fluid. Mech.*, **1**, 290-318 (1956)
12. Coulouvrat, F. , *J. Acoust. Soc. Am.*, **111**, 499-508 (2002)
13. Hodgson, J. P., *J. Fluid Mech.*, **58**, 187-196 (1973)
14. Pierce, A. D., and Kang, J., *Frontiers of Nonlinear Acoustics*, edited by M. F. Hamilton and D. T. Blackstock , Elsevier, 1990, pp. 165-170
15. Coulouvrat, F., and Auger, Th. , *Proceedings 7th Int. Symp. Long Range Sound Propagation*, Ecole Centrale de Lyon, 1996, pp. 177-191
16. Pierce, A. D., and Maglieri, D. J., *J. Acoust. Am.*, **51**, 702-721 (1972)
17. Kelly, M., Raspet, R., and Bass, H., *J. Acoust. Soc. Am.*, **107**, 3059-3064 (2000)
18. Blanc-Benon, Ph., Lipkens, B., Dallois, L., Hamilton, M. F., and Blackstock, D. T., *J. Acoust. Soc. Am.*, **111**, 487-498 (2002)
19. Guiraud, J.-P., *J. Mécanique*, **4**, 215-267 (1965)
20. Auger, Th., *Modélisation et simulation numérique de la focalisation d'ondes de choc acoustiques en milieu en mouvement. Application à la focalisation du bang sonique en accélération*, PhD Dissertation, Université Pierre et Marie Curie, 2001
21. Marchiano, R., Coulouvrat, F., and Grenon, R., *J. Acoust. Soc. Am.*, **114**, 1758-1771 (2003)
22. McCurdy, D. A., Brown, S. A., and Hilliard, R. D., *J. Acoust. Soc. Am.*, **116**, 1573-1584 (2004)
23. Gibson, R., Kallberg P., Uppala S., Hernandez A., Nomura A., and Serrano E., *ERA description. ECMWF ReAnalysis Project Report Series 1*, ECMWF (Reading, UK) , 1997

Ray theory analysis and modelling of the secondary sonic boom propagation for realistic atmosphere conditions.

Philippe Blanc-Benon*, Laurent Dallois *, Julian Scott*, Uwe Berger [†] and David Allwright**

*LMFA, UMR CNRS 5509, Ecole Centrale de Lyon, 69134 Ecully Cedex, France
[†]IAP, Schlossstrasse 6, 18225 Kühlungsborn, Germany
**OCIAM, Mathematical Institute, 24-29 St Giles', Oxford, OX1 3LB, UK

Abstract. The shock waves generated by a supersonic aircraft are reflected in the upper part of the atmosphere. Back to the ground, they are indirect sonic booms called secondary sonic booms. The recorded signals of secondary sonic booms show a low amplitude and a low frequency. They sound like rumbling noises due to amplitude bursts. These signals strongly depend on the atmospheric conditions, in particular to the amplitude and to the direction of the wind in the stratopause. In the present work, the propagation of secondary sonic booms is studied using realistic atmospheric models up to the thermosphere. The secondary carpet position is investigated by solving temporal ray equations. An amplitude equation including nonlinearity, absorption and relaxation by various chemical species is coupled to the ray solver to get the secondary boom signature at the ground level. Multipath arrivals are directly linked to wind field or 3D inhomogeneities.

Keywords: Infrasound, Ray theory, nonlinear propagation, atmosphere
PACS: 43.30.Pc , 43.30.Sf

INTRODUCTION

The atmospheric sound speed profile creates waveguides for the shock waves generated by a supersonic aeroplane. The upward shock waves is reflected back to the ground by the temperature gradients in the stratopause or in the thermosphere. The resulting noise disturbance is called secondary sonic boom. It is merely an infrasonic signal and it sounds like a rumble noise associated to bursts [1, 2]. In the present work, the propagation of secondary sonic booms is studied using realistic atmospheric models up to the thermosphere. The secondary carpet position is investigated by solving temporal ray equations. An amplitude equation including nonlinearity, absorption and relaxation by various chemical species is coupled to the ray solver to obtain the secondary boom signature at the ground level. The predicted signatures are compared to recorded signals of secondary sonic booms. A good agreement is found for the amplitude and for the time duration. The bursts seem to be related to multipath arrivals due to direct and indirect secondary sonic boom. The rumbling noise can be interpreted as the effect of finer structures of the atmosphere as gravity waves.

CP838, *Innovations in Nonlinear Acoustics: 17th International Symposium on Nonlinear Acoustics*,
edited by A. A. Atchley, V. W. Sparrow, and R. M. Keolian
© 2006 American Institute of Physics 0-7354-0330-9/06/$23.00

PREDICTION METHOD

The prediction method is based at least on two assumptions. The first one is that the atmosphere varies over length scales greater than the actual length of the supersonic boom. The second assumption is that the shocks are only weak shocks, *i.e.* the shock pressure amplitude is less than a few percent of the underlying pressure field. The model is derived from the generalized Navier-Stokes equations including earth rotation and from the state equations for the different molecular species of the atmosphere to get relaxation effects. From these equations, a two-step asymptotic development can be conducted. At the first step, a ODE system of six equations is obtained. This system provides the shock wave trajectory called boom rays in this work. The boom rays start at the aeroplane position at a given time τ and are parametrized by their launching angle θ, the azimuthal angle around the aeroplane. The boom rays are similar to the acoustical rays due to the weak shock assumption. The main difference is that their launching polar angle ϕ is fixed by the relation $\phi = \arccos(1/M)$ where M is the Mach number. At a given time, all the shock waves generated by the aeroplane along its trajectory are located on a surface called the Mach surface. The second step of the development leads to an amplitude equation which must be solved along each ray. This equation is used to models the deformation of the shock wave during its propagation into the inhomogeneous atmosphere. It is a nonlinear paraxial equation that includes dispersion, absorption and relaxation effects. This equation is no more valid when the associated ray goes through a caustic and a Hilbert transform is then applied to the shock wave to simulate the crossing of the caustic.

RESULTS FOR SECONDARY BOOMS

The initial conditions are typical of a Concorde flighing at a Mach 2 speed. The aircraft is located at a latitude of around 45 degres North for summer atmospheric conditions. As the aircraft altitude during a supersonic flight is around $15km$, two waveguides are allowed to the shock rays; the stratospheric waveguide between an altitude of few kilometers up to the stratosphere ($\sim 50km$) and the thermospheric waveguide between the ground level and the thermosphere (above $100km$)(see Fig.1).

The figure 2 shows the Mach surface and the primary and secondary carpet position for this configuration. Actually, the whole part of the secondary, the direct and the indirect one, are created by thermospheric rays which are reflected at an altitude of $120km$. Due to their long travel into the thermosphere where absorption and relaxation are the main effects, a low amplitude wave is expected. The shock rays that are restrained to the waveguide 1 never reach the ground. The secondary Mach carpet is between $250km$ and $300km$ beyond the plane (here at the origin). The interval of times between the arrivals of the direct (first secondary carpet) and the indirect (second secondary carpet) shock ray is about 2 minutes for a ground observer at the vertical of the aircraft trajectory. This is a first occurence of multipath arrivals.

The secondary sonic booms which travel in the upper part of the atmosphere before being reflected back to the ground are more dependent on the atmospheric structure and the wind velocity in altitude. Depending of the value of the maximum wind velocity

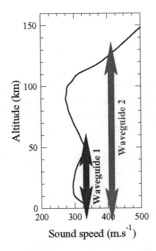

FIGURE 1. Sound speed profile.

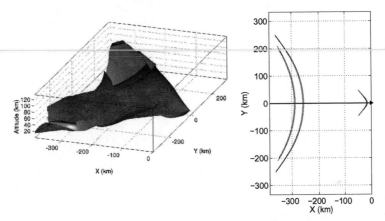

FIGURE 2. Evolution of the Mach surface. Primary and secondary boom carpets at ground .

at the stratopause altitude, the secondary sonic boom is reflected in the thermosphere or in the stratopause and this leads to two different kinds of secondary sonic boom. Considering the mean atmospheric sound speed, two waveguides exist for the boom rays (see Fig. 1). The thermosphere is the upper boundary of the first waveguide. Most of the boom rays trapped in this waveguide reach the ground. As a large part of the boom ray is at altitudes (over 50 km) where absorption and relaxation are the dominant effects, the secondary boom signature is an infrasonic signal (cf. figure 3). Due to its low frequency, this wave can travel over very long distances in the atmosphere. The peak pressure amplitude is around $0.1 Pa$ and the frequency is around $0.05 Hz$. These values are in accordance with long distance measurements of secondary boom [3, 4].

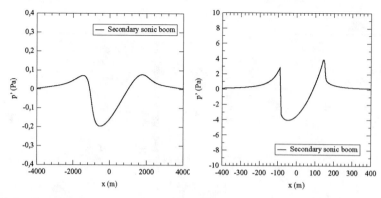

FIGURE 3. Secondary boom signatures : comparison between thermospheric ray (left graph) and stratospheric ray (right).

The second waveguide is between the stratopause and the troposphere. As the maximum sound speed in the stratopause is less than the sound speed at ground level, boom rays trapped in this waveguide reach the ground only if the shock wave is appropriately convected by wind at the reflection altitude (around $50km$). The resulting secondary sonic boom signature has a higher amplitude than the thermospheric one with a peak pressure amplitude of $4Pa$ (cf. figure 3). Its frequency is also much higher ($0.5Hz$) and shocks persist. These values are also in agreement with some measurements of secondary sonic boom [5].

INFLUENCE OF GRAVITY WAVES

The model presented here has been applied to realistic atmospheric conditions. The atmospheric fields are provided by the IAP (Leibniz-Institut für Atmosphärenphysik, Universität Rostock). They include mean pressure, density, temperature and wind data at a given latitude ($69°N$) up to an altitude of $150km$. They are discretized every $10km$ along the latitude and every $100m$ along the altitude. In addition to these mean values, the IAP provides also simulated data of atmospheric gravity waves. They correspond to finer length scale inhomogeneities of the atmosphere and their influence is investigated by adding them to the mean atmospheric fields. These data have to be interpolated over the whole propagation domain. This is performed by using third-order polynomials as continuity has to be preserved up to the second spatial derivatives of the fields.

The primary influence of the gravity waves on the secondary sonic booms concerns the secondary carpets. The figure 4 compares the secondary sonic boom carpets when the atmospheric data do or do not contain gravity waves.

In figure 4 the left plot is obtained when gravity waves are not included to atmospheric data. The leftmost carpet corresponds to the primary boom. The two carpet patches at a medium distance from the aeroplane ($x \sim 150km$) are created by boom rays reflected by the stratopause. The last carpets ($x \sim 300km$) are due to the thermospheric boom rays. Each carpet is composed of a direct and an indirect part, the indirect part being due to the rays reflected back to the atmosphere from the primary carpet position. The distance between the direct and the indirect secondary carpet is around $20km$. This is in agreeement with the time duration between the bursts of secondary sonic boom (around $30s$). The carpets are smooth and a clear difference appears between each one.

In figure 4 the right plot is obtained when gravity waves are included. The primary carpet is not influenced by the gravity waves. The gravity wave influence appears on the secondary carpet geometry. The secondary carpets look more complex and the direct and indirect secondary carpets cannot be distinguished anymore. A greater number of rays reaches a given earth location. It may be expected that it is the main reason of the rumble noise.

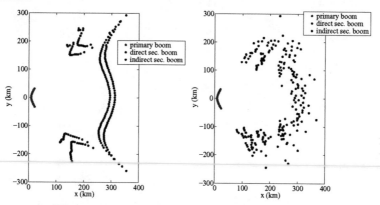

FIGURE 4.   Secondary boom carpets for realistic atmosphere (left) ; influence of gravity waves (right).

CONCLUSION

The present results show that the signature model predicts well the amplitude and the frequency of the secondary sonic booms. The rumble noise and the bursts may be explained by multipath arrivals of secondary sonic booms. The bursts seems to be linked to the arrival of direct and indirect rays and the rumble noise to the presence of gravity waves or other fine scale effects in the atmosphere.

ACKNOWLEDGMENTS

This work was supported by European Commission CONTRACT N°: G4RD-CT-2000-00398.

REFERENCES

1. ROGERS, P. H. & GARDNER, J. H., 1980, Propagation of sonic booms in the thermosphere, J. Acoust. Soc. Am. **67**, 78-91.

2. HAGLUND, GEORGE T. & POLING, HUGH W., 1995, HSCT sonic boom impact at the earth's surface due to secondary booms,Tech. Report Boeing Commercial Airplane Company.
3. GROVER F. H., 1973, Geophysical effects of Concorde sonic booms, Quart. J. Roy Astron. Soc., **14**,141-160.
4. LISZKA LUDWIK, 1978, Long-distance focusing of Concorde sonic boom, J. Acoust. Soc. Am. **64**, 631-635.
5. RICKLEY, E. J. AND PIERCE, A. D., 1980, Detection and assessment of secondary booms in New England, Technical Report FAA-AEE-80-22.

Experiment on Finite Amplitude Sound Propagation in a Fluid with a Strong Sound Speed Gradient

H. Hobæk, Å. Voll, R. Fardal and L. Calise

Department of Physics and Technology, University of Bergen, Allégt. 55, N5007 Bergen, Norway

Abstract. A closed tank of dimensions $0.5 \times 0.5 \times 2.7 \ m^3$, filled with a mixture of ethanol and water to produce an almost linear sound speed profile with a gradient near 450 (m/s)/m, served the purpose for investigating shocked sound wave propagation in a stratified environment. As the sound speed profile evolved by diffusion a number of different measurements were taken, both in areas with caustics, shadow zones, along the main beam and along the bottom. After about one year, part of the fluid was re-mixed to obtain a pronounced sound speed maximum some 20 cm above the bottom.

The high intensity sound was produced by a plane circular piston type sound source with near-field length 45 cm and half power angle 0.8^o at 1.1 MHz, placed near one end of the tank. Its tilt angle and depth could be varied. A 0.5 mm diameter PVDF needle hydrophone (Precision Acoustics) mapped the sound field in a vertical slice in the range $0.9 - 2.4$ m, remotely controlled by a PC.

We present results from measurements in a shadow zone and along the bottom. The latter, in particular, displays unexpected amplitude variations. The project was funded by the European Commission, contract number G4RD-CT-2000-00398.

Keywords: Binary fluid mixing, diffusion, shadow zones, sonic boom, sound speed profile, waveform distortion.
PACS: 43.25.Cb, 43.25.Zx, 43.28.Mw, 43.85.-e.

MOTIVATION

The sound speed in the standard atmosphere has a linear gradient between 1 and 10 km height. Above, up to some 25 km it is constant, and below 1 km highly varying. Related to sonic boom propagation it was the purpose of this experiment to measure the propagation of high amplitude sound waves in a fluid simulating this sound speed profile (SSP) scaled to laboratory dimensions. The atmospheric sound speed gradient is $g = 4 \cdot 10^{-3} \ s^{-1}$, corresponding to a radius of curvature (upwards) of $R = 75$ km. Scaling 10 km to 0.5 m in the laboratory results in need of a gradient $G \geq 400 \ s^{-1}$. The choice of ethanol and water as the components of the binary mixture was made after an extensive literature search, and is due mainly to the large range in obtainable sound speeds, low cost and low toxicity. However, it was not easy to find accurate data on such mixture. Figure 1 shows sound speed versus ethanol concentration (by weight) from two different references [2] and [1]. The curve extending to 35 % is taken at 25 °C, while the other is at $17 - 18 \,^oC$. The maximum sound speed difference is 385 m/s, i.e. $G_{max} = 770 \ s^{-1}$, almost twice what is needed. For concentrations above 40 % it is possible to obtain an almost linear SSP. For low concentrations one may obtain a profile with a clear

CP838, *Innovations in Nonlinear Acoustics: 17th International Symposium on Nonlinear Acoustics*,
edited by A. A. Atchley, V. W. Sparrow, and R. M. Keolian
© 2006 American Institute of Physics 0-7354-0330-9/06/$23.00

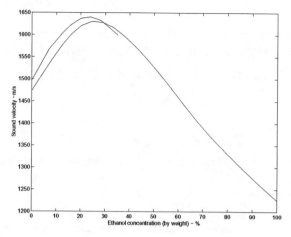

FIGURE 1. Sound velocity in ethanol/water mixture at 25 °C, (to 35%, [1]), and at 17-18 °C [2].

maximum, yielding a shadow zone at that height, and allowing for sound channels above and below. The density of the mixture varies almost linearly with ethanol concentration, C. Thus, the mixture will be gravitationally stable if C decreases monotonously from top to bottom. It should be observed, however, that it is not possible at the same time to scale the density, absorption or parameter of nonlinearity to the atmosphere.

EXPERIMENTAL ARRANGEMENT

Density gradient

In order to obtain an approximately linear density gradient (and hence SSP) the experiment tank needs to be filled in a special way. For this purpose we use two auxiliary tanks, one of which (Tank 1) is filled with the ethanol-water concentration desired at the bottom of the experiment tank, and the other (Tank 2) filled with the ethanol concentration wanted at the top. The two tanks are connected by tubes at the bottom, and the content in Tank 2 is mixed continuously during the filling process. From Tank 2 the mixed solution then enters the experiment tank, which is filled from the bottom. Analytic solutions for the resulting density gradient exists if the cross sections of the auxiliary tanks are equal. If they differ, the solution must be obtained numerically[3]. Figure 2 shows the analytical solution, which is not quite linear, and an optimized numerical solution which is almost linear. Here the ratio of the cross sections ε (Tank 2/Tank 1) is 0.788, which is almost equal to the density ratio of the pure fluids.

The mixing tanks were two cylindrical polyethylene tanks of internal diameter 78 cm and height 112 cm. Each holds a maximum of 550 l. They were equipped with tube connections and valves, so that each could be filled separately and connected later when the mixing started. The effective area of Tank 2 was reduced with a vertical cylinder of

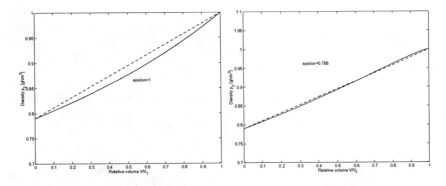

FIGURE 2. Calculated density versus height (V/V_T). Left: $\varepsilon = 1$, Right: $\varepsilon = 0.788$.

diameter 36 cm (two stacked ethanol cans). This resulted in $\varepsilon = 0.778$, which is very close to the desired value 0.788. From the outlet of Tank 2 the flow was divided between a pump - which refills the same tank in order to stir the fluid by violent turbulence - and the filling tube to the experimental tank. The flow was distributed evenly to 6 vertically aligned copper tubes along the center of the tank. The end of each of these tubes was machined to obtain the same opening between the end of the copper tubes and the glass tank bottom, with the purpose to have the same flow rate through all the tubes. Care was taken to ensure that the total length of tubing between the outlet of Tank 2 and the end of each tube was exactly the same, so that flow out of each tube should have the same concentration. The experiment tank during filling is shown in Fig. 3, with the mixing tanks barely visible in the background. Note the hermetically sealed cover of the tank, with two hatches (here open). After the tank was filled the air space was exchanged with nitrogen in order to reduce danger of explosion of the ethanol vapor. During the filling process one practical problem emerged: air bubbles in the tubes. These were probably due to dissolved air in the tap water becoming liberated because of the alcohol. Small bubbles followed the flow and became trapped at the highest points in the tubes, so that the flow rate diminished. During the filling process, which took about 4 hours, it was necessary to empty the tubes for air almost constantly. This manipulation did not influence much the final sound speed profile, which was almost like the planned one. 400 l of 96% ethanol was needed to fill the experiment tank.

Diffusion - evolution of the sound speed profile

The change in concentration in the measurement tank with time is governed by diffusion. Assuming that the concentration gradient at the start is linear, and using Landau's boundary conditions (no diffusion at the top or bottom of the tank), we find solutions of the diffusion equation as shown in Figure 3, with time steps of 10 days for a 3 month period [3]. Here we have assumed the diffusion coefficient D for ethanol-water mixture [4] as $D = 1.24 \cdot 10^{-5}$ cm^2/s at $25°C$. If the height of the tank is H this

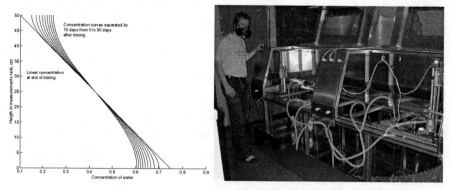

FIGURE 3. Evolution of density gradient with time - Tank during the filling process

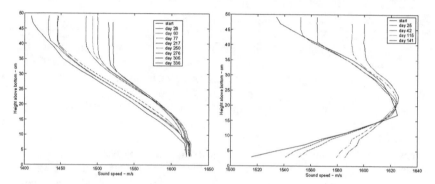

FIGURE 4. Evolution of SSP with time. Left: Linear profile. Right: Modified profile.

results in a time constant $T = H^2/(\pi^2 D) = 2 \cdot 10^7 s = 236$ days, e.g the time before the concentration has been reduced with a factor $1/e$. Diffusion of the density gradient leads to the necessity of measuring the SSP at a regular basis. It was measured directly by placing a series of acoustical reflectors in the shape of a "stair-case" with regular depth intervals (20 mm) along one tank wall. A vertically directed high frequency ultrasound beam measures the travel time to each step. A simple inversion scheme provides the SSP, to a resolution limited mainly by the size of the stairs. Calibration was made in pure water. A part of he stair case can be seen at the lower right in Figure 3. The evolution of the SSP is shown in Fig. 4. The almost linear initial profile keeps a surprisingly stable slope at middle depths over the period of almost one year. As expected, the top layer, where the sound speed is almost uniform, grows steadily. The situation at the bottom is somewhat different. The evolution here takes place at a lower pace than at the top. A probable reason for this is the evaporation from the top, but not at the bottom, so the real boundary condition at the top is not like the one assumed in the theory.

After nearly one year the mixture was modified by extracting 14 cm of the top level fluid and re-mixing it before refilling it at the bottom, with some water added. In this

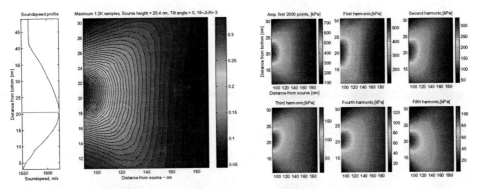

FIGURE 5. Amplitude map of horizontally oriented sound beam in a shadow region. Left: SSP and map of total amplitude. Right: Harmonics (Total and 1-5).

way we obtained a SSP with a pronounced maximum, as seen in Fig. 4 (right panel), together with its evolution over 141 days.

Equipment

Most measurements were taken with a circular high amplitude sound source at 1.1 MHz and later (after repair) 880 kHz and diameter 5 cm [5]. In some cases a low amplitude line array of cylindrical elements at 276 kHz, aligned horizontally at right angle to the sound axis, were used. The source signal bursts were delivered from an ENI 240L or an ENI A500 power amplifier. The probe was a PVDF needle hydrophone of 0.5 mm diameter (Precision Acoustics). The probe tip was covered with 15μm of parylene, to withstand high concentrations of ethanol. The received signals were captured with a LeCroy 9350 digital oscilloscope. The hydrophone had to be moved very slowly through the stratified fluid, in order to not cause extra mixing. The measurements were controlled from a PC. The measurement range was from 90 cm from the source to the end of the tank. Some measurement series took more than 8 hours to perform. Regions of particular interest were found by ray tracing using J. Bowlin's code RAY[6]. For some quantitative comparisons simulations were made with the computer code OASES[7].

RESULTS

Several different series of measurements have been taken in the stratified fluid. Space allows us to present only a few samples of these results. Fig. 5 shows a typical result obtained with the high amplitude source radiating horizontally into a shadow zone. The left panel shows the SSP, the location of the source and a map of the measured field (RMS-amplitude). The usually narrow sound beam (beam-width $\approx 4^o$) is refracted to the extent that it is nearly split into two beams. The right panel shows the map of the lowest 5 harmonics. Note the difference in the distribution of the various harmonics.

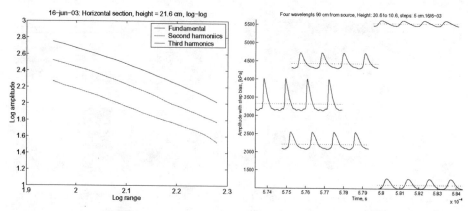

FIGURE 6. Left: Amplitudes along a horizontal section at source height, 1-3 harmonics. Right: Samples of wave forms 90 cm from the source, at 5 cm steps in height.

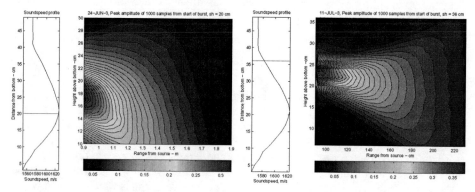

FIGURE 7. Shadow zones. Left: Horizontal beam slightly below the SSP maximum. Right: Source far above the SSP maximum tilted 12.8° down, .

The reason is not known, but it is possible that absorption plays a role here, since it is strongly dependent on concentration, and possibly has frequency anomaly due to relaxation. Figure 6 shows the amplitude of the three lowest harmonics as a function of range along a horizontal line from the source center, in a log-log plot. A best fitted line to these curves indicates a range dependence as r^{-n}, with $n \approx 2.3$; it is slightly higher for the fifth harmonics. The right panel shows sample wave forms at 90 cm range. The vertical scale is in kPa, but the wave forms have been offset for clarity. Two other typical results from shadow zone situations are shown in Figure 7. In the left panel the source is radiating horizontally, but is located slightly below the SSP maximum. Accordingly the beam is refracted downwards. The right panel shows a case where the source is far above the SSP maximum, and is tilted downwards 12.8°, resulting in an almost horizontal beam in the shadow zone.

An example of measurements close to the bottom is shown in Fig. 8 and 9. The

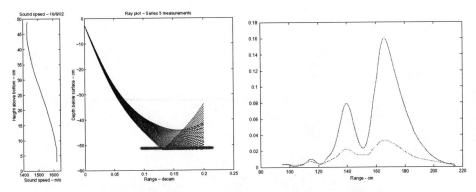

FIGURE 8. Field along the bottom. Left: SSP and ray fan. Right: Amplitude along the bottom.

SSP was almost linear, and the source was located 49 cm above the bottom, tilted 30° down. The left panel shows the sound speed profile, a ray plot and the location of the measurements (red dots). The right panel shows the amplitude at height 13 mm above the glass bottom. The two curves apply different methods to determine the amplitude. One might believe that the fluctuations reflect the main and the side lobes of the beam. They don't! The largest maximum may possibly be fitted to the main lobe (a source tilt adjustment of 0.25^o is needed), but then the side lobes fail to match the other extremes. The most likely explanation is interference with bottom reflections. This is also strengthened by inspection of wave forms taken at the maxima and minima, as shown in Figure 9, left panel. The strongly distorted wave forms are taken at the minimum amplitudes, 1.51 and 2.12 m, respectively. Such distortion can only occur by super-posing two (or more) signals rich in harmonics. The corresponding spectra are shown in the right panel. Note that the fundamental component is strongly suppressed in the most distorted wave forms. Reflections from the bottom occur far above the critical incidence angle, and the reflection coefficient is very close to -1.

The type of distorted wave forms seen here is new to us. We have no evidence that they are caused by multiple ray paths due to refraction, and therefore could also appear in homogeneous media. An inspection in previously recorded data reveals that similar wave forms may be found at locations where diffraction causes a cancellation of the fundamental frequency, but where nonlinear distortion has generated higher harmonics. Or near walls, like in the present case.

SUMMARY

The binary mixture is remarkably stable, and allows for an experimental period of substantial length. The SSP may be shaped to a linear profile, or a profile with a SSP-maximum. SSP-minima may only occur at top or bottom. Monitoring of the SSP with a reflecting stair case was very accurate and convenient, and should be performed once each day of measurements. A large amount of data has been collected, and awaits further analysis. At the moment we lack a working theoretical model, which incorporates both

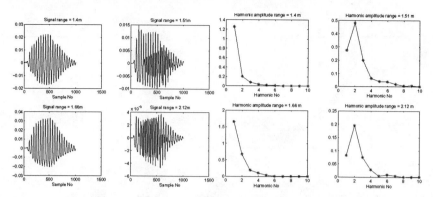

FIGURE 9. Left: Wave forms at 4 ranges (amplitude maxima and minima). Right: Harmonic amplitudes of the same.

finite amplitude sound beam propagation and a stratified medium, to compare with, but progress is being made to adapt the model by Reiso *et al.*[8] for this purpose. Some typical measurements from shadow zones and near the bottom were presented. In particular, some strangely distorted wave forms observed near the bottom seems to be caused by destructive interference due to reflections from the bottom, together with regular nonlinear deformation.

ACKNOWLEDGMENTS

This work was supported by the European Commission contract no G4RD-CT-2000-00398 (SOnic Boom European Research programme: numerical and laboratory-scale experimental simulation).

REFERENCES

1. A. Marifjæren, "Volum- og kompressibilitetsstudier av alkoholer og karbohydrater i vann og vann/etanol ved 25%", *Master thesis*, University of Bergen, 1985.
2. I.G. Mikhailov, "Velocity and absorption of ultra-acoustic waves in some binary liquid mixtures" *Comptes Rendus (Doklady) de l'Academie des Sciences de l'URSS*, **26** 145-146, (1940).
3. Å. Voll and H. Hobæk, "Sound velocity profile in a stratified ethanol-water mixture", SOBER project Task 2.2 - UiB Report 1, Department of Physics, University of Bergen, 2001.
4. D. R. Lide (ed.), *Handbook of Chemistry and Physics*, CRC Press, 81th edition, 2000-2001.
5. A. Pedersen, R. Fardal and H. Hobæk, "Design of sound source for shock waves in a stratified fluid - SOBER project WP2", Report for deliverable D7: Part 2, Department of Physics, University of Bergen, 2003.
6. J.B. Bowlin, J.L.Spiesberger, T.F. Duda, L.F. Freitag, "Ocean Acoustical Ray-Tracing Software RAY", *Tech. Rep. WHOI-93-10*, Woods Hole Oceanographic Institution, Woods Hole, Mass, USA, 1992.
7. H. Schmidt H, "OASES Version 3.1. User guide and Reference Manual". Department of Ocean Engineering, Massachusetts Institute of Technology, 2004.
8. E. Reiso, J. Naze Tjøtta and S. Tjøtta, "Nonlinear equations of acoustics in an inhomogeneous fluid", in *Frontiers of Nonlinear Acoustics, Proc. 12th ISNA*, ed. M.F. Hamilton and D.T. Blackstock, 1990.

"Once Nonlinear, Always Nonlinear"

David T. Blackstock

Applied Research Laboratories, The University of Texas at Austin, Austin, Texas 78713-8029, USA

Abstract.
   The phrase "Once nonlinear, always nonlinear" is attributed to David F. Pernet. In the 1970s he noticed that nonlinearly generated higher harmonic components (both tones and noise) don't decay as small signals, no matter how far the wave propagates. Despite being out of step with the then widespread notion that small-signal behavior is restored in "old age," Pernet's view is supported by the Burgers-equation solutions of the early 1960s. For a plane wave from a sinusoidally vibrating source in a thermoviscous fluid, the old-age decay of the nth harmonic is $e^{-n\alpha x}$, not $e^{-n^2\alpha x}$ (small-signal expectation), where α is the absorption coefficient at the fundamental frequency f and x is propagation distance. Moreover, for spherical waves (r the distance) the harmonic diminishes as $e^{-n\alpha x}/r^n$, not $e^{-n^2\alpha x}/r$. While not new, these results have special application to aircraft noise propagation, since the large propagation distances of interest imply old age. The virtual source model may be used to explain the "anomalous" decay rates. In old age most of the nth harmonic sound comes from virtual sources close to the receiver. Their strength is proportional to the nth power of the local fundamental amplitude, and that sets the decay law for the nth harmonic.

Keywords: nonlinear distortion, old age
PACS: 43.28.Kt

INTRODUCTION

The message in this paper is not new. The author was motivated to give the paper by an email message Kent Gee sent to him earlier this year:

1. I've heard the statement "once nonlinear, always nonlinear" attributed to you several times and I was wondering if (a) you said/wrote it, and (b) what was the context?

2. I'm working on the nonlinear propagation of jet noise using an Anderson-type code I wrote, and as a final point to the thesis, I was going to discuss at what distance the propagation becomes essentially linear. ...what I've been finding is that the difference between the linearly and nonlinearly predicted power spectra continues to grow as a function of range, especially at high frequencies (10-20 kHz). ... [This] surprises me because I would have expected [the difference] to diminish as overall amplitudes decay and the time scale distortion afforded by the Earnshaw solution becomes negligible. I can't decide if it's a numerical issue with the algorithm or if the result is believable, meaning "once nonlinear, always nonlinear."

   The answer to part (a) in the first paragraph is no. The person who coined the phrase is David Pernet, whose actual words in a letter to the author[1] were "once nonlinearity, always nonlinearity." However, the author agrees with Pernet's characterization and has on occasion used the phrase himself. The purpose of this paper is to show by analytical

CP838, *Innovations in Nonlinear Acoustics: 17th International Symposium on Nonlinear Acoustics*,
edited by A. A. Atchley, V. W. Sparrow, and R. M. Keolian
© 2006 American Institute of Physics 0-7354-0330-9/06/$23.00

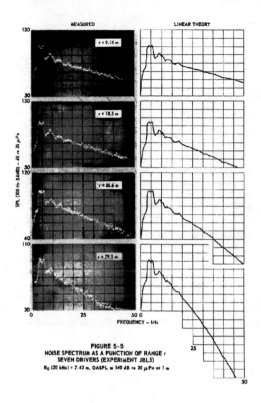

MEASURED   LINEAR THEORY

130

r = 9.14 m

50
130

r = 18.3 m

SPL (300 Hz BAND) – dB re 20 μPa

50
120

r = 36.6 m

40
110

r = 79.3 m

30
0   25   50 0
FREQUENCY – kHz

25

50

FIGURE 5-5
NOISE SPECTRUM AS A FUNCTION OF RANGE r
SEVEN DRIVERS (EXPERIMENT JBL5)
R₀ (20 kHz) = 7.42 m, OASPL = 140 dB re 20 μPa at 1 m

FIGURE 1. Measured spectra and corresponding linear-theory predictions, as a function of range, for an outdoor noise propagation experiment. Source signal was an octave band of noise centered at 4 kHz. "Source level" (overall sound pressure level, extrapolated from farfield measurement to 1 m) was 140 dB re 20 μPa. From Ref. 2.

evidence and physical explanation why the phrase is apt.

The paradox is that finite-amplitude sound, when it reaches old age, where it "ought" to behave as a small signal, still retains some of its nonlinear character. First, is it really true that small-signal theory is inadequate for old age? Second, if so, why? Gee's second paragraph gives an example from a computational study.

An example from experimental measurements is shown in Fig. 1, which is a sample from a 1978 NASA report by Webster and Blackstock[2]. They describe a series of measurements of outdoor propagation of spherically spreading high-intensity noise. In the experiment summarized by Fig. 1, the source was a seven-horn array excited by an octave band of noise centered at 4 kHz. The source level for the octave band (overall sound pressure level 1 m from the array, extrapolated from a farfield measurement) was 140 dB (re 20 μPa). All measurements were made along the axis of the array, which was pointed upward to minimize ground effects. At the closest measurement point shown, $r = 9.14$ m, the noise is seen already to have suffered considerable propagation

602

distortion. For frequencies just above the original 4-kHz octave band, the distortion is manifest as individual higher-harmonic bands. At still higher frequencies the distortion spectrum loses its band character and becomes continuous.

The "once nonlinear, always nonlinear" hypothesis may be tested by comparing the measurements in Fig. 1 with predictions based on linear theory, which are shown in the right-hand column. The 9.14-m spectrum serves as the input for the linear-theory prediction. The linear processes included are (1) spherical spreading (9.14 m is sufficiently far from the source that nearfield effects are negligible) and (2) atmospheric absorption appropriate for each spectral component. Predictions are shown for distances that match the measurement distances. While the predictions are good for the original 4-kHz octave band, they become progressively poorer (too low) as distance increases. For example, at 25 kHz the discrepancy is 0 dB at 9.14 m (as it must be), 5 dB at 18.3 m, 10 dB at 36.6 m, and 20 dB at 79.3 m. This is an accelerating trend, showing no indication that the discrepancy will ever level off and become constant. Moreover, the higher the frequency, the greater the divergence between linear theory and measured data. Growth of the divergence with range is like that found by Gee in his computational study. Note that in this case nonlinear effects are not particularly strong.

The simplistic explanation is that nonlinear propagation distortion rebuilds the high-frequency components to offset the higher atmospheric absorption they suffer. But shouldn't nonlinear distortion be so slowed by spherical spreading and ordinary absorption that eventually it disappears? In other words, shouldn't the divergence between small-signal predictions and measurements (or calculations) eventually reach a constant? The evidence is that it does not. In the next section some analytical results are presented to show that even in old age the wave appears to remember its nonlinear past.

ANALYTICAL EVIDENCE

Presented here are some known asymptotic expressions for finite-amplitude sound generated by a pure-tone source of angular frequency ω, first for plane waves and then for spherical waves. Because the fluid considered is thermoviscous, the (small-signal) absorption coefficient α is proportional to ω^2. Other absorption laws are considered in the following section.

Let the source condition for plane waves be $p(0,t) = p_{10}\sin\omega t$, where p is pressure, p_{10} is source amplitude, and t is time. The solution of Burgers' equation for this case is well known.[3] For strong waves at great distance ($\alpha x \gg 1$) the old-age (asymptotic) expression is

$$p(x,t) = \frac{4\alpha\rho_0 c_0^3}{\beta\omega} \sum_{n=1}^{\infty} e^{-n\alpha x}\sin n(\omega t - kx), \qquad (1)$$

where β is coefficient of nonlinearity, ρ_0 is static density, c_0 is small-signal sound speed, and $k = \omega/c_0$ is wave number. Besides saturation (lack of dependence of the signal on the source amplitude p_{10}), the most obvious surprise is that the harmonics decay as $e^{-n\alpha x}$, not as $e^{-n^2\alpha x}$, the latter being the decay rate that would be expected for a small signal of angular frequency $n\omega$. Although the wave has forgotten its source amplitude, it still carries remnants of its nonlinear past. One sees that the notion of "once nonlinear, always nonlinear" might have been voiced as far back as the 1950s and 1960s, when investigators were discovering the application of Burgers' equation to nonlinear acoustics. In fact, evidence for the notion is even much older: Fay's 1931 solution[4] yields Eq. (1) asymptotically. On the positive side, some investigators of the 1950s and 1960s did attempt to explain the peculiar decay rate by noting that the harmonics are generated throughout the medium, not back at the source.

The fact that the harmonics decay as $e^{-n\alpha x}$ is not limited to strong waves. The Keck-Beyer solution[5] displays the same asymptotic decay for weak waves.

For even more striking evidence, consider the case of spherical waves in a thermo-viscous medium. Although the exact solution of Burgers' equation for spherical waves is not known, solutions have been found for special cases. Let the source be a pulsating sphere of radius r_0, i.e., $p(r_0, t) = p_{10} \sin \omega t$, where r is radial distance. For weak waves the asymptotic solution ($\alpha r \gg 1$), including harmonics up through the fourth, is (see Appendix A in Ref. 2)

$$p(r,t) = p_{10} \sum_{n=1}^{4} \left(\frac{r_0}{r}\right)^n \left(\frac{\Gamma}{4}\right)^{n-1} e^{-n\alpha(r-r_0)} \sin n[\omega t - k(r-r_0)], \tag{2}$$

where $\Gamma = \beta p_{10} k / \rho_0 c_0^2 \alpha$ is the Gol'dberg number. The remarkable property displayed here is that the spreading factor for the harmonics amplitudes is $1/r^n$, not $1/r$. Despite the faster spreading loss, however, the harmonics still ultimately attenuate more slowly than would be expected for small signals. That is, $r^{-n} e^{-n\alpha(r-r_0)}$ is ultimately a weaker attenuation factor than $r^{-1} e^{-n^2 \alpha (r-r_0)}$.

It turns out that the same peculiar spreading is shown by an asymptotic solution for strong spherical waves. The following result was reported by Webster and Blackstock:[6]

$$p(r,t) = \left(\frac{4\alpha\rho_0 c_0^2}{\beta k}\right) \sum_{n=1}^{\infty} K^n \left(\frac{r_0}{r}\right)^n e^{-n\alpha(r-r_0)} \sin n[\omega t - k(r-r_0)], \tag{3}$$

where K is an undetermined constant (independent of r).

Although in striking support of "once nonlinear, always nonlinear," the solutions above are limited to fluids for which the absorption coefficient varies as ω^2. For other fluids the results can be somewhat different, as shown in the next section.

PHYSICAL EXPLANATION

Westervelt's virtual source model[7] provides one way to explain the cumulative distortion suffered by a finite-amplitude sound wave. The model may also be used to explain the apparently peculiar behavior of the harmonics in old age. The explanation is given first for plane waves.

Consider, for example, the second-harmonic component that is generated as the propagating fundamental interacts nonlinearly with itself. In effect, the interaction produces virtual sources of second-harmonic sound. Because the interaction process is quadratic, the strength of the virtual sources is proportional to the square of the local fundamental amplitude. Since the virtual sources are energized by the propagating fundamental,

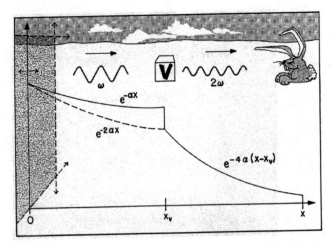

FIGURE 2. Second-harmonic sound generated in a plane wave by virtual sources.

their output is phased so that their signals add constructively in the direction the fundamental is traveling. As a result, the second-harmonic amplitude grows cumulatively with distance. The second-harmonic signal at any given point x is the sum total of all the contributions from virtual sources up to that point.

Now include the effect of absorption. For the moment do not limit the discussion to fluids for which $\alpha \propto \omega^2$. Let α_1 and α_2 be the absorption coefficients at the fundamental and second-harmonic frequencies, respectively. Let x_v be the location of a virtual source emitting second harmonic that is received at the more distant point x. First, because the amplitude of the fundamental at x_v is down by the factor $e^{-\alpha_1 x_v}$, the second-harmonic virtual source strength is proportional to $e^{-2\alpha_1 x_v}$. Second, once generated, the second harmonic decays, in traveling to the receive point x, as $e^{-\alpha_2(x-x_v)}$.

Figure 2 illustrates the process. Here a thermoviscous fluid has been assumed, i.e., $\alpha_1 = \alpha$ and $\alpha_2 = 4\alpha$. Second-harmonic sound reaches the receiver by a two-step process. First, the virtual source at x_v is produced by quadratic interaction of the fundamental with itself. Squaring the fundamental amplitude makes $e^{-2\alpha x_v}$ the appropriate decay for *generation* of the second harmonic. For *propagation* of the second harmonic, however, from x_v to x (the second step), the decay is $e^{-4\alpha(x-x_v)}$, a much faster rate. The two-step decay process discriminates against virtual sources far from the receiver. The dominant contributions come from virtual sources closest to the receiver. Since for these sources $x_v \doteq x$, the effective overall decay factor in old age is $e^{-2\alpha x}$.

The argument is easily extended to higher harmonics. Third-harmonic signals, for example, are produced by interaction of the fundamental with the second harmonic. But since the second harmonic is produced by interaction of the fundamental with itself, the third-harmonic virtual source strength is proportional to the cube of the fundamental amplitude. The appropriate decay factor associated with the third-harmonic virtual sources is therefore $e^{-3\alpha x_v}$. Again, dominance by virtual sources closest to the receiver leads to $e^{-3\alpha x}$ as the decay factor in old age.

The picture changes, however, for absorption laws for which decay associated with the generation step is less than that associated with the propagation step. For the second harmonic, for example, if α_2 is less than $2\alpha_1$, the virtual sources that contribute most to the received second-harmonic signal are those closest to the physical source ($x = 0$). The old age decay of the second harmonic is then $e^{-\alpha_2 x}$, and small-signal expectations are not overturned. However, cases of this sort are usually restricted to limited frequency ranges, such as near one of the prominent relaxation frequencies in the case of propagation in the atmosphere or the ocean. For harmonics beyond this frequency region the dominant virtual sources are again those closest to the receiver. If the plane wave is in a tube, on the other hand, absorption is due to boundary-layer dissipation, for which the absorption law is $\alpha \propto \sqrt{\omega}$. Typically, the fundamental is low enough that the dominant virtual sources for a great many harmonics are those closest to the physical source. Finally, for harmonics high enough in frequency that free-medium absorption takes over, the dominant virtual sources are again those closest to the receiver.

For spherical waves the spreading loss factor may also be understood in terms of virtual source strength. In the case of the second harmonic, for example, squaring the fundamental amplitude produces a source strength proportional to $r^{-2}e^{-2\alpha(r-r_0)}$. Similar arguments apply to the higher harmonics, and the peculiar spreading loss factors seen in Eqs. 2 and 3 follow.

CONCLUSION

The defense given here for "once nonlinear, always nonlinear" is not new, just not well known. See, for example, Refs. 6 and 8. Hamilton (see Chapter 8 in Ref. 8) has also applied the virtual-source explanation to the attenuation factor for sound beams.

In summary, Pernet's point is well taken. For many years the phenomenon of saturation for strong waves led investigators to focus on forgetfulness, that is, the wave forgets some properties of its source. We see, however, that remembrance is also powerful: the wave does not forget its nonlinear past.

REFERENCES

1. Unpublished letter, David F. Pernet to David T. Blackstock, 14 December 1977. Pernet undoubtedly arrived at the concept much earlier.
2. D. A. Webster and D. T. Blackstock, "Experimental Investigation of Outdoor Propagation of Finite-Amplitude Noise," NASA Contractor Report 2992, August 1978 (N78-31876).
3. See, for example, D. T. Blackstock, "Thermoviscous attenuation of plane, periodic, finite-amplitude sound waves," J. Acoust. Soc. Am. 36, 534-542 (1964), Eq. (31).
4. R. D. Fay, "Plane sound waves of finite amplitude," J. Acoust. Soc. Am. 3, 222-241 (1931).
5. W. Keck and R. T. Beyer, "Frequency spectrum of finite amplitude ultrasonic waves in liquids," Phys. Fluids 3, 346-352 (1960).
6. D. A. Webster and D. T. Blackstock, "Asymptotic decay of periodic spherical waves in dissipative media," J. Acoust. Soc. Am. 64, S33 (1978).
7. P. J. Westervelt, "Parametric end-fire array," J. Acoust. Soc. Am. 32, 934-935 (1960(A)). See also the full paper, P. J. Westervelt, "Parametric acoustic array," J. Acoust. Soc. Am. 35, 535-537 (1963).
8. M. F. Hamilton and D. T. Blackstock, eds., Nonlinear Acoustics (Academic Press, New York, 1998), pp. 135-136.

Computational Solution of Nonlinear Tricomi Equation for Sonic Boom Focusing and Applications

Osama Kandil[1] and Xudong Zheng[2]

Aerospace Engineering Department, Old Dominion University, Norfolk, VA 23529

Abstract. The sonic-boom focusing problem has been solved using the nonlinear Tricomi equation. A pseudo time term has been added to the equation so that marching in pseudo time will be possible. The solution is obtained by splitting the nonlinear unsteady equation into two parts: a part corresponding to a linear unsteady Tricomi equation, and another part corresponding to an unsteady nonlinear Burgers equation. The solution of the linear unsteady Tricomi equation is followed by the solution of the nonlinear unsteady Burgers equation to obtain the solution of the total nonlinear equation. The solution of the linear unsteady Tricomi equation has been accomplished using two schemes. The first is a frequency-domain (FD) spectral scheme and the second is a time-domain (TD) finite-differencing scheme. The solution of the nonlinear unsteady Burgers equation has also been accomplished using two schemes. The first is a finite-difference scheme and the second is an analytical scheme. Thus, there are four combinations to obtain the solution of the nonlinear Tricomi equation.

Keywords: Nonlinear Tricomi equation, Sonic boom focusing, Superboom
PACS: 43.25. +y, 43.28.Mw, 46.15.-x, 47.11. +j,47.15.Hg,47.40.Ki,47.40.Nm

Introduction

The most intense sonic boom is the focused sonic boom due to aircraft transonic acceleration from Mach 1 to cruise speed and due to aircraft maneuverings. It leads to amplification of ground pressures up to two or three times the carpet boom shock strength. Therefore, accurate prediction of focused sonic booms around the caustic at ground level is very important. Sonic boom focusing has been also known as sonic superboom. Focusing of shock waves occurs at surfaces called caustics. Shock wave focusing is fundamentally a nonlinear process and can be predicted using the solution of the nonlinear Tricomi equation. Analysis of weak shock focusing at a smooth caustic surface has been introduced in 1965 by Guiraud [1]. He developed a theory, which includes both diffraction and nonlinear effects up to first order, which leads to the nonlinear Tricomi equation. This result was confirmed by Hayes [2], Hunter and Keller [3], and Rosales and Tabak [4, 5]. Augar and Coulouvrat [6] have presented an algorithm using the Fast-Fourier Transform (FFT) to solve the nonlinear Tricomi equation, which was expressed in terms of the dimensionless acoustic pressure. Recently, Marchiano and Coulouvrat [7] have solved the nonlinear Tricomi equation using the FFT also, but in terms of the potential field instead of the pressure field. The nonlinear effects were treated using an "exact" solver[12]. In this paper, the nonlinear Tricomi equation is solved using a frequency-domain (FD) scheme and a time-domain (TD) scheme using a type differencing scheme and overlapping grid (OLG) scheme as well.

CP838, *Innovations in Nonlinear Acoustics: 17th International Symposium on Nonlinear Acoustics*,
edited by A. A. Atchley, V. W. Sparrow, and R. M. Keolian
© 2006 American Institute of Physics 0-7354-0330-9/06/$23.00

Nonlinear Tricomi Equation and Boundary Conditions

The sonic-boom focusing problem has been shown to be governed by the nonlinear Tricomi equation. The Tricomi equation changes its character from a hyperbolic PDE (in illumination zone) to an elliptic PDE (in the shadow zone). At the caustic surface between the illumination zone and shadow zone, the equation is parabolic. The computational solution of this equation is obtained by marching in pseudo time. This is achieved by adding a pseudo time term to the equation. This term tends to zero when the pseudo time marching scheme reaches the steady solution of $\phi (t \rightarrow \infty, \tau, z)$. The modified equation is given by

$$\frac{\partial^2 \phi}{\partial \tau \, \partial t} = \frac{\partial^2 \phi}{\partial z^2} - z \frac{\partial^2 \phi}{\partial^2 \tau} + \frac{\mu}{2} \frac{\partial}{\partial \tau} \left[(\frac{\partial \phi}{\partial \tau})^2 \right] \qquad (1)$$

Where ϕ is acoustical potential, t pseudo time variable, τ dimensionless phase variable, z dimensionless normal distance to the caustic, and μ a measure of nonlinear effects relative to diffraction effects. The boundary conditions to be satisfied are:
1. no disturbance before or after acoustic waved has passed
$$\phi (z, \tau \rightarrow \pm\infty) = 0 \qquad (2)$$
2. away from the caustic surface in the shadow zone the acoustic pressure decreases exponentially:
$$\phi (z \rightarrow -\infty, \tau) \rightarrow 0 \qquad (3)$$
3. a radiation condition is imposed at the top of the domain
$$z^{\frac{1}{4}} \frac{\partial \phi}{\partial \tau} + z^{-\frac{1}{4}} \frac{\partial \phi}{\partial z} \rightarrow 2F(\tau + \frac{2}{3} z^{\frac{3}{2}}) \qquad (4)$$

The unsteady nonlinear equation is split into two simpler equations. The first one includes the linear diffraction effects and the second one includes the nonlinear effects. The first equation is the unsteady linear Tricomi equation, which is solved using a TD scheme [8, 9] or a FD scheme [6]. The second equation is the inviscid nonlinear Burgers' equation, which is solved using a shock-capturing finite-differencing scheme [10] or by an exact shock fitting scheme [11, 12]. Therefore, four methods are developed to solve the two equations. At each time step, the total computational cycle is iterated between the linear Tricomi equation and the nonlinear Burgers equation until convergence is reached. Next, the time step is advanced and the computational cycle is repeated.

Computational Applications

First, the numerical schemes are tested for validation. The test cases are those of Ref. 6, with an incoming N-wave case or an incoming Concorde aircraft wave case, with $\mu = 0.08$. Another four tests are solved, which include symmetric and asymmetric flat-top wave cases and symmetric and asymmetric ramp-top wave case are carried out. The validation study shows very high confidence in the computational schemes. Figure 1 shows sample cases for the incoming wave; an N-wave and a Concorde aircraft wave.

Computational domain and grid

We choose a rectangular domain with $z \in [-2., 2.]$ and $\tau \in [-2.67, 3.67]$ dimensionless units. The number of grid points in the z direction is 1000 and in the τ direction is 8,000 points on using the TD scheme, or 2048 frequencies on using the FD scheme. It should be noticed that the higher the frequencies are or the points in τ the better the solutions are.

The F- function of Eq. (4) on the upper boundary z = 2.0 is used as an incoming N-wave (see Fig 1.a), which extends from $\tau = -1.386$ to $\tau = -2.386$ (duration of 1) with $p_{max} = 1.0$ and $p_{min} = -1$ (τ and p are dimensionless). With these dimensionless pressure and duration, $p = 1$ is equivalent to 2.25 psf and $\tau = 1$ is equivalent to 230 ms. The dimensionless time step for the pseudo time integration is taken as 0.001. The case has been run for 30,000 time steps until the total error of the pseudo unsteady time term was reduced to 10^{-5}.

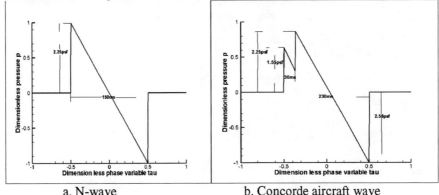

a. N-wave   b. Concorde aircraft wave

Fig. 1: Different shapes of incoming sonic boom signatures at z = 2.

Results and Discussion

Using the FD scheme, Fig. 2 shows the pressure contours of the incoming wave as it progresses toward the caustic surface, and the outgoing wave as it originates from the caustic surface. The caustic surface is also shown. Figure 3 shows p_{max} of 2.82, which is equivalent to 6.345 psf at z = 0.162. It is conclusively shown that the superboom response is predicted. It is also consistent with the results of Ref. 6. Using the TD scheme, Fig. 4 shows the pressure contours of incoming and outgoing waves. Figure 5 shows p_{max} of 2.89, which is equivalent to 6.503 psf at z = 0.15. It is observed that the TD scheme predicts a slightly higher p_{max} (2.5%) than that of the FD scheme. Figures 6 and 7 shows comparisons of the FD, TD and OLG solutions at z = 2 for an N-wave and a Concorde wave. It is observed that the width of outgoing wave of the FD solution is slightly larger than that of the TD and OLG solutions.

Conclusions

The FD, TD and OLG schemes are developed to solve the nonlinear Tricomi equation for sonic boom focusing. The three schemes produce consistent and almost matching results. The TD and OLG results are in excellent agreement, while the FD results show a slight difference.

Acknowledgement

The authors would like to acknowledge the support of Dr. Peter Coens of NASA LaRC .

References

1. Guiraud, J.-P., *J. Mecanique* **4**, pp 215-267 (1965).
2. Hayes, W. D., *Proceedings of the Sound Conference on Sonic Boom Research*, NASA **SP-180**, pp 165-171 (1968)
3. Hunter, J. K., *Wave Motion* **9**, pp 429-443 (1987)
4. Rosales, R. R. and Tabak, E. G., *Phys. Fluids* **10**, pp 206-222 (1997)
5. Tabak, E. G. and Rosales, R. R., *Phys. Fluids* **6**, pp 1874-1892(1994)
6. Augar, T. and Coulouvrat F., *AIAA J., Vol. 40*, **9**, pp 1726-1734(2002)
7. Marchiano, R. and Coulouvrat, F., *J. Acoust. Soc. Am.* **114**, pp 1758-1771 (2003)

8.   Kandil, O. A. and Zheng, X., *AIAA Paper 05-6335* (2005)
9.   Lee, Y. S. and Hamilton, M. F., *J. Acoust. Soc. Am.* **97**, pp 906-917 (1995)
10.   McDonald, B. E. and Ambrosiano, J., *J. Comput. Phys.* 56, pp 448-460 (1984)
11.   Hayes, W. D., Haefeli, R. C., and Kulsrud, H. E., *NASA CR*-1299.
12.   Gill, P. M. and Seebass, A. R., *AIAA Paper 73-103* (1973)

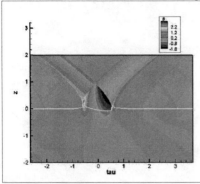

Fig. 2: Pressure contours for incoming N-wave ($\mu = 0.08$), FD.

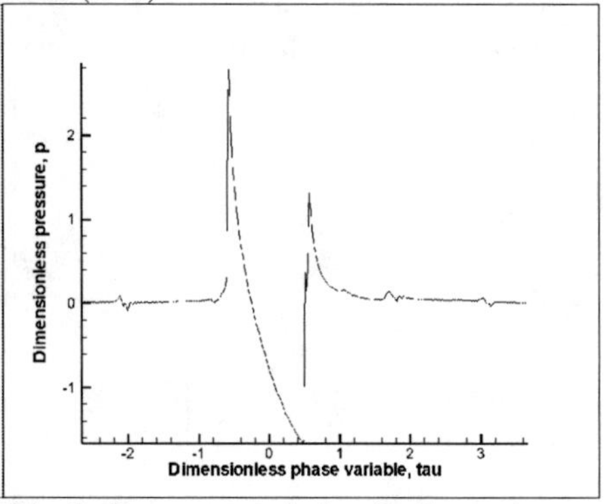

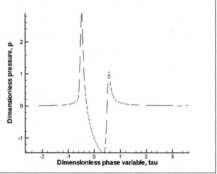

Fig. 3: Pressure variation for incoming N-wave at z of maximum pressure, FD.

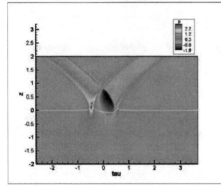

Fig. 4: Pressure contours for incoming N-wave ($\mu = 0.08$), TD.

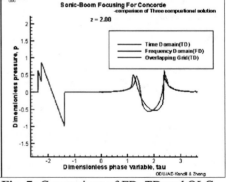

Fig. 5: Pressure variation for incoming N-wave at z of maximum pressure, TD.

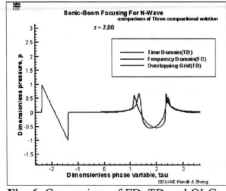

Fig. 6: Comparison of FD, TD and OLG for N-wave at z = 2.

Fig. 7: Comparison of FD, TD and OLG for Concorde-wave at z = 2.

Calculation of the front part of the sonic boom signature for a maneuvering aerofoil

S. Baskar[*] and P. Prasad[†]

[*]*Laboratoire de Modélisation en Mécanique, Université Pierre et Marie Curie & CNRS (UMR 7607), 4 place Jussieu, 75252 Paris cedex 05, France*
[†] *Department of Mathematics, Indian Insitute of Science, Bangalore - 560 012. India.*
email: prasad@math.iisc.ernet.in

Abstract. The sonic boom at a large distance from its source consists of a leading shock, a trailing shock and a one parameter family of nonlinear wavefronts in between these shocks. A new ray theoretical method using a shock ray theory and a weakly nonlinear ray theory has been used to obtain the shock fronts and wavefronts respectively, for a maneuvering aerofoil in a homogeneous medium. This method introduces a one parameter family of Cauchy problems to calculate the shock and wave fronts emerging from the surface of the aerofoil. These problems are solved numerically to obtain the leading shock front and the nonlinear wavefronts emerging from the front portion of the aerofoil.

Keywords: sonic boom, shock and wave fronts, ray theory
PACS: 43.25.Cb, 43.25.Jh

INTRODUCTION

Sonic boom had been anticipated in the early 50's of the 20th century even in early stages of supersonic flights of aircrafts. Developments in the fundamental theory of sonic boom and the implementation to practical models took almost two decades but this field of research is still active (Plotkin, 2002). When the propagation of the sonic boom is traced using linear ray theory, the rays tends to converge and form a caustic, where the amplitude becomes infinite. Experimental (Wanner *et al.*, 1972) and theoretical (Marchiano *et al.*, 2003) studies shows that the nonlinear and diffraction effects at the caustic reduces the focus intensity and keeps the amplitude finite.

We use a shock ray theory and a weakly nonlinear ray theory (see Prasad, 2001, Baskar and Prasad, 2005b) to formulate a new method involving a one parameter family of Cauchy problems for a system of conservation laws to calculate the leading and trailing shock fronts, and the nonlinear wavefronts emerging from the surface of the aerofoil. Unlike some of the earlier methods based on Whitham (1974), the nonlinear effects are incorporated from very beginning. We show that the governing systems are hyperbolic for the front part of the aerofoil and are elliptic for the trailing part. We compute the leading shock front and the nonlinear wavefronts emerging from the front portion of the aerofoil, each one of which are governed by an hyperbolic system of equations.

CP838, *Innovations in Nonlinear Acoustics: 17th International Symposium on Nonlinear Acoustics*,
edited by A. A. Atchley, V. W. Sparrow, and R. M. Keolian
© 2006 American Institute of Physics 0-7354-0330-9/06/$23.00

FORMULATION OF THE PROBLEM

Consider a two dimensional unsteady flow produced by a thin maneuvering aerofoil moving with a supersonic velocity along a curved path. We are interested in calculating the sonic boom produced by the aerofoil, the point of observation being far away say at a distance L, from the aerofoil. We use coordinates x, y and time t nondimensionalized by L and the sound velocity a_0 in the ambient medium. In a local rectangular coordinate system (x', y') with origin O' at the nose of the aerofoil and $O'x'$ axis tangential to the path of the nose, which moves along $(X_0(t), Y_0(t))$, let the upper and lower surfaces of the aerofoil be given by $(x' = \zeta, y' = b_u(\zeta))$ and $(x' = \zeta, y' = b_l(\zeta)), -d < \zeta < 0$ respectively. Here d is the nondimensional cumber length. We assume that $b_u'(-d) > 0$, $b_u'(0) < 0$, $b_l'(-d) < 0$ and $b_l'(0) > 0$, so that the nose and the tail of the aerofoil are not blunt. We further assume that $d = \bar{d}/L = O(\varepsilon)$ and $O\{\max_{-d<\zeta<0} b_u(\zeta)/d\} = O\{\max_{-d<\zeta<0}(-b_l(\zeta))/d\} = O(\varepsilon)$ where ε is a small positive number. Then the amplitude w of the perturbation in the sonic boom also satisfies $w = O(\varepsilon)$.

The sonic boom produced either by the upper or the lower surface consists of a leading shock LS: $\Omega_t^{(0)}$, a trailing shock TS: $\Omega_t^{(-d)}$ and since high frequency approximation is satisfied by the flow between the two shocks, a one parameter family of nonlinear wavefronts $\Omega_t^{(\zeta)}(-d < \zeta < 0, \zeta \neq G)$ originating from the points P_ζ on the aerofoil in between the leading and the trailing ends.

The evolution of Ω_t in the ray coordinate system (ξ, t) is governed by a pair of *Kinematical conservation laws* (KCL) (Prasad 2001)

$$(g\sin\theta)_t + (m\cos\theta)_\xi = 0, \ (g\cos\theta)_t - (m\sin\theta)_\xi = 0 \tag{1}$$

where $n = (\cos\theta, \sin\theta)$ is the normal to Ω_t, $g d\xi$ is an element of length along Ω_t and m is the Mach number of Ω_t. When Ω_t is a wavefront (weakly nonlinear ray theory), we have in addition to (1)

$$(g(m-1)^2 e^{2(m-1)})_t = 0, \ m = 1 + w(\gamma+1)/2. \tag{2}$$

When Ω_t is a shock front, we denote the corresponding variable by $M, N = (\cos\Theta, \sin\Theta)$ and G, and we have (1) and two more equations (Shock ray theory) (Baskar and Prasad, 2005a)

$$(G(M-1)^2 e^{2(M-1)})_t + 2M(M-1)^2 e^{2(M-1)} GV = 0 \tag{3}$$

$$(GV^2 e^{2(M-1)})_t + GV^3 (M+1) e^{2(M-1)} = 0 \tag{4}$$

where M and V are defined as

$$M = 1 + w|_s(\gamma+1)/4, \ V = ((\gamma+1)/4)\{\langle N, \nabla \rangle w\}|_s. \tag{5}$$

Here the normal derivative $\langle N, \nabla \rangle w$ is first obtained in the region behind the shock if the shock is moving into the undisturbed region and in the region ahead of the shock if it is moving into the disturbed region and then the limit is taken as we approach the shock. This limiting value is indicated by $|_s$. The mapping from (ξ, t)-plane to (x, y)-plane can be obtained by integrating the first part of the shock ray equations

$$x_t = M\cos\Theta, y_t = M\sin\Theta. \tag{6}$$

Let $(X(t), Y(t))$ be the parametric representation of the flight path. We introduce a ray coordinate system $(\xi = -\eta, \eta \leq t)$ and $(\xi = \eta, \eta \leq t)$ for the fronts emerging from the upper and the lower surface of the aerofoil respectively. The base point P_ζ of $\Omega_t^{(\zeta)}$ on the upper surface of the aerofoil corresponds to a point on the line $\xi + t = 0$ in the (ξ, t)-plane. The Cauchy data for the nonlinear wavefronts (system (1)-(2)) on this line is given by (Baskar and Prasad, 2005b)

$$m_0(\xi) := 1 - (\gamma + 1)(\dot{X}_0^2 + \dot{Y}_0^2)b_u'(\zeta)/(2(\dot{X}_0^2 + \dot{Y}_0^2 - 1)^{\frac{1}{2}}),$$

$$g_0(\xi) := (\dot{X}_0^2 + \dot{Y}_0^2 - 1)^{\frac{1}{2}}, \quad \theta_0(\xi) := \frac{\pi}{2} + \psi - \sin^{-1}\{1/(\dot{X}_0^2 + \dot{Y}_0^2)^{\frac{1}{2}}\}, \qquad (7)$$

where $\psi = \tan^{-1}\{\dot{Y}_0/\dot{X}_0\}$. Since $b_u'(\zeta) < 0$ and $b_u'(\zeta) > 0$ for $G < \zeta < 1$ and $-d < \zeta < G$ respectively, $m_0 > 1$ on P_ζ for $G < \zeta < 1$ and $m_0 < 1$ on P_ζ for $-d < \zeta < G$. This can be used to argue that $m > 1$ on $\Omega^{(\zeta)}, G < \zeta < 0$ and $m < 1$ on $\Omega^{(\zeta)}, -d < \zeta < G$. Since the eigenvalues of the system (1)-(2) are $\lambda_\pm = \pm((m-1)/(2g^2))^{1/2}$ and $\lambda_2 = 0$, the Cauchy problems for $\Omega_t^{(\zeta)}, G < \zeta < 0$ are hyperbolic and for $\Omega_t^{(\zeta)}, -d < \zeta < G$ are elliptic (we call it elliptic even though $\lambda_2 = 0$ is real). The same is true for LS and TS. The Cauchy data for the LS and TS ((1), (3)-(5)) are given by

$$M_0(\xi) := 1 - (\gamma + 1)(\dot{X}_0^2 + \dot{Y}_0^2)b_u'(\zeta)/(4(\dot{X}_0^2 + \dot{Y}_0^2 - 1)^{\frac{1}{2}}),$$

$$V_0(\xi) := (\gamma + 1)\{\Omega_{P_{(-d)}}w_0(\xi) - \mathscr{F}(-d, t)\}/4$$

$$\Omega_{P_{(-d)}} = \frac{(\dot{X}_0\ddot{X}_0 + \dot{Y}_0\ddot{Y}_0)}{2g(\dot{X}_0^2 + \dot{Y}_0^2)(\dot{X}_0^2 + \dot{Y}_0^2 - 1)^{1/2}} + \frac{\dot{X}_0\ddot{Y}_0 - \dot{Y}_0\ddot{X}_0}{g\dot{X}_0^2}$$

$$\mathscr{F}(\zeta, t) = \frac{(\dot{X}_0^2 + \dot{Y}_0^2)b_u''(\zeta)}{(\dot{X}_0^2 + \dot{Y}_0^2 - 1)^{1/2}}\{\ddot{\mathscr{X}}_0(t)\} - \frac{(\dot{X}_0^2 + \dot{Y}_0^2 - 2)(\dot{X}_0\ddot{X}_0 + \dot{Y}\ddot{Y}_0)}{(\dot{X}_0^2 + \dot{Y}_0^2 - 1)^{3/2}}b_u'(\zeta)$$

$$\mathscr{X}_0 = X_0 \cos\psi + Y_0 \sin\psi$$

G_0 and Θ_0 are as in (7). The Cauchy data for the lower surface are difined similarly.

NUMERICAL SIMULATION

We consider the flight path to be of two types namely, 1. a straight path where the aerofoil accelerates in the x-direction and 2. a path concave downwards. The leading nonlinear wavefronts emerging from the upper and the lower surface of the aerofoil over concave downward path are shown in FIG. 1a for various times. The nonlinear wavefronts produced from the points on the front portion of the aerofoil interacts with the LS and dissapears from the flow as shown in FIG. 1b and 2. The same should happen for the trailing part of the aerofoil. These two sets, one interacting with LS and another interacting with TS are seperated by a linear wavefront $\Omega_t^{(G)}$, which originates from a point P_G where the function $b_u(\zeta)$ $(b_l(\zeta))$ are maximum (minimum). FIG 1c shows the sonic boom wavefront (concave downward path) at $t = 5$ (caustic region) calculated from the present method and the linear theory. The linear wavefront folds whereas the nonlinear wavefront develops two kinks (called as shock-shock by Whitham, 1974), which are the images of the shocks in the ray coordinate under the transformation (6). Such kinks are also observed by Inoue *et al.* (1997) in their Euler solution for an accelerating projectile (reproduced in FIG. 3) and also some general maneuvering projectile similar to the concave downwards path. The kinks in FIG. 3 are denoted by IPE and IPS. The governing systems for the present method in $-d \leq \zeta < G$ is elliptic in nature and so we expect a geometry for the TS and the nonlinear wave fronts to be free from kinks in this region which are also observed by Inoue *et al.* (1997) (reproduced in FIG. 3). The computation in the elliptic region will be pursued separately.

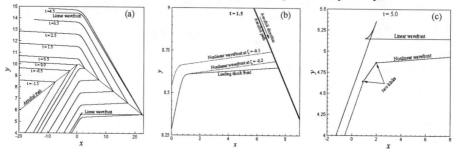

FIGURE 1. Aerofoil maneuvering over a concave downward path (a) nonlinear wavefront from the upper and the lower surface of the aerofoil (b) nonlinear wavefronts interacts with LS and disappears from the fbw (c) comparison between the linear and the nonlinear wavefronts

(d) t=2.601

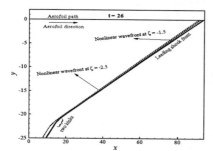

FIGURE 2: The aerofoil accelerates over a straight path. Nonlinear wavefronts interacts with LS and disappears from the fbw

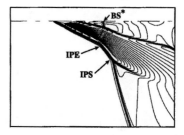

FIGURE 3: Euler solution of an accelerating projectile over a straight path (Reprinted from Inoue *et al.*, 1997, with permission from Elsevier)

REFERENCES

1. K. J. Plotkin, *J. Acoust. Soc. Am.*, **111**, 530–536 (2002).
2. J. C. L. Wanner, J. Vallee, C. Vivier, and C. Thery, *J. Acoust. Soc. Am.*, **52** (1972).
3. R. Marchiano, F. Coulouvrat, and R. Grenon, *J. Acoust. Soc. Am.*, **114**, 1758–1771 (2003).
4. P. Prasad, *Nonlinear Hyperbolic Waves in Multi-dimensions*, Monographs and Surveys in Pure and Applied Mathematics 121, Chapman and Hall/CRC, London, 2001.
5. S. Baskar, and P. Prasad, *Preprint No. 2/2005, Dept. of App. Math., Indian institute of Science* (2005b).
6. G. B. Whitham, *Linear and nonlinear waves*, John Wiley and Sons, New York, 1974.
7. S. Baskar, and P. Prasad, *J. Fluid Mech.,*, **523**, 171–198 (2005a).
8. O. Inoue, T. Sakai, and M. Nishida, *Fluid Dyn. Res.*, **21**, 403–416 (1997).

Low-Amplitude Sonic Boom From a Descending Sounding Rocket

Kenneth J. Plotkin*, Edward A. Haering, Jr.† and James E. Murray†

*Wyle Laboratories, 2001 Jefferson Davis Highway, Arlington, VA 22202, USA
†NASA Dryden Flight Research Center, Edwards, CA 93523, USA

Abstract. The target sonic boom shock strength specified in the DARPA Quiet Supersonic Platform project was 0.3 psf. The potential acceptability of that overpressure is based to a large degree on the expected relaxation-driven shock structure. It has previously been difficult to obtain sonic booms in that range from aircraft, so there had been no flight test data verifying the shock structure of such minimized booms. Low amplitudes do occur, but they are generally near the edge of the boom carpet, are distorted by ground effects, and do not represent the kind of primary boom that is expected from a low-boom aircraft. Recent flights of sounding rockets have generated low overpressure booms, providing an opportunity to examine weak boom shock structures as they occur under real atmospheric propagation. One extraordinarily clean 0.2 psf N-wave provided a shock structure which is quantitatively compared with relaxation theory.

Keywords: Sonic boom, shock wave, molecular relaxation.
PACS: 43.28.Lj, 43.28.Mw, 43.28.Rq

INTRODUCTION

The DARPA Quiet Supersonic Platform (QSP) program [1] established a goal of designing a 100,000 pound supersonic aircraft with a shaped boom [2] with maximum bow shock strength of 0.3 psf. That goal was based on human response data [3], and on the analysis by Pierce [4] that showed shock waves below about 0.4 psf would have rise times longer than might be expected from extrapolation of stronger shocks and hence would be particularly quiet. The relaxation theory upon which this is based is sufficiently well established so as to be standardized [5], has been successfully applied to sonic boom calculations [6], and agrees well with available sonic boom flight test data [7].

Shock waves in available sonic boom flight test data tend, however, to be in the 1 to 2 psf range. Shock waves below 0.4 psf are rare, and when generated by conventional aircraft tend to occur near lateral cutoff, where they are substantially distorted by turbulence and ground interference effects

In 2004, a series of sounding rocket tests (unrelated to sonic boom research) was conducted. Pre-flight analysis indicated that low-amplitude N-wave booms would occur under the reentry flight track. Accordingly, several of NASA Dryden Flight Research Center's Boom Amplitude and Shape Sensor (BASS) sonic boom recorders were deployed within the expected footprint. A 0.2 psf N-wave sonic boom was recorded. The boom is shown in Figure 1. Personnel operating the BASS array heard

CP838, *Innovations in Nonlinear Acoustics: 17th International Symposium on Nonlinear Acoustics*,
edited by A. A. Atchley, V. W. Sparrow, and R. M. Keolian
© 2006 American Institute of Physics 0-7354-0330-9/06/$23.00

the boom, but felt that they would not have noticed it had they been engaged in an unrelated activity.

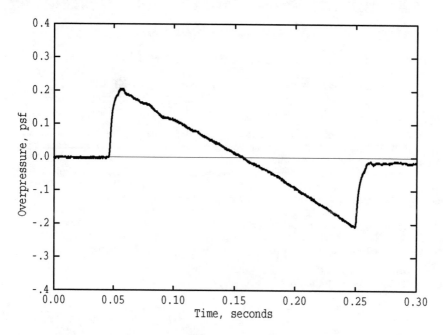

FIGURE 1. Measured 0.2 psf N-Wave Sonic Boom

Note that the front and rear shock waves have the same shape as each other, and that shape qualitatively appears similar to that of computed relaxation shocks presented in References 6 and 7. Detailed comparison of the measured shock structure with a theoretical relaxation shock is in order.

RELAXATION SHOCK STRUCTURE

The structure of a weak shock, as occurs in sonic booms, is governed by the Burgers equation. For classical absorption, the Burgers equation is written:

$$\frac{\partial P}{\partial t} + \frac{\gamma+1}{2\gamma} \frac{a_\infty}{p_\infty} P \frac{\partial P}{\partial \xi} = \frac{2}{3} m \frac{\partial^2 P}{\partial \xi^2} \tag{1}$$

where

$P(\xi,t)$ = pressure signature
ξ = x-a$_\infty$t = wave-fixed coordinate
x = length coordinate in propagation direction
t = time
a_∞ = ambient sound speed

616

p_∞ = ambient pressure
γ = ratio of specific heats
η = kinematic viscosity

Equation (1) is written in a frame moving with the wave. The wave is moving in the positive x direction, and also evolves with time. Note that t is involved in both parameters of P. As part of ξ, it represents a coordinate along the waveform. By itself, as the second parameter, it represents the time the wave has propagated. Care must be taken to not confuse the two roles of t.

The second derivative form of the term on the right is equivalent to absorption proportional to frequency squared. In the frequency domain, it can be written as $\left[bf^2 P^T\right]^{-T}$, where b is a constant, f is frequency, $[\ ]^T$ is the Fourier transform, and $[\ ]^{-T}$ is the inverse transform.

Absorption in air includes the classical absorption in Equation (1), but that is small compared to absorption due to molecular relaxation. Relaxation effects are more complex than simple f^2 behavior, but are well defined by current standards [5]. Equation (1) can be generalized to

$$\frac{\partial P}{\partial t} + \frac{\gamma+1}{2\gamma}\frac{a_\infty}{p_\infty}P\frac{\partial P}{\partial \xi} = \left[\alpha(f)P^T\right]^{-T} \tag{2}$$

where $\alpha(f)$ is frequency-dependent air absorption. The units of α must be consistent with Equation (2); as written, α should be in units of nepers per second. It should be noted that α can be complex, with the real part (generally negative) being attenuation and the complex part being dispersion.

Equation (2) may be written in a simple time difference form:

$$\Delta P = -\frac{\gamma+1}{2\gamma}\frac{a_\infty}{p_\infty}P\frac{\partial P}{\partial \xi}\cdot \Delta t + \left[\alpha(f)\cdot \Delta t \cdot P^T\right]^{-T} \tag{3}$$

where Δt is a finite time step. Evaluating Equation (3) at time t_i , P at time $t_{i+1} = t_i + \Delta t$ is $P_{i+1} = P_i + \Delta P$. Note that Equation (3) is formulated as a simple chain rule sum of two components, and avoids the (unwarranted) alternation criticism sometimes directed at the established Pestorius [8] and Anderson [9] algorithms.

RESULTS

Equation (3) was computationally implemented for a plane wave step function shock in a uniform atmosphere. Considering the high altitude of the source, modeling the boom as a plane wave is very reasonable. Temperature and humidity were taken as 20° C and 30%, the average conditions below 10,000 ft above mean sea level at the time of the boom measurement. An initial zero-thickness waveform was evolved for a 100 second propagation time, at the end of which it was asymptoting steady state.

The results are shown in Figure 2. This is the measured bow shock from the boom shown in Figure 1, with the scale expanded and the computed shock overlaid. The agreement is excellent. The expectation that relaxation shock theory applies to weak sonic booms in the real atmosphere is thus confirmed.

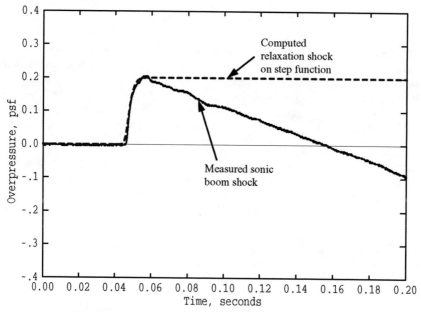

FIGURE 2. Comparison of Measured Shock With Computed Shock

REFERENCES

1. Wlezien, R., and Veitch, L., "Quiet Supersonic Platform Program," AIAA 2002-0143, Jan. 2002.
2. George. A.R., and Seebass, R, "Sonic Boom Minimization Including Both Front and Rear Shocks", *AIAA Journal*, **9** (10), 2091-2093, October 1971
3. Leatherwood, J.D., and Sullivan, B.M., "Laboratory Study of Effects of Sonic Boom Shaping on Subjective Loudness and Acceptability," NASA Technical Paper 3269, 1992
4. Pierce, A. D., "Weaker Sonic Booms May be Considerably More Quiet," presented at 118th Meeting of the Acoustical Society of America, Nov. 1989.
5. "Method for the Calculation of the Absorption of Sound by the Atmosphere," ANSI S1.26-1978, American National Standards Institute.
6. Chambers, James P., Cleveland, Robin, O., Bass, Henry E., Blackstock, David T., and Hamilton, Mark F., "Comparison of Computer Codes for the Propagation of Sonic Booms Through Realistic Atmospheres Utilizing Actual Acoustic Signatures," NASA Conference Publication 3335, pp. 151-175, July 1996.
7. Kang, J., "Nonlinear Acoustic Propagation of Shock Waves Through the Atmosphere with Molecular Relaxation," Ph.D. thesis, The Pennsylvania State University, May 1991.
8. Pestorius, F.M., "Propagation of Plane Acoustic Noise of Finite Amplitude," Technical Report ARL-TR-73-23, Applied Research Laboratories, The University of Texas at Austin, August 1973
9. Anderson, M.O., "The Propagation of a Spherical N-Wave in an Absorbing Medium and its Diffraction by a Circular Aperture," Technical Report ARL-74-25, Applied Research Laboratories, The University of Texas at Austin, August 1974

Absorption of sonic boom by clouds

Michaël Baudoin*,†, François Coulouvrat* and Jean-Louis Thomas†

*Laboratoire de Modélisation en Mécanique, UMR CNRS 7607, Université Pierre et Marie Curie,
4 place jussieu, 75252 Paris Cedex 05, France
†Institut des Nanosciences de Paris, UMR CNRS 7588, Université Pierre et Marie Curie, 140 rue
de Lourmel, 75015 Paris, France

Abstract. Atmospheric absorption may have a huge impact on the sonic boom annoyance by reducing the amplitude and increasing the rise time. However, the standard absorption due to the vibrational relaxation of molecular nitrogen and oxygen does not estimate the influence of clouds where scattering by water droplets occurs. As clouds cover more than 50 % of the Earth surface, their influence on sonic boom needs estimation. Test flights performed in the former Soviet Union in 1967-1968 indicate a strong impact. An existing model for acoustical propagation in a polydispersed air-vapor-droplet suspension [1] is reviewed. It takes into account energy and viscous momentum transfer as well as evaporation processes influenced by diffusion of vapor through the air. When applied to the conditions prevailing in atmospherical clouds, it shows a dramatical increase of sound attenuation and dispersion.

Keywords: sonic boom, cloud, scattering, nonlinear acoustics
PACS: 43.20.F, 43.25.C, 43.25.J

INTRODUCTION

Attenuation of acoustic waves by fogs has been first discussed by Sewell in 1910. In the early treatments [2] [3] of this problem, only stationary momentum transfer between phases was taken into account. Then, mass exchange and non-stationnary phenomena including heat and momentum transfer have been progressively introduced. A recently published paper [1] investigates the influence of the polydispsersion of droplets on the attenuation and dispersion in a two-fluids model which gathers most of the known damping mechanisms.

The present work aims at giving a first evaluation of such phenomena in the case of the sonic boom propagation in clouds. As shock waves cover a broad frequency spectrum, non-stationary phenomena and mass exchange cannot be neglected and last results in this area are thus of interest. A model based on Gubaidullin's approach, adapted for this study and that includes also nonlinear effects due to the high amplitude of the propagating wave, has been reviewed. Then we investigate the main parameters that drive the attenuation process. Finally we give some results obtained for typical clouds and discuss the magnitude of the attenuation.

MODEL

The wave propagation through an air-vapor-droplet mixture induces a disturbance from the thermodynamical equilibrium. Thus, some mass, energy and momentum exchange

CP838, Innovations in Nonlinear Acoustics: 17th International Symposium on Nonlinear Acoustics,
edited by A. A. Atchley, V. W. Sparrow, and R. M. Keolian
© 2006 American Institute of Physics 0-7354-0330-9/06/$23.00

may arise between the gaz and the liquid phases. The description of the evolution of every single particle is obviously impossible and that is why a continuumlike approach is embraced provided that the wavelength is much greater than the size of the particles and the interparticle distance. Once the two-fluid system of averaged equations established, one must handle with the interactions terms between the phases.

To determine their expression, a spherical droplet of fixed size is considered and the different fluxes at its interface are calculated. The momentum transfer is due to the differential velocity between the droplets and the propagating wave. As the droplet size does not exceed 50 μm in clouds, the local Reynolds number is less than unity and the Faxen formula can be used to explicit the force exerted by the continuum phase on the particles. It takes into account both Stokes drag and non-stationary forces that is to say buoyancy, virtual mass and Basset terms. The calculation of heat transfers are performed via the classical non-stationnary heat equation. Finally, the so-called Hertz-Knudsen-Langmuir formula and the mass diffusion equation of the vapor through the air are introduced to fix the evaporation rate. Once the system closed, the dispersion equation is determined as a function of the frequency.

Although this model has been originally derived for monochromatic waves, arbitrary perturbations such as N-waves can be calculated by considering them as a superposition of monochromatic waves through Fourier transforms.

Furthermore, nonlinear effects due to the amplitude of the propagating signal have been added in the numerical computation by using a split-step method. They proved to be significant especially on the evolution of the rise time.

APPLICATION TO THE SONIC BOOM

Model validity

For the sonic boom application, we are mainly interested in the acoustic annoyance and thus, the calculation must be accurate for infrasonic and audible frequencies: $[1Hz, 20kHz]$. First, the model described above is only valid if the wavelength is much greater than the size of the particle and the interparticle distance. In clouds, these conditions are satisfied if frequencies are less than about $300kHz$. Now, we can introduce the main characteristic times in order to analyse if all transfer phenomena are important in the case of the sonic boom propagation: τ_v and τ_T are the characteristic times of the stationary momentum and heat tranfer; τ_μ and τ_{λ_g} the same for non-stationary tranfers and τ_v/m the time associated with mass exchange where m is the mass content of droplets:

$$\tau_v = \frac{2}{9}\frac{\rho_l\, a^2}{\mu_g} \qquad \tau_T = \frac{\rho_l\, cp_l\, a^2}{3\, \lambda_g} \qquad \tau_\mu = \frac{\rho_g\, a^2}{\mu_g} \qquad \tau_{\lambda_g} = \frac{a^2\, \rho_g cp_g}{\lambda_g} \qquad m = \frac{\alpha_l \rho_l}{\alpha_g \rho_g}$$

where ρ_l, ρ_g, α_l, α_g, cp_l and cp_g are respectively the density, volume fraction and specific heat of the liquid (index l) and gazeous phases (index g) ; μ_g is the viscosity of the gazeous phase, λ_g its heat conductivity and a is some average radius of the droplets.

We can give some estimation of the frequencies associated with these characteristic times, considering than the mean radius of the droplets a in clouds is about 10 μm:

$$f_v \sim 120\,Hz \quad f_T \sim 30\,Hz \quad f_\mu \sim 25kHz \quad f_{\lambda_g} \sim 40kHz \quad 10^{-4} \leq m \leq 10^{-3}$$

This analysis reveals the importance of all damping mechanisms in the present study as all characteristic frequencies above are inside or close to the audible spectrum. In particular, non stationary phenomena cannot be neglected, contrarily to most litterature models. Although they occur mainly at very low frequencies (mf_v), phase changes are still significant for frequencies lower than f_v.

Characteristic parameters

Now we can analyse the main parameters that define the microstructure and macrostructure of a cloud and that will influence the sound propagation. The droplet size distribution function is typically of the following form [4]: $n(a) = A\,a^2\,exp(-B\,a)$ where A and B are related to the water liquid content $w_L = \alpha_l\,\rho_l$ in $g\,m^{-3}$ and the average droplet radius a. Ordinarily, a is included between 5 and $30\mu m$, and w_L may vary from 0.2 for early stage clouds to $3g\,m^{-3}$ for dense ones [4]. The important macroscopic parameters are of course the mean temperature related to the average altitude (from 200 m for law cumulus to 10 km for high cirrostratus) and the thickness ($\sim$ 2000 m for cumulus, $\sim$ 7000 m for cumulonimbus) of the cloud. Fig. 1 illustrates

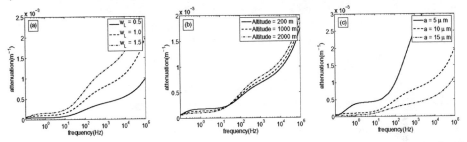

FIGURE 1. Influence of cloud characteristics on the attenuation. (a) Influence of the water liquid content w_L ($a = 10\ \mu m$, $alt. = 2000m$). (b) Influence of the altitude ($w_L = 1.0$, $a = 10\ \mu m$). (c) Influence of the mean radius a ($w_L = 1.0$, $alt. = 2000m$)

the influence of these parameters on the attenuation. First, the augmentation of w_L (fig. 1a) induces an increase of m to whom the amplitudes of momentum and heat exchanges are directly related. The variation of m will however not influence the intensity of phase exchange but will only modify its characteristic frequency. Secondly, when the altitude increases (fig. 1b), the temperature decreases and so does the mass concentration of vapour k_v. Therefore, evaporation processes will be reduced but as m concurrently increases, momentum transfer will slightly be amplified. Finally, at constant w_l, the diminution of the average droplet size (fig. 1c) corresponds to an augmentation of the interface between the liquid and gazeous phases and thus, will lead to a rise of all transfer phenomena. As f_v is directly related to a, it will also change the characteristic frequency of the attenuation effects.

Results

Now we will consider the propagation of a 0.2 s duration and 50 Pascal amplitude N-Wave representative of the sonic boom, in a typical cumulus 2000 m high, 2000 m deep and with a water liquid content and a mean radius of respectively 1 $g\ m^{-3}$ and 10 μm. To measure the impact of the sonic boom, the commonly introduced parameters

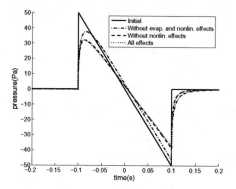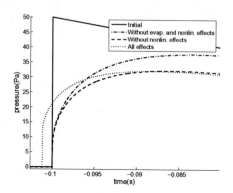

FIGURE 2. Propagation of an N-wave in a typical cumulus: overview and zoom on the first shock. The calculation is performed with all propagation mechanisms (dotted curve), without nonlinear effects (dashed curve) and without evaporation and nonlinear effects (dashdotted curve).

are the maximum overpressure P_{max}; the rise time t_m which is the time required for the pressure field to jump from 10% to 90% of the pressure maximum; and the A-weighted Sound Exposure Level in dB. After propagation, the pressure maximum is only 64% of its initial value, the rise time is increased up to 4.6 ms and there is a 12.5 dBA attenuation. Evaporation processes mainly influence the value of the pressure maximum that corresponds to low frequencies. On the contrary, nonlinear effects affect more high frequencies and so make the rise time evolve from 5.4 to 4.6 ms.

These results are in accordance with some test flights performed in 1967-1968 in the former Soviet-Union that observed a smoothing of the wave profile and a diminution of about 40 % of the pressure maximum for similar conditions.

This study demonstrates a deep impact of the presence of clouds on the sonic boom atmospherical propagation by widely reducing the amplitude and increasing the rise time. A high sensitivity to the clouds main characteristics has also been observed. Next step will be the addition of molecular relaxation and temperature stratification in the model.

REFERENCES

1. D. Gubaidullin, and R. Nigmatulin, *Int. J. Multiphase Flow*, **26**, 207–228 (2000).
2. S. Sewell, *Phil. Trans. R. Soc. Lond. A*, **210**, 239–270 (1910).
3. H. Lamb, *Hydrodynamics*, Dover, 1945.
4. H. Pruppacher, and J. D. Klett, *Microphysics of clouds and preciptation*, Kluwer Academic, 1978.

FIR Models for Sonic Boom Propagation Through Turbulence

Lance L. Locey, Victor W. Sparrow

The Pennsylvania State University
University Park, PA, lll162@psu.edu/vws1@psu.edu

Abstract. It is challenging to develop a physics-based model for propagation through atmospheric turbulence. A number of models are currently available but are typically not based on experimental results. A fresh approach might be to construct some type of "black box" filter function for producing typical distorted sonic boom waveforms on the ground. The input to the filter could be arbitrary, non-distorted sonic boom waveforms above the turbulence. As an initial model, a single input-output system was assumed. Ground-recorded waveforms were used as outputs of the system. The transfer function between the two signals was estimated using Welch's average periodogram method. Data obtained from the 2004 Shaped Sonic Boom Experiment (SSBE) was analyzed and corresponding transfer functions were obtained. With such transfer functions established, FIR filters were determined and convolved with an appropriate input waveform. Such FIR filters allow for the production of "typical" atmospherically distorted waveforms given arbitrary input waveforms.

Keywords: sonic boom, atmosphere, turbulence, SSBD, SSBE, filter
PACS: 43.28.Mw

INTRODUCTION

Atmospheric turbulence poses a challenge to researchers. Though much investigation has been conducted, little is known about how turbulence alters sound as it propagates through the atmosphere. Particularly, when a sonic boom propagates through the atmosphere, turbulence can either "spike" or round waveforms and has a tendency to increase rise times and add perturbations after the shocks[1, 2].

No simple method exists to take an undistorted waveform and modify it the way a real atmosphere does. The research presented in this paper represents a first attempt at such a model. Data taken as part of the Shaped Sonic Boom Experiment (SSBE) is used to come up with filter functions that represent simple models of the atmosphere, and those functions are used to "turbulize" clean waveforms.

This work is a component of the Partnership for AiR Transportation Noise and Emissions Reduction (PARTNER) Center of Excellence, created in 2003. Project 8, Sonic Boom Mitigation, was funded in April 2005 to study the feasibility of overland supersonic flight and to assist the FAA in making informed policy decisions concerning the possibility of permitting overland supersonic flight, currently prohibited. One would like to know if low-boom supersonic signatures remain acceptable to the public after passing through turbulence, hence this work.

CP838, *Innovations in Nonlinear Acoustics: 17th International Symposium on Nonlinear Acoustics*,
edited by A. A. Atchley, V. W. Sparrow, and R. M. Keolian
© 2006 American Institute of Physics 0-7354-0330-9/06/$23.00

APPROACH

The Shaped Sonic Boom Experiment (SSBE) occurred at NASA Dryden Flight Research Center over several days in January of 2004[3]. Research presented in this paper will focus on data taken January 15th, 2004 at approximately 10:00 (PST), referred to as test 21. In this particular flight, the Shaped Sonic Boom Demonstrator (SSBD) flew Mach 1.35 at an altitude of 9,753 meters (32,000 feet). A glider with a wingtip mounted microphone flew below the SSBD, at an altitude of 2,438.4 meters (8,000 feet), underneath the flight path of the SSBD. An array of 28 microphones was located in a linear array on the ground, separated by nominally 152 meters (500 feet) each, spanning just over 3 kilometers (about two miles). The SSBD flew directly over the array of microphones. Four different data acquisition systems were used to capture the waveforms recorded by the microphones.

A simple way to model the atmosphere is to think of it as a simple finite impulse response (FIR) filter. When an aircraft flies supersonic, the nose, tail, and various parts of the plane create disturbances which eventually coalesce in the form of a time waveform frequently referred to as a sonic boom. The waveform propagates through the atmosphere until it reaches the Planetary Boundary Layer (PBL). Once the waveform enters the PBL, it is altered by turbulence. One way to model that turbulence is to assume that the PBL is a FIR, where the output is a turbulent waveform measured on the ground and the input is a clean waveform measured above the PBL.

Several assumptions were made in this simple FIR model. First, propagation through the PBL was assumed to be linear and lossless. Additionally, all turbulence was assumed to be inside the PBL, which was assumed to be below the glider. The filter was assumed to be sufficiently accurate such that it could represent all of the waveform variation caused by the turbulence. The filter was also assumed to be long enough to accurately represent the system. Lastly, it was assumed that the energy of the waveform measured above the PBL (by the glider) was equal to the energy measured on the ground.

It was necessary to determine the temporal and spatial relationships between the SSBD, the glider, and the microphones. GPS data was used to determine these relationships, using Matlab[4]. An initial assumption was that the glider flew above the PBL. However, it was found that this was not the case in every flight. Consequently, rather than solely use the waveform measured by the glider during test 21, all the glider waveforms were compiled, sorted and averaged. Some glider waveforms were excluded from the average because they exhibited too much distortion, indicating that the glider was indeed inside the PBL when the particular measurement was made. Once an averaged glider waveform was established, it was used as the input to the filter function model.

Several steps were required to create the FIR filters. The first step was to down-sample the data to the lowest sampling frequency. That was done to assist in the comparison of the various ground measurements. The lowest sampling frequency was 8,335 Hz. The next step was to sort and average the glider booms into a single clean waveform. Once that was accomplished, a Z-domain transfer function was estimated, between the glider input and each ground measurement, using the Matlab function

TFEstimate, which estimates a Z-domain transfer function using Welch's averaged periodogram method. This yielded a Z-domain filter for each ground measurement. The parameters supplied to the TFEstimate function were: the averaged glider waveform as input, the ground measured waveform as output, 1024 point FFT length, 512 points of overlap, and a hamming window. Each Z-domain filter was then inverse transformed, yielding a four hundred point FIR filter. The filter was then applied to the original input signal, (the averaged glider waveform), and the energy of the resultant output waveform was compared to the input. If the energies were very different, the filter coefficients were rescaled by the ratio of the square root of the input to output energy, and the filter was reapplied. The filter output was compared to the ground measurement as a way to verify that the filter was accurately representing what was measured.

RESULTS

Once the filter coefficients were determined, each filter was applied to a hypothetical low boom signature supplied by one of the PARTNER Center of Excellence industry partners. The results are shown in figure 1. One can see the input waveform has been modified slightly by each of the filters.

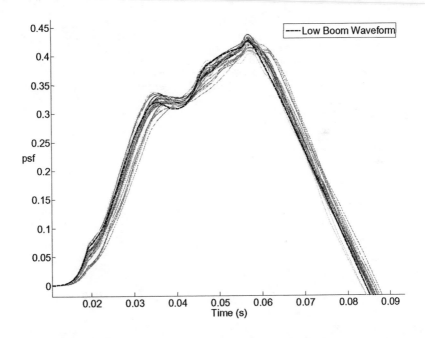

Figure 1 Positive phase portions of a clean low-boom waveform (dark-dashed) and 28 resulting "turbulized" waveforms (grey). The vertical axis is acoustic pressure in pounds per square foot (psf) where 1 psf = 47.88 Pa.

625

CONCLUSIONS

The filter functions developed represent a first attempt to incorporate atmospheric turbulence in their propagation to the ground. These filters are now available for use in perceptual testing of low boom signatures. However, more work is necessary to improve the filters by varying parameters and sampling rates. Future work will include the development of more refined models. The overall goal is to develop a statistical model of the turbulence that accurately reflects the physical processes that alter a sonic boom as it propagates through the PBL.

ACKNOWLEDGMENTS

This work was supported by the FAA and NASA. The author would like to acknowledge the PARTNER industrial partners for useful discussions.

REFERENCES

1. Plotkin, K. J., Maglieri, D. J, Sullivan, B. M., "Measured Effects of Turbulence on the Loudness and Waveforms of Conventional and Shaped Minimized Sonic Booms," AIAA-2005-2949, January 2005
2. Morgenstern, J. M., Arslan, A., Lyman, V., Vadyak, J., "F-5 Shaped Sonic Boom Demonstrator's Persistence of Boom Shaping Reduction through Turbulence," AIAA-2005-0012, January 2005
3. Pawlowski, J. W., Graham, D. H., Boccadoro, C. H., "Origins and Overview of the Shaped Sonic Boom Demonstration Program," AIAA-2005-0005, January 2005
4. Computer Program MATLAB Version 7.0.4.365 (R14) Service Pack 2, (The Math Works, 2005)

Measured Effects of Turbulence on the Waveforms and Loudness of Sonic Booms

Kenneth J. Plotkin,[*] Domenic J. Maglieri[†] and Brenda M. Sullivan[¶]

[*] *Wyle Laboratories, 2001 Jefferson Davis Highway, Arlington, VA 22202, USA*
[†] *Eagle Aeronautics, Inc., 13 West Mercury Blvd, Hampton, Virginia 23669 USA*
[¶] *NASA Langley Research Center, Hampton, Virginia, 23681*

Abstract. The recent Shaped Sonic Boom Experiment yielded a large number of digital recordings of shaped and N-wave sonic boom waveforms, all of which exhibited some degree of turbulent distortion. This digital data set provided an opportunity to test theoretical predictions of the shape of distortions, and to assess the effect of turbulence on the loudness of shaped booms. Distortions following the shocks were found to be virtually identical for the bow and tail shocks of each boom, confirming the usual explanation that the distortion is scattering of the shock waves and that turbulence may be considered to be frozen for the duration of a boom. RMS values of the distortions have been compared with the theory of S. C. Crow. The loudness reduction of shaped booms had been predicted by Plotkin to persist through turbulence, and this was found to be the case. The loudness calculations also confirmed that the benefit of reduced shock amplitude is nonlinear: there are benefits from both the reduced amplitude and from the increased shock thickness. (A full version of this paper was presented at the 2005 AIAA/CEAS Aeroacoustics Conference as AIAA-2005-2949.)

Keywords: Sonic boom, turbulence, loudness, scattering.
PACS: 43.28.Gq, 43.28.Lj, 43.28.Lv, 43.28.Mw, 43.28.Rq

TURBULENT DISTORTION OF SONIC BOOMS

Sonic booms measured from flight tests in the atmosphere always exhibit some degree of distortion. Figure 1 shows samples of N-wave sonic booms measured under two different atmospheric conditions [1]. Part (a) shows relatively clean N-waves measured under calm conditions, and part (b) shows distorted N-waves measured under turbulent, gusty conditions. Two types of distortion are seen. The first is that perturbations appear behind the shocks. Crow [2], who called these perturbations a "spiky fine structure," showed that the perturbations were caused by turbulent scattering in the lower atmosphere, and that their rms envelope has an envelope of the form $(h_c/h)^{7/12}$, where h is the distance behind the shock and h_c is a parameter related to the thickness and structure of the turbulent layer. The 7/12 power follows from the assumption of Kolmogorov turbulence. Kamali and Pierce [3] showed that Crow's 7/12 relation fit available measurements of sonic booms. The second distortion is that the shock rise times tend to be longer under turbulent conditions.

A detail seen in Figure 1 is that the distortion associated with the tail shocks strongly resembles the distortion associated with the bow shocks. This is generally

CP838, *Innovations in Nonlinear Acoustics: 17th International Symposium on Nonlinear Acoustics*,
edited by A. A. Atchley, V. W. Sparrow, and R. M. Keolian
© 2006 American Institute of Physics 0-7354-0330-9/06/$23.00

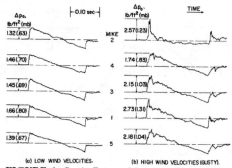

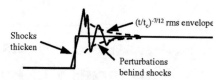

FIGURE 2. Summary of Turbulence Effects on Shock Waves

(a) LOW WIND VELOCITIES. (b) HIGH WIND VELOCITIES (GUSTY).

FIGURE 1. Sonic Booms Measured Under Calm and Turbulent Conditions

cited as evidence of frozen turbulence, i.e., the time scale of changes in the turbulence is slow compared to the time between the bow and tail shocks, so each shock sees essentially the same turbulence. Figure 2 is a sketch of the net distortion of a sonic boom shock wave: a spiky fine structure behind the shock, and a longer rise time. The same pattern is expected behind each shock of an N-wave.

Crow's theory [2] was a breakthrough in the understanding of turbulent scattering of booms. Traditional scattering theory [4] is based on continuous waves scattered by a finite volume, with solutions in the frequency domain. Application to thin shock waves was less than successful. Crow addressed this problem by formally representing the shock as a step function, and working in the time domain. He also took the primary independent variable to be the distance behind the shock, rather than the traditional scattering angle. Because the shock is a line, the scattering volume for a point a given distance behind the shock reduces to a paraboloid of revolution, with the point being the focus and the shock being the directrix. This is illustrated in Figure 3 for a receiver point a distance h behind the shock. Crow called this the "paraboloid of dependence."

The paraboloid perspective is very useful because it illustrates the scales of turbulence that matter: those that match the diameter of the paraboloid. All scales matter for a given h, but larger scales dominate at regions further from the front. For receiver points at larger h, the paraboloid is wider, so larger scales are involved. This can be used to deduce the frequency content of scattered waves [5], although Crow did not pursue that detail. This does, however, explain why the spiky fine structure becomes smoother further behind the shock.

Employing a reasonable model for the planetary boundary layer under daytime turbulent conditions, including the assumption of the Kolmogorov inertial subrange, Crow obtained the result that the rms value of the spiky fine structure is given by

$$<p^2>^{1/2} = (h_c/h)^{7/12} \tag{1}$$

with a value (for his assumptions) of $h_c = 0.7$. This value may be considered to be either feet or milliseconds (within 10% or so), corresponding to the units of h.

Kamali and Pierce [3] fit Crow's theory to data from three early afternoon supersonic flights at Edwards AFB in 1966 [6]. Their results are shown in Figure 4. They found good agreement with the 7/12 law, and obtained a value of $h_c = 1.0$ msec, slightly larger than Crow's 0.7.

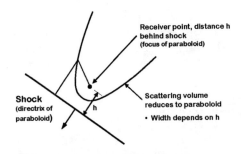

Receiver point, distance h
behind shock
(focus of paraboloid)

Shock
(directrix of
paraboloid)

Scattering volume
reduces to paraboloid
• Width depends on h

FIGURE 3. Scattering Volume for Thin Shock:
Paraboloid of Dependence

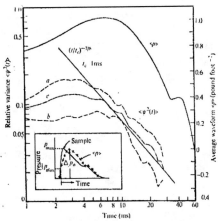

FIGURE 4. Kamali and Pierce's Fit of Crow's
Theory to Flight Test Data

Data from both the SSBD and the F-5E reference aircraft flown during the 2004 Shaped Sonic Boom Experiment [7] provided another opportunity to test Crow's theory. The "de-turbing" data cleanup procedure in that study provided a means to extract the spiky fine structure from the N-wave sonic booms. Spiky fine data from the rear shocks of both vehicles were analyzed. Figures 5 and 6 show data from each of these aircraft. Consistency with the 7/12 power law for both data sets is very good, better than that seen in Figure 4, for t less than about 20 msec. The deviation at times beyond 20 msec is probably due to the radius of the paraboloid of dependence exceeding the turbulent outer scale, as suggested by Plotkin [5].

The quantitative result of the current analysis, tc of about 0.2 msec, is smaller than Crow's calculated 0.7 and Kamali and Pierce's 1.0. Noting that the amplitude of the rms is proportional to $t_c^{7/12}$, the current measured result corresponds to rms values of about half Crow's predictions, while Kamali and Pierce's measured result corresponds to rms values about 25% higher than Crow's predictions.

Considering the general variability of atmospheric turbulence, these two results bracketing Crow's prediction are quite good. The anecdotal experience of participants

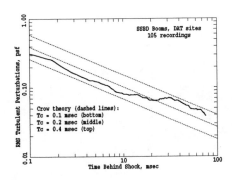

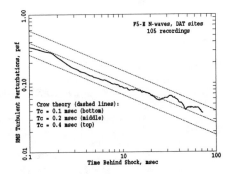

FIGURE 5. RMS Values of SSBD Rear Shock
Spiky Fine Structure, Compared With Crow's
Theory

FIGURE 6. RMS Values of F-5E Rear Shock
Spiky Fine Structure, Compared With Crow's
Theory

in the SSBE measurements was that the booms measured during these flights were unusually clean. Booms measured during the early morning calm (minimal wind, stable inversion) were generally as expected. Booms measured in the afternoon, despite the clear presence of a thermally-driven turbulent boundary layer, were generally less distorted than expected.

Loudness of Shaped Booms

The F-5E flown as a reference during SSBE generated an N-wave with 1.2 psf shocks. The SSBD generated a boom with a flat-top bow shock of 0.9 psf and a tail shock of 1.2 psf. Observers during the test did not discern any difference in loudness, nor were they expected to, since the 1.2 psf shock dominated the experience. The loudness of sonic boom shock waves is formally calculated by the procedure developed by Shepherd and Sullivan [8], based on Stevens Mark VII perceived loudness together with an auditory response time of 70 msec. From the difference in overpressure and correspondingly slower rise time, the loudness of the SSBD's bow shock was expected to be about 5 dB lower then that of the F-5E's bow shock. Those calculations were conducted and reported in Reference 9, the full version of this paper. The expected loudness benefit was confirmed. That result supports the analysis by Plotkin [5] that indicated turbulent distortion would not adversely affect the average benefit of shaped sonic booms.

ACKNOWLEDGMENTS

The Shaped Sonic Boom Experiment was funded by the National Aeronautics and Space Administration.

REFERENCES

1. Hilton, D.A., Huckel, V., and Maglieri, D.J., "Sonic Boom Measurements During Bomber Training Operations in the Chicago Area", NASA TN D-3655, 1966.
2. Crow, S.C., "Distortion of Sonic Bangs by Atmospheric Turbulence", *J.Fluid Mech.*, **37**, 529-563, 1969: also, NPL Aero Report 1260.
3. Kamali, G., and Pierce, A.D., "Time Dependence of Variances of Sonic Boom Waveforms", *Nature*, **234** (5323), 30-31, 5 November 1971.
4. Tatarskii, V.I., *Propagation of Waves Through Turbulence*, Dover, New York, 1961
5. Plotkin, K.J., "The effect of turbulent distortion on the loudness of sonic booms," *NASA High-Speed Research Sonic Boom Workshop*, NASA CP-3172, May 1992.
6. Maglieri, D.J., Huckel, V., Henderson, H.R., and McLeod, M.J., "Variability in sonic-boom ground signatures measured along an 8000-foot linear array," NASA Technical Note TN D-5040, 1969.
7. Plotkin, K.J. Haering, E.A. Jr., Murray, J.E., Maglieri, D.J., Salamone, J., Sullivan, B.M., and Schein, D, "Ground Data Collection of Shaped Sonic Boom Experiment Aircraft Pressure Signatures," AIAA-2005-0010, January 2005.
8. Shepherd, Kevin P., and Sullivan, Brenda M., "A Loudness Calculation Procedure applied to Shaped Sonic Booms," NASA TP-3134, Nov. 1991.
9. Plotkin, K.J., Maglieri, D.J., and Sullivan, B.M., "Measured Effects of Turbulence on the Loudness and Waveforms of Conventional and Shaped Minimized Sonic Booms," AIAA-2005-2949, May 2005. (You can obtain a PDF copy of this full paper at **http://hdl.handle.net/2002/15775**)

Numerical and experimental simulations of shock waves propagation through heterogeneous media

L. Ganjehi[1,2], R. Marchiano[1], J.-L. Thomas[2], F. Coulouvrat[1]

Université Pierre et Marie Curie – 4 place Jussieu – 75252 Paris cedex 05, France
(1) Laboratoire de Modélisation en Mécanique - UMR CNRS 7607
(2) Institut des Nanosciences de Paris - UMR CNRS 7588

Abstract. The influence of the planetary boundary layer on the sonic boom received at the ground level is known since the 60s to be a major importance, with a decrease of the mean pressure, an increase of the mean rise time but a huge variability. A generalized "KZ" paraxial approximation of the nonlinear wave equation is obtained in a weakly heterogeneous (small temperature fluctuations) and slowly moving medium (small Mach number), two consistent approximations for the atmosphere. For a non-moving medium, the generalized KZ equation is solved numerically using a finite differences scheme adapted from one developed for sonic boom focusing. To validate quantitatively the model in a deterministic way, experiments are conducted at a 1:100 000 scale for ultrasonic shocks in a water tank. Atmospheric heterogeneities are simulated by introducing cylindrical (sub)wavelength silicon tubes. Experiments and simulation of the nonlinear pressure fields are in very good agreements. The sensitivity of the scattered signal to the size of the heterogeneities is investigated. Also the influence of nonlinear effects on the shock wave/heterogeneity interaction is examined. Finally, the limitation of the paraxial approximation for wave scattering are discussed.

Keywords: sonic boom, nonlinear propagation, diffraction, heterogeneous and moving medium.
PACS: 43.25.43.28

GENERALIZED KZ EQUATION

The generalized KZ equation describes the propagation of nonlinear acoustic waves through a weakly heterogeneous and slowly moving medium. In Cartesian coordinates, the longitudinal and transverse space variables are respectively z and (x, y). The time is noted t and $\tau = \omega(t - z/c_0)$ is the dimensionless retarded time, where ω is a characteristic frequency and c_0 is the mean speed of sound and ρ_0 the mean density. The z coordinate is scaled by the shock distance $L_s = \omega/(c_0 \beta M)$: $\sigma = z/L_s$, where β is the nonlinear parameter of the fluid and $M = v_0/c_0$ is the acoustical Mach number, characterizing the nonlinear effects. In the case of the sonic boom it is of order $M \approx 10^{-3}$. The dimensionless acoustical pressure P is scaled by $\rho_0 c_0 v_0$. The transverse space variables are scaled by a, the transverse length of the source. In the parabolic approximation, the diffraction parameter $\eta = 1/ka$ is assumed to be small. Let

CP838, *Innovations in Nonlinear Acoustics: 17th International Symposium on Nonlinear Acoustics*,
edited by A. A. Atchley, V. W. Sparrow, and R. M. Keolian
© 2006 American Institute of Physics 0-7354-0330-9/06/$23.00

$\delta c_0(\vec{x}) = \varepsilon c^h(\vec{x}) c_0$ denote the sound speed fluctuations due to heterogeneities, where the parameter $\varepsilon = \max(\delta c_0)/c_0$ measuring the degree of heterogeneity is small $\varepsilon \ll 1$. For a slowly moving medium, the flow speed is $\vec{u}_0(\vec{x}) = \varepsilon \vec{u}^h(\vec{x}) c_0 \ u^h$, assuming the flow Mach number M_f of the same order as ε. Under these assumptions, the conservation and state equations finally leads to the generalized KZ equation:

$$\frac{\partial^2 P}{\partial \varpi \partial \sigma} - \frac{\partial}{\partial \tau}\left(P \frac{\partial P}{\partial \tau}\right) - \frac{\eta^2}{2\beta M}\Delta_{\perp}P = \frac{\varepsilon}{\beta M}\left[c_h + u_z^h\right]\frac{\partial^2 P}{\partial \tau^2} - \frac{\varepsilon}{\beta M}\left[\frac{\partial u_z^h}{\partial z} + \frac{1}{2}\left(\frac{\partial u_x^h}{\partial x} + \frac{\partial u_y^h}{\partial y}\right)\right]\frac{\partial P}{\partial \tau}. \quad (1)$$

The left hand-side term is the usual KZ equation in a homogeneous medium [1]. The first right hand-side term represents the combined effects of the heterogeneities and of the flow, involving the effective sound speed fluctuation along the propagation direction z. The last term is associated to the flow gradient.

This pressure equation is solved numerically with a finite difference scheme in a non-moving medium.

EXPERIMENTAL AND NUMERICAL RESULTS

In order to examine the influence of heterogeneities on the sonic boom, experiments were conducted at a 1:100 000 scale in a water tank (sound speed 1471 m/s), with (sub)wavelength semi-cylindrical heterogeneities made in silicon (sound speed 1083 m/s), whose diameter is accurately known.

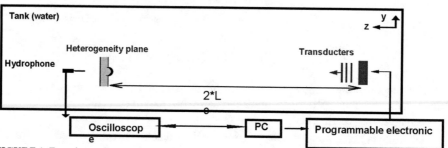

FIGURE 1. Experimental setup.

A 1-MHz frequency wave is emitted in water tank by a 2D array of 128 piezoelectric transducers electronically controlled (cf. figure 1). Each electronic channel is fully programmable. The inverse filter technique [2] allows us to synthesize a plane wave in front of the heterogeneity.

The heterogeneity is placed at twice the shock distance. The pressure field is measured with a 40-MHz bandwidth membrane hydrophone at several distances before and behind the heterogeneity. distances before and behind the heterogeneity.

Figure 2 shows the incident shock wave measured 3 mm before the heterogeneity.

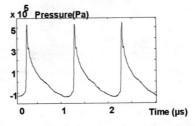

FIGURE 2. Incident shock wave. The mean rise time of the shock wave is equal to 15 ns.

The pressure field is first measured behind the 1.5-mm diameter heterogeneity, near the focus. The experimental and numerical results, represented on figure 3, are in good agreement. Clearly visible is the shock wave focusing with the swallowtail shape of the waveform, even though the heterogeneity diameter is equal to only one wavelength. Also visible are the large differences between two observations points : one (Fig. 3.c) near focus shows a "U" form, characteristic of the focusing effects, with some amplification. The other one, away from focus (Fig 3.d), shows a superposition of incident and diffracted shock waves, with reduced overall amplitude. The rise time of the shock is longer after the heterogeneity. These results agree qualitatively with sonic boom observations. However, the amplitude and the rise time of the shock are slightly different on experimental and numerical results, because the attenuation of the silicon is not taken into account yet in the numerical simulation..

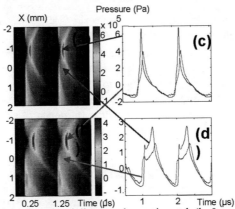

FIGURE 3 . Numerical (a & c-d red lines) and experimental (b & c-d blue lines) space/time representation of the pressure field 4 mm behind an heterogeneity Ø = 1.5 mm and time signals in front of the center (c) and the extremity (d) of the heterogeneity

In order to estimate the bias induced by the parabolic approximation, numerical and experimental results are compared near a 0.7-mm diameter heterogeneity (figure 4). The agreement is good on the axis of propagation, but the interferences between the incident plane wave and the diffracted cylindrical ones are not fully noticeable on the numerical results. Indeed, in the heterogeneity vicinity, interferences occur at large angles, even since its diameter comparable to the fundamental wavelength.

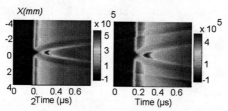

FIGURE 4. Comaprison between numerical (left) and experimental (right) space/time representation of the pressure field 4 mm behind an heterogeneity Ø = 0.7 mm

We then investigate the influence of several heterogeneities. Fig. 5 shows the experimental measurements after propagation of the shock wave through a screen of 30 heterogeneities, either near (left figure) and far beyond (right figure) the focus. We clearly observe multiple interferences leading to complex waveforms, either spiky waveforms near focus (left), or rounded signals with multiple shock fronts (right).

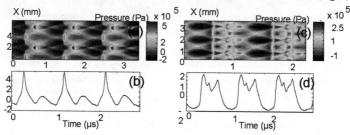

FIGURE 5. A spatio-temporel representation of the pressure field behind multi-heterogeneities d = 1.5 m, respectively near and far the focus of these heterogeneities

Again, these experimental results are qualitatively in good agreement with sonic boom observations or laboratory-scale experiments [3], and confirm the influence of focusing effects to explain spiky waveforms and multiple wavefront folding, as suggested theoretically [4] and explored numerically [5]. There remains to be explored i) the validity of paraxial approximation, ii) the influence of nonlinear effects, iii) the influence of heterogeneity size and sound speed fluctuations.

REFERENCES

1. Zabolotskaya, E. A., and Khokhlov, R. V., *Sov.Phys.Acoust.*, **15**, 35-40 (1969).
2. Tanter, M., Thomas, J.L., and Fink, M., *J.Acoust. Soc. Am.*, **108**, 223-234 (2000).
3. Lipkens, B., Blackstock, D.T., *J.Acoust. Soc. Am.*, **103**, 1301-1309 (1998b)
4. Pierce, A.D., and Maglieri, D. J., *J. Acoust. Soc. Am.*, **51**, 702-721, (1972)
5. Blanc-Benon, Ph., Lipkens, B., Dallois, L ., Hamilton, M. F., and Blackstock, D. T., *J. Acoust. Soc. Am.*, **111**, 487-498, (2002)

Time Domain Propagation of Sonic Booms

James P. Chambers Henry E. Bass, Richard Raspet, and Jack Seiner

*The Jamie L. Whitten National Center for Physical Acoustics, University of Mississippi,
University, Ms 38677*

Abstract. A number of physical processes work to modify the shape of sonic boom waveforms as the waveform propagates from the aircraft to a receiver on the ground. These include frequency-dependent absorption, nonlinear steepening, and scattering by atmospheric turbulence. In the past two decades, each of these effects has been introduced into numerical prediction algorithms and results compared to experimental measurements. There is still some disagreement between measurements and prediction, but those differences are now in the range of tens of percent. The processes seem to be fairly well understood and the present understanding of sonic boom evolution will be presented. An area of possible future development is the marrying of this propagation understanding with advances made in computational fluid dynamics (CFD) and scale model jet tests to predict the near field sonic boom waveform from aircraft shape and flight characteristics. Such a capability would allow for complete modeling from aircraft to ground impact and as such would represent a significant design tool to effectively evaluate proposed aircraft designs for reduced sonic boom impact.

Keywords: Sonic Booms, Acoustics, Non-Linear Acoustics.
PACS: 43.25, 43.28

INTRODUCTION

The effect of relaxation processes on the rise times of shock waves in the atmosphere has been acknowledged for a many years.[1] In the early 1970's, the role of vibrational relaxation of Nitrogen on absorption of sound in the atmosphere was identified and led to more accurate expressions for the vibrational relaxation times of all atmospheric constituents. In 1978, the absorption calculation developed for the American National Standard ANSI S1-26-1978, "Method for Calculation for the Absorption of Sound by the Atmosphere"[2] was incorporated into a finite wave propagation algorithm developed by Anderson[3] at the University of Texas to investigate the effect of vibrational relaxation on the propagation of explosion waves.[4] The resulting algorithm came to be known as SHOCKN. This algorithm was extended to predict the evolution of sonic boom waveforms propagating through the atmosphere.[5] Of particular interest has been the prediction of rise times of these sonic booms, since measured waveforms usually have rise times orders of magnitude greater than the rise times predicted from viscous and thermal losses. Atmospheric turbulence typically plays a major role in the increase in rise times measured at the surface of the earth. The rise time imposed by relaxation processes establish a base waveform to be perturbed by turbulence. Since turbulence most often increases the rise time, calculations based upon relaxation absorption and dispersions establish a lower limit.

CP838, *Innovations in Nonlinear Acoustics: 17ᵗʰ International Symposium on Nonlinear Acoustics*,
edited by A. A. Atchley, V. W. Sparrow, and R. M. Keolian
© 2006 American Institute of Physics 0-7354-0330-9/06/$23.00

The first predictions[4] from SHOCKN did not include dispersion which results from the same relaxation processes that give rise to atmospheric absorption. Dispersion was included in subsequent calculations of the rise times of explosion waves and sonic booms propagated through a realistic stratified atmosphere.[5] Subsequent papers compared the predictions of the SHOCKN algorithm to measurements of ballistic waves,[6] to steady state shock calculations,[7] and to time domain calculations of rise times of sonic booms.[8] Other research investigated the propagation distance for a potential shock to "heal", that is, for a artificially lengthened or shortened rise portion of a wave to return within 10% of its steady state value.[9] The rapid advancement of computation speeds have resulted in improved calculations of interest in the study of sonic boom rise times.

CALCULATION ALGORITHM

SHOCKN assumes that finite wave distortion, atmospheric attenuation and dispersion are independent over a sufficiently short distance step. The finite wave distortion is calculated in the time domain, the wave is then Fast Fourier Transformed and the attenuation, dispersion and geometrical spreading is applied to of each Fourier components in the frequency domain. The inverse Fourier transform is then used to re-form the time domain waveform and the process repeated for the next distance step.

In practice, these are the calculation steps:
1. The input wave is evenly time sampled.
2. The maximum negative slope, $(dp/dt)_{max}$ is determined and used to find the distance to shock formation,

$$d_{shock} = \rho c^2 / \beta (dp/dt)_{max},$$ (1)

β is the non-linearity parameter, c the sound speed, and ρ the ambient density.
3. The step size for calculation is chosen to be a fraction of the shock formation distance (.10<f<.25)

$$d_{step} = f\, d_{shock}$$ (2)

4. The time distortion applied to each point as a result of propagation a distance d_{step} is computed by

$$t_i' = t_i + \beta\, p(t_i)\, d_{step} / Pc^3$$ (3)

where $p(t_i)$ is the pressure at time t_i and P is the ambient pressure
5. The wave is re-sampled so that the time samples are uniform
6. The Fast Fourier Transform is applied to the discretized time waveform to calculate the complex amplitude spectrum $A(f_i)$.
7. The absorption and relative dispersion over the distance d_{step} is applied to the complex amplitude.

$$A'(f_i) = A(f_i)_e^{-\alpha(f_i)d_{step}}\, e^{i\Delta c(f_i)d_{step}}$$ (4)

8. The geometrical spreading loss is applied by keeping the intensity constant

$$I = \frac{P^2 A}{2\rho c} = const$$ (5)

9. The time waveform is reconstituted using the Inverse Fast Fourier Transform. This process is repeated until the wave has propagated the required distance.

It should be noted that studies of the rise times of steady shocks[9] has led to the development of a criteria for the sampling rate required for SHOCKN to match the results of weak shock theory for moderately strong impulses. SHOCKN was found to correctly predict the speed of propagation of the shock front of a steady state wave when there were at least 8 to 10 sample points on the rise portion of the waveform. The degradation when the number of sample points approach 5 or 6 was astonishing. This criteria was not satisfied for the shock waves with overpressures greater than 100 Pa in the earlier papers.[4-7] These calculations were subsequently repeated.[10]

TURBULENT SCATTERING

Boulanger et al.[11] reported on the use of a scattering calculation to predict the effects of turbulence on sonic boom rise times and overpressures for different atmospheric conditions. Turbulence is represented by spherical turbules and is incorporated into the equation of propagation via a first Born approximation. This method offers a closed-form expression for the pressure which is convenient for use in computer simulations. The results of this calculation were compared to data collected at the JAPE-2 tests.[12,13] These tests provided simultaneous recordings of sonic boom pressure waveforms from T-38 aircraft along with detailed meteorological data.

NEAR FIELD SIGNATURE

While the computational methodologies for Sonic Boom propagation, such as SHOCKN, are a useful, and arguably necessary tool, they do not, by themselves afford a solution to the problem at hand, namely sonic boom mitigation. The solution of that problem will require reduction or modification of the initial boom itself. Toward that end, research is suggested on near field measurements and simulations utilizing new computational tools such as computational Fluid Dynamics (CFD) and scale model analysis using newly constructed test facilities that afford more realistic and closer scale capabilities than previously available. Recent construction at the NCPA will provide a fully instrumented 12 inch cross section Mach 5 wind tunnel for testing. The facility is capable of recharging to full pressure in ½ hour to allow rapid modifications to designs as well as optical view ports on all four walls to allow full observations of shock and boundary layer effects. One additional concern in scale model (or numerical) testing is that the model is often constructed (or posed) as consisting of a well understood and uniform material such as aluminum or steel while the final flight system may be made of more exotic and complex materials such as a laminated composite. A laser vibrometer will be able to investigate the interaction of fluid boundary layers with the compliant wall effects as well. Once the near field waveform has been evaluated or estimated, the solution can be married as an input to SHOCKN to determine the waveform at the ground

CONCLUSIONS

The algorithm SHOCKN has been used to calculate rise times from weak shocks in the atmosphere. The role of vibrational relaxation attenuation and dispersion in the determination of rise times has been investigated. It has also been demonstrated that steady state calculations of rise times overestimate the rise time of finite duration waves significantly unless the rise portion of the wave is less than 2% of the positive duration of the pulse. Finally near field calculation or evaluation of sonic booms with new computational tools and high speed wind tunnels has been proposed to solve the problem of boom reduction. The output of such a measurement would provide the input to computational codes such as SHOCKN.

ACKNOWLEDGMENTS

The authors would like to acknowledge the support of NASA LaRC for their support over the years and the close collaborations of many contributors and co-authors over the years including, but not limited to, Mark Kelley, Patrice Boulanger, David Blackstock, Mark Hamilton and Robin Cleveland.

REFERENCES

1. J. P. Hodgsen, H. E. Bass and R. Raspet," Vibrational relaxation effects in weak shock wave in air and structure of sonic bangs," J. Fluid Mech., 58, pp. 187-196, (1973)
2. "Method for the calculation of the absorption of sound by the atmosphere," ANSI S1.26-1978, American National Standards Institute
3. M. O. Anderson," The propagation of a spherical N wave in an absorbing medium and its diffraction by a circular aperture," Masters Thesis, Applied Research Laboratories, The University of Texas at Austin, (1974)
4. H. E. Bass and R. Raspet," Vibrational relaxation effects on the atmospheric attenuation and rise time of explosion waves," J. Acoust. Soc. Am., 64(4), pp. 1208-1210, (1978)
5. H. E. Bass, J. Ezell, and R. Raspet," Effect of vibrational relaxation on rise times of shock waves in the atmosphere," J. Acoust. Soc. Am., 74(5), pp. 1514-1517, (1983)
6. H. E. Bass, B. A. Layton, L. N. Bolen, and R. Raspet, "Propagation of medium strength shock waves through the atmosphere," J. Acoust. Soc. Am., 82(1), pp. 306-310, (1987)
7. H. E. Bass and R. Raspet," Comparison of sonic boom rise time prediction techniques," J. Acoust. Soc. Am., 91(3), pp. 1767-1768, (1992)
8. R. O. Cleveland, J. P. Chambers, H. E. Bass, R. Raspet, D. T. Blackstock, and M. F. Hamilton," Comparison computer codes for the propagation of sonic boom waveforms though isothermal atmospheres," J. Acoust. Soc. Am., 100(5), pp. 1767-1768, (1992)
9. R. Raspet, H. E. Bass, L. Yao, and W. Wu," Steady state risetimes of shock waves in the atmosphere," Proc. Of the High Speed Research: Sonic Boom Conference, NASA Langley Research Center, (1992)
10. H.E. Bass, R. Raspet, J.P. Chambers, and M. Kelly, "Modification of Sonic Boom Waveforms during propagation from source to ground", J. Acoust. Soc. Am. 111(1), pp. 481-486 (2002)
11. P. Boulanger, R. Raspet, H. E. Bass," Sonic boom propagation through a realistic turbulent atmosphere," J. Acoust. Soc. Am., 98(3), pp. 3412-3417, (1995)
12 B. W. Kennedy, R. O. Olsen, J. R. Fox, and G. M. Mitchler, "Joint Acoustic Propagation Experiment Project Summary," The Bionetics Corporation, Las Cruces, NM, September 1991.
13. W. L. Willshire, Jr., D. P. Garber, and D. W. deVilbiss, "The effect of turbulence on the propagation of sonic booms," Proceedings of the Fifth International Symposium on Long Range Sound Propagation, The Open University, Milton Keynes, England (May 1992).

Low boom airplane design process at Dassault Aviation

Nicolas Héron, Zdenek Johan, Michel Ravachol, Franck Dagrau

DASSAULT AVIATION, Direction Générale Technique, 92552 Saint-Cloud cedex, France

Abstract. Dassault Aviation as a civil aircraft manufacturer is studying the feasibility of a supersonic business jet. One of the prerequisites for an environmentally friendly and economically viable supersonic aircraft is that its sonic boom signature is reduced to a level acceptable for flight over populated areas. The method presented here is part of the Dassault Aviation toolbox dedicated to the evaluation and/or to the design process of a low boom constrained supersonic aircraft.

Keywords: sonic boom modeling, Euler flow computations, Fourier decomposition, ray-tracing process
PACS: 47.11.+j / 43.20.Dk / 43.28.Mw

NEAR FIELD PREDICTIONS

One of the methodologies used to predict sonic boom levels on the ground consists of using CFD results as inputs for a sonic boom propagation code (see ref. 4). Surface and volumic unstructured meshes are generated starting from the CAD definition of the aircraft shape. Those meshes and flight conditions are then used as an input of an Euler flow solver. The results of the Euler computations provide the pressure variations on the body shape and around the aircraft. On the one hand lift and drag coefficients are calculated. On the other hand these pressure variations are extracted on a cylinder in the flow field around the aircraft for sonic boom prediction. It has to be performed for a radius such as this cylinder is far enough from the aircraft in order to respect the assumptions implicit in the propagation code (see ref.1).

FIGURE 1. Body mesh of a supersonic business jet, including canards and side nacelles under the wings for Euler flow computations

CP838, *Innovations in Nonlinear Acoustics: 17th International Symposium on Nonlinear Acoustics*,
edited by A. A. Atchley, V. W. Sparrow, and R. M. Keolian

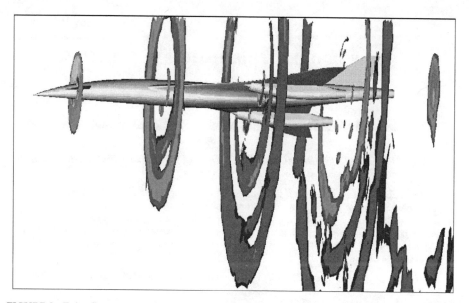

FIGURE 2. Euler flow computations around a supersonic business jet with canards and side nacelles. M=1.6, flight altitude=45000ft. From the left side to the right side, the slices of the pressure fields show the generation of wavefronts by the different parts of the aircraft shape: the first fronts are due to the nose and the canards (slice 1) then are added the impact of the wings (slice 2) and the side nacelles (slice 3).

SONIC BOOM ASSESSMENT

The pertinence of sonic boom evaluations depends strongly on what is extracted from the flow computation. CFD results have to be assessed according to the numerical scheme or mesh refinement. Trade-offs have to be made when it comes to extracting a pressure distribution from the CFD results. One would indeed want to extract the pressure as close to the airplane as possible since this is the zone where the CFD results have undergone the lesser numerical diffusion. Unfortunately, the equations solved for the propagation code (see ref. 1) are not correct in the vicinity of the aircraft: the propagation algorithm is a ray-tracing process based on equations that are valid if the source is regarded as locally axisymmetric. Respecting this criteria would imply the use of quasi-isotropic meshes from the body to the midfield(many times the length of the aircraft), that is not acceptable for reasonable computational cost.

near field to far field extrapolation

Using the spatial periodicity of the pressure variations around the aircraft (2π periodic), it is possible to write the solution as a modal distribution (see ref. 2,3), such as the general solution of the perturbation potential equation. It enables, by asymptotic development, to extrapolate from the CFD nearfield results, a solution which has the same radiation in the farfield and moreover verifies the assumptions of the propagation code.

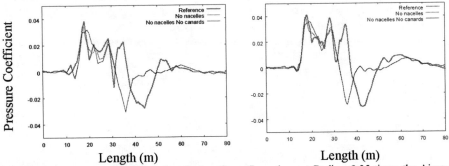

FIGURE 3. Pressure variations below 3 aircraft configurations at Radius=0.35 times the Aircraft length. M=1.6, 45000 ft, α=2°. Left curves are the extractions, Right curves are the multipoles distributions.

sonic boom prediction

The multipoles solution is used as input to the propagation code. The figure below shows the impact of side nacelles and canards on the sonic boom at the ground. This impact decreases with the growth of the angle of attack, the main lift effect generated by the wings leading the sonic boom sources.

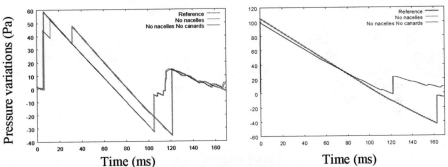

FIGURE 4. Sonic Boom at the ground for 3 Aircraft configurations. Flight Altitude = 45000 ft, Mach=1.6. Left curves are for α=2°, Right curves are for α=5°.

sonic boom minimization

In order to automate the low boom design process, two strategies are applied:
coupled approach with a combined nearfield optimization + farfield analysis. Aircraft shape optimizations (see ref.6) are performed to achieve some modifications of the pressure near field. Then the propagation model is used to check if the CFD results lead to reduce the sonic boom level at the ground as expected.

641

one-field optimization at near field relying on a pressure nearfield criteria which can be provided by inverse propagation models(starting from a low boom signature at the ground to the nearfield shape). For slender bodies, the approach relies directly to shape functions, as illustrated on the figure below. We have to notice that the drag of the optimized shape is twice this of the initial one.

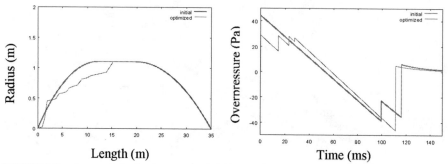

FIGURE 5. Fuselage optimization by genetic algorithm. Left curves describes the initial (grey line) and optimized (red line) shape functions. Right curves are the associated sonic boom at the ground.

CONCLUSION

The method based on CFD data to compute sonic boom is a very attractive alternative to older approach as the use of shapes function. The Euler computations are fast and more sensitive to design modifications. Moreover it allows the use of classical aerodynamic optimization methods for a low boom constrained design process. The remaining difficulties are (1) the definition of the target for the appropriate near-field pressure or for the sonic boom wave form on the ground, according to the optimization algorithm which will be performed (2) a more accurate modeling of sonic boom signatures(peak pressure(s), rise time(s), duration and how these criteria are impinging by meteorological and maneuvers effects). This second point is vital for reliable sonic boom loudness assessment.

REFERENCES

1. Thomas, Ch., Extrapolation of sonic boom pressure signatures by the waveform parameter method, NASA TN. D-6832, 1972.
2. George, A.R., Reduction of sonic boom by azimuthal redistribution of overpressure, AIAA Paper 68-159, AIAA 6th Aerospace Sciences Meeting, New-York, 1968
3. Page J.A. and Plotkin, K.J., An Efficient Method for Incorporating Computational Fluid Dynamics Into Sonic Boom Prediction, AIAA Paper 91-3275, AIAA 9th Applied Aerodynamics Conference, Baltimore, 1991.
4. Plotkin, K.J. and Page, J.A., Extrapolation of Sonic Boom signatures from CFD solutions, AIAA 2002-0922, 40th Aerospace Sciences Meeting & Exhibit, Reno, 2002.
5. Waldken F., The shock pattern of Wing-Body Combination, far from the flight path, Aeronautical Quarterly, Vol. 9, Pt.2, pp164-194, 1958.
6. Daumas, L., Dinh, Q.V., Kleinveld, S. and Rogé, G., Impact of optimization in the design process of supersonic business jet, Design Optimization International Conference, Athens, 2004.

Sonic Boom Research in Russia: Computational Methods Improvement, Aerodynamic Configuration Effects, Analysis of Meteorological Variability Influence

Sergey L. Chernyshev[*], Ludmila G. Ivanteeva[*], Andrey Ph. Kiselev[*], Victor V. Kovalenko[*], Peter P. Vorotnikov[*], Anatoly V. Rodnov[†], Leonid L.Teperin[*] and Yury A. Zavershnev[†]

[*]*Central Aerohydrodynamic Institute (TsAGI), Zhukovsky, Russia*

[†]*Flight Research Institute (LII), Zhukovsky, Russia*

Abstract. Presented paper is a brief overview of current efforts in sonic boom area taking place in the Russian Central Aerohydrodynamic Institute in cooperation with the Flight Research Institute. These efforts are focused on improvement of existing methods and development of new ones, minimization technique development, study of sonic boom atmospheric effect, search of low boom aircraft configurations in different weight categories, having good aerodynamic characteristics.

Growing interest to the development of supersonic transport (SST) requires additional efforts in studying and eventually solving the sonic boom problem. Without essential progress and acceptable solution of this problem it seems impossible to further proceed with a new generation SST. The tools available today for sonic boom research allow to determine various characteristics of this phenomenon with rather good accuracy. They can be used also for optimizing aircraft configuration to minimize shock wave impact. Particularly, quite efficient algorithms and computer codes of sonic boom calculation have been developed at TsAGI. They are based on the Euler equations integration for the near field and asymptotic quasi-linear acoustical solution for the far field [1].

The new method of designing a mean camber of SST wing surface has been recently developed. The design principle is in the construction of a mean camber line of the wing profiles, providing minimum pressure drag induced by the lift and, simultaneously, in the fulfillment of sonic boom level restrictions. The method is applied at the cruise regime of flight, which is determined by Mach number and flight altitude. The wing plan form, thickness distribution and weight of an airplane are considered to be given. The classical theory is applied to the problem of sonic boom minimization; the use of this makes it possible to determine the optimum form of a sonic boom signature for given values of airplane's length, weight and Mach number. The type of optimum signature (flat-top, ramp or hybrid type) is a parameter of the

CP838, *Innovations in Nonlinear Acoustics: 17th International Symposium on Nonlinear Acoustics*, edited by A. A. Atchley, V. W. Sparrow, and R. M. Keolian
© 2006 American Institute of Physics 0-7354-0330-9/06/$23.00

task. For the given profile of the sonic boom signature, it is possible to determine the cross section area distribution of equivalent body of revolution that gives the same as airplane the shape and intensity of shock wave on the ground. The area distribution derivative of equivalent body of revolution is an input data for the optimization task.

To connect a perturbed field near an airplane (near field) with the form of an equivalent body of revolution, the theorem of Professor Yu.L.Zhilin [2] is applied. This theorem connects the given derivative of area distribution with an integral of perturbed velocity components. The panel method of the linear theory is used in the given work to solve the problem of supersonic flow and determine the near field parameters and aerodynamic characteristics. Functional to be minimized is a function dependent on the distribution of intensity of vortexes. The condition of extremality leads to the system of linear algebraic equations (with reference to unknown vortex strengths and Lagrange multipliers). Aerodynamic coefficients and inclinations of the mean camber surface are determined by predicted vortexes intensities.

In case of "Flying wing", the signature of minimum sonic boom and corresponding distribution of derivative of cross section area of equivalent body of revolution (Figure 1a) were set for the following cruise flight regime parameters: weight 188 t, altitude 16500 m, Mach number 2, airplane length 54 m.

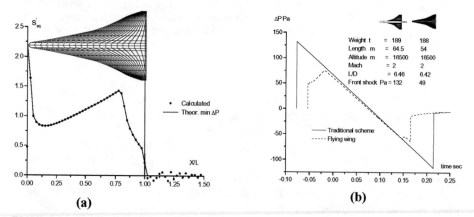

FIGURE 1. Distribution of theoretical and predicted optimum derivatives of cross-section area of equivalent body of revolution (a) and comparison of sonic boom signature for aircraft of traditional type and for "Flying wing" configuration (b).

Optimum distribution of derivative of cross section area of equivalent body of revolution (Figure 1a) was introduced into the task of designing the mean camber surface of flying wing, as a restriction. Distribution of derivative of cross section area of equivalent body of revolution obtained from optimum problem solution is shown in Figure 1a by markers. The comparison of two configurations, traditional one (like the Tu-144 aircraft) and "Flying wing", is shown in Figure 1b. The head shock intensity of sonic boom wave for the "Flying wing" is seen to be reduce up to the level being in 2.5 times lower, than for traditional configuration at the same lift-to-drag ratio. It is possible to achieve this due to a greater lift-to-drag ratio possessed by configuration "Flying wing" in case of discarded restrictions on the form of the equivalent body of revolution. Restrictions on the sonic boom level inevitably lead to an inferior lift-to-drag ratio coming to the level of the traditional configuration.

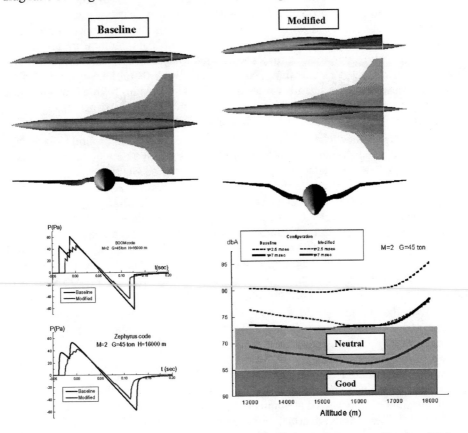

FIGURE 2. Influence of special configuration forming on sonic-boom signatures and loudness levels.

At present, the task of sonic boom exposure limit determination takes the critical importance. This task requires more thorough knowledge of the mechanism of this phenomenon as well as particular qualities of shock waves propagation and its esthesia depending on various flight regimes, weather conditions, surface geometry, etc.

Based on recently performed numerous parametric calculations the area of «design-acceptable» variations of SST configuration has been defined. Two programs were used to calculate a pressure profile in the wave of sonic boom: BOOM program that uses a classic theory and ZEPHYRUS [3] program that accounts for adsorption and dispersion of frequency components of the package of disturbances.

The choice of SST configuration is stipulated by their influence on the sonic boom signature. Some results of numerical investigations of the influence of parameters fixing the geometry (distribution of body thicknesses, angles of wing setting and coordinates of wing placing) on signatures of disturbances in the sonic boom wave are given in Figure 2. Low boom aircraft configuration means not only lower maximum overpressure in a shock wave but smoother pressure distribution $P=f(t)$ with no intermediate jumps and, preferably, with prolonged time of shock pressure raise. Besides smoother pressure signature is quieter, having lower loudness. Results of investigations show the opportunity of creation of SSBJ with mass $G \sim 45\,t$ and general length about 40 m with "admissible" sonic boom at cruise flight.

Recently some features of secondary sonic boom propagation have been investigated. Procedure to pass secondary sonic boom wave through turnaround point and to extend the problem solution continuously into secondary zones of influence have been developed for stratified atmosphere with horizontal wind for an airplane flying along an arbitrary trajectory without restriction on its flight regime. Conditions for caustic occurrence in turnaround point have been considered. Computational results have revealed some features of SSB propagation. Particularly, the essential effect of acceleration on elementary ray tube area is demonstrated. Effect of descent rate is essential and it is analogous to the effect of acceleration. In reality we used to deal with decelerated landing regime and then well-known ZEPHYRUS code [3] is not quite correct, because it operate only for steady level flight regime.

With the aim to reveal the relationships between parameters of the sonic boom wave and the atmosphere turbulence characteristics, the available experimental data concerning the influence of local atmospheric non-homogeneity on sonic boom signatures were analyzed. The following cases were considered for summer-autumn: nominal meteorological conditions (fine weather), developed thermal turbulence, dynamic turbulence and clouds. Normal logarithmic law describes the scattering of experimental data measured at nominal conditions. The maximum variations from this law occur when strong thermal turbulence is presented. The correlation between characteristics of data scattering and the time period, when the measurements were carried out, as well as atmosphere turbulence parameters was ascertained.

ACKNOWLEDGMENTS

The International Science and Technology Center (ISTC) under Project No. 2249 supported some lines of these investigations.

REFERENCES

1. Ivanteeva L.G., Kovalenko V.V., Pavlyokov E.V., Teperin L.L., Racl R.G. Validation of sonic boom propagation codes using SR-71 flight test data. *J. Acoust. Soc. Am.*, 111, No. 1, Pt 2. 2002.
2. Zhilin Yu.L., Kovalenko V.V. On relationship of near and far fields in the sonic boom task. *Uchenye Zapiski TsAGI*, Volume XXIX, № 3-4, 1998.
3. Robinson L.D. 1991, Sonic boom propagation through an inhomogeneous, windy atmosphere. *Ph.D. dissertation, University of Texas.*

Flight Demonstration Of Low Overpressure N-Wave Sonic Booms And Evanescent Waves

Edward A. Haering, Jr.*, James W. Smolka†, James E. Murray*, and Kenneth J. Plotkin¶

*Research Aerodynamics, NASA Dryden Flight Research Center, Edwards, CA 93523, USA
†Research Test Pilot, NASA Dryden Flight Research Center, Edwards, CA 93523, USA
¶Chief Scientist, Wyle Laboratories, Inc., Arlington, VA 22202, USA

Abstract. The recent flight demonstration of shaped sonic booms shows the potential for quiet overland supersonic flight, which could revolutionize air transport. To successfully design quiet supersonic aircraft, the upper limit of an acceptable noise level must be determined through quantitative recording and subjective human response measurements. Past efforts have concentrated on the use of sonic boom simulators to assess human response, but simulators often cannot reproduce a realistic sonic boom sound. Until now, molecular relaxation effects on low overpressure rise time had never been compared with flight data. Supersonic flight slower than the cutoff Mach number, which generates evanescent waves, also prevents loud sonic booms from impacting the ground. The loudness of these evanescent waves can be computed, but flight measurement validation is needed. A novel flight demonstration technique that generates low overpressure N-waves using conventional military aircraft is outlined, in addition to initial quantitative flight data. As part of this demonstration, evanescent waves also will be recorded.

Keywords: Sonic Boom, Flight Data, Evanescent Wave
PACS: 43.25.Cb, 43.28.Gq, 43.28.Mw, 43.50.Lj, 43.50.Rq, 43.50.Yw

HYPOTHESIS

The inspiration for producing low overpressure N-wave sonic booms originated with the recent measurement of a sonic boom generated by a sounding rocket upon descent[1]. This vehicle was in a very steep dive at a high altitude and low Mach number when it generated the sonic boom that hit the recorder. Sonic booms of this type were desired for recording and analysis, so the sonic boom propagation code PCBoom4[2] was used to look at similar trajectories. Because additional flights of this sounding rocket would be infrequent or nonexistent, alternative available aircraft trajectories were modeled with a multitude of PCBoom4 runs. An aircraft in a steep dive at a high supersonic Mach number was found to generate low overpressures far forward of the dive point. It is hypothesized that these low amplitude booms could be used for human acceptability studies leading to a supersonic aircraft quiet enough for overland flight.

Four attributes have been determined to contribute to low overpressure N-waves generated by conventional aircraft. The first attribute is a very long propagation distance. High vehicle altitude increases propagation distance. The maximum propagation distance occurs when the ray-path extends to near the lateral cutoff.

CP838, Innovations in Nonlinear Acoustics: 17th International Symposium on Nonlinear Acoustics, edited by A. A. Atchley, V. W. Sparrow, and R. M. Keolian
© 2006 American Institute of Physics 0-7354-0330-9/06/$23.00

The second attribute is the use of shock waves coming from the top of the vehicle. Near-field shock wave probing above and below a B-58 aircraft[3] shows that the shock waves coming from the top of the vehicle are less than those coming from the bottom because of the lift distribution around the aircraft. For the upper surface shock waves to reach the ground (as the primary sonic boom, excluding "over-the-top" booms), the vehicle must be in a dive. An additional benefit of the dive is that less lift is generated on the aircraft, causing less intense shock waves caused by lift.

The third attribute of low overpressure N-waves is the use of the smallest vehicle possible. Although an F/A-18B aircraft was used in this effort, an F-16 aircraft would produce a smaller boom, because it has less boom due to volume. A T-38 aircraft also was considered, but it has the disadvantages of less avionics to aid the pilot in the maneuver and engines that are less forgiving in the high altitude supersonic flight regime. The F/A-18B aircraft was selected, because it was readily available at the NASA Dryden Flight Research Center (Edwards, California), and it had been equipped with the appropriate instrumentation.

The fourth and last attribute is the minimization of Mach number. When a particular Mach number and dive angle was found to produce the desired boom levels on the ground, both Mach number and dive angle could be reduced while maintaining a constant shock wave direction, thus preserving the long propagation path. A lower supersonic Mach number and dive angle also created a condition in which the maneuver was easier to achieve, and more importantly, easier to recover from.

A way to prevent the sonic boom from reaching the ground is to fly slower than the Mach cutoff condition (in the range of Mach 1.1 to 1.3) such that the shock wave refracts up away from the ground. On the ground, only an evanescent wave may be heard, and if heard, would sound much like distant thunder. Flight research on evanescent waves was conducted more than 30 years ago[4], but no high-fidelity recordings were acquired that can be used in human acceptability studies. When they are acquired, such recordings then can be used to validate computational models[5].

EXPERIMENT SETUP

A flight research project called the Low Boom/No Boom Experiment has been initiated. Four flight phases are planned for this project, but the overarching objective is to aid in the determination of an acceptable noise level for certifiable overland supersonic vehicles. Table 1 presents the specific objective of each flight phase.

TABLE 1. Low Boom/No Boom Experiment Objectives.

Flight Phase	Low Boom (low overpressure N-waves)	No Boom (evanescent waves)
1	Assess feasibility and repeatability of generating low overpressure N-waves (<0.6 lbf/ft^2) in a specified geographic area	Gather limited digital high-fidelity microphone data for validation of evanescent wave loudness predictions
2	Gather high-fidelity, statistically significant data for loudness levels and operational considerations (atmospheric variability effects)	
3	Perform outdoor human response surveys with flights within the restricted airspace	
4	Perform outdoor human response surveys with flights outside the restricted airspace (populated areas)	

To implement the first phase of this experiment, a NASA Dryden F/A-18B aircraft equipped with several instrumentation packages is employed. A Research Quick Data System (RQDS) acts as the 1553 bus monitor of the aircraft parameters, including airdata and inertial navigation system parameters, which are telemetered to the ground for recording and real-time sonic boom footprint calculations. A time code generator with an embedded global positioning system (GPS) receiver is used to time-stamp the RQDS data.

In the transonic flight region the F/A-18B aircraft is known to have airdata position errors[6] on the order of 2,000 ft of altitude. Additionally, the steep dives that are performed may induce significant pneumatic lag in the aircraft airdata system. To provide the necessary additional calibration for position error and lag, a carrier-phase differential GPS receiver (that provides accurate inertial position and speeds) was installed in the aircraft. The initiation of the supersonic dive involves rolling the aircraft to an inverted attitude before diving, causing the GPS to lose data. For these maneuvers, ground-based radar tracking was obtained using a C-band radar beacon on the aircraft.

For the airdata calibrations and PCBoom4 sonic boom propagation codes, accurate measurements of the atmospheric conditions are needed. The GPS rawinsonde balloons and wind profiler data are used for atmospheric analysis. Some of these sensors will be deployed at the location of the microphone array.

Microphones were placed over many miles of relatively flat desert terrain under a supersonic corridor. The spacing of the microphones was chosen to obtain a range of overpressures while allowing for atmospheric variability in footprint location. Several NASA Dryden Boom Amplitude and Shape Sensors (BASS) and Boom Amplitude and Direction Sensors (BADS[7]) were used to record the low booms in phase 1. A few Bruel and Kjær (Denmark) 4193 microphones will be used to record evanescent waves for phase 1.

Various dive profiles first were practiced in the NASA Dryden F/A-18 flight simulator. The current dive profile involves flying at a level attitude, high subsonic speed, and altitude of nearly 50,000 ft. The aircraft is rolled to an inverted attitude; a positive g pull to the desired dive angle of 53° downward then is initiated, while the throttle is pulled to the idle position to avoid excessive speed. When the desired dive angle is reached, the aircraft is rolled to an upright attitude, and a Mach number of approximately 1.1 is achieved. At an altitude of 38,000 ft a pull-up is executed to recover the aircraft at an altitude of approximately 32,000 ft. The F/A-18B aircraft has an angle-of-attack limit in this supersonic flight regime, so angle of attack is closely monitored. The F/A-18 avionics allows a dive point to be displayed on the head-up display (HUD), which greatly aids in maintaining the proper dive angle and heading. The PCBoom4 runs of the simulation data predicted overpressures down to approximately 0.2 lbf/ft^2.

INITIAL RESULTS

The supersonic dives were practiced during a checkout flight of the RQDS system on the F/A-18B aircraft. At the same time, the BASS recorders underwent field checkouts while candidate microphone locations were sought. Although the surface weather was sometimes quite breezy, low overpressure N-wave sonic booms were recorded on multiple dives. Figure 1 shows two sets of these sonic boom measurements.

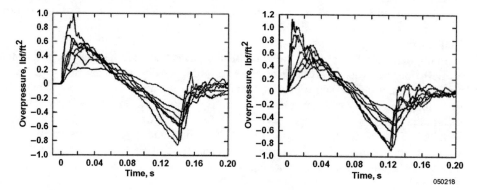

FIGURE 1. Initial ground-level sonic boom measurements from the F/A-18B aircraft in a supersonic low boom dive. Dive no. 3 is on the left, and dive no. 5 is on the right (April 1, 2005).

The microphones were spaced apart over 6 miles parallel to the ground track, and an additional microphone was placed 1 mile off of the track. The ground sensor sites located farthest from the aircraft generally show the lowest overpressure, and overpressure increases as the dive location is approached. Atmospheric turbulence is evident with repeated artifacts in each bow and tail shock. The rise time of the signature greatly increases with decreasing maximum overpressure, as much as 40 ms for one signature. The airdata calibration and lag determination has not yet been completed, and no radar tracking or local weather balloon data were obtained from this flight, but the experiment participants were very pleased to have obtained low overpressure N-waves on the first attempt, even in nonoptimal weather conditions. Such low overpressure sonic booms bracket the probable acceptable noise level and can be used for human subjective testing.

REFERENCES

1. Plotkin, K. J., Haering, E. A., Jr., and Murray, J. E., "Low-Amplitude Sonic Boom From a Descending Sounding Rocket," *Proceedings of the 17th International Symposium on Nonlinear Acoustics*, American Institute of Physics, Melville, New York, July 2005.
2. Plotkin, K. J., and Grandi, F., *Computer Models for Sonic Boom Analysis: PCBoom4, CABoom, BooMap, CORBoom*, Wyle Report WR 02-11, June 2002.
3. Maglieri, D. J., Ritchie, V. S., and Bryant, J. F., Jr., *In-Flight Shock-Wave Pressure Measurements Above and Below a Bomber Airplane at Mach Numbers from 1.42 to 1.69*, NASA TN D-1968, 1963.
4. Haglund, G. T., and Kane, E. J., *Flight Test Measurements and Analysis of Sonic Boom Phenomena Near the Shock Wave Extremity*, NASA CR-2167, 1973.
5. Fung, K. Y., "Shock Wave Formation at a Caustic," *SIAM Journal of Applied Mathematics*, Vol. 39, No. 2, pp. 355–371, Oct. 1980.
6. Haering, E. A., Jr., *Airdata Measurement and Calibration*, NASA TM-104316, 1995.
7. Plotkin, K. J., Haering, E. A., Jr., Murray, J. E., Maglieri, D. J., Salamone, J., Sullivan, B. M., and Schein, D., "Ground Data Collection of Shaped Sonic Boom Experiment Aircraft Pressure Signatures," AIAA-2005-010, Jan. 2005.

Laboratory experiments to study N-waves propagation: Effects of turbulence and/or ground roughness.

Philippe Blanc-Benon[*], Sébastien Ollivier[*],
Keith Attenborough[†], Qin Qin[†]

[*] LMFA, UMR CNRS 5509, Ecole Centrale de Lyon, 69134 Ecully Cedex, France
[†] Acoustic Research Centre, Dept. of Engineering, University of Hull, UK

Abstract. Model experiments have done in the anechoic chamber of the Ecole Centrale de Lyon in order to investigate long range propagation of short duration N-waves under various configurations. Measurements have been done in a free field configuration and in the shadow zone created by a curved surface. The surface was either smooth or rough.

Keywords: propagation, turbulence, N-wave, shadow zone.
PACS: 43.28.Mw, 43.28.En .

EXPERIMENTAL SET-UP

Model experiments have been done in the anechoic chamber of the Ecole Centrale de Lyon in order to investigate long range propagation of N-waves under various configurations, including turbulence and/or ground effects. Short duration N-waves (duration 40 µs and rise time 3 µs measured at 1 meter) are generated by electric sparks. The pressure waves are measured by using Bruel & Kjaër 1/8" microphones mounted in a baffle, and recorded by using a 5MHz sampling rate data acquisition system. Two devices are used to generated either velocity or thermal turbulence. Velocity turbulence is generated by a one meter width turbulent plane jet of variable speed. The velocity fluctuations v_{RMS} along the acoustic wave propagation path vary from 0 to 2.35 m/s. The propagation distance is 1.4 meters. Random temperature fluctuations were produced by using a 4.4m×1.1m heated grid of resistors. The distance of propagation vary from 0.6 m to 4.5 m for a fixed level of turbulence, the temperature fluctuation T_{RMS} is of the order of 1.4 K. In both cases, the outer length scale is of the order of 10cm. With turbulence 1000 pressure waveforms have been recorded for each source-receiver distance in order to perform statistical analysis. In the case of steady atmosphere (without turbulence) 100 waveforms have been recorded for reference.

CP838, *Innovations in Nonlinear Acoustics: 17th International Symposium on Nonlinear Acoustics*,
edited by A. A. Atchley, V. W. Sparrow, and R. M. Keolian
© 2006 American Institute of Physics 0-7354-0330-9/06/$23.00

EFFECT OF TURBULENCE

The peak pressure level and the rise time of initial N-waves have been analyzed after propagation through thermal or kinematic turbulence without any boundary effect. The mean peak pressure computed over the measurements at each receiver position tends to decrease both with the turbulence level and the propagation distance through turbulence. Nevertheless, in some cases the level of the maximum peak pressure is clearly amplified by turbulence [1] (up to a factor 3 in some cases). The mean rise time increases both with the turbulence level and the propagation distance. The statistical analysis of the data at each position shows that the spreading of the peak pressure and rise time at a given position is much larger than the mean variations due to the turbulence level or the propagation distance. The probability distributions are not symmetric, skewness increases with the propagation distance and with the turbulence level. No rise time is significantly shortened by turbulence, and most shortened rise times correspond to amplified waves. The change of behavior between the propagation in a steady atmosphere and the propagation through turbulence is significant only when the propagation distance is higher than the distance were focusing effects due to random caustics occur (figure 1). This is in accordance with previous numerical simulations [2].

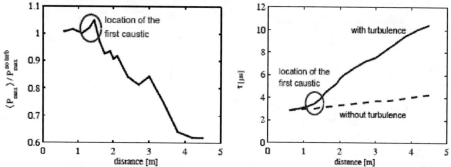

FIGURE 1. Peak pressure level and rise time as functions of the propagation distance through turbulence. Left: ratio (mean peak pressure with turbulence) / (mean peak pressure without turbulence). Right: mean front shock rise time measured with turbulence (——) and without turbulence (- - -).

PROPAGATION INTO A SHADOW ZONE

Measurements have also been done close to a curved surface boundary which was used to create a shadow zone. The radius of the cylindrical surface is 2 meters. This configuration simulates the propagation of sonic booms near a curved ground surface. It is also analogous to the effect of a temperature or wind gradient. The rigid plastic surface was smooth, additional roughness was obtained by gluing 0.8 mm height sandpaper. Measurements close to the boundary have been done with and without thermal turbulence, allowing the study of simultaneous effects of turbulence and ground roughness on weak acoustic shocks propagation in the shadow zone.

Effect of turbulence

The results obtained in the absence of turbulence over the smooth curved surface vary as expected: the peak pressure decreases and the rise time increases as the receivers are moved deeper into the shadow zone. As expected from the experiments in the free field configuration, turbulence decreases the mean level of the peak pressure in the illuminated zone and increases the mean rise time. On the contrary, in the shadow zone turbulence increases the level of the peak pressure and decreases the rise time. These effects are the results of scattering by turbulence.

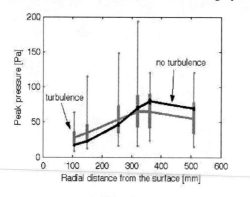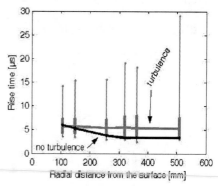

FIGURE 2. Mean peak pressure level (left) and mean rise time (right) as functions of the radial distance from the surface, with and without turbulence. The two highest positions of receivers are in the illuminated zone. The range between the maximum and minimum values are indicated by the thin vertical lines; the thick vertical lines indicate the range of values for 75% of the data.

Effect of roughness

Without turbulence, comparison of measurements made with the smooth and with the rough curved surfaces show that roughness slightly increases the level of the peak pressure in the illuminated zone and slightly decreases the level into the shadow zone. The effect of roughness is more significant on the rise time, which is clearly increased by roughness (figure 3). Spectral analysis and comparison with residue series prediction shows that high frequencies are more influenced by scattering from the surface roughness [3].

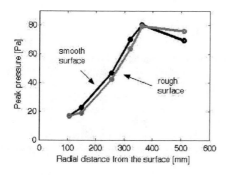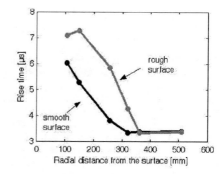

FIGURE 3. Measurements into the shadow zone without turbulence, close to the rough and the smooth curved surfaces. Peak pressure level (left) and rise time (right) as functions of the radial distance from the surface. The two highest positions of receivers are in the illuminated zone.

Combined effects of roughness and turbulence

Close to the rough curved surface, comparison of measurements with and without turbulence show that roughness decreases the mean peak pressure and increases the mean rise time. The spreading of the peak pressure and rise time values are not modified. The decrease of the mean peak pressure induced by roughness is enhanced by turbulence. The increase of the mean rise time induced by roughness is of the same order of magnitude as the increase measured without turbulence. (figures 3 & 4).

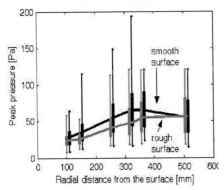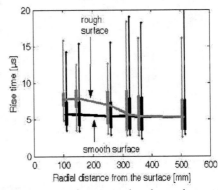

FIGURE 4. Measurement with turbulence into the shadow zone, close to rough and smooth curved surfaces. Mean peak pressure level (left) and rise time (right) as functions of the radial distance from the surface. The two highest positions of receivers are in the illuminated zone. The range between the maximum and minimum values are indicated by the thin vertical lines; the thick vertical lines indicate the range of values for 75% of the data.

ACKNOWLEDGMENTS

This work is partly supported by the European Community (SOBER project, Contract No. G4RD-CT-2000-00398) and by the French government (decision No. 00 T0116).

REFERENCES

1. Ph. Blanc-Benon & S. Ollivier, Model experiments to study acoustic N-waves propagation through turbulence, *11th Long Range Sound Propagation Symposium*, Lake Morey Resort, VT, 1-3 June 2004.
2. Ph. Blanc-Benon, B. Lipkens, L. Dallois, M.F. Hamilton & D.T. Blackstock, Propagation of finite amplitude sound through turbulence: modeling with geometrical acoustics and the parabolic approximation, *J. Acoust. Soc. Am.*, 111(1) Pt 2, 2002, pp 487-498.
3. Qin Q., K. Attenborough, S. Ollivier, & Ph. Blanc-Benon, Effects of surface roughness and turbulence on propagation of shock waves above a curved surface, *11th Long Range Sound Propagation Symposium*, Lake Morey Resort, VT, 1-3 June 2004.

Sonic Booms And Building Vibration Revisited

Louis C. Sutherland, Consultant in Acoustics, 27803 Longhill Dr., Rancho Palos Verdes, CA. 90275, Karl D. Kryter, Adjunct Research Professor, San Diego State University, 3969 Maricopa Dr., Santa Barbara, CA 93110 and Joseph Czech, Wyle Laboratories, 128 Maryland St. El Segundo, CA 90245

Abstract. Lessons learned from the 1960's sonic boom tests at St. Louis, Oklahoma City and at Edwards Air Force Base (EAFB) and more recently in communities near EAFB and Nellis AFB are briefly reviewed from the standpoint of building vibration and rattle response induced by the sonic boom signature. Available data on the vibro-acoustic threshold of rattle are considered along with the principal sonic boom signature parameters, peak overpressure and duration, which drive the low frequency vibration response of buildings to sonic booms. Implications for the current effort to develop an acceptable sonic boom signature are considered with this overview of current understanding of building vibration response to sonic booms. Possible gaps in this current knowledge for current technology boom signatures are considered.

Keywords: Sonic Boom, Building Vibration, Rattle
PACS: 43.28 Mw, 43.50 Qp

Introduction

In a 1971 paper discussing the 51- 46 Senate vote to cancel support of the SST, a spokesman for the National Center for Atmospheric Research stated: "Although the SST was defeated in Congress for economic as well as environmental reasons, it was really the prospect of environmental damage that brought the public into the fight"

Thus, it seems appropriate at this conference to ask if we have the information needed to assure the public (and Congress) that an environmentally-acceptable, economically-viable, supersonic commercial aircraft is now possible. This paper focuses on just one aspect of this question – the problem of residential building vibration or rattle induced by sonic booms.

Prior Assessments of Subjective Response to Sonic Booms

Most of the studies of subjective or community response to sonic boom carried out in the U.S. involved survey respondents either living in their own homes[1,2,4] or interviewed while experiencing sonic booms inside a fully furnished home or while outdoors[3]. The surveys given to those living in-home asked what was the <u>primary</u>

CP838, *Innovations in Nonlinear Acoustics: 17th International Symposium on Nonlinear Acoustics*,
edited by A. A. Atchley, V. W. Sparrow, and R. M. Keolian
© 2006 American Institute of Physics 0-7354-0330-9/06/$23.00

source of annoyance with the sonic booms[1,2,4] and, for the more recent Western USA Sonic Boom study [4], also asked what was the most disturbing aspect of their sonic boom exposure.

A summary of these data is shown in Table 1(a) (Annoyance) and 1(b) (Disturbing). The txable shows that the dominant source of annoyance was building vibration from all three surveys asking this question but for the Western USA study, the most disturbing factor was startle. (As Kryter has observed,[5] "startle response" (to a noise) depends jointly on the suddenness of the onset and intensity of the noise..". Since these are also the controlling factors for loudness of an impulsive sound, it is reasonable to assume that loudness is a suitable surrogate for startle and is thus customarily used as a basic measure of the subjective impact of sonic booms. NASA, Langley has been one of the leaders in studies of loudness of sonic booms as exemplified by the next paper in this program covering loudness studies of low boom wave forms. This research clearly demonstrates that the ability exists to accurately measure the loudness of real or simulated sonic booms.

However, "percent highly annoyed" is also a universally recognized descriptor for assessment of community noise exposure. Thus, the authors believe that the dominance of building vibration as the strongest source of annoyance, dictates the need to place more emphasis on this aspect of defining an acceptable sonic boom.

Table 1   Percent Responses to Survey Questions:" What Is Very Much or Moderately Annoying" or "What Is Most Disturbing" To You About Sonic Booms? (Respondents)

Response Factor	--------- a) Very Much or Moderately Annoying --------				b) Disturbing
	Oklahoma City (3,000)	St. Louis (2,300)	Western USA Study (1,573)	Avg of all three studies	Western USA Study
Vibration	53.7 %	37.7 %	45.2 %	45.5 %	24.0 %
Startle	27.7	30.7	34.8	31.1	42.0
Sleep/Rest	14.0	18.9	NA	16.5	NA
Talking/Radio/TV	8.1	8.1	NA	8.1	NA
Building Damage					21.0
Noisiness					10.4

(Note that data from the two survey areas (A and B) in the Western USA Study[4] showed marked differences in acceptance of sonic booms of the same magnitude but relative differences between "vibration" and "startle" annoyance responses were similar for each area.)

Further possible support for the significance of building vibration and rattle can be inferred from the 1966 study by Kryter, et al[3] at Edwards AFB. The sonic booms or aircraft noise heard indoors were judged more acceptable than when heard outdoors as expected due to the building noise attenuation. However, this higher acceptability corresponded to only about a 5 dB reduction in the perceived sound indoors in contrast to an expected 10 dB building noise reduction at the dominant low frequencies. This inconsistency is believed due to the added influence of rattle noise inside the test residences. Kryter has thus proposed that high intensity, low frequency sounds such as sonic booms, heard indoors, should be rated as about 5 dB "less acceptable" due to building vibration than sounds without vibration effects.[5]

Predicting Building Vibration and Rattle from Sonic Booms.

The vibro-acoustic response of a building wall can be defined by the product $|P(j\omega)|$ x $|R(j\omega)|$ where $R(j\omega)$ is of the complex Frequency Response Function defining the ratio of acceleration response to an acoustic excitation and $P(j\omega)$ is the complex Fourier Transform, of the acoustic excitation.[6] $|P(j\omega)|$ can be derived from the Sound Exposure Spectral Density, $E_s(f) = 2|P(j\omega)|^2$. For an ideal sonic boom N-wave with a peak pressure P_f and duration T, at frequency f, $E_s(f)$ is given by:

$$E_s(f) = 2\{P_f/(\pi f)| \ [\sin(\pi f T)/\pi f T - \cos(\pi f T)]\}^2 \tag{1}$$

The asymptotic value for $fT \to 0$ is $E_s(f) \to 2 \ [\pi P_f fT/3]^2$ and for $fT \to \infty$, $E_s(f) \to 2 \ [P_f/(\pi f)]^2$. (Note that the high frequency portion of this ideal N wave spectrum is <u>independent of the duration.</u>). These two asymptotes meet at spectrum peak frequency, $f_{max} = [\sqrt{3}/(\pi T)]$. Figure 1 shows the <u>envelope</u> of maximum values for the Sound Exposure Spectral Density for the ideal N-wave and for five other idealized low boom wave forms, all with a peak pressure of 1 psf and a duration, T of 200 ms.

Figure 1 Sound Exposure Spectral Density of Booms

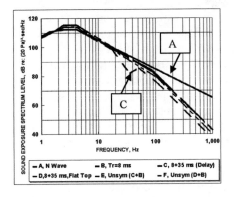

The alternate 1 psf wave shapes were

A – Ideal N- wave
B - N-wave with 8 ms rise time
C. - 8 ms rise to 0.5 psf, 35 rise to 1 psf (symmetric)
D - 8 ms rise to 1 psf flat top with 35 ms duration (symmetric)
E - Front shock as C, back as B.
F - Front shock as D, back as B.

Note the modest change in level for all but wave form C at building resonance frequencies in the range of 15 to 50 Hz. For C, loudness is 21 dB lower than A.

Next consider prediction of building vibration based on extensive NASA tests summarized in Table 2. Applying the dimensionless empirical expression for the peak acceleration A_{pk} as a function of f_oT at the bottom of the table, the estimated A_{pk} for $P_{pk} = 0.4$ psf and $T = 200$ ms indicates a value for A_{pk} for a wall with $w = 5$ psf and typical resonance frequency, f_o of 20 Hz to be about 0.048 g $\pm$ 100 %. This compares to an estimated wall vibration rattle threshold of about 0.045 g $\pm$ 50% developed in Ref. 6 from the available experimental (primarily NASA) data[6]. However, the above estimate of wall vibration does not take into account the possible modest reduction in the effective acoustical excitation due to modified boom shapes suggested by Fig. 1.

Table 2 Summary of Building Vibration Data Measured at EAFB in 1966 [3,6]

Building Element	Type of A/C	No. of Tests	Avg T	f_0	$f_0 T$	Surface Wt. (w)	Measured $-A_{pk} \cdot w / P_{pk}$ (g's psf/psf)	Regression [a]
			ms	Hz	-	Psf	-	-
Ceiling	B58	149	164	16.0	2.6	9.8	1.17	0.92
Ceiling	F104	59	79	16.0	1.3	9.8	2.09	1.84
Floor	B58	343	187	20.8	3.9	5.0	0.26	0.60
Wall	B58	170	165	21.0	3.5	4.8	0.86	0.64
Wall	F104	158	79	21.0	1.7	4.8	1.06	1.51
All	XB70	3	267	19.0	5.1	6.8	0.80	0.59

[a] $A_{pk} \cdot w / P_{pk} = 3.24 - 1.31(f_0 T) + 0.162(f_0 T)^2$ for $f_0 T \leq 4$ or 0.60 for $f_0 T > 4$
where A_{pk} = peak g's, w = surface weight and P_{pk} = peak pressure, both in psf.
Estimated accuracy $\approx \pm 100$ % (± 6 dB) for conventional boom shapes.

Conclusions

The key points for this overview of vibration in sonic boom environment are:
1) Previous studies consistently show people being "most annoyed" by building vibration or rattle by sonic boom but the recent Western USA study[4] showed people were "most disturbed" by startle from the sonic boom.
2) The acoustical spectra of possible "reduced boom" shapes show very little reduction in anticipated building vibration due only to the shape of the boom with the exception of booms with a delayed secondary rise and fall of, say, 35 ms after, and before an initial rise and final fall of, say, 8 ms. (Type C in Fig.1)
3) The vibration of building walls can be predicted from both analytical and empirical, experimentally-derived models with sufficient accuracy to estimate the approximate potential for onset of rattle in building elements.
4) Reduction in peak sonic boom pressure is the most reliable means of minimizing building rattle and accompanying annoyance response.
5) Non-flight testing of sonic boom acceptance should incorporate realistic simulation of building vibration and rattle along with measurement of loudness (or startle) to help define the trade-off between building vibration and rattle vs loudness and startle in determining the subjective acceptance of sonic booms.

REFERENCES

1. Nixon, C.W. and Borsky, P.N., "Effects of Sonic Boom on People: St. Louis, Missouri, *J. Acoust. Soc. Am.* 39, 551-558, (1964).
2. Borksy, P.N., "Community Reactions to Sonic Booms in the Oklahoma City Area, (Parts I & II), Wright-Patterson AFB, AMRL-TR-65-37, (1965)
3. Kryter, K.D., Johnson, D.J. and Young, J.R: "Psychological Experiments on Sonic Booms Conducted at Edwards AF Base". Contract AF 49 (638)-1758, Stanford Res. Inst., August (1968).
4. Fields, J..M., "Reactions of Residents to Long-Term Sonic Boom Noise Environments", NASA CR 2 01704 by Wyle Laboratories, June, (1997).
5. Kryter, K.D., *The Effects of Noise on Man*, Second Edition, Academic Press, San Diego, (1985).
6. Sutherland, L.C. and Czech, J., "Evaluation of Human Response to Structural Vibrations Induced by Sonic Booms", Wyle Labs, NASA CR 189584, (1992).

Research On Subjective Response To Simulated Sonic Booms At NASA Langley Research Center

Brenda M. Sullivan

NASA Langley Research Center, Hampton, Virginia, 23681

Abstract. Over the past 15 years, NASA Langley Research Center has conducted many tests investigating subjective response to simulated sonic booms. Most tests have used the Sonic Boom Booth, an airtight concrete booth fitted with loudspeakers that play synthesized sonic booms pre-processed to compensate for the response of the booth/loudspeaker system. Tests using the Booth have included investigations of shaped booms, booms with simulated ground reflections, recorded booms, outdoor and indoor booms, booms with differing loudness for bow and tail shocks, and comparisons of aircraft flyover recordings with sonic booms. Another study used loudspeakers placed inside people's houses, so that they could experience the booms while in their own homes. This study investigated the reactions of people to different numbers of booms heard within a 24-hour period. The most recent Booth test used predicted boom shapes from candidate low-boom aircraft. At present, a test to compare the Booth with boom simulators constructed by Gulfstream Aerospace Corporation and Lockheed Martin Aeronautics Company is underway. The Lockheed simulator is an airtight booth similar to the Langley booth; the Gulfstream booth uses a traveling wave method to create the booms. Comparison of "realism" as well as loudness and other descriptors is to be studied.

Keywords: Sonic boom, human response, simulation.
PACS: 43.66

INTRODUCTION

NASA's High Speed Research (HSR) program in the 1990s was intended to develop a technology base for a future 300-passenger High-Speed Civil Transport (HSCT). As part of this program, the NASA Langley Research Center sonic boom simulator (SBS) was built and used for a series of tests on subjective response to sonic booms. At the end of the HSR program, an HSCT was deemed impractical, but since then interest in supersonic flight has reawakened, this time focusing on a smaller aircraft. The demonstrated ability to design a "low-boom" aircraft has encouraged re-appraisal of the FAA ruling that supersonic flight should be banned overland, because of the impact of sonic boom on the community.

This paper presents an overview of work of human response to sonic boom completed at NASA in the 1990s, and a summary of studies performed since then on the impact of low-intensity sonic booms, including work performed as part of Project 8 of the FAA/NASA/Transport Canada Center of Excellence Partnership for Air Transportation Noise and Emissions Reduction (PARTNER). Details of PARTNER can be found at http://web.mit.edu/aeroastro/www/partner/mission.htm.

CP838, *Innovations in Nonlinear Acoustics: 17th International Symposium on Nonlinear Acoustics*,
edited by A. A. Atchley, V. W. Sparrow, and R. M. Keolian
2006 American Institute of Physics 0-7354-0330-9/06/$23.00

RESPONSE TO SONIC BOOM 1990-1996

During the 1990s, NASA Langley Research Center conducted three groups of studies on sonic booms: laboratory, "inhome" and field. These three complementary parts consisted of (a) laboratory studies, which have very good control over the sound stimuli that the subjects hear but require a very abnormal listening environment; (b) an "inhome" study where sounds are played through loudspeakers in people's homes, thus improving the realism of the environment but reducing the control over the sound field; and (c) field studies with completely normal environment but very poor knowledge of the precise details of the sound exposure. The "acceptability" of or "annoyance" caused by a sound is affected by many factors. In a laboratory situation, while some of these factors are under control, others may be missing. Thus ratings of sounds can be considered as accurate relative to one another, but not in an absolute sense. The inhome study moves closer to absolute measurements and the field studies measure absolute, real reactions. A compilation of these studies, with details of the findings, is given in Leatherwood *et al.* [1].

Laboratory Studies

The laboratory experiments were designed to (a) develop an improved understanding of sonic boom subjective effects; (b) quantify in a systematic and comprehensive manner the loudness and annoyance benefits due to intentional sonic boom shaping as well as distortions due to passage through walls, ground reflections, and atmospheric propagation; and (c) assess various noise descriptors as predictors of sonic boom subjective effects. To study these factors, NASA Langley built a Sonic Boom Simulator (SBS). Construction details, performance capabilities, and operating procedures of the simulator were described by Leatherwood *et al.* [2]. The simulator is a person-rated, airtight, loudspeaker driven booth, carpeted and lined with foam to reduce acoustic resonances, and free of loose objects capable of creating rattles. Input waveforms are computer generated and preprocessed to compensate for non-uniformities in the magnitude and phase characteristics of the frequency response of the booth and sound reproduction system. Preprocessing is accomplished by the use of a digital broadband equalization filter as described by Brown and Sullivan [3]. One significant finding from the series of 14 studies conducted in the SBS were that subjective reactions to simulated sonic booms were predicted well by calculated loudness metrics, such as Steven's Perceived Level (PL). The metric and its application to sonic boom are described in Shepherd and Sullivan [4]. Such metrics as unweighted or C-weighted dB or peak overpressure were poor predictors of human response.

Inhome Study

One question affecting assessment of a sound exposure is the combination of single events into a long-term multiple-event situation. Within the laboratory, such combining of reactions is unrepresentative of a realistic situation. To enable the investigation of more realistic multiple-event environments, Langley Research Center completed an "inhome" study (McCurdy et al. [5]) in which simulated sonic booms were played through loudspeaker systems in people's homes. Various scenarios involving different numbers of booms at different levels were played, with participants giving annoyance judgments after a day's exposure. The study confirmed that the increase in annoyance resulting from multiple occurrences can be modeled by the addition of the term "10×(log(Number of Occurrences)" to the sonic boom level.

Field Studies

The third part of NASA Langley Research Center's program to study subjective response to sonic boom was a series of field studies in which the responses of community residents experiencing supersonic overflights were measured, together with their boom exposures. This study was unique in that no other study has investigated the reactions of people routinely exposed to sonic booms over a long time period. As reported in Fields [6], the study found that sonic boom annoyance increased as the number and/or level of the booms increased. Large differences noted in responses from two localities were not attributable to sonic boom exposure, but were explained in part by differences in attitudes towards the "noise makers" and differences in exposure to low altitude, subsonic aircraft flyovers.

RESPONSE TO LOW-INTENSITY SONIC BOOM

Recent interest in "low-boom" aircraft prompted NASA Langley Research Center to complete a study in 2003 (Sullivan [7]) that used twenty-four predicted low-intensity boom waveforms resulting from designs of candidate aircraft supplied by aircraft designers. Participants rated these sounds presented in the SBS at the levels expected for the aircraft operating under cruise conditions. As was expected, unweighted and C-weighted Sound Exposure Levels, and overpressure were not good predictors of human response as measured in the SBS; PL and A-weighted Sound Exposure Level were good predictors. These metrics predicted the loudness of complex, multishocked booms and simple booms equally well.

SIMULATOR ASSESSMENT

As part of Project 8 of the Partnership for AiR Transportation Noise and Emissions Reduction (PARTNER) (a FAA/NASA/Transport Canada-Sponsored Center of Excellence) an evaluation of three sonic boom simulators to determine if they can reproduce realistic sonic booms is under way. This consists of a series of three experiments at NASA Langley, NASA Dryden and Lockheed Martin's Palmdale facility in California. The Lockheed Martin simulator is a very similar design to the SBS, both being airtight booths that achieve the characteristic sonic boom N-wave shape by compressing the air within the booth. The Gulfstream simulator (Salamone [8]) is a folded-horn design which creates a pressure wave that travels past the listener into an anechoic termination. This "traveling wave" simulation is closer to the pressure waveform of a real sonic boom than the booths' compression simulation.

However the Gulfstream simulator cannot recreate the very low frequencies (below 7 Hz) that are present in a sonic boom, but which can be created in the Langley simulator.

In the first test, participants experienced in listening to sonic booms evaluated the realism of booms played in the NASA and the Gulfstream simulators. In the second test subjects heard supersonic military jet flyovers and afterwards rated the realism of recordings of some of those flyovers reproduced in the Gulfstream simulator. In the third test in this realism series, expert listeners will evaluate the Lockheed Martin and the Gulfstream simulators.

In the NASA/Gulfstream back-to-back comparison test, the sounds used were all created from recordings of F-18s and F-5Es made during the Shaped Sonic Boom Experiment at Edwards AFB in January 2004. In order to identify important characteristics of sonic boom sounds, some booms were modified in various ways, including (1) high pass filtering to remove the low frequencies so the booms in the NASA LaRC simulator would have the same frequency content as those in the Gulfstream simulator; (2) the addition of a synthesized "ground reflection" to simulate what a listener standing above hard ground would hear; (3) modification of the length of the recording to vary the amount of noise that was heard after the boom. Preliminary results indicate that the absence of the very low frequencies was less important in the sensation of realism than the "feeling" imparted by the traveling wave. Also evident is the importance of the "post-boom" noise in giving an impression of realism to the simulations. This post-boom noise is not fully understood, but is considered to be a combination of engine noise and boom sounds that have been disturbed by turbulence and have traveled by complex paths through the atmosphere to reach the listener after the direct boom.

REFERENCES

1. Leatherwood, J. D., Sullivan, B. M., Shepherd, K. P., McCurdy, D. A., and Brown, S. A., "A summary of recent NASA studies of human response to sonic booms", *J. Acoust. Soc. Am.*, **111**(1) pt. 2, 586-598 (2002).
2. Leatherwood, J. D., Shepherd, K. P., and Sullivan, B. M., "A New Simulator for Assessing Subjective Effects of Sonic Booms", *NASA Tech. Memo.* **104150**, 1–35 (1991) http://ntrs.nasa.gov/archive/nasa/casi.ntrs.nasa.gov/19920002541_1992002541.pdf.
3. Brown, D. E., and Sullivan, B. M., "Adaptive Equalization of the Acoustic Response in the NASA Langley Sonic Boom Chamber" *Proceedings, International Conference on Recent Advances in Active Control of Sound and Vibration*, VPI & SU, Blacksburg, Virginia, 360-371 (1991)
4. Shepherd, K. P., and Sullivan, B. M., "A Loudness Calculation Procedure applied to Shaped Sonic Booms", *NASA Technical Paper* **3134**, 1–10 (1991) http://hdl.handle.net/2002/14923
5. McCurdy, D. A., Brown, S. A., and Hilliard, R. D., "Subjective Response of People to Simulated Sonic Booms in their Homes" *J. Acoust. Soc. Am.* **116**(3) 1573-1584 (2004)
6. Fields, J. A.: "Reactions of Residents to Long-Term Sonic Boom Noise Environments" *NASA Contractor Report* 201704, 1-157 (1997) http://ntrs.nasa.gov/archive/nasa/casi.ntrs.nasa.gov/19970023685_1997038340.pdf
7. Sullivan, B. M., "Human Response to Simulated Low-Intensity Sonic Booms" *Proceedings of NOISE-CON* **2004**, Baltimore, Maryland, 541-550 (2004) http://techreports.larc.nasa.gov/ltrs/PDF/2004/mtg/NASA-2004-noisecon-bms.pdf
8. Salamone, J., "Portable Sonic Boom Simulation" *Proceedings of the 17th International Symposium on Nonlinear Acoustics*, American Institute of Physics Press (2005)

Preliminary work about the reproduction of sonic boom signals for perception studies

N. Epain*, P. Herzog*, G. Rabau* and E. Friot*

*Laboratoire de Mécanique et d'Acoustique,
31 chemin Joseph Aiguier
13402 Marseille cedex 20 France

Abstract. As part of a French research program, a sound restitution cabin was designed for investigating the annoyance of sonic boom signals. The first goal was to reproduce the boom spectrum and temporal waveform: this required linear generation of high pressure levels at infrasonic frequencies (110 SPL dB around 3 Hz), and response equalization over the full frequency range (1 Hz–20 kHz). At this stage the pressure inside the cabin was almost uniform around the listener, emulating an outdoor situation. A psychoacoustic study was then conducted which confirmed that the loudness (related to annoyance) of N-waves is roughly governed by the peak pressure, the rise/fall time, and the wave duration. A longer-term goal is to reproduce other aspects of an indoor situation including rattle noise, ground vibrations, and a more realistic spatial repartition of pressure. This latter point has been addressed through an Active Noise Control study aiming at monitoring the low-frequency acoustic pressure on a surface enclosing a listener. Frequency and time-domain numerical simulations of boom reproduction via ANC are given, including a sensitivity study of the coupling between a listener's head and the incident boom wave which combine into the effective sound-field to be reproduced.

Keywords: Sonic boom, Reproduction, Psychoacoustics, Loudness
PACS: 43.38.Hz, 43.66.Lj, 43.66.Yw

INTRODUCTION

A French research project about future supersonic planes has risen the need for assessing the loudness of a sonic boom. Previous studies had been done for the Concorde project or military fighter many years ago [1, 2], and did not provide definitive conclusions about the possibility to use supersonic planes above or near populated areas.

Conducting psychoacoustic tests [3, 4] rises two main difficulties: the sonic boom spectrum has its maximum at subaudio frequencies, and it is a transient signal, so higher frequencies play a significant role. As it is not yet known, how such signals are perceived, reliable tests require a dedicated reproduction tool which must ensure that the listener is immersed into a sound field as close as possible to the actual one.

Similarly to previous studies, a "closed simulator" (pressure chamber) has thus been designed. Embedded electroacoustic sources have been optimized for the reproduction of the subsonic and audio spectrum of sonic booms. A series of psychoacoustic tests has been done, which essentially confirmed the results already provided by similar studies, concerning the case of free-field sonic booms.

However, final conclusions cannot be given because annoyance of sonic booms may be quite different for indoor situations, where the soundfield is much more complicated and also involve vibration and rattle. Numerical simulations have therefore been con-

CP838, *Innovations in Nonlinear Acoustics: 17th International Symposium on Nonlinear Acoustics*,
edited by A. A. Atchley, V. W. Sparrow, and R. M. Keolian
© 2006 American Institute of Physics 0-7354-0330-9/06/$23.00

ducted to design a system able to reproduce an arbitrary soundfield, using active noise control techniques to control each source separately.

AXIAL REPRODUCTION OF SONIC BOOMS

An example of sonic boom signal is given by figure 1. It is an "N-wave", with a quite high peak pressure. Its spectrum may be maximum around 3Hz, and extends over the full audio range, even if most of the energy is below 30 Hz. This signal is scaled down for the psychoacoustic tests: for annoyance assessment, our goal is to reproduce sonic boom peaks with a good accuracy up to 110 dB.

At very low frequencies, monopole subwoofers radiating inside an almost sealed room ("pressure chamber") create a spatially uniform pressure. For increasing frequencies, the structure of the field is more complex, involving several degrees of freedom, but it may also be almost uniform with a suitable repartition and control of speakers. Conversely, at higher frequencies, a suitable lining on the walls allows to consider the room as close to semi-anechoic. The whole frequency range is therefore driven by a triple control system, with different constraints for each frequency band.

The simulation cabin has been built inside a room with concrete walls ensuring a very good insulation from outside noise. Its dimensions have been kept to a minimum, thus reducing the requirement on the sources, and shifting modal resonance toward higher frequencies. Cabin dimensions of 3 x 2 x 2 m have been considered adequate, and leave enough room for fitting 16 subwoofers at appropriate locations on the walls. More details about the cabin and loudspeaker design can be found in [5].

In a first step, we designed a reproduction system for simulating a plane wave coming from the front direction. This control system is a pre-processing, using inverse filters computed from pressure measurements close to the position of the listener's head. Fig. 1 shows that this equalization helps greatly to achieve the fidelity of the waveform.

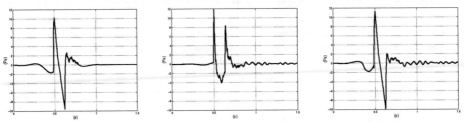

FIGURE 1. Temporal waveforms of a boom signal : recorded filtered N-wave (left), its reproduction without equalization (center), and with a dual channel system (right)

A first series of psychoacoustic tests has been conducted to evaluate the influence of different physical parameters on the perceived loudness of synthesized sonic-booms. The studied parameters were the rise time, the interpeak duration, and the peak over-pressure. More details about these tests can be found in [6].

Fig. 2 shows some results obtained for the assessment of N-waves signals. As in previous studies [4], the results show that the rise time is an important parameter in the

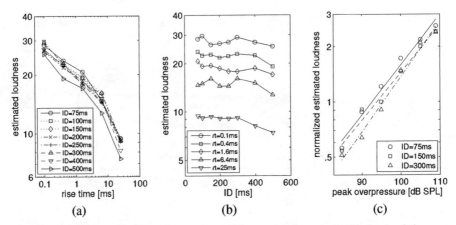

FIGURE 2. Results of the perceptual assessment of N-wave signals synthesized in the simulation room: (a) estimated loudness as a function of the rise time (rt) for various values of the interpeak durations (ID), with a peak overpressure of 110 dB SPL; (b) estimated loudness as a function of the interpeak duration for various values of the rise time, with a peak overpressure of 110 dB SPL; (c) loudness function for various values of the interpeak duration, with a rise time of 2 ms.

perception of N-waves (Fig. 2a), whereas the interpeak duration has a little influence (Fig. 2b). The most important parameter is the peak overpressure as shown on Fig. 2c.

IMPROVEMENT OF THE SOUND FIELD REPRODUCTION USING ACTIVE NOISE CONTROL TECHNIQUES

An almost spatially uniform acoustic pressure has been obtained in the room by filtering the signal fed to the subwoofers. However, this method does not enable to reproduce the spatio-temporal properties of actual sonic booms, such as the travel of the wavefront past the listener. In the aim of reproducing more accurately the sound in the room, each subwoofer must be separately driven with an optimized signal.

A sound field reproduction strategy called Boundary Pressure Control (BPC) has been proposed [7], which is based on active noise control techniques. BPC intends to reproduce the required sound field inside a given volume via the active control of the acoustic pressure measured at some microphones placed over the whole boundary surface of the volume. Unlike other sound reproduction strategies, BPC is able to compensate efficiently for the room reflections, using adaptive filtering methods. Besides, the control microphones allow to monitor the performances of the system in real time.

The reproduction of plane waves by the room's subwoofers using BPC has been simulated in both the frequency and time domains, in the case where the reproduction area is a cylinder large enough to surround a listener, with 32 virtual control microphones distributed on the surface. As shown in Fig. 3, harmonic plane waves may be very accurately reproduced everywhere inside the cylinder below 100 Hz. Above this frequency, however, the error increases strongly, due to spatial aliasing. Besides, time-domain sim-

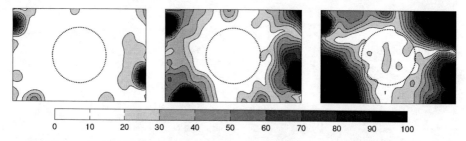

FIGURE 3. Frequency-domain simulation of optimal reproduction via the control of the acoustic pressure at the surface of a microphone cylinder: contour of equal sound pressure error level (in percents) measured at an horizontal plane in the room for plane waves at 50 Hz (left), 100 Hz (middle), and 150 Hz (right). The dashed circle represents the boundary of the cylinder.

ulations have shown that the subwoofers may efficiently reconstruct low-pass filtered sonic booms using BPC, including the travel of the wavefront through the control area. More details about these simulations can be found in [8].

In theory, the use of a particular sound field for each listener is required as a reference for the control, since the incident wave is diffracted in a different way by each body. This problem has been addressed via frequency domain simulations in free field, which have shown that BPC allows to reproduce around a listener the soundfield recorded around another person if the control microphones are far enough from them. However, even at low frequency, the reproduction is inaccurate if only the incident wave is reproduced at the level of the control microphones, which suggests that a dummy head should be used for recording the sound field to be reproduced.

In conclusion, numerical simulations have shown that low-frequency components of sonic booms can be reproduced more accurately using BPC. Sound field synthesis using BPC will be soon implemented in the reproduction room, using an off-line filter design in a first step, and a new series of psychoacoustic tests will be conducted.

REFERENCES

1. L. Bobin and Al, "Les ondes acoustiques et leurs effets : résumé de 25 années d'activité à l'ISL," Notice biblio NB 401/85 (1985).
2. J.D. Leatherwood, K.P. Shepherd, B.M. Sullivan, "A new simulator for assessing subjective effects of sonic booms," NASA TM 104150 (1991).
3. K. Buck and Al, "Etude du sursaut provoqué par le bang sonique chez l'homme," ISL, Rapport R 114/77.
4. J.D Leatherwood and Al., "Summary of recent NASA studies of human response to sonic boom," *J. Acoust. Soc. Am.*, **111** (1), Pt. 2 (2002).
5. G. Rabau and P. Herzog, "A specific cabin for restitution of sonic boom : application for perceptive tests," in *Proceedings of the Joint Congress CFA/DAGA'04* (2004), on CD-ROM (2 pages).
6. G. Rabau, S. Meunier and P. Herzog, "Evaluation psychoacoustique de la gêne induite au sol par un bang supersonique", Technical Report, French Ministry of Research (2004).
7. S. Takane, Y. Suzuki and T. Sone, "A New Method for Global Sound Field Reproduction Based on the Kirchhoff's Integral Equation," *ACUSTICA - Acta Acustica* **85** (1999) 250–257.
8. N. Epain, E. Friot and G. Rabau, "Indoor Sonic Boom Reproduction Using ANC," in *Proceedings of ACTIVE 2004* (2004), on CD-ROM (12 pages).

Portable Sonic Boom Simulation

Joe Salamone

Gulfstream Aerospace Corporation, P.O. Box 2206 M/S C-215, Savannah, Georgia 31402, USA

Abstract. A method is presented to simulate sonic booms using high fidelity and custom-built audio equipment that output to an acoustically treated listening environment, all of which is contained in a portable vehicle. The audio system has inherent low and high frequency performance limitations and also introduces distortion due to the frequency response of the system. The limitations of the system are compensated for by band-pass filtering a full-fidelity sonic boom signature and applying a system equalization filter. The purpose of the band-pass filter is to remove frequency content above and below the capabilities of the system yet retain the audible and felt characteristics of the full-fidelity waveform. The equalization filter, computed from time-domain Wiener filtering, compensates for the frequency-dependent system response of the audio system at several listening positions. The system performance is evaluated by comparing the PLdB, SEL(A) and SEL(C) of the measured system output to the full-fidelity waveform. Results show good agreement between the loudness levels of the full-fidelity waveform and the corresponding measured system output.

Keywords: Sonic boom simulation.
PACS: 43.28.Mw, 43.50.Pn

INTRODUCTION

Gulfstream Aerospace has a unique, portable facility that can simulate both measured and synthesized sonic booms. This acoustic simulation enables a design team to listen to the results of their efforts toward sonic boom suppression [1]. A significant advantage of a portable sonic boom simulator is its capability to extend to an international listening audience. The simulator has already been used for human subject research pertaining to simulation realism [2] and will be used in future research for acceptability of shaped sonic boom signatures. This paper will discuss the vehicle and its contents along with the method and results of the sonic boom simulation.

SYSTEM DESCRIPTION

The simulator is contained in a towable trailer 32 feet long, 8.5 feet wide and approximately 11.5 feet tall. The trailer can receive electrical power from two gas generators or by building electricity. Some safety features include emergency lighting and fire extinguishers. The interior of the trailer is air-conditioned. There are two sections inside the trailer: the operator area and the listening area. The audio speakers are located between the two main sections. A video camera and microphone are

CP838, *Innovations in Nonlinear Acoustics: 17th International Symposium on Nonlinear Acoustics*,
edited by A. A. Atchley, V. W. Sparrow, and R. M. Keolian
© 2006 American Institute of Physics 0-7354-0330-9/06/$23.00

installed in the listener section so that activity can be monitored from the operator area. A top-down view of the trailer and its interior is shown in Figure 1.

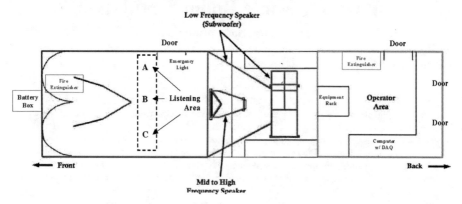

FIGURE 1. Top-down view of the interior layout in the Gulfstream sonic boom simulator.

The operator area contains the computing and audio hardware which produce the simulation. A laptop contains discrete-time data sampled at 24kHz that is output to the audio system via a data acquisition system. That hardware then sends an analog signal to the digital audio equalizer that is pre-programmed with the gains, filtering and delays necessary for each power amplifier. Two power amplifiers supply signals to the four drivers in the waveguided subwoofer and two power amplifiers supply signals to the four drivers in the mid/high frequency speaker. The full performance bandwidth of the system is 8 Hz to 18 kHz, but for sonic boom simulations the upper end of the system is reduced to 5 kHz.

The listening section in the simulator has interior dimensions of 6 feet wide, 7 feet high and approximately 12 feet long. One to four people can stand in the designated listener area, which is roughly 5 feet wide and 2 feet deep (see Fig. 1). The sound travels from the speakers in the middle of the trailer, passes through the listener area and into the anechoic termination at the front of the trailer. The anechoic termination, filled from floor to ceiling with fiberglass insulation, was constructed to absorb sound down to nearly 50 Hz. The walls and ceiling are thermally and acoustically treated with fiberglass insulation and absorptive open-cell foam. A thin sheet of perforated aluminum separates the foam and fiberglass. The foam has a thin, acoustically transparent cloth glued to its surface for aesthetics. The floor is not acoustically treated but is thermally insulated and carpeted.

SYSTEM EQUALIZATION

There are several reasons to perform a discrete-time equalization of the system. The frequency response of the system is not perfectly flat in magnitude nor is it linear or zero phase. The bandwidth of the system does not reach DC (zero Hz) and has a finite upper limit. It is also desirable to establish a transfer function between the input to the system and its output at several locations in the listener area. A multi-point equalization using time-domain Wiener filtering was computed that addresses the above mentioned items.

The Wiener filtering scheme and equations used to compute the FIR equalization filter are similar to those found in [3] except for changes due to multiple microphone locations and system frequency bandwidth. An elliptic band-pass filter with 8 Hz to 5 kHz bandwidth has replaced the filters in the desired path. The delay in the desired path was different for each microphone location and adjusted for the different propagation distances to their respective measurement positions. Ten different microphone locations (instead of one) were "randomly" chosen for equalization within the designated listener area. Ten different training signals, one for each microphone position, consisting of 15 seconds of white noise, were input to the system and the corresponding acoustic outputs were measured. The optimum FIR filter, w_{opt}, for system equalization was computed as:

$$ w_{opt} = \left(\sum_{i=1}^{10} R_i \right)^{-1} \sum_{i=1}^{10} p_i \qquad (1) $$

That is, the autocorrelation matrix inverted is the sum of the autocorrelation matrices from each microphone measurement and likewise for the cross-correlation vectors. A filter length of 4096 samples at a sample rate of 24 kHz was computed for w_{opt}. Input waveforms can then be pre-filtered with the equalization filter and sent through the system.

SYSTEM PERFORMANCE

The simulator performance is evaluated by comparing the loudness of the full-fidelity sonic boom waveform to the loudness of the corresponding measurements taken at the three main positions in the listening area. Table 1 compares the loudness calculations for a full-fidelity N-wave as measured in the field to the output levels measured in the trailer at positions A, B and C in the listening area. The loudness levels for the listening locations are, in general, within 1 dB or less of the full-fidelity waveform at the three measured positions.

TABLE 1. Loudness calculations for a measured N-wave, full-fidelity vs. simulated levels.

Location	PLdB	SEL(C)	SEL(A)
Full-fidelity	106.1	109.1	91.5
Position A	105.1	108.8	91.0
Position B	106.3	108.7	92.7
Position C	105.2	108.4	91.2

Because the system has a frequency bandwidth constraint, the desired or target acoustic output is a band-pass filtered version of the full-fidelity waveform. Simulator performance can then be evaluated by how well the system can match the desired waveform in the time domain. Figure 2 compares the pressure versus time data for the full-fidelity waveform (dashed line), the band-pass filtered waveform (dotted line) and the simulator output (solid line) at listener position B for a measured N-wave. The simulator acoustic output closely matches the filtered version of the full-fidelity waveform.

The Gulfstream simulator has also been involved with research regarding human subjective response to simulated sonic booms [2]. Feedback from those who have experienced "real-world" sonic booms have indicated realistic sound and felt perception are present in the Gulfstream sonic boom simulation.

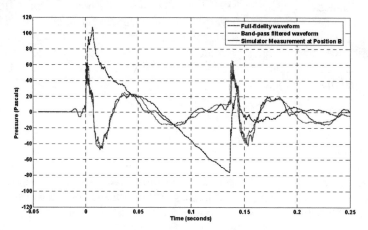

FIGURE 2. Time-domain comparison of the simulator acoustic output at listener position B to the desired output and full-fidelity waveform.

CONCLUSION

A means and method to simulate sonic booms via a portable system and listening environment have been presented. This system can simulate the audible and felt characteristics as well as the loudness and realism of a full-fidelity waveform. The Gulfstream sonic boom simulator has the capability to simulate sonic booms for one to four people simultaneously and the potential for travel to demonstrations at venues worldwide.

REFERENCES

1. Howe, Donald C., "Improved Sonic Boom Minimization with Extendable Nose Spike", AIAA-2005-1014, 43rd AIAA Aerospace Sciences Meeting and Exhibit, Reno, NV, January 10-13, 2005.
2. Sullivan, Brenda M., "Research on Subjective Response to Simulated Sonic Booms at NASA Langley Research Center," 17th International Symposium on Nonlinear Acoustics and International Sonic Boom Forum, State College, PA, July 18-22, 2005.
3. Salamone, Joe, "Sonic Boom Simulation Using Conventional Audio Equipment," Noise-Con 04, Baltimore, MD, July 12-14, 2004.

PARTICIPANTS IN ISNA17

E. Agrest	S. Garrett	A. Maksimov	I. Shkolnik
F. Akhmedzhanov	K. Gee	R. Marchiano	I. Shreiber
O. Al-Bataineh	X. Gong	T. Marston	B. Smith
V. Aleshin	D. Graham	P. Marston	R. Smith III
W. Arthur	Y. Guo	T. Mast	J. So
A. Atchley	R. Gutierrez	I. Mastikhin	I. Solodov
T. Auger	E. Haering	B. McDonald	V. Sparrow
C. Babcock	K. Haller	J. Morgenstern	M. Stone
M. Bailey	M. Hamilton	A. Mortlock	N. Sugimoto
S. Baskar	A. Hanford	T. Muir	B. Sullivan
M. Baudoin	L. Hargrove	A. Murakami	L. Sutherland
T. Biwa	T. Hartmann	S. Nakagawa	A. Sutin
D. Blackstock	T. Hay	K. Naugolnykh	G. Swift
P. Blanc-Benon	N. Heron	V. Nikolaevskiy	C. Tao
C. Bradley	M. Hilpert	P. O'Connor	J. Tarkington
J. Brady	H. Hobaek	W. Ohm	Y. Tashiro
M. Breazeale	K. Hodgdon	B. O'Neill	J. Thomas
H. Camin	R. Holt	K. Orth	B. Tittmann
D. Carpenter	E. Horan	I. Ostrovski	M. Toda
J. Chambers	J. Huijssen	L. Ostrovsky	V. Tournat
S. Chernyshev	Y. Hwang	J. Page	R. Tracy
E. Chesnokov	Y. Ilinskii	F. Pearce	B. Tuttle
S. Chi	P. Johnson	C. Pecorari	T. Ulrich
C. Church	K. Jung	B. Petitjean	M. Verweij
L. Clark	T. Kamakura	A. Pierce	N. Vykhodtseva
R. Cleveland	O. Kandil	A. Pilon	K. Wallace
P. Coen	H. Kao	Y. Pishchalnikov	Y. Wang
E. Cosharek	R. Keolian	K. Plotkin	Y. Watanabe
F. Coulouvrat	N. Kharin	M. Postema	R. Waxler
F. Curra	S. Klausmeyer	A. Puckett	N. Weiland
T. Darling	T. Kling	D. Reiner	M. Wochner
S. Dos Santos	M. Knuth	P. Rendon	X. Yan
M. Downing	M. Korman	P. Roberts	T. Yano
G. Duval	W. Kreider	R. Roy	K. Yoshida
K. Elmer	R. Kumon	O. Rudenko	R. Young
B. Enflo	E. Kurihara	S. Sakai	E. Zabolotskaya
N. Epain	C. Lafon	S. Sakamoto	J. Zeegers
I. Esipov	H. Lee	J. Salamone	Y. Zhang
V. Espinosa	G. Lilis	A. Samsonov	D. Zhang
L. Falco	L. Locey	V. Sanchez-Morcillo	X. Zheng
D. Fenneman	S. Lopatnikov	G. Sankin	Z. M. Zhu
L. Fisher	A. Loubeau	K. Sarkar	Z. Zhu
S. Fogg	J. Louis	D. Sastrapradja	A. Zyryanova
B. Forssmann	Z. Lu	D. Schein	
L. Ganjehi	J. Lyons	P. Shapiro	

AUTHOR INDEX

A

Abdel-Fattah, A. I., 186, 191
Agrest, E. M., 483
Akhmedzhanov, F. R., 143
Akiyama, M., 233
Aleshin, V., 104
Allwright, D., 587
Anderson, J. B., 556
Araki, Y., 359
Arthur, W., 135
Atchley, A. A., 544, 572
Attenborough, K., 651

B

Backhaus, S., 371, 399
Bai, Y., 345
Bailey, M. R., 299, 315, 319, 323
Baskar, S., 536, 611
Bass, H. E., 635
Baudoin, M., 619
Berger, U., 587
Birkin, P. R., 512
Biwa, T., 363, 395
Blackstock, D. T., 601
Blanc-Benon, P., 355, 587, 651
Blumrich, R., 579
Bou Matar, O., 79, 95, 284
Bradley, C., 247
Bras, N., 504
Breazeale, M. A., 11
Busse, G., 35, 108

C

Calise, L., 593
Camarena, F., 131, 237
Campos-Pozuelo, C., 449
Castagnède, B., 67
Chambers, J. P., 635
Chatterjee, D., 271
Cheng, J., 99
Chernyshev, S. L., 643
Chi, S., 203

Church, C. C., 217
Cleveland, R. O., 263, 299, 333
Coulouvrat, F., 536, 579, 619, 631
Covington, A. M., 19
Cruañes, J., 237
Crum, L. A., 299, 315, 323
Cui, J., 229
Culligan, S., 469
Czech, J., 655

D

Dagrau, F., 639
Daley, T., 167
Dallois, L., 587
Danforth-Hanford, A., 556
Darling, T. W., 19, 71
Dean, A. J., 387
de Jong, N., 275
Dos Santos, S., 79, 95
Downing, M., 560
Dreiden, G. V., 147
Driscoll, D., 267

E

Emmons, E., 19
Enflo, B. O., 457
Epain, N., 663
Esipov, I., 195
Espinosa, V., 131, 237, 453
Evan, A. P., 299, 319

F

Faidi, W., 209
Falco, L. E., 572
Fan, Z., 99
Fardal, R., 593
Farny, C. H., 225
Fenneman, D., 59
Fogg, S. L., 439
Forssmann, B., 291
Fortineau, J., 79, 95

673

674